NONEQUILIBRIUM MANY-BODY THEORY OF QUANTUM SYSTEMS

The Green's function method is one of the most powerful and versatile formalisms in physics, and its nonequilibrium version has proved invaluable in many research fields. This book provides a unique, self-contained introduction to nonequilibrium many-body theory.

Starting with basic quantum mechanics, the authors introduce the equilibrium and nonequilibrium Green's function formalisms within a unified framework called the contour formalism. The physical content of the contour Green's functions and the diagrammatic expansions are explained with a focus on the time-dependent aspect. Every result is derived step-by-step, critically discussed and then applied to different physical systems, ranging from molecules and nanostructures to metals and insulators. With an abundance of illustrative examples, this accessible book is ideal for graduate students and researchers who are interested in excited state properties of matter and nonequilibrium physics.

GIANLUCA STEFANUCCI is a Researcher at the Physics Department of the University of Rome Tor Vergata, Italy. His current research interests are in quantum transport through nanostructures and nonequilibrium open systems.

ROBERT VAN LEEUWEN is Professor of Physics at the University of Jyväskylä in Finland. His main areas of research are time-dependent quantum systems, many-body theory, and quantum transport through nanostructures.

NONEQUILIBRIUM MANY-BODY THEORY OF QUANTUM SYSTEMS

A Modern Introduction

GIANLUCA STEFANUCCI

University of Rome Tor Vergata

ROBERT VAN LEEUWEN

University of Jyväskylä

CAMBRIDGE
UNIVERSITY PRESS

Contents

Appendices

Preface

This textbook contains a pedagogical introduction to the theory of Green's functions *in* and *out* of equilibrium, and is accessible to students with a standard background in basic quantum mechanics and complex analysis. Two main motivations prompted us to write a monograph for beginners on this topic.

The first motivation is research oriented. With the advent of nanoscale physics and ultrafast lasers it became possible to probe the correlation between particles in excited quantum states. New fields of research like, e.g., molecular transport, nanoelectronics, Josephson nanojunctions, attosecond physics, nonequilibrium phase transitions, ultracold atomic gases in optical traps, optimal control theory, kinetics of Bose condensates, quantum computation, etc. added to the already existing fields in mesoscopic physics and nuclear physics. The Green's function method is probably one of the most powerful and versatile formalisms in physics, and its nonequilibrium version has already proven to be extremely useful in several of the aforementioned contexts. Extending the method to deal with the new emerging nonequilibrium phenomena holds promise to facilitate and quicken our comprehension of the excited state properties of matter. At present, unfortunately, to learn the nonequilibrium Green's function formalism requires more effort than learning the equilibrium (zero-temperature or Matsubara) formalism, despite the fact that *nonequilibrium Green's functions are not more difficult*. This brings us to the second motivation.

The second motivation is educational in nature. As students we had to learn the method of Green's functions at zero temperature, with the normal-orderings and contractions of Wick's theorem, the adiabatic switching-on of the interaction, the Gell–Mann–Low theorem, the Feynman diagrams, etc. Then we had to learn the finite-temperature or Matsubara formalism where there is no need of normal-orderings to prove Wick's theorem, and where it is possible to prove a diagrammatic expansion without the adiabatic switching-on and the Gell–Mann–Low theorem. The Matsubara formalism is often taught as a disconnected topic but the diagrammatic expansion is exactly the same as that of the zero-temperature formalism. Why do the two formalisms look the same? Why do we need more "assumptions" in the zero-temperature formalism? And isn't it enough to study the finite-temperature formalism? After all zero temperature is just one possible temperature. When we became post-docs we bumped into yet another version of Green's functions, the nonequilibrium Green's functions or the so called Keldysh formalism. And again this was another different way to prove Wick's theorem and the diagrammatic expansion. Furthermore, while several excellent textbooks on the equilibrium formalisms are available, here the learning process is considerably slowed down by the absence of introductory textbooks. There exist few review articles on the Keldysh formalism and they are scattered over the years and the journals. Students have to face different jargons and different notations, dig out original papers (not all downloadable from the web), and have to find the answer to lots of typical newcomer questions like, e.g., why is the diagrammatic expansion of the Keldysh formalism again the same as that of the zero-temperature and Matsubara formalisms? How do we see that the Keldysh formalism reduces to the zero-temperature formalism in equilibrium? How do we introduce the temperature in the Keldysh formalism? It is easy to imagine the frustration of many students during their early days of study of nonequilibrium Green's functions. In this book we introduce only *one* formalism, which we call the *contour formalism*, and we do it using a very pedagogical style. The contour formalism is not more difficult than the zero-temperature, Matsubara or Keldysh formalism and we explicitly show how it reduces to those under special conditions. Furthermore, the contour formalism provides a natural answer to all previous questions. Thus the message is: *there is no need to learn the same thing three times*.

Starting from basic quantum mechanics we introduce the contour Green's function formalism step by step. The physical content of the Green's function is discussed with particular attention to the time-dependent aspect and applied to different physical systems ranging from molecules and nanostructures to metals and insulators. With this powerful tool at our disposal we then go through the Feynman diagrams, the theory of conserving approximations, the Kadanoff–Baym equations, the Luttinger–Ward variational functionals, the Bethe–Salpeter equation, and the Hedin equations.

This book is not a collection of chapters on different applications but a self-contained introduction to mathematical and physical concepts of general use. As such, we have preferred to refer to books, reviews and classical articles rather than to recent research papers whenever this was possible. We have made a serious effort in organizing apparently disconnected topics in a *logical* instead of *chronological* way, and in filling many small gaps. The adjective "modern" in the title refers to the presentation more than to specific applications. The overall goal of the present book is to derive a set of kinetic equations governing the quantum dynamics of many identical particles and to develop perturbative as well as nonperturbative approximation schemes for their solution.

About 600 pages may seem too many for a textbook on Green's functions, so let us justify this voluminousness. First of all *there is not a single result which is not derived*. This means that we have inserted several intermediate steps to guide the reader through every calculation. Secondly, for every formal development or new mathematical quantity we present carefully selected examples which illustrate the physical content of what we are doing. Sometimes the reader will find further supplementary discussion or explanations printed in smaller type; these can be skipped at a first reading. Without examples and illustrations (more than 250 figures) this book would be half the size but the actual understanding would probably be much less. The large number of examples compensates for the moderate number of exercises. Thirdly, in the effort of writing a comprehensive presentation of the various topics we came across several small subtleties which, if not addressed and properly explained, could give rise to serious misunderstandings. We have therefore added many remarks and clarifying discussions throughout the text.

The structure of the book is illustrated in Fig. 1 and can be roughly partitioned in three parts: mathematical tools, approximation schemes, and applications. For a detailed list of

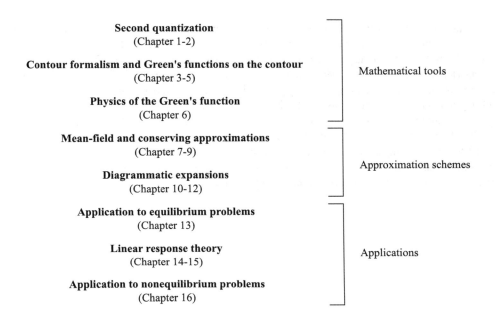

Figure 1 Structure of the book

topics the reader can look at the table of contents. Of course the choice of topics reflects our personal background and preferences. However, we feel reasonably confident to have covered all fundamental aspects of Green's function theory in and out of equilibrium. We have tried to create a self-contained and self-study book capable of bringing the undergraduate or PhD student to the level of approaching modern literature and enabling him/her to model or solve new problems with physically justified approximations. If we are successful in this endeavor it will be due to the enthusiastic and motivated students in Rome and Jyväskylä to whom we had the privilege to teach part of this book. We thank them for their feedback from which we indeed benefited enormously.

Speaking of thanks: our first and biggest thank you goes to Carl-Olof Almbladh and Ulf von Barth who introduced us to the wonderful world of many-body perturbation theory and Green's function theory during our post-doc years in Lund. Only now that we have been forced to deepen our understanding in order to explain these methods can we fully appreciate all their "of-course-I-don't-need-to-tell-you" or "you-probably-already-know" answers to our questions. We are also thankful to Evert Jan Baerends, Michele Cini, and Hardy Gross from whom we learned a large part of what today is our background in physics and chemistry and with whom we undertook many exciting research projects. We wish to express our gratitude to our PhD students, post-docs and local colleagues Klaas Giesbertz, Petri Myöhänen, Enrico Perfetto, Michael Ruggenthaler, Niko Säkkinen, Adrian Stan, Riku Tuovinen, and Anna-Maija Uimonen, for providing us with many valuable suggestions and for helping out in generating several figures. The research on the Kadanoff–Baym equations and their implementation which forms the last chapter of the book would not have been possible without the enthusiasm and the excellent numerical work of Nils Erik Dahlen. We are indebted to Heiko Appel, Karsten Balzer, Michael Bonitz, Raffaele Filosofi, Ari Harju, Maria Hellgren, Stefan Kurth, Matti Manninen, Kristian Thygesen, and Claudio Verdozzi with whom we had many inspiring and insightful discussions which either directly or indirectly influenced part of the contents of the book. We further thank the Department of Physics and the Nanoscience Center of the University of Jyväskylä and the Department of Physics of the University of Rome Tor Vergata for creating a very pleasant and supportive environment for the writing of the book. Finally we would like to thank a large number of people, too numerous to mention, in the research community who have shaped our view on many scientific topics in and outside of many-body theory.

Abbreviations and acronyms

a.u. : atomic units

BvK : Born-von Karman

e.g. : exempli gratia

HOMO : highest occupied molecular orbital

i.e. : id est

KMS : Kubo–Martin–Schwinger

l.h.s. : left hand side

LUMO : lowest unoccupied molecular orbital

LW : Luttinger–Ward

MBPT : Many-body perturbation theory

PPP : Pariser–Parr–Pople

QMC : Quantum Monte Carlo

r.h.s. : right hand side

RPA : Random Phase Approximation

WBLA : Wide Band Limit Approximation

XC : Exchange-Correlation

Fundamental constants and basic relations

Fundamental constants

Electron charge: $e = -1$ a.u. $= 1.60217646 \times 10^{-19}$ Coulomb

Electron mass: $m_e = 1$ a.u. $= 9.10938188 \times 10^{-31}$ kg

Planck constant: $\hbar = 1$ a.u. $= 1.054571 \times 10^{-34}$ Js $= 6.58211 \times 10^{-16}$ eVs

Speed of light: $c = 137$ a.u. $= 3 \times 10^5$ km/s

Boltzmann constant: $K_{\mathrm{B}} = 8.3 \times 10^{-5}$ eV/K

Basic quantities and relations

Bohr radius: $a_{\mathrm{B}} = \frac{\hbar^2}{m_e e^2} = 1$ a.u. $= 0.5$ Å

Electron gas density: $n = \frac{(\hbar p_{\mathrm{F}})^3}{3\pi^2} = $ (p_{F} being the Fermi momentum)

Electron gas radius: $\frac{1}{n} = \frac{4\pi}{3}(a_{\mathrm{B}} r_s)^3$, $r_s = \frac{(9\pi/4)^{1/3}}{\hbar a_{\mathrm{B}} p_{\mathrm{F}}}$

Plasma frequency: $\omega_{\mathrm{p}} = \sqrt{\frac{4\pi e^2 n}{m_e}}$ (n being the electron gas density)

Rydberg $R = \frac{e^2}{2a_{\mathrm{B}}} = 0.5$ a.u. $\simeq 13.6$ eV

Bohr magneton $\mu_{\mathrm{B}} = \frac{e\hbar}{2m_e c} = 3.649 \times 10^{-3}$ a.u. $= 5.788 \times 10^{-5}$ eV/T

Room temperature ($T \sim 300$ K) energy: $K_B T \sim \frac{1}{40}$ eV

$\hbar c \sim 197$ MeV fm (1 fm $= 10^{-15}$ m)

$m_e c^2 = 0.5447$ MeV

Fundamental constants and basic relations

Fundamental constants

Electron charge: $e = -1$ a.u. $= 1.60217646 \times 10^{-19}$ Coulomb

Electron mass: $m_e = 1$ a.u. $= 9.10938188 \times 10^{-31}$ kg

Planck constant: $\hbar = 1$ a.u. $= 1.054571 \times 10^{-34}$ Js $= 6.58211 \times 10^{-16}$ eVs

Speed of light: $c = 137$ a.u. $= 3 \times 10^5$ km/s

Boltzmann constant: $K_B = 8.3 \times 10^{-5}$ eV/K

Basic quantities and relations

Bohr radius: $a_B = \frac{\hbar^2}{m_e e^2} = 1$ a.u. $= 0.5$ Å

Electron gas density: $n = \frac{(\hbar p_F)^3}{3\pi^2} = $ (p_F being the Fermi momentum)

Electron gas radius: $\frac{1}{n} = \frac{4\pi}{3}(a_B r_s)^3$, $r_s = \frac{(9\pi/4)^{1/3}}{\hbar a_B p_F}$

Plasma frequency: $\omega_p = \sqrt{\frac{4\pi e^2 n}{m_e}}$ (n being the electron gas density)

Rydberg $R = \frac{e^2}{2a_B} = 0.5$ a.u. $\simeq 13.6$ eV

Bohr magneton $\mu_B = \frac{e\hbar}{2m_e c} = 3.649 \times 10^{-3}$ a.u. $= 5.788 \times 10^{-5}$ eV/T

Room temperature ($T \sim 300$ K) energy: $K_B T \sim \frac{1}{40}$ eV

$\hbar c \sim 197$ MeV fm (1 fm $= 10^{-15}$ m)

$m_e c^2 = 0.5447$ MeV

1

Second quantization

1.1 Quantum mechanics of one particle

In quantum mechanics the physical state of a particle is described in terms of a *ket* $|\Psi\rangle$. This ket belongs to a *Hilbert space* which is nothing but a vector space endowed with an inner product. The dimension of the Hilbert space is essentially fixed by our physical intuition; it is we who decide which kets are relevant for the description of the particle. For instance, if we want to describe how a laser works we can choose those energy eigenkets that get populated and depopulated and discard the rest. This selection of states leads to the well-known description of a laser in terms of a three-level system, four-level system, etc. A fundamental property following from the vector nature of the Hilbert space is that any linear superposition of kets is another ket in the Hilbert space. In other words we can make a linear superposition of physical states and the result is another physical state. In quantum mechanics, however, it is only the "direction" of the ket that matters, so $|\Psi\rangle$ and $C|\Psi\rangle$ represent the same physical state for all complex numbers C. This redundancy prompts us to work with *normalized* kets. What do we mean by that? We said before that there is an inner product in the Hilbert space. Let us denote by $\langle\Phi|\Psi\rangle = \langle\Psi|\Phi\rangle^*$ the inner product between two kets $|\Psi\rangle$ and $|\Phi\rangle$ of the Hilbert space. Then every ket has a real positive inner product with itself

$$0 < \langle\Psi|\Psi\rangle < \infty.$$

A ket is said to be normalized if the inner product with itself is 1. Throughout this book we always assume that a ket is normalized unless otherwise stated. Every ket can be normalized by choosing the complex constant $C = e^{i\alpha}/\sqrt{\langle\Psi|\Psi\rangle}$ with α an arbitrary real number. Thus, the normalization fixes the ket of a physical state only modulo a phase factor. As we see in Section 1.3, this freedom is at the basis of a fundamental property about the nature of elementary particles. The notion of inner product also allows us to define the *dual space* as the vector space of linear operators $\langle\Phi|$ which deliver the complex number $\langle\Phi|\Psi\rangle$ when acting on the ket $|\Psi\rangle$. The elements of the dual space are called *bra* and we can think of the inner product as the action of a bra on a ket. The formulation of quantum mechanics in terms of bras and kets is due to Dirac [1,2] and turns out to be extremely useful.

According to the basic principles of quantum mechanics [2]:

- With every physical observable is associated a Hermitian operator whose eigenvalues λ represent the outcome of an experimental measurement of the observable.

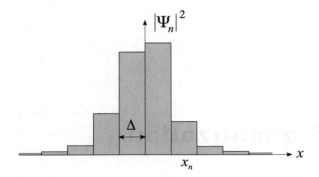

Figure 1.1 Histogram of the normalized number of clicks of the detector in $x_n = n\Delta$. The height of the function corresponds to the probability $|\Psi_n|^2$.

- If the particle is described by the ket $|\Psi\rangle$ then the probability of measuring λ is given by
$$P(\lambda) = |\langle\lambda|\Psi\rangle|^2,$$
  where $|\lambda\rangle$ is the eigenket of the operator with eigenvalue λ.

- The experimental measurement is so invasive that just after measurement the particle *collapses* in the ket $|\lambda\rangle$.

Let us discuss the implications of these principles with an example. Suppose that we want to measure the position of a particle living in a one-dimensional world. Then we can construct a detector with the property that it clicks whenever the particle is no further away than, say, $\Delta/2$ from the position of the detector. We distribute these detectors on a uniform grid $x_n = n\Delta$, with n integers, so as to cover the entire one-dimensional world. The experiment consists in preparing the particle in a state $|\Psi\rangle$ and in taking note of which detector clicks. After the click we know for sure that the particle is in the interval $x_n \pm \Delta/2$, where x_n is the position of the detector that clicked. Repeating the experiment $N \gg 1$ times, counting the number of times that a given detector clicks and dividing the result by N we obtain the probability that the particle is in the interval $x_n \pm \Delta/2$, see histogram of Fig. 1.1. Quantum mechanics tells us that this probability is

$$P(n) = |\langle n|\Psi\rangle|^2,$$

where $|n\rangle$ is the ket describing the particle in the interval $x_n \pm \Delta/2$. The experimental setup does not allow us to say where exactly the particle is within this interval. In fact, it does not make sense to speak about the exact position of the particle since it cannot be measured. From the experimental output we could even argue that the one-dimensional world is discrete! What we want to say is that in our experiment the "exact position" of the particle is a mere speculative concept, like the gender, color or happiness of the particle. These degrees of freedom may also exist but if they cannot be measured then we should not include them in the description of the physical world. As scientists we can only assign a ket $|n\rangle$ to the state of the particle just after measurement, and we can interpret this ket as describing the particle in some discrete position. The probability of finding the particle in $|n'\rangle$ just after the nth detector has clicked is zero for all $n' \neq n$ and unity for $n' = n$, and hence

$$\langle n'|n\rangle = \delta_{n'n}. \tag{1.1}$$

The kets $|n\rangle$ are orthonormal and it is easy to show that they form a basis of our Hilbert space. Suppose by *reductio ad absurdum* that there exists another ket $|\chi\rangle$ orthogonal to all the $|n\rangle$. If the particle is described by this ket then the probability that the nth detector clicks is $|\langle n|\chi\rangle|^2 = 0$ for all n. This cannot be unless the particle is somewhere else outside the one-dimensional world, i.e., in a state not included in our original description.

Let us continue to elaborate on the example of the particle in a one-dimensional world. We said before that the kets $|n\rangle$ form a basis. Therefore any ket $|\Psi\rangle$ can be expanded in terms of them

$$|\Psi\rangle = \sum_n \Psi_n |n\rangle. \tag{1.2}$$

Since the basis is orthonormal the coefficient Ψ_n is simply

$$\Psi_n = \langle n|\Psi\rangle, \tag{1.3}$$

and its square modulus is exactly the probability $P(n)$

$$|\Psi_n|^2 = \left(\begin{array}{c} \text{probability of finding the particle in} \\ \text{volume element } \Delta \text{ around } x_n \end{array} \right).$$

It is important to appreciate the advantage of working with normalized kets. Since $\langle \Psi|\Psi\rangle = 1$ then

$$\sum_n |\Psi_n|^2 = 1, \tag{1.4}$$

according to which the probability of finding the particle anywhere is unity. The interpretation of the $|\Psi_n|^2$ as probabilities would not be possible if $|\Psi\rangle$ and $|n\rangle$ were not normalized.

> *Given an orthonormal basis the inner product of a normalized ket $|\Psi\rangle$ with a basis ket gives the probability amplitude of having the particle in that ket.*

Inserting (1.3) back into (1.2) we find the interesting relation

$$|\Psi\rangle = \sum_n \langle n|\Psi\rangle \, |n\rangle = \sum_n |n\rangle\langle n|\Psi\rangle.$$

This relation is interesting because it is true for all $|\Psi\rangle$ and hence

$$\sum_n |n\rangle\langle n| = \hat{\mathbb{1}}, \tag{1.5}$$

with $\hat{1}$ the identity operator. Equation (1.5) is known as the *completeness relation* and expresses the fact that the set $\{|n\rangle\}$ is an orthonormal basis. Vice versa, any orthonormal basis satisfies the completeness relation.

We now assume that we can construct more and more precise detectors and hence reduce the range Δ. Then we can also refine the description of our particle by putting the detectors closer and closer. In the limit $\Delta \to 0$ the probability $|\Psi_n|^2$ approaches zero and it makes more sense to reason in terms of the *probability density* $|\Psi_n|^2/\Delta$ of finding the particle in x_n. Let us rewrite (1.2) as

$$|\Psi\rangle = \Delta \sum_n \frac{\Psi_n}{\sqrt{\Delta}} \frac{|n\rangle}{\sqrt{\Delta}}. \tag{1.6}$$

We now define the continuous function $\Psi(x_n)$ and the continuous ket $|x_n\rangle$ as

$$\Psi(x_n) \equiv \lim_{\Delta \to 0} \frac{\Psi_n}{\sqrt{\Delta}}, \qquad |x_n\rangle = \lim_{\Delta \to 0} \frac{|n\rangle}{\sqrt{\Delta}}.$$

In this definition the limiting function $\Psi(x_n)$ is well defined while the limiting ket $|x_n\rangle$ makes *mathematical sense* only under an integral sign since the norm $\langle x_n|x_n\rangle = \infty$. However, we can still give to $|x_n\rangle$ a precise *physical meaning* since in quantum mechanics only the "direction" of a ket matters.[1] With these definitions (1.6) can be seen as the Riemann sum of $\Psi(x_n)|x_n\rangle$. In the limit $\Delta \to 0$ the sum becomes an integral over x and we can write

$$|\Psi\rangle = \int dx \, \Psi(x)|x\rangle.$$

The function $\Psi(x)$ is usually called the *wavefunction* or the *probability amplitude* and its square modulus $|\Psi(x)|^2$ is the probability density of finding the particle in x, or equivalently

$$|\Psi(x)|^2 \, dx \; = \; \left(\begin{array}{c} \text{probability of finding the particle} \\ \text{in volume element } dx \text{ around } x \end{array} \right).$$

In the continuum formulation the orthonormality relation (1.1) becomes

$$\langle x_{n'}|x_n\rangle = \lim_{\Delta \to 0} \frac{\delta_{n'n}}{\Delta} = \delta(x_{n'} - x_n),$$

where $\delta(x)$ is the Dirac δ-function, see Appendix A. Similarly the completeness relation becomes

$$\int dx \, |x\rangle\langle x| = \hat{1}.$$

The entire discussion can now easily be generalized to particles with spin in three (or any other) dimensions. Let us denote by $\mathbf{x} = (\mathbf{r}\sigma)$ the collective index for the position $\mathbf{r}$ and

---

[1]The formulation of quantum mechanics using non-normalizable states requires the extension of Hilbert spaces to *rigged Hilbert spaces*. Readers interested in the mathematical foundations of this extension can consult, e.g., Ref. [3]. Here we simply note that in a rigged Hilbert space everything works just as in the more familiar Hilbert space. We simply have to keep in mind that every divergent quantity comes from some continuous limit and that in all physical quantities the divergency is cancelled by an infinitesimally small quantity.

the spin projection (say along the z axis) σ of the particle. If in every point of space we put a spin-polarized detector which clicks only if the particle has spin σ then $|\mathbf{x}\rangle$ is the state of the particle just after the spin-polarized detector in $\mathbf{r}$ has clicked. The *position-spin kets* $|\mathbf{x}\rangle$ are orthonormal

$$\langle\mathbf{x}'|\mathbf{x}\rangle = \delta_{\sigma'\sigma}\delta(\mathbf{r}' - \mathbf{r}) \equiv \delta(\mathbf{x}' - \mathbf{x}), \qquad (1.7)$$

and form a basis. Hence they satisfy the completeness relation which in this case reads

$$\boxed{\int d\mathbf{x}\, |\mathbf{x}\rangle\langle\mathbf{x}| = \hat{\mathbb{1}}} \qquad (1.8)$$

Here and in the remainder of the book we use the symbol

$$\int d\mathbf{x} \equiv \sum_\sigma \int d\mathbf{r}$$

to signify a sum over spin and an integral over space. The expansion of a ket in this continuous Hilbert space follows directly from the completeness relation

$$|\Psi\rangle = \hat{\mathbb{1}}|\Psi\rangle = \int d\mathbf{x}\, |\mathbf{x}\rangle\langle\mathbf{x}|\Psi\rangle,$$

and the square modulus of the wavefunction $\Psi(\mathbf{x}) \equiv \langle\mathbf{x}|\Psi\rangle$ is the probability density of finding the particle in $\mathbf{x} = (\mathbf{r}\sigma)$,

$$|\Psi(\mathbf{x})|^2\, d\mathbf{r} = \left(\begin{array}{c} \text{probability of finding the particle with spin } \sigma \\ \text{in volume element } d\mathbf{r} \text{ around } \mathbf{r} \end{array} \right).$$

So far we have only discussed the possible states of the particle and the physical interpretation of the expansion coefficients. To say something about the dynamics of the particle we must know the Hamiltonian operator $\hat{h}$. A knowledge of the Hamiltonian in quantum mechanics is analogous to a knowledge of the forces in Newtonian mechanics. In Newtonian mechanics the dynamics of the particle is completely determined by the position and velocity at a certain time and by the forces. In quantum mechanics the dynamics of the wavefunction is completely determined by the wavefunction at a certain time and by $\hat{h}$. The Hamiltonian operator $\hat{h} \equiv h(\hat{\mathbf{r}}, \hat{\mathbf{p}}, \hat{\mathbf{S}})$ will, in general, depend on the position operator $\hat{\mathbf{r}}$, the momentum operator $\hat{\mathbf{p}}$ and the spin operator $\hat{\mathbf{S}}$. An example is the Hamiltonian for a particle of mass m, charge q, and gyromagnetic ratio g moving in an external scalar potential V, vector potential $\mathbf{A}$ and whose spin is coupled to the magnetic field $\mathbf{B} = \nabla \times \mathbf{A}$,

$$\hat{h} = \frac{1}{2m}\left(\hat{\mathbf{p}} - \frac{q}{c}\mathbf{A}(\hat{\mathbf{r}})\right)^2 + qV(\hat{\mathbf{r}}) - g\mu_{\mathrm{B}}\mathbf{B}(\hat{\mathbf{r}}) \cdot \hat{\mathbf{S}}, \qquad (1.9)$$

with c the speed of light and μ_{B} the Bohr magneton. Unless otherwise stated, in this book we use atomic units so that $\hbar = 1$, $c \sim 1/137$, the electron charge $e = -1$ and the electron mass $m_e = 1$. Thus in (1.9) the Bohr magneton $\mu_{\mathrm{B}} = \frac{e\hbar}{2m_e c} \sim 3.649 \times 10^{-3}$, and the charge and mass of the particles are measured in units of e and m_e respectively. To distinguish operators from scalar or matrix quantities we always put the symbol " ˆ " (read "hat") on them. The position–spin kets are eigenstates of the position operator and of the z-component of the spin operator

$$\hat{r}|\mathbf{x}\rangle = \mathbf{r}|\mathbf{x}\rangle, \qquad \hat{S}_z|\mathbf{x}\rangle = \sigma|\mathbf{x}\rangle,$$

with $\sigma = -S, -S+1, \ldots, S-1, S$ for spin S particles. The eigenstates of the momentum operator are instead the *momentum–spin kets* $|\mathbf{p}\sigma\rangle$

$$\hat{p}|\mathbf{p}\sigma\rangle = \mathbf{p}|\mathbf{p}\sigma\rangle.$$

These kets are also eigenstates of $\hat{S}_z$ with eigenvalue σ. The momentum–spin kets form an orthonormal basis like the position–spin kets. The inner product between $|\mathbf{x}\rangle = |\mathbf{r}\sigma\rangle$ and $|\mathbf{p}\sigma'\rangle$ is proportional to $\delta_{\sigma\sigma'}$ times the plane wave $e^{i\mathbf{p}\cdot\mathbf{r}}$. In this book we choose the constant of proportionality to be unity so that

$$\boxed{\langle\mathbf{x}|\mathbf{p}\sigma'\rangle = \delta_{\sigma\sigma'}\langle\mathbf{r}|\mathbf{p}\rangle \qquad \text{with} \qquad \langle\mathbf{r}|\mathbf{p}\rangle = e^{i\mathbf{p}\cdot\mathbf{r}}} \tag{1.10}$$

This inner product fixes uniquely the form of the completeness relation for the kets $|\mathbf{p}\sigma\rangle$. We have

$$\langle\mathbf{p}'\sigma'|\mathbf{p}\sigma\rangle = \delta_{\sigma'\sigma}\langle\mathbf{p}'|\mathbf{p}\rangle = \delta_{\sigma'\sigma}\int d\mathbf{r}\,\langle\mathbf{p}'|\mathbf{r}\rangle\langle\mathbf{r}|\mathbf{p}\rangle = \delta_{\sigma'\sigma}\int d\mathbf{r}\, e^{i(\mathbf{p}-\mathbf{p}')\cdot\mathbf{r}}$$

$$= (2\pi)^3\delta_{\sigma'\sigma}\delta(\mathbf{p}'-\mathbf{p}),$$

and therefore

$$\boxed{\sum_\sigma \int \frac{d\mathbf{p}}{(2\pi)^3}|\mathbf{p}\sigma\rangle\langle\mathbf{p}\sigma| = \hat{\mathbb{1}}} \tag{1.11}$$

as can easily be verified by acting with (1.11) on the ket $|\mathbf{p}'\sigma'\rangle$ or on the bra $\langle\mathbf{p}'\sigma'|$.

Before moving to the quantum mechanical description of many particles let us briefly recall how to calculate the matrix elements of the Hamiltonian $\hat{h}$ in the position–spin basis. If $|\Psi\rangle$ is the ket of the particle then

$$\langle\mathbf{x}|\hat{p}|\Psi\rangle = -i\boldsymbol{\nabla}\langle\mathbf{x}|\Psi\rangle \quad \Rightarrow \quad \langle\Psi|\hat{p}|\mathbf{x}\rangle = i\langle\Psi|\mathbf{x}\rangle\overleftarrow{\boldsymbol{\nabla}},$$

where the arrow over the gradient specifies that $\boldsymbol{\nabla}$ acts on the quantity to its left. It follows from these identities that

$$\langle\mathbf{x}|\hat{p}|\mathbf{x}'\rangle = -i\delta_{\sigma\sigma'}\boldsymbol{\nabla}\delta(\mathbf{r}-\mathbf{r}') = i\delta_{\sigma\sigma'}\delta(\mathbf{r}-\mathbf{r}')\overleftarrow{\boldsymbol{\nabla}}', \tag{1.12}$$

where $\boldsymbol{\nabla}'$ means that the gradient acts on the primed variable. Therefore, the matrix element $\langle\mathbf{x}|\hat{h}|\mathbf{x}'\rangle$ with $\hat{h} = h(\hat{r}, \hat{p}, \hat{\boldsymbol{S}})$ can be written as

$$\boxed{\langle\mathbf{x}|\hat{h}|\mathbf{x}'\rangle = h_{\sigma\sigma'}(\mathbf{r}, -i\boldsymbol{\nabla}, \mathbf{S})\delta(\mathbf{r}-\mathbf{r}') = \delta(\mathbf{r}-\mathbf{r}')h_{\sigma\sigma'}(\mathbf{r}', i\overleftarrow{\boldsymbol{\nabla}}', \mathbf{S})} \tag{1.13}$$

where $\mathbf{S}$ is the matrix of the spin operator with elements $\langle\sigma|\hat{\boldsymbol{S}}|\sigma'\rangle = \mathbf{S}_{\sigma\sigma'}$. For example, for the one-particle Hamiltonian in (1.9) we have

$$h_{\sigma\sigma'}(\mathbf{r}, -i\boldsymbol{\nabla}, \mathbf{S}) = \frac{\delta_{\sigma\sigma'}}{2m}\left(-i\boldsymbol{\nabla} - \frac{q}{c}\mathbf{A}(\mathbf{r})\right)^2 + \delta_{\sigma\sigma'}qV(\mathbf{r}) - g\mu_\mathrm{B}\mathbf{B}(\mathbf{r})\cdot\mathbf{S}_{\sigma\sigma'}.$$

We use (1.13) over and over in the following chapters to recognize the matrix structure of several equations.

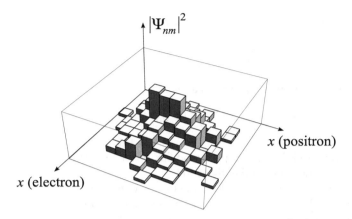

Figure 1.2 Histogram of the normalized number of simultaneous clicks of the electron and positron detectors in $x_n = n\Delta$ and $x_m = m\Delta$ respectively. The height of the function corresponds to the probability $|\Psi_{nm}|^2$.

1.2 Quantum mechanics of many particles

We want to generalize the concepts of the previous section to many particles. Let us first discuss the case of *distinguishable particles*. Particles are called distinguishable if they differ in one or more of their properties, like mass, charge, spin, etc. Let us consider, for instance, an electron and a positron in one dimension. These particles are distinguishable since the charge of the positron is opposite to the charge of the electron. To measure the position of the electron and the position of the positron at a certain time we put an electron-detector and a positron-detector in every point $x_n = n\Delta$ of the real axis and perform a *coincidence experiment*. This means that we take note of the position of the electron-detector and of the positron-detector only if they click *at the same time*. The result of the experiment is the pair of points (x_n, x_m) where the first entry x_n refers to the electron whereas the second entry x_m refers to the positron. Performing the experiment $N \gg 1$ times, counting the number of times that the pair (x_n, x_m) is measured and dividing the result by N we obtain the probability that the electron is in x_n and the positron in x_m, see the histogram of Fig. 1.2. According to quantum mechanics the electron–positron pair collapses in the ket $|n\rangle|m\rangle$ just after measurement. This ket describes an electron in the interval $x_n \pm \Delta/2$ and a positron in the interval $x_m \pm \Delta/2$. Therefore the probability of finding the electron–positron pair in $|n'\rangle|m'\rangle$ is zero unless $n' = n$ and $m' = m$, i.e.

$$(\langle n'|\langle m'|) (|n\rangle|m\rangle) = \delta_{n'n}\delta_{m'm}.$$

The kets $|n\rangle|m\rangle$ are orthonormal and form a basis since if there was a ket $|\chi\rangle$ orthogonal to all of them then the electron–positron pair described by $|\chi\rangle$ would not be on the real axis. The orthonormality of the basis is expressed by the completeness relation

$$\sum_{nm} (\, |n\rangle|m\rangle \,) (\, \langle n|\langle m| \,) = \hat{\mathbb{1}}.$$

This relation can be used to expand any ket as

$$|\Psi\rangle = \hat{\mathbb{1}}|\Psi\rangle = \sum_{nm} (\, |n\rangle|m\rangle \,) (\, \langle n|\langle m| \,) \, |\Psi\rangle,$$

and if $|\Psi\rangle$ is normalized then the square modulus of the coefficients $\Psi_{nm} \equiv (\, \langle n|\langle m| \,) \, |\Psi\rangle$ is the probability represented in the histogram.

As in the previous section, we could refine the experiment by putting the detectors closer and closer. We could also rethink the entire experiment in three (or any other) dimensions and use spin-polarized detectors. We then arrive at the position–spin kets $|\mathbf{x}_1\rangle|\mathbf{x}_2\rangle$ for the electron–positron pair with inner product

$$(\, \langle \mathbf{x}_1'|\langle \mathbf{x}_2'| \,) (\, |\mathbf{x}_1\rangle|\mathbf{x}_2\rangle \,) = \delta(\mathbf{x}_1' - \mathbf{x}_1)\delta(\mathbf{x}_2' - \mathbf{x}_2),$$

from which we deduce the completeness relation

$$\int d\mathbf{x}_1 d\mathbf{x}_2 \,(\, |\mathbf{x}_1\rangle|\mathbf{x}_2\rangle \,) (\, \langle \mathbf{x}_1|\langle \mathbf{x}_2| \,) = \hat{\mathbb{1}}.$$

The expansion of a generic ket is

$$|\Psi\rangle = \int d\mathbf{x}_1 d\mathbf{x}_2 \,(\, |\mathbf{x}_1\rangle|\mathbf{x}_2\rangle \,) (\, \langle \mathbf{x}_1|\langle \mathbf{x}_2| \,) \, |\Psi\rangle,$$

and if $|\Psi\rangle$ is normalized then the square modulus of the wavefunction $\Psi(\mathbf{x}_1, \mathbf{x}_2) \equiv (\, \langle \mathbf{x}_1|\langle \mathbf{x}_2| \,) \, |\Psi\rangle$ yields the probability density of finding the electron in $\mathbf{x}_1 = (\mathbf{r}_1\sigma_1)$ and the positron in $\mathbf{x}_2 = (\mathbf{r}_2\sigma_2)$:

$$|\Psi(\mathbf{x}_1, \mathbf{x}_2)|^2 \, d\mathbf{r}_1 d\mathbf{r}_2 \; = \; \begin{pmatrix} \text{probability of finding the electron with spin } \sigma_1 \\ \text{in volume element } d\mathbf{r}_1 \text{ around } \mathbf{r}_1 \text{ and the positron} \\ \text{with spin } \sigma_2 \text{ in volume element } d\mathbf{r}_2 \text{ around } \mathbf{r}_2 \end{pmatrix}.$$

The generalization to N distinguishable particles is now straightforward. The position–spin ket $|\mathbf{x}_1\rangle \ldots |\mathbf{x}_N\rangle$ describes the physical state in which the first particle is in $\mathbf{x}_1$, the second particle is in $|\mathbf{x}_2\rangle$ etc. These kets form an orthonormal basis with inner product

$$(\, \langle \mathbf{x}_1'| \ldots \langle \mathbf{x}_N'| \,) (\, |\mathbf{x}_1\rangle \ldots |\mathbf{x}_N\rangle \,) = \delta(\mathbf{x}_1' - \mathbf{x}_1) \ldots \delta(\mathbf{x}_N' - \mathbf{x}_N), \qquad (1.14)$$

and therefore the completeness relation reads

$$\int d\mathbf{x}_1 \ldots d\mathbf{x}_N \,(\, |\mathbf{x}_1\rangle \ldots |\mathbf{x}_N\rangle \,) (\, \langle \mathbf{x}_1| \ldots \langle \mathbf{x}_N| \,) = \hat{\mathbb{1}}.$$

Having discussed the Hilbert space for N distinguishable particles we now consider the operators acting on the N-particle kets. We start with an example and consider again the electron–positron pair. Suppose that we are interested in measuring the center-of-mass position of the pair. The center-of-mass position is an observable quantity and hence, associated with it, there exists an operator $\hat{\mathcal{R}}_{\mathrm{CM}}$. By definition the eigenstates of this operator are the position–spin kets $|\mathbf{x}_1\rangle|\mathbf{x}_2\rangle$ and the corresponding eigenvalues are $(\mathbf{r}_1 + \mathbf{r}_2)/2$, independent of the spin of the particles. The operator $\hat{\mathcal{R}}_{\mathrm{CM}}$ is then the sum of the position operator acting on the first particle and doing nothing to the second particle and the position operator acting on the second particle and doing nothing to the first particle, i.e.,

$$\hat{\mathcal{R}}_{\mathrm{CM}} = \frac{1}{2}\left(\hat{\boldsymbol{r}} \otimes \hat{\mathbb{1}} + \hat{\mathbb{1}} \otimes \hat{\boldsymbol{r}}\right). \tag{1.15}$$

The symbol "$\otimes$" denotes the *tensor product* of operators acting on different particles. For instance

$$\hat{\mathcal{R}}_{\mathrm{CM}}|\mathbf{x}_1\rangle|\mathbf{x}_2\rangle = \frac{1}{2}\left(\hat{\boldsymbol{r}}|\mathbf{x}_1\rangle\hat{\mathbb{1}}|\mathbf{x}_2\rangle + \hat{\mathbb{1}}|\mathbf{x}_1\rangle\hat{\boldsymbol{r}}|\mathbf{x}_2\rangle\right) = \frac{1}{2}(\mathbf{r}_1 + \mathbf{r}_2)|\mathbf{x}_1\rangle|\mathbf{x}_2\rangle.$$

The generalization of the center-of-mass operator to N particles is rather voluminous,

$$\hat{\mathcal{R}}_{\mathrm{CM}} = \frac{1}{N}\left(\hat{\boldsymbol{r}} \otimes \underbrace{\hat{\mathbb{1}} \otimes \ldots \otimes \hat{\mathbb{1}}}_{N-1 \text{ times}} + \hat{\mathbb{1}} \otimes \hat{\boldsymbol{r}} \otimes \underbrace{\ldots \otimes \hat{\mathbb{1}}}_{N-2 \text{ times}} + \ldots + \underbrace{\hat{\mathbb{1}} \otimes \hat{\mathbb{1}} \otimes \ldots \otimes \hat{\boldsymbol{r}}}_{N-1 \text{ times}}\right), \tag{1.16}$$

and it is typically shortened as

$$\hat{\mathcal{R}}_{\mathrm{CM}} = \frac{1}{N}\sum_{j=1}^{N}\hat{\boldsymbol{r}}_j,$$

where $\hat{\boldsymbol{r}}_j$ is the position operator acting on the jth particle and doing nothing to the other particles. Similarly the noninteracting part of the Hamiltonian of N particles is typically written as

$$\hat{\mathcal{H}}_0 = \sum_{j=1}^{N}\hat{h}_j = \sum_{j=1}^{N}h(\hat{\boldsymbol{r}}_j, \hat{\boldsymbol{p}}_j, \hat{\boldsymbol{S}}_j), \tag{1.17}$$

while the interaction part is written as

$$\hat{\mathcal{H}}_{\mathrm{int}} = \frac{1}{2}\sum_{i \neq j}^{N}v(\hat{\boldsymbol{r}}_i, \hat{\boldsymbol{r}}_j), \tag{1.18}$$

with $v(\mathbf{r}_1, \mathbf{r}_2)$ the interparticle interaction. We observe that these operators depend explicitly on the number of particles and are therefore difficult to manipulate in problems where the number of particles can fluctuate, as in systems at finite temperature. As we see later in this chapter, another disadvantage is that the evaluation of their action on kets describing *identical* particles is very lengthy. Fortunately, an incredible simplification occurs for identical particles and the expressions for operators and kets become much lighter and easier to manipulate. To appreciate this simplification, however, we first have to understand how the quantum-mechanical formulation changes when the particles are identical.

1.3   Quantum mechanics of many identical particles

Two particles are called *identical particles* or *indistinguishable particles* if they have the same internal properties, i.e., the same mass, charge, spin etc. For example two electrons are two identical particles. To understand the qualitative difference between distinguishable and identical particles let us perform the coincidence experiment of the previous section for two electrons both with spin projection $1/2$ and again in one dimension. At every point $x_n = n\Delta$ we put a spin-polarized electron-detector, and since the particles are identical we need only one kind of detector. If the detectors in x_n and x_m click at the same time then we can be sure that just after this time there is one electron around x_n and another electron around x_m. Let us denote by $|nm\rangle$ the ket describing the physical state in which the two electrons collapse after measurement.[2] As the electrons are identical the natural question to ask is: do the kets $|nm\rangle$ and $|mn\rangle$ correspond to two different physical states? If the answer were positive then we should be able to hear a difference in the clicks corresponding to $|nm\rangle$ and $|mn\rangle$. For example in the case of the electron–positron pair we could make the positron-click louder than the electron-click and hence distinguish the state $|n\rangle|m\rangle$ from the state $|m\rangle|n\rangle$. However, in this case we only have electron-detectors and it is impossible to distinguish which electron has made a given detector click. We therefore must assign to $|mn\rangle$ the *same physical state* as to $|nm\rangle$. We would like to emphasize that the kets $|nm\rangle$ are not given by nature. It is we who decide to represent nature in terms of them. For our representation of nature to make sense we *must* impose that $|nm\rangle$ and $|mn\rangle$ correspond to the same physical state. In Section 1.1 we observed that the normalized ket of a physical state is uniquely defined up to a phase factor and hence

$$|nm\rangle = e^{i\alpha}|mn\rangle \qquad \text{for all } n, m.$$

Using the above relation twice we find that $e^{2i\alpha} = 1$, or equivalently $e^{i\alpha} = \pm 1$. Consequently the ket

$$|nm\rangle = \pm|mn\rangle \tag{1.19}$$

is either symmetric or antisymmetric under the interchange of the electron positions. This is a fundamental property of nature: all particles can be grouped in two main classes. Particles described by a symmetric ket are called *bosons* while those described by an antisymmetric ket are called *fermions*. The electrons of our example are fermions. Here and in the rest of the book the upper sign always refers to bosons whereas the lower sign refers to fermions. In the case of fermions (1.19) implies $|nn\rangle = -|nn\rangle$ and hence $|nn\rangle$ must be the *null ket* $|\emptyset\rangle$, i.e., it is not possible to create two fermions in the same position and with the same spin. This peculiarity of fermions is known as the *Pauli exclusion principle*.

   If we now repeat the coincidence experiment $N \gg 1$ times, count the number of times that the detectors click simultaneously in x_n and x_m and divide the result by N we can draw the histograms of Fig. 1.3 for bosons and fermions. The probability is symmetric under the interchange $n \leftrightarrow m$ due to property (1.19). The fermions are easily recognizable since the probability of finding them in the same place is zero.

---

[2]Note the different notation with respect to the previous section where we used the ket $|n\rangle|m\rangle$ to describe the first particle around x_n and the second particle around x_m.

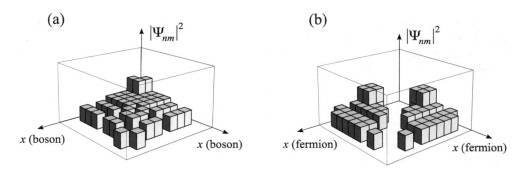

Figure 1.3 Histogram of the normalized number of simultaneous clicks of the detector in $x_n = n\Delta$ and in $x_m = m\Delta$ for (a) two bosons and (b) two fermions. The height of the function corresponds to the probability $|\Psi_{nm}|^2$.

In this book we learn how to deal with systems of many identical particles, like molecules or solids, and therefore we do not always repeat that the particles are identical. By particles we mean identical particles unless otherwise stated. Unlike the case of the electron–positron pair the probability of measuring a particle in $x_{n'}$ and the other in $x_{m'}$ just after the detectors in x_n and x_m have simultaneously clicked is zero unless $n = n'$ and $m = m'$ or $n = m'$ and $m = n'$, and hence

$$\langle n'm'|nm\rangle = c_1\delta_{n'n}\delta_{m'm} + c_2\delta_{m'n}\delta_{n'm}. \tag{1.20}$$

To fix the constants c_1 and c_2 we observe that

$$\langle n'm'|nm\rangle = \pm\langle n'm'|mn\rangle = \pm c_1\delta_{n'm}\delta_{m'n} \pm c_2\delta_{m'm}\delta_{n'n},$$

from which it follows that $c_1 = \pm c_2$. Furthermore, since the kets are normalized we must have for all $n \neq m$

$$1 = \langle nm|nm\rangle = c_1.$$

For $n = m$ the ket $|nn\rangle$ exists only for bosons and one finds $\langle nn|nn\rangle = 2c_1$. It is therefore more convenient to work with a non-normalized ket $|nn\rangle$ so that $c_1 = 1$ in all cases. We choose the normalization of the bosonic ket $|nn\rangle$ to be 2:

$$\langle nn|nn\rangle = 2. \tag{1.21}$$

Putting everything together we can rewrite the inner product (1.20) as

$$\langle n'm'|nm\rangle = \delta_{n'n}\delta_{m'm} \pm \delta_{m'n}\delta_{n'm}.$$

The inner product for the fermionic ket $|nn\rangle$ is automatically zero, in agreement with the fact that $|nn\rangle = |\emptyset\rangle$.

Let us now come to the completeness relation in the Hilbert space of two particles. Since $|nm\rangle = \pm|mn\rangle$ a basis in this space is given by the set $\{|nm\rangle\}$ with $n \geq m$. In other words the basis comprises only *inequivalent* configurations, meaning configurations not related by a permutation of the coordinates. The elements of this set are orthogonal and normalized except for the bosonic ket $|nn\rangle$ whose normalization is 2. Therefore, the completeness relation reads

$$\sum_{n>m} |nm\rangle\langle nm| + \frac{1}{2}\sum_{n} |nn\rangle\langle nn| = \hat{1},$$

where the second sum does not contribute in the fermionic case. We can rewrite the completeness relation as an unrestricted sum over all n and m using the (anti)symmetry property (1.19). The resulting expression is

$$\frac{1}{2}\sum_{nm} |nm\rangle\langle nm| = \hat{1},$$

which is much more elegant. The completeness relation can be used to expand any other ket in the same Hilbert space,

$$|\Psi\rangle = \hat{1}|\Psi\rangle = \frac{1}{2}\sum_{nm} |nm\rangle\langle nm|\Psi\rangle, \qquad (1.22)$$

and if $|\Psi\rangle$ is normalized then the square modulus of the coefficients of the expansion $\Psi_{nm} \equiv \langle nm|\Psi\rangle$ has the standard probabilistic interpretation

$$|\Psi_{nm}|^2 = \left(\begin{array}{c}\text{probability of finding one particle in volume}\\\text{element } \Delta \text{ around } x_n \text{ and the other particle}\\\text{in volume element } \Delta \text{ around } x_m\end{array}\right)$$

for all $n \neq m$. For $n = m$ we must remember that the normalization of the ket $|nn\rangle$ is 2 and therefore $|\Psi_{nn}|^2$ gives *twice* the probability of finding two particles in the same place (since the proper normalized ket is $|nn\rangle/\sqrt{2}$). Consequently

$$\frac{|\Psi_{nn}|^2}{2} = \left(\begin{array}{c}\text{probability of finding two particles in}\\\text{volume element } \Delta \text{ around } x_n\end{array}\right).$$

We can now refine the experiment by putting the detectors closer and closer. The continuum limit works in exactly the same manner as in the previous two sections. We rewrite the expansion (1.22) as

$$|\Psi\rangle = \frac{1}{2}\Delta^2 \sum_{nm} \frac{|nm\rangle}{\Delta}\frac{\Psi_{nm}}{\Delta}, \qquad (1.23)$$

and define the continuous wavefunction $\Psi(x_n, x_m)$ and the continuous ket $|x_n x_m\rangle$ according to

$$\Psi(x_n, x_m) = \lim_{\Delta\to 0} \frac{\Psi_{nm}}{\Delta}, \qquad |x_n x_m\rangle = \lim_{\Delta\to 0} \frac{|nm\rangle}{\Delta}.$$

The expansion (1.23) can then be seen as the Riemann sum of $\Psi(x_n, x_m)|x_n x_m\rangle$, and in the limit $\Delta \to 0$ the sum becomes the integral

$$|\Psi\rangle = \frac{1}{2} \int dx dx' \; \Psi(x, x')|xx'\rangle.$$

We can also derive the continuous representation of the completeness relation and the continuous representation of the inner product between two basis kets. We have

$$\lim_{\Delta \to 0} \frac{1}{2} \Delta^2 \sum_{nm} \frac{|nm\rangle}{\Delta} \frac{\langle nm|}{\Delta} = \frac{1}{2} \int dx dx' |xx'\rangle\langle xx'| = \hat{\mathbb{1}}, \tag{1.24}$$

and

$$\lim_{\Delta \to 0} \frac{\langle n'm'|nm\rangle}{\Delta^2} = \langle x_{n'} x_{m'}|x_n x_m\rangle$$
$$= \delta(x_{n'} - x_n)\delta(x_{m'} - x_m) \pm \delta(x_{m'} - x_n)\delta(x_{n'} - x_m). \tag{1.25}$$

The generalization to higher dimensions and to particles with different spin projections is now straightforward. We define the position–spin ket $|\mathbf{x}_1 \mathbf{x}_2\rangle$ as the ket of the physical state in which the particles collapse after the simultaneous clicking of a spin-polarized detector for particles of spin-projection σ_1 placed in $\mathbf{r}_1$ and a spin-polarized detector for particles of spin-projection σ_2 placed in $\mathbf{r}_2$. The set of inequivalent configurations $|\mathbf{x}_1 \mathbf{x}_2\rangle$ forms a basis of the Hilbert space of two identical particles. In the following we refer to this space as $\mathcal{H}_2$. In analogy with (1.25) the continuous kets have inner product

$$\langle \mathbf{x}_1' \mathbf{x}_2'|\mathbf{x}_1 \mathbf{x}_2\rangle = \delta(\mathbf{x}_1' - \mathbf{x}_1)\delta(\mathbf{x}_2' - \mathbf{x}_2) \pm \delta(\mathbf{x}_1' - \mathbf{x}_2)\delta(\mathbf{x}_2' - \mathbf{x}_1)$$

$$= \sum_P (\pm)^P \delta(\mathbf{x}_1' - \mathbf{x}_{P(1)})\delta(\mathbf{x}_2' - \mathbf{x}_{P(2)}), \tag{1.26}$$

where the upper/lower sign refers to bosons/fermions. The second line of this equation is an equivalent way of rewriting the (anti)symmetric product of δ-functions. The sum runs over the permutations P of $(1, 2)$ which are the identity permutation $(P(1), P(2)) = (1, 2)$ and the interchange $(P(1), P(2)) = (2, 1)$. The quantity $(\pm)^P$ is equal to $+1$ if the permutation requires an even number of interchanges and ± 1 if the permutation requires an odd number of interchanges. In the fermionic case all position–spin kets have the same norm since (1.26) implies

$$\langle \mathbf{x}_1 \mathbf{x}_2|\mathbf{x}_1 \mathbf{x}_2\rangle = \delta(0)^2 \qquad \text{for fermions.}$$

Due to the possibility in the bosonic case that two coordinates are identical the norms of the position–spin kets are instead not all the same since

$$\langle \mathbf{x}_1 \mathbf{x}_2|\mathbf{x}_1 \mathbf{x}_2\rangle = \delta(0)^2 \times \begin{cases} 1 & \text{if } \mathbf{x}_1 \neq \mathbf{x}_2 \\ 2 & \text{if } \mathbf{x}_1 = \mathbf{x}_2 \end{cases} \qquad \text{for bosons,}$$

in agreement with (1.21).

In complete analogy with (1.24) we can also write the completeness relation according to

$$\frac{1}{2} \int d\mathbf{x}_1 d\mathbf{x}_2 |\mathbf{x}_1 \mathbf{x}_2\rangle \langle \mathbf{x}_1 \mathbf{x}_2| = \hat{\mathbb{1}}. \tag{1.27}$$

Then, any ket $|\Psi\rangle \in \mathcal{H}_2$ can be expanded in the position–spin basis as

$$|\Psi\rangle = \hat{\mathbb{1}}|\Psi\rangle = \frac{1}{2} \int d\mathbf{x}_1 d\mathbf{x}_2 \, |\mathbf{x}_1 \mathbf{x}_2\rangle \underbrace{\langle \mathbf{x}_1 \mathbf{x}_2 |\Psi\rangle}_{\Psi(\mathbf{x}_1, \mathbf{x}_2)}. \tag{1.28}$$

If $|\Psi\rangle$ is normalized we can give a probability interpretation to the square modulus of the wavefunction $\Psi(\mathbf{x}_1, \mathbf{x}_2)$,

$$|\Psi(\mathbf{x}_1, \mathbf{x}_2)|^2 d\mathbf{r}_1 d\mathbf{r}_2 = \begin{pmatrix} \text{probability of finding one particle with spin } \sigma_1 \text{ in} \\ \text{volume element } d\mathbf{r}_1 \text{ around } \mathbf{r}_1 \text{ and the other particle with} \\ \text{spin } \sigma_2 \text{ in volume element } d\mathbf{r}_2 \text{ around a } \textit{different} \text{ point } \mathbf{r}_2 \end{pmatrix}.$$

However, in the case $\mathbf{x}_1 = \mathbf{x}_2$ the above formula needs to be replaced by

$$\frac{|\Psi(\mathbf{x}_1, \mathbf{x}_1)|^2}{2} d\mathbf{r}_1 d\mathbf{r}_2 = \begin{pmatrix} \text{probability of finding one particle with spin } \sigma_1 \text{ in} \\ \text{volume element } d\mathbf{r}_1 \text{ around } \mathbf{r}_1 \text{ and the other particle} \\ \text{with the } \textit{same} \text{ spin in volume element } d\mathbf{r}_2 \text{ around} \\ \text{the } \textit{same} \text{ point } \mathbf{r}_1 \end{pmatrix},$$

due to the different normalization of the diagonal kets. We stress again that the above probability interpretation follows from the normalization $\langle \Psi|\Psi\rangle = 1$, which in the continuum case reads [see (1.28)]

$$1 = \frac{1}{2} \int d\mathbf{x}_1 d\mathbf{x}_2 \, |\Psi(\mathbf{x}_1, \mathbf{x}_2)|^2.$$

It should now be clear how to extend the above relations to the case of N identical particles. We say that if the detector for a particle of spin-projection σ_1 placed in $\mathbf{r}_1$, the detector for a particle of spin-projection σ_2 placed in $\mathbf{r}_2$, etc. all click at the same time then the N-particle state collapses into the position-spin ket $|\mathbf{x}_1 \ldots \mathbf{x}_N\rangle$. Due to the nature of identical particles this ket must have the symmetry property (as usual upper/lower sign refers to bosons/fermions)

$$\boxed{|\mathbf{x}_{P(1)} \ldots \mathbf{x}_{P(N)}\rangle = (\pm)^P |\mathbf{x}_1 \ldots \mathbf{x}_N\rangle} \tag{1.29}$$

where P is a permutation of the labels $(1, \ldots, N)$, and $(\pm)^P = 1$ for even permutations and ± 1 for odd permutations (thus for bosons is always 1). A permutation is even/odd if the number of interchanges is even/odd.[3] Therefore, given the ket $|\mathbf{x}_1 \ldots \mathbf{x}_N\rangle$ with all different coordinates there are $N!$ equivalent configurations that represent the same physical state. More generally, if the ket $|\mathbf{x}_1 \ldots \mathbf{x}_N\rangle$ has m_1 coordinates equal to $\mathbf{y}_1$, m_2 coordinates equal to $\mathbf{y}_2 \neq \mathbf{y}_1$, ..., m_M coordinates equal to $\mathbf{y}_M \neq \mathbf{y}_1, \ldots, \mathbf{y}_{M-1}$, with $m_1 + \ldots + m_M = N$,

---

[3]The reader can learn more on how to calculate the sign of a permutation in Appendix B.

then the number of equivalent configurations is $N!/(m_1! \ldots m_M!)$. In the fermionic case, if two or more coordinates are the same then the ket $|\mathbf{x}_1 \ldots \mathbf{x}_N\rangle$ is the null ket $|\emptyset\rangle$. The set of position–spin kets corresponding to inequivalent configurations forms a basis in the Hilbert space of N identical particles; we refer to this space as $\mathcal{H}_N$.

The inner product between two position–spin kets is

$$\langle \mathbf{x}'_1 \ldots \mathbf{x}'_N | \mathbf{x}_1 \ldots \mathbf{x}_N \rangle = \sum_P c_P \prod_{j=1}^{N} \delta(\mathbf{x}'_j - \mathbf{x}_{P(j)}),$$

where the c_Ps are numbers depending on the permutation P. As in the two-particle case, the (anti)symmetry (1.29) of the position–spin kets requires that $c_P = c(\pm)^P$, and the normalization of $|\mathbf{x}_1 \ldots \mathbf{x}_N\rangle$ with all *different* coordinates fixes the constant $c = 1$. Hence

$$\boxed{\langle \mathbf{x}'_1 \ldots \mathbf{x}'_N | \mathbf{x}_1 \ldots \mathbf{x}_N \rangle = \sum_P (\pm)^P \prod_{j=1}^{N} \delta(\mathbf{x}'_j - \mathbf{x}_{P(j)})} \qquad (1.30)$$

This is the familiar expression for the permanent/determinant $|A|_\pm$ of a $N \times N$ matrix A (see Appendix B),

$$|A|_\pm \equiv \sum_P (\pm)^P A_{1P(1)} \ldots A_{NP(N)}.$$

Choosing the matrix elements of A to be $A_{ij} = \delta(\mathbf{x}'_i - \mathbf{x}_j)$ we can rewrite (1.30) as

$$\langle \mathbf{x}'_1 \ldots \mathbf{x}'_N | \mathbf{x}_1 \ldots \mathbf{x}_N \rangle = \begin{vmatrix} \delta(\mathbf{x}'_1 - \mathbf{x}_1) & \ldots & \delta(\mathbf{x}'_1 - \mathbf{x}_N) \\ \cdot & \cdots & \cdot \\ \cdot & \cdots & \cdot \\ \delta(\mathbf{x}'_N - \mathbf{x}_1) & \ldots & \delta(\mathbf{x}'_N - \mathbf{x}_N) \end{vmatrix}_\pm . \qquad (1.31)$$

As in the two-particle case, these formulas are so elegant because we took the bosonic kets at equal coordinates with a slightly different normalization. Consider N bosons in M *different* coordinates of which m_1 have coordinate $\mathbf{y}_1$, ..., m_M have coordinate $\mathbf{y}_M$ (hence $m_1 + \ldots + m_M = N$). Then the norm is given by

$$\langle \overbrace{\mathbf{y}_1 \cdots \mathbf{y}_1}^{m_1} \ldots \overbrace{\mathbf{y}_M \cdots \mathbf{y}_M}^{m_M} | \overbrace{\mathbf{y}_1 \cdots \mathbf{y}_1}^{m_1} \ldots \overbrace{\mathbf{y}_M \cdots \mathbf{y}_M}^{m_M} \rangle = \delta(0)^N m_1! m_2! \ldots m_M!,$$

as follows directly from (1.30).[4] In the case of fermions, instead, all position–spin kets have norm $\delta(0)^N$ since it is not possible for two or more fermions to have the same coordinate.

Given the norm of the position–spin kets, the completeness relation for N particles is a straightforward generalization of (1.27) and reads

$$\boxed{\frac{1}{N!} \int d\mathbf{x}_1 \ldots d\mathbf{x}_N \, |\mathbf{x}_1 \ldots \mathbf{x}_N\rangle\langle \mathbf{x}_1 \ldots \mathbf{x}_N| = \hat{\mathbb{1}}} \qquad (1.32)$$

---

[4]According to (1.29) the order of the arguments in the inner product does not matter.

Therefore the expansion of a ket $|\Psi\rangle \in \mathcal{H}_N$ can be written as

$$|\Psi\rangle = \hat{\mathbb{1}}|\Psi\rangle = \frac{1}{N!} \int d\mathbf{x}_1 \ldots d\mathbf{x}_N |\mathbf{x}_1 \ldots \mathbf{x}_N\rangle \underbrace{\langle \mathbf{x}_1 \ldots \mathbf{x}_N|\Psi\rangle}_{\Psi(\mathbf{x}_1,\ldots,\mathbf{x}_N)},$$

which generalizes the expansion (1.28) to the case of N particles. The wavefunction $\Psi(\mathbf{x}_1,\ldots,\mathbf{x}_N)$ is totally symmetric for bosons and totally antisymmetric for fermions due to (1.29). If $|\Psi\rangle$ is normalized then the normalization of the wavefunction reads

$$1 = \langle\Psi|\Psi\rangle = \frac{1}{N!} \int d\mathbf{x}_1 \ldots d\mathbf{x}_N |\Psi(\mathbf{x}_1,\ldots,\mathbf{x}_N)|^2. \tag{1.33}$$

The probabilistic interpretation of the square modulus of the wavefunction can be extracted using the same line of reasoning as for the two-particle case

$$\frac{|\Psi(\overbrace{\mathbf{y}_1 \ldots \mathbf{y}_1}^{m_1} \ldots \overbrace{\mathbf{y}_M \ldots \mathbf{y}_M}^{m_M})|^2}{m_1! \ldots m_M!} \prod_{j=1}^{M} d\mathbf{R}_j = \begin{pmatrix} \text{probability of finding} \\ m_1 \text{ particles in } d\mathbf{R}_1 \text{ around } \mathbf{y}_1 \\ \vdots \\ m_M \text{ particles in } d\mathbf{R}_M \text{ around } \mathbf{y}_M \end{pmatrix}, \tag{1.34}$$

where $d\mathbf{R}_j$ is the product of volume elements

$$d\mathbf{R}_j \equiv \prod_{i=m_1+\ldots+m_{j-1}+1}^{m_1+\ldots+m_j} d\mathbf{r}_i.$$

When all coordinates are different (1.34) tells us that the quantity $|\Psi(\mathbf{x}_1,\ldots,\mathbf{x}_N)|^2 \, d\mathbf{r}_1 \ldots d\mathbf{r}_N$ is the probability of finding one particle in volume element $d\mathbf{r}_1$ around $\mathbf{x}_1$, $\ldots$, and one particle in volume element $d\mathbf{r}_N$ around $\mathbf{x}_N$. We could have absorbed the prefactor $1/N!$ in (1.33) in the wavefunction (as is commonly done) but then we could not interpret the quantity $|\Psi(\mathbf{x}_1,\ldots,\mathbf{x}_N)|^2 d\mathbf{r}_1 \ldots d\mathbf{r}_N$ as the r.h.s. of (1.34) since this would amount to regarding equivalent configurations as distinguishable and consequently the probability would be overestimated by a factor of $N!$.

The reader might wonder why we have been so punctilious about the possibility of having more than one boson with the same position–spin coordinate, since these configurations are of zero measure in the space of all configurations. However, such configurations are the physically most relevant in bosonic systems at low temperature. Indeed bosons can condense in states in which certain (continuum) quantum numbers are *macroscopically* occupied and hence have a finite probability. A common example is the zero momentum state of a free boson gas in three dimensions.

We close this section by illustrating a practical way to construct the N-particle position–spin kets using the N-particle position–spin kets of distinguishable particles. The procedure simply consists in forming (anti)symmetrized products of one-particle position–spin kets. For instance, we have for the case of two particles

$$|\mathbf{x}_1\mathbf{x}_2\rangle = \frac{|\mathbf{x}_1\rangle|\mathbf{x}_2\rangle \pm |\mathbf{x}_2\rangle|\mathbf{x}_1\rangle}{\sqrt{2}}, \tag{1.35}$$

and more generally, for N particles

$$|\mathbf{x}_1 \ldots \mathbf{x}_N\rangle = \frac{1}{\sqrt{N!}} \sum_P (\pm)^P |\mathbf{x}_{P(1)}\rangle \ldots |\mathbf{x}_{P(N)}\rangle. \tag{1.36}$$

Using the inner product (1.14) one can check directly that these states have inner product (1.30). We refer to the above representation of the position–spin kets as kets in *first quantization* since it is the representation usually found in basic quantum mechanics books. Using (1.36) we could proceed to calculate matrix elements of operators such as the center-of-mass position, total energy, spin, angular momentum, density, etc. However, this involves rather cumbersome expressions with a large number of terms differing only in the sign and the order of the coordinates. In the next section we describe a formalism, known as *second quantization*, that makes it easy to do such calculations efficiently, as the position–spin ket is represented by a *single* ket rather than by $N!$ products of one-particle kets as in (1.36). As we shall see the merits of second quantization are the compactness of the expressions and an enormous simplification in the calculation of the action of operators over states in $\mathcal{H}_N$. This formalism further treats systems with different numbers of identical particles on the same footing and it is therefore well suited to study ionization processes, transport phenomena, and finite temperature effects within the grand canonical ensemble of quantum statistical physics.

1.4 Field operators

The advantage of the bra-and-ket notation invented by Dirac is twofold. First of all, it provides a geometric interpretation of the physical states in Hilbert space as abstract kets independent of the basis in which they are expanded. For example, it does not matter whether we expand $|\Psi\rangle$ in terms of the position–spin kets or momentum–spin kets; $|\Psi\rangle$ remains the same although the expansion coefficients in the two bases are different. The second advantage is that the abstract kets can be systematically generated by repeated applications of a *creation operator* on the empty or zero-particle state. This approach forms the basis of an elegant formalism known as *second quantization*, which we describe in detail in this section.

To deal with many arbitrary identical particles we define a collection $\mathcal{F}$ of Hilbert spaces, also known as *Fock space*, according to

$$\mathcal{F} = \{\mathcal{H}_0, \mathcal{H}_1, \ldots, \mathcal{H}_N, \ldots\},$$

with $\mathcal{H}_N$ the Hilbert space for N identical particles. An arbitrary element of the Fock space is a ket that can be written as

$$|\Psi\rangle = \sum_{N=0}^{\infty} c_N |\Psi_N\rangle, \tag{1.37}$$

where $|\Psi_N\rangle$ belongs to $\mathcal{H}_N$. The inner product between the ket (1.37) and another element in the Fock space

$$|\chi\rangle = \sum_{N=0}^{\infty} d_N |\chi_N\rangle$$

is defined as

$$\langle \chi | \Psi \rangle \equiv \sum_{N=0}^{\infty} d_N^* c_N \langle \chi_N | \Psi_N \rangle,$$

where $\langle \chi_N | \Psi_N \rangle$ is the inner product in $\mathcal{H}_N$. This definition is dictated by common sense: the probability of having $M \neq N$ particles in a N-particle ket is zero and therefore kets with different number of particles are orthogonal, i.e., have zero overlap.

The Hilbert space $\mathcal{H}_0$ is the space with *zero* particles. Since an empty system has no degrees of freedom, $\mathcal{H}_0$ is a one-dimensional space and we denote by $|0\rangle$ the only normalized ket in $\mathcal{H}_0$,

$$\langle 0 | 0 \rangle = 1.$$

According to the expansion (1.37) the ket $|0\rangle$ has all $c_N = 0$ except for c_0. This state should not be confused with the null ket $|\emptyset\rangle$ which is defined as the state in Fock space with all $c_N = 0$ and, therefore, is *not* a physical state. The *empty ket* $|0\rangle$ is a physical state; indeed the normalization $\langle 0 | 0 \rangle = 1$ means that the probability of finding nothing in an empty space is 1.

The goal of this section is to find a clever way to construct a basis for each Hilbert space $\mathcal{H}_1$, $\mathcal{H}_2$, ... To accomplish this goal the central idea of the second quantization formalism is to define a *field operator* $\hat{\psi}^{\dagger}(\mathbf{x}) = \hat{\psi}^{\dagger}(\mathbf{r}\sigma)$ that generates the position-spin kets by repeated action on the empty ket, i.e.,

$$
\begin{aligned}
|\mathbf{x}_1\rangle &= \hat{\psi}^{\dagger}(\mathbf{x}_1)|0\rangle \\
|\mathbf{x}_1 \mathbf{x}_2\rangle &= \hat{\psi}^{\dagger}(\mathbf{x}_2)|\mathbf{x}_1\rangle = \hat{\psi}^{\dagger}(\mathbf{x}_2)\hat{\psi}^{\dagger}(\mathbf{x}_1)|0\rangle \\
|\mathbf{x}_1 \dots \mathbf{x}_N\rangle &= \hat{\psi}^{\dagger}(\mathbf{x}_N)|\mathbf{x}_1 \dots \mathbf{x}_{N-1}\rangle = \hat{\psi}^{\dagger}(\mathbf{x}_N) \dots \hat{\psi}^{\dagger}(\mathbf{x}_1)|0\rangle
\end{aligned}
\tag{1.38}
$$

Since an operator is uniquely defined from its action on a complete set of states in the Hilbert space (the Fock space in our case), the above relations *define* the field operator $\hat{\psi}^{\dagger}(\mathbf{x})$ for all $\mathbf{x}$. The field operator $\hat{\psi}^{\dagger}(\mathbf{x})$ transforms a ket of $\mathcal{H}_N$ into a ket of $\mathcal{H}_{N+1}$ for all N, see Fig. 1.4(a). We may say that the field operator $\hat{\psi}^{\dagger}(\mathbf{x})$ creates a particle in $\mathbf{x}$ and it is therefore called the *creation operator*. Since the position–spin kets change a plus or minus sign under interchange of any two particles it follows that

$$
\begin{aligned}
\hat{\psi}^{\dagger}(\mathbf{x})\hat{\psi}^{\dagger}(\mathbf{y})|\mathbf{x}_1 \dots \mathbf{x}_N\rangle &= |\mathbf{x}_1 \dots \mathbf{x}_N\,\mathbf{y}\,\mathbf{x}\rangle = \pm|\mathbf{x}_1 \dots \mathbf{x}_N\,\mathbf{x}\,\mathbf{y}\rangle \\
&= \pm\hat{\psi}^{\dagger}(\mathbf{y})\hat{\psi}^{\dagger}(\mathbf{x})|\mathbf{x}_1 \dots \mathbf{x}_N\rangle,
\end{aligned}
$$

where we recall that the upper sign in $\pm$ refers to bosons and the lower sign to fermions. This identity is true for all $\mathbf{x}_1, \dots, \mathbf{x}_N$ and for all N, i.e., for all states in $\mathcal{F}$, and hence

$$\hat{\psi}^{\dagger}(\mathbf{x})\hat{\psi}^{\dagger}(\mathbf{y}) = \pm\hat{\psi}^{\dagger}(\mathbf{y})\hat{\psi}^{\dagger}(\mathbf{x}).$$

If we define the (anti)commutator between two generic operators $\hat{A}$ and $\hat{B}$ according to

$$\left[\hat{A}, \hat{B}\right]_{\mp} = \hat{A}\hat{B} \mp \hat{B}\hat{A},$$

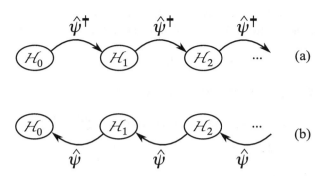

Figure 1.4 Action of the creation operator $\hat{\psi}^{\dagger}$ in (a), and of the annihilation operator $\hat{\psi}$ in (b).

we can rewrite the above relation as

$$\left[\hat{\psi}^{\dagger}(\mathbf{x}),\hat{\psi}^{\dagger}(\mathbf{y})\right]_{\mp}=0 \qquad (1.39)$$

Corresponding to the operator $\hat{\psi}^{\dagger}(\mathbf{x})$ there is the adjoint operator $\hat{\psi}(\mathbf{x})$ [or equivalently $\hat{\psi}^{\dagger}(\mathbf{x})$ is the adjoint of $\hat{\psi}(\mathbf{x})$]. Let us recall the definition of adjoint operators. An operator $\hat{O}^{\dagger}$ with the superscript "†" (read *dagger*) is the adjoint of the operator $\hat{O}$ if

$$\langle\chi|\hat{O}|\Psi\rangle=\langle\Psi|\hat{O}^{\dagger}|\chi\rangle^{*}$$

for all $|\chi\rangle$ and $|\Psi\rangle$, which implies $(\hat{O}^{\dagger})^{\dagger}=\hat{O}$. In particular, when $\hat{O}=\hat{\psi}(\mathbf{x})$ we have

$$\langle\chi|\hat{\psi}(\mathbf{x})|\Psi\rangle=\langle\Psi|\hat{\psi}^{\dagger}(\mathbf{x})|\chi\rangle^{*}.$$

Since for any $|\Psi\rangle\in\mathcal{H}_{N+1}$ the quantity $\langle\Psi|\hat{\psi}^{\dagger}(\mathbf{x})|\chi\rangle$ is zero for all $|\chi\rangle$ with no components in $\mathcal{H}_{N}$, the above equation implies that $\hat{\psi}(\mathbf{x})|\Psi\rangle\in\mathcal{H}_{N}$, i.e., the operator $\hat{\psi}(\mathbf{x})$ maps the elements of $\mathcal{H}_{N+1}$ into elements of $\mathcal{H}_{N}$, see Fig. 1.4(b). Thus, whereas the operator $\hat{\psi}^{\dagger}(\mathbf{x})$ adds a particle its adjoint operator $\hat{\psi}(\mathbf{x})$ removes a particle and, for this reason, it is called the *annihilation operator*. Below we study its properties and how it acts on the position–spin kets.

By taking the adjoint of the identity (1.39) we immediately obtain the (anti) commutation relation

$$\left[\hat{\psi}(\mathbf{x}),\hat{\psi}(\mathbf{y})\right]_{\mp}=0 \qquad (1.40)$$

The action of $\hat{\psi}(\mathbf{x})$ on any state can be deduced from its definition as the adjoint of $\hat{\psi}^{\dagger}(\mathbf{x})$ together with the inner product (1.31) between the position–spin kets. Let us illustrate this first for the action on the empty ket $|0\rangle$. For any $|\Psi\rangle\in\mathcal{F}$,

$$\langle\Psi|\hat{\psi}(\mathbf{x})|0\rangle=\langle0|\hat{\psi}^{\dagger}(\mathbf{x})|\Psi\rangle^{*}=0,$$

since $\hat{\psi}^\dagger(\mathbf{x})|\Psi\rangle$ contains at least one particle and is therefore orthogonal to $|0\rangle$. We conclude that $\hat{\psi}(\mathbf{x})|0\rangle$ is orthogonal to all $|\Psi\rangle$ in $\mathcal{F}$ and hence it must be equal to the null ket

$$\hat{\psi}(\mathbf{x})|0\rangle = |\emptyset\rangle. \tag{1.41}$$

The action of $\hat{\psi}(\mathbf{x})$ on the one-particle ket $|\mathbf{y}\rangle$ can be inferred from (1.30) and (1.38); we have

$$\delta(\mathbf{y} - \mathbf{x}) = \langle\mathbf{y}|\mathbf{x}\rangle = \langle\mathbf{y}|\hat{\psi}^\dagger(\mathbf{x})|0\rangle = \langle0|\hat{\psi}(\mathbf{x})|\mathbf{y}\rangle^*.$$

Since $\hat{\psi}(\mathbf{x})|\mathbf{y}\rangle \in \mathcal{H}_0$ it follows that

$$\hat{\psi}(\mathbf{x})|\mathbf{y}\rangle = \delta(\mathbf{y} - \mathbf{x})|0\rangle. \tag{1.42}$$

We see from this relation that the operator $\hat{\psi}(\mathbf{x})$ removes a particle from the state $|\mathbf{y}\rangle$ when $\mathbf{x} = \mathbf{y}$ and otherwise yields zero.

The derivation of the action of $\hat{\psi}(\mathbf{x})$ on the empty ket and on the one-particle ket was rather elementary. Let us now derive the action of $\hat{\psi}(\mathbf{x})$ on the general N-particle ket $|\mathbf{y}_1 \ldots \mathbf{y}_N\rangle$. For this purpose we consider the matrix element

$$\langle\mathbf{x}_1 \ldots \mathbf{x}_{N-1}|\hat{\psi}(\mathbf{x}_N)|\mathbf{y}_1 \ldots \mathbf{y}_N\rangle = \langle\mathbf{x}_1 \ldots \mathbf{x}_N|\mathbf{y}_1 \ldots \mathbf{y}_N\rangle. \tag{1.43}$$

The overlap on the r.h.s. is given in (1.31); expanding the permanent/determinant along row N (see Appendix B) we get

$$\langle\mathbf{x}_1 \ldots \mathbf{x}_{N-1}|\hat{\psi}(\mathbf{x}_N)|\mathbf{y}_1 \ldots \mathbf{y}_N\rangle$$
$$= \sum_{k=1}^{N}(\pm)^{N+k}\delta(\mathbf{x}_N - \mathbf{y}_k)\langle\mathbf{x}_1 \ldots \mathbf{x}_{N-1}|\mathbf{y}_1 \ldots \mathbf{y}_{k-1}\mathbf{y}_{k+1} \ldots \mathbf{y}_N\rangle.$$

This expression is valid for any $|\mathbf{x}_1 \ldots \mathbf{x}_{N-1}\rangle$ and since $\hat{\psi}(\mathbf{x})$ maps from $\mathcal{H}_N$ only to $\mathcal{H}_{N-1}$ we conclude that

$$\boxed{\hat{\psi}(\mathbf{x})|\mathbf{y}_1 \ldots \mathbf{y}_N\rangle = \sum_{k=1}^{N}(\pm)^{N+k}\delta(\mathbf{x} - \mathbf{y}_k)|\mathbf{y}_1 \ldots \mathbf{y}_{k-1}\mathbf{y}_{k+1} \ldots \mathbf{y}_N\rangle} \tag{1.44}$$

We have just derived an important equation for the action of the annihilation operator on a position–spin ket. It correctly reduces to (1.42) when $N = 1$ and for $N > 1$ yields, for example,

$$\hat{\psi}(\mathbf{x})|\mathbf{y}_1\mathbf{y}_2\rangle = \delta(\mathbf{x} - \mathbf{y}_2)|\mathbf{y}_1\rangle \pm \delta(\mathbf{x} - \mathbf{y}_1)|\mathbf{y}_2\rangle,$$
$$\hat{\psi}(\mathbf{x})|\mathbf{y}_1\mathbf{y}_2\mathbf{y}_3\rangle = \delta(\mathbf{x} - \mathbf{y}_3)|\mathbf{y}_1\mathbf{y}_2\rangle \pm \delta(\mathbf{x} - \mathbf{y}_2)|\mathbf{y}_1\mathbf{y}_3\rangle + \delta(\mathbf{x} - \mathbf{y}_1)|\mathbf{y}_2\mathbf{y}_3\rangle.$$

So the annihilation operator removes sequentially a particle from every position–spin coordinate while keeping the final result totally symmetric or antisymmetric in all $\mathbf{y}$ variables by adjusting the signs of the prefactors. With the help of (1.44) we can derive a fundamental (anti)commutation relation between the annihilation and creation operators. Acting on both sides of (1.44) with $\hat{\psi}^\dagger(\mathbf{y})$ and denoting by $|R\rangle$ the ket on the r.h.s. we have

$$\hat{\psi}^\dagger(\mathbf{y})\hat{\psi}(\mathbf{x})|\mathbf{y}_1 \ldots \mathbf{y}_N\rangle = \hat{\psi}^\dagger(\mathbf{y})|R\rangle. \tag{1.45}$$

Exchanging the order of the field operators in the l.h.s. of the above identity and using (1.44) we find

$$\hat{\psi}(\mathbf{x})\hat{\psi}^\dagger(\mathbf{y})|\mathbf{y}_1 \ldots \mathbf{y}_N\rangle = \hat{\psi}(\mathbf{x})|\mathbf{y}_1 \ldots \mathbf{y}_N \mathbf{y}\rangle = \delta(\mathbf{x} - \mathbf{y})\,|\mathbf{y}_1 \ldots \mathbf{y}_N\rangle$$
$$+ \sum_{k=1}^{N}(\pm)^{N+1+k}\delta(\mathbf{x} - \mathbf{y}_k)\,|\mathbf{y}_1 \ldots \mathbf{y}_{k-1}\mathbf{y}_{k+1}\ldots\mathbf{y}_N\mathbf{y}\rangle$$
$$= \delta(\mathbf{x} - \mathbf{y})\,|\mathbf{y}_1 \ldots \mathbf{y}_N\rangle \pm \hat{\psi}^\dagger(\mathbf{y})|R\rangle. \tag{1.46}$$

Subtraction and addition of (1.45) and (1.46) for bosons and fermions respectively then gives

$$\left[\hat{\psi}(\mathbf{x}), \hat{\psi}^\dagger(\mathbf{y})\right]_{\mp}|\mathbf{y}_1 \ldots \mathbf{y}_N\rangle = \delta(\mathbf{x} - \mathbf{y})|\mathbf{y}_1 \ldots \mathbf{y}_N\rangle,$$

which must be valid for all position–spin kets and for all N, and therefore

$$\boxed{\left[\hat{\psi}(\mathbf{x}), \hat{\psi}^\dagger(\mathbf{y})\right]_{\mp} = \delta(\mathbf{x} - \mathbf{y})} \tag{1.47}$$

The (anti)commutation relations (1.39), (1.40), and (1.47) are the main results of this section and form the basis of most derivations in this book. As we shall see in Section 1.6, all many-particle operators, like total energy, density, current, spin, etc., consist of simple expressions in terms of the field operators $\hat{\psi}$ and $\hat{\psi}^\dagger$, and the calculation of their averages can easily be performed with the help of the (anti)commutation relations. It is similar to the harmonic oscillator of quantum mechanics: both the eigenstates and the operators are expressed in terms of the raising and lowering operators $\hat{a}^\dagger$ and $\hat{a}$, and to calculate all sorts of average it is enough to know the commutation relations $[\hat{a}, \hat{a}]_- = [\hat{a}^\dagger, \hat{a}^\dagger]_- = 0$ and $[\hat{a}, \hat{a}^\dagger]_- = 1$. The difference with second quantization is that we have a "harmonic oscillator" for every $\mathbf{x}$. Using the (anti)commutation properties we can manipulate directly the kets and never have to deal with the rather cumbersome expressions of the wavefunctions; the field operators take care of the symmetry of the kets automatically. The great achievement of second quantization is comparable to that of a programming language. When we program we use a nice friendly text-editor to write a code which tells the computer what operations to do, and we do not bother if the instructions given through the text-editor are correctly executed by the machine. A bug in the code is an error in the text of the program (the way we manipulate the field operators) and not an erroneous functioning of some logic gate (the violation of the symmetry properties of the many-particle kets).

Exercise 1.1. We define the *density operator*

$$\hat{n}(\mathbf{x}) \equiv \hat{\psi}^\dagger(\mathbf{x})\hat{\psi}(\mathbf{x}).$$

Using the identities $[\hat{A}\hat{B}, \hat{C}]_- = \hat{A}[\hat{B}, \hat{C}]_- + [\hat{A}, \hat{C}]_-\hat{B} = \hat{A}[\hat{B}, \hat{C}]_+ - [\hat{A}, \hat{C}]_+\hat{B}$ prove the following relations for fermionic and bosonic field operators:

$$[\hat{n}(\mathbf{x}), \hat{\psi}(\mathbf{x}')]_- = -\delta(\mathbf{x} - \mathbf{x}')\hat{\psi}(\mathbf{x}), \tag{1.48}$$

$$[\hat{n}(\mathbf{x}), \hat{\psi}^\dagger(\mathbf{x}')]_- = \delta(\mathbf{x} - \mathbf{x}')\hat{\psi}^\dagger(\mathbf{x}). \tag{1.49}$$

1.5   General basis states

In the previous section we learned how to construct states of many identical particles with a given spin and position. The position–spin is, however, just one possible choice of quantum number to characterize every single particle. We now show how the field operators can be used to construct states of many identical particles in which every particle is labeled by general quantum numbers, such as momentum, energy, etc.

Let us consider a normalized one-particle ket $|n\rangle$. The quantum number $n = (s\tau)$ comprises an orbital quantum number s and the spin projection τ along some quantization axis. Choosing the quantization axis of the spin to be the same as that of the position–spin ket $|\mathbf{x}\rangle = |\mathbf{r}\sigma\rangle$ the overlap between $|n\rangle$ and $|\mathbf{x}\rangle$ is

$$\langle\mathbf{x}|n\rangle \equiv \varphi_n(\mathbf{x}) = \varphi_s(\mathbf{r})\delta_{\tau\sigma}. \tag{1.50}$$

The one-particle ket $|n\rangle$ can be expanded in the position–spin kets using the completeness relation (1.8)

$$|n\rangle = \int d\mathbf{x}\, |\mathbf{x}\rangle\langle\mathbf{x}|n\rangle = \int d\mathbf{x}\, \varphi_n(\mathbf{x})|\mathbf{x}\rangle = \int d\mathbf{x}\, \varphi_n(\mathbf{x})\hat{\psi}^\dagger(\mathbf{x})|0\rangle. \tag{1.51}$$

One can easily check that the normalization $\langle n|n\rangle = 1$ is equivalent to saying that $\int d\mathbf{x}|\varphi_n(\mathbf{x})|^2 = 1$. From (1.51) we see that $|n\rangle$ is obtained by applying to the empty ket $|0\rangle$ the operator

$$\boxed{\hat{d}_n^\dagger \equiv \int d\mathbf{x}\, \varphi_n(\mathbf{x})\hat{\psi}^\dagger(\mathbf{x})} \tag{1.52}$$

i.e., $\hat{d}_n^\dagger|0\rangle = |n\rangle$. We may say that $\hat{d}_n^\dagger$ creates a particle with quantum number n. Similarly, if we take the adjoint of (1.52),

$$\boxed{\hat{d}_n \equiv \int d\mathbf{x}\, \varphi_n^*(\mathbf{x})\hat{\psi}(\mathbf{x})} \tag{1.53}$$

we obtain an operator that destroys a particle with quantum number n since

$$\hat{d}_n|n\rangle = \hat{d}_n\hat{d}_n^\dagger|0\rangle = \int d\mathbf{x}d\mathbf{x}'\varphi_n^*(\mathbf{x})\varphi_n(\mathbf{x}') \underbrace{\hat{\psi}(\mathbf{x})\hat{\psi}^\dagger(\mathbf{x}')|0\rangle}_{\delta(\mathbf{x}-\mathbf{x}')|0\rangle} = \int d\mathbf{x}\, |\varphi_n(\mathbf{x})|^2|0\rangle = |0\rangle.$$

The operators $\hat{d}_n$ and $\hat{d}_n^\dagger$, being linear combinations of field operators at different $\mathbf{x}$, can act on states with arbitrarily many particles. Below we derive some important relations for the $\hat{d}$-operators when the set $\{|n\rangle\}$ forms an orthonormal basis in the one-particle Hilbert space.

We can easily derive the important (anti)commutation relations using the corresponding relations for the field operators

$$\left[\hat{d}_n, \hat{d}_m^\dagger\right]_\mp = \int d\mathbf{x}d\mathbf{x}' \, \varphi_n^*(\mathbf{x})\varphi_m(\mathbf{x}') \underbrace{\left[\hat{\psi}(\mathbf{x}), \hat{\psi}^\dagger(\mathbf{x}')\right]_\mp}_{\delta(\mathbf{x}-\mathbf{x}')} = \langle n|m\rangle = \delta_{nm}, \qquad (1.54)$$

and the more obvious ones

$$\left[\hat{d}_n, \hat{d}_m\right]_\mp = \left[\hat{d}_n^\dagger, \hat{d}_m^\dagger\right]_\mp = 0, \qquad (1.55)$$

that follow similarly. It is worth noting that the $\hat{d}$-operators obey the same (anti)commutation relations as the field operators with the index n playing the role of $\mathbf{x}$. This is a very important observation since the results of the previous section relied only on the (anti)commutation relations of $\hat{\psi}$ and $\hat{\psi}^\dagger$, and hence remain valid in this more general basis. To convince the reader of this fact we derive some of the results of the previous section directly from the (anti)commutation relations. We define the N-particle ket

$$|n_1 \ldots n_N\rangle \equiv \hat{d}_{n_N}^\dagger \ldots \hat{d}_{n_1}^\dagger |0\rangle = \hat{d}_{n_N}^\dagger |n_1 \ldots n_{N-1}\rangle, \qquad (1.56)$$

which has the symmetry property

$$|n_{P(1)} \ldots n_{P(N)}\rangle = (\pm)^P |n_1 \ldots n_N\rangle,$$

as follows immediately from (1.55). Like the position–spin kets, the kets $|n_1 \ldots n_N\rangle$ span the N-particle Hilbert space $\mathcal{H}_N$. The action of $\hat{d}_n$ on $|n_1 \ldots n_N\rangle$ is similar to the action of $\hat{\psi}(\mathbf{x})$ on $|\mathbf{x}_1 \ldots \mathbf{x}_N\rangle$. Using the (anti)commutation relation (1.54) we can move the $\hat{d}_n$-operator through the string of $d^\dagger$-operators[5]

$$\hat{d}_n|n_1 \ldots n_N\rangle = \left(\left[\hat{d}_n, \hat{d}_{n_N}^\dagger\right]_\mp \pm \hat{d}_{n_N}^\dagger \hat{d}_n\right)|n_1 \ldots n_{N-1}\rangle$$

$$= \delta_{nn_N}|n_1 \ldots n_{N-1}\rangle \pm \hat{d}_{n_N}^\dagger \left(\left[\hat{d}_n, \hat{d}_{n_{N-1}}^\dagger\right]_\mp \pm \hat{d}_{n_{N-1}}^\dagger \hat{d}_n\right)|n_1 \ldots n_{N-2}\rangle$$

$$= \delta_{nn_N}|n_1 \ldots n_{N-1}\rangle \pm \delta_{nn_{N-1}}|n_1 \ldots n_{N-2}n_N\rangle$$

$$(\pm)^2 \hat{d}_{n_N}^\dagger \hat{d}_{n_{N-1}}^\dagger \left(\left[\hat{d}_n, \hat{d}_{n_{N-2}}^\dagger\right]_\mp \pm \hat{d}_{n_{N-2}}^\dagger \hat{d}_n\right)|n_1 \ldots n_{N-3}\rangle$$

$$= \sum_{k=1}^N (\pm)^{N+k} \delta_{nn_k}|n_1 \ldots n_{k-1}n_{k+1} \ldots n_N\rangle. \qquad (1.57)$$

This result can also be used to calculate directly the overlap between two states of the general basis. For example, for the case of two particles we have

$$\langle n_1'n_2'|n_1n_2\rangle = \langle n_1'|\hat{d}_{n_2'}|n_1n_2\rangle = \langle n_1'|\left(\delta_{n_2'n_2}|n_1\rangle \pm \delta_{n_2'n_1}|n_2\rangle\right)$$

$$= \delta_{n_1'n_1}\delta_{n_2'n_2} \pm \delta_{n_1'n_2}\delta_{n_2'n_1},$$

---

[5]Alternatively (1.57) can be derived from (1.44) together with the definitions of the $\hat{d}$-operators.

which is the analog of (1.26). More generally, for N particles we have

$$\langle n'_1 \ldots n'_N | n_1 \ldots n_N \rangle = \sum_P (\pm)^P \prod_{j=1}^N \delta_{n'_j\, n_{P(j)}}, \tag{1.58}$$

which should be compared with the overlap $\langle \mathbf{x}'_1 \ldots \mathbf{x}'_N | \mathbf{x}_1 \ldots \mathbf{x}_N \rangle$ in (1.30).

The states $|n_1 \ldots n_N\rangle$ are orthonormal (with the exception of the bosonic kets with two or more equal quantum numbers) and can be used to construct a basis. In analogy with (1.32) the completeness relation is

$$\frac{1}{N!} \sum_{n_1,\ldots,n_N} |n_1 \ldots n_N\rangle\langle n_1 \ldots n_N| = \hat{\mathbb{1}},$$

and hence the expansion of a ket $|\Psi\rangle$ belonging to $\mathcal{H}_N$ reads

$$|\Psi\rangle = \hat{\mathbb{1}}|\Psi\rangle = \frac{1}{N!} \sum_{n_1,\ldots,n_N} |n_1 \ldots n_N\rangle \underbrace{\langle n_1 \ldots n_N|\Psi\rangle}_{\Psi(n_1,\ldots,n_N)}. \tag{1.59}$$

If $|\Psi\rangle$ is normalized then the coefficients $\Psi(n_1,\ldots,n_N)$ have the following probabilistic interpretation

$$\frac{|\Psi(\overbrace{n_1 \ldots n_1}^{m_1} \ldots \overbrace{n_M \ldots n_M}^{m_M})|^2}{m_1! \ldots m_M!} = \left(\begin{array}{c} \text{probability of finding} \\ m_1 \text{ particles with quantum number } n_1 \\ \vdots \\ m_M \text{ particles with quantum number } n_M \end{array} \right).$$

We have already observed that the $\hat{d}$-operators obey the same (anti)commutation relations as the field operators provided that $\{|n\rangle\}$ is an orthonormal basis in $\mathcal{H}_1$. Likewise we can construct linear combinations of the $\hat{d}$-operators that preserve the (anti)commutation relations. It is left as an exercise for the reader to prove that the operators

$$\hat{c}_\alpha = \sum_n U_{\alpha n} \hat{d}_n, \qquad \hat{c}_\alpha^\dagger = \sum_n U_{\alpha n}^* \hat{d}_n^\dagger$$

obey

$$\left[\hat{c}_\alpha, \hat{c}_\beta^\dagger \right]_\mp = \delta_{\alpha\beta},$$

provided that

$$U_{\alpha n} \equiv \langle \alpha | n \rangle$$

is the inner product between the elements of the original orthonormal basis $\{|n\rangle\}$ and the elements of another orthonormal basis $\{|\alpha\rangle\}$. Indeed in this case the $U_{\alpha n}$ are the matrix elements of a unitary matrix since

$$\sum_n U_{\alpha n} U_{n\beta}^\dagger = \sum_n \langle \alpha | n \rangle \langle n | \beta \rangle = \langle \alpha | \beta \rangle = \delta_{\alpha\beta},$$

where we use the completeness relation. In particular, when $\alpha = \mathbf{x}$ we have $U_{\mathbf{x}n} = \langle \mathbf{x}|n\rangle = \varphi_n(\mathbf{x})$ and we find that $\hat{c}_{\mathbf{x}} = \hat{\psi}(\mathbf{x})$. We thus recover the field operators as

$$\hat{\psi}(\mathbf{x}) = \sum_n \varphi_n(\mathbf{x})\hat{d}_n, \qquad \hat{\psi}^\dagger(\mathbf{x}) = \sum_n \varphi_n^*(\mathbf{x})\hat{d}_n^\dagger. \tag{1.60}$$

These relations tell us that the expansion of the position–spin kets in terms of the kets $|n_1 \ldots n_N\rangle$ is simply

$$\boxed{|\mathbf{x}_1 \ldots \mathbf{x}_N\rangle = \sum_{n_1 \ldots n_N} \varphi_{n_1}^*(\mathbf{x}_1) \ldots \varphi_{n_N}^*(\mathbf{x}_N)|n_1 \ldots n_N\rangle} \tag{1.61}$$

Conversely, using (1.52) we can expand the general basis kets in terms of the position–spin kets as

$$\boxed{|n_1 \ldots n_N\rangle = \int d\mathbf{x}_1 \ldots d\mathbf{x}_N\, \varphi_{n_1}(\mathbf{x}_1) \ldots \varphi_{n_N}(\mathbf{x}_N)|\mathbf{x}_1 \ldots \mathbf{x}_N\rangle} \tag{1.62}$$

If we are given a state $|\Psi\rangle$ that is expanded in a general basis and we subsequently want to calculate properties in position–spin space, such as the particle density or the current density, we need to calculate the overlap between $|n_1 \ldots n_N\rangle$ and $|\mathbf{x}_1 \ldots \mathbf{x}_N\rangle$. This overlap is the wavefunction for N particles with quantum numbers $n_1, \ldots, n_N$

$$\Psi_{n_1 \ldots n_N}(\mathbf{x}_1, \ldots, \mathbf{x}_N) = \langle \mathbf{x}_1 \ldots \mathbf{x}_N|n_1 \ldots n_N\rangle.$$

The explicit form of the wavefunction follows directly from the inner product (1.30) and from the expansion (1.62), and reads

$$\Psi_{n_1 \ldots n_N}(\mathbf{x}_1, \ldots, \mathbf{x}_N) = \sum_P (\pm)^P \varphi_{n_1}(\mathbf{x}_{P(1)}) \ldots \varphi_{n_N}(\mathbf{x}_{P(N)})$$

$$= \begin{vmatrix} \varphi_{n_1}(\mathbf{x}_1) & \cdots & \varphi_{n_1}(\mathbf{x}_N) \\ \cdot & \cdots & \cdot \\ \cdot & \cdots & \cdot \\ \varphi_{n_N}(\mathbf{x}_1) & \cdots & \varphi_{n_N}(\mathbf{x}_N) \end{vmatrix}_\pm. \tag{1.63}$$

Since for any matrix A we have $|A|_\mp = |A^T|_\mp$ with A^T the transpose of A we can equivalently write

$$\Psi_{n_1 \ldots n_N}(\mathbf{x}_1, \ldots, \mathbf{x}_N) = \begin{vmatrix} \varphi_{n_1}(\mathbf{x}_1) & \cdots & \varphi_{n_N}(\mathbf{x}_1) \\ \cdot & \cdots & \cdot \\ \cdot & \cdots & \cdot \\ \varphi_{n_1}(\mathbf{x}_N) & \cdots & \varphi_{n_N}(\mathbf{x}_N) \end{vmatrix}_\pm.$$

In the case of fermions the determinant is also known as the *Slater determinant*. For those readers already familiar with Slater determinants we note that the absence on the r.h.s. of the prefactor $1/\sqrt{N!}$ is a consequence of forcing on the square modulus of the wavefunction a probability interpretation, as discussed in detail in Section 1.3. The actions of the $\hat{d}$-operators have a simple algebraic interpretation in terms of permanents or determinants.

The action of the creation operator $\hat{d}_n^\dagger$ on $|n_1 \ldots n_N\rangle$ in the position–spin representation, i.e., $\langle \mathbf{x}_1 \ldots \mathbf{x}_{N+1} | \hat{d}_n^\dagger | n_1 \ldots n_N \rangle$ simply amounts to adding a column with coordinate $\mathbf{x}_{N+1}$ and a row with wavefunction φ_n in (1.63). For the annihilation operator we have a similar algebraic interpretation. Taking the inner product with $\langle \mathbf{x}_1 \ldots \mathbf{x}_{N-1} |$ of both sides of (1.57) we get

$$\langle \mathbf{x}_1 \ldots \mathbf{x}_{N-1} | \hat{d}_n | n_1 \ldots n_N \rangle = \sum_{k=1}^{N} (\pm 1)^{N+k} \delta_{n n_k}$$

$$\times \begin{vmatrix} \varphi_{n_1}(\mathbf{x}_1) & \cdots & \cdots & \varphi_{n_1}(\mathbf{x}_N - 1) & \varphi_{n_1}(\mathbf{x}_N) \\ \vdots & \vdots & \vdots & \vdots & \\ \varphi_{n_{k-1}}(\mathbf{x}_1) & \cdots & \cdots & \varphi_{n_{k-1}}(\mathbf{x}_{N-1}) & \varphi_{n_{k-1}}(\mathbf{x}_N) \\ \varphi_{n_k}(\mathbf{x}_1) & \cdots & \cdots & \varphi_{n_k}(\mathbf{x}_{N-1}) & \varphi_{n_k}(\mathbf{x}_N) \\ \varphi_{n_{k+1}}(\mathbf{x}_1) & \cdots & \cdots & \varphi_{n_{k+1}}(\mathbf{x}_{N-1}) & \varphi_{n_{k+1}}(\mathbf{x}_N) \\ \vdots & \vdots & \vdots & \vdots & \\ \varphi_{n_N}(\mathbf{x}_1) & \cdots & \cdots & \varphi_{n_N}(\mathbf{x}_{N-1}) & \varphi_{n_N}(\mathbf{x}_N) \end{vmatrix}_\pm ,$$

i.e., the action of the $\hat{d}_n$-operator amounts to deleting the last column and, if present, the row with quantum number n from the permanent/determinant of (1.63), and otherwise yields zero. Already at this stage the reader can appreciate how powerful it is to work with the field operators and not to have anything to do with Slater determinants.

Exercise 1.2. Prove the inverse relations (1.60).

Exercise 1.3. Let $|n\rangle = |\mathbf{p}\tau\rangle$ be a momentum–spin ket so that $\langle \mathbf{x} | \mathbf{p}\tau \rangle = e^{i\mathbf{p}\cdot\mathbf{r}} \delta_{\sigma\tau}$, see (1.10). Show that the (anti)commutation relation in (1.54) then reads

$$\left[\hat{d}_{\mathbf{p}\tau}, \hat{d}_{\mathbf{p}'\tau'}^\dagger \right]_\mp = (2\pi)^3 \delta(\mathbf{p} - \mathbf{p}') \delta_{\tau\tau'}, \tag{1.64}$$

and that the expansion (1.60) of the field operators in terms of the $\hat{d}$-operators is

$$\hat{\psi}(\mathbf{x}) = \int \frac{d\mathbf{p}}{(2\pi)^3} e^{i\mathbf{p}\cdot\mathbf{r}} \hat{d}_{\mathbf{p}\sigma}, \qquad \hat{\psi}^\dagger(\mathbf{x}) = \int \frac{d\mathbf{p}}{(2\pi)^3} e^{-i\mathbf{p}\cdot\mathbf{r}} \hat{d}_{\mathbf{p}\sigma}^\dagger. \tag{1.65}$$

1.6 Hamiltonian in second quantization

The field operators are useful not only to construct the kets of N identical particles but also the operators acting on them. Let us consider again two identical particles and the center-of-mass operator (1.15). In first quantization the ket $|\mathbf{x}_1 \mathbf{x}_2\rangle$ is represented by the (anti)symmetrized product (1.35) of one-particle kets. It is instructive to calculate the action of $\hat{\mathcal{R}}_{CM}$ on $|\mathbf{x}_1\mathbf{x}_2\rangle$ to later appreciate the advantages of second quantization. We have

$$\hat{\mathcal{R}}_{CM}\frac{|\mathbf{x}_1\rangle|\mathbf{x}_2\rangle \pm |\mathbf{x}_2\rangle|\mathbf{x}_1\rangle}{\sqrt{2}} = \frac{1}{2}\frac{\mathbf{r}_1|\mathbf{x}_1\rangle|\mathbf{x}_2\rangle \pm \mathbf{r}_2|\mathbf{x}_2\rangle|\mathbf{x}_1\rangle + \mathbf{r}_2|\mathbf{x}_1\rangle|\mathbf{x}_2\rangle \pm \mathbf{r}_1|\mathbf{x}_2\rangle|\mathbf{x}_1\rangle}{\sqrt{2}}$$

$$= \frac{1}{2}(\mathbf{r}_1 + \mathbf{r}_2)\frac{|\mathbf{x}_1\rangle|\mathbf{x}_2\rangle \pm |\mathbf{x}_2\rangle|\mathbf{x}_1\rangle}{\sqrt{2}}. \tag{1.66}$$

Throughout this book we use calligraphic letters for operators acting on kets written in first quantization as opposed to operators (like the field operators) acting on kets written in second quantization (like $\hat{\psi}^\dagger(\mathbf{x}_N)\ldots\hat{\psi}^\dagger(\mathbf{x}_1)|0\rangle$). We refer to the former as *operators in first quantization* and to the latter as *operators in second quantization*. We now show that the very same result (1.66) can be obtained if we write the center-of-mass operator as

$$\hat{\mathbf{R}}_{CM} = \frac{1}{\hat{N}}\int d\mathbf{x}\,\mathbf{r}\,\hat{n}(\mathbf{x}),$$

where $1/\hat{N}$ is the inverse of the *operator of the total number of particles*

$$\boxed{\hat{N} \equiv \int d\mathbf{x}\,\hat{n}(\mathbf{x})}$$

in which

$$\boxed{\hat{n}(\mathbf{x}) = \hat{\psi}^\dagger(\mathbf{x})\hat{\psi}(\mathbf{x})} \tag{1.67}$$

is the so called *density operator* already introduced in Exercise 1.1. The origin of these names for the operators $\hat{n}(\mathbf{x})$ and $\hat{N}$ stems from the fact that $|\mathbf{x}_1\ldots\mathbf{x}_N\rangle$ is an eigenket of the density operator whose eigenvalue is exactly the density of N particles in the position–spin coordinates $\mathbf{x}_1,\ldots,\mathbf{x}_N$. Indeed

$$\hat{n}(\mathbf{x})\underbrace{\hat{\psi}^\dagger(\mathbf{x}_N)\hat{\psi}^\dagger(\mathbf{x}_{N-1})\ldots\hat{\psi}^\dagger(\mathbf{x}_1)|0\rangle}_{|\mathbf{x}_1\ldots\mathbf{x}_N\rangle} = \left[\hat{n}(\mathbf{x}),\hat{\psi}^\dagger(\mathbf{x}_N)\right]_{-}\hat{\psi}^\dagger(\mathbf{x}_{N-1})\ldots\hat{\psi}^\dagger(\mathbf{x}_1)|0\rangle$$

$$+ \hat{\psi}^\dagger(\mathbf{x}_N)\left[\hat{n}(\mathbf{x}),\hat{\psi}^\dagger(\mathbf{x}_{N-1})\right]_{-}\ldots\hat{\psi}^\dagger(\mathbf{x}_1)|0\rangle$$

$$\vdots$$

$$+ \hat{\psi}^\dagger(\mathbf{x}_N)\hat{\psi}^\dagger(\mathbf{x}_{N-1})\ldots\left[\hat{n}(\mathbf{x}),\hat{\psi}^\dagger(\mathbf{x}_1)\right]_{-}|0\rangle$$

$$= \underbrace{\left(\sum_{i=1}^N \delta(\mathbf{x}-\mathbf{x}_i)\right)}_{\substack{\text{density of }N\text{ particles}\\\text{in }\mathbf{x}_1,\ldots,\mathbf{x}_N}}|\mathbf{x}_1\ldots\mathbf{x}_N\rangle, \tag{1.68}$$

where we repeatedly use (1.49). This result tells us that any ket with N particles is an eigenket of $\hat{N}$ with eigenvalue N. By acting with $\hat{\mathbf{R}}_{CM}$ on $|\mathbf{x}_1\mathbf{x}_2\rangle = \hat{\psi}^\dagger(\mathbf{x}_2)\hat{\psi}^\dagger(\mathbf{x}_1)|0\rangle$ and taking into account (1.68) we find

$$\hat{\mathbf{R}}_{\mathrm{CM}}|\mathbf{x}_1\mathbf{x}_2\rangle = \int d\mathbf{x}\,\mathbf{r}\,\frac{1}{2}\left(\sum_{i=1}^{2}\delta(\mathbf{x}-\mathbf{x}_i)\right)|\mathbf{x}_1\mathbf{x}_2\rangle = \frac{1}{2}(\mathbf{r}_1+\mathbf{r}_2)|\mathbf{x}_1\mathbf{x}_2\rangle.$$

Simple and elegant! Both the operator and the ket are easy to manipulate and their expressions are undoubtedly shorter than the corresponding expressions in first quantization. A further advantage of second quantization is that the operator $\hat{\mathbf{R}}_{\mathrm{CM}}$ keeps the very same form independently of the number of particles; using (1.68) it is straightforward to verify that

$$\hat{\mathbf{R}}_{\mathrm{CM}}|\mathbf{x}_1\ldots\mathbf{x}_N\rangle = \frac{1}{N}\left(\sum_{i=1}^{N}\mathbf{r}_i\right)|\mathbf{x}_1\ldots\mathbf{x}_N\rangle.$$

On the contrary, $\hat{\boldsymbol{\mathcal{R}}}_{\mathrm{CM}}$ in (1.15) acts only on kets belonging to $\mathcal{H}_2$. For kets in $\mathcal{H}_N$ the center-of-mass operator in first quantization is given by (1.16). Thus, when working in Fock space it would be more rigorous to specify on which Hilbert space $\hat{\boldsymbol{\mathcal{R}}}_{\mathrm{CM}}$ acts. Denoting by $\hat{\boldsymbol{\mathcal{R}}}_{\mathrm{CM}}(N)$ the operator in (1.16) we can write down the relation between operators in first and second quantization as

$$\hat{\mathbf{R}}_{\mathrm{CM}} = \sum_{N=0}^{\infty}\hat{\boldsymbol{\mathcal{R}}}_{\mathrm{CM}}(N),$$

with the extra rule that $\hat{\boldsymbol{\mathcal{R}}}_{\mathrm{CM}}(N)$ yields the null ket when acting on a state of $\mathcal{H}_{M\neq N}$. In this book, however, we are not so meticulous with the notation. The Hilbert space on which operators in first quantization act is clear from the context.

The goal of this section is to extend the above example to general operators and in particular to derive an expression for the many-particle Hamiltonian $\hat{H} = \hat{H}_0 + \hat{H}_{\mathrm{int}}$. According to (1.17) the matrix element of the noninteracting Hamiltonian $\hat{H}_0$ between a position–spin ket and a generic ket $|\Psi\rangle$ is

$$\langle\mathbf{x}_1\ldots\mathbf{x}_N|\hat{H}_0|\Psi\rangle = \sum_{j=1}^{N}\sum_{\sigma'}h_{\sigma_j\sigma'}(\mathbf{r}_j,-\mathrm{i}\boldsymbol{\nabla}_j,\mathbf{S})\,\Psi(\mathbf{x}_1,\ldots,\mathbf{x}_{j-1},\mathbf{r}_j\sigma',\mathbf{x}_{j+1},\ldots,\mathbf{x}_N).$$

$$(1.69)$$

It is worth observing that for $N=1$ this expression reduces to

$$\langle\mathbf{x}|\hat{H}_0|\Psi\rangle = \sum_{\sigma'}h_{\sigma\sigma'}(\mathbf{r},-\mathrm{i}\boldsymbol{\nabla},\mathbf{S})\Psi(\mathbf{r}\sigma'),$$

which agrees with (1.13) when $|\Psi\rangle = |\mathbf{x}''\rangle$ since in this case $\Psi(\mathbf{r}\sigma') = \langle\mathbf{r}\sigma'|\mathbf{r}''\sigma''\rangle = \delta(\mathbf{r}-\mathbf{r}'')\delta_{\sigma'\sigma''}$. Similarly, we see from (1.18) that the matrix element of the interaction Hamiltonian $\hat{H}_{\mathrm{int}}$ between a position–spin ket and a generic ket $|\Psi\rangle$ is

$$\langle\mathbf{x}_1\ldots\mathbf{x}_N|\hat{H}_{\mathrm{int}}|\Psi\rangle = \frac{1}{2}\sum_{i\neq j}^{N}v(\mathbf{x}_i,\mathbf{x}_j)\,\Psi(\mathbf{x}_1,\ldots,\mathbf{x}_N). \qquad (1.70)$$

In (1.70) we considered the more general case of spin-dependent interactions $v(\mathbf{x}_1,\mathbf{x}_2)$, according to which the interaction energy between a particle in $\mathbf{r}_1$ and a particle in $\mathbf{r}_2$

depends also on the spin orientation σ_1 and σ_2 of these particles. Now the question is: how can we express $\hat{H}_0$ and $\hat{H}_{\text{int}}$ in terms of field operators?

We start our discussion with the noninteracting Hamiltonian. For pedagogical purposes we derive the operator $\hat{H}_0$ in second quantization in two different ways.

Derivation 1: In first quantization the noninteracting Hamiltonian $\hat{\mathcal{H}}_0$ of a system of N particles each described by $\hat{h}$ is given in (1.17). The first-quantization eigenkets of $\hat{\mathcal{H}}_0$ are obtained by forming (anti)symmetrized products of one-particle eigenkets of $\hat{h}$ and look like

$$|n_1 \ldots n_N\rangle = \frac{1}{\sqrt{N!}} \sum_P (\pm)^P |n_{P(1)}\rangle \ldots |n_{P(N)}\rangle, \qquad (1.71)$$

with

$$\hat{h}|n\rangle = \epsilon_n |n\rangle.$$

We leave as an exercise for the reader to show that

$$\hat{\mathcal{H}}_0 |n_1 \ldots n_N\rangle = (\epsilon_{n_1} + \ldots + \epsilon_{n_N}) |n_1 \ldots n_N\rangle.$$

The proof of this identity involves the same kinds of manipulation used to derive (1.66). To carry them out is useful to appreciate the simplicity of second quantization. We show below that in second quantization the noninteracting Hamiltonian $\hat{H}_0$ takes the compact form

$$\boxed{\hat{H}_0 = \int d\mathbf{x} d\mathbf{x}' \, \hat{\psi}^\dagger(\mathbf{x}) \langle \mathbf{x}|\hat{h}|\mathbf{x}'\rangle \hat{\psi}(\mathbf{x}')} \qquad (1.72)$$

independently of the number of particles. We prove (1.72) by showing that the second quantization ket $|n_1 \ldots n_N\rangle$ is an eigenket of $\hat{H}_0$ with eigenvalue $\epsilon_{n_1} + \ldots + \epsilon_{n_N}$. In second quantization $|n_1 \ldots n_N\rangle = \hat{d}^\dagger_{n_N} \ldots \hat{d}^\dagger_{n_1}|0\rangle$, with the $\hat{d}$-operators defined in (1.52) and (1.53). It is then natural to express $\hat{H}_0$ in terms of the $\hat{d}$-operators. Inserting a completeness relation between $\hat{h}$ and $|\mathbf{x}'\rangle$ we find

$$\hat{H}_0 = \sum_n \int d\mathbf{x} d\mathbf{x}' \, \hat{\psi}^\dagger(\mathbf{x}) \langle \mathbf{x}|\hat{h}|n\rangle \langle n|\mathbf{x}'\rangle \hat{\psi}(\mathbf{x}')$$

$$= \sum_n \epsilon_n \int d\mathbf{x} \, \hat{\psi}^\dagger(\mathbf{x}) \underbrace{\langle \mathbf{x}|n\rangle}_{\varphi_n(\mathbf{x})} \int d\mathbf{x}' \, \underbrace{\langle n|\mathbf{x}'\rangle}_{\varphi_n^*(\mathbf{x}')} \hat{\psi}(\mathbf{x}') = \sum_n \epsilon_n \hat{d}^\dagger_n \hat{d}_n, \qquad (1.73)$$

where we use $\hat{h}|n\rangle = \epsilon_n |n\rangle$. The $\hat{d}$-operators bring the Hamiltonian into a diagonal form, i.e., none of the off-diagonal combinations $\hat{d}^\dagger_n d_m$ with $m \neq n$ appears in $\hat{H}_0$. The *occupation operator*

$$\hat{n}_n \equiv \hat{d}^\dagger_n \hat{d}_n \qquad (1.74)$$

is the analog of the density operator $\hat{n}(\mathbf{x})$ in the position–spin basis; it counts how many particles have quantum number n. Using the (anti)commutation relations (1.54) and (1.55) it is easy to prove that

$$\left[\hat{n}_n, \hat{d}^\dagger_m\right]_- = \delta_{nm} \hat{d}^\dagger_m, \qquad \left[\hat{n}_n, \hat{d}_m\right]_- = -\delta_{nm} \hat{d}_m, \qquad (1.75)$$

which should be compared with the relations (1.48) and (1.49). The action of $\hat{H}_0$ on $|n_1 \dots n_N\rangle$ is then

$$\hat{H}_0 \underbrace{\hat{d}^\dagger_{n_N} \hat{d}^\dagger_{n_{N-1}} \cdots \hat{d}^\dagger_{n_1} |0\rangle}_{|n_1\dots n_N\rangle} = \sum_n \epsilon_n \left(\left[\hat{n}_n, \hat{d}^\dagger_{n_N}\right]_- \hat{d}^\dagger_{n_{N-1}} \cdots \hat{d}^\dagger_{n_1} |0\rangle \right.$$

$$+ \hat{d}^\dagger_{n_N} \left[\hat{n}_n, \hat{d}^\dagger_{n_{N-1}}\right]_- \cdots \hat{d}^\dagger_{n_1} |0\rangle$$

$$\vdots$$

$$\left. + \hat{d}^\dagger_{n_N} \hat{d}^\dagger_{n_{N-1}} \cdots \left[\hat{n}_n, \hat{d}^\dagger_{n_1}\right]_- |0\rangle \right)$$

$$= (\epsilon_{n_1} + \dots + \epsilon_{n_N}) \, \hat{d}^\dagger_{n_N} \hat{d}^\dagger_{n_{N-1}} \cdots \hat{d}^\dagger_{n_1} |0\rangle. \qquad (1.76)$$

This is exactly the result we wanted to prove: the Hamiltonian $\hat{H}_0$ is the correct second-quantized form of $\hat{\mathcal{H}}_0$. We can write $\hat{H}_0$ in different ways using the matrix elements (1.13) of $\hat{h}$. For instance

$$\hat{H}_0 = \sum_{\sigma\sigma'} \int d\mathbf{r} \, \hat{\psi}^\dagger(\mathbf{r}\sigma) h_{\sigma\sigma'}(\mathbf{r}, -i\boldsymbol{\nabla}, \mathbf{S})\hat{\psi}(\mathbf{r}\sigma'), \qquad (1.77)$$

or, equivalently,

$$\hat{H}_0 = \sum_{\sigma\sigma'} \int d\mathbf{r} \, \hat{\psi}^\dagger(\mathbf{r}\sigma) h_{\sigma\sigma'}(\mathbf{r}, i\overleftarrow{\boldsymbol{\nabla}}, \mathbf{S})\hat{\psi}(\mathbf{r}\sigma'). \qquad (1.78)$$

In these expressions the action of the gradient $\boldsymbol{\nabla}$ on a field operator is a formal expression which makes sense only when we sandwich $\hat{H}_0$ with a bra and a ket. For instance

$$\langle\chi|\hat{\psi}^\dagger(\mathbf{r}\sigma)\boldsymbol{\nabla}\hat{\psi}(\mathbf{r}\sigma')|\Psi\rangle \equiv \lim_{\mathbf{r}'\to\mathbf{r}} \boldsymbol{\nabla}'\langle\chi|\hat{\psi}^\dagger(\mathbf{r}\sigma)\hat{\psi}(\mathbf{r}'\sigma')|\Psi\rangle, \qquad (1.79)$$

where $\boldsymbol{\nabla}'$ is the gradient with respect to the primed variable. It is important to observe that for any arbitrary large but finite system the physical states have no particles at infinity. Therefore, if $|\chi\rangle$ and $|\Psi\rangle$ are physical states then (1.79) vanishes when $|\mathbf{r}| \to \infty$. More generally, the sandwich of a string of field operators $\hat{\psi}^\dagger(\mathbf{x}_1)\dots\hat{\psi}^\dagger(\mathbf{x}_N)\hat{\psi}(\mathbf{y}_1)\dots\hat{\psi}(\mathbf{y}_M)$ with two physical states vanishes when one of the coordinates of the field operators approaches infinity. The equivalence between (1.77) and (1.78) has to be understood as an equivalence between the sandwich of the corresponding r.h.s. with physical states. Consider for example $\hat{h} = \hat{p}^2/2m$. Equating the r.h.s. of (1.77) and (1.78) we get

$$\sum_\sigma \int d\mathbf{r} \, \hat{\psi}^\dagger(\mathbf{r}\sigma) \left[-\frac{\nabla^2}{2m} \hat{\psi}(\mathbf{r}\sigma) \right] = \sum_\sigma \int d\mathbf{r} \left[-\frac{\nabla^2}{2m} \hat{\psi}^\dagger(\mathbf{r}\sigma) \right] \hat{\psi}(\mathbf{r}\sigma).$$

This is an equality only provided that the integration by part produces a vanishing boundary term, i.e., only provided that for any two physical states $|\chi\rangle$ and $|\Psi\rangle$ the quantity in (1.79) vanishes when $|\mathbf{r}| \to \infty$.

Derivation 2: The second derivation consists in showing that the matrix elements of (1.77) or (1.78) are given by (1.69). Using (1.44) we find

$$\langle \mathbf{x}_1 \ldots \mathbf{x}_N | \hat{\psi}^\dagger(\mathbf{r}\sigma) h_{\sigma\sigma'}(\mathbf{r}, -i\boldsymbol{\nabla}, \mathbf{S})\hat{\psi}(\mathbf{r}\sigma') | \Psi \rangle$$

$$= \lim_{\mathbf{r}' \to \mathbf{r}} h_{\sigma\sigma'}(\mathbf{r}', -i\boldsymbol{\nabla}', \mathbf{S}) \sum_{j=1}^{N} (\pm)^{N+j} \delta(\mathbf{x}_j - \mathbf{x}) \langle \mathbf{x}_1 \ldots \mathbf{x}_{j-1}\mathbf{x}_{j+1} \ldots \mathbf{x}_N | \hat{\psi}(\mathbf{x}') | \Psi \rangle$$

$$= \lim_{\mathbf{r}' \to \mathbf{r}} h_{\sigma\sigma'}(\mathbf{r}', -i\boldsymbol{\nabla}', \mathbf{S}) \sum_{j=1}^{N} \delta(\mathbf{x}_j - \mathbf{x}) \Psi(\mathbf{x}_1, \ldots \mathbf{x}_{j-1}, \mathbf{x}', \mathbf{x}_{j+1}, \ldots, \mathbf{x}_N),$$

where we use the fact that it requires $N - j$ interchanges to put $\mathbf{x}'$ at position between $\mathbf{x}_{j-1}$ and $\mathbf{x}_{j+1}$. Summing over σ, σ' and integrating over $\mathbf{r}$ we get

$$\langle \mathbf{x}_1 \ldots \mathbf{x}_N | \hat{H}_0 | \Psi \rangle = \sum_{j=1}^{N} \sum_{\sigma'} \lim_{\mathbf{r}' \to \mathbf{r}_j} h_{\sigma_j\sigma'}(\mathbf{r}', -i\boldsymbol{\nabla}', \mathbf{S}) \Psi(\mathbf{x}_1, \ldots \mathbf{x}_{j-1}, \mathbf{x}', \mathbf{x}_{j+1}, \ldots, \mathbf{x}_N),$$

which coincides with the matrix element (1.69). Here and in the following we call *one-body operators* those operators in second quantization that can be written as a quadratic form of the field operators. The Hamiltonian $\hat{H}_0$ as well as the center-of-mass position operator $\hat{\mathbf{R}}_{\mathrm{CM}}$ are one-body operators.

From (1.76) it is evident that one-body Hamiltonians can only describe noninteracting systems since the eigenvalues are the sum of one-particle eigenvalues, and the latter do not depend on the position of the other particles. If there is an interaction $v(\mathbf{x}_1, \mathbf{x}_2)$ between one particle in $\mathbf{x}_1$ and another particle in $\mathbf{x}_2$ the corresponding interaction energy operator $\hat{H}_{\mathrm{int}}$ cannot be a one-body operator. The energy to put a particle in a given point depends on where the other particles are located. Suppose that there is a particle in $\mathbf{x}_1$. Then if we want to put a particle in $\mathbf{x}_2$ we must pay an energy $v(\mathbf{x}_1, \mathbf{x}_2)$. The addition of another particle in $\mathbf{x}_3$ will cost an energy $v(\mathbf{x}_1, \mathbf{x}_3) + v(\mathbf{x}_2, \mathbf{x}_3)$. In general if we have N particles in $\mathbf{x}_1, \ldots, \mathbf{x}_N$ the total interaction energy is $\sum_{i<j} v(\mathbf{x}_i, \mathbf{x}_j) = \frac{1}{2} \sum_{i \neq j} v(\mathbf{x}_i, \mathbf{x}_j)$. To derive the form of $\hat{H}_{\mathrm{int}}$ in second quantization we simply note that the ket $|\mathbf{x}_1 \ldots \mathbf{x}_N\rangle$ is an eigenket of $\hat{H}_{\mathrm{int}}$ with eigenvalue $\frac{1}{2} \sum_{i \neq j} v(\mathbf{x}_i, \mathbf{x}_j)$, i.e.,

$$\hat{H}_{\mathrm{int}} | \mathbf{x}_1 \ldots \mathbf{x}_N \rangle = \left(\frac{1}{2} \sum_{i \neq j} v(\mathbf{x}_i, \mathbf{x}_j) \right) | \mathbf{x}_1 \ldots \mathbf{x}_N \rangle. \tag{1.80}$$

Equivalently (1.80) follows directly from the matrix element (1.70), which is valid for all $|\Psi\rangle$. Due to the presence of a double sum in (1.80) the operator $\hat{H}_{\mathrm{int}}$ must be a *quartic* form in the field operators. In (1.68) we proved that $|\mathbf{x}_1 \ldots \mathbf{x}_N\rangle$ is an eigenket of the density operator $\hat{n}(\mathbf{x})$ with eigenvalue $\sum_i \delta(\mathbf{x} - \mathbf{x}_i)$. This implies that $|\mathbf{x}_1 \ldots \mathbf{x}_N\rangle$ is also an eigenket of the operator $\hat{n}(\mathbf{x})\hat{n}(\mathbf{x}')$ with eigenvalue $\sum_{i,j} \delta(\mathbf{x} - \mathbf{x}_i)\delta(\mathbf{x}' - \mathbf{x}_j)$. Thus, taking into account that the double sum in (1.80) does not contain terms with $i = j$, the interaction energy operator is given by

$$\hat{H}_{\text{int}} = \frac{1}{2} \int d\mathbf{x}\, d\mathbf{x}'\, v(\mathbf{x}, \mathbf{x}') \hat{n}(\mathbf{x}) \hat{n}(\mathbf{x}') - \frac{1}{2} \int d\mathbf{x}\, v(\mathbf{x}, \mathbf{x}) \hat{n}(\mathbf{x})$$

$$= \frac{1}{2} \int d\mathbf{x}\, d\mathbf{x}'\, v(\mathbf{x}, \mathbf{x}') \left(\hat{\psi}^\dagger(\mathbf{x}) \hat{\psi}(\mathbf{x}) \hat{\psi}^\dagger(\mathbf{x}') \hat{\psi}(\mathbf{x}') - \delta(\mathbf{x} - \mathbf{x}') \hat{\psi}^\dagger(\mathbf{x}) \hat{\psi}(\mathbf{x}) \right)$$

$$= \frac{1}{2} \int d\mathbf{x}\, d\mathbf{x}'\, v(\mathbf{x}, \mathbf{x}') \hat{\psi}^\dagger(\mathbf{x}) \hat{\psi}^\dagger(\mathbf{x}') \hat{\psi}(\mathbf{x}') \hat{\psi}(\mathbf{x}). \qquad (1.81)$$

In the last equality we first use the (anti)commutation relation (1.47) to cancel the term proportional to $\delta(\mathbf{x} - \mathbf{x}')$, and then (1.40) to exchange the operators $\hat{\psi}(\mathbf{x})$ and $\hat{\psi}(\mathbf{x}')$. It is easy to verify that the action of $\hat{H}_{\text{int}}$ on $|\mathbf{x}_1 \ldots \mathbf{x}_N\rangle$ yields (1.80). Like the one-body Hamiltonian $\hat{H}_0$, the interaction energy operator keeps the very same form independently of the number of particles. We call *two-body operators* those operators that can be written as a quartic form of the field operators and, in general, *n-body operators* those operators that contain a string of n field operators $\hat{\psi}^\dagger$ followed by a string of n field operators $\hat{\psi}$.

The total Hamiltonian of a system of interacting identical particles is the sum of $\hat{H}_0$ and $\hat{H}_{\text{int}}$ and reads

$$\boxed{\hat{H} = \int d\mathbf{x}\, d\mathbf{x}'\, \hat{\psi}^\dagger(\mathbf{x}) \langle \mathbf{x} | \hat{h} | \mathbf{x}' \rangle \hat{\psi}(\mathbf{x}') + \frac{1}{2} \int d\mathbf{x}\, d\mathbf{x}'\, v(\mathbf{x}, \mathbf{x}') \hat{\psi}^\dagger(\mathbf{x}) \hat{\psi}^\dagger(\mathbf{x}') \hat{\psi}(\mathbf{x}') \hat{\psi}(\mathbf{x})} \qquad (1.82)$$

Equation (1.82) is the main result of this section. To calculate the action of $\hat{H}$ on a ket $|\Psi\rangle$ we only need to know the (anti)commutation relations since $|\Psi\rangle$ can always be expanded in terms of $\hat{\psi}^\dagger(\mathbf{x}_1) \ldots \hat{\psi}^\dagger(\mathbf{x}_N)|0\rangle$. Equivalently, given a convenient one-body basis $\{|n\rangle\}$ we may work with the $\hat{d}$-operators. This is done by expressing $\hat{H}$ in terms of the $\hat{d}$-operators, expanding $|\Psi\rangle$ on the basis $\hat{d}_{n_1}^\dagger \ldots \hat{d}_{n_N}^\dagger |0\rangle$ and then using the (anti)commutation relations (1.54) and (1.55). To express $\hat{H}$ in terms of the $\hat{d}$-operators we simply substitute the expansion (1.60) in (1.82) and find

$$\hat{H} = \underbrace{\sum_{ij} h_{ij} \hat{d}_i^\dagger \hat{d}_j}_{\hat{H}_0} + \underbrace{\frac{1}{2} \sum_{ijmn} v_{ijmn} \hat{d}_i^\dagger \hat{d}_j^\dagger \hat{d}_m \hat{d}_n}_{\hat{H}_{\text{int}}}, \qquad (1.83)$$

with

$$h_{ij} = \langle i | \hat{h} | j \rangle = \sum_{\sigma\sigma'} \int d\mathbf{r}\, \varphi_i^*(\mathbf{r}\sigma) h_{\sigma\sigma'}(\mathbf{r}, -i\boldsymbol{\nabla}, \mathbf{S}) \varphi_j(\mathbf{r}\sigma') = h_{ji}^*, \qquad (1.84)$$

and the so called *Coulomb integrals*[6]

$$v_{ijmn} = \int d\mathbf{x}\, d\mathbf{x}'\, \varphi_i^*(\mathbf{x}) \varphi_j^*(\mathbf{x}') v(\mathbf{x}, \mathbf{x}') \varphi_m(\mathbf{x}') \varphi_n(\mathbf{x}). \qquad (1.85)$$

In the new basis the single-particle Hamiltonian in first quantization can be written in the ket-bra form

$$\hat{h} = \sum_{ij} h_{ij} |i\rangle\langle j|, \qquad (1.86)$$

as can easily be checked by taking the matrix element $\langle i | \hat{h} | j \rangle$ and comparing with (1.84).

---

[6]In fact the nomenclature Coulomb integral is appropriate only if v is the Coulomb interaction.

We recall that the quantum numbers of the general basis comprise an orbital and a spin quantum number. For later purposes it is instructive to highlight the spin structure in (1.83). We write the quantum numbers i, j, m, n as

$$i = s_1\sigma_1, \quad j = s_2\sigma_2, \quad m = s_3\sigma_3, \quad n = s_4\sigma_4.$$

Then the one-body part reads

$$\hat{H}_0 = \sum_{\substack{s_1 s_2 \\ \sigma_1\sigma_2}} h_{s_1\sigma_1 \, s_2\sigma_2} \hat{d}^\dagger_{s_1\sigma_1} \hat{d}_{s_2\sigma_2}.$$

In the absence of magnetic fields or spin-orbit coupling h does not depend on $\mathbf{S}$ and hence its matrix elements are diagonal in spin space $h_{ij} = \delta_{\sigma_1\sigma_2} h_{s_1 s_2}$. In this case $\hat{H}_0$ takes the simpler form

$$\hat{H}_0 = \sum_{s_1 s_2} \sum_\sigma h_{s_1 s_2} \hat{d}^\dagger_{s_1\sigma} \hat{d}_{s_2\sigma}, \tag{1.87}$$

where $h_{s_1 s_2}$ is the spatial integral in (1.84) with the functions $\varphi_s(\mathbf{r})$ defined in (1.50). For interparticle interactions $v(\mathbf{x}_1, \mathbf{x}_2) = v(\mathbf{r}_1, \mathbf{r}_2)$ which are independent of spin the interaction Hamiltonian can be manipulated in a similar manner. From (1.85) we see that v_{ijmn} vanishes if j and m have different spin projection ($\sigma_2 \neq \sigma_3$) or if i and n have different spin projection ($\sigma_1 \neq \sigma_4$), i.e.,

$$v_{ijmn} = \delta_{\sigma_2\sigma_3} \delta_{\sigma_1\sigma_4} v_{s_1 s_2 s_3 s_4},$$

where $v_{s_1 s_2 s_3 s_4}$ is the spatial integral in (1.85) with the functions $\varphi_s(\mathbf{r})$. Inserting this form of the interaction into $\hat{H}_{\text{int}}$ we find

$$\hat{H}_{\text{int}} = \frac{1}{2} \sum_{\substack{s_1 s_2 s_3 s_4 \\ \sigma\sigma'}} v_{s_1 s_2 s_3 s_4} \hat{d}^\dagger_{s_1\sigma} \hat{d}^\dagger_{s_2\sigma'} \hat{d}_{s_3\sigma'} \hat{d}_{s_4\sigma}. \tag{1.88}$$

We propose below a few simple exercises to practice with operators in second quantization. In the next chapter we illustrate physically relevant examples and use some of the identities from the exercises to acquire familiarity with this new formalism.

Exercise 1.4. Let $\hat{n}_n \equiv \hat{d}^\dagger_n \hat{d}_n$ be the occupation operator for particles with quantum number n, see (1.74). Prove that in the fermionic case

$$\hat{n}_n^2 = \hat{n}_n, \tag{1.89}$$

and hence that the eigenvalues of $\hat{n}_n$ are either 0 or 1, i.e., it is not possible to create two fermions in the same state $|n\rangle$. This is a direct consequence of the Pauli exclusion principle.

Exercise 1.5. Prove that the total number of particle operator $\hat{N} = \int d\mathbf{x}\, \hat{\psi}^\dagger(\mathbf{x})\hat{\psi}(\mathbf{x})$ can also be written as $\hat{N} = \sum_n \hat{d}^\dagger_n \hat{d}_n$ for any orthonormal basis $|n\rangle$. Calculate the action of $\hat{N}$ on a generic ket $|\Psi_N\rangle$ with N particles ($|\Psi_N\rangle \in \mathcal{H}_N$) and prove that

$$\hat{N}|\Psi_N\rangle = N|\Psi_N\rangle.$$

Exercise 1.6. Prove that $\hat{N}$ commutes with $\hat{H}_0$ and $\hat{H}_{\text{int}}$, i.e.,

$$[\hat{N}, \hat{H}_0]_- = [\hat{N}, \hat{H}_{\text{int}}]_- = 0. \tag{1.90}$$

This means that the eigenkets of $\hat{H}$ can be chosen as kets with a fixed number of particles.

Exercise 1.7. Let $n = s\sigma$ and $\sigma = \uparrow, \downarrow$ be the spin projection for fermions of spin $1/2$. We consider the operators

$$\hat{S}_s^z \equiv \frac{1}{2}(\hat{n}_{s\uparrow} - \hat{n}_{s\downarrow}), \qquad \hat{S}_s^+ \equiv \hat{d}_{s\uparrow}^\dagger \hat{d}_{s\downarrow}, \qquad \hat{S}_s^- \equiv \hat{d}_{s\downarrow}^\dagger \hat{d}_{s\uparrow} = (\hat{S}_s^+)^\dagger. \tag{1.91}$$

Using the anticommutation relations prove that the action of the above operators on the kets $|s\sigma\rangle \equiv \hat{d}_{s\sigma}^\dagger|0\rangle$ is

$$\hat{S}_s^z|s\uparrow\rangle = \frac{1}{2}|s\uparrow\rangle, \qquad \hat{S}_s^+|s\uparrow\rangle = |\emptyset\rangle, \qquad \hat{S}_s^-|s\uparrow\rangle = |s\downarrow\rangle,$$

and

$$\hat{S}_s^z|s\downarrow\rangle = -\frac{1}{2}|s\downarrow\rangle, \qquad \hat{S}_s^+|s\downarrow\rangle = |s\uparrow\rangle, \qquad \hat{S}_s^-|s\downarrow\rangle = |\emptyset\rangle.$$

To what operators do $\hat{S}_s^z$, $\hat{S}_s^+$, $\hat{S}_s^-$ correspond?

Exercise 1.8. Let us define the *spin operators* along the x and y directions as

$$\hat{S}_s^x \equiv \frac{1}{2}(\hat{S}_s^+ + \hat{S}_s^-), \qquad \hat{S}_s^y \equiv \frac{1}{2i}(\hat{S}_s^+ - \hat{S}_s^-),$$

and the spin–density operator $\hat{S}_s^z$ along the z direction as in (1.91). Prove that these operators can also be written as

$$\hat{S}_s^j = \frac{1}{2}\sum_{\sigma\sigma'}\hat{d}_{s\sigma}^\dagger\sigma_{\sigma\sigma'}^j\hat{d}_{s\sigma'}, \qquad j = x, y, z, \tag{1.92}$$

with

$$\sigma^x = \begin{pmatrix} 0 & 1 \\ 1 & 0 \end{pmatrix}, \qquad \sigma^y = \begin{pmatrix} 0 & -i \\ i & 0 \end{pmatrix}, \qquad \sigma^z = \begin{pmatrix} 1 & 0 \\ 0 & -1 \end{pmatrix},$$

the *Pauli matrices*. Using the anticommutation relations verify also that

$$[\hat{S}_s^i, \hat{S}_{s'}^j]_- = i\delta_{ss'}\sum_{k=x,y,z}\varepsilon_{ijk}\hat{S}_s^k,$$

where ε_{ijk} is the *Levi–Civita tensor*.[7]

---

[7]The Levi-Civita tensor is zero if at least two indices are equal and otherwise

$$\varepsilon_{P(1)P(2)P(3)} = (-)^P,$$

where P is an arbitrary permutation of the indices 1, 2, 3.

1.7 Density matrices and quantum averages

We have already stressed several times that if we know the ket $|\Psi\rangle$, which means if we know the wavefunction $\Psi(\mathbf{x}_1, \ldots, \mathbf{x}_N) = \langle \mathbf{x}_1 \ldots \mathbf{x}_N | \Psi \rangle$, then we can use the (anti)commutation rules of the field operators to calculate any quantum average $\langle \Psi | \hat{O} | \Psi \rangle$, where $\hat{O}$ is an operator in second quantization. Most operators of physical interest are one- or two-body operators (or in any case n-body operators with $n \ll N$). For the averages of these operators a full knowledge of the wavefunction is redundant, as we now show. Consider for instance the most general one-body operator

$$\hat{O} = \int d\mathbf{x} d\mathbf{y} \; O(\mathbf{x}, \mathbf{y}) \, \hat{\psi}^\dagger(\mathbf{x}) \hat{\psi}(\mathbf{y}).$$

To calculate the quantum average over the state $|\Psi\rangle$ we must evaluate the quantum average $\langle \Psi | \hat{\psi}^\dagger(\mathbf{x}) \hat{\psi}(\mathbf{y}) | \Psi \rangle$ for all $\mathbf{x}$ and $\mathbf{y}$. This quantity can be seen as the inner product between the states $\hat{\psi}(\mathbf{x}) | \Psi \rangle$ and $\hat{\psi}(\mathbf{y}) | \Psi \rangle$ which both contain $(N-1)$ particles. We therefore have

$$\begin{aligned}
\langle \Psi | \hat{\psi}^\dagger(\mathbf{x}) \hat{\psi}(\mathbf{y}) | \Psi \rangle &= \langle \Psi | \hat{\psi}^\dagger(\mathbf{x}) \hat{\mathbb{1}} \hat{\psi}(\mathbf{y}) | \Psi \rangle \\
&= \frac{1}{(N-1)!} \int d\mathbf{x}_1 \ldots d\mathbf{x}_{N-1} \langle \Psi | \mathbf{x}_1 \ldots \mathbf{x}_{N-1} \mathbf{x} \rangle \langle \mathbf{x}_1 \ldots \mathbf{x}_{N-1} \mathbf{y} | \Psi \rangle \\
&= \frac{1}{(N-1)!} \int d\mathbf{x}_1 \ldots d\mathbf{x}_{N-1} \Psi^*(\mathbf{x}_1, \ldots, \mathbf{x}_{N-1}, \mathbf{x}) \Psi(\mathbf{x}_1, \ldots, \mathbf{x}_{N-1}, \mathbf{y}), \quad (1.93)
\end{aligned}$$

where in the second line we used the completeness relation (1.32) as well as the definition of the field operators, according to which

$$\hat{\psi}^\dagger(\mathbf{x}) | \mathbf{x}_1 \ldots \mathbf{x}_{N-1} \rangle = | \mathbf{x}_1 \ldots \mathbf{x}_{N-1} \mathbf{x} \rangle.$$

We thus see that to calculate $\langle \Psi | \hat{O} | \Psi \rangle$ it is enough to know the integral over all coordinates except one of the product of the wavefunction with itself. The quantity

$$\boxed{\Gamma_1(\mathbf{y}; \mathbf{x}) \equiv \langle \Psi | \hat{\psi}^\dagger(\mathbf{x}) \hat{\psi}(\mathbf{y}) | \Psi \rangle}$$

is called the *one-particle density matrix*. Sometimes we use the alternative notation

$$n(\mathbf{y}, \mathbf{x}) = \Gamma_1(\mathbf{y}; \mathbf{x}),$$

since the one-particle density matrix is the generalization of the quantum average of the density operator

$$n(\mathbf{x}) = \langle \Psi | \hat{n}(\mathbf{x}) | \Psi \rangle = \langle \Psi | \hat{\psi}^\dagger(\mathbf{x}) \hat{\psi}(\mathbf{x}) | \Psi \rangle = n(\mathbf{x}, \mathbf{x}).$$

From a knowledge of the one-particle density matrix we can, in particular, calculate the quantum average of the noninteracting Hamiltonian $\hat{H}_0$

$$\langle \Psi | \hat{H}_0 | \Psi \rangle = \int d\mathbf{x} d\mathbf{y} \langle \mathbf{x} | \hat{h} | \mathbf{y} \rangle \Gamma_1(\mathbf{y}; \mathbf{x}).$$

Taking for example the Hamiltonian $\hat{h}$ with matrix elements (1.13) we find

$$\langle\Psi|\hat{H}_0|\Psi\rangle = \sum_{\sigma'\sigma} \int d\mathbf{r}\, h_{\sigma'\sigma}(\mathbf{r}, -i\mathbf{\nabla}, \mathbf{S})\, \Gamma_1(\mathbf{r}\sigma; \mathbf{r}'\sigma')\bigg|_{\mathbf{r}'=\mathbf{r}}.$$

The one-particle density matrix is Hermitian in the position–spin indices, i.e.,

$$\boxed{\Gamma_1(\mathbf{y}; \mathbf{x}) = \Gamma_1^*(\mathbf{x}; \mathbf{y})}$$

and its trace equals the number of particles in the state $|\Psi\rangle$

$$\boxed{\int d\mathbf{x}\, \Gamma_1(\mathbf{x}, \mathbf{x}) = \int d\mathbf{x}\, n(\mathbf{x}) = N} \tag{1.94}$$

The eigenfunctions ϕ_k of Γ_1 are defined as the solution of the eigenvalue problem

$$\int d\mathbf{x}\, \Gamma_1(\mathbf{y}; \mathbf{x})\phi_k(\mathbf{x}) = n_k\phi_k(\mathbf{y}) \tag{1.95}$$

and are called the *natural orbitals*. The natural orbitals can always be chosen orthonormal and their set forms a basis in the one-particle Hilbert space. The eigenvalues n_k can be interpreted as the occupation of the natural orbitals for the following reasons. First, they sum up to N due to (1.94), i.e.,

$$\sum_k n_k = N.$$

Second, if we multiply both sides of (1.95) by $\phi_k^*(\mathbf{y})$ and integrate over $\mathbf{y}$ we find

$$n_k = \int d\mathbf{x} d\mathbf{y}\, \phi_k^*(\mathbf{y}) \underbrace{\langle\Psi|\hat{\psi}^\dagger(\mathbf{x})\hat{\psi}(\mathbf{y})|\Psi\rangle}_{\Gamma_1(\mathbf{y};\mathbf{x})} \phi_k(\mathbf{x}) = \langle\Psi|\hat{d}_k^\dagger \hat{d}_k|\Psi\rangle,$$

where the $\hat{d}$-operators are defined as in (1.52) and (1.53). Thus n_k is the quantum average of the occupation operator of the kth natural orbital. In fermionic systems the eigenvalues of $\hat{d}_k^\dagger \hat{d}_k$ are either 0 or 1, see (1.89), and therefore[8]

$$0 \leq n_k \leq 1.$$

Finally, if the state $|\Psi\rangle = |m_1 \ldots m_N\rangle$ is the permanent/determinant of N functions belonging to some orthonormal basis $\varphi_m(\mathbf{x}) = \langle\mathbf{x}|m\rangle$ then the natural orbitals are simply the functions of the basis, i.e., $\phi_k = \varphi_k$, as can easily be checked. If the quantum number k appears N_k times in the string $m_1 \ldots m_N$ then the eigenvalues n_k are equal to N_k.

Let us continue our discussion on quantum averages by considering two-body operators. The most general form of a two-body operator is

$$\hat{O} = \int d\mathbf{x} d\mathbf{x}' d\mathbf{y} d\mathbf{y}'\, O(\mathbf{x}, \mathbf{x}', \mathbf{y}, \mathbf{y}')\, \hat{\psi}^\dagger(\mathbf{x})\hat{\psi}^\dagger(\mathbf{x}')\hat{\psi}(\mathbf{y}')\hat{\psi}(\mathbf{y}).$$

---

[8]Any fermionic ket $|\Psi\rangle$ can be written as the linear combination of two eigenkets $|\Psi_0\rangle$, $|\Psi_1\rangle$ of the occupation operator $\hat{d}_k^\dagger \hat{d}_k$ with eigenvalues 0 and 1 respectively: $|\Psi\rangle = \alpha|\Psi_0\rangle + \beta|\Psi_1\rangle$. Then $\langle\Psi|\hat{d}_k^\dagger \hat{d}_k|\Psi\rangle = |\beta|^2 \leq 1$ since the normalization of $|\Psi\rangle$ requires that $|\alpha|^2 + |\beta|^2 = 1$.

To evaluate $\langle\Psi|\hat{O}|\Psi\rangle$ we need to know the *two-particle density-matrix*

$$\Gamma_2(\mathbf{y},\mathbf{y}';\mathbf{x},\mathbf{x}') \equiv \langle\Psi|\hat{\psi}^\dagger(\mathbf{x})\hat{\psi}^\dagger(\mathbf{x}')\hat{\psi}(\mathbf{y}')\hat{\psi}(\mathbf{y})|\Psi\rangle$$

for all $\mathbf{y},\mathbf{y}',\mathbf{x},\mathbf{x}'$. Note that for notational convenience the order of the first two arguments of Γ_2 is reversed with respect to the string of the $\hat{\psi}$ operators. As before we can see Γ_2 as the inner product between two states with $(N-2)$ particles and therefore

$$\Gamma_2(\mathbf{y},\mathbf{y}';\mathbf{x},\mathbf{x}') = \frac{1}{(N-2)!}\int d\mathbf{x}_1\ldots d\mathbf{x}_{N-2}\,\Psi^*(\mathbf{x}_1,\ldots,\mathbf{x}_{N-2},\mathbf{x}',\mathbf{x})$$

$$\times\Psi(\mathbf{x}_1,\ldots,\mathbf{x}_{N-2},\mathbf{y}',\mathbf{y}). \qquad (1.96)$$

Thus, to calculate $\langle\Psi|\hat{O}|\Psi\rangle$ we do not need the full N-particle wavefunction; it is enough to know the integral over all coordinates except two of the product of the wavefunction with itself. In particular, the quantum average of the interaction Hamiltonian over the state $|\Psi\rangle$ reads

$$\langle\Psi|\hat{H}_{\text{int}}|\Psi\rangle = \frac{1}{2}\int d\mathbf{x}d\mathbf{x}'\,v(\mathbf{x},\mathbf{x}')\Gamma_2(\mathbf{x},\mathbf{x}';\mathbf{x},\mathbf{x}'). \qquad (1.97)$$

It is instructive to derive a few properties of the density matrices. We start by introducing a generalization of Γ_1 and Γ_2 which is the n-particle density matrix

$$\Gamma_n(\mathbf{y}_1,\ldots,\mathbf{y}_n;\mathbf{x}_1,\ldots,\mathbf{x}_n) = \langle\Psi|\hat{\psi}^\dagger(\mathbf{x}_1)\ldots\hat{\psi}^\dagger(\mathbf{x}_n)\hat{\psi}(\mathbf{y}_n)\ldots\hat{\psi}(\mathbf{y}_1)|\Psi\rangle.$$

If the ket $|\Psi\rangle$ describes N particles then $\Gamma_n = 0$ for all $n > N$. For $n = N$ we have

$$\Gamma_N(\mathbf{y}_1,\ldots,\mathbf{y}_N;\mathbf{x}_1,\ldots,\mathbf{x}_N) = \langle\Psi|\hat{\psi}^\dagger(\mathbf{x}_1)\ldots\hat{\psi}^\dagger(\mathbf{x}_N)|0\rangle\langle 0|\hat{\psi}(\mathbf{y}_N)\ldots\hat{\psi}(\mathbf{y}_1)|\Psi\rangle$$

$$= \langle\Psi|\mathbf{x}_N\ldots\mathbf{x}_1\rangle\langle\mathbf{y}_N\ldots\mathbf{y}_1|\Psi\rangle$$

$$= \Psi^*(\mathbf{x}_1,\ldots,\mathbf{x}_N)\,\Psi(\mathbf{y}_1,\ldots,\mathbf{y}_N),$$

where we use the fact that acting with N annihilation operators on $|\Psi\rangle$ produces the zero-particle ket $|0\rangle$ and we further use the (anti)symmetry of the wavefunction in the last equality. So the N-particle density matrix is simply the product of two wavefunctions. The $(N-1)$-particle density matrix can be obtained by integrating out one coordinate. Setting $\mathbf{y}_N = \mathbf{x}_N$ we have

$$\int d\mathbf{x}_N\Gamma_N(\mathbf{y}_1,\ldots,\mathbf{y}_{N-1},\mathbf{x}_N;\mathbf{x}_1,\ldots,\mathbf{x}_N)$$

$$= \langle\Psi|\hat{\psi}^\dagger(\mathbf{x}_1)\ldots\hat{\psi}^\dagger(\mathbf{x}_{N-1})\underbrace{\left(\int d\mathbf{x}_N\,\hat{\psi}^\dagger(\mathbf{x}_N)\hat{\psi}(\mathbf{x}_N)\right)}_{\hat{N}}\hat{\psi}(\mathbf{y}_{N-1})\ldots\hat{\psi}(\mathbf{y}_1)|\Psi\rangle$$

$$= \langle\Psi|\hat{\psi}^\dagger(\mathbf{x}_1)\ldots\hat{\psi}^\dagger(\mathbf{x}_{N-1})\hat{\psi}(\mathbf{y}_{N-1})\ldots\hat{\psi}(\mathbf{y}_1)|\Psi\rangle$$

$$= \Gamma_{N-1}(\mathbf{y}_1,\ldots,\mathbf{y}_{N-1};\mathbf{x}_1,\ldots,\mathbf{x}_{N-1}),$$

where we use the fact that the operator $\hat{N}$ acts on a one-particle state. We can continue this procedure and integrate out coordinate $\mathbf{y}_{N-1} = \mathbf{x}_{N-1}$. We then obtain again an expression involving the operator $\hat{N}$ that now acts on a two-particle state finally yielding $2\Gamma_{N-2}$. Subsequently integrating out more coordinates we find

$$\Gamma_k(\mathbf{y}_1,\ldots,\mathbf{y}_k;\mathbf{x}_1,\ldots,\mathbf{x}_k)$$
$$= \frac{1}{(N-k)!}\int d\mathbf{x}_{k+1}\ldots d\mathbf{x}_N\,\Gamma_N(\mathbf{y}_1,\ldots,\mathbf{y}_k,\mathbf{x}_{k+1},\ldots,\mathbf{x}_N;\mathbf{x}_1,\ldots,\mathbf{x}_N).$$

We see that the k-particle density matrices are simply obtained by integrating $N-k$ coordinates out of the N-particle density matrix. In particular for $k=1,\,2$ the above results coincide with (1.93) and (1.96). In Appendix C we derive further properties of the density matrices and discuss how one can give them a probability interpretation.

Before concluding the section we would like to draw the attention of the reader to an important point. In many physical situations one is interested in calculating the ground-state energy of the Hamiltonian $\hat{H} = \hat{H}_0 + \hat{H}_{\text{int}}$. From basic courses of quantum mechanics we know that this energy can be obtained, in principle, by minimizing the quantity $\langle\Psi|\hat{H}|\Psi\rangle$ over *all* possible (normalized) N-particle states $|\Psi\rangle$. This is in general a formidable task since the wavefunction $\Psi(\mathbf{x}_1,\ldots,\mathbf{x}_N)$ depends on N coordinates. We have just learned, however, that

$$\langle\Psi|\hat{H}|\Psi\rangle = \int d\mathbf{x}d\mathbf{y}\,\langle\mathbf{x}|\hat{h}|\mathbf{y}\rangle\,\Gamma_1(\mathbf{y};\mathbf{x}) + \frac{1}{2}\int d\mathbf{x}d\mathbf{x}'\,v(\mathbf{x},\mathbf{x}')\Gamma_2(\mathbf{x},\mathbf{x}';\mathbf{x},\mathbf{x}').$$

Could we not minimize the quantum average of the Hamiltonian with respect to Γ_1 and Γ_2? This would be a great achievement since Γ_1 and Γ_2 depend only on 2 and 4 coordinates respectively. Unfortunately the answer to the question is, at present, negative. While we know the constraints to construct physical one-particle density matrices (for instance for fermions Γ_1 must be Hermitian with eigenvalues between 0 and 1 that sum up to N) we still do not know the constraints for Γ_2 [4]. In this book we learn how to *approximate* Γ_1 and Γ_2. As we shall see the density matrices are not the most natural objects to work with. We therefore introduce other quantities called *Green's functions* from which to extract the density matrices and much more.

2

Getting familiar with second quantization: model Hamiltonians

2.1 Model Hamiltonians

In all practical calculations the properties of a many-particle system are extracted by using a finite number of (physically relevant) single-particle basis functions. For instance, in systems like crystals or molecules the electrons are attracted by the positive charge of the nuclei and it is reasonable to expect that a few localized orbitals around each nucleus provide a good-enough basis set. If we think of the H_2 molecule the simplest description consists in taking one basis function $\varphi_{n=1\sigma}$ for an electron of spin σ localized around the first nucleus and another one $\varphi_{n=2\sigma}$ for an electron of spin σ localized around the second nucleus, see the schematic representation below.

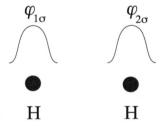

If the set $\{\varphi_n\}$ is not complete in $\mathcal{H}_1$ then the expansion (1.60) is an approximation for the field operators. The approximate field operators satisfy approximate (anti)commutation relations. Let us show this with an example. We consider a space grid in three-dimensional space with uniform grid spacing Δ. To each grid point $\mathbf{r}_s = (x_s, y_s, z_s)$ we assign a basis function that is constant in a cube of linear dimension Δ centered in $\mathbf{r}_s$, see Fig. 2.1(a). These basis functions are orthonormal and have the following mathematical structure

$$\varphi_{n=s\sigma_s}(\mathbf{r}\sigma) = \delta_{\sigma\sigma_s} \frac{\theta_{x_s}(x)\theta_{y_s}(y)\theta_{z_s}(z)}{\Delta^{3/2}}, \tag{2.1}$$

with the Heaviside function $\theta_a(x) = \theta(\frac{1}{2}\Delta - |a - x|)$, see Fig. 2.1(b), and the prefactor $1/\Delta^{3/2}$ which guarantees the correct normalization. For any finite grid spacing Δ the set

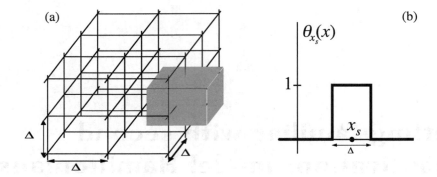

Figure 2.1 (a) Space grid with vertices (grid points) in $\mathbf{r}_s$. To each grid point is assigned a basis function as described in the main text. (b) The piecewise function used to construct the approximate basis.

$\{\varphi_n\}$ is not a complete set. However, if the physical properties of the system vary on a length scale much larger than Δ then the use of the approximate field operators,

$$\hat{\psi}^\dagger(\mathbf{r}\sigma) \sim \hat{\psi}_\Delta^\dagger(\mathbf{r}\sigma) = \sum_s \frac{\theta_{x_s}(x)\theta_{y_s}(y)\theta_{z_s}(z)}{\Delta^{3/2}}\hat{d}_{s\sigma}^\dagger,$$

is expected to work fine. The (anti)commutation relation for these approximate field operators is

$$\left[\hat{\psi}_\Delta(\mathbf{x}),\hat{\psi}_\Delta^\dagger(\mathbf{x}')\right]_\mp = \delta_{\sigma\sigma'}\sum_{ss'}\frac{\theta_{x_s}(x)\theta_{y_s}(y)\theta_{z_s}(z)\theta_{x_{s'}}(x')\theta_{y_{s'}}(y')\theta_{z_{s'}}(z')}{\Delta^3}\underbrace{\left[\hat{d}_{s,\sigma},\hat{d}_{s',\sigma}^\dagger\right]_\mp}_{\delta_{ss'}}.$$

Thus we see that the (anti)commutator is zero if $\mathbf{r}$ and $\mathbf{r}'$ belong to different cubes and is equal to $\delta_{\sigma\sigma'}/\Delta^3$ otherwise. The accuracy of the results can be checked by increasing the number of basis functions. In our example this corresponds to reducing the spacing Δ. Indeed, in the limit $\Delta \to 0$ the product $\theta_{x_s}(x)\theta_{x_s}(x')/\Delta \to \delta(x - x')$ and similarly for y and z, and hence the (anti)commutator approaches the exact result

$$\left[\hat{\psi}_\Delta(\mathbf{x}),\hat{\psi}_\Delta^\dagger(\mathbf{x}')\right]_\mp \xrightarrow{\Delta\to0} \delta_{\sigma\sigma'}\delta(x - x')\delta(y - y')\delta(z - z') = \delta(\mathbf{x} - \mathbf{x}').$$

In this chapter we discuss how to choose a proper set of basis functions for some relevant physical systems, construct the corresponding Hamiltonians, and derive a few elementary results. The aim of the following sections is not to provide an exhaustive presentation of these *model Hamiltonians*, something that would require a monograph for each model, but rather to become familiar with the formalism of second quantization. We think that the best way to learn how to manipulate the field operators is by seeing them at work.

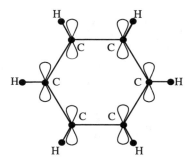

Figure 2.2 Representation of the benzene molecule C_6H_6 with a p_z orbital for each carbon atom. We recall that the electronic configuration of carbon is $1s^2 2s^2 2p^2$ and that in the benzene geometry the $2s$, $2p_x$, $2p_y$ orbitals hybridize to form three sp^2 orbitals [8]. The latter share an electron with the nearest hydrogen as well as with the two nearest neighbour carbons.

2.2 Pariser–Parr–Pople model

A popular model Hamiltonian often employed to describe organic molecules is the so called *Pariser–Parr–Pople model* or simply the PPP model [5, 6]. We give here an elementary derivation of the PPP model and refer the reader to more specialized textbooks for a careful justification of the simplifications involved [7]. As a concrete example we consider an atomic ring like, e.g., the benzene molecule C_6H_6 of Fig. 2.2, but the basic ideas can be used for other molecular geometries as well. If we are interested in the low energy physics of the system, such as for instance its ground-state properties, we can assume that the inner shell electrons are "frozen" in their molecular orbitals while the outer shell electrons are free to wander around the molecule. In the case of benzene we may consider as frozen the two $1s$ electrons of each carbon atom C as well as the three electrons of the in-plane sp^2 orbitals which form σ-bonds with the hydrogen atom H and with the two nearest carbon atoms. For the description of the remaining six electrons (one per C–H unit) we could limit ourselves to use a p_z orbital for each carbon atom. In general the problem is always to find a minimal set of functions to describe the dynamics of the "free" (also called valence) electrons responsible for the low-energy excitations.

Let us assign a single orbital to each atomic position $\mathbf{R}_s$ of the ring

$$\tilde{\varphi}_{s\tau}(\mathbf{r}\sigma) = \delta_{\sigma\tau} f(\mathbf{r} - \mathbf{R}_s),$$

where $s = 1, \ldots, N$ and N is the number of atoms in the ring. The function $f(\mathbf{r})$ is localized around $\mathbf{r} = 0$; an example could be the exponential function $e^{-\alpha|\mathbf{r}|}$, see Fig. 2.3.

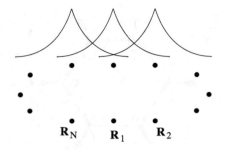

Figure 2.3 Orbitals $\tilde{\varphi}_{s\tau}$ localized around the atomic position $\mathbf{R}_s$ of the atomic ring.

The set of functions $\{\tilde{\varphi}_{s\tau}\}$ is, in general, not an orthonormal set since the *overlap matrix*

$$S_{ss'} = \int d\mathbf{x}\ \tilde{\varphi}_{s\tau}^*(\mathbf{x})\tilde{\varphi}_{s'\tau}(\mathbf{x}) = \int d\mathbf{r}\ f^*(\mathbf{r} - \mathbf{R}_s)f(\mathbf{r} - \mathbf{R}_{s'})$$

may have nonvanishing off-diagonal elements. A simple way to orthonormalize the set $\{\tilde{\varphi}_{s\tau}\}$ without losing the local character of the functions is the following. The overlap matrix is Hermitian and positive-definite,[1] meaning that all its eigenvalues λ_k are larger than zero. Let $D = \text{diag}(\lambda_1, \lambda_2, \ldots)$ and U be the unitary matrix which brings S into its diagonal form, i.e., $S = UDU^\dagger$. We define the square root of the matrix S according to

$$S^{1/2} = UD^{1/2}U^\dagger \qquad \text{with} \qquad D^{1/2} = \text{diag}(\sqrt{\lambda_1}, \sqrt{\lambda_2}, \ldots).$$

The matrix $S^{1/2}$ is also Hermitian and positive-definite and can easily be inverted

$$S^{-1/2} = UD^{-1/2}U^\dagger \qquad \text{with} \qquad D^{-1/2} = \text{diag}\left(\frac{1}{\sqrt{\lambda_1}}, \frac{1}{\sqrt{\lambda_2}}, \ldots\right).$$

We then construct the new set of functions

$$\varphi_{s\tau}(\mathbf{x}) = \sum_{s'} \tilde{\varphi}_{s'\tau}(\mathbf{x})S_{s's}^{-1/2}$$

whose overlap is

$$\int d\mathbf{x}\ \varphi_{s_1\tau_1}^*(\mathbf{x})\varphi_{s_2\tau_2}(\mathbf{x}) = \sum_{s's''}\int d\mathbf{x}\ S_{s_1s'}^{-1/2}\tilde{\varphi}_{s'\tau_1}^*(\mathbf{x})\tilde{\varphi}_{s''\tau_2}(\mathbf{x})S_{s''s_2}^{-1/2}$$

$$= \delta_{\tau_1\tau_2}(S^{-1/2}SS^{-1/2})_{s_1s_2} = \delta_{\tau_1\tau_2}\delta_{s_1s_2}.$$

---

[1]The positive definiteness of S follows from its definition. Given an arbitrary vector with components v_s we have

$$\sum_{ss'} v_s^* S_{ss'} v_{s'} = \int d\mathbf{x}\ \left(\sum_s v_s^*\tilde{\varphi}_{s\tau}^*(\mathbf{x})\right)\left(\sum_{s'} v_{s'}\tilde{\varphi}_{s'\tau}(\mathbf{x})\right) > 0.$$

The set $\{\varphi_{s\tau}\}$ is therefore orthonormal. If the off-diagonal elements of the overlap matrix are small compared to unity then the new functions are only slightly delocalized. Consider for instance an overlap matrix of the form

$$
S = \begin{pmatrix}
1 & \delta & 0 & 0 & \cdots & \delta \\
\delta & 1 & \delta & 0 & & 0 \\
0 & \delta & 1 & \ddots & & 0 \\
0 & 0 & \ddots & \ddots & \ddots & \vdots \\
\vdots & & & \ddots & 1 & \delta \\
\delta & 0 & 0 & \cdots & \delta & 1
\end{pmatrix} = \hat{\mathbb{1}} + \Delta,
$$

according to which we only have an overlap of amount $\delta \ll 1$ between nearest neighbor atoms (note that the matrix element $S_{1N} = S_{N1} = \delta$ since atom N is the nearest neighbor of atom 1). To first order in δ the inverse of the square root of S is

$$
S^{-1/2} = (\hat{\mathbb{1}} + \Delta)^{-1/2} \sim \hat{\mathbb{1}} - \frac{1}{2}\Delta,
$$

and therefore the new functions are slightly spread over the nearest neighbor atoms

$$
\varphi_{s\tau}(\mathbf{x}) = \tilde{\varphi}_{s\tau}(\mathbf{x}) - \frac{\delta}{2}\tilde{\varphi}_{s+1\tau}(\mathbf{x}) - \frac{\delta}{2}\tilde{\varphi}_{s-1\tau}(\mathbf{x}), \tag{2.2}
$$

where it is understood that the index $s \pm N$ must be identified with s. The orthonormal functions $\{\varphi_{s\tau}\}$ are our set over which to expand the field operators

$$
\hat{\psi}^\dagger(\mathbf{x}) \sim \sum_{s\tau} \varphi_{s\tau}^*(\mathbf{x})\hat{d}_{s\tau}^\dagger,
$$

with

$$
\hat{d}_{s\tau}^\dagger = \int d\mathbf{x}\, \varphi_{s\tau}(\mathbf{x})\hat{\psi}^\dagger(\mathbf{x}),
$$

and similar relations for the adjoint operators. The $\hat{d}$-operators satisfy the anti-commutation relations (1.54) since $\{\varphi_{s\tau}\}$ is an orthonormal set. Inserting the approximate expansion of the field operators into the Hamiltonian $\hat{H}$ we obtain an approximate Hamiltonian. This Hamiltonian looks like (1.83) but the sums are restricted to the incomplete set of quantum numbers s. In this way the original field operators $\hat{\psi}(\mathbf{x})$ and $\hat{\psi}^\dagger(\mathbf{x})$ get replaced by a finite (or at most countable) number of $\hat{d}$-operators. The parameters h_{ij} and v_{ijmn} of the approximate Hamiltonian depend on the *specific choice of basis functions* and on *the microscopic details of the system* such as, e.g., the mass and charge of the particles (we remind the reader that in this book we work in atomic units so that for electrons $m_e = -e = 1$), the scalar and vector potentials, the interparticle interaction, etc. Once these parameters are given, the approximate Hamiltonian is fully specified and we refer to it as the *model Hamiltonian*. Below we estimate the parameters h_{ij} and v_{ijmn} for the set $\{\varphi_{s\tau}\}$ in (2.2) and for an atomic ring like benzene.

Let us write the functions (2.2) as a product of an orbital and a spin part: $\varphi_{s\tau}(\mathbf{r}\sigma) = \delta_{\sigma\tau}\varphi_s(\mathbf{r})$. Since the $\varphi_s(\mathbf{r})$s are localized around the atomic position $\mathbf{R}_s$ the dominant Coulomb integrals $v_{s_1 s_2 s_3 s_4}$ are those with $s_1 = s_4$ and $s_2 = s_3$, see (1.85). We therefore make the approximation

$$v_{s_1 s_2 s_3 s_4} \sim \delta_{s_1 s_4}\delta_{s_2 s_3}\underbrace{\int d\mathbf{r}d\mathbf{r}'|\varphi_{s_1}(\mathbf{r})|^2 v(\mathbf{r},\mathbf{r}')|\varphi_{s_2}(\mathbf{r}')|^2}_{v_{s_1 s_2}}. \tag{2.3}$$

The quantity $v_{s_1 s_2}$ equals the classical interaction energy between the charge distributions $|\varphi_{s_1}|^2$ and $|\varphi_{s_2}|^2$. For the Coulomb interaction $v(\mathbf{r},\mathbf{r}') = 1/|\mathbf{r} - \mathbf{r}'|$ we can further approximate the integral $v_{s_1 s_2}$ when $s_1 \neq s_2$ as

$$v_{s_1 s_2} \sim \frac{1}{|\mathbf{R}_{s_1} - \mathbf{R}_{s_2}|}, \tag{2.4}$$

since in the neighborhood of $|\mathbf{R}_{s_1} - \mathbf{R}_{s_2}|$ the function $1/r$ is slowly varying and can be considered constant. Inserting these results into (1.88) we obtain the model form of the interaction operator

$$\hat{H}_{\text{int}} = \frac{1}{2}\sum_{\substack{ss'\\ \sigma\sigma'}} v_{ss'}\hat{d}^\dagger_{s\sigma}\hat{d}^\dagger_{s'\sigma'}\hat{d}_{s'\sigma'}\hat{d}_{s\sigma} = \frac{1}{2}\sum_{s\neq s'} v_{ss'}\hat{n}_s\hat{n}_{s'} + \sum_s v_{ss}\hat{n}_{s\uparrow}\hat{n}_{s\downarrow}, \tag{2.5}$$

with $n_{s\sigma} = \hat{d}^\dagger_{s\sigma}\hat{d}_{s\sigma}$ the occupation operator that counts how many electrons (0 or 1) are in the spin-orbital $\varphi_{s\sigma}$ and $\hat{n}_s \equiv \hat{n}_{s\uparrow} + \hat{n}_{s\downarrow}$.[2]

By the same overlap argument we can neglect all matrix elements $h_{ss'}$ between atomic sites that are not nearest neighbors (we are implicitly assuming that $h_{s\sigma s'\sigma'} = \delta_{\sigma\sigma'}h_{ss'}$). Let $\langle ss'\rangle$ denote the couples of nearest neighbor atomic sites. Then the noninteracting part (1.87) of the Hamiltonian takes the form

$$\hat{H}_0 = \sum_s h_{ss}\hat{n}_s + \sum_{\langle ss'\rangle}\sum_\sigma h_{ss'}\hat{d}^\dagger_{s\sigma}\hat{d}_{s'\sigma}.$$

In the present case the coefficients $h_{ss'}$ are given by

$$h_{ss'} = \langle s\sigma|\hat{h}|s'\sigma\rangle = \int d\mathbf{r}\,\varphi_s^*(\mathbf{r})\left[-\frac{\nabla^2}{2} - V(\mathbf{r})\right]\varphi_{s'}(\mathbf{r}).$$

The potential $V(\mathbf{r})$ in this expression is the sum of the electrostatic potentials between an electron in $\mathbf{r}$ and the atomic nuclei in $\mathbf{R}_s$, i.e.,

$$V(\mathbf{r}) = +\sum_s \frac{Z_s}{|\mathbf{r} - \mathbf{R}_s|},$$

---

[2]In writing the first term on the r.h.s. of (2.5) we use the fact that for $s\sigma \neq s'\sigma'$ the operator $\hat{d}_{s\sigma}$ commutes with $\hat{d}^\dagger_{s'\sigma'}\hat{d}_{s'\sigma'}$. For the second term we further took into account that for $s = s'$ and $\sigma = \sigma'$ the product $\hat{d}_{s\sigma}\hat{d}_{s\sigma} = 0$.

where Z_s is the effective nuclear positive charge of atom s, i.e., the sum of the bare nuclear charge and the screening charge of the frozen electrons. Let us manipulate the diagonal elements h_{ss}. Using the explicit form of the potential we can write

$$h_{ss} = \epsilon_s + \beta_s,$$

with

$$\epsilon_s = \int d\mathbf{r}\, \varphi_s^*(\mathbf{r}) \left[-\frac{\nabla^2}{2} - \frac{Z_s}{|\mathbf{r} - \mathbf{R}_s|} \right] \varphi_s(\mathbf{r}),$$

and

$$\beta_s = -\sum_{s' \neq s} \int d\mathbf{r}\, |\varphi_s(\mathbf{r})|^2 \frac{Z_{s'}}{|\mathbf{r} - \mathbf{R}_{s'}|}.$$

Since $|\varphi_s|^2$ is the charge distribution of an electron localized in $\mathbf{R}_s$ we can, as before, approximately write

$$\beta_s \sim -\sum_{s' \neq s} \frac{Z_{s'}}{|\mathbf{R}_s - \mathbf{R}_{s'}|} = -\sum_{s' \neq s} Z_{s'} v_{ss'},$$

where we use (2.4). Inserting these results into $\hat{H}_0$ and adding $\hat{H}_{\text{int}}$ we find the PPP model Hamiltonian

$$\hat{H} = \hat{H}_0 + \hat{H}_{\text{int}} = \sum_s \epsilon_s \hat{n}_s + \sum_{\langle ss' \rangle} \sum_\sigma h_{ss'} \hat{d}_{s\sigma}^\dagger \hat{d}_{s'\sigma}$$

$$+ \frac{1}{2} \sum_{s \neq s'} v_{ss'} (\hat{n}_s - Z_s)(\hat{n}_{s'} - Z_{s'}) + \sum_s v_{ss} \hat{n}_{s\uparrow} \hat{n}_{s\downarrow},$$

where we have also added the constant $\frac{1}{2} \sum_{s \neq s'} v_{ss'} Z_s Z_{s'}$ corresponding to the electrostatic energy of the screened nuclei. If the ground state $|\Psi_0\rangle$ has exactly Z_s electrons on atom s, i.e., $\hat{n}_s |\Psi_0\rangle = Z_s |\Psi_0\rangle$ for all s, then the only interaction energy comes from the last term. In general, however, this is not the case since for $|\Psi_0\rangle$ to be an eigenstate of all $\hat{n}_s$ it must be $|\Psi_0\rangle = |s_1\sigma_1 s_2\sigma_2 \ldots\rangle$. This state is an eigenstate of $\hat{H}$ only provided that the off-diagonal elements $h_{ss'} = 0$, which is satisfied when the atoms are infinitely far apart from each other.

The purpose of deriving the PPP model was mainly pedagogical. Every model Hamiltonian in the scientific literature has similar underlying assumptions and approximations. Most of these models cannot be solved exactly despite the fact that the Hamiltonian is undoubtedly simpler than the original continuum Hamiltonian. In the next sections we discuss other examples of model Hamiltonians suitable for describing other physical systems.

2.3 Noninteracting models

We discussed in Section 1.6 how to find eigenvalues and eigenvectors of the noninteracting Hamiltonian $\hat{H}_0$. In the absence of interactions the many-particle problem reduces to a single-particle problem. The interparticle interaction makes our life much more complicated

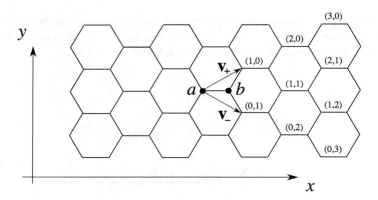

Figure 2.4 Crystal structure of graphene: a unit cell with two carbon atoms (a and b) repeated periodically along the directions $\mathbf{v}_+$ and $\mathbf{v}_-$ to form a honeycomb lattice. For illustrative purposes some unit cells have been labeled with the two-dimensional vector of integers $\mathbf{n} = (n_1, n_2)$.

(and interesting) and we have to resort to approximative methods in order to make progress. The zeroth order approximation consists of neglecting the interparticle interaction altogether; how much physics can we capture in this way? The answer to this question clearly depends on the system at hand and on the physical properties that interest us. It turns out that in nature *some* physical properties of *some* systems are not so sensitive to the interparticle interaction. For instance, a noninteracting treatment of crystals is, in many cases, enough to assess whether the crystal is a metal or an insulator. In this section we consider two paradigmatic examples of noninteracting models.

2.3.1 Bloch theorem and band structure

A crystal consists of a unit cell repeated periodically in three, two, or one dimensions. Each unit cell contains the same (finite) number of atoms arranged in the same geometry. An example is *graphene*, i.e., a planar structure of sp^2-bonded carbon atoms arranged in a honeycomb lattice as illustrated in Fig. 2.4. In this case the unit cell consists of two carbon atoms, a and b in the figure, repeated periodically along the directions $\mathbf{v}_+$ and $\mathbf{v}_-$. The unit cells can be labeled with a vector of integers $\mathbf{n} = (n_1, \ldots, n_d)$ where d is the dimension of the crystal, see again Fig. 2.4. The expansion of the vector $\mathbf{n}$ over the orthonormal basis $\{\mathbf{e}_i\}$ with $(\mathbf{e}_i)_j = \delta_{ij}$ reads

$$\mathbf{n} = \sum_{i=1}^{d} n_i \mathbf{e}_i.$$

Two unit cells with labels $\mathbf{n}$ and $\mathbf{n}'$ are nearest neighbor if $|\mathbf{n} - \mathbf{n}'| = 1$. As for the PPP model we assign to each unit cell a set of localized orbitals $\{\varphi_{\mathbf{n}s\sigma}\}$ which we assume already orthonormal and denote by $\hat{d}^\dagger_{\mathbf{n}s\sigma}$ and $\hat{d}_{\mathbf{n}s\sigma}$, the creation and annihilation operators for electrons in the orbital $\varphi_{\mathbf{n}s\sigma}$. The index s runs over all orbitals of a given unit cell. In the case of graphene we may assign a single p_z orbital to each carbon, so $s = 1, 2$.

In the absence of magnetic fields or other couplings that break the spin symmetry, the matrix elements of $\hat{h}$ are

$$h_{\mathbf{n}s\sigma\,\mathbf{n}'s'\sigma'} = \delta_{\sigma\sigma'} h_{\mathbf{n}s\,\mathbf{n}'s'}.$$

If we choose orbitals localized around the atomic positions then the matrix elements $h_{\mathbf{n}s\,\mathbf{n}'s'}$ are very small for $|\mathbf{n} - \mathbf{n}'| \gg 1$. We then discard $h_{\mathbf{n}s\,\mathbf{n}'s'}$ unless $\mathbf{n}$ and $\mathbf{n}'$ are nearest neighbors. The periodicity of a crystal is reflected in the fact that the unit cell Hamiltonian $h_{\mathbf{n}s\,\mathbf{n}s'}$, as well as the Hamiltonian connecting two nearest neighbor cells $h_{\mathbf{n}s\,\mathbf{n}\pm\mathbf{e}_is'}$, does not depend on $\mathbf{n}$. We therefore define the matrices

$$h_{ss'} \equiv h_{\mathbf{n}s\,\mathbf{n}s'}, \tag{2.6}$$

and

$$T_{i,ss'} \equiv h_{\mathbf{n}+\mathbf{e}_is\,\mathbf{n}s'}, \quad \Rightarrow \quad T_{i,ss'}^\dagger = h_{\mathbf{n}+\mathbf{e}_is'\,\mathbf{n}}^* = h_{\mathbf{n}s\,\mathbf{n}+\mathbf{e}_is'}$$
$$= h_{\mathbf{n}-\mathbf{e}_is\,\mathbf{n}s'}. \tag{2.7}$$

With these definitions our model for the noninteracting Hamiltonian of a crystal takes the form

$$\hat{H}_0 = \sum_{\mathbf{n}\sigma} \sum_{ss'} \left(h_{ss'} \hat{d}_{\mathbf{n}s\sigma}^\dagger \hat{d}_{\mathbf{n}s'\sigma} + \sum_{i=1}^d T_{i,ss'} \hat{d}_{\mathbf{n}+\mathbf{e}_is\sigma}^\dagger \hat{d}_{\mathbf{n}s'\sigma} + T_{i,ss'}^\dagger \hat{d}_{\mathbf{n}-\mathbf{e}_is\sigma}^\dagger \hat{d}_{\mathbf{n}s'\sigma} \right).$$

To better visualize the matrix structure of $\hat{H}_0$ we introduce the vector of $\hat{d}$-operators

$$\hat{\mathbf{d}}_{\mathbf{n}\sigma}^\dagger \equiv (\hat{d}_{\mathbf{n}1\sigma}^\dagger, \hat{d}_{\mathbf{n}2\sigma}^\dagger, \ldots), \qquad \hat{\mathbf{d}}_{\mathbf{n}\sigma} = (\hat{d}_{\mathbf{n}1\sigma}^\dagger, \hat{d}_{\mathbf{n}2\sigma}^\dagger, \ldots)^\dagger,$$

and rewrite $\hat{H}_0$ in terms of the product between these vectors and the matrices h, T_i and $T_i^\dagger$,

$$\hat{H}_0 = \sum_{\mathbf{n}\sigma} \left(\hat{\mathbf{d}}_{\mathbf{n}\sigma}^\dagger h\, \hat{\mathbf{d}}_{\mathbf{n}\sigma} + \sum_{i=1}^d \hat{\mathbf{d}}_{\mathbf{n}+\mathbf{e}_i\sigma}^\dagger T_i\, \hat{\mathbf{d}}_{\mathbf{n}\sigma} + \hat{\mathbf{d}}_{\mathbf{n}-\mathbf{e}_i\sigma}^\dagger T_i^\dagger\, \hat{\mathbf{d}}_{\mathbf{n}\sigma} \right). \tag{2.8}$$

The strategy to find the one-particle eigenvalues of $\hat{H}_0$ consists of considering a finite block of the crystal and then letting the volume of the block go to infinity. This block is, for convenience, chosen as a parallelepiped with edges given by d linearly independent vectors of integers $\mathbf{N}_1, \ldots, \mathbf{N}_d$ each radiating from a given unit cell. Without loss of generality we can choose the radiating unit cell at the origin. The set V_b of all unit cells contained in the block is then

$$\mathsf{V}_b = \left\{ \mathbf{n} : \frac{\mathbf{n} \cdot \mathbf{N}_i}{\mathbf{N}_i \cdot \mathbf{N}_i} < 1 \text{ for all } i \right\}. \tag{2.9}$$

In the case of a one-dimensional crystal we can choose only one vector $\mathbf{N}_1 = N$ and the unit cells of the block $\mathbf{n} = n \in \mathsf{V}_b$ are those with $n = 0, 1, \ldots, N-1$. For the two-dimensional graphene, instead, we may choose, e.g., $\mathbf{N}_1 = (N, 0)$ and $\mathbf{N}_2 = (0, M)$ or $\mathbf{N}_1 = (N, N)$ and $\mathbf{N}_2 = (M, -M)$ or any other couple of linearly independent vectors. When the volume of the block tends to infinity the eigenvalues of the crystal are independent of the choice of the block. Interestingly, however, the procedure below allows us to know the eigenvalues of $\hat{H}_0$ for any *finite* $\mathbf{N}_1, \ldots, \mathbf{N}_d$, and hence to have access to the eigenvalues of blocks of different shapes. We come back to this aspect at the end of the section. To simplify the mathematical treatment we impose periodic boundary conditions along $\mathbf{N}_1, \ldots, \mathbf{N}_d$, i.e., we "wrap" the block onto itself forming a ring in $d = 1$, a torus in $d = 2$, etc. This choice of boundary conditions is known as the *Born–von Karman (BvK) boundary condition* and turns out to be very convenient. Other kinds of boundary condition would lead to the same results in the limit of large V_b. The BvK condition implies that the cells $\mathbf{n}$ and $\mathbf{n} + \mathbf{N}_i$ are actually the same cell and we therefore make the identifications

$$\hat{d}^\dagger_{\mathbf{n}+\mathbf{N}_i\sigma} \equiv \hat{d}^\dagger_{\mathbf{n}\sigma}, \qquad \hat{d}_{\mathbf{n}+\mathbf{N}_i\sigma} \equiv \hat{d}_{\mathbf{n}\sigma}, \tag{2.10}$$

for all $\mathbf{N}_i$.

We are now in the position to show how the diagonalization procedure works. Consider the Hamiltonian (2.8) in which the sum over $\mathbf{n}$ is restricted to unit cells in V_b and the identification (2.10) holds for all boundary terms with one $\hat{d}$-operator outside V_b. To bring $\hat{H}_0$ into a diagonal form we must find a suitable linear combination of the $\hat{d}$-operators that preserves the anti-commutation relations, as discussed in Section 1.6. For this purpose we construct the matrix U with elements

$$U_{\mathbf{n}\mathbf{k}} = \frac{1}{\sqrt{|\mathsf{V}_b|}} e^{i\mathbf{k}\cdot\mathbf{n}}, \tag{2.11}$$

where $|\mathsf{V}_b|$ is the number of unit cells in V_b. In (2.11) the row index runs over all $\mathbf{n} \in \mathsf{V}_b$ whereas the column index runs over all vectors $\mathbf{k} = (k_1, \ldots, k_d)$ with components $-\pi < k_i \leq \pi$ and, more importantly, fulfilling

$$\mathbf{k} \cdot \mathbf{N}_i = 2\pi m_i, \qquad \text{with } m_i \text{ integers}. \tag{2.12}$$

We leave it as an exercise for the reader to prove that the number of $\mathbf{k}$ vectors with these properties is exactly $|\mathsf{V}_b|$ and hence that U is a square matrix. Due to property (2.12) the quantity $U_{\mathbf{n}\mathbf{k}}$ is periodic in $\mathbf{n}$ with periods $\mathbf{N}_1, \ldots, \mathbf{N}_d$. It is this periodicity that we now exploit to prove that U is unitary. Consider the following set of equalities

$$e^{ik'_1} \sum_{\mathbf{n}\in\mathsf{V}_b} U^*_{\mathbf{n}\mathbf{k}} U_{\mathbf{n}\mathbf{k}'} = \sum_{\mathbf{n}\in\mathsf{V}_b} U^*_{\mathbf{n}\mathbf{k}} U_{\mathbf{n}+\mathbf{e}_1\mathbf{k}'}$$

$$= \sum_{\mathbf{n}\in\mathsf{V}_b} U^*_{\mathbf{n}-\mathbf{e}_1\mathbf{k}} U_{\mathbf{n}\mathbf{k}'}$$

$$= e^{ik_1} \sum_{\mathbf{n}\in\mathsf{V}_b} U^*_{\mathbf{n}\mathbf{k}} U_{\mathbf{n}\mathbf{k}'},$$

and the likes with $k'_2, \ldots, k'_d$. In the second line of the above identities we use the fact that the sum over all $\mathbf{n} \in \mathsf{V}_b$ of a periodic function $f(\mathbf{n})$ is the same as the sum of $f(\mathbf{n} - \mathbf{e}_1)$. For the left and the right hand side of these equations to be the same we must either have $\mathbf{k} = \mathbf{k}'$ or

$$\sum_{\mathbf{n}} U^*_{\mathbf{n}\mathbf{k}} U_{\mathbf{n}\mathbf{k}'} = 0 \qquad \text{for } \mathbf{k} \neq \mathbf{k}'.$$

Since for $\mathbf{k} = \mathbf{k}'$ the sum $\sum_{\mathbf{n}} U_{\mathbf{nk}}^{*} U_{\mathbf{nk}} = 1$, the matrix U is unitary and hence the operators

$$\hat{c}_{\mathbf{k}\sigma} = \frac{1}{\sqrt{|V_b|}} \sum_{\mathbf{n} \in V_b} e^{-i\mathbf{k}\cdot\mathbf{n}} \, \hat{d}_{\mathbf{n}\sigma}, \qquad \hat{c}_{\mathbf{k}\sigma}^{\dagger} = \frac{1}{\sqrt{|V_b|}} \sum_{\mathbf{n} \in V_b} e^{i\mathbf{k}\cdot\mathbf{n}} \, \hat{d}_{\mathbf{n}\sigma}^{\dagger},$$

preserve the anti-commutation relations. The inverse relations read

$$\hat{d}_{\mathbf{n}\sigma} = \frac{1}{\sqrt{|V_b|}} \sum_{\mathbf{k}} e^{i\mathbf{k}\cdot\mathbf{n}} \, \hat{c}_{\mathbf{k}\sigma}, \qquad \hat{d}_{\mathbf{n}\sigma}^{\dagger} = \frac{1}{\sqrt{|V_b|}} \sum_{\mathbf{k}} e^{-i\mathbf{k}\cdot\mathbf{n}} \, \hat{c}_{\mathbf{k}\sigma}^{\dagger},$$

and the reader can easily check that due to property (2.12) the $\hat{d}$-operators satisfy the BvK boundary conditions (2.10). Inserting these inverse relations into (2.8) (in which the sum is restricted to $\mathbf{n} \in V_b$) we get the Hamiltonian

$$\hat{H}_0 = \sum_{\mathbf{k}\sigma} \hat{c}_{\mathbf{k}\sigma}^{\dagger} \underbrace{\left(h + \sum_{i=1}^{d} (T_i e^{-ik_i} + T_i^{\dagger} e^{ik_i}) \right)}_{h_{\mathbf{k}}} \hat{c}_{\mathbf{k}\sigma}.$$

In this expression the matrix $h_{\mathbf{k}}$ is Hermitian and can be diagonalized. Let $\epsilon_{\mathbf{k}\nu}$ be the eigenvalues of $h_{\mathbf{k}}$ and $u_{\mathbf{k}}$ be the unitary matrix that brings $h_{\mathbf{k}}$ into the diagonal form, i.e., $h_{\mathbf{k}} = u_{\mathbf{k}} \mathrm{diag}(\epsilon_{\mathbf{k}1}, \epsilon_{\mathbf{k}2}, \ldots) u_{\mathbf{k}}^{\dagger}$. The unitary matrix $u_{\mathbf{k}}$ has the dimension of the number of orbitals in the unit cell and should not be confused with the matrix U whose dimension is $|V_b|$. We now perform a further change of basis and construct the following linear combinations of the $\hat{c}$-operators with fixed $\mathbf{k}$ vector

$$\hat{b}_{\mathbf{k}\sigma} = u_{\mathbf{k}}^{\dagger} \hat{c}_{\mathbf{k}\sigma}, \qquad \hat{b}_{\mathbf{k}\sigma}^{\dagger} = \hat{c}_{\mathbf{k}\sigma}^{\dagger} u_{\mathbf{k}}.$$

Denoting by $u_{\mathbf{k}\nu}(s) = (u_{\mathbf{k}})_{s\nu}$ the (s, ν) matrix element of $u_{\mathbf{k}}$ the explicit form of the $\hat{b}$-operators is

$$\hat{b}_{\mathbf{k}\nu\sigma} = \sum_{s} u_{\mathbf{k}\nu}^{*}(s) \hat{c}_{\mathbf{k}s\sigma}, \qquad \hat{b}_{\mathbf{k}\nu\sigma}^{\dagger} = \sum_{s} u_{\mathbf{k}\nu}(s) \hat{c}_{\mathbf{k}s\sigma}^{\dagger}.$$

With these definitions the Hamiltonian $\hat{H}_0$ takes the desired form since

$$\hat{H}_0 = \sum_{\mathbf{k}\sigma} \hat{c}_{\mathbf{k}\sigma}^{\dagger} u_{\mathbf{k}} \, \mathrm{diag}(\epsilon_{\mathbf{k}1}, \epsilon_{\mathbf{k}2}, \ldots) \, u_{\mathbf{k}}^{\dagger} \hat{c}_{\mathbf{k}\sigma}$$

$$= \sum_{\mathbf{k}\nu\sigma} \epsilon_{\mathbf{k}\nu} \hat{b}_{\mathbf{k}\nu\sigma}^{\dagger} \hat{b}_{\mathbf{k}\nu\sigma}.$$

We have derived the *Bloch theorem*: the one-particle eigenvalues of $\hat{H}_0$ are obtained by diagonalizing $h_{\mathbf{k}}$ for all $\mathbf{k}$ and the corresponding one-particle eigenkets $|\mathbf{k}\nu\sigma\rangle = \hat{b}_{\mathbf{k}\nu\sigma}^{\dagger}|0\rangle$ have overlap with the original basis functions $|\mathbf{n}s\sigma'\rangle = \hat{d}_{\mathbf{n}s\sigma'}^{\dagger}|0\rangle$ given by

$$\psi_{\mathbf{k}\nu\sigma}(\mathbf{n}s\sigma') = \langle \mathbf{n}s\sigma' | \mathbf{k}\nu\sigma \rangle = \frac{\delta_{\sigma\sigma'}}{\sqrt{|V_b|}} e^{i\mathbf{k}\cdot\mathbf{n}} u_{\mathbf{k}\nu}(s), \qquad (2.13)$$

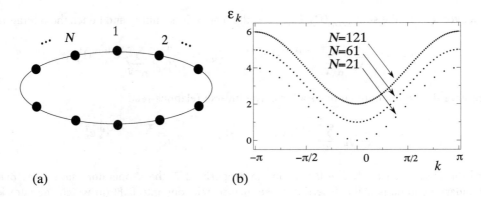

(a)   (b)

Figure 2.5 (a) A ring with N unit cells. (b) Eigenvalues of the ring Hamiltonian with $t = -1$ and $\epsilon = 2$ for different numbers N of unit cells. The eigenvalues with $N = 61$ and 121 are shifted upward by 1 and 2 respectively.

which is a plane-wave with different amplitudes on different atoms of the same unit cell. When the volume of the block $|V_b| \to \infty$ the eigenvalues $\epsilon_{\mathbf{k}\nu}$ become a continuum called a *band*.[3] We then have a band for each ν and the total number of bands coincides with the number of localized orbitals per unit cell. Each crystal is characterized by its *band structure* and, as we see in Section 6.3.4, the band structure can be experimentally measured. If we choose the localized basis functions to be of the form (2.1) and subsequently we want to increase the accuracy of the calculations by reducing the spacing Δ, then in the limit $\Delta \to 0$ the quantities $u_{\mathbf{k}\nu}(s) \to u_{\mathbf{k}\nu}(\mathbf{r})$ become continuous functions of the position $\mathbf{r}$ in the cell. These functions can be periodically extended to all space by imposing a requirement that they assume the same value in the same point of all unit cells. In the literature the periodic functions $u_{\mathbf{k}\nu}(\mathbf{r})$ are called *Bloch functions*. Below we illustrate some elementary applications of this general framework.

One-band model: The simplest example is a one-dimensional crystal with one orbital per unit cell, see Fig. 2.5(a). Then the matrices $h \equiv \epsilon$ and $T \equiv t$ are 1×1 matrices and the one-particle eigenvalues are

$$\epsilon_k = \epsilon + t e^{-ik} + t e^{ik} = \epsilon + 2t \cos k. \tag{2.14}$$

If the number N of unit cells is, e.g., odd, then the one-dimensional k vector takes the values $k = 2\pi m/N$ with $m = -\frac{(N-1)}{2}, \ldots, \frac{(N-1)}{2}$, as follows from (2.12). The eigenvalues (2.14) are displayed in Fig. 2.5(b) for different N. It is clear that when $N \to \infty$ they become a continuum and form one band.

Two-band model: Another example of a one-dimensional crystal is shown in Fig. 2.6(a). Each unit cell consists of two different atoms, a and b in the figure. We assign a single orbital per atom and neglect all off-diagonal matrix elements except those connecting atoms

---

[3]For $|V_b| \to \infty$ the index $\mathbf{k}$ becomes a continuous index in accordance with (2.12).

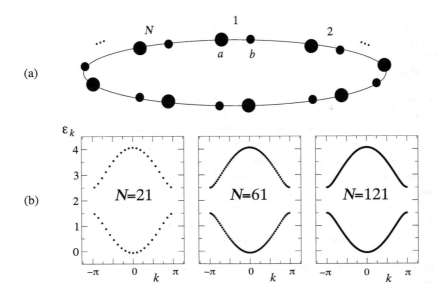

Figure 2.6 (a) A ring with N unit cells and two atoms per cell. (b) Eigenvalues of the ring Hamiltonian with $t = -1$, $\epsilon = 2$ and $\Delta = 1/2$ for different numbers N of unit cells.

of type a to atoms of type b. For simplicity we also consider the case that the distance between two nearest neighbor atoms is everywhere the same along the crystal. Then the matrices h and T are 2×2 matrices with the following structure

$$h = \begin{pmatrix} \epsilon + \Delta & t \\ t & \epsilon - \Delta \end{pmatrix}, \qquad T = \begin{pmatrix} 0 & t \\ 0 & 0 \end{pmatrix},$$

where Δ is an energy parameter that takes into account the different nature of the atoms. The structure of $T_{ss'} = h_{n+1s\,ns'}$, see (2.7), can be derived by checking the overlap between atom s in cell $n+1$ and atom s' in cell n. For instance, if we take cell 2 we see from Fig. 2.6(a) that atom a has overlap with atom b in cell 1, hence $T_{ab} = t$, but it does not have overlap with atom a in cell 1, hence $T_{aa} = 0$. On the other hand atom b in cell 2 does not have overlap with any atom in cell 1, and hence $T_{ba} = T_{bb} = 0$. The eigenvalues of

$$h_k = h + Te^{-ik} + T^\dagger e^{ik} = \begin{pmatrix} \epsilon + \Delta & t(1 + e^{-ik}) \\ t(1 + e^{ik}) & \epsilon - \Delta \end{pmatrix}$$

are

$$\epsilon_{k\pm} = \epsilon \pm \sqrt{\Delta^2 + 2t^2(1 + \cos k)}, \qquad (2.15)$$

and are displayed in Fig. 2.6(b) for different numbers N of unit cells. As in the previous example, the values of k have the form $2\pi m/N$ with $m = -\frac{(N-1)}{2}, \ldots, \frac{(N-1)}{2}$ for odd N. When $N \to \infty$ the $\epsilon_{k\pm}$ become a continuum and form two bands separated by an *energy gap* of width 2Δ. If the crystal contains two electrons per unit cell then the lower band is fully occupied whereas the upper band is empty. In this situation we must provide a minimum energy 2Δ to excite one electron in an empty state; the crystal behaves like an insulator if the gap is large and like a semiconductor if the gap is small. It is also worth noting that for $\Delta = 0$ we recover the previous example but with $2N$ atoms rather than N.

Graphene: Another important example that we wish to discuss is the two-dimensional crystal with which we opened this section. Graphene is a *single layer* of graphite and, as such, not easy to isolate. It was not until 2004 that graphene was experimentally realized by transferring a single layer of graphite onto a silicon dioxide substrate [9], an achievement which was rewarded with the Nobel prize in physics in 2010. The coupling between the graphene and the substrate is very weak and does not alter the electrical properties of the graphene. Before 2004 experimentalists were already able to produce tubes of graphene with a diameter of few nanometers. These *carbon nanotubes* can be considered as an infinitely long graphene strip wrapped onto a cylinder. Due to their mechanical (strong and stiff) and electrical properties, carbon nanotubes are among the most studied systems at the time of writing this book. There exist several kinds of nanotube depending on how the strip is wrapped. The wrapping is specified by a pair of integers (P, Q) so that atoms separated by $P\mathbf{v}_+ + Q\mathbf{v}_-$ are identified. Below we consider the *armchair nanotubes* for which $(P, Q) = (N, N)$ and hence the axis of the tube is parallel to $\mathbf{v}_+ - \mathbf{v}_-$.

Let us define the vector $\mathbf{N}_1 = (N, N)$ and the vector parallel to the tube axis $\mathbf{N}_2 = (M, -M)$. The armchair nanotube is recovered when $M \to \infty$. For simplicity we assign a single p_z orbital to each carbon atom and consider only those off-diagonal matrix elements that connect atoms a to atoms b. From Fig. 2.4 we then see that the matrices h, T_1 and T_2 must have the form[4]

$$h = \begin{pmatrix} 0 & t \\ t & 0 \end{pmatrix}, \qquad T_1 = \begin{pmatrix} 0 & t \\ 0 & 0 \end{pmatrix}, \qquad T_2 = \begin{pmatrix} 0 & t \\ 0 & 0 \end{pmatrix}.$$

The eigenvalues of

$$h_{\mathbf{k}} = h + \sum_{i=1}^{2} (T_i e^{-ik_i} + T_i^\dagger e^{ik_i}) = \begin{pmatrix} 0 & t(1 + e^{-ik_1} + e^{-ik_2}) \\ t(1 + e^{ik_1} + e^{ik_2}) & 0 \end{pmatrix}$$

can easily be calculated and read

$$\epsilon_{\mathbf{k}\pm} = \pm t \sqrt{1 + 4\cos\frac{k_1 - k_2}{2}\left(\cos\frac{k_1 - k_2}{2} + \cos\frac{k_1 + k_2}{2}\right)}.$$

The possible values of $\mathbf{k}$ belong to a square with vertices in $(\pm\pi, \pm\pi)$ and fulfill (2.12), i.e.,

$$\mathbf{k} \cdot \mathbf{N}_1 = N(k_1 + k_2) = 2\pi m_1,$$
$$\mathbf{k} \cdot \mathbf{N}_2 = M(k_1 - k_2) = 2\pi m_2. \tag{2.16}$$

---

[4]The addition of a constant energy to the diagonal elements of h simply leads to a rigid shift of the eigenvalues.

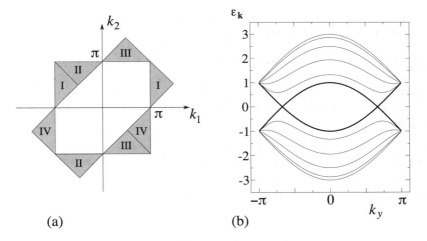

Figure 2.7 (a) Equivalent domain of **k** vectors. The grey areas with the same roman numbers are separated either by the vector $(2\pi, 0)$ or by the vector $(0, 2\pi)$. (b) Eigenvalues of a $(5, 5)$ armchair nanotube. The thick lines correspond to the valence band (below zero) and the conduction band (above zero). All bands in between the top and the conduction band and in between the bottom and the valence band are doubly degenerate.

From these relations it is evident that an equivalent domain of **k** vectors, like the one illustrated in Fig. 2.7(a), is more convenient for our analysis. This equivalent domain is a rectangle tilted by $\pi/4$, with the short edge equal to $\sqrt{2}\pi$ and the long edge equal to $2\sqrt{2}\pi$. If we define $k_x = k_1 + k_2$ and $k_y = k_1 - k_2$ then $-2\pi < k_x \le 2\pi$ and $-\pi < k_y \le \pi$ and hence

$$k_x = 2\pi \frac{m_1}{N}, \quad \text{with } m_1 = -N + 1, \ldots, N$$

$$k_y = 2\pi \frac{m_2}{M}, \quad \text{with } m_2 = -\frac{(M-1)}{2}, \ldots, \frac{M-1}{2},$$

for odd M. The reader can easily check that the number of **k** is the same as the number of atoms in the graphene block. In the limit $M \to \infty$ the quantum number k_y becomes a continuous index and we obtain $2 \times 2N$ one-dimensional bands corresponding to the number of carbon atoms in the transverse direction,[5]

$$\epsilon_{k_y k_x \pm} = \pm t \sqrt{1 + 4 \cos \frac{k_y}{2} \left(\cos \frac{k_y}{2} + \cos \frac{k_x}{2} \right)}.$$

The set of all carbon atoms in the transverse direction can be considered as the unit cell of the one-dimensional crystal, which is the nanotube.

In Fig. 2.7(b) we plot the eigenvalues $\epsilon_{k_y k_x \pm}$ for the $(N, N) = (5, 5)$ armchair nanotube. Since the neutral nanotube has one electron per atom, all bands with negative energy are

---

[5]In this example the quantum number k_x together with the sign $\pm$ of the eigenvalues plays the role of the band index ν.

fully occupied. It is noteworthy that the highest occupied band (valence band) and the lowest unoccupied band (conduction band) touch *only in two points*. Around these points the energy dispersion is linear and the electrons behave as if they were relativistic [10]. Increasing the diameter of the nanotube also, k_x becomes a continuous index and the $\epsilon_{\mathbf{k}\pm}$ become the eigenvalues of the two-dimensional graphene. In this limit the $\epsilon_{\mathbf{k}\pm}$ form two two-dimensional bands which still touch in only two points. Due to its so peculiar band structure graphene is classified as a *semi-metal*, that is a crystal with properties between a metal and a semiconductor.

Exercise 2.1. Show that the eigenvalues (2.15) with $\Delta = 0$ coincide with the eigenvalues (2.14) of the simple ring with $2N$ atoms.

Exercise 2.2. Consider the one-dimensional crystal below

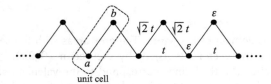

with matrices $h = \begin{pmatrix} \epsilon & \sqrt{2}t \\ \sqrt{2}t & \epsilon \end{pmatrix}$ and $T = \begin{pmatrix} t & \sqrt{2}t \\ 0 & 0 \end{pmatrix}$. Show that for $t > 0$ the two bands are

$$\epsilon_{k1} = \epsilon - 2t,$$
$$\epsilon_{k2} = \epsilon + 2t + 2t\cos k,$$

with $k \in (-\pi, \pi)$. The first band is therefore perfectly flat. If we have half an electron per unit cell then the ground state is highly degenerate. For instance, the states obtained by occupying each k-level of the flat band with an electron of either spin up or down all have the same energy. This degeneracy is lifted by the electron–electron interaction and the ground state turns out to be the one in which all electrons have parallel spins. The crystal is then a *ferromagnet*. The ferromagnetism in flat-band crystals has been proposed by Mielke and Tasaki [11–13] and is usually called *flat-band ferromagnetism*.

2.3.2 Fano model

The Fano model is ubiquitous in condensed matter physics as it represents the simplest schematization of a discrete state "interacting" with a continuum of states. It was originally introduced by Fano to explain a universal asymmetric line-shape observed in the absorption spectrum of several physical systems [14]. We introduce it here to model an atom or a molecule adsorbed on the surface of a metal, see the schematic illustration below.

Let $|\epsilon Q\rangle$ be the single-particle energy eigenket of the metal, where Q is a quantum number (continuous or discrete) that accounts for the degeneracy of the eigenvalue ϵ. In this section the spin index carried by states and operators is understood since everything is diagonal in spin space. Equivalently, we can think of fermions with spin zero. For the atom we consider only a single localized orbital and denote by $|\epsilon_0\rangle$ the corresponding ket. For large distances between the atom and the surface the Hamiltonian of the system is the sum of the Hamiltonian $\hat{H}_{\mathrm{met}}$ of the metal and the Hamiltonian $\hat{H}_{\mathrm{at}}$ of the atom

$$\hat{H}_{\mathrm{met}} + \hat{H}_{\mathrm{at}} = \int d\epsilon \, dQ \, \epsilon \, \hat{d}_{\epsilon Q}^{\dagger} \hat{d}_{\epsilon Q} + \epsilon_0 \, \hat{d}_0^{\dagger} \hat{d}_0,$$

where the $\hat{d}$-operators create or annihilate electrons and hence satisfy the anti-commutation relations

$$\left[\hat{d}_{\epsilon Q}, \hat{d}_{\epsilon' Q'}^{\dagger} \right]_+ = \delta(\epsilon - \epsilon')\delta(Q - Q'), \qquad \left[\hat{d}_0, \hat{d}_0^{\dagger} \right]_+ = 1.$$

When the atom approaches the surface the overlap $T_{\epsilon Q} \equiv \langle \epsilon Q | \hat{h} | \epsilon_0 \rangle$ increases and cannot be neglected any longer. In this regime the Hamiltonian of the system becomes

$$\hat{H}_0 = \hat{H}_{\mathrm{met}} + \hat{H}_{\mathrm{at}} + \int d\epsilon \, dQ \left(T_{\epsilon Q} \hat{d}_{\epsilon Q}^{\dagger} \hat{d}_0 + T_{\epsilon Q}^* \hat{d}_0^{\dagger} \hat{d}_{\epsilon Q} \right). \tag{2.17}$$

This is the general structure of the *Fano model*. It is common (and for later purposes also instructive) to discretize the continuum of states by retaining only one state $|\epsilon_k Q_k\rangle$ in the volume element $\Delta\epsilon\Delta Q$. Then, for small volume elements the Fano Hamiltonian can be approximated as

$$\hat{H}_0 = \sum_k \epsilon_k \hat{d}_k^{\dagger} \hat{d}_k + \epsilon_0 \hat{d}_0^{\dagger} \hat{d}_0 + \sum_k \left(T_k \hat{d}_k^{\dagger} \hat{d}_0 + T_k^* \hat{d}_0^{\dagger} \hat{d}_k \right), \tag{2.18}$$

where the discrete $\hat{d}$-operators satisfy the anti-commutation relations $\left[\hat{d}_k, \hat{d}_{k'} \right]_+ = \delta_{kk'}$. To recover the continuum limit we establish the correspondence

$$T_k = \sqrt{\Delta\epsilon\Delta Q} \, T_{\epsilon_k Q_k} \tag{2.19}$$

for the off-diagonal matrix elements of $\hat{h}$, and

$$\hat{d}_k = \sqrt{\Delta\epsilon\Delta Q} \, \hat{d}_{\epsilon_k Q_k}, \qquad \hat{d}_k^{\dagger} = \sqrt{\Delta\epsilon\Delta Q} \, \hat{d}_{\epsilon_k Q_k}^{\dagger},$$

for the fermionic operators. Letting $\Delta\epsilon\Delta Q \to 0$ the discrete Hamiltonian (2.18) reduces to the continuum Hamiltonian (2.17) and

$$\left[\hat{d}_{\epsilon_k Q_k}, \hat{d}_{\epsilon_{k'} Q_{k'}}^{\dagger} \right]_+ = \frac{\delta_{kk'}}{\Delta\epsilon\Delta Q} \to \delta(\epsilon - \epsilon')\delta(Q - Q').$$

We now calculate the atomic occupation n_0 for a given Fermi energy ϵ_F, which is the energy of the highest occupied level of the metal. Let $|\lambda\rangle$ be the single-particle eigenkets of

$\hat{H}_0$ with eigenenergies ϵ_λ. The corresponding annihilation and creation operators $\hat{c}_\lambda$ and $\hat{c}_\lambda^\dagger$ are a linear combination of the original $\hat{d}$-operators as discussed in Section 1.5,

$$\hat{c}_\lambda^\dagger = \langle\epsilon_0|\lambda\rangle\,\hat{d}_0^\dagger + \sum_k \langle k|\lambda\rangle\,\hat{d}_k^\dagger,$$

and with a similar equation for the adjoint. The ground state $|\Phi_0\rangle$ of the system is obtained by occupying all the $|\lambda\rangle$s with energies $\epsilon_\lambda \leq \epsilon_F$ and reads

$$|\Phi_0\rangle = \prod_{\lambda:\epsilon_\lambda\leq\epsilon_F} \hat{c}_\lambda^\dagger\,|0\rangle. \tag{2.20}$$

To calculate the atomic occupation $n_0 = \langle\Phi_0|\hat{d}_0^\dagger\hat{d}_0|\Phi_0\rangle$ we evaluate the ket $\hat{d}_0|\Phi_0\rangle$ by moving the operator $\hat{d}_0$ through the string of $\hat{c}^\dagger$-operators, as we did in (1.57). The difference here is that the anti-commutator is $[\hat{d}_0,\hat{c}_\lambda^\dagger]_+ = \langle\epsilon_0|\lambda\rangle$ rather than a Kronecker delta. We then find

$$\hat{d}_0|\Phi_0\rangle = \sum_\lambda (-)^{p_\lambda}\langle\epsilon_0|\lambda\rangle \prod_{\lambda'\neq\lambda} \hat{c}_{\lambda'}^\dagger\,|0\rangle,$$

where the sum and the product are restricted to states below the Fermi energy. The integer p_λ in the above equation refers to the position of $\hat{c}_\lambda^\dagger$ in the string of operators (2.20), in agreement with (1.57). The atomic occupation n_0 is the inner product of the many-particle state $\hat{d}_0|\Phi_0\rangle$ with itself. In this inner product all cross terms vanish since they are proportional to the inner product of states with different strings of $\hat{c}_\lambda^\dagger$ operators, see (1.58). Taking into account that $\prod_{\lambda'\neq\lambda} \hat{c}_{\lambda'}^\dagger\,|0\rangle$ is normalized to 1 for every λ we obtain the intuitive result

$$n_0 = \sum_{\lambda:\epsilon_\lambda\leq\epsilon_F} |\langle\epsilon_0|\lambda\rangle|^2; \tag{2.21}$$

the atomic occupation is the sum over all occupied states of the probability of finding an electron in $|\epsilon_0\rangle$.

From (2.21) it seems that we need to know the eigenkets $|\lambda\rangle$ and eigenenergies ϵ_λ in order to determine n_0. We now show that this is not strictly the case. Let us rewrite n_0 as

$$n_0 = \int_{-\infty}^{\epsilon_F} \frac{d\omega}{2\pi} \sum_\lambda 2\pi\delta(\omega - \epsilon_\lambda)|\langle\epsilon_0|\lambda\rangle|^2. \tag{2.22}$$

The eigenkets $|\lambda\rangle$ satisfy $\hat{h}|\lambda\rangle = \epsilon_\lambda|\lambda\rangle$ where $\hat{h}$ is the one-particle Hamiltonian in first quantization. For the Fano model $\hat{h}$ has the following ket-bra form:

$$\hat{h} = \sum_k \epsilon_k|k\rangle\langle k| + \epsilon_0|\epsilon_0\rangle\langle\epsilon_0| + \sum_k \left(T_k|k\rangle\langle\epsilon_0| + T_k^*|\epsilon_0\rangle\langle k| \right),$$

where we use (1.86). Then

$$\sum_\lambda \delta(\omega - \epsilon_\lambda)|\langle\epsilon_0|\lambda\rangle|^2 = \sum_\lambda \langle\epsilon_0|\delta(\omega - \hat{h})|\lambda\rangle\langle\lambda|\epsilon_0\rangle = \langle\epsilon_0|\delta(\omega - \hat{h})|\epsilon_0\rangle,$$

where we use the completeness relation $\sum_\lambda |\lambda\rangle\langle\lambda| = \hat{\mathbb{1}}$. Inserting this result into (2.22) we get

$$n_0 = \int_{-\infty}^{\epsilon_F} \frac{d\omega}{2\pi} \, \langle\epsilon_0|2\pi\delta(\omega - \hat{h})|\epsilon_0\rangle.$$

This is our first encounter with the *spectral function* (first quantization) operator

$$\hat{\mathcal{A}}(\omega) \equiv 2\pi\delta(\omega - \hat{h}) = i\left[\frac{1}{\omega - \hat{h} + i\eta} - \frac{1}{\omega - \hat{h} - i\eta}\right].$$

In the second equality η is an infinitesimally small positive constant and we used the *Cauchy relation*

$$\frac{1}{\omega - \epsilon \pm i\eta} = P\frac{1}{\omega - \epsilon} \mp i\pi\delta(\omega - \epsilon) \qquad (2.23)$$

where P denotes the principal part. Thus we can calculate n_0 if we find a way to determine the matrix element $A_{00}(\omega) = \langle\epsilon_0|\hat{\mathcal{A}}(\omega)|\epsilon_0\rangle$ of the spectral function. As we see in Chapter 6, this matrix element can be interpreted as the probability that an electron in $|\epsilon_0\rangle$ has energy ω.[6]

To determine $A_{00}(\omega)$ we separate $\hat{h} = \hat{\mathcal{E}} + \hat{\mathcal{T}}$ into a sum of the metal+atom Hamiltonian $\hat{\mathcal{E}}$ and the off-diagonal part $\hat{\mathcal{T}}$, and use the identity

$$\frac{1}{\zeta - \hat{h}} = \frac{1}{\zeta - \hat{\mathcal{E}}} + \frac{1}{\zeta - \hat{h}}\,\hat{\mathcal{T}}\,\frac{1}{\zeta - \hat{\mathcal{E}}}, \qquad (2.24)$$

where ζ is an arbitrary complex number. This identity can easily be verified by multiplying both sides from the left by $(\zeta - \hat{h})$. Sandwiching of (2.24) between $\langle\epsilon_0|$ and $|\epsilon_0\rangle$ and between $\langle\epsilon_0|$ and $|k\rangle$ we find

$$\langle\epsilon_0|\frac{1}{\zeta - \hat{h}}|\epsilon_0\rangle = \frac{1}{\zeta - \epsilon_0} + \sum_k T_k \langle\epsilon_0|\frac{1}{\zeta - \hat{h}}|k\rangle \frac{1}{\zeta - \epsilon_0},$$

$$\langle\epsilon_0|\frac{1}{\zeta - \hat{h}}|k\rangle = T_k^* \langle\epsilon_0|\frac{1}{\zeta - \hat{h}}|\epsilon_0\rangle \frac{1}{\zeta - \epsilon_k}.$$

Substituting the second of these equations into the first we arrive at the following important result

$$\langle\epsilon_0|\frac{1}{\zeta - \hat{h}}|\epsilon_0\rangle = \frac{1}{\zeta - \epsilon_0 - \Sigma_{\text{em}}(\zeta)}, \qquad \text{with} \quad \Sigma_{\text{em}}(\zeta) = \sum_k \frac{|T_k|^2}{\zeta - \epsilon_k}. \qquad (2.25)$$

The *embedding self-energy* $\Sigma_{\text{em}}(\zeta)$ appears because the atom is not isolated; we can think of it as a correction to the atomic level ϵ_0 induced by the presence of the metal. Taking $\zeta = \omega + i\eta$ we can separate $\Sigma_{\text{em}}(\zeta)$ into a real and an imaginary part,

$$\Sigma_{\text{em}}(\omega + i\eta) = \sum_k \frac{|T_k|^2}{\omega - \epsilon_k + i\eta} = \underbrace{P\sum_k \frac{|T_k|^2}{\omega - \epsilon_k}}_{\Lambda(\omega)} - \frac{i}{2}\underbrace{2\pi\sum_k |T_k|^2\delta(\omega - \epsilon_k)}_{\Gamma(\omega)}. \qquad (2.26)$$

---

[6]For the time being we observe that this interpretation is supported by the normalization condition $\int \frac{d\omega}{2\pi} A_{00}(\omega) = 1$, i.e., the probability that the electron has energy between $-\infty$ and ∞ is 1.

The real and imaginary parts are not independent but instead related by a *Hilbert transformation*

$$\Lambda(\omega) = P \int \frac{d\omega'}{2\pi} \frac{\Gamma(\omega')}{\omega - \omega'},$$

as can be verified by inserting the explicit expression for $\Gamma(\omega)$ into the r.h.s.. In conclusion, we have obtained an expression for n_0 in terms of the quantity $\Gamma(\omega)$ only

$$
\begin{aligned}
n_0 &= \int_{-\infty}^{\epsilon_F} \frac{d\omega}{2\pi} A_{00}(\omega) \\
&= \int_{-\infty}^{\epsilon_F} \frac{d\omega}{2\pi} \, i \left[\frac{1}{\omega + i\eta - \epsilon_0 - \Sigma_{\text{em}}(\omega + i\eta)} - \frac{1}{\omega - i\eta - \epsilon_0 - \Sigma_{\text{em}}(\omega - i\eta)} \right] \\
&= -2 \int_{-\infty}^{\epsilon_F} \frac{d\omega}{2\pi} \, \text{Im} \frac{1}{\omega - \epsilon_0 - \Lambda(\omega) + \frac{i}{2}\Gamma(\omega) + i\eta}.
\end{aligned}
\tag{2.27}
$$

If the coupling T_k between the atom and the metal is weak then Λ and Γ are small and the dominant contribution to the above integral comes from a region around ϵ_0. In the continuum limit the function $\Gamma(\omega)$ is a smooth function since

$$\Gamma(\omega) \to 2\pi \int d\epsilon \, dQ \, |T_{\epsilon Q}|^2 \delta(\omega - \epsilon) = 2\pi \int dQ \, |T_{\omega Q}|^2, \tag{2.28}$$

and in the integral (2.27) we can approximate $\Gamma(\omega)$ by a constant $\Gamma = \Gamma(\epsilon_0)$. This approximation is known as the *Wide Band Limit Approximation* (WBLA) since for the r.h.s. of (2.28) to be ω independent the spectrum of the metal must extend from $-\infty$ and $+\infty$. For a $\Gamma(\omega) = \Gamma$ independent of ω the Hilbert transform $\Lambda(\omega)$ vanishes and the formula for the atomic occupation simplifies to[7]

$$n_0 = \int_{-\infty}^{\epsilon_F} \frac{d\omega}{2\pi} \frac{\Gamma}{(\omega - \epsilon_0)^2 + \Gamma^2/4}.$$

We thus see that in the WBLA n_0 is the integral up to the Fermi energy of a Lorentzian of width Γ centered at ϵ_0. The atomic level is occupied ($n_0 \sim 1$) if $\epsilon_0 < \epsilon_F + \Gamma$, whereas it is empty ($n_0 \sim 0$) if $\epsilon_0 > \epsilon_F + \Gamma$. This result should be compared with the atomic limit, $\Gamma = 0$, corresponding to the isolated atom. In this limit the Lorentzian becomes a δ-function since

$$\lim_{\Gamma \to 0} \frac{1}{\pi} \frac{\Gamma/2}{(\omega - \epsilon_0)^2 + \Gamma^2/4} = \delta(\omega - \epsilon_0),$$

as follows immediately from the Cauchy relation (2.23). In the atomic limit the atomic occupation is exactly 1 for $\epsilon_0 < \epsilon_F$ and zero otherwise. We say that the presence of the metal broadens the sharp atomic level and transforms it into a resonance of finite width as illustrated in Fig. 2.8. This broadening is a general feature of discrete levels "interacting" or "in contact" with a continuum and it is observed in interacting systems as well.

---

[7]Since $\Gamma(\omega) = \Gamma$ for all ω we can discard the infinitesimal η in the denominator of (2.27).

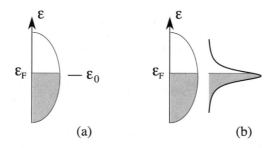

Figure 2.8 (a) The metallic band and the sharp level in the atomic limit. (b) The metallic band and the spectral function A_{00} for finite Γ.

2.4 Hubbard model

The *Hubbard model* was originally introduced to describe transition metals and rare-earth metals, i.e., solids composed of atoms with very localized outer electrons (d or f shells). In these materials the degree of localization of the outer-electron orbitals is so high that the Coulomb integrals $v_{s_1 s_2 s_3 s_4}$ can be approximated as in (2.3) with the diagonal elements $U_s \equiv v_{ss}$ about an order of magnitude larger than the off-diagonal ones. For this reason Hubbard, in a milestone paper from 1963 [15], included only the diagonal interaction U_s in his treatment and wrote the model Hamiltonian

$$\hat{H} = \hat{H}_0 + \hat{H}_{\text{int}} = \sum_\sigma \sum_{ss'} h_{ss'} \hat{d}^\dagger_{s\sigma} \hat{d}_{s'\sigma} + \sum_s U_s \hat{n}_{s\uparrow} \hat{n}_{s\downarrow}, \qquad (2.29)$$

that today carries his name. In this Hamiltonian the sums run over the lattice position s of the nuclei, the $\hat{d}$-operators are fermionic annihilation and creation operators for electrons with the usual anti-commutation relations

$$\left[\hat{d}_{s\sigma}, \hat{d}^\dagger_{s'\sigma'} \right]_+ = \delta_{\sigma\sigma'} \delta_{ss'}, \qquad \left[\hat{d}_{s\sigma}, \hat{d}_{s'\sigma'} \right]_+ = 0,$$

and $\hat{n}_{s\sigma} = \hat{d}^\dagger_{s\sigma} \hat{d}_{s\sigma}$ is the occupation operator for electrons in s with spin σ. The matrix elements $h_{ss'}$ are typically set to zero if the distance between s and s' exceeds a few lattice spacings. In the context of model Hamiltonians the off-diagonal matrix elements $h_{ss'}$ are also called *hopping integrals* or simply *hoppings* since they multiply the operator $\hat{d}^\dagger_{s\sigma} \hat{d}_{s'\sigma}$, which destroys an electron in s' and creates an electron in s, an operation that can be pictured as the hopping of an electron from s' to s. It is interesting to note that in the Hubbard model two electrons interact only if they occupy the same atomic site (and of course have opposite spins).

To gain some insight into the physics of the Hubbard model we compare two limiting situations, i.e., $\hat{H}_{\text{int}} = 0$ and $\hat{H}_0 = 0$. We see that electrons behave as "waves" if $\hat{H}_{\text{int}} = 0$ and as "particles" if $\hat{H}_0 = 0$; how they behave when both $\hat{H}_0$ and $\hat{H}_{\text{int}}$ are nonzero is a fascinating problem intimately related to wave–particle dualism. To readers interested in the physics of the Hubbard model we suggest the review article in Ref. [16] and the books in Refs. [17, 18].

For $\hat{H}_{\text{int}} = 0$ the Hubbard Hamiltonian describes noninteracting electrons on a lattice. Let $|k\sigma\rangle = c^\dagger_{k\sigma}|0\rangle$ be the one-particle eigenkets with energies ϵ_k. If we order the energies $\epsilon_k \leq \epsilon_{k+1}$, the state of lowest energy with $N_{\uparrow/\downarrow}$ electrons of spin up/down is

$$|\Phi\rangle = \prod_{k=1}^{N_\uparrow} \prod_{k'=1}^{N_\downarrow} \hat{c}^\dagger_{k\uparrow} \hat{c}^\dagger_{k'\downarrow}|0\rangle \quad \text{with} \quad \hat{H}_0|\Phi\rangle = \left(\sum_{k=1}^{N_\uparrow} \epsilon_k + \sum_{k'=1}^{N_\downarrow} \epsilon_{k'} \right)|\Phi\rangle. \quad (2.30)$$

We can say that in $|\Phi\rangle$ the electrons behave as "waves" since they have probability 0 or 1 of being in a delocalized state $|k\sigma\rangle$. The N-particle ground state of $\hat{H}_0$ is obtained by minimizing the eigenvalue in (2.30) with respect to $N_\uparrow$ and $N_\downarrow$ under the constraint $N_\uparrow + N_\downarrow = N$. For nondegenerate energies $\epsilon_k < \epsilon_{k+1}$, the ground state of $\hat{H}_0$ with an even number N of electrons is unique and reads

$$|\Phi_0\rangle = \prod_{k=1}^{N/2} \hat{c}^\dagger_{k\uparrow} \hat{c}^\dagger_{k\downarrow}|0\rangle. \quad (2.31)$$

We leave it as an exercise for the reader to show that $|\Phi_0\rangle$ is also an eigenstate of the total spin operators [see (1.92)],

$$\hat{\mathbf{S}} = (\hat{S}^x, \hat{S}^y, \hat{S}^z) \equiv \sum_s \hat{\mathbf{S}}_s = \sum_s (\hat{S}^x_s, \hat{S}^y_s, \hat{S}^z_s)$$

with vanishing eigenvalue, i.e., it is a singlet.

If we perturb the system with a weak external magnetic field $\mathbf{B}$ along, say, the z axis and discard the coupling to the orbital motion, the noninteracting part of the Hamiltonian changes to

$$\hat{H}_0 \rightarrow \hat{H}_0 - g\mu_B\hat{\mathbf{S}} \cdot \mathbf{B} = \hat{H}_0 - \frac{1}{2}g\mu_B B(\hat{N}_\uparrow - \hat{N}_\downarrow), \quad (2.32)$$

with g the electron gyromagnetic ratio, μ_B the Bohr magneton, and $\hat{N}_\sigma = \sum_s \hat{n}_{s\sigma}$ the operator for the total number of particles of spin σ. The eigenkets (2.30) are also eigenkets of this new Hamiltonian but with a different eigenvalue

$$(\hat{H}_0 - g\mu_B\hat{\mathbf{S}} \cdot \mathbf{B})|\Phi\rangle = \left(\sum_{k=1}^{N_\uparrow}(\epsilon_k - \frac{1}{2}g\mu_B B) + \sum_{k'=1}^{N_\downarrow}(\epsilon_{k'} + \frac{1}{2}g\mu_B B) \right)|\Phi\rangle.$$

Thus, in the presence of an external magnetic field the state (2.31) is no longer the lowest in energy since for, e.g., $B > 0$, it becomes energetically convenient to have more electrons of spin up than of spin down. This is the typical behavior of a *Pauli paramagnet*, that is a system whose total spin is zero for $B = 0$ and grows parallel to $\mathbf{B}$ for $B \neq 0$.[8]

Next we discuss the case $\hat{H}_0 = 0$. The Hamiltonian $\hat{H}_{\text{int}}$ is already in a diagonal form and the generic eigenket can be written as

$$|\Phi_{XY}\rangle = \prod_{s \in X} \prod_{s' \in Y} \hat{d}^\dagger_{s\uparrow} \hat{d}^\dagger_{s'\downarrow}|0\rangle,$$

---

[8]The Pauli paramagnetic behavior is due to the spin degrees of freedom and it is therefore distinct from the paramagnetic behavior due to the orbital degrees of freedom, see Section 3.4.

where X and Y are two arbitrary collections of atomic sites. We can say that in $|\Phi_{XY}\rangle$ the electrons behave like "particles" since they have probability 0 or 1 of being on a given atomic site. The energy eigenvalue of $|\Phi_{XY}\rangle$ is

$$E_{XY} = \sum_{s \in X \cap Y} U_s.$$

The ground state(s) for a given number $N = N_\uparrow + N_\downarrow$ of electrons can be constructed by choosing X and Y that minimize the energy E_{XY}. Denoting by N_V the total number of atomic sites, we have that for $N \leq N_V$ all states with $X \cap Y = \emptyset$ are ground states (with zero energy) and, again, the system behaves like a Pauli paramagnet.

In conclusion, neither $\hat{H}_0$ nor $\hat{H}_{\text{int}}$ favors any kind of magnetic order. However, their sum $\hat{H}_0 + \hat{H}_{\text{int}}$ sometimes does. In Section 2.5 we illustrate an example of this phenomenon.

2.4.1 Particle–hole symmetry: application to the Hubbard dimer

In this section we solve explicitly the Hubbard model with only two sites, the so called *Hubbard dimer*, with Hamiltonian

$$\hat{H} = T \sum_\sigma (\hat{d}_{1\sigma}^\dagger \hat{d}_{2\sigma} + \hat{d}_{2\sigma}^\dagger \hat{d}_{1\sigma}) + U\left(\hat{n}_{1\uparrow}\hat{n}_{1\downarrow} + \hat{n}_{2\uparrow}\hat{n}_{2\downarrow}\right).$$

This Hamiltonian belongs to a general class of Hubbard Hamiltonians defined on *bipartite lattices*, i.e., lattices that can be divided into two sets of sites A and B with hoppings only from sites of A to sites of B and vice versa, i.e.,

$$\hat{H}_{\text{bip}} = \sum_\sigma \sum_{\substack{s \in A \\ s' \in B}} \left(h_{ss'}\hat{d}_{s\sigma}^\dagger \hat{d}_{s'\sigma} + h_{s's}\hat{d}_{s'\sigma}^\dagger \hat{d}_{s\sigma} \right) + \sum_s U_s \hat{n}_{s\uparrow}\hat{n}_{s\downarrow}.$$

In the Hubbard dimer A is site 1 and B is site 2 and $h_{12} = T$. If the lattice is bipartite and if $h_{ss'} = h_{s's}$ and $U_s = U$ is independent of s, the Hubbard Hamiltonian enjoys an interesting *particle–hole symmetry* which can be used to simplify the calculation of eigenvalues and eigenkets. Let us first explain what this symmetry is. The fermionic operators

$$\hat{b}_{s\sigma} \equiv \begin{cases} \hat{d}_{s\sigma}^\dagger & s \in A \\ -\hat{d}_{s\sigma}^\dagger & s \in B \end{cases}$$

satisfy the same anti-commutation relations as the original $\hat{d}$ operators, and in terms of them $\hat{H}_{\text{bip}}$ becomes

$$\begin{aligned} \hat{H}_{\text{bip}} &= -\sum_\sigma \sum_{\substack{s \in A \\ s' \in B}} \left(h_{ss'}\hat{b}_{s\sigma}\hat{b}_{s'\sigma}^\dagger + h_{s's}\hat{b}_{s'\sigma}\hat{b}_{s\sigma}^\dagger \right) + U\sum_s \hat{b}_{s\uparrow}\hat{b}_{s\uparrow}^\dagger \hat{b}_{s\downarrow}\hat{b}_{s\downarrow}^\dagger \\ &= \sum_\sigma \sum_{\substack{s \in A \\ s' \in B}} \left(h_{ss'}\hat{b}_{s\sigma}^\dagger \hat{b}_{s'\sigma} + h_{s's}\hat{b}_{s'\sigma}^\dagger \hat{b}_{s\sigma} \right) + U\sum_s \hat{n}_{s\uparrow}^{(b)}\hat{n}_{s\downarrow}^{(b)} - U\hat{N}^{(b)} + UN_V, \end{aligned}$$

where $\hat{n}_{s\sigma}^{(b)} = \hat{b}_{s\sigma}^\dagger \hat{b}_{s\sigma}$, $\hat{N}^{(b)} = \sum_{s\sigma} \hat{n}_{s\sigma}^{(b)}$ and N_{V} is the total number of sites. Except for the last two terms, which both commute with the first two terms, the Hamiltonian written with the $\hat{b}$-operators is identical to the Hamiltonian written with the $\hat{d}$-operators. This means that if

$$|\Psi\rangle = \sum_{s_1 \ldots s_{N_\uparrow}} \sum_{s_1' \ldots s_{N_\downarrow}'} \Psi(s_1\uparrow, \ldots, s_{N_\uparrow}\uparrow, s_1'\downarrow, \ldots, s_{N_\downarrow}'\downarrow)\, \hat{d}_{s_1\uparrow}^\dagger \cdots \hat{d}_{s_{N_\uparrow}\uparrow}^\dagger \hat{d}_{s_1'\downarrow}^\dagger \cdots \hat{d}_{s_{N_\downarrow}'\downarrow}^\dagger |0\rangle$$

is an eigenstate of $\hat{H}_{\mathrm{bip}}$ with N_σ electrons of spin σ and energy E, then the state $|\Psi^{(b)}\rangle$ obtained from $|\Psi\rangle$ by replacing the $\hat{d}$-operators with the $\hat{b}$-operators *and* the empty ket $|0\rangle$ with the empty ket $|0^{(b)}\rangle$ of the $\hat{b}$-operators is also an eigenstate of $\hat{H}_{\mathrm{bip}}$ but with energy $E - U(N_\uparrow + N_\downarrow) + U N_{\mathrm{V}}$. The empty ket $|0^{(b)}\rangle$ is by definition the ket for which

$$\hat{b}_{s\sigma}|0^{(b)}\rangle = 0 \qquad \text{for all } s \text{ and } \sigma.$$

Recalling the definition of the $\hat{b}$-operators, this is equivalent to saying that $\hat{d}_{s\sigma}^\dagger|0^{(b)}\rangle = 0$ for all s and σ, i.e., the ket $|0^{(b)}\rangle$ is the full ket with one electron of spin up and down on every site. Thus the state $|\Psi^{(b)}\rangle$ can alternatively be written as a linear combination of products of annihilation $\hat{d}$-operators acting on the full ket. The number of electrons of spin σ in $|\Psi^{(b)}\rangle$ is therefore $N_{\mathrm{V}} - N_\sigma$. In conclusion we can say that if E is an eigenvalue of $\hat{H}_{\mathrm{bip}}$ with N_σ electrons of spin σ then $E - U(N_\uparrow + N_\downarrow) + U N_{\mathrm{V}}$ is also an eigenvalue of $\hat{H}_{\mathrm{bip}}$ but with $N_{\mathrm{V}} - N_\sigma$ electrons of spin σ. The relation between the corresponding eigenkets is that the empty ket is replaced by the full ket and the creation operators are replaced by the annihilation operators.

Let us now return to the Hubbard dimer. The Fock space $\mathcal{F} = \mathcal{H}_0 \oplus \mathcal{H}_1 \oplus \mathcal{H}_2 \oplus \mathcal{H}_3 \oplus \mathcal{H}_4$ is in this case finite because we cannot have more than four electrons. The zero-particle Hilbert space $\mathcal{H}_0$ contains only the empty ket which is an eigenket of $\hat{H}$ with eigenvalue 0. In $\mathcal{H}_1$ we have four possible kets $|i\sigma\rangle = \hat{d}_{i\sigma}^\dagger|0\rangle$ with $i = 1, 2$. They are all eigenkets of the interaction operator with eigenvalue 0. Furthermore the matrix $\langle i\sigma|\hat{H}|j\sigma'\rangle = \delta_{\sigma\sigma'} h_{ij}$ is spin diagonal. The eigenvalues of $h = \begin{pmatrix} 0 & T \\ T & 0 \end{pmatrix}$ are $\pm T$ and the corresponding eigenvectors are $\frac{1}{\sqrt{2}}(1, \pm 1)$. Thus, the one-particle eigenkets of the Hubbard dimer are $\frac{1}{\sqrt{2}}(\hat{d}_{1\sigma}^\dagger \pm \hat{d}_{2\sigma}^\dagger)|0\rangle$, see also Table 2.1. Let us now consider the two-particle Hilbert space $\mathcal{H}_2$. If both electrons have spin σ then the only possible ket is $\hat{d}_{1\sigma}^\dagger \hat{d}_{2\sigma}^\dagger|0\rangle$ due to the Pauli exclusion principle. The reader can easily verify that this ket is an eigenket of $\hat{H}$ with eigenvalue 0. In particular it is also an eigenket of the interaction Hamiltonian with eigenvalue 0 since the electrons have parallel spin. On the other hand, if the electrons have opposite spin then we have four possible kets:

$$|\Psi_1\rangle = \hat{d}_{1\uparrow}^\dagger \hat{d}_{1\downarrow}^\dagger|0\rangle, \quad |\Psi_2\rangle = \hat{d}_{2\uparrow}^\dagger \hat{d}_{2\downarrow}^\dagger|0\rangle, \quad |\Psi_3\rangle = \hat{d}_{1\uparrow}^\dagger \hat{d}_{2\downarrow}^\dagger|0\rangle, \quad |\Psi_4\rangle = \hat{d}_{1\downarrow}^\dagger \hat{d}_{2\uparrow}^\dagger|0\rangle.$$

Space	Eigenvalues	Eigenkets
$\mathcal{H}_0$	0	$\|0\rangle$
$\mathcal{H}_1$	$\pm T$	$\frac{1}{\sqrt{2}}(\hat{d}_{1\sigma}^\dagger \pm \hat{d}_{2\sigma}^\dagger)\|0\rangle$
$\mathcal{H}_2$	0	$\hat{d}_{1\uparrow}^\dagger \hat{d}_{2\uparrow}^\dagger \|0\rangle$
	0	$\hat{d}_{1\downarrow}^\dagger \hat{d}_{2\downarrow}^\dagger \|0\rangle$
	0	$\dfrac{\hat{d}_{1\uparrow}^\dagger \hat{d}_{2\downarrow}^\dagger + \hat{d}_{1\downarrow}^\dagger \hat{d}_{2\uparrow}^\dagger}{\sqrt{2}}\|0\rangle$
	U	$\dfrac{\hat{d}_{1\uparrow}^\dagger \hat{d}_{1\downarrow}^\dagger - \hat{d}_{2\uparrow}^\dagger \hat{d}_{2\downarrow}^\dagger}{\sqrt{2}}\|0\rangle$
	$E_3 = \frac{1}{2}(U - \Delta)$	$\dfrac{E_3 \hat{d}_{1\uparrow}^\dagger \hat{d}_{1\downarrow}^\dagger + E_3 \hat{d}_{2\uparrow}^\dagger \hat{d}_{2\downarrow}^\dagger + 2T\hat{d}_{1\uparrow}^\dagger \hat{d}_{2\downarrow}^\dagger - 2T\hat{d}_{1\downarrow}^\dagger \hat{d}_{2\uparrow}^\dagger}{\sqrt{2E_3^2 + 8T^2}}\|0\rangle$
	$E_4 = \frac{1}{2}(U + \Delta)$	$\dfrac{E_4 \hat{d}_{1\uparrow}^\dagger \hat{d}_{1\downarrow}^\dagger + E_4 \hat{d}_{2\uparrow}^\dagger \hat{d}_{2\downarrow}^\dagger + 2T\hat{d}_{1\uparrow}^\dagger \hat{d}_{2\downarrow}^\dagger - 2T\hat{d}_{1\downarrow}^\dagger \hat{d}_{2\uparrow}^\dagger}{\sqrt{2E_4^2 + 8T^2}}\|0\rangle$
$\mathcal{H}_3$	$\pm T + U$	$\frac{1}{\sqrt{2}}(\hat{d}_{1\sigma}^\dagger \hat{d}_{2\uparrow}^\dagger \hat{d}_{2\downarrow}^\dagger \pm \hat{d}_{1\uparrow}^\dagger \hat{d}_{1\downarrow}^\dagger \hat{d}_{2\sigma}^\dagger)\|0\rangle$
$\mathcal{H}_4$	$2U$	$\hat{d}_{1\uparrow}^\dagger \hat{d}_{1\downarrow}^\dagger \hat{d}_{2\uparrow}^\dagger \hat{d}_{2\downarrow}^\dagger \|0\rangle$

Table 2.1 Eigenvalues and normalized eigenkets of the Hubbard dimer in the different Hilbert spaces. In the table the quantity $\Delta = \sqrt{16T^2 + U^2}$.

The Hamiltonian does not couple these states to those with parallel spin. Therefore, we can calculate the remaining eigenvalues and eigenkets in $\mathcal{H}_2$ by diagonalizing the matrix h_2 with elements $(h_2)_{ij} = \langle \Psi_i | \hat{H} | \Psi_j \rangle$. After some simple algebra one finds

$$h_2 = \begin{pmatrix} U & 0 & T & -T \\ 0 & U & T & -T \\ T & T & 0 & 0 \\ -T & -T & 0 & 0 \end{pmatrix}.$$

This matrix has eigenvalues $E_1 = 0$, $E_2 = U$, $E_3 = \frac{1}{2}(U - \Delta)$, and $E_4 = \frac{1}{2}(U + \Delta)$ where $\Delta = \sqrt{16T^2 + U^2}$. Let us take, e.g., $U > 0$ so that $E_3 < 0$ and $E_4 > 0$. The normalized eigenkets corresponding to these eigenvalues are reported in Table 2.1. The occurrence of the zero eigenvalue E_1 should not come as a surprise. The Hamiltonian commutes with the total spin operator $\hat{S}^2 = (\hat{\mathbf{S}}_1 + \hat{\mathbf{S}}_2) \cdot (\hat{\mathbf{S}}_1 + \hat{\mathbf{S}}_2)$, as well as with $\hat{S}^z = \hat{S}_1^z + \hat{S}_2^z$ (the spin operators were defined in (1.92)). Therefore the degenerate eigenkets of $\hat{H}$ must belong to spin multiplets. It is easy to see that the three eigenkets with vanishing eigenvalue belong to a triplet whereas the eigenkets with eigenvalues E_2, E_3 and E_4 are singlets. To conclude our analysis we must calculate the eigenkets and eigenvalues with three and four particles. This can be done using the particle–hole symmetry and the results are reported in Table 2.1.

It is interesting to observe that when $T \to 0$ the eigenvalues in $\mathcal{H}_2$ become 0 (4 times degenerate) and U (two times degenerate). This is the *dissociation limit* which corresponds to pulling the two atomic sites infinitely far apart. The results agree with our physical intuition; if two isolated (Hubbard) atoms have one electron each (either with spin up or down) then the energy is zero whereas if both electrons are on the same atom then the energy is U.

Exercise 2.3. Consider a chain of N sites, $s = 1, \ldots, N$, with $T_{ss'} = T$ if s and s' are nearest neighbor and zero otherwise. The one-body Hamiltonian is

$$\hat{H}_0 = T \sum_\sigma \sum_{s=1}^{N-1} (\hat{d}_{s\sigma}^\dagger \hat{d}_{s+1\sigma} + \hat{d}_{s+1\sigma}^\dagger \hat{d}_{s\sigma}).$$

Prove that the single particle eigenkets of $\hat{H}_0$ are

$$|k\sigma\rangle = \hat{c}_{k\sigma}^\dagger |0\rangle = \sum_s \sqrt{\frac{2}{N+1}} \sin\left(\frac{\pi k s}{N+1}\right) \hat{d}_{s\sigma}^\dagger |0\rangle,$$

with $k = 1, \ldots, N$, and have eigenvalues $\epsilon_k = 2T \cos(\frac{\pi k}{N+1})$.

Exercise 2.4. Let $\hat{N}_\sigma = \sum_s \hat{d}_{s\sigma}^\dagger \hat{d}_{s\sigma}$ be the operator for the total number of particles of spin σ and $\hat{S}^j = \sum_s \hat{S}_s^j$ be the *total spin operators*, where the spin density operators are defined in (1.92). Show that these operators commute with both $\hat{H}_0$ and $\hat{H}_{\text{int}}$ in (2.29).

2.5 Heisenberg model

A model system in which the number N of electrons is identical to the number N_V of space-orbitals basis-functions is said to be half-filled, since the maximum possible value of N is $2N_V$. In the half-filled Hubbard model, when the U_s are much larger than the $h_{ss'}$, states with two electrons on the same site have very high energy. It would then be useful to construct an effective theory in the truncated space of states with one single electron per atomic site

$$|\Phi_{\{\sigma\}}\rangle = \prod_s \hat{d}_{s\sigma(s)}^\dagger |0\rangle, \tag{2.33}$$

where $\sigma(s) = \uparrow, \downarrow$ is a collection of spin indices. The kets (2.33) are eigenkets of $\hat{H}_{\text{int}}$ in (2.29) with eigenvalue zero and of $\hat{N}$ with eigenvalue N_V. Their number is equal to the number of possible spin configurations, that is 2^{N_V}. Since $\hat{H}_{\text{int}}$ is positive semidefinite the kets $|\Phi_{\{\sigma\}}\rangle$ form a basis in the ground (lowest energy) subspace of $\hat{H}_{\text{int}}$. To construct an effective low-energy theory we need to understand how the states (2.33) are mixed by the one-body part $\hat{H}_0$ of the Hubbard Hamiltonian (2.29). Let us start by separating the matrix h into a diagonal part, $\epsilon_s = h_{ss}$, and an off-diagonal part $T_{ss'} = h_{ss'}$ for $s \neq s'$. For simplicity, in the discussion below we set the diagonal elements $\epsilon_s = 0$. If $|\Psi\rangle$ is an eigenket of the full Hamiltonian $\hat{H} = \hat{H}_0 + \hat{H}_{\text{int}}$ with energy E then the eigenvalue equation can be written as

$$|\Psi\rangle = \frac{1}{E - \hat{H}_{\text{int}}} \hat{H}_0 |\Psi\rangle. \tag{2.34}$$

We now approximate $|\Psi\rangle$ as a linear combination of the $|\Phi_{\{\sigma\}}\rangle$. Then, the action of $(E - \hat{H}_{\text{int}})^{-1} \hat{H}_0$ on $|\Psi\rangle$ yields a linear combination of kets with a doubly occupied site.[9] Since these kets are orthogonal to $|\Psi\rangle$ we iterate (2.34) so as to generate an eigenvalue problem in the subspace of the $|\Phi_{\{\sigma\}}\rangle$

$$\langle\Phi_{\{\sigma\}}|\Psi\rangle = \langle\Phi_{\{\sigma\}}| \frac{1}{E - \hat{H}_{\text{int}}} \hat{H}_0 \frac{1}{E - \hat{H}_{\text{int}}} \hat{H}_0 |\Psi\rangle, \qquad |\Psi\rangle = \sum_{\{\sigma\}} \alpha_{\{\sigma\}} |\Phi_{\{\sigma\}}\rangle. \tag{2.35}$$

This eigenvalue equation tells us how $\hat{H}_0$ lifts the ground-state degeneracy to second order in $\hat{H}_0$. To evaluate the r.h.s. we consider a generic term in $\hat{H}_0|\Psi\rangle$, say, $\hat{d}^{\dagger}_{s\uparrow}\hat{d}_{s'\uparrow}|\Phi_{\{\sigma\}}\rangle$ with $s \neq s'$ (recall that $\epsilon_s = 0$). This term corresponds to removing an electron of spin up from s' and creating an electron of spin up in s. If the spin configuration $\{\sigma\}$ is such that $\sigma(s') = \uparrow$ and $\sigma(s) = \downarrow$ the ket $\hat{d}^{\dagger}_{s\uparrow}\hat{d}_{s'\uparrow}|\Phi_{\{\sigma\}}\rangle$ has the site s occupied by two electrons (of opposite spin) and the site s' empty; in all other cases $\hat{d}^{\dagger}_{s\uparrow}\hat{d}_{s'\uparrow}|\Phi_{\{\sigma\}}\rangle = |\emptyset\rangle$ since we cannot remove an electron of spin up from s' if $\sigma(s') = \downarrow$ and we cannot create an electron of spin up in s if there is one already, i.e., $\sigma(s) = \uparrow$. From this observation we conclude that $\hat{d}^{\dagger}_{s\uparrow}\hat{d}_{s'\uparrow}|\Phi_{\{\sigma\}}\rangle$ is either an eigenket of $\hat{H}_{\text{int}}$ with eigenvalue U_s or the null ket. Similar considerations apply to $\hat{d}^{\dagger}_{s\downarrow}\hat{d}_{s'\downarrow}|\Phi_{\{\sigma\}}\rangle$. Therefore, we can write

$$\frac{1}{E - \hat{H}_{\text{int}}} \hat{H}_0 |\Psi\rangle = \sum_{\sigma} \sum_{ss'} \frac{T_{ss'}}{E - U_s} \hat{d}^{\dagger}_{s\sigma}\hat{d}_{s'\sigma}|\Psi\rangle \sim -\sum_{\sigma}\sum_{ss'} \frac{T_{ss'}}{U_s} \hat{d}^{\dagger}_{s\sigma}\hat{d}_{s'\sigma}|\Psi\rangle,$$

where in the last step we use the fact that the U_s are much larger than E.[10] Inserting this result into (2.35) and taking into account that $\langle\Phi_{\{\sigma\}}|(E - \hat{H}_{\text{int}})^{-1} = \langle\Phi_{\{\sigma\}}|/E$ we find the eigenvalue equation

$$E\langle\Phi_{\{\sigma\}}|\Psi\rangle = \langle\Phi_{\{\sigma\}}| \underbrace{-\sum_{\sigma\sigma'}\sum_{ss'}\sum_{rr'} \frac{T_{rr'}T_{ss'}}{U_s} \hat{d}^{\dagger}_{r\sigma'}\hat{d}_{r'\sigma'}\hat{d}^{\dagger}_{s\sigma}\hat{d}_{s'\sigma}}_{\hat{H}_{\text{eff}}} |\Psi\rangle. \tag{2.36}$$

---

[9] If the reader does not see that this is the case, it becomes clear in a few lines.

[10] We remind the reader that E is the correction to the degenerate ground-state energy due to the presence of $\hat{H}_0$ and therefore $E \to 0$ for $T_{ss'} \to 0$.

The eigenvalues E of the *effective Hamiltonian* $\hat{H}_{\text{eff}}$ in the subspace of the $|\Phi_{\{\sigma\}}\rangle$ are the second-order corrections of the highly degenerate ground-state energy. We now show that in the subspace of the $|\Phi_{\{\sigma\}}\rangle$ the effective Hamiltonian takes a very elegant form.

The terms of $\hat{H}_{\text{eff}}|\Psi\rangle$ that have a nonvanishing inner product with the $|\Phi_{\{\sigma\}}\rangle$ are those generated by removing an electron from s' (which then remains empty), creating an electron of the same spin in s (which becomes doubly occupied), removing the same electron or the one with opposite spin from $r' = s$ and creating it back in $r = s'$, see the illustration below.

In (2.36) we can therefore restrict the sums to $r' = s$ and $r = s'$. In doing so the effective Hamiltonian simplifies to

$$
\hat{H}_{\text{eff}} \equiv -\sum_{\sigma\sigma'}\sum_{ss'} \frac{T_{s's}T_{ss'}}{U_s} \hat{d}^{\dagger}_{s'\sigma'}\hat{d}_{s\sigma'}\hat{d}^{\dagger}_{s\sigma}\hat{d}_{s'\sigma}
$$

$$
= -\sum_{\sigma\sigma'}\sum_{ss'} \frac{|T_{ss'}|^2}{U_s} (\delta_{\sigma\sigma'}\hat{n}_{s'\sigma} - \hat{d}^{\dagger}_{s'\sigma'}\hat{d}_{s'\sigma}\hat{d}^{\dagger}_{s\sigma}\hat{d}_{s\sigma'}).
$$

Using the definition (1.91) of the spin density operators we can perform the sum over σ,σ' and rewrite the effective Hamiltonian as

$$
\hat{H}_{\text{eff}} = -\sum_{ss'} \frac{|T_{ss'}|^2}{U_s} \left(\hat{n}_{s'\uparrow} + \hat{n}_{s'\downarrow} - \hat{n}_{s'\uparrow}\hat{n}_{s\uparrow} - \hat{n}_{s'\downarrow}\hat{n}_{s\downarrow} - \hat{S}^+_{s'}\hat{S}^-_s - \hat{S}^-_{s'}\hat{S}^+_s \right).
$$

The above equation can be further manipulated to obtain a physically transparent formula. In the subspace of the $|\Phi_{\{\sigma\}}\rangle$ the operator $\hat{n}_{s'\uparrow} + \hat{n}_{s'\downarrow}$ is always 1. This implies that $\hat{S}^z_s = \frac{1}{2}(\hat{n}_{s\uparrow} - \hat{n}_{s\downarrow}) = \hat{n}_{s\uparrow} - \frac{1}{2} = -\hat{n}_{s\downarrow} + \frac{1}{2}$, and hence $\hat{n}_{s'\uparrow}\hat{n}_{s\uparrow} + \hat{n}_{s'\downarrow}\hat{n}_{s\downarrow} = 2\hat{S}^z_{s'}\hat{S}^z_s + \frac{1}{2}$. In conclusion, $\hat{H}_{\text{eff}}$ can be expressed solely in terms of the spin density operators

$$
\hat{H}_{\text{eff}} = \sum_{ss'} J_{ss'}(\hat{\mathbf{S}}_s \cdot \hat{\mathbf{S}}_{s'} - \frac{1}{4}), \quad \text{with} \quad J_{ss'} = |T_{ss'}|^2 \left(\frac{1}{U_s} + \frac{1}{U_{s'}} \right).
$$

This effective Hamiltonian is known as the *Heisenberg model* [19] and tells us that the interaction between two electrons frozen in their atomic positions has the same form as the interaction between two magnetic moments. There is, however, an important difference. The coupling constants $J_{ss'}$ are not proportional to μ^2_{B} (square of the Bohr magneton). The origin of the Heisenberg spin–spin interaction is purely quantum mechanical and has no classical analogue; it stems from virtual transitions where an electron hops to another occupied site and then hops back with the same or opposite spin.

Since $J_{ss'} > 0$ the Heisenberg Hamiltonian favors those configurations in which the electron spins in s and s' point in opposite directions. For instance, in a one-dimensional chain with sites $s = -\infty, \ldots, -1, 0, 1, \ldots \infty$ and $T_{ss'} = T$ for $|s - s'| = 1$ and zero otherwise, the ground-state average $\langle \Psi | \hat{\mathbf{S}}_s \cdot \hat{\mathbf{S}}_{s'} | \Psi \rangle = (-)^{s-s'} f(s - s')$ with $f(s - s') > 0$. This means that the spin magnetization is *staggered* and the system behaves like an *antiferromagnet*. The Heisenberg model is a clear example of how the subtle interplay between the tendency towards localization brought about by $\hat{H}_{\text{int}}$ and delocalization due to $\hat{H}_0$ favors some kind of magnetic order.

Exercise 2.5. Calculate the ground state average $\langle \Psi | \hat{\mathbf{S}}_s \cdot \hat{\mathbf{S}}_{s'} | \Psi \rangle$ of the Heisenberg dimer $\hat{H}_{\text{eff}} = J\hat{\mathbf{S}}_1 \cdot \hat{\mathbf{S}}_2$ for $s, s' = 1, 2$.

2.6 BCS model and the exact Richardson solution

The BCS model was introduced by Bardeen, Cooper, and Schrieffer in 1957 [20] to explain the phenomenon of superconductivity in metals at low enough temperatures. In the simplest version of the BCS model two electrons in the energy eigenstates $\varphi_{k\uparrow}(\mathbf{r}\sigma) = \delta_{\sigma\uparrow}\varphi_k(\mathbf{r})$ and $\varphi_{k\downarrow}(\mathbf{r}\sigma) = \delta_{\sigma\downarrow}\varphi_k^*(\mathbf{r})$ interact and after scattering end up in another couple of states $\varphi_{k'\uparrow}$ and $\varphi_{k'\downarrow}$.[11] The probability for this process is assumed to be independent of k and k' and, furthermore, all other scattering processes are discarded. We refer the reader to the original paper for the microscopic justification of these assumptions. Denoting by ϵ_k the single particle eigenenergies, the BCS Hamiltonian reads

$$\hat{H} = \hat{H}_0 + \hat{H}_{\text{int}} = \sum_{k\sigma} \epsilon_k \hat{c}_{k\sigma}^\dagger \hat{c}_{k\sigma} - v \sum_{kk'} \hat{c}_{k\uparrow}^\dagger \hat{c}_{k\downarrow}^\dagger \hat{c}_{k'\downarrow} \hat{c}_{k'\uparrow}, \tag{2.37}$$

where v is the so called *scattering amplitude* and the sum runs over a set of one-particle states like, e.g., the Bloch states of Section 2.3.1. The operators $\hat{c}_{k\sigma}^\dagger$ create an electron in $\varphi_{k\sigma}$ and satisfy the usual anti-commutation relations $\left[\hat{c}_{k\sigma}, \hat{c}_{k'\sigma'}^\dagger \right]_+ = \delta_{\sigma\sigma'} \delta_{kk'}$. The BCS model is usually treated within the so called BCS approximation, according to which the ground state is a superposition of states with different numbers of electrons. That this is certainly an approximation follows from the fact that $\hat{H}$ commutes with the total number of particle operator $\hat{N}$, and hence the *exact* ground state has a well defined number of electrons. Nevertheless, in macroscopic systems the fluctuations around the average value of $\hat{N}$ tend to zero in the thermodynamic limit, and the BCS approximation is, in these cases, very accurate. The BCS approximation breaks down in small systems, like, e.g., in superconducting aluminum nanograins [21]. Starting from the mid-nineties, the experimental progress in characterizing superconducting nanostructures has renewed interest in the BCS model beyond the BCS approximation. In the effort of constructing new approximation schemes, a series of old papers in nuclear physics containing the *exact solution* of the BCS model came to light. It just happens that the Hamiltonian (2.37) also describes a system of nucleons with an effective nucleon–nucleon interaction v; in this context it is known

---

[11] The state $\varphi_{k\downarrow}(\mathbf{r}\sigma)$ is the time-reverse of $\varphi_{k\uparrow}(\mathbf{r}\sigma)$.

as the *pairing model*. In the mid-sixties Richardson derived a set of algebraic equations to calculate all eigenstates and eigenvalues of the pairing model [22]. His work is today enjoying an increasing popularity. Below we derive the Richardson solution of the BCS model.

The first important observation is that $\hat{H}_{int}$ acts only on doubly occupied k-states. This means that given a many-particle ket $|\Psi\rangle$ in which a k-state is occupied by a single electron (either with spin up or down), the ket $\hat{H}|\Psi\rangle$ also has the k-state occupied by a single electron. In other words the singly occupied states remain blocked from participating in the dynamics. The labels of these states are therefore good quantum numbers. Let B be a subset of k-states whose number is $|B|$. Then, a generic eigenket of $\hat{H}$ has the form

$$|\Psi_B^{(N)}\rangle = \prod_{k \in B} \hat{c}_{k\,\sigma(k)}^\dagger |\Psi^{(N)}\rangle,$$

with $\sigma(k) = \uparrow, \downarrow$ a collection of spin indices and $|\Psi^{(N)}\rangle$ a $2N$-electron ket whose general form is

$$|\Psi^{(N)}\rangle = \sum_{k_1 \ldots k_N \notin B} \alpha_{k_1 \ldots k_N} \hat{b}_{k_1}^\dagger \ldots \hat{b}_{k_N}^\dagger |0\rangle, \tag{2.38}$$

where $\hat{b}_k^\dagger \equiv \hat{c}_{k\uparrow}^\dagger \hat{c}_{k\downarrow}^\dagger$ are electron-pair creation operators or simply pair operators. Since $\hat{b}_k^\dagger \hat{b}_k^\dagger = 0$ we can restrict the sum in (2.38) to $k_1 \neq k_2 \neq \ldots \neq k_N$. Furthermore $\hat{b}_k^\dagger \hat{b}_{k'}^\dagger = \hat{b}_{k'}^\dagger \hat{b}_k^\dagger$ and hence the amplitudes $\alpha_{k_1 \ldots k_N}$ are symmetric under a permutation of the indices $k_1, \ldots, k_N$. The eigenket $|\Psi_B^{(N)}\rangle$ describes $2N + |B|$ electrons, $|B|$ of which are in singly-occupied k-states and contribute $E_B = \sum_{k \in B} \epsilon_k$ to the energy, and the remaining N pairs of electrons are distributed among the remaining unblocked k-states. Since the dynamics of the blocked electrons is trivial, in the following we shall assume $B = \emptyset$.

The key to solving the BCS model is the commutation relation between the pair operators

$$\left[\hat{b}_k, \hat{b}_{k'}^\dagger\right]_- = \delta_{kk'}(1 - \hat{n}_{k\uparrow} - \hat{n}_{k\downarrow}), \qquad \left[\hat{b}_k, \hat{b}_{k'}\right]_- = \left[\hat{b}_k^\dagger, \hat{b}_{k'}^\dagger\right]_- = 0,$$

with $\hat{n}_{k\sigma} = \hat{c}_{k\sigma}^\dagger \hat{c}_{k\sigma}$ the occupation operator. In the Hilbert space of states (2.38) one can discard the term $\hat{n}_{k\uparrow} + \hat{n}_{k\downarrow}$ in the commutator since $k_1 \neq k_2 \neq \ldots \neq k_N$ and hence

$$\hat{b}_k \hat{b}_{k_1}^\dagger \hat{b}_{k_2}^\dagger \ldots \hat{b}_{k_N}^\dagger |0\rangle = \left[\hat{b}_k, \hat{b}_{k_1}^\dagger\right]_- \hat{b}_{k_2}^\dagger \ldots \hat{b}_{k_N}^\dagger |0\rangle + \hat{b}_{k_1}^\dagger \left[\hat{b}_k, \hat{b}_{k_2}^\dagger\right]_- \ldots \hat{b}_{k_N}^\dagger |0\rangle$$

$$+ \ldots + \hat{b}_{k_1}^\dagger \hat{b}_{k_2}^\dagger \ldots \left[\hat{b}_k, \hat{b}_{k_N}^\dagger\right]_- |0\rangle$$

$$= \delta_{kk_1} \hat{b}_{k_2}^\dagger \hat{b}_{k_3}^\dagger \ldots \hat{b}_{k_N}^\dagger |0\rangle + \delta_{kk_2} \hat{b}_{k_1}^\dagger \hat{b}_{k_3}^\dagger \ldots \hat{b}_{k_N}^\dagger |0\rangle$$

$$+ \ldots + \delta_{kk_N} \hat{b}_{k_1}^\dagger \hat{b}_{k_2}^\dagger \ldots \hat{b}_{k_{N-1}}^\dagger |0\rangle, \tag{2.39}$$

which is the same result as (1.57) for bosonic operators. This is not, a posteriori, surprising since $\hat{b}_k^\dagger$ creates two electrons in a *singlet state*. These pair singlets are known as the *Cooper pairs* in honour of Cooper who first recognized the formation of bound states in a fermionic system with attractive interactions [23]. The analogy between Cooper pairs and bosons is, however, not so strict: we cannot create two Cooper pairs in the same state while we can certainly do it for two bosons. We may say that Cooper pairs are *hard core bosons* since we cannot find two or more of them in the same state.

To understand the logic of the Richardson solution we start with the simplest nontrivial case, i.e., a number $N = 2$ of pairs

$$|\Psi^{(2)}\rangle = \sum_{p \neq q} \alpha_{pq} \hat{b}_p^\dagger \hat{b}_q^\dagger |0\rangle,$$

with $\alpha_{pq} = \alpha_{qp}$. Using (2.39) we find

$$\hat{H}|\Psi^{(2)}\rangle = \sum_{p \neq q} \alpha_{pq} \left((2\epsilon_p + 2\epsilon_q) \hat{b}_p^\dagger \hat{b}_q^\dagger - v \sum_{k \neq q} \hat{b}_k^\dagger \hat{b}_q^\dagger - v \sum_{k \neq p} \hat{b}_p^\dagger \hat{b}_k^\dagger \right) |0\rangle.$$

Renaming the indices $k \leftrightarrow p$ in the first sum and $k \leftrightarrow q$ in the second sum, we can easily extract from $\hat{H}|\Psi^{(2)}\rangle = E|\Psi^{(2)}\rangle$ an eigenvalue equation for the amplitudes α_{pq},

$$(2\epsilon_p + 2\epsilon_q)\alpha_{pq} - v \sum_{k \neq q} \alpha_{kq} - v \sum_{k \neq p} \alpha_{pk} = E\alpha_{pq}. \tag{2.40}$$

The constraint in the sums is reminiscent of the fact that there cannot be more than one pair in a k-state. It is this constraint that renders the problem complicated since for a pair to stay in k no other pair must be there. Nevertheless (2.40) still admits a simple solution! The idea is to reduce (2.40) to two coupled eigenvalue equations. We therefore make the ansatz

$$\alpha_{pq} = \alpha_p^{(1)}\alpha_q^{(2)} + \alpha_q^{(1)}\alpha_p^{(2)},$$

which entails the symmetry property $\alpha_{pq} = \alpha_{qp}$, and write the energy E as the sum of two pair energies $E = E_1 + E_2$. Then, the eigenvalue equation (2.40) becomes

$$\{\alpha_q^{(2)}(2\epsilon_p - E_1)\alpha_p^{(1)} + \alpha_q^{(1)}(2\epsilon_p - E_2)\alpha_p^{(2)}\} + \{p \leftrightarrow q\}$$
$$= v \left(\{\alpha_q^{(2)} \sum_k \alpha_k^{(1)} + \alpha_q^{(1)} \sum_k \alpha_k^{(2)}\} + \{p \leftrightarrow q\} \right) - 2v(\alpha_p^{(1)}\alpha_p^{(2)} + \alpha_q^{(1)}\alpha_q^{(2)}), \tag{2.41}$$

where on the r.h.s. we have extended the sums to all ks and subtracted the extra term. Without this extra term (2.41) is solved by the amplitudes

$$\alpha_p^{(1)} = \frac{1}{2\epsilon_p - E_1}, \qquad \alpha_p^{(2)} = \frac{1}{2\epsilon_p - E_2}, \tag{2.42}$$

with E_i a root of

$$\sum_k \frac{1}{2\epsilon_k - E_i} = \frac{1}{v}.$$

This solution would correspond to two *independent* pairs since $\alpha_p^{(1)}$ does not depend on $\alpha_p^{(2)}$. The inclusion of the extra term does not change the structure (2.42) of the αs but, instead, changes the equation that determines the E_i. Indeed, from (2.42) it follows that

$$\alpha_p^{(1)}\alpha_p^{(2)} = \frac{\alpha_p^{(2)} - \alpha_p^{(1)}}{E_2 - E_1},$$

and therefore (2.41) can be rewritten as

$$\left\{ \alpha_q^{(2)} \left((2\epsilon_p - E_1)\alpha_p^{(1)} - v \sum_k \alpha_k^{(1)} + \frac{2v}{E_2 - E_1} \right) \right\} + \{p \leftrightarrow q\}$$

$$+ \left\{ \alpha_q^{(1)} \left((2\epsilon_p - E_2)\alpha_p^{(2)} - v \sum_k \alpha_k^{(2)} - \frac{2v}{E_2 - E_1} \right) \right\} + \{p \leftrightarrow q\} = 0.$$

We have shown that (2.42) is a solution, provided that the pair energies E_1 and E_2 are the roots of the coupled system of algebraic equations

$$\sum_k \frac{1}{2\epsilon_k - E_1} = \frac{1}{v} + \frac{2}{E_2 - E_1},$$

$$\sum_k \frac{1}{2\epsilon_k - E_2} = \frac{1}{v} + \frac{2}{E_1 - E_2}.$$

The generalization to N pairs is tedious but straightforward. (In the remaining part of this section we do not introduce concepts or formulas needed for the following chapters. Therefore, the reader can, if not interested, move forward to the next section.) Let $|\Psi^{(N)}\rangle$ be the N-pair eigenket of $\hat{H}$ and $\{k\}_i$ be the set $\{k_1, .., k_{i-1}, k_{i+1}, .., k_N\}$ in which the k_i state is removed. The eigenvalue equation for the symmetric tensor $\alpha_{k_1...k_N}$ reads

$$\left(\sum_{i=1}^N 2\epsilon_{k_i} \right) \alpha_{k_1...k_N} - v \sum_{i=1}^N \sum_{p \neq \{k\}_i} \alpha_{k_1...k_{i-1}\,p\,k_{i+1}...k_N} = E\alpha_{k_1...k_N}.$$

As in the case of two pairs, the sum over p on the l.h.s. is constrained and some extra work must be done to make the equation separable. The ansatz is

$$\alpha_{k_1...k_N} = \sum_P \alpha_{k_{P(1)}}^{(1)} \cdots \alpha_{k_{P(N)}}^{(N)} = \sum_P \prod_{i=1}^N \alpha_{k_{P(i)}}^{(i)},$$

where the sum is over all permutations of $(1, \ldots, N)$, and $E = E_1 + \ldots + E_N$. Substitution into the eigenvalue equation leads to

$$\sum_P \sum_{i=1}^N \left\{ \left(\prod_{l \neq i}^N \alpha_{k_{P(l)}}^{(l)} \right) \left[(2\epsilon_{k_{P(i)}} - E_i)\alpha_{k_{P(i)}}^{(i)} - v \sum_p \alpha_p^{(i)} \right] \right.$$

$$\left. + v \sum_{j \neq i}^N \left(\prod_{l \neq i,j}^N \alpha_{k_{P(l)}}^{(l)} \right) \alpha_{k_{P(j)}}^{(i)} \alpha_{k_{P(j)}}^{(j)} \right\} = 0, \qquad (2.43)$$

where, as in the example with $N = 2$, the last term is what we have to subtract to perform the unconstrained sum over p. We look for solutions of the form

$$\alpha_k^{(i)} = \frac{1}{2\epsilon_k - E_i}, \qquad (2.44)$$

from which it follows that the product of two pair amplitudes with the same k can be written as

$$\alpha_k^{(i)}\alpha_k^{(j)} = \frac{\alpha_k^{(j)} - \alpha_k^{(i)}}{E_j - E_i}.$$

We use this identity for some manipulation of the last term in (2.43)

$$v\sum_P\sum_{i=1}^N\sum_{j\neq i}^N\left(\prod_{l\neq i,j}^N\alpha_{k_{P(l)}}^{(l)}\right)\frac{\alpha_{k_{P(j)}}^{(j)} - \alpha_{k_{P(j)}}^{(i)}}{E_j - E_i} = 2v\sum_P\sum_{i=1}^N\left(\prod_{l\neq i}^N\alpha_{k_{P(l)}}^{(l)}\right)\sum_{j\neq i}^N\frac{1}{E_j - E_i},$$

a result which allows us to decouple the eigenvalue equation

$$\sum_P\sum_{i=1}^N\left(\prod_{l\neq i}^N\alpha_{k_{P(l)}}^{(l)}\right)\left[(2\epsilon_{k_{P(i)}} - E_i)\alpha_{k_{P(i)}}^{(i)} - v\sum_p\alpha_p^{(i)} + 2v\sum_{j\neq i}\frac{1}{E_j - E_i}\right] = 0.$$

Therefore the amplitudes (2.44) are solutions provided that the pair-energies E_i are the roots of the coupled system of algebraic equations

$$\boxed{\frac{1}{v_i} \equiv \frac{1}{v} + 2\sum_{j\neq i}\frac{1}{E_j - E_i} = \sum_k\frac{1}{2\epsilon_k - E_i}} \qquad (2.45)$$

The system (2.45) can be regarded as a generalization of the eigenvalue equation for a single pair, where the effective scattering amplitude v_i depends on the relative distribution of all other pairs. Numerical solutions show that with increasing v some of the E_i become complex; however, they always occur in complex conjugate pairs, so that the total energy E remains real.

2.7 Holstein model

The motion of an electron in a crystal (in particular an ionic crystal) causes a displacement of the nuclei and, therefore, differs quite substantially from the motion through a rigid nuclear structure. Consider an electron in some point of the crystal: as a result of the attractive interaction the nuclei move toward new positions and create a potential well for the electron. If the well is deep enough and if the electron is sufficiently slow this effect causes self-trapping of the electron: the electron cannot move unless accompanied by the well, also called a *polarization cloud*. The *quasi-particle* which consists of the electron and of the surrounding polarization cloud is called the *polaron*.

Depending on the nature of the material, the electron–nuclear interaction gives rise to different kinds of polaron, small or large, heavy or light, etc. In ionic crystals (like sodium chloride) the electron–nuclear interaction is long ranged and a popular model to describe the polaron features is the Fröhlich model [24]. In 1955 Feynman proposed a variational solution in which the polaron was considered as an electron bound to a particle of mass M

by a spring of elastic constant K [25]. The Feynman solution is in good agreement with the numerical results for both strong and weak electron–nuclear coupling, and hence it provides a good physical picture of the polaron behavior in these regimes. Other physically relevant models leading to the formation of polarons are the Su–Schrieffer–Heeger model [26] for conducting polymers (originally polyacetylene) and the Holstein model [27] for molecular crystals (originally one-dimensional). Below we discuss the Holstein model.

Let us consider electrons moving along a chain of diatomic molecules whose center of mass and orientations are fixed whereas the intra-nuclear distance can vary. The Hamiltonian of the system can be regarded as the sum of the molecular chain Hamiltonian $\hat{H}_C$, the electron Hamiltonian $\hat{H}_{el}$ and the Hamiltonian $\hat{H}_{int}$ that describes the interaction between the electrons and the nuclei. Assuming that the potential energy of a single molecule in the chain is not too different from that of the isolated molecule, and approximating the latter with the energy of a harmonic oscillator of frequency ω_0, the Hamiltonian of the molecular chain takes the form

$$\hat{H}_C = \sum_s \left(\frac{\hat{p}_s^2}{2M} + \frac{1}{2} M\omega_0^2 \hat{x}_s^2 \right),$$

with $\hat{x}_s$ the intra-molecular distance (measured with respect to the equilibrium position) operator of the sth molecule, $\hat{p}_s$ the corresponding conjugate momentum, and M the relative mass. Introducing the lowering operators $\hat{a}_s = \sqrt{\frac{M\omega_0}{2}}(\hat{x}_s + \frac{i}{M\omega_0}\hat{p}_s)$ and raising operators $\hat{a}_s^\dagger = \sqrt{\frac{M\omega_0}{2}}(\hat{x}_s - \frac{i}{M\omega_0}\hat{p}_s)$ with *bosonic* commutation relations $[\hat{a}_s, \hat{a}_{s'}^\dagger]_- = \delta_{ss'}$ and $[\hat{a}_s, \hat{a}_{s'}]_- = 0$, the chain Hamiltonian can be rewritten as

$$\hat{H}_C = \sum_s \omega_0(\hat{a}_s^\dagger \hat{a}_s + \frac{1}{2}).$$

We wish here to include a few words on the physical meaning of the bosonic operators $\hat{a}_s^\dagger$ and $\hat{a}_s$. First of all it is important to stress that *they do not create or destroy the molecules but quanta of vibrations*. These quanta can be labeled with their energy (in our case the energy is the same for all oscillators) or with their position (the center of mass x_s of the molecules) and have, therefore, the *same degrees of freedom as a "particle" of spin zero*. In solid-state physics these "particles" are called *phonons* or, if the molecule is isolated, *vibrons*. Let us elaborate on this "particle" interpretation by considering a one-dimensional harmonic oscillator. In quantum mechanics the energy eigenket $|j\rangle$ describes *one particle* (either a fermion or a boson) in the jth energy level. Within the above interpretation of the vibrational quanta the ket $|j\rangle$ corresponds to a state with j phonons since $|j\rangle = \frac{1}{\sqrt{j!}}(\hat{a}^\dagger)^j|0\rangle$; in particular the ground state $|0\rangle$, which describes *one* particle in the lowest energy eigenstate, corresponds to *zero* phonons. Even more "exotic" is the new interpretation of ket $|x\rangle$ that in quantum mechanics describes *one* particle in the position x. Indeed, $|x\rangle = \sum_j |j\rangle\langle j|x\rangle$ and therefore it is a linear combination of kets *with different numbers of phonons*. Accordingly, an eigenstate of the position operator is obtained by acting with the operator $\hat{A}^\dagger(x) \equiv \sum_j \langle j|x\rangle \frac{1}{\sqrt{j!}}(\hat{a}^\dagger)^j$ on the zero phonons ket $|0\rangle$. The interpretation of the quantum of vibration as a bosonic particle has been and continues to be a very fruitful idea. After the scientific revolutions of special relativity and quantum mechanics, scientists started to look for a quantum theory consistent with relativistic mechanics.[12] In the attempt to construct such a theory Dirac proposed a relativistic equation for the wavefunction [28]. The Dirac equation, however, led to a new interpretation of the electrons. The electrons, and more generally the fermions, can be seen as the vibrational quanta of some *fermionic oscillator* in the same way as the phonons, and more generally the bosons, are the vibrational quanta of some *bosonic oscillator*. Of course to describe a particle in three-dimensional space we need many oscillators, as in the Holstein model. In a continuum description the operator $\hat{a}_s$ is labeled with the continuous variable $\mathbf{r}$, $\hat{a}_s \to \hat{a}(\mathbf{r})$, and it becomes what is called a *quantum field*. The origin of the name quantum field stems from the fact that to every point in space is assigned a quantity, as is done with the classical electric field $\mathbf{E}(\mathbf{r})$ or magnetic field $\mathbf{B}(\mathbf{r})$, but this quantity is an operator. Similarly, the fermions are described by some fermionic quantum field. For these reasons the relativistic generalization of quantum mechanics is called quantum field theory [29–34].

Let us continue with the description of the Holstein model and derive an approximation for the interaction between the electrons and the vibrational quanta of the molecules. We denote by $E(x)$ the ground state energy of the isolated molecule with one more electron (molecular ion). Due to the presence of the extra electron, $E(x)$ is not stationary at the equilibrium position $x = 0$ of the isolated charge-neutral molecule. For small x we can expand the energy to linear order and write $E(x) = -g\sqrt{2M\omega_0}\,x$,[13] where the constant $g > 0$ governs the strength of the coupling. Due to the minus sign the presence of the electron causes a force $F = -dE/dx > 0$ which tends to increase the interatomic distance in the molecule, as shown in the schematic representation below.

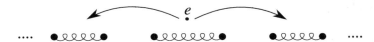

If we discard the corrections induced by the presence of the other molecules the interaction Hamiltonian can then be modeled as

$$\hat{H}_{\text{int}} = -g\sqrt{2M\omega_0}\sum_{s\sigma}\hat{d}^\dagger_{s\sigma}\hat{d}_{s\sigma}\hat{x}_s = -g\sum_{s\sigma}\hat{d}^\dagger_{s\sigma}\hat{d}_{s\sigma}(\hat{a}_s + \hat{a}^\dagger_s),$$

where the $\hat{d}_{s\sigma}$ operator destroys an electron of spin σ on the sth molecule and therefore $\hat{d}^\dagger_{s\sigma}\hat{d}_{s\sigma}$ is simply the occupation operator that counts how many electrons (0 or 1) of spin σ sit on the sth molecule. It is also important to say that the fermionic operators $\hat{d}$ *commute* with the bosonic operators $\hat{a}$.

Finally we need to model the Hamiltonian $\hat{H}_{\text{el}}$ which describes "free" electrons moving along the chain. Since the orthonormal functions $\varphi_{s\sigma}$ that define the $\hat{d}$-operators are localized around the sth molecule we neglect all matrix elements $T_{ss'} \equiv \langle s\sigma|\hat{h}|s'\sigma\rangle$ except those for which $|s - s'| = 1$ (nearest neighbor molecules). The diagonal matrix elements

---

[12]We recall that the original Schrödinger equation is invariant under Galilean transformation but not under the more fundamental Lorentz transformations.

[13]The zeroth order term leads to a trivial shift of the total energy for any fixed number of electrons and it is therefore ignored.

can be ignored as well since $\epsilon_s \equiv \langle s\sigma|\hat{h}|s\sigma \rangle$ does not depend on s; these matrix elements simply give rise to a constant energy shift for any fixed number of electrons. In conclusion we have

$$\hat{H}_{\mathrm{el}} = T \sum_{s\sigma}(\hat{d}^{\dagger}_{s\sigma}\hat{d}_{s+1\sigma} + \hat{d}^{\dagger}_{s+1\sigma}\hat{d}_{s\sigma}),$$

with $T = T_{ss\pm1}$. The Hamiltonian that defines the Holstein model is

$$\hat{H} = \hat{H}_{\mathrm{C}} + \hat{H}_{\mathrm{int}} + \hat{H}_{\mathrm{el}}. \tag{2.46}$$

This Hamiltonian commutes with the operator $\hat{N}_{\mathrm{el}} = \sum_{s\sigma}\hat{d}^{\dagger}_{s\sigma}\hat{d}_{s\sigma}$ of the total number of electrons but it does not commute with the operator $\hat{N}_{\mathrm{ph}} = \sum_s \hat{a}^{\dagger}_s\hat{a}_s$ of the total number of phonons due to the presence of $\hat{H}_{\mathrm{int}}$. The eigenstates of $\hat{H}$ are therefore linear combinations of kets with a fixed number of electrons and different number of phonons. Despite its simplicity the Holstein model cannot be solved exactly and one has to resort to approximations.

2.7.1 Peierls instability

A common approximation is the so called *adiabatic approximation* which considers the mass M infinitely large. Nuclei of large mass move so slowly that the electrons have ample time to readjust their wavefunctions in the configuration of minimum energy, i.e., the ground state. In terms of the elastic constant $K = M\omega_0^2$ and the coupling $\tilde{g} = g\sqrt{2M\omega_0}$ the Holstein model in the adiabatic approximation reads

$$\hat{H}_{\mathrm{ad}} = T \sum_{s\sigma}(\hat{d}^{\dagger}_{s\sigma}\hat{d}_{s+1\sigma} + \hat{d}^{\dagger}_{s+1\sigma}\hat{d}_{s\sigma}) - \tilde{g}\sum_{s\sigma}\hat{d}^{\dagger}_{s\sigma}\hat{d}_{s\sigma}\hat{x}_s + \frac{1}{2}K\sum_s \hat{x}_s^2.$$

The adiabatic Hamiltonian is much simpler than the original one since $[\hat{H}_{\mathrm{ad}}, \hat{x}_s]_- = 0$ for all s, and hence the eigenvalues of the position operators are good quantum numbers. We then introduce the operators

$$\hat{A}^{\dagger}_s(x) \equiv \sum_j \langle j|x\rangle \frac{1}{\sqrt{j!}}(\hat{a}^{\dagger}_s)^j$$

that create an eigenket of $\hat{x}_s$ with eigenvalue x when acting on the empty ket $|0\rangle$. The general eigenket of $\hat{H}_{\mathrm{ad}}$ has the form

$$\hat{A}^{\dagger}_1(x_1)\hat{A}^{\dagger}_2(x_2)\ldots|\Phi_{\mathrm{el}}, \{x_s\}\rangle,$$

where $|\Phi_{\mathrm{el}}, \{x_s\}\rangle$ is a pure electronic ket (obtained by acting with the $\hat{d}^{\dagger}$-operators over the empty ket) obeying the eigenvalue equation

$$\left(\hat{H}_{\mathrm{el}} - \tilde{g}\sum_{s\sigma}x_s\hat{d}^{\dagger}_{s\sigma}\hat{d}_{s\sigma}\right)|\Phi_{\mathrm{el}}, \{x_s\}\rangle = E(\{x_s\})|\Phi_{\mathrm{el}}, \{x_s\}\rangle. \tag{2.47}$$

If $|\Phi_{el}, \{x_s\}\rangle$ is the ground state of (2.47) with energy $E_0(\{x_s\})$, the lowest energy of $\hat{H}_{ad}$ is obtained by minimizing

$$E_0(\{x_s\}) + \frac{1}{2}K\sum_s x_s^2$$

over the space of all possible configurations $x_1, x_2, \ldots$ Even though the Hamiltonian in (2.47) is a one-body operator, and hence it does not take long to calculate $E_0(\{x_s\})$ for a given set of x_s, the minimization of a function of many variables is, in general, a complicated task. The best we can do to gain some physical insight is to calculate $E_0(\{x_s\})$ for some "reasonable" configuration and then compare the results.

Let us consider, for instance, a molecular ring with $2N$ molecules and $2N$ electrons (half-filling), N of spin up and N of spin down. It is intuitive to expect that in the ground state the nuclear displacement is uniform, $x_s = x$ for all s, since the ring is invariant under (discrete) rotations. If so, we could calculate the ground-state energy by finding the x that minimize the total energy. For a uniform displacement the Hamiltonian in (2.47) becomes

$$\hat{H}_{el} - \tilde{g}x\sum_{s=1}^{2N}\sum_\sigma \hat{d}_{s\sigma}^\dagger \hat{d}_{s\sigma} = \hat{H}_{el} - \tilde{g}x\hat{N}_{el},$$

and from the result (2.14) we know that the single particle eigenenergies of this Hamiltonian are $\epsilon_k = -\tilde{g}x + 2T\cos k$ with $k = 2\pi m/2N$. Therefore, the lowest energy of $\hat{H}_{ad}$ (with $2N$ electrons) *in the subspace of uniform displacements* is the minimum of

$$E_{unif}(x) = 2\times\left[\sum_{m=-\frac{N}{2}+1}^{\frac{N}{2}} 2T\cos\frac{2\pi m}{2N}\right] - 2N\tilde{g}x + \frac{2NKx^2}{2},$$

where, for simplicity, we have assumed that N is even and that $T < 0$. In this formula the factor of 2 multiplying the first term on the r.h.s. comes from spin.

Intuition is not always a good guide. Below we show that the lowest energy of $\hat{H}_{ad}$ in the subspace of *dimerized configurations* $x_s = y+(-)^s x$ is always lower than the minimum of $E_{unif}(x)$. In the dimerized case the electronic operator in (2.47) can be written as

$$\hat{H}_{el} - \tilde{g}y\hat{N}_{el} - \tilde{g}x\sum_{s=1}^{2N}\sum_\sigma (-)^s \hat{d}_{s\sigma}^\dagger \hat{d}_{s\sigma}.$$

We have already encountered this Hamiltonian in Section 2.3.1. It describes the ring of Fig. 2.6(a) with $\epsilon = -\tilde{g}y$, $\Delta = \tilde{g}x$, $T = t$ and N unit cells. Using the one-particle eigenvalues given in (2.15) we find that the lowest energy of $\hat{H}_{ad}$ (with $2N$ electrons) *in the subspace of dimerized configurations* is the minimum of

$$E_{dim}(x, y) = -2\times\left[\sum_{m=-\frac{N}{2}+1}^{\frac{N}{2}}\sqrt{(\tilde{g}x)^2 + 2T^2(1 + \cos\frac{2\pi m}{N})}\right] - 2N\tilde{g}y$$

$$+ \frac{NK}{2}\left[(y+x)^2 + (y-x)^2\right].$$

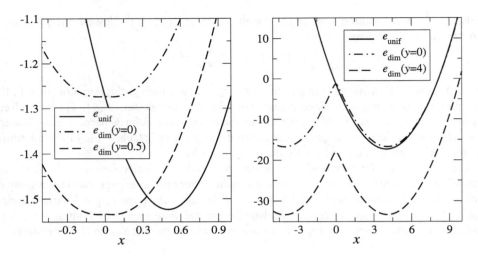

Figure 2.9 Uniform and dimerized energy densities $e_{\mathrm{unif}}(x) = E_{\mathrm{unif}}(x)/2N$ and $e_{\mathrm{dim}}(x,y) = E_{\mathrm{dim}}(x,y)/2N$ for a ring of 100 molecules and $T = -1$, $K = 2$. In the left panel $\tilde{g} = 1$ while in the right panel $\tilde{g} = 8$. The minimum of the dimerized energy $e_{\mathrm{dim}}(x,y)$ is always the lowest when y is close to the minimum of the uniform energy $e_{\mathrm{unif}}(y)$.

In Fig. 2.9 we show $E_{\mathrm{unif}}(x)/2N$ and $E_{\mathrm{dim}}(x,y)/2N$ as a function of x and for different values of y. In all cases there exist values of y for which the *dimerized configuration has lower energy*. This phenomenon is called the *Peierls instability* [35]: a periodic one-dimensional molecular crystal is unstable towards dimerization since the energy gain in opening a gap in the electronic spectrum is always larger than the elastic energy loss.

2.7.2　Lang–Firsov transformation: the heavy polaron

Let us go back to the original Holstein Hamiltonian and consider another limiting case, i.e., $g \gg T$. In this strong coupling regime the low-energy states have localized electrons. Indeed, if a state has a localized electron of spin σ on site s then the average of the occupation operator $\hat{n}_{s\sigma}$ is close to 1 and hence the energy gain in stretching the molecule is maximized. In order to treat $\hat{H}_{\mathrm{el}}$ as a perturbation it is convenient to perform a unitary transformation of the electron and phonon operators so to bring $\hat{H}_{\mathrm{C}} + \hat{H}_{\mathrm{int}}$ into a diagonal form. This unitary transformation is known as the *Lang–Firsov transformation* [36] and reads

$$\hat{p}_{s\sigma} = e^{i\hat{S}}\hat{d}_{s\sigma}e^{-i\hat{S}}, \qquad \hat{b}_s = e^{i\hat{S}}\hat{a}_s e^{-i\hat{S}}, \tag{2.48}$$

where $\hat{S}$ is the following Hermitian operator

$$\hat{S} = -i\frac{g}{\omega_0}\sum_{s\sigma}\hat{n}_{s\sigma}(\hat{a}_s^\dagger - \hat{a}_s).$$

Since the transformation is unitary the new operators obey the same (anti)commutation relations as the old ones, which in our case means $\left[\hat{p}_{s\sigma}, \hat{p}^\dagger_{s'\sigma'}\right]_+ = \delta_{ss'}\delta_{\sigma\sigma'}$ and $\left[\hat{b}_s, \hat{b}^\dagger_{s'}\right]_- = \delta_{ss'}$. To express the old operators in terms of the new ones we use the following trick. Let $\hat{O}$ be a generic operator and $\hat{F}_\alpha(\hat{O}) \equiv e^{i\alpha\hat{S}}\hat{O}e^{-i\alpha\hat{S}}$. This transformation is unitary for every real α and has the properties:

(1) $\hat{F}_0(\hat{O}) = \hat{O}$,

(2) $\hat{F}_\alpha(\hat{O}^\dagger) = \hat{F}^\dagger_\alpha(\hat{O})$,

(3) $\hat{F}_\alpha(\hat{O}_1 + \hat{O}_2) = \hat{F}_\alpha(\hat{O}_1) + \hat{F}_\alpha(\hat{O}_2)$,

(4) $\hat{F}_\alpha(\hat{O}_1\hat{O}_2) = \hat{F}_\alpha(\hat{O}_1)\hat{F}_\alpha(\hat{O}_2)$.

Furthermore, the derivative with respect to α is simply

$$\hat{F}'_\alpha(\hat{O}) = i\,[\hat{S}, \hat{F}_\alpha(\hat{O})] = i\,e^{i\alpha\hat{S}}\,[\hat{S}, \hat{O}]\,e^{-i\alpha\hat{S}}.$$

We now show how to invert (2.48) by calculating $\hat{F}'_\alpha(\hat{O})$ with $\hat{O}$ either the fermionic $\hat{d}$-operators or the bosonic $\hat{a}$-operators.

The commutators we are interested in are

$$[\hat{S}, \hat{d}_{s\sigma}] = i\frac{g}{\omega_0}\hat{d}_{s\sigma}(\hat{a}^\dagger_s - \hat{a}_s), \quad \Rightarrow \quad [\hat{S}, \hat{d}^\dagger_{s\sigma}] = -i\frac{g}{\omega_0}\hat{d}^\dagger_{s\sigma}(\hat{a}^\dagger_s - \hat{a}_s), \tag{2.49}$$

and

$$[\hat{S}, \hat{a}_s] = i\frac{g}{\omega_0}\sum_\sigma \hat{n}_{s\sigma}, \quad \Rightarrow \quad [\hat{S}, \hat{a}^\dagger_s] = i\frac{g}{\omega_0}\sum_\sigma \hat{n}_{s\sigma}. \tag{2.50}$$

From (2.50) it follows that [property (3)] $\hat{F}'_\alpha(\hat{a}^\dagger_s - \hat{a}_s) = 0$ and hence $\hat{F}_\alpha(\hat{a}^\dagger_s - \hat{a}_s) = \hat{F}_0(\hat{a}^\dagger_s - \hat{a}_s) = \hat{a}^\dagger_s - \hat{a}_s$ [property (1)] does not depend on α. This independence allows us to integrate the equation for the electron operators. From the first relation in (2.49) we find

$$\hat{F}'_\alpha(\hat{d}_{s\sigma}) = -\frac{g}{\omega_0}\hat{F}_\alpha(\hat{d}_{s\sigma}(\hat{a}^\dagger_s - \hat{a}_s)) = -\frac{g}{\omega_0}\hat{F}_\alpha(\hat{d}_{s\sigma})(\hat{a}^\dagger_s - \hat{a}_s),$$

where in the last equality we use property (4). Taking into account that $\hat{F}_0(\hat{d}_{s\sigma}) = \hat{d}_{s\sigma}$ the solution of the differential equation is

$$\hat{F}_\alpha(\hat{d}_{s\sigma}) = \hat{d}_{s\sigma}e^{-\alpha\frac{g}{\omega_0}(\hat{a}^\dagger_s - \hat{a}_s)} \quad \overset{\text{property (2)}}{\Longrightarrow} \quad \hat{F}_\alpha(\hat{d}^\dagger_{s\sigma}) = \hat{d}^\dagger_{s\sigma}e^{\alpha\frac{g}{\omega_0}(\hat{a}^\dagger_s - \hat{a}_s)}.$$

Having the transformed electron operators we can calculate the transformed phonon operators. We use the result in (2.50) and the fact that $\hat{F}_\alpha(\hat{n}_{s\sigma}) = \hat{F}_\alpha(\hat{d}^\dagger_{s\sigma})\hat{F}_\alpha(\hat{d}_{s\sigma}) = \hat{n}_{s\sigma}$ does not depend on α; then we find $\hat{F}'_\alpha(\hat{a}_s) = -\frac{g}{\omega_0}\sum_\sigma \hat{n}_{s\sigma}$. Integrating over α we eventually obtain [property (1)]

$$\hat{F}_\alpha(\hat{a}_s) = \hat{a}_s - \alpha\frac{g}{\omega_0}\sum_\sigma \hat{n}_{s\sigma} \quad \overset{\text{property (2)}}{\Longrightarrow} \quad \hat{F}_\alpha(\hat{a}^\dagger_s) = \hat{a}^\dagger_s - \alpha\frac{g}{\omega_0}\sum_\sigma \hat{n}_{s\sigma}.$$

The transformed operators in (2.48) follow from the above identities with $\alpha = 1$ and the inverse transformation reads

$$\hat{d}_{s\sigma} = \hat{p}_{s\sigma} e^{\frac{g}{\omega_0}(\hat{b}_s^\dagger - \hat{b}_s)}, \qquad \hat{a}_s = \hat{b}_s + \frac{g}{\omega_0}\sum_\sigma \hat{n}_{s\sigma}. \qquad (2.51)$$

These results are physically sound. The unitary transformation $\hat{S}$ is the exponential of an operator proportional to the sum of the momentum operators, $\hat{p}_s \propto (\hat{a}_s^\dagger - \hat{a}_s)$, and therefore it is similar to a translation operator. As a consequence the new position operator is shifted; this follows from the second relation in (2.51) which implies

$$\hat{F}_1(\hat{x}_s) = \frac{\hat{b}_s^\dagger + \hat{b}_s}{\sqrt{2M\omega_0}} = \underbrace{\frac{\hat{a}_s^\dagger + \hat{a}_s}{\sqrt{2M\omega_0}}}_{\hat{x}_s} - 2\frac{g}{\omega_0}\frac{\sum_\sigma \hat{n}_{s\sigma}}{\sqrt{2M\omega_0}}.$$

Upon substitution of (2.51) in the Holstein Hamiltonian (2.46) we find

$$\hat{H}_{\mathrm{C}} + \hat{H}_{\mathrm{int}} = \omega_0 \sum_s (\hat{b}_s^\dagger \hat{b}_s + \frac{1}{2}) - \frac{g^2}{\omega_0}\sum_s \left(\sum_\sigma \hat{n}_{s\sigma}\right)^2,$$

and

$$\hat{H}_{\mathrm{el}} = T \sum_{s\sigma}\left(\hat{B}_s^\dagger \hat{B}_{s+1}\hat{p}_{s\sigma}^\dagger \hat{p}_{s+1\sigma} + \hat{B}_{s+1}^\dagger \hat{B}_s \hat{p}_{s+1\sigma}^\dagger \hat{p}_{s\sigma}\right),$$

where we defined $\hat{B}_s \equiv e^{\frac{g}{\omega_0}(\hat{b}_s^\dagger - \hat{b}_s)}$. As anticipated the zeroth order Hamiltonian $\hat{H}_{\mathrm{C}} + \hat{H}_{\mathrm{int}}$ is diagonalized by the Lang–Firsov transformation. The eigenstates consist of a given number of transformed electrons and of transformed phonons. For instance, if X is the set of molecules hosting a transformed electron of spin $\uparrow$ and Y is the set of molecules hosting a transformed electron of spin $\downarrow$, then the ket

$$|\Phi\rangle = \prod_{s\in X}\hat{p}_{s\uparrow}^\dagger \prod_{s'\in Y}\hat{p}_{s'\downarrow}^\dagger \prod_r (\hat{b}_r^\dagger)^{m_r}|0\rangle$$

is an eigenket of $\hat{H}_{\mathrm{C}} + \hat{H}_{\mathrm{int}}$ with eigenvalue

$$E = \omega_0 \sum_r (m_r + \frac{1}{2}) - \frac{g^2}{\omega_0}(|X| + |Y| + 2|X \cap Y|),$$

where $|X|$, $|Y|$ is the number of elements in the set X, Y. We refer to the transformed electrons as the *polarons* since the action of $\hat{p}_{s\sigma}^\dagger$ on the empty ket $|0\rangle$ generates an electron of spin σ in the sth molecule surrounded by a cloud of phonons.

In the subspace with only one polaron, say of spin up, all kets with no transformed phonons, $|\Phi_s\rangle \equiv \hat{p}_{s\uparrow}^\dagger|0\rangle$, have the same energy $-g^2/\omega_0$, which is also the lowest possible energy. This means that the ground state is degenerate; a ket of the ground-state multiplet describes a still electron trapped in s by the polarization cloud. The ground state degeneracy is lifted by the perturbation $\hat{H}_{\mathrm{el}}$. To first order the splitting is given by the eigenvalues of the matrix $\langle \Phi_s | \hat{H}_{el} | \Phi_{s'} \rangle$. Since the polarons can only hop to nearest neighbor molecules the matrix elements are zero unless $s' = s \pm 1$, in which case

$$\langle \Phi_s | \hat{H}_{el} | \Phi_{s\pm 1} \rangle = T \langle 0 | \hat{B}_s^\dagger \hat{B}_{s\pm 1} | 0 \rangle.$$

To evaluate the matrix element on the r.h.s. we make use of the Baker–Campbell–Hausdorff formula.[14] Recalling the definition of the $\hat{B}$ operators we have

$$\hat{B}_s^\dagger \hat{B}_{s\pm 1} | 0 \rangle = e^{-\frac{g}{\omega_0} \hat{b}_s^\dagger} e^{\frac{g}{\omega_0} \hat{b}_s} e^{-\frac{g^2}{2\omega_0^2}} e^{\frac{g}{\omega_0} \hat{b}_{s\pm 1}^\dagger} e^{-\frac{g}{\omega_0} \hat{b}_{s\pm 1}} e^{-\frac{g^2}{2\omega_0^2}} | 0 \rangle$$

$$= e^{-\frac{g^2}{\omega_0^2}} e^{-\frac{g}{\omega_0} \hat{b}_s^\dagger} e^{\frac{g}{\omega_0} \hat{b}_{s\pm 1}^\dagger} | 0 \rangle, \tag{2.52}$$

where in the last step we use the fact that the $\hat{b}$ operators with different indices commute and that for any constant α the ket $e^{\alpha \hat{b}_s} | 0 \rangle = \sum_k \frac{1}{k!} (\alpha \hat{b}_s)^k | 0 \rangle = | 0 \rangle$, since $\hat{b}_s | 0 \rangle = | \emptyset \rangle$. Taylor expanding the exponentials in (2.52) and multiplying by the bra $\langle 0 |$ it is evident that only the zeroth order terms remain. Hence $\langle 0 | \hat{B}_s^\dagger \hat{B}_{s\pm 1} | 0 \rangle = \exp(-g^2/\omega_0^2)$ which, as expected, does not depend on s. We conclude that

$$\langle \Phi_s | \hat{H}_{el} | \Phi_{s'} \rangle = T_{eff} (\delta_{s,s'+1} + \delta_{s,s'-1}), \qquad T_{eff} \equiv T e^{-\frac{g^2}{\omega_0^2}}. \tag{2.53}$$

For rings with N molecular sites the eigenvalues of the matrix (2.53) are $\epsilon_k = 2T_{eff} \cos k$, as we saw in (2.14). In the large N limit the variable $k = 2\pi m/N$ as well as the eigenvalues ϵ_k become a continuum and we can extract the *effective mass* of the polaron. This is typically done by comparing the energy dispersion ϵ_k (at low energies) with the energy dispersion $p^2/(2m^*)$ of a free particle of mass m^* and momentum p. In our case the momentum is the crystal momentum $p = k/a \in (-\pi/a, \pi/a)$, with a the equilibrium distance between two neighboring molecules. Assuming, e.g., $T < 0$, the lowest energies are those with $k \ll 1$ and hence we can approximate ϵ_k as $\epsilon_{k=ap} \sim 2T_{eff} + |T_{eff}| a^2 p^2$, from which it follows that the polaron mass is

$$m^* = \frac{e^{g^2/\omega_0^2}}{2|T| a^2}.$$

Thus, in the strong coupling regime the polaron mass increases exponentially with the coupling g. The polaron is heavy and the molecular deformations are localized around it.

---

[14] Given two operators $\hat{A}$ and $\hat{B}$ the Baker–Campbell–Hausdorff formula is the solution to $\hat{C} = \ln(e^{\hat{A}} e^{\hat{B}})$ or, equivalently, $e^{\hat{C}} = e^{\hat{A}} e^{\hat{B}}$. There is no closed expression for the solution of this problem. However, if the commutator $[\hat{A}, \hat{B}] = c \hat{\mathbb{1}}$ then

$$e^{\hat{A}+\hat{B}} = e^{\hat{A}} e^{\hat{B}} e^{-\frac{c}{2}}.$$

3

Time-dependent problems and equations of motion

3.1 Introduction

After the excursion of Chapter 2 on model systems we now go back to the general second-quantized Hamiltonian (1.82) and specialize the discussion to common physical situations. The operator $\hat{h}$ typically describes a particle of mass m and charge q in an external electromagnetic field and reads

$$\hat{h} = \frac{1}{2m}\left(\hat{\boldsymbol{p}} - \frac{q}{c}\mathbf{A}(\hat{\boldsymbol{r}})\right)^2 + qV(\hat{\boldsymbol{r}}) - g\mu_{\mathrm{B}}\mathbf{B}(\hat{\boldsymbol{r}})\cdot\hat{\boldsymbol{S}}, \qquad (3.1)$$

with V the scalar potential, $\mathbf{A}$ the vector potential, and $\mathbf{B} = \boldsymbol{\nabla}\times\mathbf{A}$ the magnetic field. In (3.1) the constant c is the speed of light while g and μ_{B} are the gyromagnetic ratio and the Bohr magneton respectively, see also (1.9). The many-body Hamiltonian $\hat{H}$ in (1.82) with $\hat{h}$ from (3.1) describes a system of interacting identical particles in some external *static* field. If we consider a molecule, in the absence of a magnetic field a standard choice for the potentials is $\mathbf{A}(\mathbf{r}) = 0$ and $V(\mathbf{r}) = \sum_i Z_i/|\mathbf{r} - \mathbf{R}_i|$, with Z_i being the charge of the ith nucleus at position $\mathbf{R}_i$. In the next chapters we develop approximation schemes to calculate equilibrium properties, like total energies, ionization energies, spectral functions, charge and current distributions, etc. of these systems. We are, however, especially interested in situations where at some time t_0 a *time-dependent* perturbation is switched on. For instance, one could change the electromagnetic field and study how the particles move under the influence of a time-dependent electric field $\mathbf{E}(\mathbf{r}, t)$ and magnetic field $\mathbf{B}(\mathbf{r}, t)$ with

$$\mathbf{E}(\mathbf{r}, t) = -\boldsymbol{\nabla}V(\mathbf{r}, t) - \frac{1}{c}\frac{\partial}{\partial t}\mathbf{A}(\mathbf{r}, t),$$
$$\mathbf{B}(\mathbf{r}, t) = \boldsymbol{\nabla}\times\mathbf{A}(\mathbf{r}, t),$$

and, of course, $V(\mathbf{r}, t \leq t_0) = V(\mathbf{r})$ and $\mathbf{A}(\mathbf{r}, t \leq t_0) = \mathbf{A}(\mathbf{r})$. In this case $\hat{h} \to \hat{h}(t)$ becomes time-dependent through its dependence on the external potentials. Consequently the full Hamiltonian (1.82) also acquires a time dependence, $\hat{H} \to \hat{H}(t) = \hat{H}_0(t) + \hat{H}_{\mathrm{int}}$.

The time evolution of the system is governed by the *Schrödinger equation*

$$i\frac{d}{dt}|\Psi(t)\rangle = \hat{H}(t)|\Psi(t)\rangle \tag{3.2}$$

with $|\Psi(t)\rangle$ the ket of the system at time t. The Schrödinger equation is a first-order differential equation in time and, therefore, $|\Psi(t)\rangle$ is uniquely determined once the initial ket $|\Psi(t_0)\rangle$ is given. For time-independent Hamiltonians $\hat{H}(t) = \hat{H}(t_0)$ for all times t and (3.2) is solved by

$$|\Psi(t)\rangle = e^{-i\hat{H}(t_0)(t-t_0)}|\Psi(t_0)\rangle. \tag{3.3}$$

How does this solution change when $\hat{H}(t)$ is time dependent?

3.2 Evolution operator

To generalize (3.3) we look for an operator $\hat{U}(t, t_0)$ which maps $|\Psi(t_0)\rangle$ into $|\Psi(t)\rangle$:

$$|\Psi(t)\rangle = \hat{U}(t, t_0)|\Psi(t_0)\rangle. \tag{3.4}$$

The operator $\hat{U}(t, t_0)$ must be unitary since the time-dependent Schrödinger equation preserves the norm of the states, i.e., $\langle\Psi(t)|\Psi(t)\rangle = \langle\Psi(t_0)|\Psi(t_0)\rangle$.

Let us discuss first the case $t > t_0$. We start by considering a Hamiltonian $\hat{H}(t)$ which is piecewise constant, i.e., $\hat{H}(t) = \hat{H}(t_p)$ for $t_p < t \leq t_{p+1}$, where the t_p are times at which the Hamiltonian changes suddenly. If we know the ket at time t_0 then we can calculate the ket at time $t \in (t_n, t_{n+1})$ by using (3.3) repeatedly

$$\begin{aligned}|\Psi(t)\rangle &= e^{-i\hat{H}(t_n)(t-t_n)}|\Psi(t_n)\rangle = e^{-i\hat{H}(t_n)(t-t_n)}e^{-i\hat{H}(t_{n-1})(t_n-t_{n-1})}|\Psi(t_{n-1})\rangle \\ &= e^{-i\hat{H}(t_n)(t-t_n)}e^{-i\hat{H}(t_{n-1})(t_n-t_{n-1})}\dots e^{-i\hat{H}(t_0)(t_1-t_0)}|\Psi(t_0)\rangle.\end{aligned} \tag{3.5}$$

As expected, the operator acting on $|\Psi(t_0)\rangle$ is unitary since it is the product of unitary operators. It is important to observe the order of the operators: the exponential calculated with Hamiltonian $\hat{H}(t_p)$ is on the left of all exponentials calculated with $\hat{H}(t_q)$ if $t_p > t_q$ (t_p later than t_q). Equation (3.5) takes a simpler form when the times t_p are equally spaced, i.e., $t_p = t_0 + p\Delta_t$ with Δ_t some given time interval. Then, $t_p - t_{p-1} = \Delta_t$ for all p and hence

$$|\Psi(t_{n+1})\rangle = e^{-i\hat{H}(t_n)\Delta_t}e^{-i\hat{H}(t_{n-1})\Delta_t}\dots e^{-i\hat{H}(t_0)\Delta_t}|\Psi(t_0)\rangle. \tag{3.6}$$

At this point it is natural to introduce the so called *chronological ordering operator* or *time ordering operator* T. Despite its name T is not an operator in the usual sense,[1] like the Hamiltonian or the density operator, but rather a rule which establishes how to rearrange products of operators. We define the chronologically ordered product of m Hamiltonians at times $t_m \geq \dots \geq t_1$ as

$$T\left\{\hat{H}(t_{P(m)})\hat{H}(t_{P(m-1)})\dots\hat{H}(t_{P(1)})\right\} = \hat{H}(t_m)\hat{H}(t_{m-1})\dots\hat{H}(t_1), \tag{3.7}$$

---

[1] Note that the symbol T for the chronological ordering operator does not have a "hat."

for all permutations P of $(1, \ldots, m-1, m)$. The action of T is to rearrange the Hamiltonians according to the *two ls rule*: later times go to the left. Since the order of the Hamiltonians in the T product is irrelevant, under the T sign they can be treated as commuting operators. This observation can be used to rewrite (3.6) in a more compact and useful form. Indeed, in (3.6) the Hamiltonians are already chronologically ordered and hence nothing changes if we act with the T operator on the product of the exponentials, i.e.,

$$|\Psi(t_{n+1})\rangle = T\left\{e^{-i\hat{H}(t_n)\Delta_t}e^{-i\hat{H}(t_{n-1})\Delta_t} \ldots e^{-i\hat{H}(t_0)\Delta_t}\right\}|\Psi(t_0)\rangle. \tag{3.8}$$

The equivalence between (3.6) and (3.8) can easily be checked by expanding the exponentials in power series and by taking into account that the action of the T operator on Hamiltonians with the same time argument is unambiguously defined since these Hamiltonians commute. For instance, in the case of only two exponentials

$$e^{-i\hat{H}(t_1)\Delta_t}e^{-i\hat{H}(t_0)\Delta_t} = \sum_{m,n=0}^{\infty} \frac{(-i\Delta_t)^n}{n!}\frac{(-i\Delta_t)^m}{m!}\hat{H}(t_1)^n\hat{H}(t_0)^m,$$

and we see that the action of T does not change this expression. We now recall that according to the Baker–Campbell–Hausdorff formula for two commuting operators $\hat{A}$ and $\hat{B}$ the product $\exp(\hat{A})\exp(\hat{B})$ is equal to $\exp(\hat{A} + \hat{B})$ (see also footnote 14 in Section 2.7.2). Under the T sign $\hat{H}(t_p)$ commutes with $\hat{H}(t_q)$ for all p and q and therefore

$$|\Psi(t_{n+1})\rangle = T\left\{e^{-i\Delta_t \sum_{p=0}^{n}\hat{H}(t_p)}\right\}|\Psi(t_0)\rangle, \tag{3.9}$$

which is a very nice result.

We are now in the position to answer the general question of how to solve (3.2) for time-dependent Hamiltonians. Let $t > t_0$ be the time at which we want to know the time-evolved ket $|\Psi(t)\rangle$. We divide the interval (t_0, t) into $n+1$ equal sub-intervals $(t_p, t_{p+1} = t_p + \Delta_t)$ with $\Delta_t = (t - t_0)/(n+1)$ as shown in the figure below.

If n is large enough the Hamiltonian $\hat{H}(t) \sim \hat{H}(t_p)$ for $t \in (t_p, t_{p+1})$ and $|\Psi(t)\rangle$ is approximately given by (3.9). Increasing n, and hence reducing Δ_t, the approximated ket $|\Psi(t)\rangle$ approaches the exact one and eventually coincides with it when $n \to \infty$. We conclude that the general solution of (3.2) can be written as

$$|\Psi(t)\rangle = \lim_{n\to\infty} T\left\{e^{-i\Delta_t \sum_{p=0}^{n}\hat{H}(t_p)}\right\}|\Psi(t_0)\rangle = T\left\{e^{-i\int_{t_0}^{t} d\bar{t}\,\hat{H}(\bar{t})}\right\}|\Psi(t_0)\rangle. \tag{3.10}$$

The operator in (3.4) is therefore

$$\hat{U}(t, t_0) \equiv T\left\{e^{-i\int_{t_0}^{t} d\bar{t}\,\hat{H}(\bar{t})}\right\}, \tag{3.11}$$

and is usually called the *evolution operator*. The evolution operator obeys a simple differential equation. Inserting the solution (3.10) into the time-dependent Schrödinger equation (3.2) we find $i\frac{d}{dt}\hat{U}(t,t_0)|\Psi(t_0)\rangle = \hat{H}(t)\hat{U}(t,t_0)|\Psi(t_0)\rangle$ and since this must be true for all initial states $|\Psi(t_0)\rangle$ we find

$$i\frac{d}{dt}\hat{U}(t,t_0) = \hat{H}(t)\hat{U}(t,t_0), \qquad \hat{U}(t_0,t_0) = \hat{1}, \tag{3.12}$$

with $\hat{1}$ the identity operator. The differential equation in (3.12) together with the initial condition at t_0 determines uniquely $\hat{U}(t,t_0)$. We can use (3.12) for an alternative proof of (3.11). We integrate (3.12) between t_0 and t and take into account that $\hat{U}(t_0,t_0) = \hat{1}$,

$$\hat{U}(t,t_0) = \hat{1} - i\int_{t_0}^{t} dt_1\, \hat{H}(t_1)\hat{U}(t_1,t_0).$$

This integral equation contains the same information as (3.12). This is a completely general fact: a first order differential equation endowed with a boundary condition can always be written as an integral equation which automatically incorporates the boundary condition. We can now replace the evolution operator under the integral sign with the whole r.h.s. and iterate. In doing so we find

$$\hat{U}(t,t_0) = \hat{1} - i\int_{t_0}^{t} dt_1\, \hat{H}(t_1) + (-i)^2 \int_{t_0}^{t} dt_1 \int_{t_0}^{t_1} dt_2\, \hat{H}(t_1)\hat{H}(t_2)\hat{U}(t_2,t_0)$$

$$= \sum_{k=0}^{\infty} (-i)^k \int_{t_0}^{t} dt_1 \int_{t_0}^{t_1} dt_2 \ldots \int_{t_0}^{t_{k-1}} dt_k\, \hat{H}(t_1)\hat{H}(t_2)\ldots\hat{H}(t_k).$$

The product of the Hamiltonians is chronologically time-ordered since we integrate t_j between t_0 and t_{j-1} for all j. Therefore nothing changes if we act with T on the r.h.s.. We now remind the reader that under the T sign the Hamiltonians can be treated as commuting operators and hence

$$\hat{U}(t,t_0) = \sum_{k=0}^{\infty} \frac{(-i)^k}{k!} \int_{t_0}^{t} dt_1 \int_{t_0}^{t} dt_2 \ldots \int_{t_0}^{t} dt_k\, T\left\{\hat{H}(t_1)\hat{H}(t_2)\ldots\hat{H}(t_k)\right\}$$

$$= T\left\{e^{-i\int_{t_0}^{t} d\bar{t}\,\hat{H}(\bar{t})}\right\}, \tag{3.13}$$

where in the first equality we have extended the domain of integration from (t_0, t_{j-1}) to (t_0, t) for all t_js and divided by $k!$, i.e., the number of identical contributions generated by the extension of the domain.[2]

---

[2]For the k-dimensional integral on a hypercube we have

$$\int_{t_0}^{t} dt_1 \int_{t_0}^{t} dt_2 \ldots \int_{t_0}^{t} dt_k\, f(t_1,\ldots,t_k) = \sum_{P} \int_{t_0}^{t} dt_1 \int_{t_0}^{t_1} dt_2 \ldots \int_{t_0}^{t_{k-1}} dt_k\, f(t_{P(1)},\ldots,t_{P(k)}),$$

where the sum over P runs over all permutations of $(1,\ldots,k)$. In the special case of a totally symmetric function f, i.e., $f(t_{P(1)},\ldots,t_{P(k)}) = f(t_1,\ldots,t_k)$, the above identity reduces to

$$\int_{t_0}^{t} dt_1 \int_{t_0}^{t} dt_2 \ldots \int_{t_0}^{t} dt_k\, f(t_1,\ldots,t_k) = k! \int_{t_0}^{t} dt_1 \int_{t_0}^{t_1} dt_2 \ldots \int_{t_0}^{t_{k-1}} dt_k\, f(t_1,\ldots,t_k).$$

This is the case of the time-ordered product of Hamiltonians in (3.13).

Next we consider the evolution operator $\hat{U}(t, t_0)$ for $t < t_0$, or equivalently $\hat{U}(t_0, t)$ for $t > t_0$ (since t and t_0 are two arbitrary times, $\hat{U}(t < t_0, t_0)$ is given by $\hat{U}(t_0, t > t_0)$ after renaming the times $t \to t_0$ and $t_0 \to t$). By definition the evolution operator has the property

$$\hat{U}(t_3, t_2)\hat{U}(t_2, t_1) = \hat{U}(t_3, t_1), \tag{3.14}$$

from which it follows that $\hat{U}(t_0, t)\hat{U}(t, t_0) = \hat{1}$. This result can be used to find the explicit expression of $\hat{U}(t_0, t)$. Taking into account (3.10) we have

$$\lim_{n \to \infty} \hat{U}(t_0, t)e^{-i\hat{H}(t_n)\Delta_t}e^{-i\hat{H}(t_{n-1})\Delta_t}\dots e^{-i\hat{H}(t_0)\Delta_t} = \hat{1}.$$

We then see by inspection that

$$\hat{U}(t_0, t) = \lim_{n \to \infty} e^{i\hat{H}(t_0)\Delta_t}\dots e^{i\hat{H}(t_{n-1})\Delta_t}e^{i\hat{H}(t_n)\Delta_t}$$
$$= \bar{T}\left\{e^{i\int_{t_0}^{t} d\bar{t}\,\hat{H}(\bar{t})}\right\},$$

where we have introduced the *anti-chronological ordering operator* or *anti-time ordering operator* $\bar{T}$ whose action is to move the operators with later times to the right. The operator $\hat{U}(t_0, t)$ can be interpreted as the operator that evolves a ket backward in time from t to t_0.

To summarize, the main result of this section is

$$\hat{U}(t_2, t_1) = \begin{cases} T\left\{e^{-i\int_{t_1}^{t_2} d\bar{t}\hat{H}(\bar{t})}\right\} & t_2 > t_1 \\ \bar{T}\left\{e^{+i\int_{t_2}^{t_1} d\bar{t}\hat{H}(\bar{t})}\right\} & t_2 < t_1 \end{cases} \tag{3.15}$$

Exercise 3.1. Show that

$$\bar{T}\left\{e^{i\int_{t_0}^{t} d\bar{t}\,\hat{H}(\bar{t})}\right\} = T\left\{e^{-i\int_{t}^{2t-t_0} d\bar{t}\,\hat{H}'(\bar{t})}\right\},$$

where $\hat{H}'(\bar{t}) = -\hat{H}(2t - \bar{t})$.

Exercise 3.2. Consider the Hamiltonian of a forced harmonic oscillator

$$\hat{H}(t) = \omega\hat{d}^\dagger\hat{d} + f(t)(\hat{d}^\dagger + \hat{d}),$$

where $f(t)$ is a real function of time and $[\hat{d}, \hat{d}^\dagger]_- = 1$. Let $|\Phi_n\rangle = \frac{(\hat{d}^\dagger)^n}{\sqrt{n!}}|0\rangle$ be the normalized eigenstates of $\hat{d}^\dagger\hat{d}$ with eigenvalues n. Show that the ket

$$|\Psi_n(t)\rangle = e^{-i\alpha(t)}e^{y(t)\hat{d}^\dagger - y^*(t)\hat{d}}|\Phi_n\rangle$$

is normalized to 1 for any real function $\alpha(t)$ and for any complex function $y(t)$. Then show that this ket also satisfies the time-dependent Schrödinger equation $i\frac{d}{dt}|\Psi_n(t)\rangle =$

$\hat{H}(t)|\Psi_n(t)\rangle$ with boundary condition $|\Psi_n(t_0)\rangle = |\Phi_n\rangle$, provided that the functions $\alpha(t)$ and $y(t)$ satisfy the differential equations

$$\frac{dy}{dt} + i\omega y = -if, \qquad \frac{d\alpha}{dt} = (n + |y|^2)\omega + (y + y^*)f + \frac{i}{2}\left(y\frac{dy^*}{dt} - \frac{dy}{dt}y^*\right),$$

with boundary conditions $y(t_0) = 0$ and $\alpha(t_0) = 0$. From these results prove that the evolution operator is given by

$$\hat{U}(t, t_0) = e^{i\int_{t_0}^t dt_1 \int_{t_0}^{t_1} dt_2\, f(t_1)f(t_2)\sin[\omega(t_1-t_2)]}\, e^{y(t)\hat{d}^\dagger - y^*(t)\hat{d}}\, e^{-i\omega\hat{d}^\dagger\hat{d}(t-t_0)},$$

with $y(t) = -ie^{-i\omega t}\int_{t_0}^t dt'\, f(t')e^{i\omega t'}$. Useful relations to solve this exercise are

$$\hat{d}\, e^{y\hat{d}^\dagger} = e^{y\hat{d}^\dagger}(\hat{d} + y), \qquad \hat{d}^\dagger e^{-y^*\hat{d}} = e^{-y^*\hat{d}}(\hat{d}^\dagger + y^*), \qquad e^{y\hat{d}^\dagger - y^*\hat{d}} = e^{y\hat{d}^\dagger}e^{-y^*\hat{d}}e^{-|y|^2/2}.$$

3.3 Equations of motion for operators in the Heisenberg picture

In quantum mechanics we associate with any observable quantity O a Hermitian operator $\hat{O}$, and an experimental measurement of O yields one of the eigenvalues of $\hat{O}$. The probability of measuring one of these eigenvalues depends on the state of the system at the time of measurement. If $|\Psi(t)\rangle$ is the normalized ket describing the system at time t then the probability of measuring the eigenvalue λ_i is $P_i(t) = |\langle\Psi_i|\Psi(t)\rangle|^2$, with $|\Psi_i\rangle$ the normalized eigenket of $\hat{O}$ with eigenvalue λ_i, i.e., $\hat{O}|\Psi_i\rangle = \lambda_i|\Psi_i\rangle$. The same is true if the observable quantity depends explicitly on time, as, e.g., the time-dependent Hamiltonian introduced in the previous sections. In this case the eigenvalues and eigenkets of $\hat{O}(t)$ depend on t, $\lambda_i \to \lambda_i(t)$ and $|\Psi_i\rangle \to |\Psi_i(t)\rangle$, and the probability of measuring $\lambda_i(t)$ at time t becomes $P_i(t) = |\langle\Psi_i(t)|\Psi(t)\rangle|^2$. Note that $|\Psi_i(t)\rangle$ *is not* the time evolved ket $|\Psi_i\rangle$ but instead the ith eigenket of the operator $\hat{O}(t)$ which can have any time dependence. The knowledge of all probabilities can be used to construct a dynamical *quantum average* of the observable O according to $\sum_i \lambda_i(t)P_i(t)$. This quantity represents the average of the outcomes of an experimental measurement performed in $N \to \infty$ independent systems all in the same state $|\Psi(t)\rangle$. Using the completeness relation $\sum_i |\Psi_i(t)\rangle\langle\Psi_i(t)| = \hat{1}$ (which is valid for all t) the dynamical quantum average can be rewritten as

$$\sum_i \lambda_i(t)P_i(t) = \sum_i \lambda_i(t)\langle\Psi(t)|\Psi_i(t)\rangle\langle\Psi_i(t)|\Psi(t)\rangle$$

$$= \sum_i \langle\Psi(t)|\hat{O}(t)|\Psi_i(t)\rangle\langle\Psi_i(t)|\Psi(t)\rangle$$

$$= \langle\Psi(t)|\hat{O}(t)|\Psi(t)\rangle.$$

The dynamical quantum average is the expectation value of $\hat{O}(t)$ over the state of the system at time t. The enormous amount of information that can be extracted from the dynamical averages prompts us to develop mathematical techniques for their calculation. In the remainder of this chapter we introduce some fundamental concepts and derive a few important identities to lay down the basis of a very powerful mathematical apparatus.

From the results of the previous section we know that the time-evolved ket $|\Psi(t)\rangle = \hat{U}(t, t_0)|\Psi(t_0)\rangle$ and therefore the expectation value $\langle\Psi(t)|\hat{O}(t)|\Psi(t)\rangle$ can also be written as $\langle\Psi(t_0)|\hat{U}(t_0, t)\hat{O}(t)\hat{U}(t, t_0)|\Psi(t_0)\rangle$. This leads us to introduce the notion of operators in the *Heisenberg picture*. An operator $\hat{O}(t)$ in the Heisenberg picture is denoted by $\hat{O}_H(t)$ and is defined according to

$$\boxed{\hat{O}_H(t) \equiv \hat{U}(t_0, t)\hat{O}(t)\hat{U}(t, t_0)} \tag{3.16}$$

We will apply the above definition not only to those operators associated with observable quantities but to *all* operators, including the field operators. For instance, the density operator $\hat{n}(\mathbf{x}) = \hat{\psi}^\dagger(\mathbf{x})\hat{\psi}(\mathbf{x})$ is associated with an observable quantity and is written in terms of two field operators which also admit a Heisenberg picture

$$\hat{n}_H(\mathbf{x}, t) = \hat{U}(t_0, t)\hat{\psi}^\dagger(\mathbf{x})\hat{\psi}(\mathbf{x})\hat{U}(t, t_0) = \hat{U}(t_0, t)\hat{\psi}^\dagger(\mathbf{x})\underbrace{\hat{U}(t, t_0)\hat{U}(t_0, t)}_{\hat{1}}\hat{\psi}(\mathbf{x})\hat{U}(t, t_0)$$

$$= \hat{\psi}_H^\dagger(\mathbf{x}, t)\hat{\psi}_H(\mathbf{x}, t). \tag{3.17}$$

From (3.17) it is evident that the Heisenberg picture of the product of two operators $\hat{O} = \hat{O}_1\hat{O}_2$ is simply the product of the two operators in the Heisenberg picture, i.e., $\hat{O}_H(t) = \hat{O}_{1,H}(t)\hat{O}_{2,H}(t)$. An important consequence of this fact is that *operators in the Heisenberg picture at equal times satisfy the same (anti)commutation relations as the original operators*. In particular for the field operators we have

$$\left[\hat{\psi}_H(\mathbf{x}, t), \hat{\psi}_H^\dagger(\mathbf{x}', t)\right]_\mp = \delta(\mathbf{x} - \mathbf{x}').$$

Operators in the Heisenberg picture obey a simple *equation of motion*. Taking into account (3.12) we find

$$i\frac{d}{dt}\hat{O}_H(t) = -\hat{U}(t_0, t)\hat{H}(t)\hat{O}(t)\hat{U}(t, t_0) + \hat{U}(t_0, t)\hat{O}(t)\hat{H}(t)\hat{U}(t, t_0)$$

$$+ \hat{U}(t_0, t)\left(i\frac{d}{dt}\hat{O}(t)\right)\hat{U}(t, t_0). \tag{3.18}$$

For operators independent of time the last term vanishes. In most textbooks the last term is also written as $i\frac{\partial}{\partial t}\hat{O}_H(t)$ where the symbol of partial derivative signifies that only the derivative with respect to the explicit time dependence of $\hat{O}(t)$ must be taken. In this book we use the same notation. The first two terms can be rewritten in two equivalent ways. The first, and most obvious, one is $\hat{U}(t_0, t)[\hat{O}(t), \hat{H}(t)]_-\hat{U}(t, t_0)$. We could, alternatively, insert the identity operator $\hat{1} = \hat{U}(t, t_0)\hat{U}(t_0, t)$ between the Hamiltonian and $\hat{O}(t)$ in order to have both operators in the Heisenberg picture. Thus

$$i\frac{d}{dt}\hat{O}_H(t) = \hat{U}(t_0, t)\left[\hat{O}(t), \hat{H}(t)\right]_-\hat{U}(t, t_0) + i\frac{\partial}{\partial t}\hat{O}_H(t)$$

$$= \left[\hat{O}_H(t), \hat{H}_H(t)\right]_- + i\frac{\partial}{\partial t}\hat{O}_H(t). \tag{3.19}$$

In second quantization all operators are expressed in terms of field operators $\hat{\psi}(\mathbf{x})$, $\hat{\psi}^\dagger(\mathbf{x})$ and hence the equations of motion for $\hat{\psi}_H(\mathbf{x},t)$ and $\hat{\psi}^\dagger_H(\mathbf{x},t)$ play a very special role; they actually constitute the seed of the formalism that we develop in the next chapters. Due to their importance we derive them below. Let us consider the many-body Hamiltonian (1.82) with some time-dependent one-body part $\hat{h}(t)$

$$\hat{H}(t) = \underbrace{\sum_{\sigma\sigma'} \int d\mathbf{r}\, \hat{\psi}^\dagger(\mathbf{r}\sigma) h_{\sigma\sigma'}(\mathbf{r}, -i\boldsymbol{\nabla}, \mathbf{S}, t)\hat{\psi}(\mathbf{r}\sigma')}_{\hat{H}_0(t)}$$

$$+ \frac{1}{2}\int d\mathbf{x}\, d\mathbf{x}'\, v(\mathbf{x},\mathbf{x}')\hat{\psi}^\dagger(\mathbf{x})\hat{\psi}^\dagger(\mathbf{x}')\hat{\psi}(\mathbf{x}')\hat{\psi}(\mathbf{x}). \qquad (3.20)$$

Since the field operators have no explicit time dependence, the last term in (3.19) vanishes and we only need to evaluate the commutators between $\hat{\psi}(\mathbf{x})$, $\hat{\psi}^\dagger(\mathbf{x})$ and the Hamiltonian. Let us start with the fermionic field operator $\hat{\psi}(\mathbf{x})$. Using the identity

$$\left[\hat{\psi}(\mathbf{x}), \hat{A}\hat{B}\right]_- = \left[\hat{\psi}(\mathbf{x}), \hat{A}\right]_+ \hat{B} - \hat{A}\left[\hat{\psi}(\mathbf{x}), \hat{B}\right]_+$$

with

$$\hat{A} = \hat{\psi}^\dagger(\mathbf{r}'\sigma'), \qquad \hat{B} = \sum_{\sigma''} h_{\sigma'\sigma''}(\mathbf{r}', -i\boldsymbol{\nabla}', \mathbf{S}, t)\hat{\psi}(\mathbf{r}'\sigma''),$$

and taking into account that $\left[\hat{\psi}(\mathbf{x}), \hat{B}\right]_+ = 0$ since $\left[\hat{\psi}(\mathbf{x}), \hat{\psi}(\mathbf{x}')\right]_+ = 0$, we find

$$\left[\hat{\psi}(\mathbf{r}\sigma), \hat{H}_0(t)\right]_- = \sum_{\sigma'} h_{\sigma\sigma'}(\mathbf{r}, -i\boldsymbol{\nabla}, \mathbf{S}, t)\hat{\psi}(\mathbf{r}\sigma').$$

The evaluation of $\left[\hat{\psi}(\mathbf{x}), \hat{H}_{\text{int}}\right]_-$ is also a simple exercise. Indeed

$$\left[\hat{\psi}(\mathbf{x}), \hat{\psi}^\dagger(\mathbf{x}')\hat{\psi}^\dagger(\mathbf{x}'')\hat{\psi}(\mathbf{x}'')\hat{\psi}(\mathbf{x}')\right]_- = \left[\hat{\psi}(\mathbf{x}), \hat{\psi}^\dagger(\mathbf{x}')\hat{\psi}^\dagger(\mathbf{x}'')\right]_- \hat{\psi}(\mathbf{x}'')\hat{\psi}(\mathbf{x}')$$

$$= \left(\delta(\mathbf{x}-\mathbf{x}')\hat{\psi}^\dagger(\mathbf{x}'') - \delta(\mathbf{x}-\mathbf{x}'')\hat{\psi}^\dagger(\mathbf{x}')\right)\hat{\psi}(\mathbf{x}'')\hat{\psi}(\mathbf{x}'),$$

and therefore

$$\left[\hat{\psi}(\mathbf{x}), \hat{H}_{\text{int}}\right]_- = \frac{1}{2}\int d\mathbf{x}''\, v(\mathbf{x},\mathbf{x}'')\hat{\psi}^\dagger(\mathbf{x}'')\hat{\psi}(\mathbf{x}'')\hat{\psi}(\mathbf{x}) - \frac{1}{2}\int d\mathbf{x}'\, v(\mathbf{x}',\mathbf{x})\hat{\psi}^\dagger(\mathbf{x}')\hat{\psi}(\mathbf{x})\hat{\psi}(\mathbf{x}')$$

$$= \int d\mathbf{x}'\, v(\mathbf{x},\mathbf{x}')\hat{n}(\mathbf{x}')\hat{\psi}(\mathbf{x}),$$

where in the last step we use the symmetry property $v(\mathbf{x},\mathbf{x}') = v(\mathbf{x}',\mathbf{x})$. Substituting these results in (3.19), the equation of motion for the field operator reads

$$\boxed{i\frac{d}{dt}\hat{\psi}_H(\mathbf{x},t) = \sum_{\sigma'} h_{\sigma\sigma'}(\mathbf{r}, -i\boldsymbol{\nabla}, \mathbf{S}, t)\hat{\psi}_H(\mathbf{r}\sigma',t) + \int d\mathbf{x}'\, v(\mathbf{x},\mathbf{x}')\hat{n}_H(\mathbf{x}',t)\hat{\psi}_H(\mathbf{x},t)}$$

$$(3.21)$$

The equation of motion for $\hat{\psi}_H^\dagger(\mathbf{x}, t)$ can be obtained from the adjoint of the equation above and reads

$$
\mathrm{i}\frac{d}{dt}\hat{\psi}_H^\dagger(\mathbf{x}, t) = -\sum_{\sigma'} \hat{\psi}_H^\dagger(\mathbf{r}\sigma', t)h_{\sigma'\sigma}(\mathbf{r}, \mathrm{i}\overleftarrow{\boldsymbol{\nabla}}, \mathbf{S}, t) - \int d\mathbf{x}'\, v(\mathbf{x}, \mathbf{x}')\hat{\psi}_H^\dagger(\mathbf{x}, t)\hat{n}_H(\mathbf{x}', t)
$$

(3.22)

In the next sections we use (3.21) and (3.22) to derive the equation of motion of other physically relevant operators like the density and the total momentum of the system. These results are exact and hence provide important benchmarks to check the quality of an approximation.

Exercise 3.3. Show that the same equations of motion (3.21) and (3.22) are valid for the bosonic field operators.

Exercise 3.4. Show that for a system of identical particles in an external time-dependent electromagnetic field with

$$
h_{\sigma\sigma'}(\mathbf{r}, -\mathrm{i}\boldsymbol{\nabla}, \mathbf{S}, t) = \delta_{\sigma\sigma'}\left[\frac{1}{2m}\left(-\mathrm{i}\boldsymbol{\nabla} - \frac{q}{c}\mathbf{A}(\mathbf{r}, t)\right)^2 + qV(\mathbf{r}, t)\right],
$$

(3.23)

the equations of motion of the field operators are invariant under the gauge transformation

$$
\begin{aligned}
\mathbf{A}(\mathbf{r}, t) &\to \mathbf{A}(\mathbf{r}, t) + \boldsymbol{\nabla}\Lambda(\mathbf{r}, t) \\
V(\mathbf{r}, t) &\to V(\mathbf{r}, t) - \frac{1}{c}\frac{\partial}{\partial t}\Lambda(\mathbf{r}, t) \\
\hat{\psi}_H(\mathbf{x}, t) &\to \hat{\psi}_H(\mathbf{x}, t)\exp\left[\mathrm{i}\frac{q}{c}\Lambda(\mathbf{r}, t)\right]
\end{aligned}
$$

(3.24)

3.4 Continuity equation: paramagnetic and diamagnetic currents

Let us consider a system of interacting and identical particles under the influence of an external time-dependent electromagnetic field. The one-body part of the Hamiltonian is given by the first term of (3.20) with $h(\mathbf{r}, -\mathrm{i}\boldsymbol{\nabla}, \mathbf{S}, t)$ as in (3.23). The equation of motion for the density operator $\hat{n}_H(\mathbf{x}, t) = \hat{\psi}_H^\dagger(\mathbf{x}, t)\hat{\psi}_H(\mathbf{x}, t)$ can easily be obtained using (3.21) and (3.22) since[3]

$$
\mathrm{i}\frac{d}{dt}\hat{n}_H = \left(\mathrm{i}\frac{d}{dt}\hat{\psi}_H^\dagger\right)\hat{\psi}_H + \hat{\psi}_H^\dagger\left(\mathrm{i}\frac{d}{dt}\hat{\psi}_H\right).
$$

On the r.h.s. the terms containing the scalar potential V and the interaction v cancel. This cancellation is a direct consequence of the fact that the operator $\int d\mathbf{x}'\, V(\mathbf{r}', t)\hat{n}(\mathbf{x}')$ in

---

[3]In the rest of this section we omit the arguments of the operators as well as of the scalar and vector potentials if there is no ambiguity.

$\hat{H}_0(t)$ as well as $\hat{H}_{\text{int}}$ are expressed in terms of the density only, see (1.81), and therefore commute with $\hat{n}(\mathbf{x})$. To calculate the remaining terms we write

$$\left(\pm i\boldsymbol{\nabla} - \frac{q}{c}\mathbf{A}\right)^2 = -\nabla^2 \mp \frac{iq}{c}\left(\boldsymbol{\nabla}\cdot\mathbf{A}\right) \mp \frac{2iq}{c}\mathbf{A}\cdot\boldsymbol{\nabla} + \frac{q^2}{c^2}A^2, \tag{3.25}$$

from which it follows that the term proportional to A^2 also cancels (like the scalar potential, this term is coupled to the density operator).[4] Collecting the remaining terms we find

$$\frac{d}{dt}\hat{n}_H = \frac{1}{2mi}\left[\left(\nabla^2\hat{\psi}_H^\dagger\right)\hat{\psi}_H - \hat{\psi}_H^\dagger\left(\nabla^2\hat{\psi}_H\right)\right]$$
$$+ \frac{q}{mc}\left[\hat{n}_H\boldsymbol{\nabla} + \left(\boldsymbol{\nabla}\hat{\psi}_H^\dagger\right)\hat{\psi}_H + \hat{\psi}_H^\dagger\left(\boldsymbol{\nabla}\hat{\psi}_H\right)\right]\cdot\mathbf{A}.$$

The first term on the r.h.s can be written as minus the divergence of the *paramagnetic current density operator*

$$\boxed{\hat{\mathbf{j}}(\mathbf{x}) \equiv \frac{1}{2mi}\left[\hat{\psi}^\dagger(\mathbf{x})\left(\boldsymbol{\nabla}\hat{\psi}(\mathbf{x})\right) - \left(\boldsymbol{\nabla}\hat{\psi}^\dagger(\mathbf{x})\right)\hat{\psi}(\mathbf{x})\right]} \tag{3.26}$$

in the Heisenberg picture, while the second term can be written as minus the divergence of the *diamagnetic current density operator*

$$\boxed{\hat{\mathbf{j}}_d(\mathbf{x},t) \equiv -\frac{q}{mc}\hat{n}(\mathbf{x})\mathbf{A}(\mathbf{r},t)} \tag{3.27}$$

also in the Heisenberg picture. Note that the diamagnetic current depends explicitly on time through the time dependence of the vector potential. The resulting equation of motion for the density $\hat{n}_H(\mathbf{x},t)$ is known as the *continuity equation*

$$\boxed{\frac{d}{dt}\hat{n}_H(\mathbf{x},t) = -\boldsymbol{\nabla}\cdot\left[\hat{\mathbf{j}}_H(\mathbf{x},t) + \hat{\mathbf{j}}_{d,H}(\mathbf{x},t)\right]} \tag{3.28}$$

The origin of the names paramagnetic and diamagnetic current stems from the behavior of the orbital magnetic moment that these currents generate when the system is exposed to a magnetic field. To make these definitions less abstract let us a consider a system of noninteracting particles and rewrite the Hamiltonian $\hat{H}_0$ using the identity (3.25) together with the rearrangement

$$\frac{1}{mi}\hat{\psi}^\dagger(\boldsymbol{\nabla}\hat{\psi})\cdot\mathbf{A} = \hat{\mathbf{j}}\cdot\mathbf{A} + \frac{1}{2mi}(\boldsymbol{\nabla}\hat{n})\cdot\mathbf{A}.$$

We find

$$\hat{H}_0 = -\frac{1}{2m}\int\hat{\psi}^\dagger\nabla^2\hat{\psi} - \frac{q}{c}\int\hat{\mathbf{j}}\cdot\mathbf{A} + \int\hat{n}\left(qV + \frac{q^2}{2mc^2}A^2\right) + \frac{iq}{2mc}\int\boldsymbol{\nabla}\cdot(\hat{n}\mathbf{A}). \tag{3.29}$$

---

[4]Equation (3.25) can easily be verified by applying the differential operator to a test function.

Equation (3.29) is an exact manipulation of the original $\hat{H}_0$.[5] We now specialize the discussion to situations with no electric field, $\mathbf{E} = 0$, and with a static and uniform magnetic field $\mathbf{B}$. A possible gauge for the scalar and vector potentials is $V = 0$ and $\mathbf{A}(\mathbf{r}) = -\frac{1}{2}\mathbf{r} \times \mathbf{B}$. Substituting these potentials in (3.29) we find a coupling between $\mathbf{B}$ and the currents

$$\hat{H}_0 = -\frac{1}{2m}\int \hat{\psi}^\dagger \nabla^2 \hat{\psi} \;-\; \int \left(\hat{\mathbf{m}} + \frac{1}{2}\hat{\mathbf{m}}_d\right)\cdot\mathbf{B} \;+\; \frac{iq}{2mc}\int \nabla\cdot(\hat{n}\mathbf{A}), \qquad (3.30)$$

where we define the magnetic moments

$$\hat{\mathbf{m}}(\mathbf{x}) \equiv \frac{q}{2c}\mathbf{r}\times\hat{\mathbf{j}}(\mathbf{x}), \qquad\qquad \hat{\mathbf{m}}_d(\mathbf{x},t) \equiv \frac{q}{2c}\mathbf{r}\times\hat{\mathbf{j}}_d(\mathbf{x},t),$$

generated by the paramagnetic and diamagnetic currents. In (3.30) we use the identity $\mathbf{a}\cdot(\mathbf{b}\times\mathbf{c}) = -(\mathbf{b}\times\mathbf{a})\cdot\mathbf{c}$, with $\mathbf{a}$, $\mathbf{b}$, and $\mathbf{c}$ three arbitrary vectors. The negative sign in front of the coupling $\hat{\mathbf{m}}\cdot\mathbf{B}$ energetically favors configurations in which the paramagnetic moment is large and aligned along $\mathbf{B}$, a behavior similar to the total spin magnetization in a Pauli paramagnet. This explains the name "paramagnetic current." On the contrary the diamagnetic moment tends, by definition, to be aligned in the opposite direction since

$$\begin{aligned}
\hat{\mathbf{m}}_d = \frac{q}{2c}\mathbf{r}\times\hat{\mathbf{j}}_d &= -\frac{q^2}{2mc^2}\,\hat{n}\,[\mathbf{r}\times\mathbf{A}] \\
&= \frac{q^2}{4mc^2}\,\hat{n}\,[\mathbf{r}\times(\mathbf{r}\times\mathbf{B})] \\
&= \frac{q^2}{4mc^2}\,\hat{n}\,[\mathbf{r}(\mathbf{r}\cdot\mathbf{B}) - r^2\mathbf{B}],
\end{aligned}$$

and hence, denoting by θ the angle between $\mathbf{r}$ and $\mathbf{B}$,

$$\hat{\mathbf{m}}_d\cdot\mathbf{B} = -\frac{q^2}{4mc^2}\hat{n}\,(rB)^2(1-\cos^2\theta) \le 0.$$

This explains the name "diamagnetic current." The minus sign in front of the coupling $\hat{\mathbf{m}}_d\cdot\mathbf{B}$ in (3.30) favors configurations in which the diamagnetic moment is small. It should be said, however, that there are important physical situations in which the diamagnetic contribution is the dominant one. For instance, in solids composed of atoms with filled shells the average of $\hat{\mathbf{m}}$ over the ground state is zero and the system is called a *Larmor diamagnet*. Furthermore, interactions among particles may change drastically the relative contribution between the paramagnetic and diamagnetic moments. A system is said to exhibit perfect diamagnetism if the magnetic field generated by the total magnetic moment $\mathbf{M} = \mathbf{m} + \mathbf{m}_d$ cancels exactly the external magnetic field $\mathbf{B}$. An example perfect diamagnet is a bulk superconductor in

---

[5]A brief comment about the last term is in order. For any arbitrary large but finite system all many-body kets $|\Psi_i\rangle$ relevant to its description yield matrix elements of the density operator $\langle\Psi_i|\hat{n}(\mathbf{r})|\Psi_j\rangle$ that vanish at large $\mathbf{r}$ (no particles at infinity). Then, the total divergence in the last term of (3.29) does not contribute to the quantum averages and can be discarded.

which $\mathbf{B}$ is expelled outside the bulk by the diamagnetic currents, a phenomenon known as the *Meissner effect*. As a final remark we observe that neither the paramagnetic nor the diamagnetic current density operator is invariant under the gauge transformation (3.24). On the contrary the *current density operator*

$$\boxed{\hat{\mathbf{J}}(\mathbf{x},t) = \hat{\mathbf{j}}(\mathbf{x}) + \hat{\mathbf{j}}_d(\mathbf{x},t)}$$

(3.31)

in the Heisenberg picture, $\hat{\mathbf{J}}_H(\mathbf{x},t)$, is gauge invariant and hence observable.

Exercise 3.5. Show that

$$\left[\hat{\mathbf{J}}(\mathbf{x},t), \hat{n}(\mathbf{x}')\right]_- = -\frac{i}{m}\hat{n}(\mathbf{x})\boldsymbol{\nabla}\delta(\mathbf{x}-\mathbf{x}').$$

Exercise 3.6. Show that under the gauge transformation (3.24)

$$\hat{\mathbf{j}}_H(\mathbf{x},t) \to \hat{\mathbf{j}}_H(\mathbf{x},t) + \frac{q}{mc}\hat{n}_H(\mathbf{x},t)\boldsymbol{\nabla}\Lambda(\mathbf{r},t),$$

$$\hat{\mathbf{j}}_{d,H}(\mathbf{x},t) \to \hat{\mathbf{j}}_{d,H}(\mathbf{x},t) - \frac{q}{mc}\hat{n}_H(\mathbf{x},t)\boldsymbol{\nabla}\Lambda(\mathbf{r},t).$$

3.5 Lorentz Force

From the continuity equation it follows that the dynamical quantum average of the current density operator $\hat{\mathbf{J}}(\mathbf{x},t)$ is the particle flux, and hence the operator

$$\boxed{\hat{\mathbf{P}}(t) \equiv m \int d\mathbf{x}\,\hat{\mathbf{J}}(\mathbf{x},t)}$$

(3.32)

is the *total momentum operator*. Consequently, the time derivative of $\hat{\mathbf{P}}_H(t)$ is the total force acting on the system. For particles of charge q in an external electromagnetic field this force is the Lorentz force. In this section we see how the Lorentz force comes out from our equations.

The calculation of $i\frac{d}{dt}\hat{\mathbf{P}}_H$ can be performed along the same lines as the continuity equation. We proceed by first evaluating $i\frac{d}{dt}\hat{\mathbf{j}}_{d,H}$, then $i\frac{d}{dt}\hat{\mathbf{j}}_H$ and finally integrating over all space and spin the sum of the two. The equation of motion for the diamagnetic current density follows directly from the continuity equation and reads

$$i\frac{d}{dt}\hat{\mathbf{j}}_{d,H} = \frac{iq}{mc}\mathbf{A}(\boldsymbol{\nabla}\cdot\hat{\mathbf{J}}_H) - \frac{iq}{mc}\hat{n}_H\frac{d}{dt}\mathbf{A}.$$

(3.33)

The equation of motion for the paramagnetic current density is slightly more complicated. It is convenient to work in components and define $f_{k=x,y,x}$ as the components of a generic vector function $\mathbf{f}$ and $\partial_{k=x,y,z}$ as the partial derivative with respect to x, y, and z. The kth component of the paramagnetic current density (3.26) in the Heisenberg picture is

$\hat{j}_{k,H} = \frac{1}{2mi}[\hat{\psi}_H^\dagger(\partial_k\hat{\psi}_H) - (\partial_k\hat{\psi}_H^\dagger)\hat{\psi}_H]$. To calculate its time derivative we rewrite the equations of motion (3.21) and (3.22) for the field operators as

$$i\frac{d}{dt}\hat{\psi}_H = i\frac{d}{dt}\hat{\psi}_H\Big|_{V=\mathbf{A}=0} + \frac{iq}{mc}\sum_p\left[A_p\partial_p + \frac{1}{2}(\partial_p A_p)\right]\hat{\psi}_H + w\,\hat{\psi}_H,$$

$$i\frac{d}{dt}\hat{\psi}_H^\dagger = i\frac{d}{dt}\hat{\psi}_H^\dagger\Big|_{V=\mathbf{A}=0} + \frac{iq}{mc}\sum_p\left[A_p\partial_p + \frac{1}{2}(\partial_p A_p)\right]\hat{\psi}_H^\dagger - w\,\hat{\psi}_H^\dagger,$$

where the first term is defined as the contribution to the derivative which does not *explicitly* depend on the scalar and vector potentials (there is, of course, an implicit dependence through the evolution operator $\hat{U}$ in the field operators $\hat{\psi}_H$ and $\hat{\psi}_H^\dagger$). Furthermore the function w is defined as

$$w(\mathbf{r},t) = qV(\mathbf{r},t) + \frac{q^2}{2mc^2}A^2(\mathbf{r},t). \tag{3.34}$$

We can evaluate $i\frac{d}{dt}\hat{j}_{k,H}$ in a systematic way by collecting the terms which do not depend on V and $\mathbf{A}$, the terms linear in $\mathbf{A}$ and the terms proportional to w. We find

$$i\frac{d}{dt}\hat{j}_{k,H} = i\frac{d}{dt}\hat{j}_{k,H}\Big|_{V=\mathbf{A}=0} + \frac{iq}{mc}\sum_p\left[\partial_p(A_p\hat{j}_{k,H}) + (\partial_k A_p)\hat{j}_{p,H}\right] - \frac{i}{m}\hat{n}_H\partial_k w. \tag{3.35}$$

It is not difficult to show that

$$i\frac{d}{dt}\hat{j}_{k,H}\Big|_{V=\mathbf{A}=0} = -i\sum_p\partial_p\hat{T}_{pk,H} - i\hat{W}_{k,H},$$

where the *momentum–stress tensor operator* $\hat{T}_{pk} = \hat{T}_{kp}$ reads

$$\hat{T}_{pk} = \frac{1}{2m^2}\left[(\partial_k\hat{\psi}^\dagger)(\partial_p\hat{\psi}) + (\partial_p\hat{\psi}^\dagger)(\partial_k\hat{\psi}) - \frac{1}{2}\partial_k\partial_p\hat{n}\right], \tag{3.36}$$

while the operator $\hat{W}_k(\mathbf{x})$ is given by

$$\hat{W}_k(\mathbf{x},t) = \frac{1}{m}\int d\mathbf{x}'\hat{\psi}^\dagger(\mathbf{x})\hat{\psi}^\dagger(\mathbf{x}')(\partial_k v(\mathbf{x},\mathbf{x}'))\hat{\psi}(\mathbf{x}')\hat{\psi}(\mathbf{x})$$

$$= i\left[\hat{j}_k(\mathbf{x}), \hat{H}_{\text{int}}\right]_-.$$

The equation of motion for the current density operator $\hat{\mathbf{J}}_H$ follows by adding (3.33) to (3.35). After some algebra one finds

$$i\frac{d}{dt}\hat{j}_{k,H} = \frac{iq}{m}\hat{n}_H\left[-\partial_k V - \frac{1}{c}\frac{d}{dt}A_k\right] + \frac{iq}{mc}\sum_p\hat{j}_{p,H}\left[\partial_k A_p - \partial_p A_k\right]$$

$$-i\sum_p\partial_p\left[\hat{T}_{pk,H} - \frac{q}{mc}(A_k\hat{j}_{p,H} + A_p\hat{j}_{k,H})\right] - i\hat{W}_{k,H}. \tag{3.37}$$

In the first row we recognize the kth component of the electric field $\mathbf{E}$ as well as the kth component of the vector product $\hat{\mathbf{J}}_H \times (\boldsymbol{\nabla} \times \mathbf{A}) = \hat{\mathbf{J}}_H \times \mathbf{B}$.[6] The equation of motion for $\hat{\mathbf{P}}_H$ is obtained by integrating (3.37) over all space and spin. The first two terms give $\int (q\,\hat{n}_H \mathbf{E} + \frac{q}{c}\,\hat{\mathbf{J}}_H \times \mathbf{B})$ which is exactly the operator of the Lorentz force, as expected. What about the remaining terms? Let us start by discussing $\hat{W}_{k,H}$. The interparticle interaction $v(\mathbf{x}, \mathbf{x}')$ is symmetric under the interchange $\mathbf{x} \leftrightarrow \mathbf{x}'$. Moreover, to not exert a net force on the system v must depend only on the difference $\mathbf{r} - \mathbf{r}'$, a requirement that guarantees the conservation of $\hat{\mathbf{P}}_H$ in the absence of external fields.[7] Taking into account these properties we have $\partial_k v(\mathbf{x}, \mathbf{x}') = -\partial'_k v(\mathbf{x}, \mathbf{x}') = -\partial'_k v(\mathbf{x}', \mathbf{x})$ where ∂'_k is the partial derivative with respect to x', y', z'. Therefore $\partial_k v(\mathbf{x}, \mathbf{x}')$ is antisymmetric under the interchange $\mathbf{x} \leftrightarrow \mathbf{x}'$ and consequently the integral over all space of $\hat{W}_{k,H}$ is zero. No further simplifications occur upon integration and hence the Lorentz force operator is recovered only modulo the total divergence of another operator [first term in the second row of (3.37)]. This fact should not worry the reader. Operators themselves are not measurable; it is the quantum average of an operator that can be compared to an experimental result. As already mentioned in footnote 5 of the previous section, for any large but finite system the probability of finding a particle far away from the system is vanishingly small for all physical states. Therefore, *in an average sense* the total divergence does not contribute and we can write

$$\boxed{\frac{d}{dt}\langle \hat{\mathbf{P}}_H \rangle = \int \left(q\langle \hat{n}_H \rangle \mathbf{E} + \frac{q}{c}\langle \hat{\mathbf{J}}_H \rangle \times \mathbf{B} \right)} \tag{3.38}$$

where $\langle \hat{O} \rangle$ denotes the quantum average of the operator $\hat{O}$ over some physical state.

---

[6] Given three arbitrary vectors $\mathbf{a}$, $\mathbf{b}$, and $\mathbf{c}$ the kth component of $\mathbf{a} \times (\mathbf{b} \times \mathbf{c})$ can be obtained as follows

$$\begin{aligned}
[\mathbf{a} \times (\mathbf{b} \times \mathbf{c})]_k &= \sum_{pq} \varepsilon_{kpq} a_p (\mathbf{b} \times \mathbf{c})_q \\
&= \sum_{pq} \sum_{lm} \varepsilon_{kpq} \varepsilon_{qlm} a_p b_l c_m = \sum_{plm} (\delta_{kl}\delta_{pm} - \delta_{km}\delta_{pl}) a_p b_l c_m \\
&= \sum_p (a_p b_k c_p - a_p b_p c_k),
\end{aligned}$$

where we use the identity $\sum_q \varepsilon_{kpq}\varepsilon_{qlm} = (\delta_{kl}\delta_{pm} - \delta_{km}\delta_{pl})$ for the contraction of two Levi–Civita tensors.

[7] These properties do not imply a dependence on $|\mathbf{r} - \mathbf{r}'|$ only. For instance $v(\mathbf{x}, \mathbf{x}') = \delta_{\sigma\sigma'}/(y - y')^2$ is symmetric, depends only on $\mathbf{r} - \mathbf{r}'$ but cannot be written as a function of $|\mathbf{r} - \mathbf{r}'|$ only. In the special (and physically relevant) case that $v(\mathbf{x}, \mathbf{x}')$ depends only on $|\mathbf{r} - \mathbf{r}'|$ then the internal torque is zero and also the angular momentum is conserved.

4

The contour idea

4.1 Time-dependent quantum averages

In the previous chapter we discussed how to calculate the time-dependent quantum average of an operator $\hat{O}(t)$ at time t when the system is prepared in the state $|\Psi(t_0)\rangle \equiv |\Psi_0\rangle$ at time t_0: if the time evolution is governed by the Schrödinger equation (3.2) the expectation value, $O(t)$, is given by

$$O(t) = \langle\Psi(t)|\hat{O}(t)|\Psi(t)\rangle = \langle\Psi_0|\hat{U}(t_0,t)\hat{O}(t)\hat{U}(t,t_0)|\Psi_0\rangle,$$

with $\hat{U}$ the evolution operator (3.15). We may say that $O(t)$ is the overlap between the initial bra $\langle\Psi_0|$ and a ket obtained by evolving $|\Psi_0\rangle$ from t_0 to t, after which the operator $\hat{O}(t)$ acts, and then evolving the ket backward from t to t_0.

For $t > t_0$ the evolution operator $\hat{U}(t,t_0)$ is expressed in terms of the chronological ordering operator while $\hat{U}(t_0,t)$ is in terms of the anti-chronological ordering operator. Inserting their explicit expressions in $O(t)$ we find

$$O(t) = \langle\Psi_0|\bar{T}\left\{e^{-i\int_t^{t_0} d\bar{t}\,\hat{H}(\bar{t})}\right\}\,\hat{O}(t)\,T\left\{e^{-i\int_{t_0}^t d\bar{t}\,\hat{H}(\bar{t})}\right\}|\Psi_0\rangle. \tag{4.1}$$

The structure of the r.h.s. is particularly interesting. Reading the operators from left to right we note that inside the T operator all Hamiltonians are ordered with later time arguments to the left, the latest time being t. Then the operator $\hat{O}(t)$ appears and to its left all Hamiltonians inside the $\bar{T}$ operator ordered with earlier time arguments to the left, the earliest time being t_0. The main purpose of this section is to elucidate the mathematical structure of (4.1) and to introduce a convenient notation to manipulate chronologically and anti-chronologically ordered products of operators.

If we expand the exponentials in (4.1) in powers of the Hamiltonian then a generic term of the expansion consists of integrals over time of operators like

$$\bar{T}\left\{\hat{H}(t_1)\ldots\hat{H}(t_n)\right\}\hat{O}(t)\,T\left\{\hat{H}(t_1')\ldots\hat{H}(t_m')\right\}, \tag{4.2}$$

where all $\{t_i\}$ and $\{t_i'\}$ have values between t_0 and t.[1] This quantity can be rewritten in a

---

[1] We remind the reader that *by construction* the (anti-)time-ordered operator acting on the multiple integral of Hamiltonians is the multiple integral of the (anti)time-ordered operator acting on the Hamiltonians, see for instance (3.13). In other words $T\int\ldots \equiv \int T\ldots$ and $\bar{T}\int\ldots \equiv \int\bar{T}\ldots$

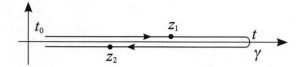

Figure 4.1 The oriented contour γ in the complex time plane as described in the main text. The contour consists of a forward and a backward branch along the real axis between t_0 and t. The branches are displaced from the real axis only for graphical purposes. According to the orientation the point z_2 is later than the point z_1.

more convenient way by introducing a few new definitions. We start by defining the oriented "contour"

$$\gamma \equiv \underbrace{(t_0, t)}_{\gamma_-} \oplus \underbrace{(t, t_0)}_{\gamma_+}, \tag{4.3}$$

which goes from t_0 to t and then back to t_0. The contour γ consists of two paths: a forward branch γ_- and a backward branch γ_+ as shown in Fig. 4.1. A generic point z' of γ can lie either on γ_- or on γ_+ and once the branch is specified it can assume any value between t_0 and t. We denote by $z' = t'_-$ the point of γ lying on the branch γ_- with value t' and by $z' = t'_+$ the point of γ lying on the branch γ_+ with value t'. Having defined γ we introduce operators with arguments on the contour according to

$$\hat{A}(z') \equiv \begin{cases} \hat{A}_-(t') & \text{if } z' = t'_- \\ \hat{A}_+(t') & \text{if } z' = t'_+ \end{cases}. \tag{4.4}$$

In general, the operator $\hat{A}(z')$ on the forward branch ($\hat{A}_-(t')$) can be different from the operator on the backward branch ($\hat{A}_+(t')$). We further define a suitable ordering operator for the product of many operators with arguments on γ. Let $\mathcal{T}$ be the *contour ordering operator* which moves operators with "later" contour-arguments to the left. Then for every permutation P of the times z_m later than z_{m-1} later than $z_{m-2} \ldots$ later than z_1 we have

$$\mathcal{T}\left\{ \hat{A}_m(z_{P(m)})\hat{A}_{m-1}(z_{P(m-1)}) \ldots \hat{A}_1(z_{P(1)}) \right\} = \hat{A}_m(z_m)\hat{A}_{m-1}(z_{m-1}) \ldots \hat{A}_1(z_1),$$

which should be compared with (3.7). A point z_2 is later than a point z_1 if z_1 is closer to the starting point, see again Fig. 4.1. In particular a point on the backward branch is always later than a point on the forward branch. Furthermore, due to the orientation, if $t_1 > t_2$ then t_{1-} is later than t_{2-} while t_{1+} is earlier than t_{2+}. Thus, $\mathcal{T}$ acts like the chronological ordering operator for arguments on γ_- and like the anti-chronological ordering operator for arguments on γ_+. The definition of $\mathcal{T}$, however, also allows us to consider other cases. For example, given two operators $\hat{A}(z_1)$ and $\hat{B}(z_2)$ with argument on the contour we have the following possibilities:

$$
\mathcal{T}\left\{\hat{A}(z_1)\hat{B}(z_2)\right\} =
\begin{cases}
T\left\{\hat{A}_-(t_1)\hat{B}_-(t_2)\right\} & \text{if } z_1 = t_{1-} \text{ and } z_2 = t_{2-} \\
\hat{A}_+(t_1)\hat{B}_-(t_2) & \text{if } z_1 = t_{1+} \text{ and } z_2 = t_{2-} \\
\hat{B}_+(t_2)\hat{A}_-(t_1) & \text{if } z_1 = t_{1-} \text{ and } z_2 = t_{2+} \\
\bar{T}\left\{\hat{A}_+(t_1)\hat{B}_+(t_2)\right\} & \text{if } z_1 = t_{1+} \text{ and } z_2 = t_{2+}
\end{cases}
\tag{4.5}
$$

Operators on the contour and the contour ordering operator can be used to rewrite (4.2) in a compact form. We define the Hamiltonian and the operator $\hat{O}$ with arguments on γ according to

$$
\hat{H}(z' = t'_\pm) \equiv \hat{H}(t'), \qquad \hat{O}(z' = t'_\pm) \equiv \hat{O}(t').
\tag{4.6}
$$

Both the Hamiltonian and $\hat{O}$ are the same on the forward and backward branches, and they equal the corresponding operators with real-time argument; this is a special case of (4.4) with $\hat{A}_- = \hat{A}_+$. In this book the field operators, and hence all operators associated with observable quantities (like the density, current, energy, etc.), with argument on the contour are defined as in (4.6), i.e., they are the same on the two branches of γ and they equal the corresponding operators with real-time argument. Since the field operators carry no dependence on time we have

$$
\boxed{\hat{\psi}(\mathbf{x}, z = t_\pm) \equiv \hat{\psi}(\mathbf{x}), \qquad \hat{\psi}^\dagger(\mathbf{x}, z = t_\pm) \equiv \hat{\psi}^\dagger(\mathbf{x})}
\tag{4.7}
$$

and hence, for instance, the density with argument on the contour is

$$
\hat{n}(\mathbf{x}, z) = \hat{\psi}^\dagger(\mathbf{x}, z)\hat{\psi}(\mathbf{x}, z) = \hat{\psi}^\dagger(\mathbf{x})\hat{\psi}(\mathbf{x}) = \hat{n}(\mathbf{x}),
$$

the diamagnetic current with arguments on the contour is

$$
\hat{\mathbf{j}}_d(\mathbf{x}, z) = -\frac{q}{mc}\hat{n}(\mathbf{x}, z)\mathbf{A}(\mathbf{x}, z) = -\frac{q}{mc}\hat{n}(\mathbf{x})\mathbf{A}(\mathbf{x}, t),
$$

etc. Examples of operators which take different values on the forward and backward branch can be constructed as in (4.5): let $\hat{O}_1(z_1)$ and $\hat{O}_2(z_2)$ be such that $\hat{O}_1(z_1 = t_{1\pm}) = \hat{O}_1(t_1)$ and $\hat{O}_2(z_2 = t_{2\pm}) = \hat{O}_2(t_2)$. Then, for any fixed value of z_2 the operator

$$
\hat{A}(z_1) \equiv \mathcal{T}\left\{\hat{O}_1(z_1)\hat{O}_2(z_2)\right\}
$$

is, in general, different on the two branches, see again (4.5). Notice the slight abuse of notation in (4.6) and (4.7). The same symbol $\hat{H}$ (or $\hat{O}$) is used for the operator with argument on the contour and for the operator with real times. There is, however, no risk of ambiguity as long as we always specify the argument: from now on we use the letter z for variables on γ. With the definition (4.6) we can rewrite (4.2) as

$$
\mathcal{T}\left\{\hat{H}(t_{1+})\ldots\hat{H}(t_{n+})\hat{O}(t_\pm)\hat{H}(t'_{1-})\ldots\hat{H}(t'_{m-})\right\},
\tag{4.8}
$$

where the argument of the operator $\hat{O}$ can be either t_+ or t_-. This result is only the first of a series of simplifications entailed by our new definitions.

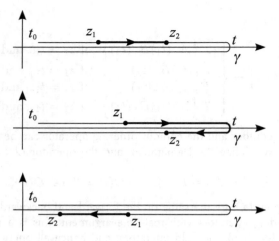

Figure 4.2 The three possible locations of z_2 and z_1 for z_2 later than z_1. The domain of integration is highlighted with bold lines.

We can proceed further by introducing the contour integral between two points z_1 and z_2 on γ in the same way as the standard integral along any contour. If z_2 is later than z_1, see Fig. 4.2, then we have

$$
\int_{z_1}^{z_2} d\bar{z}\, \hat{A}(\bar{z}) = \begin{cases} \int_{t_1}^{t_2} d\bar{t}\, \hat{A}_-(\bar{t}) & \text{if } z_1 = t_{1-} \text{ and } z_2 = t_{2-} \\[2mm] \int_{t_1}^{t} d\bar{t}\, \hat{A}_-(\bar{t}) + \int_{t}^{t_2} d\bar{t}\, \hat{A}_+(\bar{t}) & \text{if } z_1 = t_{1-} \text{ and } z_2 = t_{2+}, \\[2mm] \int_{t_1}^{t_2} d\bar{t}\, \hat{A}_+(\bar{t}) & \text{if } z_1 = t_{1+} \text{ and } z_2 = t_{2+} \end{cases}
$$

while if z_2 is earlier than z_1

$$
\int_{z_1}^{z_2} d\bar{z}\, \hat{A}(\bar{z}) = -\int_{z_2}^{z_1} d\bar{z}\, \hat{A}(\bar{z}).
$$

In this definition $\bar{z}$ is the integration variable along γ and *not* the complex conjugate of z. The latter is denoted by z^*. The generic term of the expansion (4.1) is obtained by integrating the operator (4.2) over all $\{t_i\}$ between t and t_0 and over all $\{t'_i\}$ between t_0 and t. Taking into account that (4.2) is equivalent to (4.8) and using the definition of the contour integral we can write

$$
\int_t^{t_0} dt_1 \dots dt_n \int_{t_0}^{t} dt'_1 \dots dt'_m\, \bar{T}\left\{\hat{H}(t_1)\dots\hat{H}(t_n)\right\} \hat{O}(t)\, T\left\{\hat{H}(t'_1)\dots\hat{H}(t'_m)\right\}
$$

$$
= \int_{\gamma_+} dz_1 \dots dz_n \int_{\gamma_-} dz'_1 \dots dz'_m\, \mathcal{T}\left\{\hat{H}(z_1)\dots\hat{H}(z_n)\hat{O}(t_\pm)\hat{H}(z'_1)\dots\hat{H}(z'_m)\right\},
$$

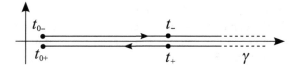

Figure 4.3 The extended oriented contour γ described in the main text with a forward and a backward branch between t_0 and ∞. For any physical time t we have two points $t_\pm$ on γ at the same distance from t_0.

where the symbol $\int_{\gamma_+}$ signifies that the integral is between t_+ and t_{0+} while the symbol $\int_{\gamma_-}$ signifies that the integral is between t_{0-} and t_-.[2] Using this general result for all the terms of the expansion we can rewrite the time-dependent quantum average (4.1) as

$$O(t) = \langle \Psi_0 | \mathcal{T} \left\{ e^{-i \int_{\gamma_+} d\bar{z}\, \hat{H}(\bar{z})} \hat{O}(t_\pm) e^{-i \int_{\gamma_-} d\bar{z}\, \hat{H}(\bar{z})} \right\} | \Psi_0 \rangle.$$

Next we use the fact that operators inside the $\mathcal{T}$ sign can be treated as commuting operators (as for the chronological or anti-chronological ordering operators), and hence

$$O(t) = \langle \Psi_0 | \mathcal{T} \left\{ e^{-i \int_{\gamma} d\bar{z}\, \hat{H}(\bar{z})} \hat{O}(t_\pm) \right\} | \Psi_0 \rangle, \tag{4.9}$$

where $\int_\gamma = \int_{\gamma_-} + \int_{\gamma_+}$ is the contour integral between t_{0-} and t_{0+}. Equation (4.9) is, at the moment, no more than a compact way to rewrite $O(t)$. As we shall see, however, the new notation is extremely useful to manipulate more complicated quantities. We emphasize that in (4.9) $\hat{O}(t)$ is *not* the operator in the Heisenberg picture [the latter is denoted by $\hat{O}_H(t)$].

The contour γ has an aesthetically unpleasant feature: its length depends on t. It would be desirable to have a formula similar to (4.9) but in terms of a universal contour that does not change when we vary the time. Let us explore the implications of extending γ up to infinity, as shown in Fig. 4.3. We evaluate the contour ordered product in (4.9), with γ the extended contour, when the operator $\hat{O}$ is placed in the position $t_\pm$

$$\mathcal{T} \left\{ e^{-i \int_\gamma d\bar{z}\, \hat{H}(\bar{z})} \hat{O}(t_-) \right\} = \hat{U}(t_0, \infty)\hat{U}(\infty, t)\hat{O}(t)\hat{U}(t, t_0) = \hat{U}(t_0, t)\hat{O}(t)\hat{U}(t, t_0),$$

and similarly

$$\mathcal{T} \left\{ e^{-i \int_\gamma d\bar{z}\, \hat{H}(\bar{z})} \hat{O}(t_+) \right\} = \hat{U}(t_0, t)\hat{O}(t)\hat{U}(t, \infty)\hat{U}(\infty, t_0) = \hat{U}(t_0, t)\hat{O}(t)\hat{U}(t, t_0).$$

Thus, the expectation value $O(t)$ in (4.9) does not change if we extend the contour γ as in Fig. 4.3. The extended contour γ is commonly referred to as the *Keldysh contour* in honour of Keldysh who developed the "contour idea" in the context of nonequilibrium Green's functions in a classic paper from 1964 [37]. It should be said, however, that the idea

---

[2]Considering the orientation of the contour the notation is very intuitive.

of using the contour γ of Fig. 4.3 (which over the years has been named in several different ways such as "round trip contour", "closed loop contour", etc.) was first presented by Schwinger a few years earlier in 1961 [38]. For this reason the contour γ is sometimes called the *Schwinger–Keldysh contour*. Schwinger's paper [38] deals with a study of the Brownian motion of a quantum oscillator in external fields and even though his idea is completely general (like most of his ideas) the modern nonequilibrium Green's function formalism is much closer to that developed independently by Keldysh.[3] In the following we simply use the name *contour* since, as we see in the next section, other authors also came up with the "contour idea" independently and more or less at the same time. Furthermore, we do not give a *chronological* presentation of the formalism but, rather, a *logical* presentation. As we see, *all Green's function formalisms usually treated as independent naturally follow from a single one.*

Let us continue and go back to (4.9). An important remark about this equation concerns the explicit time dependence of $\hat{O}(t)$. If the operator does not depend on time we can safely write $\hat{O}(t) = \hat{O}$ in (4.1). However, if we do so in (4.9) it is not clear where to place the operator $\hat{O}$ when acted upon by $\mathcal{T}$. The reason to keep the contour argument even for operators that do not have an explicit time dependence (like the field operators) stems from the need to specify their position along the contour, thus rendering unambiguous the action of $\mathcal{T}$. Once the operators are ordered we can omit the time arguments if there is no time dependence.

Finally, we observe that the l.h.s. in (4.9) contains the physical time t while the r.h.s. contains operators with arguments on γ. We can transform (4.9) into an identity between quantities on the contour if we define $O(t_\pm) \equiv O(t)$. In this way (4.9) takes the elegant form

$$O(z) = \langle \Psi_0 | \mathcal{T} \left\{ e^{-i \int_\gamma d\bar{z}\, \hat{H}(\bar{z})} \, \hat{O}(z) \right\} | \Psi_0 \rangle \qquad (4.10)$$

In (4.10) the contour argument z can be either t_- or t_+ and according to (4.9) our definition is consistent since $O(t_-) = O(t_+) = O(t)$.

4.2 Time-dependent ensemble averages

So far we have used the term *system* to denote an isolated system of particles. In reality, however, it is not possible to completely isolate the system from the surrounding environment; the isolated system is an idealization. The interaction, no matter how weak, between the system and the environment renders a description in terms of one single many-body state impossible. The approach of quantum statistical physics to this problem consists in assigning a probability $w_n \in [0, 1]$ of finding the system at time t_0 in the state $|\chi_n\rangle$, with $\sum_n w_n = 1$. The states $|\chi_n\rangle$ are normalized, $\langle \chi_n | \chi_n \rangle = 1$, but they may not be orthogonal and they may have different energies, momentum, spin, and also different numbers of particles. The underlying idea is to describe the system+environment in terms of the isolated

---

[3]For an interesting historical review on the status of Russian science in nonequilibrium physics in the fifties and early sixties see the article by Keldysh in Ref. [39].

system only, and to account for the interaction with the environment through the probability distribution w_n. The latter, of course, depends on the features of the environment itself. It is important to stress that such probabilistic description is not a consequence of the quantum limitations imposed by the Heisenberg principle. There are no *theoretical* limitations to how well a system can be isolated. Furthermore, for a perfectly isolated system there are no *theoretical* limitations to the accuracy with which one can determine its quantum state. In quantum mechanics a state is uniquely characterized by a complete set of quantum numbers and these quantum numbers are the eigenvalues of a complete set of *commuting* *operators*. Thus, it is in principle possible to measure *at the same time* the value of all these operators and to determine the exact state of the system. In the language of statistical physics we then say that the system is in a *pure state* since the probabilities w_n are all zero except for a single w_n which is 1.

The *ensemble average* of an operator $\hat{O}(t)$ at time t_0 is defined in the most natural way as

$$O(t_0) = \sum_n w_n \langle \chi_n | \hat{O}(t_0) | \chi_n \rangle, \tag{4.11}$$

and reduces to the quantum average previously introduced in the case of pure states. If we imagine an ensemble of identical *and* isolated systems each in a different pure state $|\chi_n\rangle$, then the ensemble average is the result of calculating the weighted sum of the quantum averages $\langle \chi_n | \hat{O}(t_0) | \chi_n \rangle$ with weights w_n. Ensemble averages incorporate the interaction between the system and the environment.

The ensemble average leads us to introduce an extremely useful quantity called the *density matrix operator* $\hat{\rho}$ which contains all the statistical information

$$\hat{\rho} = \sum_n w_n |\chi_n\rangle\langle\chi_n|. \tag{4.12}$$

The density matrix operator is self-adjoint, $\hat{\rho} = \hat{\rho}^\dagger$, and positive-semidefinite since

$$\langle \Psi | \hat{\rho} | \Psi \rangle = \sum_n w_n |\langle \Psi | \chi_n \rangle|^2 \geq 0,$$

for all states $|\Psi\rangle$. Denoting by $|\Psi_k\rangle$ a generic basis of orthonormal states, we can rewrite the ensemble average (4.11) in terms of $\hat{\rho}$ as[4]

$$O(t_0) = \sum_k \sum_n w_n \langle \chi_n | \Psi_k \rangle \langle \Psi_k | \hat{O}(t_0) | \chi_n \rangle = \sum_k \langle \Psi_k | \hat{O}(t_0) \hat{\rho} | \Psi_k \rangle$$
$$= \mathrm{Tr}\left[\hat{O}(t_0)\, \hat{\rho} \right] = \mathrm{Tr}\left[\hat{\rho}\, \hat{O}(t_0) \right], \tag{4.13}$$

where the symbol Tr denotes a trace over all many-body states, i.e., a trace in the Fock space $\mathcal{F}$. If $\hat{O}(t_0) = \hat{1}$ then from (4.11) $O(t_0) = 1$ since the $|\chi_n\rangle$s are normalized and we have the property

$$\mathrm{Tr}\left[\hat{\rho}\right] = 1.$$

---

[4]We recall that the states $|\chi_n\rangle$ may not be orthogonal.

Choosing the kets $|\Psi_k\rangle$ to be the eigenkets of $\hat{\rho}$ with eigenvalues ρ_k, we can write $\hat{\rho} = \sum_k \rho_k |\Psi_k\rangle\langle\Psi_k|$. The eigenvalues ρ_k are non-negative (since $\hat{\rho}$ is positive-semidefinite) and sum up to 1, meaning that $\rho_k \in [0,1]$ and hence that $\mathrm{Tr}\left[\hat{\rho}^2\right] \leq 1$. The most general expression for the ρ_k which incorporates the above constraints is

$$\rho_k = \frac{e^{-x_k}}{\sum_p e^{-x_p}},$$

where x_k are real (positive or negative) numbers. In particular if $\rho_k = 0$ then $x_k = \infty$. For reasons that soon become clear we write $x_k = \beta E_k^{\mathrm{M}}$, where β is a real positive constant, and construct the operator $\hat{H}^{\mathrm{M}}$ according to[5]

$$\hat{H}^{\mathrm{M}} = \sum_k E_k^{\mathrm{M}} |\Psi_k\rangle\langle\Psi_k|.$$

The density matrix operator can then be written as

$$\hat{\rho} = \sum_k \frac{e^{-\beta E_k^{\mathrm{M}}}}{Z} |\Psi_k\rangle\langle\Psi_k| = \frac{e^{-\beta \hat{H}^{\mathrm{M}}}}{Z} \tag{4.14}$$

with the *partition function*

$$Z \equiv \sum_k e^{-\beta E_k^{\mathrm{M}}} = \mathrm{Tr}\left[e^{-\beta \hat{H}^{\mathrm{M}}}\right] \tag{4.15}$$

For example if we number the $|\Psi_k\rangle$ with an integer $k = 0, 1, 2, \ldots$ and if $\hat{\rho} = \rho_1 |\Psi_1\rangle\langle\Psi_1| + \rho_5 |\Psi_5\rangle\langle\Psi_5|$ then

$$\hat{H}^{\mathrm{M}} = E_1^{\mathrm{M}} |\Psi_1\rangle\langle\Psi_1| + E_5^{\mathrm{M}} |\Psi_5\rangle\langle\Psi_5| + \lim_{E\to\infty} E \sum_{k\neq 1,5} |\Psi_k\rangle\langle\Psi_k|, \qquad \begin{cases} E_1^{\mathrm{M}} = -\frac{1}{\beta}\ln(\rho_1 Z) \\[2mm] E_5^{\mathrm{M}} = -\frac{1}{\beta}\ln(\rho_5 Z) \end{cases}.$$

In the special case of pure states, $\hat{\rho} = |\Psi_0\rangle\langle\Psi_0|$, we could either take $E_k^{\mathrm{M}} \to \infty$ for all $k \neq 0$ or alternatively we could take E_k^{M} finite but larger than E_0^{M} and $\beta \to \infty$ since

$$\lim_{\beta\to\infty} \hat{\rho} = \lim_{\beta\to\infty} \frac{\sum_k e^{-\beta E_k^{\mathrm{M}}} |\Psi_k\rangle\langle\Psi_k|}{\sum_k e^{-\beta E_k^{\mathrm{M}}}} = |\Psi_0\rangle\langle\Psi_0|. \tag{4.16}$$

For operators $\hat{H}^{\mathrm{M}}$ with degenerate ground states (4.16) reduces instead to an equally weighted ensemble of degenerate ground states.

In general the expression of $\hat{H}^{\mathrm{M}}$ in terms of field operators is a complicated linear combination of one-body, two-body, three-body, etc. operators. However, in most physical

---

[5]The superscript "^{M}" stands for "Matsubara" since quantities with this superscript have to do with the initial preparation of the system. Matsubara put forward a perturbative formalism, described in the next chapter, to evaluate the ensemble averages of operators at the initial time t_0.

situations the density matrix $\hat{\rho}$ is chosen to describe a system in thermodynamic equilibrium at a given temperature T and chemical potential μ. This density matrix can be determined by maximizing the entropy of the system with the constraints that the average energy and number of particles are fixed, see Appendix D. The resulting $\hat{\rho}$ can be written as in (4.14) with

$$\hat{H}^{\mathrm{M}} = \hat{H} - \mu\hat{N} \qquad \text{and} \qquad \beta = \frac{1}{K_{\mathrm{B}}T},$$

where $\hat{H}$ is the Hamiltonian of the system and K_{B} is the Boltzmann constant. Thus for Hamiltonians as in (1.82) the operator $\hat{H}^{\mathrm{M}}$ is the sum of a one-body and two-body operators.

Let us now address the question how the ensemble averages evolve in time. According to the statistical picture outlined above we have to evolve each system of the ensemble and then calculate the weighted sum of the time-dependent quantum averages $\langle\chi_n(t)|\hat{O}(t)|\chi_n(t)\rangle$ with weights w_n. The systems of the ensemble are all identical and hence described by the same Hamiltonian $\hat{H}(z)$. Using the same logic that led to (4.10) we then find

$$O(z) = \sum_n w_n \langle\chi_n|\hat{U}(t_0,t)\hat{O}(t)\hat{U}(t,t_0)|\chi_n\rangle = \mathrm{Tr}\left[\hat{\rho}\,\hat{U}(t_0,t)\hat{O}(t)\hat{U}(t,t_0)\right]$$

$$= \mathrm{Tr}\left[\hat{\rho}\,\mathcal{T}\left\{e^{-\mathrm{i}\int_\gamma d\bar{z}\,\hat{H}(\bar{z})}\,\hat{O}(z)\right\}\right], \tag{4.17}$$

and taking into account the representation (4.14) for the density matrix operator

$$O(z) = \frac{\mathrm{Tr}\left[e^{-\beta\hat{H}^{\mathrm{M}}}\,\mathcal{T}\left\{e^{-\mathrm{i}\int_\gamma d\bar{z}\,\hat{H}(\bar{z})}\,\hat{O}(z)\right\}\right]}{\mathrm{Tr}\left[e^{-\beta\hat{H}^{\mathrm{M}}}\right]}. \tag{4.18}$$

Two observations are now in order:
(1) The contour ordered product

$$\mathcal{T}\left\{e^{-\mathrm{i}\int_\gamma d\bar{z}\,\hat{H}(\bar{z})}\right\} = \hat{U}(t_0,\infty)\hat{U}(\infty,t_0) = \hat{\mathbb{1}}, \tag{4.19}$$

and can therefore be inserted inside the trace in the denominator of (4.18).
(2) The exponential of $\hat{H}^{\mathrm{M}}$ can be written as

$$e^{-\beta\hat{H}^{\mathrm{M}}} = e^{-\mathrm{i}\int_{\gamma^{\mathrm{M}}} d\bar{z}\,\hat{H}^{\mathrm{M}}},$$

where γ^{M} is any contour in the complex plane starting in z_a and ending in z_b with the only constraint that

$$\boxed{z_b - z_a = -\mathrm{i}\beta}$$

Using the observations (1) and (2) in (4.18) we find

$$O(z) = \frac{\mathrm{Tr}\left[e^{-\mathrm{i}\int_{\gamma^{\mathrm{M}}} d\bar{z}\hat{H}^{\mathrm{M}}}\,\mathcal{T}\left\{e^{-\mathrm{i}\int_\gamma d\bar{z}\,\hat{H}(\bar{z})}\,\hat{O}(z)\right\}\right]}{\mathrm{Tr}\left[e^{-\mathrm{i}\int_{\gamma^{\mathrm{M}}} d\bar{z}\hat{H}^{\mathrm{M}}}\,\mathcal{T}\left\{e^{-\mathrm{i}\int_\gamma d\bar{z}\,\hat{H}(\bar{z})}\right\}\right]}. \tag{4.20}$$

This is a very interesting formula. Performing a statistical average is similar to performing a time propagation as they are both described by the exponential of a "Hamiltonian" operator.

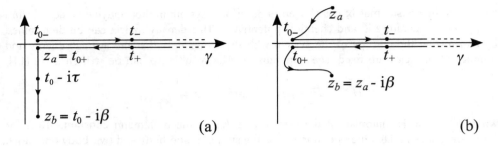

Figure 4.4 Two examples of the extension of the original contour. In (a) a vertical track going from t_0 to $t_0 - i\beta$ has been added and, according to the orientation, any point on this track is later than a point on the forward or backward branch. In (b) any point between z_a and t_{0-} (t_{0+} and z_b) is earlier (later) than a point on the forward or backward branch.

In particular the statistical average is equivalent to a time propagation along the complex path γ^M. The complex time evolution can be incorporated inside the contour ordering operator provided that we connect γ^M to the original contour and define $\hat{H}(z) = \hat{H}^M$ for any z on γ^M. Two examples of such contours are given in Fig. 4.4.[6] The idea of a contour with a complex path γ^M which incorporates the information on how the system is initially prepared was proposed for the first time in 1960 by Konstantinov and Perel' [40] and subsequently developed by several authors, see e.g. Refs. [41, 42]. According to the orientation displayed in the figure, a point on γ^M lying between z_a and t_{0-} is *earlier* than any point lying on the forward or backward branch (there are no such points for the contour of Fig. 4.4(a)). Similarly, a point on γ^M lying between t_{0+} and z_b is *later* than any point lying on the forward or backward branch. We use this observation and the cyclic property of the trace to rewrite the numerator of (4.20) as

$$\mathrm{Tr}\left[e^{-i\int_{t_{0+}}^{z_b} d\bar{z}\hat{H}^M}\mathcal{T}\left\{e^{-i\int_\gamma d\bar{z}\,\hat{H}(\bar{z})}\hat{O}(z)\right\}e^{-i\int_{z_a}^{t_{0-}}d\bar{z}\hat{H}^M}\right]=\mathrm{Tr}\left[\mathcal{T}\left\{e^{-i\int_{\gamma^M\oplus\gamma}d\bar{z}\,\hat{H}(\bar{z})}\hat{O}(z)\right\}\right],$$

where $\gamma^M\oplus\gamma$ denotes a Konstantinov–Perel' contour of Fig. 4.4 and $\mathcal{T}$ is the contour ordering operator along $\gamma^M\oplus\gamma$. From now on we simply denote by γ the Konstantinov–Perel' contour, i.e.,

$$\gamma = \gamma_- \oplus \gamma_+ \oplus \gamma^M,$$

with γ_- the forward branch and γ_+ the backward branch. Furthermore, in the remainder of the book we simply refer to γ as the *contour*. Performing the same manipulations for the

---

[6]Strictly speaking it is not necessary to connect γ^M to the horizontal branches since we can define an ordering also for disconnected contours by saying, e.g., that all points on a piece are earlier or later than all points on the other piece. The fundamental motivation for us to connect γ^M is explained at the end of Section 5.1. For the time being let us say that it is an aesthetically appealing choice.

denominator of (4.20) we get

$$O(z) = \frac{\text{Tr}\left[\mathcal{T}\left\{e^{-i\int_\gamma d\bar{z}\,\hat{H}(\bar{z})}\,\hat{O}(z)\right\}\right]}{\text{Tr}\left[\mathcal{T}\left\{e^{-i\int_\gamma d\bar{z}\,\hat{H}(\bar{z})}\right\}\right]} \tag{4.21}$$

Equation (4.21) is the main result of this section.

We have already shown that if z lies on the forward/backward branch then (4.21) yields the time-dependent ensemble average of the observable $O(t)$. Does (4.21) make any sense if z lies on γ^M? In general it does not since the operator $\hat{O}(z)$ itself is not defined in this case. In the absence of a definition for $\hat{O}(z)$ when $z \in \gamma^M$ we can always invent one. A natural definition would be

$$\hat{O}(z \in \gamma^M) \equiv \hat{O}^M, \tag{4.22}$$

which is the same operator for every point $z \in \gamma^M$ and hence it is compatible with our definition of $\hat{H}(z \in \gamma^M) = \hat{H}^M$. Throughout this book we use the superscript "M" to indicate the constant value of the Hamiltonian or of any other operator (both in first and second quantization) along the path γ^M,

$$\hat{H}^M \equiv \hat{H}(z \in \gamma^M), \qquad \hat{h}^M \equiv \hat{h}(z \in \gamma^M), \qquad \hat{O}^M \equiv \hat{O}(z \in \gamma^M). \tag{4.23}$$

We often consider the case $\hat{O}^M = \hat{O}(t_0)$ except for the Hamiltonian which we take as $\hat{H}^M = \hat{H}(t_0) - \mu\hat{N}$. However, the formalism itself is not restricted to these situations. In some of the applications of the following chapters we use this freedom for studying systems initially prepared in an excited configuration. It is important to stress again that the definition (4.22) is operative only after the operators have been ordered along the contour. Inside the $\mathcal{T}$-product all operators must have a contour argument, also those operators with no explicit time dependence. Having a definition for $\hat{O}(z)$ for all z on the contour we can calculate what (4.21) yields for $z \in \gamma^M$. Taking into account (4.19) we find

$$O(z \in \gamma^M) = \frac{\text{Tr}\left[e^{-i\int_z^{z_b} d\bar{z}\hat{H}(\bar{z})}\hat{O}^M e^{-i\int_{z_a}^z d\bar{z}\hat{H}(\bar{z})}\right]}{\text{Tr}\left[e^{-\beta\hat{H}^M}\right]} = \frac{\text{Tr}\left[e^{-\beta\hat{H}^M}\hat{O}^M\right]}{Z}, \tag{4.24}$$

where we have used the cyclic property of the trace. The r.h.s. is independent of z and for systems in thermodynamic equilibrium coincides with the thermal average of the observable O^M.

Let us summarize what we have derived so far. In (4.21) the variable z lies on the contour γ of Fig. 4.4; the r.h.s. gives the time-dependent ensemble average of $\hat{O}(t)$ when $z = t_\pm$ lies on the forward or backward branch, and the ensemble average of $\hat{O}^M$ when z lies on γ^M.

We conclude this section with an observation. We have already explained that the system-environment interaction is taken into account by assigning a density matrix $\hat{\rho}$ or equivalently an operator $\hat{H}^M$. This corresponds to having an ensemble of identical and *isolated* systems with probabilities w_n. In calculating a time-dependent ensemble average, however, each of these systems evolves as an *isolated* system. To understand the implications of this procedure we consider an environment at zero temperature, so that the density matrix is simply $\hat{\rho} = |\Psi_0\rangle\langle\Psi_0|$ with $|\Psi_0\rangle$ the ground state of the system. Suppose that we switch on a perturbation which we switch off after a time T. In the real world, the average of any observable quantity will go back to its original value for $t \gg T$ due to the interaction with the environment. Strictly speaking this is *not* what our definition of time-dependent ensemble average predicts. The average $\langle\Psi_0(t)|\hat{O}|\Psi_0(t)\rangle$ corresponds to the following thought experiment: the system is initially in contact with the environment at zero temperature (and hence in its ground state), then it is *disconnected* from the environment and it is left to evolve as an isolated system [40]. In other words the effects of the system-environment interaction are included only through the initial configuration but are discarded during the time propagation. In particular, if we inject energy in the system (as we do when we perturb it) there is no way to dissipate it. It is therefore important to keep in mind that the results of our calculations are valid up to times much shorter than the typical relaxation time of the system-environment interaction. To overcome this limitation one should explicitly include the system-environment interaction in the formalism. This leads to a stochastic formulation of the problem [43] which is outside the scope of the present book.

4.3   Initial equilibrium and adiabatic switching

The actual evaluation of (4.21) is, in general, a very difficult task. The formula involves the trace over the full Fock space $\mathcal{F}$ of the product of several operators.[7] As we shall see, we can make progress whenever $\hat{H}(z)$ is the sum of a one-body operator $\hat{H}_0(z)$, which is typically easy to deal with, and an interaction energy operator $\hat{H}_{\text{int}}(z)$ which is, in some sense, "small" and can therefore be treated perturbatively. The derivation of the perturbative scheme to calculate (4.21) is the topic of the next chapter. In this section we look for alternative ways of including $\hat{H}_{\text{int}}(z)$ along the contour without altering the exact result. Depending on the problem at hand it can be advantageous to use one formula or the other when dealing with $\hat{H}_{\text{int}}$ perturbatively. The alternative formulas, however, are valid only under some extra assumptions.

We consider a system initially in equilibrium at a given temperature and chemical potential so that

$$\hat{H}^{\text{M}} = \hat{H}_0^{\text{M}} + \hat{H}_{\text{int}} \qquad \text{with} \qquad \hat{H}_0^{\text{M}} = \hat{H}_0 - \mu\hat{N}.$$

The curious reader can generalize the following discussion to more exotic initial preparations. As usual we take t_0 to be the time at which the Hamiltonian

$$\hat{H} \to \hat{H}(t) = \hat{H}_0(t) + \hat{H}_{\text{int}}$$

acquires some time dependence. To be concrete we also choose the contour of Fig. 4.4(a). In Fig. 4.5(a) we illustrate how the Hamiltonian $\hat{H}(z)$ appearing in (4.21) changes along the contour.

The *adiabatic assumption* is based on the idea that one can generate the density matrix $\hat{\rho}$ with Hamiltonian $\hat{H}^{\text{M}}$ starting from the density matrix $\hat{\rho}_0$ with Hamiltonian $\hat{H}_0^{\text{M}}$ and then

---

[7]The product of several operators stems from the Taylor expansion of the exponential.

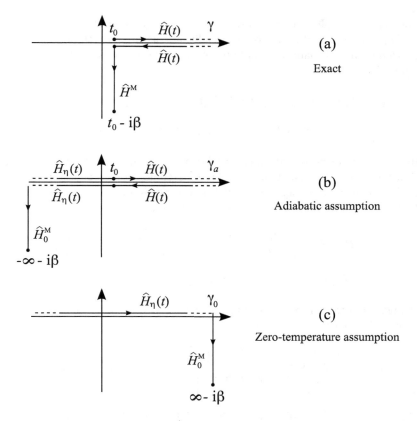

Figure 4.5 Contours and Hamiltonian $\hat{H}(z)$ for: (a) the exact formula, (b) the adiabatic formula, and (c) the zero-temperature formula with $\beta \to \infty$.

switching on the interaction adiabatically, i.e.,

$$\hat{\rho} = \frac{e^{-\beta \hat{H}^{\mathrm{M}}}}{Z} = \hat{U}_{\eta}(t_0, -\infty) \frac{e^{-\beta \hat{H}_0^{\mathrm{M}}}}{Z_0} \hat{U}_{\eta}(-\infty, t_0) = \hat{U}_{\eta}(t_0, -\infty) \hat{\rho}_0 \hat{U}_{\eta}(-\infty, t_0), \quad (4.25)$$

where $\hat{U}_{\eta}$ is the real-time evolution operator with Hamiltonian

$$\hat{H}_{\eta}(t) = \hat{H}_0 + e^{-\eta|t-t_0|} \hat{H}_{\mathrm{int}},$$

and η is an infinitesimally small positive constant. This Hamiltonian coincides with the noninteracting Hamiltonian when $t \to -\infty$ and with the full interacting Hamiltonian when $t = t_0$. Mathematically the adiabatic assumption is supported by the *Gell-Mann–Low theorem* [44, 45] according to which if $\beta \to \infty$ and $\hat{H}_0^{\mathrm{M}}$ has a nondegenerate ground state

$|\Phi_0\rangle$ (hence $\hat{\rho}_0 = |\Phi_0\rangle\langle\Phi_0|$) then $\hat{U}_\eta(t_0, -\infty)|\Phi_0\rangle \equiv |\Psi_0\rangle$ is an eigenstate[8] of $\hat{H}^M$ (hence $\hat{\rho} = |\Psi_0\rangle\langle\Psi_0|$). In general the validity of the adiabatic assumption should be checked case by case. Under the adiabatic assumption we can rewrite the time-dependent ensemble average in (4.17) as

$$O(z = t_\pm) = \text{Tr}\left[\hat{\rho}\,\hat{U}(t_0, t)\hat{O}(t)\hat{U}(t, t_0)\right]$$
$$= \text{Tr}\left[\hat{\rho}_0\,\hat{U}_\eta(-\infty, t_0)\,\hat{U}(t_0, t)\hat{O}(t)\hat{U}(t, t_0)\hat{U}_\eta(t_0, -\infty)\right], \qquad (4.26)$$

where the cyclic property of the trace has been used. Similarly to what we did in the previous sections we can cast (4.26) in terms of a contour-ordered product of operators. Consider the contour γ_a of Fig. 4.5(b) which is essentially the same contour of Fig. 4.5(a) where $t_0 \to -\infty$. If the Hamiltonian changes along the contour as

$$\hat{H}(t_\pm) = \begin{cases} \hat{H}_\eta(t) = \hat{H}_0 + e^{-\eta|t-t_0|}\hat{H}_{\text{int}} & \text{for } t < t_0 \\ \hat{H}(t) = \hat{H}_0(t) + \hat{H}_{\text{int}} & \text{for } t > t_0 \end{cases}$$

$$\hat{H}(z \in \gamma^M) = \hat{H}_0^M = \hat{H}_0 - \mu\hat{N},$$

then (4.26) takes the same form as (4.21) in which $\gamma \to \gamma_a$ and the Hamiltonian goes into the Hamiltonian of Fig. 4.5(b)

$$O(z) = \frac{\text{Tr}\left[\mathcal{T}\left\{e^{-i\int_{\gamma_a} d\bar{z}\,\hat{H}(\bar{z})}\,\hat{O}(z)\right\}\right]}{\text{Tr}\left[\mathcal{T}\left\{e^{-i\int_{\gamma_a} d\bar{z}\,\hat{H}(\bar{z})}\right\}\right]} \qquad \text{(adiabatic assumption)}. \qquad (4.27)$$

We refer to this way of calculating time-dependent ensemble averages as the *adiabatic formula*. This is exactly the formula used by Keldysh in his original paper [37]. The adiabatic formula is correct only provided that the adiabatic assumption is fulfilled. The adiabatic formula gives the noninteracting ensemble average of the operator $\hat{O}$ if $z \in \gamma^M$.

We can derive yet another expression of the ensemble average of the operator $\hat{O}$ for systems unperturbed by external driving fields and hence described by a Hamiltonian $\hat{H}(t > t_0) = \hat{H}$ independent of time. In this case, for any finite time t we can approximate $\hat{U}(t, t_0)$ with $\hat{U}_\eta(t, t_0)$ since we can always choose $\eta \ll 1/|t - t_0|$. If we do so in (4.26) we get

$$O(z = t_\pm) = \text{Tr}\left[\hat{\rho}_0\,\hat{U}_\eta(-\infty, t)\hat{O}(t)\hat{U}_\eta(t, -\infty)\right]. \qquad (4.28)$$

According to the adiabatic assumption we can generate the interacting $\hat{\rho}$ starting from the noninteracting $\hat{\rho}_0$ and then propagating *forward* in time from $-\infty$ to t_0 using the evolution operator $\hat{U}_\eta$. If so then ρ could also be generated starting from $\hat{\rho}_0$ and then propagating *backward* in time from ∞ to t_0 using the same evolution operator $\hat{U}_\eta$ since $\hat{H}_\eta(t_0 - \Delta_t) = \hat{H}_\eta(t_0 + \Delta_t)$. In other words

$$\hat{\rho} = \hat{U}_\eta(t_0, \infty)\,\hat{\rho}_0\,\hat{U}_\eta(\infty, t_0).$$

---

[8]The state $|\Psi_0\rangle$ is not necessarily the ground state.

Comparing this equation with (4.25) we conclude that

$$\hat{\rho}_0 = \hat{U}_\eta(-\infty, \infty)\, \hat{\rho}_0\, \hat{U}_\eta(\infty, -\infty). \tag{4.29}$$

Now it may be reasonable to expect that this identity is fulfilled because every eigenstate $|\Phi_k\rangle$ of $\hat{\rho}_0$ goes back to $|\Phi_k\rangle$ by switching on and off the interaction adiabatically, i.e.,

$$\langle\Phi_k|\hat{U}_\eta(\infty, -\infty) = e^{i\alpha_k}\langle\Phi_k|. \tag{4.30}$$

This expectation is generally wrong due to the occurrence of level crossings and degeneracies in the spectrum of $\hat{H}_\eta$.[9] However, if $\hat{\rho}_0 = |\Phi_0\rangle\langle\Phi_0|$ is a pure state then (4.29) implies that (4.30) with $k = 0$ is satisfied.[10] For $\hat{\rho}_0$ to be a pure state we must take the zero-temperature limit $\beta \to \infty$ and have a nondegenerate ground state $|\Phi_0\rangle$ of $\hat{H}_0^{\mathrm{M}}$. We refer to the adiabatic assumption in combination with equilibrium at zero temperature and with the condition of no ground-state degeneracy as the *zero-temperature assumption*. The zero-temperature assumption can be used to manipulate (4.28) a little more. Since $|\Phi_0\rangle$ is nondegenerate we have

$$\lim_{\beta \to \infty} \hat{\rho}_0 = |\Phi_0\rangle\langle\Phi_0| = \frac{|\Phi_0\rangle\langle\Phi_0|\hat{U}_\eta(\infty, -\infty)}{\langle\Phi_0|\hat{U}_\eta(\infty, -\infty)|\Phi_0\rangle} = \lim_{\beta \to \infty} \frac{e^{-\beta\hat{H}_0^{\mathrm{M}}}\hat{U}_\eta(\infty, -\infty)}{\mathrm{Tr}\left[e^{-\beta\hat{H}_0^{\mathrm{M}}}\hat{U}_\eta(\infty, -\infty)\right]}.$$

Inserting this result into (4.28) we find

$$\lim_{\beta \to \infty} O(z = t_\pm) = \lim_{\beta \to \infty} \frac{\mathrm{Tr}\left[e^{-\beta\hat{H}_0^{\mathrm{M}}}\hat{U}_\eta(\infty, t)\hat{O}(t)\hat{U}_\eta(t, -\infty)\right]}{\mathrm{Tr}\left[e^{-\beta\hat{H}_0^{\mathrm{M}}}\hat{U}_\eta(\infty, -\infty)\right]}. \tag{4.31}$$

If we now rewrite the exponential $e^{-\beta\hat{H}_0^{\mathrm{M}}}$ as $\exp[-i\int_{\gamma_{\mathrm{M}}} d\bar{z}\, \hat{H}_0^{\mathrm{M}}]$ and construct the contour γ_0 which starts at $-\infty$, goes all the way to ∞ and then down to $\infty - i\beta$ we see that (4.31) again has the same mathematical structure as (4.21) in which $\gamma \to \gamma_0$ and the Hamiltonian along the contour changes as illustrated in Fig. 4.5(c). It is worth noting that the contour γ_0 has the special property of having only a forward branch. We refer to this way of calculating ensemble averages as the *zero-temperature formula*. The zero-temperature formula is correct only provided that the zero-temperature assumption is fulfilled. This is certainly not the case if the Hamiltonian of the system is time dependent. There is indeed no reason to expect that by switching on and off the interaction the system goes back to the same state

---

[9]Consider for instance a density matrix $\hat{\rho}_0 = \frac{1}{2}(|\Phi_1\rangle\langle\Phi_1| + |\Phi_2\rangle\langle\Phi_2|)$. Then the most general solution of (4.29) is not (4.30) but

$$\langle\Phi_1|\hat{U}_\eta(\infty, -\infty) = e^{i\alpha}\cos\theta\langle\Phi_1| + \sin\theta\langle\Phi_2|,$$
$$\langle\Phi_2|\hat{U}_\eta(\infty, -\infty) = \sin\theta\langle\Phi_1| - e^{-i\alpha}\cos\theta\langle\Phi_2|,$$

with α and θ two arbitrary real numbers.

[10]For completeness we should mention that the phase factor $\alpha_0 \sim 1/\eta$. This has no consequence in the calculation of observable quantities but it may lead to instabilities when solving the time-dependent Schrödinger equation numerically, see also Exercise 4.1.

in the presence of external driving fields. Finally, we observe that the zero-temperature formula, like the adiabatic formula, gives the noninteracting ensemble average of the operator $\hat{O}$ if $z \in \gamma^{\mathrm{M}}$.

> *To summarize, the average of the operator $\hat{O}$ can be calculated using the very same formula (4.21) in which the contour and the Hamiltonian along the contour are each one of those illustrated in Fig. 4.5. The equivalence between these different flavours of (4.21) relies on the validity of the adiabatic assumption or the zero-temperature assumption.*

Exercise 4.1. Consider the Hamiltonian of a bosonic harmonic oscillator, $\hat{H} = \omega \hat{d}^\dagger \hat{d}$, with eigenstates $|\Phi_n\rangle = \frac{(\hat{d}^\dagger)^n}{\sqrt{n!}}|0\rangle$ and eigenvalues $n\omega$. Consider switching on adiabatically the perturbation $\hat{H}' = \lambda(\hat{d}^\dagger + \hat{d})$ so that the total adiabatic Hamiltonian reads

$$\hat{H}_\eta(t) = \omega \hat{d}^\dagger \hat{d} + e^{-\eta|t|}\lambda(\hat{d}^\dagger + \hat{d}),$$

where η is an infinitesimally small energy. Using the results of Exercise 3.2 show that the state $\hat{U}_\eta(0, \pm\infty)|\Phi_n\rangle$ is, up to an infinite phase factor, the normalized eigenstate of the shifted harmonic oscillator, $\hat{H}_\eta(0) = \omega \hat{d}^\dagger \hat{d} + \lambda(\hat{d}^\dagger + \hat{d})$, with eigenvalue $n\omega - \lambda^2/\omega$. We remind the reader that the eigenstates of the shifted harmonic oscillator are $|\Psi_n\rangle = e^{-\frac{\lambda}{\omega}(\hat{d}^\dagger - \hat{d})}|\Phi_n\rangle$.

Exercise 4.2. Consider the noninteracting Hubbard model at zero temperature in the presence of a magnetic field $\mathbf{B}$ along the z axis. The Hamiltonian is therefore $\hat{H}_0 - \frac{1}{2}g\mu_{\mathrm{B}}B(\hat{N}_\uparrow - \hat{N}_\downarrow)$, see (2.32). Suppose that $B > 0$ is large enough such that the ground state has more electrons of spin up than electrons of spin down. Show that starting from the ground state of $\hat{H}_0$, hence with the same number of spin up and down electrons, and then switching on the magnetic field adiabatically we never generate the spin-polarized ground state.

4.4   Equations of motion on the contour

The interpretation of the density matrix operator as an evolution operator along a complex path prompts us to extend the definition of the evolution operator. We define the *contour evolution operator* $\hat{U}(z_2, z_1)$ with z_1, z_2 belonging to some of the contours discussed in the previous section as

$$\hat{U}(z_2, z_1) = \begin{cases} \mathcal{T}\left\{ e^{-i\int_{z_1}^{z_2} d\bar{z}\, \hat{H}(\bar{z})} \right\} & z_2 \text{ later than } z_1 \\ \\ \bar{\mathcal{T}}\left\{ e^{+i\int_{z_2}^{z_1} d\bar{z}\, \hat{H}(\bar{z})} \right\} & z_2 \text{ earlier than } z_1 \end{cases}, \tag{4.32}$$

where we introduce the anti-chronological contour ordering operator $\bar{\mathcal{T}}$ which rearranges operators with later contour variables to the right. The contour evolution operator is unitary only for z_1 and z_2 on the horizontal branches: if z_1 and/or z_2 lie on γ^M then $\hat{U}$ is proportional to the exponential of a Hermitian operator (like the density matrix operator). In particular $\hat{U}(z_b, z_a) = e^{-\beta \hat{H}^M}$. The contour evolution operator has properties very similar to the real-time evolution operator, namely:

(1) $\hat{U}(z, z) = \hat{\mathbb{1}}$,

(2) $\hat{U}(z_3, z_2)\hat{U}(z_2, z_1) = \hat{U}(z_3, z_1)$,

(3) $\hat{U}$ satisfies a simple differential equation: if z is later than z_0

$$
\mathrm{i}\frac{d}{dz}\hat{U}(z, z_0) = \mathcal{T}\left\{\mathrm{i}\frac{d}{dz}e^{-\mathrm{i}\int_{z_0}^{z} d\bar{z}\, \hat{H}(\bar{z})}\right\} = \mathcal{T}\left\{\hat{H}(z)e^{-\mathrm{i}\int_{z_0}^{z} d\bar{z}\, \hat{H}(\bar{z})}\right\} = \hat{H}(z)\hat{U}(z, z_0),
$$
(4.33)

and

$$
\mathrm{i}\frac{d}{dz}\hat{U}(z_0, z) = \bar{\mathcal{T}}\left\{\mathrm{i}\frac{d}{dz}e^{+\mathrm{i}\int_{z_0}^{z} d\bar{z}\, \hat{H}(\bar{z})}\right\} = -\bar{\mathcal{T}}\left\{\hat{H}(z)e^{+\mathrm{i}\int_{z_0}^{z} d\bar{z}\, \hat{H}(\bar{z})}\right\} = -\hat{U}(z_0, z)\hat{H}(z).
$$
(4.34)

In (4.33) we use the fact that $\mathcal{T}\{\hat{H}(z)\ldots\} = \hat{H}(z)\mathcal{T}\{\ldots\}$ since the operators in $\{\ldots\}$ are calculated at earlier times. A similar property has been used to obtain (4.34).

It goes without saying that the derivative with respect to z of an operator $\hat{A}(z)$ with argument on the contour is defined in the same way as the standard contour derivative. Let z' be a point on γ infinitesimally later than z. If $z = t_-$ (forward branch) we can write $z' = (t + \varepsilon)_-$ and the distance between these two points is simply $(z' - z) = \varepsilon$. Then

$$
\frac{d}{dz}\hat{A}(z) = \lim_{z' \to z}\frac{\hat{A}(z') - \hat{A}(z)}{z' - z} = \lim_{\varepsilon \to 0}\frac{\hat{A}_-(t + \varepsilon) - \hat{A}_-(t)}{\varepsilon} = \frac{d}{dt}\hat{A}_-(t).
$$
(4.35)

On the other hand, if $z = t_+$ (backward branch) then a point infinitesimally later than z can be written as $z' = (t - \varepsilon)_+$ and the distance is in this case $(z' - z) = -\varepsilon$. Therefore

$$
\frac{d}{dz}\hat{A}(z) = \lim_{z' \to z}\frac{\hat{A}(z') - \hat{A}(z)}{z' - z} = \lim_{\varepsilon \to 0}\frac{\hat{A}_+(t - \varepsilon) - \hat{A}_+(t)}{-\varepsilon} = \frac{d}{dt}\hat{A}_+(t).
$$
(4.36)

From this result it follows that operators which are the same on the forward and backward branch have a derivative which is also the same on the forward and backward branch. Finally we consider the case $z \in \gamma^M$. For simplicity let γ^M be the vertical track of Fig. 4.4(a) so that $z = t_0 - \mathrm{i}\tau$ (no extra complications arise from more general paths). A point infinitesimally later than z can be written as $z' = t_0 - \mathrm{i}(\tau + \varepsilon)$ so that the distance between these two points is $(z' - z) = -\mathrm{i}\varepsilon$. The derivative along the imaginary track is then

$$
\frac{d}{dz}\hat{A}(z) = \lim_{z' \to z}\frac{\hat{A}(z') - \hat{A}(z)}{z' - z} = \lim_{\varepsilon \to 0}\frac{\hat{A}(t_0 - \mathrm{i}(\tau + \varepsilon)) - \hat{A}(t_0 - \mathrm{i}\tau)}{-\mathrm{i}\varepsilon} = \mathrm{i}\frac{d}{d\tau}\hat{A}(t_0 - \mathrm{i}\tau).
$$

Equations (4.33) and (4.34) are differential equations along a contour. They have been derived without using the property of unitarity and they should be compared with the differential equation (3.12) and its adjoint for the real-time evolution operator. The conclusion is that $\hat{U}(t_2, t_1)$ and $\hat{U}(z_2, z_1)$ are closely related. In particular, taking into account that on the forward/backward branch $\mathcal{T}$ orders the operators as the chronological/anti-chronological ordering operator while $\bar{\mathcal{T}}$ orders them as the anti-chronological/chronological ordering operator the reader can easily verify that

$$\boxed{\hat{U}(t_2, t_1) = \hat{U}(t_{2-}, t_{1-}) = \hat{U}(t_{2+}, t_{1+})} \tag{4.37}$$

The contour evolution operator can be used to rewrite the ensemble average (4.21) in an alternative way. Let us denote by z_i the initial point of the contour and by z_f the final point of the contour.[11] Then

$$O(z) = \frac{\text{Tr}\left[\hat{U}(z_f, z)\,\hat{O}(z)\,\hat{U}(z, z_i)\right]}{\text{Tr}\left[\hat{U}(z_f, z_i)\right]} = \frac{\text{Tr}\left[\hat{U}(z_f, z_i)\hat{U}(z_i, z)\,\hat{O}(z)\,\hat{U}(z, z_i)\right]}{\text{Tr}\left[\hat{U}(z_f, z_i)\right]}.$$

Looking at this result it is natural to introduce the *contour Heisenberg picture* according to

$$\boxed{\hat{O}_H(z) \equiv \hat{U}(z_i, z)\,\hat{O}(z)\,\hat{U}(z, z_i)} \tag{4.38}$$

If z lies on the horizontal branches then the property (4.37) implies a simple relation between the contour Heisenberg picture and the standard Heisenberg picture

$$\hat{O}_H(t_+) = \hat{O}_H(t_-) = \hat{O}_H(t),$$

where $\hat{O}_H(t)$ is the operator in the standard Heisenberg picture.

The equation of motion for an operator in the contour Heisenberg picture is easily derived from (4.33) and (4.34) and reads

$$i\frac{d}{dz}\hat{O}_H(z) = \hat{U}(z_i, z)\,[\hat{O}(z), \hat{H}(z)]\,\hat{U}(z, z_i) + i\frac{\partial}{\partial z}\hat{O}_H(z)$$

$$= [\hat{O}_H(z), \hat{H}_H(z)] + i\frac{\partial}{\partial z}\hat{O}_H(z), \tag{4.39}$$

where the partial derivative is with respect to the explicit z-dependence of the operator $\hat{O}(z)$. Equation (4.39) has exactly the same structure as the real-time equation of motion (3.19). In particular for $z = t_\pm$ the r.h.s. of (4.39) is identical to the r.h.s. of (3.19) in which t is replaced by z. This means that the equation of motion for the field operators on the contour is given by (3.21) and (3.22) in which $\hat{\psi}_H(\mathbf{x}, t) \to \hat{\psi}_H(\mathbf{x}, z)$, $\hat{\psi}_H^\dagger(\mathbf{x}, t) \to \hat{\psi}_H^\dagger(\mathbf{x}, z)$. For $z \in \gamma^M$ the r.h.s. of (4.39) contains the commutator $[\hat{O}^M, \hat{H}^M]$, see (4.22) and (4.23). We define the field operators with argument on γ^M as

$$\boxed{\hat{\psi}(\mathbf{x}, z \in \gamma^M) \equiv \hat{\psi}(\mathbf{x}), \qquad \hat{\psi}^\dagger(\mathbf{x}, z \in \gamma^M) \equiv \hat{\psi}^\dagger(\mathbf{x})} \tag{4.40}$$

---

[11]The contour can be, e.g., that of Fig. 4.4(a), in which case $z_i = t_{0-}$ and $z_f = t_0 - i\beta$, or that of Fig. 4.4(b), in which case $z_i = z_a$ and $z_f = z_b$, or also that of Fig. 4.5(c), in which case $z_i = -\infty$ and $z_f = \infty - i\beta$ with $\beta \to \infty$.

which together with (4.7) implies that the field operators are constant over the entire contour. In order to keep the presentation suitable to that of an introductory book we only consider systems prepared with an $\hat{H}^{\mathrm{M}}$ of the form

$$\hat{H}^{\mathrm{M}} = \underbrace{\int d\mathbf{x}d\mathbf{x}'\hat{\psi}^{\dagger}(\mathbf{x})\langle\mathbf{x}|\hat{h}^{\mathrm{M}}|\mathbf{x}'\rangle\hat{\psi}(\mathbf{x}')}_{\hat{H}_{0}^{\mathrm{M}}} + \underbrace{\frac{1}{2}\int d\mathbf{x}d\mathbf{x}'v^{\mathrm{M}}(\mathbf{x},\mathbf{x}')\hat{\psi}^{\dagger}(\mathbf{x})\hat{\psi}^{\dagger}(\mathbf{x}')\hat{\psi}(\mathbf{x}')\hat{\psi}(\mathbf{x})}_{\hat{H}_{\mathrm{int}}^{\mathrm{M}}}.$$

(4.41)

The generalization to more complicated $\hat{H}^{\mathrm{M}}$ with three-body or higher order interactions is simply more tedious but it does not require more advanced mathematical tools. For our purposes it is instructive enough to show that no complication arises when $v^{\mathrm{M}} \neq v$ and $\hat{h}^{\mathrm{M}} \neq \hat{h} - \mu$.[12] With an $\hat{H}^{\mathrm{M}}$ of the form (4.41) the equation of motion for the field operators on the entire contour can be written as [compare with (3.21) and (3.22)]

$$i\frac{d}{dz}\hat{\psi}_{H}(\mathbf{x},z) = \sum_{\sigma'}h_{\sigma\sigma'}(\mathbf{r},-i\boldsymbol{\nabla},\mathbf{S},z)\hat{\psi}_{H}(\mathbf{r}\sigma',z) + \int d\mathbf{x}'v(\mathbf{x},\mathbf{x}',z)\hat{n}_{H}(\mathbf{x}',z)\hat{\psi}_{H}(\mathbf{x},z),$$

(4.42)

$$-i\frac{d}{dz}\hat{\psi}_{H}^{\dagger}(\mathbf{x},z) = \sum_{\sigma'}\hat{\psi}_{H}^{\dagger}(\mathbf{r}\sigma',z)h_{\sigma'\sigma}(\mathbf{r},i\overleftarrow{\boldsymbol{\nabla}},\mathbf{S},z) + \int d\mathbf{x}'v(\mathbf{x},\mathbf{x}',z)\hat{\psi}_{H}^{\dagger}(\mathbf{x},z)\hat{n}_{H}(\mathbf{x}',z),$$

(4.43)

with

$$\begin{cases} \hat{h}(z=t_{\pm}) = \hat{h}(t) \\ \hat{h}(z\in\gamma^{\mathrm{M}}) = \hat{h}^{\mathrm{M}} \end{cases} \qquad \begin{cases} v(\mathbf{x},\mathbf{x}',t_{\pm}) = v(\mathbf{x},\mathbf{x}',t) \\ v(\mathbf{x},\mathbf{x}',z\in\gamma^{\mathrm{M}}) = v^{\mathrm{M}}(\mathbf{x},\mathbf{x}') \end{cases}.$$

In this equation we include an explicit time-dependence in the interparticle interaction $v(\mathbf{x},\mathbf{x}',t_{\pm}) = v(\mathbf{x},\mathbf{x}',t)$. The reader can easily verify that in the derivation of (3.21) and (3.22) we do not make any use of the time-independence of v. This extension is useful to deal with situations like those of the previous section in which the interaction was switched on adiabatically.

Even though simple, the equations of motion for the field operators still look rather complicated. We can unravel the underlying "matrix structure" using (1.13), which we rewrite below for convenience

$$\langle\mathbf{x}|\hat{h}(z)|\mathbf{x}'\rangle = h_{\sigma\sigma'}(\mathbf{r},-i\boldsymbol{\nabla},\mathbf{S},z)\delta(\mathbf{r}-\mathbf{r}') = \delta(\mathbf{r}-\mathbf{r}')h_{\sigma\sigma'}(\mathbf{r}',i\overleftarrow{\boldsymbol{\nabla}}',\mathbf{S},z).$$

Then we see by inspection that (4.42) and (4.43) are equivalent to

$$\boxed{i\frac{d}{dz}\hat{\psi}_{H}(\mathbf{x},z) = \int d\mathbf{x}'\langle\mathbf{x}|\hat{h}(z)|\mathbf{x}'\rangle\hat{\psi}_{H}(\mathbf{x}',z) + \int d\mathbf{x}'v(\mathbf{x},\mathbf{x}',z)\hat{n}_{H}(\mathbf{x}',z)\hat{\psi}_{H}(\mathbf{x},z)}$$

(4.44)

---

[12]If $v^{\mathrm{M}} = v$ and $\hat{h}^{\mathrm{M}} = \hat{h} - \mu$ then $\hat{H}^{\mathrm{M}} = \hat{H}_{0} + \hat{H}_{\mathrm{int}} - \mu\hat{N}$ yields the density matrix of thermodynamic equilibrium.

$$-i\frac{d}{dz}\hat{\psi}_H^\dagger(\mathbf{x}, z) = \int d\mathbf{x}'\hat{\psi}_H^\dagger(\mathbf{x}', z)\langle\mathbf{x}'|\hat{h}(z)|\mathbf{x}\rangle + \int d\mathbf{x}'v(\mathbf{x}, \mathbf{x}', z)\hat{\psi}_H^\dagger(\mathbf{x}, z)\hat{n}_H(\mathbf{x}', z)$$

(4.45)

These equations are valid for systems initially in equilibrium and then perturbed by external fields as well as for more exotic situations like, for instance, an interacting system ($v \neq 0$) prepared in a noninteracting configuration ($v^M = 0$) or a noninteracting system ($v = 0$) prepared in an interacting configuration ($v^M \neq 0$) and then perturbed by external fields. Finally we wish to emphasize that the operator $\hat{\psi}_H^\dagger(\mathbf{x}, z)$ is not the adjoint of $\hat{\psi}_H(\mathbf{x}, z)$ if $z \in \gamma^M$ since $\hat{U}(z, z_i)$ is not unitary in this case.

Exercise 4.3. Prove (4.37).

Exercise 4.4. Suppose that the interaction Hamiltonian contains also a three-body operator

$$\hat{H}_{\text{int}}(z) = \frac{1}{2}\int d\mathbf{x}_1 d\mathbf{x}_2\, v(\mathbf{x}_1, \mathbf{x}_2, z)\hat{\psi}^\dagger(\mathbf{x}_1)\hat{\psi}^\dagger(\mathbf{x}_2)\hat{\psi}(\mathbf{x}_2)\hat{\psi}(\mathbf{x}_1)$$

$$+ \frac{1}{3}\int d\mathbf{x}_1 d\mathbf{x}_2 d\mathbf{x}_3\, v(\mathbf{x}_1, \mathbf{x}_2, \mathbf{x}_3, z)\hat{\psi}^\dagger(\mathbf{x}_1)\hat{\psi}^\dagger(\mathbf{x}_2)\hat{\psi}^\dagger(\mathbf{x}_3)\hat{\psi}(\mathbf{x}_3)\hat{\psi}(\mathbf{x}_2)\hat{\psi}(\mathbf{x}_1),$$

(4.46)

where $v(\mathbf{x}_1, \mathbf{x}_2, \mathbf{x}_3, z)$ is totally symmetric under a permutation of $\mathbf{x}_1, \mathbf{x}_2, \mathbf{x}_3$. Show that the equation of motion for the field operator is

$$i\frac{d}{dz}\hat{\psi}_H(\mathbf{x}, z) = \int d\mathbf{x}_1\langle\mathbf{x}|\hat{h}(z)|\mathbf{x}_1\rangle\hat{\psi}_H(\mathbf{x}_1, z) + \int d\mathbf{x}_1\, v(\mathbf{x}, \mathbf{x}_1, z)\hat{n}_H(\mathbf{x}_1, z)\hat{\psi}_H(\mathbf{x}, z)$$

$$+ \int d\mathbf{x}_1 d\mathbf{x}_2\, v(\mathbf{x}_1, \mathbf{x}_2, \mathbf{x}, z)\hat{\psi}_H^\dagger(\mathbf{x}_1, z)\hat{\psi}_H^\dagger(\mathbf{x}_2, z)\hat{\psi}_H(\mathbf{x}_2, z)\hat{\psi}_H(\mathbf{x}_1, z)\hat{\psi}_H(\mathbf{x}, z).$$

4.5 Operator correlators on the contour

In the previous sections we showed how to rewrite the time-dependent ensemble average of an operator $\hat{O}(t)$ and derived the equation of motion for operators in the contour Heisenberg picture. The main question that remains to be answered is: how can we calculate $O(t)$? We have already mentioned that the difficulty lies in evaluating the exponential in (4.21) and in taking the trace over the Fock space. For evaluation of the exponential a natural way to proceed is to expand it in a Taylor series. This would lead to traces of time-ordered strings of operators on the contour. These strings are of the general form

$$\hat{k}(z_1, \ldots, z_n) = \mathcal{T}\left\{\hat{O}_1(z_1)\ldots\hat{O}_n(z_n)\right\},$$

(4.47)

in which $\hat{O}_k(z_k)$ are operators located at position z_k on the contour. Alternatively, we could calculate $O(t)$ by tracing with $\hat{\rho}$ the equation of motion for $\hat{O}_H(z)$ and then solving the resulting differential equation. However, as clearly shown in (4.44) and (4.45), the time derivative generates new operators whose time derivative generates yet other and more complex operators, and so on and so forth. As a result we are again back to calculating the trace of strings of operators like (4.47), but this time with operators in the contour Heisenberg picture. For instance the r.h.s. of (4.44) contains the operator $\hat{n}_H(\mathbf{x}', z)\hat{\psi}_H(\mathbf{x}, z) = \mathcal{T}\left\{ n_H(\mathbf{x}', z^+)\hat{\psi}_H(\mathbf{x}, z) \right\}$ where z^+ is a contour time infinitesimally later than z.

From the above discussion we conclude that if we want to calculate $O(t)$ we must be able to manipulate objects like (4.47). We refer to these strings of operators as the *operator correlators*. As we shall see they play an important part in the subsequent development of the perturbative scheme. The aim of this section is to discuss some of their basic properties. At this stage it is not important what the origin of the time-dependence of the operators is. We can take operators in the contour Heisenberg picture or consider any other arbitrary time-dependence. It will not even be important what the actual shape of the contour is. The only thing that matters in the derivations below is that the operators are under the contour-ordered sign $\mathcal{T}$.

The most natural way to find relations for the operator correlators is to differentiate them with respect to their contour arguments. As we shall see this leads to a very useful set of hierarchy equations. In order not to overcrowd our equations we introduce the abbreviation

$$\hat{O}_j \equiv O_j(z_j).$$

The simplest example of an operator correlator is the contour ordered product of just two operators [see also (4.5)]

$$\mathcal{T}\left\{\hat{O}_1\hat{O}_2\right\} = \theta(z_1, z_2)\hat{O}_1\hat{O}_2 + \theta(z_2, z_1)\hat{O}_2\hat{O}_1, \tag{4.48}$$

where $\theta(z_1, z_2) = 1$ if z_1 is later than z_2 on the contour and zero otherwise. This θ-function can be thought of as the Heaviside function on the contour. If we differentiate (4.48) with respect to the contour variable z_1 we obtain

$$\frac{d}{dz_1}\mathcal{T}\left\{\hat{O}_1\hat{O}_2\right\} = \delta(z_1, z_2)\left[\hat{O}_1, \hat{O}_2\right]_- + \mathcal{T}\left\{\left(\frac{d}{dz_1}\hat{O}_1\right)\hat{O}_2\right\}. \tag{4.49}$$

In this formula we defined the Dirac δ-function on the contour in the obvious way

$$\delta(z_1, z_2) \equiv \frac{d}{dz_1}\theta(z_1, z_2) = -\frac{d}{dz_2}\theta(z_1, z_2).$$

By definition $\delta(z_1, z_2)$ is zero everywhere except in $z_1 = z_2$ where it is infinite. Furthermore, for any operator $\hat{A}(z)$ we have[13]

$$\int_{z_i}^{z_f} d\bar{z}\, \delta(z, \bar{z})\hat{A}(\bar{z}) = \hat{A}(z).$$

---

[13] This identity can easily be checked with, e.g., an integration by parts. We have

$$\int_{z_i}^{z_f} d\bar{z}\, \delta(z, \bar{z})\hat{A}(\bar{z}) = -\int_{z_i}^{z_f} d\bar{z}\, [\frac{d}{d\bar{z}}\theta(z, \bar{z})]\hat{A}(\bar{z})$$

$$= \hat{A}(z_i) + \int_{z_i}^{z_f} d\bar{z}\, \theta(z, \bar{z})\frac{d}{d\bar{z}}\hat{A}(\bar{z}) = \hat{A}(z_i) + \hat{A}(z) - \hat{A}(z_i).$$

A nice feature of (4.49) is the equal time commutator between the two operators. Since we can build any many-body operator from the field operators one of the most important cases to consider is when $\hat{O}_1$ and $\hat{O}_2$ are field operators. In this case the commutator has a simple value for bosons [either 0 or a δ-function, see (1.39), (1.40), and (1.47)] but has no simple expression for fermions. A simple expression for fermions would be obtained if we could replace the commutator with the anti-commutator. This can be readily achieved by defining the time ordering of two fermionic field operators to be given by

$$\mathcal{T}\{\hat{O}_1\hat{O}_2\} \equiv \theta(z_1, z_2)\hat{O}_1\hat{O}_2 - \theta(z_2, z_1)\hat{O}_2\hat{O}_1,$$

where we introduce a minus sign in front of the last term. If we differentiate this expression with respect to the time z_1 we get

$$\frac{d}{dz_1}\mathcal{T}\{\hat{O}_1\hat{O}_2\} = \delta(z_1, z_2)\left[\hat{O}_1, \hat{O}_2\right]_+ + \mathcal{T}\left\{\left(\frac{d}{dz_1}\hat{O}_1\right)\hat{O}_2\right\},$$

where now the anti-commutator appears. In order for this nice property to be present in general n-point correlators also we introduce the following generalized definition of the contour-ordered product

$$\mathcal{T}\{\hat{O}_1\ldots\hat{O}_n\} = \sum_P (\pm)^P \theta_n(z_{P(1)}, \ldots, z_{P(n)})\hat{O}_{P(1)}\ldots\hat{O}_{P(n)}, \tag{4.50}$$

where we sum over all permutations P of n variables and where the $+$ sign refers to a string of bosonic field operators and the $-$ sign refers to a string of fermionic field operators. In (4.50) we further define the n-time theta function θ_n to be

$$\theta_n(z_1, \ldots, z_n) = \begin{cases} 1 & \text{if } z_1 > z_2 > \ldots > z_n \\ 0 & \text{otherwise} \end{cases},$$

or equivalently

$$\theta_n(z_1, \ldots, z_n) = \theta(z_1, z_2)\theta(z_2, z_3)\ldots\theta(z_{n-1}, z_n).$$

From now on we interchangeably write "$z_1 > z_2$" or "z_1 later than z_2" as well as "$z_1 < z_2$" or "z_1 earlier than z_2." The definition (4.50) considers all possible orderings of the contour times through the sum over P, and for a given set of times $z_1, \ldots, z_n$ only one term of the sum survives. It follows from the definition that

$$\mathcal{T}\{\hat{O}_1\ldots\hat{O}_n\} = (\pm)^P \mathcal{T}\{\hat{O}_{P(1)}\ldots\hat{O}_{P(n)}\}.$$

In particular it follows that bosonic field operators commute within the contour-ordered product whereas fermionic field operators anti-commute.

At this point we should observe that the contour ordering operator is ambiguously defined if the operators have the same time variable. To cure this problem we introduce the further rule that operators at equal times do not change their relative order after the contour-ordering. Thus, for instance

$$\mathcal{T}\{\hat{\psi}(\mathbf{x}_1, z)\hat{\psi}(\mathbf{x}_2, z)\} = \hat{\psi}(\mathbf{x}_1)\hat{\psi}(\mathbf{x}_2).$$

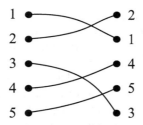

Figure 4.6 Graphical way to calculate the sign of a permutation.

or

$$\mathcal{T}\left\{\hat{\psi}(\mathbf{x}_1,t_-)\hat{\psi}(\mathbf{x}_2,t'_+)\hat{\psi}^\dagger(\mathbf{x}_3,t_-)\right\} = \pm\hat{\psi}(\mathbf{x}_2)\hat{\psi}(\mathbf{x}_1)\hat{\psi}^\dagger(\mathbf{x}_3)$$

or

$$\mathcal{T}\left\{\hat{\psi}(\mathbf{x}_1,z)\hat{\psi}(\mathbf{x}_2,z)\hat{\psi}^\dagger(\mathbf{x}_2,z^+)\hat{\psi}^\dagger(\mathbf{x}_1,z^+)\right\} = \hat{\psi}^\dagger(\mathbf{x}_2)\hat{\psi}^\dagger(\mathbf{x}_1)\hat{\psi}(\mathbf{x}_1)\hat{\psi}(\mathbf{x}_2),$$

where z^+ is a time infinitesimally later than z. From the last example we easily deduce that a composite operator which consists of M equal-time field operators, like the density or the current or $\hat{n}(\mathbf{x})\hat{\psi}(\mathbf{y})$, behaves like a bosonic/fermionic field operator for even/odd M under the $\mathcal{T}$ sign. We use the general nomenclature bosonic/fermionic operators for these kinds of composite operator. In particular, a string of Hamiltonians behaves like a set of bosonic operators. Therefore, our generalized definition (4.50) is consistent with the earlier definition in Section 4.1.

According to our new definition the derivative of the two-operator correlator is given by

$$\frac{d}{dz_1}\mathcal{T}\left\{\hat{O}_1\hat{O}_2\right\} = \delta(z_1,z_2)\left[\hat{O}_1,\hat{O}_2\right]_\mp + \mathcal{T}\left\{\left(\frac{d}{dz_1}\hat{O}_1\right)\hat{O}_2\right\}, \qquad (4.51)$$

where the upper/lower sign refers to bosonic/fermionic operators. Let us generalize this result to higher order operator correlators. For long strings of fermionic operators, finding the sign of the prefactor in (4.50) can be awkward. However, there exists a very nice and elegant graphical way to find the sign by the simple drawing of one diagram. Let us illustrate it with an example. Consider the case of five fermionic operators $\hat{O}_1 \ldots \hat{O}_5$ with the contour variables $z_2 > z_1 > z_4 > z_5 > z_3$. Then

$$\mathcal{T}\left\{\hat{O}_1\hat{O}_2\hat{O}_3\hat{O}_4\hat{O}_5\right\} = -\hat{O}_2\hat{O}_1\hat{O}_4\hat{O}_5\hat{O}_3.$$

The reordering of these operators corresponds to the permutation

$$P(1,2,3,4,5) = (2,1,4,5,3)$$

and has sign -1. This permutation can be drawn graphically as in Fig. 4.6. The operators $\hat{O}_1 \ldots \hat{O}_5$ are denoted by dots and are ordered from top to bottom on a vertical line on the left of the figure. The permuted operators $\hat{O}_{P(1)} \ldots \hat{O}_{P(5)}$ are similarly drawn on a vertical line on the right of the figure with the latest time at the top and the earliest time at the bottom. We then connect the left dot i with the right dot i by a line. This line does not have to be a straight line, it can be any curve as long as it does not extend beyond the left and right boundaries of the graph. If we count the number of crossings n_c then the sign of the permutation is simply given by $(-1)^{n_c}$. Since in our case there are three crossings the sign of the permutation is $(-1)^3 = -1$. This graphical trick is explained in more detail in Appendix B where it is also used to derive some useful relations for determinants and permanents. It is readily seen that any interchange of neighboring operators on the left or right vertical line increases or decreases the number of crossings by one. An interchange on the left vertical line corresponds to transposing operators under the contour ordering sign. For example, if we interchange the operators $\hat{O}_3$ and $\hat{O}_4$

$$\mathcal{T}\left\{\hat{O}_1\hat{O}_2\hat{O}_3\hat{O}_4\hat{O}_5\right\} = -\mathcal{T}\left\{\hat{O}_1\hat{O}_2\hat{O}_4\hat{O}_3\hat{O}_5\right\} \tag{4.52}$$

in agreement with the fact that the number of crossings for the operator correlator on the r.h.s. is 2. On the other hand, if we interchange two operators on the right vertical line we change the contour-ordering. For instance, interchanging operators $\hat{O}_1$ and $\hat{O}_2$ on the right will change the contour-ordering from $z_2 > z_1$ to $z_1 > z_2$. These operations and the graphs as in Fig. 4.6 are very useful in proving relations involving contour-ordered products. We can use these pictures to derive a generalization of (4.51).

We consider strings of only fermionic or only bosonic operators as the generalization to mixed strings is straightforward. From (4.50) we see that the derivative of a time-ordered product consists of a part where the Heaviside functions are differentiated and a part where an operator is differentiated, i.e.,

$$\frac{d}{dz_k}\mathcal{T}\left\{\hat{O}_1 \ldots \hat{O}_n\right\} = \partial^\theta_{z_k}\mathcal{T}\left\{\hat{O}_1 \ldots \hat{O}_n\right\} + \mathcal{T}\left\{\hat{O}_1 \ldots \hat{O}_{k-1}\left(\frac{d}{dz_k}\hat{O}_k\right)\hat{O}_{k+1} \ldots \hat{O}_n\right\}, \tag{4.53}$$

where we define [46]

$$\partial^\theta_{z_k}\mathcal{T}\left\{\hat{O}_1 \ldots \hat{O}_n\right\} \equiv \sum_P (\pm)^P \left(\frac{d}{dz_k}\theta_n(z_{P(1)}, \ldots, z_{P(n)})\right)\hat{O}_{P(1)} \ldots \hat{O}_{P(n)}. \tag{4.54}$$

It remains to give a more explicit form to (4.54). Imagine that we have a given contour-ordering of the operators as in Fig. 4.6 and that we subsequently change the time z_k. As previously explained, this corresponds to moving a dot on the right vertical line. When z_k moves along the contour the Heaviside function leads to a sudden change in the time-ordering whenever z_k passes another time z_l. In such a case the derivative of the Heaviside function gives a contribution proportional to $\delta(z_k, z_l)$. We only need to find what the prefactor of this δ-function is and for this purpose we need to know how the correlator behaves when z_k is very close to z_l. Let us start by moving the operator $\hat{O}_l$ directly after $\hat{O}_k$ inside the contour ordering. This requires $l - k - 1$ interchanges when $l > k$ and $k - l$ interchanges when $k > l$ and therefore

$$\mathcal{T}\left\{\hat{O}_1 \ldots \hat{O}_n\right\} = (\pm)^{l-k-1}\mathcal{T}\left\{\hat{O}_1 \ldots \hat{O}_k\hat{O}_l\hat{O}_{k+1} \ldots \hat{O}_{l-1}\hat{O}_{l+1} \ldots \hat{O}_n\right\},$$

when $l > k$, and

$$\mathcal{T}\left\{\hat{O}_1 \ldots \hat{O}_n\right\} = (\pm)^{k-l}\mathcal{T}\left\{\hat{O}_1 \ldots \hat{O}_{l-1}\hat{O}_{l+1} \ldots \hat{O}_k\hat{O}_l\hat{O}_{k+1} \ldots \hat{O}_n\right\},$$

when $k > l$. The key observation is now that the operators $\hat{O}_k$ and $\hat{O}_l$ stay next to each other after the contour-ordering since z_k has been assumed to be very close to z_l, and therefore there is no other time z_j between them. The sign of the permutation that achieves this contour-ordering is therefore equal to the sign of the permutation Q of the subset of operators not including $\hat{O}_k$ and $\hat{O}_l$. This is easily seen graphically since the pair of lines connecting the operators $\hat{O}_k$ and $\hat{O}_l$ to their images is always crossed an even number of times. For instance, if we move 4 just below 1 in the left vertical line of Fig. 4.6 and then shift the pair 1 and 4 in the right vertical line upward or downward we see that the number of crossings n_c always changes by an even number. Then for $l > k$ and for z_k just above and below z_l we can write

$$\mathcal{T}\left\{\hat{O}_1\ldots\hat{O}_n\right\} = (\pm)^{l-k-1}(\pm)^Q\left[\theta(z_k,z_l)\hat{O}_{Q(1)}\ldots\hat{O}_k\hat{O}_l\ldots\hat{O}_{Q(n)}\right.$$
$$\left.\pm\,\theta(z_l,z_k)\hat{O}_{Q(1)}\ldots\hat{O}_l\hat{O}_k\ldots\hat{O}_{Q(n)}\right].$$

We can now differentiate the Heaviside function with respect to z_k and find that

$$\partial^\theta_{z_k}\mathcal{T}\left\{\hat{O}_1\ldots\hat{O}_n\right\} = (\pm)^{l-k-1}\delta(z_k,z_l)(\pm)^Q\hat{O}_{Q(1)}\ldots\left[\hat{O}_k,\hat{O}_l\right]_{\mp}\ldots\hat{O}_{Q(n)}$$
$$= (\pm)^{l-k-1}\delta(z_k,z_l)\mathcal{T}\left\{\hat{O}_1\ldots\left[\hat{O}_k,\hat{O}_l\right]_{\mp}\hat{O}_{k+1}\ldots\hat{O}_{l-1}\hat{O}_{l+1}\ldots\hat{O}_n\right\}.$$

Due to the presence of the δ-function the (anti)commutator under the contour-ordering sign can be regarded as a function of one time variable only and behaves like a bosonic operator. Similarly for $k > l$ we can, in exactly the same way, derive the equation

$$\partial^\theta_{z_k}\mathcal{T}\left\{\hat{O}_1\ldots\hat{O}_n\right\} = (\pm)^{k-l}\delta(z_k,z_l)\mathcal{T}\left\{\hat{O}_1\ldots\hat{O}_{l-1}\hat{O}_{l+1}\ldots\left[\hat{O}_k,\hat{O}_l\right]_{\mp}\hat{O}_{k+1}\ldots\hat{O}_n\right\}.$$

These two equations are valid only for z_k very close to z_l. The general result for the derivative (4.54) is obtained by summing over all possible values of $l \neq k$. This yields the final expression

$$\partial^\theta_{z_k}\mathcal{T}\left\{\hat{O}_1\ldots\hat{O}_n\right\}$$
$$= \sum_{l=1}^{k-1}(\pm)^{k-l}\delta(z_k,z_l)\mathcal{T}\left\{\hat{O}_1\ldots\hat{O}_{l-1}\hat{O}_{l+1}\ldots\left[\hat{O}_k,\hat{O}_l\right]_{\mp}\hat{O}_{k+1}\ldots\hat{O}_n\right\}$$
$$+ \sum_{l=k+1}^{n}(\pm)^{l-k-1}\delta(z_k,z_l)\mathcal{T}\left\{\hat{O}_1\ldots\left[\hat{O}_k,\hat{O}_l\right]_{\mp}\hat{O}_{k+1}\ldots\hat{O}_{l-1}\hat{O}_{l+1}\ldots\hat{O}_n\right\}. \quad (4.55)$$

This expression can be rewritten in various other ways since the (anti)commutator behaves as a bosonic operator and can therefore be placed anywhere we like under the $\mathcal{T}$ sign. Equation (4.53) together with (4.55) represents the generalization of (4.51). For example, if we differentiate the five-operator correlator in (4.52) with respect to z_3 we get

$$
\frac{d}{dz_3} \mathcal{T} \left\{ \hat{O}_1 \hat{O}_2 \hat{O}_3 \hat{O}_4 \hat{O}_5 \right\} = \delta(z_3, z_1) \mathcal{T} \left\{ \hat{O}_2 \left[\hat{O}_3, \hat{O}_1 \right]_{\mp} \hat{O}_4 \hat{O}_5 \right\}
$$

$$
\pm \delta(z_3, z_2) \mathcal{T} \left\{ \hat{O}_1 \left[\hat{O}_3, \hat{O}_2 \right]_{\mp} \hat{O}_4 \hat{O}_5 \right\} + \delta(z_3, z_4) \mathcal{T} \left\{ \hat{O}_1 \hat{O}_2 \left[\hat{O}_3, \hat{O}_4 \right]_{\mp} \hat{O}_5 \right\}
$$

$$
\pm \delta(z_3, z_5) \mathcal{T} \left\{ \hat{O}_1 \hat{O}_2 \left[\hat{O}_3, \hat{O}_5 \right]_{\mp} \hat{O}_4 \right\} + \mathcal{T} \left\{ \hat{O}_1 \hat{O}_2 \left(\frac{d}{dz_3} \hat{O}_3 \right) \hat{O}_4 \hat{O}_5 \right\},
$$

where the sign in front of each δ-function is simply given by the number of interchanges required to shift the operators $\hat{O}_l$ for $l = 1, 2, 4, 5$ directly after $\hat{O}_3$ inside the $\mathcal{T}$ product.

We have mentioned that an important application of (4.55) is when the operators $\hat{O}_l$ are the field operators in the contour Heisenberg picture. In this case the equal-time (anti)commutator is a real number equal to either zero or a δ-function. Let us therefore investigate how (4.55) simplifies if

$$
\left[\hat{O}_k(z), \hat{O}_l(z) \right]_{\mp} = c_{kl}(z) \hat{\mathbb{1}}, \tag{4.56}
$$

where $c_{kl}(z)$ is a scalar function of z and where on the l.h.s. we reinsert the explicit time dependence of the operators to stress that (4.56) is valid only for the equal-time (anti)commutator. Since the unit operator $\hat{\mathbb{1}}$ commutes with all operators in Fock space the (anti)commutators in (4.55) can be moved outside the contour-ordered product and we obtain

$$
\partial^\theta_{z_k} \mathcal{T} \left\{ \hat{O}_1 \dots \hat{O}_n \right\} = \sum_{l=1}^{k-1} (\pm)^{k-l} \delta(z_k, z_l) \left[\hat{O}_k, \hat{O}_l \right]_{\mp} \mathcal{T} \{ \hat{O}_1 \dots \overset{\sqcap}{\hat{O}_l} \dots \overset{\sqcap}{\hat{O}_k} \dots \hat{O}_n \}
$$

$$
+ \sum_{l=k+1}^{n} (\pm)^{l-k-1} \delta(z_k, z_l) \left[\hat{O}_k, \hat{O}_l \right]_{\mp} \mathcal{T} \{ \hat{O}_1 \dots \overset{\sqcap}{\hat{O}_k} \dots \overset{\sqcap}{\hat{O}_l} \dots \hat{O}_n \}, \tag{4.57}
$$

where the symbol $\sqcap$ above an operator $\hat{O}_k$ signifies that this operator is missing from the list, for example

$$
\overset{\sqcap}{\hat{O}_1} \hat{O}_2 \hat{O}_3 \overset{\sqcap}{\hat{O}_4} \hat{O}_5 = \hat{O}_1 \hat{O}_3 \hat{O}_5.
$$

Equation (4.57) shows very clearly why it is useful to introduce the definition (4.50) of the contour-ordering. If we had stuck to our original version of contour-ordering, then (4.55) would still have been valid but with all prefactors $+$ and, more importantly, with a commutator also for fermionic operators. The unpleasant consequence of this fact would be that for the fermionic field operators ψ and $\psi^\dagger$ no simplification as in (4.57) can be made. This is a nice example of something to keep in mind when learning new topics: definitions are always introduced to simplify the calculations and never fall from the sky. To appreciate the simplification entailed by the definition (4.50) we work out the time derivative of a string of four field operators in the contour Heisenberg picture. Let us introduce a notation which will be used throughout the book

$$
i = \mathbf{x}_i, z_i, \quad j = \mathbf{x}_j, z_j, \quad \dots
$$

as well as

$$i' = \mathbf{x}'_i, z'_i, \quad j' = \mathbf{x}'_j, z'_j, \quad \ldots$$

and

$$\bar{i} = \bar{\mathbf{x}}_i, \bar{z}_i, \quad \bar{j} = \bar{\mathbf{x}}_j, \bar{z}_j, \quad \ldots$$

etc., to denote the collective position–spin–time coordinate, and

$$\delta(j; k) \equiv \delta(z_j, z_k)\delta(\mathbf{x}_j - \mathbf{x}_k)$$

to denote the space–spin–time δ-function. Then (4.57) together with (4.53) tells us, for instance, that

$$\frac{d}{dz_2}\mathcal{T}\left\{\hat{\psi}_H(1)\hat{\psi}_H(2)\hat{\psi}^\dagger_H(3)\hat{\psi}^\dagger_H(4)\right\} = \mathcal{T}\left\{\hat{\psi}_H(1)\left(\frac{d}{dz_2}\hat{\psi}_H(2)\right)\hat{\psi}^\dagger_H(3)\hat{\psi}^\dagger_H(4)\right\}$$
$$+ \delta(2;3)\mathcal{T}\left\{\hat{\psi}_H(1)\hat{\psi}^\dagger_H(4)\right\} \pm \delta(2;4)\mathcal{T}\left\{\hat{\psi}_H(1)\hat{\psi}^\dagger_H(3)\right\},$$

where we take into account that the equal-time (anti)commutator between $\hat{\psi}_H(2)$ and $\hat{\psi}_H(1)$ vanishes.

At the beginning of the section we showed that the operator correlators $\hat{k}$ appear naturally in the expansion of the contour-ordered exponential and in the equations of motion. Thus, the operators $\hat{O}_k$ are typically composite operators corresponding to observable quantities and, as such, with an equal number of creation and annihilation field operators. Let us therefore define the special correlators

$$\hat{G}_n(1,\ldots,n;1',\ldots,n') \equiv \frac{1}{\mathrm{i}^n}\mathcal{T}\left\{\hat{\psi}_H(1)\ldots\hat{\psi}_H(n)\hat{\psi}^\dagger_H(n')\ldots\hat{\psi}^\dagger_H(1')\right\}, \qquad (4.58)$$

where the primed $j' = \mathbf{x}'_j, z'_j$ and unprimed $j = \mathbf{x}_j, z_j$ coordinates label creation and annihilation field operators respectively (which can be either all bosonic or all fermionic). As we shall see the prefactor $1/\mathrm{i}^n$ in this equation is a useful convention. For the case $n = 0$ we define $\hat{G}_0 \equiv \hat{\mathbb{1}}$. We can derive an extremely useful set of coupled equations for these operator correlators using (4.53) and (4.57) with the identification

$$\hat{O}_j = \begin{cases} \hat{\psi}_H(j) & \text{for } j = 1,\ldots,n \\ \hat{\psi}^\dagger_H((2n - j + 1)') & \text{for } j = n+1,\ldots,2n \end{cases}.$$

After some relabeling we easily find

$$\mathrm{i}\frac{d}{dz_k}\hat{G}_n(1,\ldots,n;1',\ldots,n')$$
$$= \frac{1}{\mathrm{i}^n}\mathcal{T}\left\{\hat{\psi}_H(1)\ldots\left(\mathrm{i}\frac{d}{dz_k}\hat{\psi}_H(k)\right)\ldots\hat{\psi}_H(n)\hat{\psi}^\dagger_H(n')\ldots\hat{\psi}^\dagger_H(1')\right\}$$
$$+ \sum_{j=1}^{n}(\pm)^{k+j}\,\delta(k;j')\,\hat{G}_{n-1}(1,\ldots\overset{\sqcap}{k}\ldots,n;1',\ldots\overset{\sqcap}{j'}\ldots,n') \qquad (4.59)$$

and

$$-\mathrm{i}\frac{d}{dz'_k}\hat{G}_n(1,\ldots,n;1',\ldots,n')$$

$$= \frac{1}{\mathrm{i}^n}\mathcal{T}\left\{\hat{\psi}_H(1)\ldots\hat{\psi}_H(n)\hat{\psi}_H^\dagger(n')\ldots\left(-\mathrm{i}\frac{d}{dz'_k}\hat{\psi}_H^\dagger(k')\right)\ldots\hat{\psi}_H^\dagger(1')\right\}$$

$$+\sum_{j=1}^{n}(\pm)^{k+j}\,\delta(j;k')\,\hat{G}_{n-1}(1,\ldots\overset{\sqcap}{j}\ldots,n;1',\ldots\overset{\sqcap}{k'}\ldots,n'). \qquad (4.60)$$

Indeed the shift of, e.g., $\hat{\psi}_H^\dagger(j')$ just after $\hat{\psi}_H(k)$ in (4.59) requires $(n-j)+(n-k)$ interchanges and $(\pm)^{(n-j)+(n-k)} = (\pm)^{k+j}$. The terms involving the time-derivatives of the field operators can be worked out from (4.44) and (4.45). In order not to generate formulas which are too voluminous, we assume that $\hat{h}$ is diagonal in spin space so that

$$\langle\mathbf{x}_1|\hat{h}(z_1)|\mathbf{x}_2\rangle = h(1)\delta(\mathbf{x}_1 - \mathbf{x}_2) = \delta(\mathbf{x}_1 - \mathbf{x}_2)h(2), \qquad (4.61)$$

with $h(1) = h(\mathbf{r}_1, -\mathrm{i}\boldsymbol{\nabla}_1, \mathbf{S}_1, z_1)$ when it acts on quantities to its right and $h(1) = h(\mathbf{r}_1, \mathrm{i}\overset{\leftarrow}{\boldsymbol{\nabla}}_1, \mathbf{S}_1, z_1)$ when it acts on quantities to its left. The extension to arbitrary matrix elements $\langle\mathbf{x}_1|\hat{h}(z_1)|\mathbf{x}_2\rangle$ is straightforward. Using (4.61) the equations of motion (4.44) and (4.45) can be rewritten in the following compact form

$$\mathrm{i}\frac{d}{dz_k}\hat{\psi}_H(k) = h(k)\hat{\psi}_H(k) + \int d\bar{1}\,v(k,\bar{1})\hat{n}_H(\bar{1})\hat{\psi}_H(k), \qquad (4.62)$$

$$-\mathrm{i}\frac{d}{dz'_k}\hat{\psi}_H^\dagger(k') = \hat{\psi}_H^\dagger(k')h(k') + \int d\bar{1}\,v(k',\bar{1})\hat{\psi}_H^\dagger(k')\hat{n}_H(\bar{1}), \qquad (4.63)$$

where we introduce the definition

$$v(i;j) \equiv \delta(z_i, z_j)v(\mathbf{x}_i, \mathbf{x}_j, z_i).$$

The r.h.s. of the equations of motion contains three field operators with the same time argument. For instance in (4.62) we have

$$\hat{n}_H(\bar{1})\hat{\psi}_H(k) = \hat{\psi}_H^\dagger(\bar{1})\hat{\psi}_H(\bar{1})\hat{\psi}_H(k) = \pm\hat{\psi}_H^\dagger(\bar{1})\hat{\psi}_H(k)\hat{\psi}_H(\bar{1}),$$

where we use the fact that field operators in the Heisenberg picture at equal times satisfy the same (anti)commutation relations as the field operators. Inserting this composite operator inside the contour ordering we would like to move the field operator $\hat{\psi}_H^\dagger(\bar{1})$ to the right in order to form a $\hat{G}_n$. To make sure that after the reordering $\hat{\psi}_H^\dagger(\bar{1})$ ends up to the left of $\hat{\psi}_H(\bar{1})$ and $\hat{\psi}_H(k)$ we calculate it in $\bar{1}^+ = \bar{\mathbf{x}}_1, \bar{z}_1^+$, where $\bar{z}_1^+$ is infinitesimally later than $\bar{z}_1$. Inside the $\mathcal{T}$ sign we can then write

$$\mathcal{T}\left\{\ldots\hat{n}_H(\bar{1})\hat{\psi}_H(k)\ldots\right\} = \pm\mathcal{T}\left\{\ldots\hat{\psi}_H(k)\hat{\psi}_H(\bar{1})\hat{\psi}_H^\dagger(\bar{1}^+)\ldots\right\},$$

As $\hat{\psi}_H(\bar{1})\hat{\psi}_H^\dagger(\bar{1}^+)$ is composed of an even number of field operators it behaves like a bosonic operator inside the contour ordering and can therefore be placed wherever we like without caring about the sign. With this trick the first term on the r.h.s. of (4.59) can be written as

$$\frac{1}{i^n} \mathcal{T} \left\{ \hat{\psi}_H(1) \dots \left(i \frac{d}{dz_k} \hat{\psi}_H(k) \right) \dots \hat{\psi}_H(n) \hat{\psi}_H^\dagger(n') \dots \hat{\psi}_H^\dagger(1') \right\}$$

$$= h(k) \hat{G}_n(1, \dots, n; 1', \dots, n')$$

$$\pm \frac{1}{i^n} \int d\bar{1}\, v(k; \bar{1})\, \mathcal{T} \left\{ \hat{\psi}_H(1) \dots \hat{\psi}_H(n) \hat{\psi}_H(\bar{1}) \hat{\psi}_H^\dagger(\bar{1}^+) \hat{\psi}_H^\dagger(n') \dots \hat{\psi}_H^\dagger(1') \right\}$$

$$= h(k) \hat{G}_n(1, \dots, n; 1', \dots, n') \pm i \int d\bar{1}\, v(k; \bar{1})\, \hat{G}_{n+1}(1, \dots, n, \bar{1}; 1', \dots, n', \bar{1}^+).$$

A similar equation can be derived in a completely analogous way for the contour-ordered product in (4.60). This time we write

$$\mathcal{T} \left\{ \dots \hat{\psi}_H^\dagger(k') \hat{n}_H(\bar{1}) \dots \right\} = \pm \mathcal{T} \left\{ \dots \hat{\psi}_H(\bar{1}^-) \hat{\psi}_H^\dagger(\bar{1}) \hat{\psi}_H^\dagger(k') \dots \right\}$$

where the time coordinate $\bar{z}_1^-$ is infinitesimally earlier than $\bar{z}_1$. We again use that $\hat{\psi}_H(\bar{1}^-)\hat{\psi}_H^\dagger(\bar{1})$ is a bosonic operator, and hence it can be placed between $\hat{\psi}_H(n)$ and $\hat{\psi}_H^\dagger(n')$ without caring about the sign. Inserting these results into the equations of motion for the $\hat{G}_n$ we find

$$\left[i \frac{d}{dz_k} - h(k) \right] \hat{G}_n(1, \dots, n; 1', \dots, n')$$

$$= \pm i \int d\bar{1}\, v(k; \bar{1})\, \hat{G}_{n+1}(1, \dots, n, \bar{1}; 1', \dots, n', \bar{1}^+) \tag{4.64}$$

$$+ \sum_{j=1}^n (\pm)^{k+j} \delta(k; j')\, \hat{G}_{n-1}(1, \dots \overset{\sqcap}{k} \dots, n; 1', \dots \overset{\sqcap}{j'} \dots, n')$$

and

$$\hat{G}_n(1, \dots, n; 1', \dots, n') \left[-i \frac{\overleftarrow{d}}{dz_k'} - h(k') \right]$$

$$= \pm i \int d\bar{1}\, v(k'; \bar{1})\, \hat{G}_{n+1}(1, \dots, n, \bar{1}^-; 1', \dots, n', \bar{1}) \tag{4.65}$$

$$+ \sum_{j=1}^n (\pm)^{k+j} \delta(j; k')\, \hat{G}_{n-1}(1, \dots \overset{\sqcap}{j} \dots, n; 1', \dots \overset{\sqcap}{k'} \dots, n')$$

where in the last equation the arrow over d/dz_k' specifies that the derivative acts on the quantity to its left. We have thus derived an infinite hierarchy of operator equations in Fock space in which the derivative of $\hat{G}_n$ is expressed in terms of $\hat{G}_{n-1}$ and $\hat{G}_{n+1}$. These equations are very general as we made no assumptions on the particular shape of the contour, which can be any of the contours encountered in the previous sections. We only used an identity for the contour derivative of a string of contour ordered operators and the equations of motion for $\hat{\psi}_H$ and $\hat{\psi}_H^\dagger$. The equations (4.64) and (4.65) play a pivotal role in the development of the theory. As we explain in the subsequent chapters the whole structure of diagrammatic perturbation theory is encoded in them.

Exercise 4.5. Let $\hat{O}_j \equiv \hat{O}_j(z_j)$ be a set of composite operators consisting of an even or odd number of fermion field operators. Show that a bosonic operator (even number) always commutes within the contour-ordered product and that a fermionic operator (odd number) anti-commutes with another fermionic operator. Let further $z_1 > z_2 > \ldots > z_n$. Show that

$$\mathcal{T}\left\{\hat{O}_{P(1)} \ldots \hat{O}_{P(n)}\right\} = (-1)^F \hat{O}_1 \ldots \hat{O}_n,$$

where P is a permutation of the labels and where F is the number of interchanges of fermion operators in the permutation P that puts all the operators in the right order.

Exercise 4.6. Write down (4.64) and (4.65) in the case that the one-particle Hamiltonian $\langle \mathbf{x}|\hat{h}(z)|\mathbf{x}'\rangle$ is not diagonal in spin space, and even more generally in the case that $\langle \mathbf{x}|\hat{h}(z)|\mathbf{x}'\rangle$ is not diagonal in position space.

Exercise 4.7. How do (4.64) and (4.65) change if the interaction Hamiltonian contains a three-body operator as in (4.46)?

Exercise 4.8. Consider the Hamiltonian on the contour

$$\hat{H}(z) = \sum_{i=1}^N \left(\omega_i \hat{d}_i^\dagger \hat{d}_i + f_i(z)\hat{x}_i\right),$$

where $\hat{x}_i = \frac{1}{\sqrt{2}}(\hat{d}_i^\dagger + \hat{d}_i)$ and the $\hat{d}$-operators satisfy the bosonic commutation relations $[\hat{d}_i, \hat{d}_j^\dagger]_- = \delta_{ij}$ and $[\hat{d}_i, \hat{d}_j]_- = [\hat{d}_i^\dagger, \hat{d}_j^\dagger]_- = 0$. Show that the operator correlator

$$\mathcal{T}\{\hat{x}_{1,H}(z_1) \ldots \hat{x}_{n,H}(z_n)\}$$

satisfies the equation of motion

$$\left[\frac{d^2}{dz_k^2} + \omega_k^2\right] \mathcal{T}\{\hat{x}_{1,H}(z_1) \ldots \hat{x}_{n,H}(z_n)\}$$

$$= -\omega_k f_k(z)\mathcal{T}\left\{\hat{x}_{1,H}(z_1) \ldots \overset{\sqcap}{\hat{x}}_{k,H}(z_k) \ldots \hat{x}_{n,H}(z_n)\right\}$$

$$- i\omega_k \sum_{j \neq k}^n \delta(j;k)\, \mathcal{T}\left\{\hat{x}_{1,H}(z_1) \ldots \overset{\sqcap}{\hat{x}}_{k,H}(z_k) \ldots \overset{\sqcap}{\hat{x}}_{j,H}(z_j) \ldots \hat{x}_{n,H}(z_n)\right\},$$

where $\delta(j;k) = \delta_{jk}\delta(z_j, z_k)$. Why do we need a second order equation to obtain a closed set of equations for these types of operator correlator?

5

Many-particle Green's functions

5.1 Martin–Schwinger hierarchy

In the previous chapter we derived the differential equations (4.64) and (4.65) for the operator correlators $\hat{G}_n$ which form the building blocks to construct any other operator correlator. This set of operator equations can be turned into a coupled set of differential equations by taking the average with $\hat{\rho} = e^{-\beta \hat{H}^{\mathrm{M}}}/Z$. If we discuss a system that is initially in thermodynamic equilibrium then $\hat{\rho}$ is the grand canonical density matrix. More generally we define the *n-particle Green's function* G_n to be

$$
\begin{aligned}
G_n(1,\ldots,n;1',\ldots,n') &\equiv \frac{\mathrm{Tr}\left[e^{-\beta\hat{H}^{\mathrm{M}}}\hat{G}_n(1,\ldots,n;1',\ldots,n')\right]}{\mathrm{Tr}\left[e^{-\beta\hat{H}^{\mathrm{M}}}\right]} \\
&= \frac{1}{\mathrm{i}^n}\frac{\mathrm{Tr}\left[\mathcal{T}\left\{e^{-\mathrm{i}\int_\gamma d\bar{z}\hat{H}(\bar{z})}\hat{\psi}(1)\ldots\hat{\psi}(n)\hat{\psi}^\dagger(n')\ldots\hat{\psi}^\dagger(1')\right\}\right]}{\mathrm{Tr}\left[\mathcal{T}\left\{e^{-\mathrm{i}\int_\gamma d\bar{z}\hat{H}(\bar{z})}\right\}\right]}.
\end{aligned}
\tag{5.1}
$$

In the second equality we use the definition of operators in the contour Heisenberg picture. Consider, for instance, G_1 with $z_1 < z_1'$. Then

$$
\begin{aligned}
e^{-\beta\hat{H}^{\mathrm{M}}}\mathcal{T}\left\{\hat{\psi}_H(1)\hat{\psi}_H^\dagger(1')\right\} &= \pm\hat{U}(z_{\mathrm{f}},z_{\mathrm{i}})\hat{U}(z_{\mathrm{i}},z_1')\hat{\psi}^\dagger(1')\hat{U}(z_1',z_1)\hat{\psi}(1)\hat{U}(z_1,z_{\mathrm{i}}) \\
&= \pm\mathcal{T}\left\{e^{-\mathrm{i}\int_\gamma d\bar{z}\hat{H}(\bar{z})}\hat{\psi}^\dagger(1')\hat{\psi}(1)\right\} \\
&= \mathcal{T}\left\{e^{-\mathrm{i}\int_\gamma d\bar{z}\hat{H}(\bar{z})}\hat{\psi}(1)\hat{\psi}^\dagger(1')\right\},
\end{aligned}
$$

where we take into account that $e^{-\beta\hat{H}^{\mathrm{M}}} = \hat{U}(z_{\mathrm{f}},z_{\mathrm{i}})$ and that under the $\mathcal{T}$ sign the field operators (anti)commute. Similar considerations apply to G_n. The Green's function G_1 at times $z_1 = z$ and $z_1' = z^+$ infinitesimally later than z_1 is the time-dependent ensemble average of the operator $\hat{\psi}^\dagger(\mathbf{x}_1')\hat{\psi}(\mathbf{x}_1)$ (modulo a factor of i) from which we can calculate the time-dependent ensemble average of *any* one-body operator. More generally, by choosing the contour arguments of G_n to be $z_i = z$ and $z'_i = z^+$ for all i we obtain the time-dependent ensemble average of the operator $\hat{\psi}^\dagger(\mathbf{x}'_n)...\hat{\psi}^\dagger(\mathbf{x}'_1)\hat{\psi}(\mathbf{x}_1)...\hat{\psi}(\mathbf{x}_n)$ (modulo a factor i^n) from which we can calculate the time-dependent ensemble average of *any* n-body operator. At the initial time $z = t_0$ these averages are the generalization to ensembles of the n-particle density matrices introduced in Section 1.7.

It is worth observing that under the adiabatic assumption, see Section 4.3, we can calculate the Green's functions G_n from the second line of (5.1) in which $\gamma \to \gamma_a$ is the adiabatic contour of Fig. 4.5(b) and the "adiabatic Hamiltonian" changes along the contour as illustrated in the same figure. This is the same as saying that in the first line of (5.1) we replace $\hat{H}^M \to \hat{H}_0^M$, take the arguments of $\hat{G}_n$ on γ_a and the field operators in the Heisenberg picture with respect to the adiabatic Hamiltonian. These adiabatic Green's functions coincide with the exact ones when all contour arguments are larger than t_0. The proof of this statement is the same as the proof of the adiabatic formula (4.27). If we further have a system in equilibrium at zero temperature then we can use the zero-temperature assumption. The Green's functions can then be calculated from the second line of (5.1) in which $\gamma \to \gamma_0$ is the zero-temperature contour of Fig. 4.5(c) and the "zero-temperature Hamiltonian" changes along the contour as illustrated in the same figure. Equivalently, the zero-temperature Green's functions can be calculated from the first line of (5.1) in which $\hat{H}^M \to \hat{H}_0^M$, the arguments of $\hat{G}_n$ are on the contour γ_0 and the field operators are in the Heisenberg picture with respect to the zero-temperature Hamiltonian.[1] These zero-temperature Green's functions coincide with the exact ones for contour arguments on the forward branch, as can readily be verified by following the same logic as Section 4.3. Using the (equal-time) adiabatic or zero-temperature Green's functions corresponds to using the adiabatic or zero-temperature formula for the time-dependent averages. The reader is strongly encouraged to pause a while and get convinced of this fact. A useful exercise is, e.g., to write the explicit expression of $G_1(\mathbf{x}, z; \mathbf{x}', z^+)$ from both the first and second line of (5.1) with γ one of the three contours of Fig. 4.5 (and, of course, with the Hamiltonian $\hat{H}(z)$ as illustrated in these contours) and to show that they are the same under the corresponding assumptions. The special appealing feature of writing the Green's functions as in the second line of (5.1) is that this formula nicely embodies all possible cases (exact, adiabatic and zero-temperature ones), once we assign the form of γ and the way the Hamiltonian changes along the contour.

Let us now come back to the differential equations (4.64) and (4.65). They have been derived without any assumption on the shape of the contour and without any assumption on the time-dependence of $\hat{h}(z)$ and $v(\mathbf{x}, \mathbf{x}', z)$ along the contour. Therefore, no matter how we define the Green's functions G_n (to be the exact, the adiabatic, or the zero-temperature) they satisfy the same kind of differential equations. To show it we simply multiply (4.64) and (4.65) by the appropriate density matrix $\hat{\rho}$ (equal to $e^{-\beta \hat{H}^M}/Z$ in the exact case and equal to $e^{-\beta \hat{H}_0^M}/Z_0$ in the adiabatic and zero-temperature case), take the trace and use the

---

[1] We recall that under the zero-temperature assumption $\hat{U}_\eta(-\infty, \infty)|\Phi_0\rangle = e^{-i\alpha_0}|\Phi_0\rangle$, where $|\Phi_0\rangle$ is the ground state of $\hat{H}_0^M$.

definition (5.1) to obtain the following system of coupled differential equations:

$$\left[i\frac{d}{dz_k} - h(k)\right] G_n(1,\ldots,n;1',\ldots,n')$$

$$= \pm i \int d\bar{1}\, v(k;\bar{1})\, G_{n+1}(1,\ldots,n,\bar{1};1',\ldots,n',\bar{1}^+)$$

$$+ \sum_{j=1}^{n} (\pm)^{k+j}\, \delta(k;j')\, G_{n-1}(1,\ldots\overset{\sqcap}{k}\ldots,n;1',\ldots\overset{\sqcap}{j'}\ldots,n')$$

(5.2)

$$G_n(1,\ldots,n;1',\ldots,n')\left[-i\frac{\overleftarrow{d}}{dz'_k} - h(k')\right]$$

$$= \pm i \int d\bar{1}\, v(k';\bar{1})\, G_{n+1}(1,\ldots,n,\bar{1}^-;1',\ldots,n',\bar{1})$$

$$+ \sum_{j=1}^{n} (\pm)^{k+j}\, \delta(j;k')\, G_{n-1}(1,\ldots\overset{\sqcap}{j}\ldots,n;1',\ldots\overset{\sqcap}{k'}\ldots,n')$$

(5.3)

This set of equations is known as the *Martin–Schwinger hierarchy* [47] for the Green's functions. If we could solve it we would be able to calculate any time-dependent ensemble average of any observable that we want. For example, the average of the density operator and paramagnetic current density operator are given in terms of the *one-particle Green's function* G_1 as

$$n(\mathbf{x},z) = \frac{\mathrm{Tr}\left[e^{-\beta\hat{H}^{\mathrm{M}}}\hat{\psi}_H^\dagger(\mathbf{x},z)\hat{\psi}_H(\mathbf{x},z)\right]}{\mathrm{Tr}\left[e^{-\beta\hat{H}^{\mathrm{M}}}\right]} = \pm i G_1(\mathbf{x},z;\mathbf{x},z^+)$$

(5.4)

and [see (3.26)]

$$\mathbf{j}(\mathbf{x},z) = \frac{1}{2mi} \frac{\mathrm{Tr}\left[e^{-\beta\hat{H}^{\mathrm{M}}}\left(\hat{\psi}_H^\dagger(\mathbf{x},z)(\boldsymbol{\nabla}\hat{\psi}_H(\mathbf{x},z)) - (\boldsymbol{\nabla}\hat{\psi}_H^\dagger(\mathbf{x},z))\hat{\psi}_H(\mathbf{x},z)\right)\right]}{\mathrm{Tr}\left[e^{-\beta\hat{H}^{\mathrm{M}}}\right]}$$

$$= \pm\left(\frac{\boldsymbol{\nabla}-\boldsymbol{\nabla}'}{2m} G(\mathbf{x},z;\mathbf{x}',z^+)\right)_{\mathbf{x}'=\mathbf{x}},$$

(5.5)

while, for instance, the interaction energy is given in terms of the *two-particle Green's function* as[2]

$$E_{\mathrm{int}}(z) = \frac{1}{2}\int d\mathbf{x}d\mathbf{x}'\, v(\mathbf{x},\mathbf{x}';z)\frac{\mathrm{Tr}\left[e^{-\beta\hat{H}^{\mathrm{M}}}\hat{\psi}_H^\dagger(\mathbf{x},z)\hat{\psi}_H^\dagger(\mathbf{x}',z)\hat{\psi}_H(\mathbf{x}',z)\hat{\psi}_H(\mathbf{x},z)\right]}{\mathrm{Tr}\left[e^{-\beta\hat{H}^{\mathrm{M}}}\right]}$$

$$= -\frac{1}{2}\int d\mathbf{x}d\mathbf{x}'\, v(\mathbf{x},\mathbf{x}';z)G_2(\mathbf{x}',z,\mathbf{x},z;\mathbf{x}',z^+,\mathbf{x},z^+).$$

---

[2]Recall that under the $\mathcal{T}$ sign the order of operators with the same contour argument is preserved, see also examples in Section 4.5.

Choosing z on the vertical track we have the initial value of these observable quantities while for $z = t_\pm$ on the horizontal branches we have their ensemble average at time t. Moreover, according to the previous discussion, we can state that:

> *The exact, adiabatic, and zero-temperature formula corresponds to using the G_n which solve the Martin–Schwinger hierarchy on the contours of Fig. 4.5 with a one-particle Hamiltonian h and interaction v as specified in the same figure.*

In all cases the task is therefore to solve the hierarchy. As with any set of differential equations their solution is unique provided we pose appropriate spatial and temporal boundary conditions. These boundary conditions obviously depend on the physical problem at hand. The spatial boundary conditions on the Green's functions are determined directly by the corresponding spatial boundary conditions on the many-body states of the system. For instance, if we describe a finite system such as an isolated molecule we know that the many-body wave-functions vanish at spatial infinity and hence we require the same for the Green's functions. For an infinite periodic crystal the many-body wave-functions are invariant (modulo a phase factor) under translations over a lattice vector. For this case we thus demand that the Green's functions obey the same lattice-periodic symmetry in each of their spatial coordinates since the physics of adding or removing a particle cannot depend on the particular unit cell where we add or remove the particle, see Appendix E. Since the differential equations are first order in the time-derivatives we also need one condition per time-argument for each Green's function G_n. From the definition (5.1) it follows that (we derive these relations below)

$$
\begin{aligned}
G_n(1,\ldots,\mathbf{x}_k,z_\mathrm{i},\ldots,n;1',\ldots,n') &= \pm G_n(1,\ldots,\mathbf{x}_k,z_\mathrm{f},\ldots,n;1',\ldots,n') \\
G_n(1,\ldots,n;1',\ldots,\mathbf{x}'_k,z_\mathrm{i},\ldots,n') &= \pm G_n(1,\ldots,n;1',\ldots,\mathbf{x}'_k,z_\mathrm{f},\ldots,n')
\end{aligned}
\tag{5.6}
$$

with sign $+/-$ for the bosonic/fermionic case. The Green's functions are therefore (anti)periodic along the contour γ. The boundary conditions (5.6) are known as the *Kubo–Martin–Schwinger* (KMS) *relations* [47,48]. To derive these relations we consider only the numerator of (5.1). Let us insert the value z_f in the kth contour argument. Since z_f is the latest time on the contour, we have, using the cyclic property of the trace,[3] the set of identities illustrated in Fig. 5.1. In the figure the arrows indicate how the operator $\hat\psi(\mathbf{x}_k)$ moves from one step to the next. The final result is that we have replaced the contour-argument z_f by z_i and gained a sign $(\pm)^{k-1}(\pm)^{2n-k} = \pm 1$. One can similarly derive the second KMS relation for a time-argument z'_k. As we see in Section 5.3, these boundary conditions are sufficient to determine a unique solution for the differential equations (5.2) and (5.3) provided that *the boundary points belong to the same connected contour on which the differential equations are solved*. It is indeed very important that the Martin–Schwinger hierarchy is valid on the vertical track and that the vertical track is connected to the horizontal branches. If the vertical track were not attached then there would be a unique solution on the vertical track,

---

[3]Since in the bosonic case $[\psi(\mathbf{x}),\psi^\dagger(\mathbf{x}')]_- = \delta(\mathbf{x}-\mathbf{x}')$ it seems dubious to use the property $\mathrm{Tr}[\hat A\hat B] = \mathrm{Tr}[\hat B\hat A]$ for field operators. However, under the trace the field operators are always multiplied by $e^{-\beta\hat H^\mathrm{M}}$ and the use of the cyclic property is allowed. For a mathematical discussion see Ref. [49].

$$\text{Tr}\left[\mathcal{T}\left\{e^{-\mathrm{i}\int_\gamma d\bar{z}\hat{H}(\bar{z})}\hat{\psi}(1)\dots\hat{\psi}(k-1)\hat{\psi}(\mathbf{x}_k,z_{\mathrm{f}})\hat{\psi}(k+1)\dots\hat{\psi}^\dagger(1')\right\}\right]$$

$$=(\pm)^{k-1}\text{Tr}\left[\hat{\psi}(\mathbf{x}_k)\mathcal{T}\left\{e^{-\mathrm{i}\int_\gamma d\bar{z}\hat{H}(\bar{z})}\hat{\psi}(1)\dots\hat{\psi}(k-1)\hat{\psi}(k+1)\dots\hat{\psi}^\dagger(1')\right\}\right]$$

$$=(\pm)^{k-1}\text{Tr}\left[\mathcal{T}\left\{e^{-\mathrm{i}\int_\gamma d\bar{z}\hat{H}(\bar{z})}\hat{\psi}(1)\dots\hat{\psi}(k-1)\hat{\psi}(k+1)\dots\hat{\psi}^\dagger(1')\right\}\hat{\psi}(\mathbf{x}_k)\right]$$

$$=(\pm)^{k-1}\text{Tr}\left[\mathcal{T}\left\{e^{-\mathrm{i}\int_\gamma d\bar{z}\hat{H}(\bar{z})}\hat{\psi}(1)\dots\hat{\psi}(k-1)\hat{\psi}(k+1)\dots\hat{\psi}^\dagger(1')\hat{\psi}(\mathbf{x}_k,z_{\mathrm{i}})\right\}\right]$$

$$=(\pm)^{k-1}(\pm)^{2n-k}\text{Tr}\left[\mathcal{T}\left\{e^{-\mathrm{i}\int_\gamma d\bar{z}\hat{H}(\bar{z})}\hat{\psi}(1)\dots\hat{\psi}(k-1)\hat{\psi}(\mathbf{x}_k,z_{\mathrm{i}})\hat{\psi}(k+1)\dots\hat{\psi}^\dagger(1')\right\}\right]$$

Figure 5.1 Proof of the KMS relations.

but there would be no unique solution on the horizontal branches, see Exercise 5.1. This provides a posteriori yet another and more fundamental reason to attach the vertical track.

In the next section we show a few examples on how to calculate some approximate Green's functions using the hierarchy equations together with the boundary conditions. Already at this early stage it is possible to introduce several quantities and concepts which are used and further discussed in the following chapters. Following the derivations with pencil and paper will facilitate understanding of Section 5.3 where *we give an exact formal solution of the hierarchy for the one- and two-particle Green's function as an infinite expansion in powers of the interparticle interaction v.*

Exercise 5.1. In order to illustrate the importance of the KMS boundary conditions and of connected domains when solving the Martin–Schwinger hierarchy consider the differential equation

$$\mathrm{i}\frac{df(z)}{dz}-\epsilon f(z)=g(z)$$

valid for $z\in(0,-\mathrm{i}\beta)$ and $z\in(1,\infty)$ and with boundary condition $f(0)=\pm f(-\mathrm{i}\beta)$. This equation has the same structure as the Martin–Schwinger hierarchy but the real domain $(1,\infty)$ is disconnected from the imaginary domain $(0,-\mathrm{i}\beta)$. Show that for any complex ϵ with a nonvanishing real part the solution is unique in $(0,-\mathrm{i}\beta)$ but not in $(1,\infty)$. Are there imaginary values of ϵ for which the solution is not unique in $(0,-\mathrm{i}\beta)$?

5.2 Truncation of the hierarchy

The Martin–Schwinger hierarchy together with the KMS boundary conditions completely define the many-body problem. It therefore remains to find ways to solve the hierarchy. What we want to show here is that we can already get some useful approximations for the one- and two-particle Green's functions by making a few physically inspired approximations in the hierarchy equations. This leads to new equations for the Green's functions that depend on the Green's functions themselves in a nonlinear way. The solutions are therefore nonperturbative in the interaction strength v. Let us start by writing down the lowest order equations of the hierarchy. For the one-particle Green's function (or simply the Green's function), which we denote by $G(1;2) \equiv G_1(1;2)$, we have the two equations

$$\left[i\frac{d}{dz_1} - h(1) \right] G(1;1') = \delta(1;1') \pm i \int d2\, v(1;2) G_2(1,2;1',2^+), \qquad (5.7)$$

$$G(1;1') \left[-i\frac{\overleftarrow{d}}{dz_1'} - h(1') \right] = \delta(1;1') \pm i \int d2\, v(1';2) G_2(1,2^-;1',2). \qquad (5.8)$$

In the following we often refer to this couple of equations as the *equations of motion* for the Green's function. Similarly for the two-particle Green's function we have the four equations of motion:

$$\left[i\frac{d}{dz_1} - h(1) \right] G_2(1,2;1',2') = \delta(1;1')G(2;2') \pm \delta(1;2')G(2;1')$$

$$\pm i \int d3\, v(1;3) G_3(1,2,3;1',2',3^+), \qquad (5.9)$$

$$\left[i\frac{d}{dz_2} - h(2) \right] G_2(1,2;1',2') = \pm\delta(2;1')G(1;2') + \delta(2;2')G(1;1')$$

$$\pm i \int d3\, v(2;3) G_3(1,2,3;1',2',3^+), \qquad (5.10)$$

$$G_2(1,2;1',2') \left[-i\frac{\overleftarrow{d}}{dz_1'} - h(1') \right] = \delta(1;1')G(2;2') \pm \delta(2;1')G(1;2')$$

$$\pm i \int d3\, v(1';3) G_3(1,2,3^-;1',2',3), \qquad (5.11)$$

$$G_2(1,2;1',2') \left[-i\frac{\overleftarrow{d}}{dz_2'} - h(2') \right] = \pm\delta(1;2')G(2;1') + \delta(2;2')G(1;1')$$

$$\pm i \int d3\, v(2';3) G_3(1,2,3^-;1',2',3). \qquad (5.12)$$

The presence of the δ-functions on the right hand side of the equations of motion suggests the following decomposition of the two-particle Green's function:

$$G_2(1,2;1',2') = G(1;1')G(2;2') \pm G(1;2')G(2;1') + \Upsilon(1,2;1',2'), \qquad (5.13)$$

which implicitly defines the so called *correlation function* Υ. It is easy to see that if the interaction $v = 0$ then the above G_2 with $\Upsilon = 0$ satisfies the equations of motion. Consider for instance (5.9); inserting G_2 with $\Upsilon = 0$ and using the equations of motion for G with $v = 0$, the l.h.s. becomes

$$\left[i\frac{d}{dz_1} - h(1) \right] [G(1;1')G(2;2') \pm G(1;2')G(2;1')] = \delta(1;1')G(2;2') \pm \delta(1;2')G(2;1'),$$

which is the same as the r.h.s. of (5.9) for $v = 0$. Furthermore this G_2 satisfies the KMS boundary conditions whenever G does and therefore it is the exact noninteracting G_2 if we use the exact noninteracting G. This suggests that for weak interactions $\Upsilon \approx 0$. The corresponding approximation

$$G_2(1,2;1',2') = G(1;1')G(2;2') \pm G(1;2')G(2;1') \tag{5.14}$$

is the so-called *Hartree–Fock approximation* to G_2. This expression can be given a nice physical interpretation if we choose a contour ordering. For instance, if $z_1, z_2 > z_1', z_2'$ (here the inequality symbol ">" signifies "later than") the two-particle Green's function is an object that describes the time-evolution of a many-body state to which we add two particles at times z_1' and z_2' and remove them at times z_1 and z_2. For this reason G_2 is also sometimes called the two-particle propagator. Equation (5.14) then tells us that, if the interaction between the particles is weak, the two-particle propagator can be approximately written as the product of functions that describe how a single particle propagates when added to the system. Due to the indistinguishability of the particles we do not know which of the particles that we added at space–spin–time points 1 and 2 end up at points $1'$ and $2'$ and we therefore need to (anti)symmetrize the product for bosons/fermions. This explains the form of (5.14). We come back to the Hartree–Fock approximation for G_2 in Chapter 7. We can now insert the Hartree–Fock G_2 into the equations of motion (5.7) and (5.8) for G and obtain

$$\left[i\frac{d}{dz_1} - h(1) \right] G(1;1') = \delta(1;1') \pm i\int d2\, v(1;2)[G(1;1')G(2;2^+) \pm G(1;2^+)G(2;1')]$$

$$= \delta(1;1') + \int d2\, \Sigma(1;2)G(2;1'), \tag{5.15}$$

$$G(1;1')\left[-i\frac{\overleftarrow{d}}{dz_1'} - h(1') \right] = \delta(1;1') \pm i\int d2\, v(1';2)[G(1;1')G(2^-;2) \pm G(1;2)G(2^-;1')]$$

$$= \delta(1;1') + \int d2\, G(1;2)\Sigma(2;1'), \tag{5.16}$$

where the second equalities implicitly define the integral kernel Σ. The reader can easily verify that

$$\Sigma(1;2) = \delta(1;2)\, V_H(1) + iv(1;2)\, G(1;2^+)$$

with

$$V_H(1) = \pm i\int d3\, v(1;3)G(3;3^+) = \int d\mathbf{x}_3\, v(\mathbf{x}_1, \mathbf{x}_3, z_1)n(\mathbf{x}_3, z_1)$$

by inserting Σ back in the equations of motion. In the equation above we take into account that $v(1;3) = \delta(z_1, z_3)v(\mathbf{x}_1, \mathbf{x}_3, z_1)$ and that $\pm iG(\mathbf{x}_3, z_1; \mathbf{x}_3, z_1^+) = n(\mathbf{x}_3, z_1)$ is the density at space–spin–time point 3. The quantity $V_{\rm H}$ is called the *Hartree potential* and it should be an old acquaintance of the reader. The Hartree potential $V_{\rm H}(\mathbf{x}_1, z_1)$ has a clear physical interpretation as the classical potential that a particle in $\mathbf{x}_1$ experiences from a density distribution n of all the particles in the system (for Coulombic interactions $v(\mathbf{x}_1, \mathbf{x}_3, z_1) = 1/|\mathbf{r}_1 - \mathbf{r}_3|$ the Hartree potential is the potential of classical electrostatics). The Hartree potential is the first term of the integral kernel Σ which is known as the *self-energy*. The second term in Σ is often called the *Fock* or *exchange potential*. This quantity is local in time (due to the δ-function in the definition of $v(1;2)$) but nonlocal in space and therefore cannot be interpreted as a classical potential.

The equations (5.15) and (5.16) need to be solved *self-consistently* as the self-energy depends on the Green's function itself (nonlinear equations). The resulting approximated Green's function is therefore *nonperturbative* in v. In fact, we can write the solution in integral form if we use the KMS boundary conditions. To do this we define the noninteracting Green's function G_0 as the solution of the equations of motion with $v = 0$,

$$\left[\mathrm{i}\frac{d}{dz_1} - h(1) \right] G_0(1;1') = \delta(1;1'), \tag{5.17}$$

$$G_0(1;1') \left[-\mathrm{i}\frac{\overleftarrow{d}}{dz'_1} - h(1') \right] = \delta(1;1'). \tag{5.18}$$

As usual we need to solve these equations together with the KMS boundary conditions. We then have

$$\int d1\, G_0(2;1) \left[\mathrm{i}\frac{d}{dz_1} - h(1) \right] G(1;1')$$

$$= \int d1\, G_0(2;1) \left[-\mathrm{i}\frac{\overleftarrow{d}}{dz_1} - h(1) \right] G(1;1') + \mathrm{i} \int d\mathbf{x}_1 G_0(2;\mathbf{x}_1, z_1) G(\mathbf{x}_1, z_1; 1') \Big|_{z_1=z_i}^{z_1=z_f}$$

$$= \int d1\, \delta(2;1) G(1;1')$$

$$= G(2;1'), \tag{5.19}$$

where in the partial integration we use the fact that both G_0 and G satisfy the KMS relations so that the boundary term vanishes. In a similar manner one can show that

$$\int d1'\, G(1;1') \left[-\mathrm{i}\frac{\overleftarrow{d}}{dz'_1} - h(1') \right] G_0(1';2) = G(1;2). \tag{5.20}$$

If we now multiply (5.15) from the right and (5.16) from the left with G_0 and use the above identities we obtain the following two equivalent equations for G

$$G(1;2) = G_0(1;2) + \int d3d4\, G_0(1;3)\Sigma(3;4)G(4;2),$$

$$G(1;2) = G_0(1;2) + \int d3d4\, G(1;3)\Sigma(3;4)G_0(4;2).$$

These equations are known as *Dyson equations*. The difference between the Dyson equations and the integro-differential equations (5.15) and (5.16) is that in the Dyson equations the boundary conditions are incorporated through G_0. As we see later in the book, the Dyson equation is not only valid for the approximate case that we derived here but also the exact Green's function satisfies an equation of this form in which Σ depends in a much more complicated but well-defined way on the Green's function G. The Dyson equation is therefore the formal solution of the Martin–Schwinger hierarchy for the one-particle Green's function.

Let us now see how to go beyond the Hartree–Fock approximation (5.14) for the two-particle Green's function. For this purpose we necessarily have to consider the correlation function Υ. From the equations of motion for the one- and two-particle Green's functions (5.7) and (5.9) we find that if we act on (5.13) with $[i\frac{d}{dz_1} - h(1)]$ we get

$$\left[i\frac{d}{dz_1} - h(1)\right] \Upsilon(1,2;1',2') = \pm i \int d3\, v(1;3) \left[G_3(1,2,3;1',2',3^+)\right.$$
$$\left. - G_2(1,3;1',3^+)G(2;2') \mp G_2(1,3;2',3^+)G(2;1')\right]. \quad (5.21)$$

Now we use from the Martin–Schwinger hierarchy the following equation of motion for the three-particle Green's function:

$$\left[i\frac{d}{dz_2} - h(2)\right] G_3(1,2,3;1',2',3') = \pm\delta(2;1')G_2(1,3;2,3') + \delta(2;2')G_2(1,3;1',3')$$
$$\pm \delta(2;3')G_2(1,3;1',2') \pm i \int d4\, v(2;4)\, G_4(1,2,3,4;1',2',3',4^+).$$

Acting on (5.21) with $[i\frac{d}{dz_2} - h(2)]$ and further taking into account the equations of motion for G we find

$$\left[i\frac{d}{dz_1} - h(1)\right]\left[i\frac{d}{dz_2} - h(2)\right] \Upsilon(1,2;1',2') = iv(1;2)G_2(1,2;1',2')$$
$$- \int d4\, v(1;3)v(2;4)\left[G_4(1,2,3,4;1',2',3^+,4^+) - G_2(1,3;1',3^+)G_2(2,4;2',4^+)\right.$$
$$\left. \mp G_2(1,3;2',3^+)G_2(2,4;1',4^+)\right].$$

Since the last integral is at least one order higher in the interaction than the term directly after the equal sign we neglect it. Our first approximation beyond Hartree–Fock is thus given by

$$\left[i\frac{d}{dz_1} - h(1)\right]\left[i\frac{d}{dz_2} - h(2)\right] \Upsilon(1,2;1',2') = iv(1;2)G_2(1,2;1',2').$$

By definition Υ satisfies the KMS boundary conditions and therefore we can solve this equation doing the same integration as in (5.19). We then obtain the following explicit expression

$$\Upsilon(1,2;1',2') = i \int d3d4\, G_0(1;3)G_0(2;4)v(3;4)G_2(3,4;1',2').$$

This result together with (5.13) gives us an approximate (integral) equation for the two-particle Green's function beyond Hartree–Fock:

$$G_2(1,2;1',2') = G(1;1')G(2;2') \pm G(1;2')G(2;1')$$
$$+ i \int d3d4 \, G_0(1;3)G_0(2;4)v(3;4)G_2(3,4;1',2'). \qquad (5.22)$$

To solve this equation we note that if we define the function S satisfying

$$S(1,2;3,4) = \delta(1;3)\delta(2;4) + i \int d5d6 \, G_0(1;5)G_0(2;6)v(5;6)S(5,6;3,4),$$

then the solution of (5.22) can be written as[4]

$$G_2(1,2;1',2') = \int d3d4 \, S(1,2;3,4) \left[G(3;1')G(4;2') \pm G(3;2')G(4;1') \right]. \qquad (5.23)$$

In this expression S can be calculated from G_0 only. The Green's function G, instead, still depends on G_2 through the equations of motion (5.7) and (5.8). In these equations G_2 is always multiplied by the interaction and it is therefore convenient to define the function

$$T_0(1,2;1',2') \equiv v(1;2)S(1,2;1',2'),$$

which is also known as the T-*matrix* . Then (5.7) attains the form

$$\left[i\frac{d}{dz_1} - h(1) \right] G(1;1') = \delta(1;1')$$
$$\pm i \int d2d3d4 \, T_0(1,2;3,4) \left[G(3;1')G(4;2^+) \pm G(3;2^+)G(4;1') \right]$$
$$= \delta(1;1') + \int d3 \, \Sigma(1;3)G(3;1'),$$

where this time the self-energy is defined as

$$\Sigma(1;3) = \pm i \int d2d4 \, \left[T_0(1,2;3,4) \pm T_0(1,2;4,3) \right] G(4;2^+). \qquad (5.24)$$

We have thus generated another self-consistent equation for G, the solution of which goes beyond the Hartree–Fock approximation.

We could develop even more sophisticated approximations by truncating the hierarchy at a higher level. It is clear, however, that this procedure is rather cumbersome and not so systematic. In the next section we lay down the basic results to construct a powerful approximation method to solve the Martin–Schwinger hierarchy. As we shall see the method is not just a mathematical technique but also a new way of thinking of the many-body problem.

---

[4]To prove this result one can iterate (5.22) to get an expansion of G_2 in powers of v. Iterating in a similar way the equation for S and inserting the resulting expansion in (5.23) one can verify that the terms are equal to those of the expansion of G_2 order by order in v. We do not give further details here since we work out the same solution using diagrammatic techniques in Section 13.6.

5.3 Exact solution of the hierarchy from Wick's theorem

It is clear from the previous section that the main difficulty in solving the hierarchy is due to the coupling of the equation for G_n to those for G_{n+1} and G_{n-1}. However, we note that in the noninteracting case, $v = 0$, we have only a coupling to lower order Green's functions given by the equations

$$
\left[i\frac{d}{dz_k} - h(k) \right] G_n(1, \ldots, n; 1', \ldots, n')
$$
$$
= \sum_{j=1}^{n} (\pm)^{k+j}\, \delta(k; j')\, G_{n-1}(1, \ldots \overset{\sqcap}{k} \ldots, n; 1', \ldots \overset{\sqcap}{j'} \ldots, n'), \tag{5.25}
$$

$$
G_n(1, \ldots, n; 1', \ldots, n') \left[-i\frac{\overleftarrow{d}}{dz_k'} - h(k') \right]
$$
$$
= \sum_{j=1}^{n} (\pm)^{k+j}\, \delta(j; k')\, G_{n-1}(1, \ldots \overset{\sqcap}{j} \ldots, n; 1', \ldots \overset{\sqcap}{k'} \ldots, n'). \tag{5.26}
$$

In this case the hierarchy can be solved exactly. Below we prove that the solution is simply [46, 47]

$$
G_{0,n}(1, \ldots, n; 1', \ldots, n') =
\begin{vmatrix}
G_0(1; 1') & \ldots & G_0(1; n') \\
\vdots & & \vdots \\
G_0(n; 1') & \ldots & G_0(n; n')
\end{vmatrix}_{\pm} \tag{5.27}
$$

where we add the subscript zero to denote noninteracting Green's functions and define $G_0 \equiv G_{0,1}$. As in Section 1.5, the quantity $|A|_{\pm}$, where A is an $n \times n$ matrix with elements A_{ij}, is the permanent/determinant for bosons/fermions

$$
|A|_{\pm} \equiv \sum_{P} (\pm)^{P} A_{1P(1)} \ldots A_{nP(n)}.
$$

In (5.27) the matrix A has elements $A_{ij} = G_0(i; j')$. To show that (5.27) is a solution of the noninteracting Martin–Schwinger hierarchy we expand the permanent/determinant along row k (see Appendix B) and find

$$
G_{0,n}(1, \ldots, n; 1', \ldots, n') = \sum_{j=1}^{n} (\pm)^{k+j} G_0(k; j')\, G_{0,n-1}(1, \ldots \overset{\sqcap}{k} \ldots n; 1', \ldots \overset{\sqcap}{j'} \ldots, n').
$$

By acting from the left with $[i\frac{d}{dz_k} - h(k)]$ we see immediately that this $G_{0,n}$ satisfies (5.25) since the first equation of the hierarchy tells us that G_0 satisfies (5.17). On the other hand, expanding along column k we find

$$
G_{0,n}(1, \ldots, n; 1', \ldots, n') = \sum_{j=1}^{n} (\pm)^{k+j} G_0(j; k')\, G_{0,n-1}(1, \ldots \overset{\sqcap}{j} \ldots, n; 1', \ldots \overset{\sqcap}{k'} \ldots, n'),
$$

which can similarly be seen to satisfy the Martin–Schwinger hierarchy equations (5.26) using (5.18). Therefore we conclude that (5.27) is a solution to the noninteracting Martin–Schwinger hierarchy. However, since this is just a particular solution of the set of differential equations we still need to check that it satisfies the KMS boundary conditions (5.6). This is readily done. From (5.27) $G_{0,n}$ satisfies these conditions whenever G_0 satisfies them as multiplying G_0 by a ± 1 in a row or a column the permanent/determinant is multiplied by the same factor. It is therefore sufficient to solve (5.17) and (5.18) with KMS boundary conditions to obtain all $G_{0,n}$s!

The result (5.27) is also known as *Wick's theorem* [50] and it is exceedingly useful in deriving diagrammatic perturbation theory. It seems somewhat of an exaggeration to give the simple statement summarized in (5.27) the status of a theorem. The Wick theorem, however, is usually derived in textbooks in a different way which requires much more effort and a separate discussion depending on whether we are working with the exact Green's function in equilibrium (also known as finite temperature Matsubara formalism) or in nonequilibrium or with the adiabatic Green's function or with the zero-temperature Green's function. The above derivation, apart from being shorter, is valid in all cases and highlights the physical content of Wick's theorem as a solution to a boundary value problem for the Martin–Schwinger hierarchy (or equations of motion for the Green's functions). To recover the Wick theorem of preference it is enough to assign the corresponding contour and the Hamiltonian along the contour. We comment more on how to recover the various formalisms in Section 5.4.

The Wick theorem suggests that we can calculate the interacting n-particle Green's function by a *brute force* expansion in powers of the interaction. From the definition (5.1) it follows that (omitting the arguments of the Green's function)

$$
G_n = \frac{1}{i^n} \frac{\mathrm{Tr}\left[\mathcal{T}\left\{ e^{-i\int_\gamma d\bar{z}\,\hat{H}_0(\bar{z})} e^{-i\int_\gamma d\bar{z}\,\hat{H}_{\mathrm{int}}(\bar{z})} \hat{\psi}(1)\ldots\hat{\psi}^\dagger(1')\right\}\right]}{\mathrm{Tr}\left[\mathcal{T}\left\{ e^{-i\int_\gamma d\bar{z}\,\hat{H}_0(\bar{z})} e^{-i\int_\gamma d\bar{z}\,\hat{H}_{\mathrm{int}}(\bar{z})}\right\}\right]},
$$

where we split the Hamiltonian into a one-body part $\hat{H}_0$ and a two-body part $\hat{H}_{\mathrm{int}}$ and we used the property that $\hat{H}_0$ and $\hat{H}_{\mathrm{int}}$ commute under the $\mathcal{T}$ sign. Expanding in powers of $\hat{H}_{\mathrm{int}}$ we get

$$
G_n = \frac{1}{i^n} \frac{\sum_{k=0}^{\infty} \frac{(-i)^k}{k!} \int_\gamma d\bar{z}_1\ldots d\bar{z}_k \,\langle \mathcal{T}\{\hat{H}_{\mathrm{int}}(\bar{z}_1)\ldots\hat{H}_{\mathrm{int}}(\bar{z}_k)\hat{\psi}(1)\ldots\hat{\psi}^\dagger(1')\}\rangle_0}{\sum_{k=0}^{\infty} \frac{(-i)^k}{k!} \int_\gamma d\bar{z}_1\ldots d\bar{z}_k \,\langle \mathcal{T}\{\hat{H}_{\mathrm{int}}(\bar{z}_1)\ldots\hat{H}_{\mathrm{int}}(\bar{z}_k)\}\rangle_0}, \tag{5.28}
$$

where we introduce the short-hand notation

$$
\mathrm{Tr}\left[\mathcal{T}\left\{ e^{-i\int_\gamma d\bar{z}\,\hat{H}_0(\bar{z})}\ldots\right\}\right] = \langle \mathcal{T}\{\ldots\}\rangle_0,
$$

according to which any string of operators can be inserted for the dots. We now recall that $\hat{H}_{\mathrm{int}}(z)$ is a two-body operator both on the horizontal branches and on the vertical track of the contour, see Section 4.4. Therefore we can write

$$
\hat{H}_{\mathrm{int}}(z) = \frac{1}{2}\int dz' \int d\mathbf{x}d\mathbf{x}'\, v(\mathbf{x},z;\mathbf{x}',z')\hat{\psi}^\dagger(\mathbf{x},z^+)\hat{\psi}^\dagger(\mathbf{x}',z'^+)\hat{\psi}(\mathbf{x}',z')\hat{\psi}(\mathbf{x},z), \tag{5.29}
$$

where

$$
v(\mathbf{x}, z; \mathbf{x}', z') = \delta(z, z') \begin{cases} v(\mathbf{x}, \mathbf{x}', t) & \text{if } z = t_\pm \text{ is on the horizontal branches of } \gamma \\ v^{\mathrm{M}}(\mathbf{x}, \mathbf{x}') & \text{if } z \text{ is on the vertical track of } \gamma \end{cases}.
$$

The reason for shifting the contour arguments of the $\hat{\psi}^\dagger$ operators from z, z' to z^+, z'^+ in (5.29) stems from the possibility of moving the field operators freely under the $\mathcal{T}$ sign without loosing the information that after the reordering the $\hat{\psi}^\dagger$ operators must be placed to the left of the $\hat{\psi}$ operators, see also below. Taking into account (5.29) and reordering the field operators, (5.28) provides an expansion of G_n in terms of noninteracting Green's functions $G_{0,m}$. Below we work out explicitly the case of the one- and two-particle Green's functions $G \equiv G_1$ and G_2 which are the ones most commonly used.

Let us start with the one-particle Green's function. For notational convenience we denote $a = (\mathbf{x}_a, z_a)$ and $b = (\mathbf{x}_b, z_b)$ and rename all time-integration variables $\bar{z}_k$ to z_k. Then (5.28) for $n = 1$ yields

$$
G(a; b) = \frac{1}{i} \frac{\sum_{k=0}^{\infty} \frac{(-i)^k}{k!} \int_\gamma dz_1 \ldots dz_k \, \langle \mathcal{T} \{ \hat{H}_{\mathrm{int}}(z_1) \ldots \hat{H}_{\mathrm{int}}(z_k) \hat{\psi}(a) \hat{\psi}^\dagger(b) \} \rangle_0}{\sum_{k=0}^{\infty} \frac{(-i)^k}{k!} \int_\gamma dz_1 \ldots dz_k \, \langle \mathcal{T} \{ \hat{H}_{\mathrm{int}}(z_1) \ldots \hat{H}_{\mathrm{int}}(z_k) \} \rangle_0}. \tag{5.30}
$$

Using the explicit form (5.29) of the interaction Hamiltonian the numerator of this equation can be rewritten as

$$
\sum_{k=0}^{\infty} \frac{1}{k!} \left(-\frac{i}{2} \right)^k \int d1 \ldots dk \, d1' \ldots dk' \, v(1; 1') \ldots v(k; k')
$$
$$
\times \langle \mathcal{T} \{ \hat{\psi}^\dagger(1^+) \hat{\psi}^\dagger(1'^+) \hat{\psi}(1') \hat{\psi}(1) \ldots \hat{\psi}^\dagger(k^+) \hat{\psi}^\dagger(k'^+) \hat{\psi}(k') \hat{\psi}(k) \hat{\psi}(a) \hat{\psi}^\dagger(b) \} \rangle_0.
$$

We reorder the quantity in the bracket as follows

$$
\langle \mathcal{T} \{ \hat{\psi}(a) \hat{\psi}(1) \hat{\psi}(1') \ldots \hat{\psi}(k) \hat{\psi}(k') \hat{\psi}^\dagger(k'^+) \hat{\psi}^\dagger(k^+) \ldots \hat{\psi}^\dagger(1'^+) \hat{\psi}^\dagger(1^+) \hat{\psi}^\dagger(b) \} \rangle_0,
$$

which requires an even number of interchanges so there is no sign change. In this expression we recognize the noninteracting $(2k+1)$-particle Green's function

$$
G_{0,2k+1}(a, 1, 1', \ldots, k, k'; b, 1^+, 1'^+, \ldots k^+, k'^+)
$$

multiplied by $i^{2k+1} Z_0$.[5] Similarly, the kth order bracket of the denominator can be written as the noninteracting $2k$-particle Green's function

$$
G_{0,2k}(1, 1', \ldots, k, k'; 1^+, 1'^+, \ldots k^+, k'^+)
$$

---

[5] Recall that $Z_0 = e^{-\beta \hat{H}_0^{\mathrm{M}}} = \mathrm{Tr}[\mathcal{T}\{e^{-i \int_\gamma d\bar{z} \hat{H}_0(\bar{z})}\}]$ is the noninteracting partition function.

multiplied by $i^{2k}Z_0$. We therefore find that (5.30) is equivalent to

$$G(a;b) = \frac{\sum_{k=0}^{\infty} \frac{1}{k!} \left(\frac{i}{2}\right)^k \int v(1;1')\ldots v(k;k') G_{0,2k+1}(a,1,1',\ldots;b,1^+,1'^+,\ldots)}{\sum_{k=0}^{\infty} \frac{1}{k!} \left(\frac{i}{2}\right)^k \int v(1;1')\ldots v(k;k') G_{0,2k}(1,1',\ldots;1^+,1'^+,\ldots)}, \quad (5.31)$$

where the integrals are over $1,1',\ldots,k,k'$. Next we observe that the $G_{0,n}$ can be decomposed in products of noninteracting one-particle Green's functions G_0 using the Wick's theorem (5.27). We thus arrive at the following important formula

$$G(a;b) = \frac{\sum_{k=0}^{\infty} \frac{1}{k!} \left(\frac{i}{2}\right)^k \int v(1;1')..v(k;k') \begin{vmatrix} G_0(a;b) & G_0(a;1^+) & \cdots & G_0(a;k'^+) \\ G_0(1;b) & G_0(1;1^+) & \cdots & G_0(1;k'^+) \\ \vdots & \vdots & \ddots & \vdots \\ G_0(k';b) & G_0(k';1^+) & \cdots & G_0(k';k'^+) \end{vmatrix}_{\pm}}{\sum_{k=0}^{\infty} \frac{1}{k!} \left(\frac{i}{2}\right)^k \int v(1;1')..v(k;k') \begin{vmatrix} G_0(1;1^+) & G_0(1;1'^+) & \cdots & G_0(1;k'^+) \\ G_0(1';1^+) & G_0(1';1'^+) & \cdots & G_0(1';k'^+) \\ \vdots & \vdots & \ddots & \vdots \\ G_0(k';1^+) & G_0(k';1'^+) & \cdots & G_0(k';k'^+) \end{vmatrix}_{\pm}}$$

$$(5.32)$$

which is an exact expansion of the interacting G in terms of the noninteracting G_0.

We can derive the expansion for the two-particle Green's function in a similar way. We have

$$G_2(a,b;c,d)$$
$$= \frac{1}{i^2} \frac{\sum_{k=0}^{\infty} \frac{(-i)^k}{k!} \int_\gamma dz_1 \ldots dz_k \langle \mathcal{T}\{\hat{H}_{\text{int}}(z_1)\ldots\hat{H}_{\text{int}}(z_k)\hat{\psi}(a)\hat{\psi}(b)\hat{\psi}^\dagger(d)\hat{\psi}^\dagger(c)\}\rangle_0}{\sum_{k=0}^{\infty} \frac{(-i)^k}{k!} \int_\gamma dz_1 \ldots dz_k \langle \mathcal{T}\{\hat{H}_{\text{int}}(z_1)\ldots\hat{H}_{\text{int}}(z_k)\}\rangle_0}.$$

$$(5.33)$$

Using again the explicit form (5.29) of the interaction Hamiltonian the numerator of (5.33) can be rewritten as

$$\sum_{k=0}^{\infty} \frac{1}{k!} \left(-\frac{i}{2}\right)^2 \int d1\ldots dk d1' \ldots dk' v(1;1')\ldots v(k;k')$$

$$\times \langle \mathcal{T}\{\hat{\psi}^\dagger(1^+)\hat{\psi}^\dagger(1'^+)\hat{\psi}(1')\hat{\psi}(1)\ldots\hat{\psi}^\dagger(k^+)\hat{\psi}^\dagger(k'^+)\hat{\psi}(k')\hat{\psi}(k)$$

$$\times \hat{\psi}(a)\hat{\psi}(b)\hat{\psi}^\dagger(d)\hat{\psi}^\dagger(c)\}\rangle_0.$$

Reordering the quantity in the bracket as

$$\langle \mathcal{T}\{\hat{\psi}(a)\hat{\psi}(b)\hat{\psi}(1)\hat{\psi}(1')\ldots\hat{\psi}(k)\hat{\psi}(k')\hat{\psi}^\dagger(k'^+)\hat{\psi}^\dagger(k^+)\ldots\hat{\psi}^\dagger(1'^+)\hat{\psi}^\dagger(1^+)\hat{\psi}^\dagger(d)\hat{\psi}^\dagger(c)\}\rangle_0$$

requires an even number of interchanges and we recognize the noninteracting $(2k + 2)$-particle Green's function

$$G_{0,2k+2}(a, b, 1, 1' \ldots, k, k'; c, d, 1^+, 1'^+, \ldots, k^+, k'^+)$$

multiplied by $i^{2k+2} Z_0$. The denominator in (5.33) is the same as that of the one-particle Green's function in (5.30) and therefore G_2 becomes

$$G_2(a, b; c, d) = \frac{\sum\limits_{k=0}^{\infty} \frac{1}{k!} \left(\frac{i}{2}\right)^k \int v(1; 1') \ldots v(k; k') G_{0,2k+2}(a, b, 1, 1' \ldots; c, d, 1^+, 1'^+, \ldots)}{\sum\limits_{k=0}^{\infty} \frac{1}{k!} \left(\frac{i}{2}\right)^k \int v(1; 1') \ldots v(k; k') G_{0,2k}(1, 1', \ldots; 1^+, 1'^+, \ldots)},$$

where the integrals are over $1, 1', \ldots, k, k'$. Using Wick's theorem we can now transform G_2 into an exact perturbative expansion in terms of the noninteracting Green's function G_0:

$$
\begin{aligned}
&G_2(a, b; c, d) \\
&= \frac{\sum\limits_{k=0}^{\infty} \frac{1}{k!} \left(\frac{i}{2}\right)^k \int v(1; 1') \ldots v(k; k') \begin{vmatrix} G_0(a; c) & G_0(a; d) & \ldots & G_0(a; k'^+) \\ G_0(b; c) & G_0(b; d) & \ldots & G_0(b; k'^+) \\ \vdots & \vdots & \ddots & \vdots \\ G_0(k'; c) & G_0(k'; d) & \ldots & G_0(k'; k'^+) \end{vmatrix}_{\pm}}{\sum\limits_{k=0}^{\infty} \frac{1}{k!} \left(\frac{i}{2}\right)^k \int v(1; 1') \ldots v(k; k') \begin{vmatrix} G_0(1; 1^+) & G_0(1; 1'^+) & \ldots & G_0(1; k'^+) \\ G_0(1'; 1^+) & G_0(1'; 1'^+) & \ldots & G_0(1'; k'^+) \\ \vdots & \vdots & \ddots & \vdots \\ G_0(k'; 1^+) & G_0(k'; 1'^+) & \ldots & G_0(k'; k'^+) \end{vmatrix}_{\pm}}
\end{aligned}
$$

$$(5.34)$$

In a similar way the reader can work out the equations for the higher order Green's functions as well as for the partition function Z which we write below:[6]

$$\frac{Z}{Z_0} = \sum\limits_{k=0}^{\infty} \frac{1}{k!} \left(\frac{i}{2}\right)^k \int v(1; 1') \ldots v(k; k') \begin{vmatrix} G_0(1; 1^+) & G_0(1; 1'^+) & \ldots & G_0(1; k'^+) \\ G_0(1'; 1^+) & G_0(1'; 1'^+) & \ldots & G_0(1'; k'^+) \\ \vdots & \vdots & \ddots & \vdots \\ G_0(k'; 1^+) & G_0(k'; 1'^+) & \ldots & G_0(k'; k'^+) \end{vmatrix}_{\pm}$$

$$(5.35)$$

Many-body perturbation theory (MBPT) is now completely defined. The evaluation of (5.32) or (5.34) is a well-defined mathematical problem and there exist many useful tricks to carry on the calculations in an efficient way. In his memorable talk at the Pocono Manor Inn in 1948 Feynman showed how to represent the cumbersome Wick expansion in terms

---

[6]The ratio Z/Z_0 is simply the denominator of the formulas (5.32) and (5.34) for G and G_2.

of physically insightful diagrams [51], and since then the Feynman diagrams have become an invaluable tool in many areas of physics. The Feynman diagrams give us a deep insight into the microscopic scattering processes, help us to evaluate (5.32) or (5.34) more efficiently, and also provide us with techniques to resum certain classes of diagram to infinite order in the interaction strength. These resummations are particularly important in bulk systems, where diagrams can be divergent (as, e.g., in the case of the long-range Coulomb interaction), as well as in finite systems. The expansion in powers of the interaction is indeed an *asymptotic* expansion, meaning that the inclusion of diagrams beyond a certain order brings the sum further and further away from the exact result, see e.g. Ref. [52] (for a brief introduction to asymptotic expansions see Appendix F).

5.4 Finite and zero-temperature formalism from the exact solution

In this section we discuss how different versions of MBPT that are commonly used follow as special cases of the perturbation theory developed in the previous section. Examples of these formalisms are the finite temperature Matsubara formalism, the zero-temperature formalism, the Keldysh formalism and the Konstantinov–Perel' formalism. In most textbooks these different flavors of MBPT appear as disconnected topics even though the perturbative terms have the same mathematical structure. This is not a coincidence but has its origin in the Martin–Schwinger hierarchy as defined on the contour. We now demonstrate this fact by discussing these formalisms one by one.

Konstantinov–Perel' formalism

In the Konstantinov–Perel' formalism the Green's functions satisfy the Martin–Schwinger hierarchy on the contour of Fig. 4.5(a) and, therefore, they are given by (5.32) and (5.34) where the z-integrals run on the same contour. The averages calculated from these Green's functions at equal time, i.e., $z_i = z$ and $z_i' = z^+$ for all i, correspond to the exact formula (4.21). We remind the reader that this formula yields the initial ensemble average for z on the vertical track and the time-dependent ensemble average for z on the horizontal branches. It is therefore clear that if we are interested in calculating time-dependent ensemble averages up to a maximum time T we only need the Green's functions with real time contour arguments up to T. *In this case it is sufficient to solve the Martin–Schwinger hierarchy over a shrunken contour like the one illustrated below.*

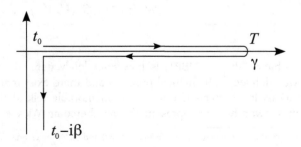

The solution is again given by (5.32) and (5.34) but the z-integrals run over the shrunken contour. This observation tells us that if the contour is longer than T then the terms with integrals after T cancel off. The cancellation is analogous to the one discussed in Section 4.2 where we showed that nothing changes in (4.9) if the contour extends all the way to $+\infty$.

Matsubara formalism

The Matsubara formalism is used to calculate the initial ensemble average (4.24) and it is typically applied to systems in equilibrium at finite temperature. For this reason the Matsubara formalism is sometimes called the "finite-temperature formalism." The crucial observation here is that in order to calculate the initial averages we only need the Green's function with times on the vertical track. Then, according to the previous discussion, we can take the time $T = t_0$ and hence shrink the horizontal branches to a point leaving only the vertical track. The corresponding Matsubara Green's functions are given by the solution of the Martin–Schwinger hierarchy on γ^{M}. The Matsubara formalism therefore consists of expanding the Green's functions as in (5.32) and (5.34) with the z-integrals restricted to the vertical track. It is important to realize that no assumptions, like the adiabatic or the zero temperature assumption, are made in this formalism. The Matsubara formalism is exact but limited to initial (or equilibrium) averages. Note that not all equilibrium properties can be extracted from the Matsubara Green's functions. As we shall see, quantities like photo-emission currents, hyper-polarizabilities and more generally high-order response properties require a knowledge of equilibrium Green's functions with *real* time contour arguments.

Keldysh formalism

The formalism originally used by Keldysh was based on the adiabatic assumption according to which the interacting density matrix $\hat{\rho}$ can be obtained from the noninteracting $\hat{\rho}_0$ by an adiabatic switch-on of the interaction.[7] Under this assumption we can calculate time-dependent ensemble averages from the adiabatic Green's functions, i.e., from the solution of the Martin–Schwinger hierarchy on the contour γ_a of Fig. 4.5(b) with the adiabatic Hamiltonian shown in the same figure. These adiabatic Green's functions are again given by (5.32) and (5.34) but the z-integrals run over γ_a and the Hamiltonian is the adiabatic Hamiltonian. The important simplification entailed by the adiabatic assumption is that the interaction v is zero on the vertical track, see again Fig. 4.5(b). Consequently in (5.32) and (5.34) we can restrict the z-integrals to the horizontal branches. Like the exact formalism, the adiabatic formalism can be used to deal with nonequilibrium situations in which the external perturbing fields are switched on after time t_0. In the special case of no external fields we can calculate interacting *equilibrium* Green's functions at any finite temperature with *real*-time contour arguments. In contrast with the exact and Matsubara formalism, however, the Green's functions with imaginary time contour arguments are noninteracting since on the vertical track we have $\hat{H}_0^{\mathrm{M}}$ and not $\hat{H}^{\mathrm{M}}$.

---

[7]As already observed this assumption is supported by the Gell-Mann–Low theorem [44].

To summarize: *the adiabatic Green's functions can be expanded as in (5.32) or (5.34) where the z-integrals run over the contour below:*

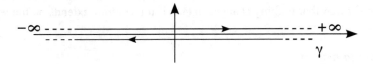

Zero-temperature formalism

The zero-temperature formalism relies on the zero-temperature assumption according to which the ground state of $\hat{H}_0^M$ evolves into the ground state of $\hat{H}^M$ by adiabatically switching on the interaction in the remote past or in the distant future. As already discussed this assumption makes sense only *in the absence of external fields*. The corresponding zero-temperature Green's functions are the solution of the Martin–Schwinger hierarchy on the contour γ_0 of Fig. 4.5(c) where the Hamiltonian changes along the contour as shown in the same figure. As in the adiabatic case, the interaction v of the zero-temperature Hamiltonian vanishes along the vertical track and hence the z-integrals in the expansions (5.32) and (5.34) can be restricted to a contour that goes from $-\infty$ to ∞. The contour ordering on the horizontal branch of γ_0 is the same as the standard time-ordering. For this reason the zero-temperature Green's functions are also called time-ordered Green's functions. The zero-temperature formalism allows us to calculate interacting Green's functions in equilibrium at zero temperature with *real*-time contour arguments. It cannot, however, be used to study systems out of equilibrium and/or at finite temperature. In some cases, however, the zero-temperature formalism is used also at finite temperatures (finite β) as the finite temperature corrections are small.[8] This approximated formalism is sometimes referred to as the *real-time finite temperature formalism* [45]. It is worth noting that in the real-time finite-temperature formalism (as in the Keldysh formalism) the temperature enters in (5.32) and (5.34) only through G_0, which satisfies the KMS relations. In the Konstantinov–Perel' formalism, on the other hand, the temperature enters through G_0 and through the contour integrals since the interaction is nonvanishing along the vertical track.

To summarize: *the zero-temperature Green's functions can be expanded as in (5.32) and (5.34) where the z-integrals run over the contour below.*

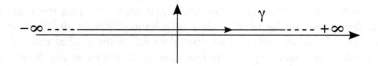

In conclusion, if one does not want to make the assumptions implicit in the last two formalisms then there is no alternative but to use the full contour of Fig. 4.5(a). This is the contour that we use in the remainder of the book. However, all derivations in the book rely

---

[8]At finite temperature the use of the contour γ_0 implies the assumption that *all* eigenstates of $\hat{H}_0^M$ evolve into themselves by adiabatically switching on and then off the interaction, i.e., the satisfaction of (4.30).

exclusively on the solution of the Martin–Schwinger hierarchy and therefore can be directly adapted to any formalism preferred by the reader by simply changing

$$\int_{\gamma_{\text{book}}} \rightarrow \int_{\gamma_{\text{preferred}}}$$

in every integral with a contour symbol attached.

As a final remark we observe that the explicit form (5.2) and (5.3) of the Martin–Schwinger hierarchy assumes the fact that $\hat{H}^M$ and $\hat{H}(t)$ contain at most two-body operators. In Appendix G we present an alternative approach to deal with general density matrices $\hat{\rho}$ (or equivalently with general $\hat{H}^M$). We also want to point out that if we are only interested in the equal-time G_ns, then we can easily derive an exact hierarchy for these quantities starting from the Martin–Schwinger hierarchy. This is the so called BBGKY hierarchy and we refer the reader to Appendix H for its derivation and further discussion.

5.5 Langreth rules

In the expansion of the interacting Green's functions, see (5.32) or (5.34), convolutions like, e.g., $\int d\bar{1} G_0(1,\bar{1}) G_0(\bar{1},2)$, or products like, e.g., $G_0(1;2) G_0(2;1)$ appear. In order to evaluate these formal expressions we must convert contour integrals into standard real-time integrals, and products of functions on the contour into products of functions with real-time arguments. The purpose of this section is to derive a set of identities to transform contour convolutions and contour products into convolutions and products that can be calculated analytically or implemented numerically. Some of these identities are known as the *analytic continuation rules* or *Langreth rules* [53], while others, equally important, are often ignored.[9] We think that the name "analytic continuation rules" misleads the reader into thinking that these rules are applicable only to analytic functions of the complex time. The analyticity property is never used in the derivations below and, therefore, we prefer to use the name "generalized Langreth rules" or simply "Langreth rules." In this section we present a comprehensive and self-contained derivation of all of them. Table 5.1 at the end of the section summarizes the main results. For those readers who are not already familiar with the formalism, we strongly recommend following this section with pencil and paper.

We specialize the discussion to two-point correlators (like the one-particle Green's function)

$$k(z, z') = \text{Tr}\left[\hat{\rho}\, \hat{k}(z, z')\right] = \text{Tr}\left[\hat{\rho}\, \mathcal{T}\left\{\hat{O}_1(z)\hat{O}_2(z')\right\}\right],$$

which are the ensemble average of the two-point operator correlators (4.47). Higher order correlators are conceptually no more complicated and can be treated in a similar manner. As in Section 4.5 we do not specify the origin of the z-dependence of the operators; they could be operators with an explicit z-dependence or operators in the Heisenberg picture. The important thing is that $\hat{O}(t_+) = \hat{O}(t_-)$.[10] Unless otherwise stated, here and in the remainder of the book we always consider the contour of Fig. 4.5(a).

---

[9] This is due to the fact that MBPT is mostly used within the Keldysh (adiabatic) formalism, see Section 5.4.

[10] This property is, by definition, satisfied by the field operators $\hat{\psi}$ and $\hat{\psi}^\dagger$ (and hence by arbitrary products of field operators) as well as by the field operators $\hat{\psi}_H$ and $\hat{\psi}_H^\dagger$ in the contour Heisenberg picture. See also the discussion around equations (4.6) and (4.7).

Due to the contour ordering $k(z, z')$ has the following structure:

$$k(z, z') = \theta(z, z')k^>(z, z') + \theta(z', z)k^<(z, z'),$$

with $\theta(z, z')$ the Heaviside function on the contour and

$$k^>(z, z') = \text{Tr}\left[\hat{\rho}\,\hat{O}_1(z)\hat{O}_2(z')\right], \qquad k^<(z, z') = \pm\text{Tr}\left[\hat{\rho}\,\hat{O}_2(z')\hat{O}_1(z)\right],$$

where the $\pm$ sign in $k^<$ is for bosonic/fermionic operators $\hat{O}_1$ and $\hat{O}_2$. It is important to observe that the functions $k^>$ and $k^<$ are well defined for *all* z and z' on the contour. Furthermore, these functions have the important property that their value is independent of whether z, z' lie on the forward or backward branch

$$\boxed{k^\lessgtr(t_+, z') = k^\lessgtr(t_-, z'), \qquad k^\lessgtr(z, t'_+) = k^\lessgtr(z, t'_-)} \qquad (5.36)$$

since $\hat{O}_i(t_+) = \hat{O}_i(t_-)$ for both $i = 1, 2$. We say that a function $k(z, z')$ belongs to the *Keldysh space* if it can be written as

$$k(z, z') = k^\delta(z)\delta(z, z') + \theta(z, z')k^>(z, z') + \theta(z', z)k^<(z, z'),$$

with $k^\lessgtr(z, z')$ satisfying the properties (5.36), $k^\delta(t_+) = k^\delta(t_-) \equiv k^\delta(t)$ and $\delta(z, z')$ the δ-function on the contour. We here observe that the δ-function on the contour is zero if z and z' lie on different branches, $\delta(t_\pm, t_\mp) = 0$, and that, due to the orientation of the contour, $\delta(t_-, t'_-) = \delta(t - t')$ whereas $\delta(t_+, t'_+) = -\delta(t - t')$, with $\delta(t - t')$ the δ-function on the real axis. The precise meaning of these identities is the following. From the definition of δ-function on the contour, i.e., $\int_\gamma dz'\delta(z, z')f(z') = f(z)$, and from the definition (4.4) of functions on the contour we have for $z = t_-$

$$f_-(t) = f(t_-) = \int_\gamma dz'\delta(t_-, z')f(z') = \int_{t_0}^\infty dt'\,\delta(t_-, t'_-)f(t'_-) = \int_{t_0}^\infty dt'\,\delta(t - t')f_-(t'),$$

whereas for $z = t_+$

$$f_+(t) = f(t_+) = \int_\gamma dz'\delta(t_+, z')f(z') = \int_\infty^{t_0} dt'\,\delta(t_+, t'_+)f(t'_+) = -\int_\infty^{t_0} dt'\,\delta(t - t')f_+(t').$$

In order to extract physical information from $k(z, z')$ we must evaluate this function for all possible positions of z and z' on the contour: both arguments on the horizontal branches, one argument on the vertical track and the other on the horizontal branches or both arguments on the vertical track. We define the *greater* and *lesser* Keldysh components as the following functions on the real time axis

$$\boxed{\begin{aligned} k^>(t, t') &\equiv k(t_+, t'_-) \\ k^<(t, t') &\equiv k(t_-, t'_+) \end{aligned}} \qquad (5.37)$$

In other words $k^>(t, t')$ is the real-time function with the values of the contour function $k^>(z, z')$.[11] Similarly $k^<(t, t') = k^<(z, z')$. From these considerations we see that the *equal-time* greater and lesser functions can also be written as

$$k^>(t, t) = k(z^+, z) = k(z, z^-), \qquad k^<(t, t) = k(z, z^+) = k(z^-, z), \qquad (5.38)$$

where z^+ (z^-) is a contour point infinitesimally later (earlier) than the contour point $z = t_\pm$, which can lie either on the forward or backward branch. We sometimes use these alternative expressions to calculate time-dependent ensemble averages.

We also define the *left* and *right* Keldysh components from $k(z, z')$ with one real time t and one imaginary time $t_0 - i\tau$

$$\begin{aligned} k^\lceil(\tau, t) &\equiv k(t_0 - i\tau, t_\pm) \\ k^\rceil(t, \tau) &\equiv k(t_\pm, t_0 - i\tau) \end{aligned} \qquad (5.39)$$

The names "left" and "right" refer to the position of the vertical segment of the hooks "$\lceil$" and "$\rceil$" with respect to the horizontal segment. In the definition of $k^\lceil$ and $k^\rceil$ we can arbitrarily choose t_+ or t_- since $t_0 - i\tau$ is later than both of them and $k^\lessgtr(z, z')$ fulfills (5.36). In (5.39) the notation has been chosen to help visualization of the contour arguments [54]. For instance, the symbol "$\rceil$" has a horizontal segment followed by a vertical one; accordingly, $k^\rceil$ has a first argument which is real (and thus lies on the horizontal axis) and a second argument which is imaginary (and lies on the vertical track). In a similar way we can explain the use of the left symbol "$\lceil$". In the definition of the left and right functions we also introduce a convention of denoting the real times with latin letters and the imaginary times with greek letters. This convention is adopted throughout the book.

As we shall see it is also useful to define the *Matsubara* component $k^M(\tau, \tau')$ with both contour arguments on the vertical track:

$$\begin{aligned} k^M(\tau, \tau') &\equiv k(t_0 - i\tau, t_0 - i\tau') \\ &= \delta(t_0 - i\tau, t_0 - i\tau')k^\delta(t_0 - i\tau) + k_r^M(\tau, \tau') \end{aligned} \qquad (5.40)$$

where

$$k_r^M(\tau, \tau') = \theta(\tau - \tau')k^>(t_0 - i\tau, t_0 - i\tau') + \theta(\tau' - \tau)k^<(t_0 - i\tau, t_0 - i\tau')$$

is the regular part of the function.[12] To convert the δ-function on the vertical track into a standard δ-function we again use the definition

$$f(t_0 - i\tau) = \int_\gamma dz' \delta(t_0 - i\tau, z')f(z') \qquad \text{(setting } z' = t_0 - i\tau')$$

$$= -i \int_0^\beta d\tau' \delta(t_0 - i\tau, t_0 - i\tau')f(t_0 - i\tau'),$$

---

[11]As already emphasized, $k^>(z, z')$ is independent of whether z and z' are on the forward or backward branch.
[12]In writing k_r^M we take into account that $\theta(t_0 - i\tau, t_0 - i\tau')$ is 1 if $\tau > \tau'$ and zero otherwise, hence it is equal to $\theta(\tau - \tau')$.

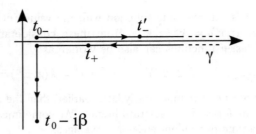

Figure 5.2 Contour with t on the backward branch and t' on the forward branch.

from which it follows that $\delta(t_0 - i\tau, t_0 - i\tau') = i\delta(\tau - \tau')$. Thus, if we introduce the short-hand notation $k^\delta(\tau) \equiv k^\delta(t_0 - i\tau)$, we can rewrite the Matsubara component as

$$k^{\mathrm{M}}(\tau, \tau') = i\delta(\tau - \tau')k^\delta(\tau) + k_r^{\mathrm{M}}(\tau, \tau'). \tag{5.41}$$

The advantage of working with the Keldysh components is that $k^{\lessgtr}$, $k^{\lceil,\rceil}$ and k^{M} are functions of real variables (as opposed to contour variables) and therefore can be numerically implemented, visualized, plotted, etc.; in other words the Keldysh components are easier to handle. Furthermore, they encode all the information to reconstruct $k^{\lessgtr}(z, z')$.

We now consider the convolution

$$c(z, z') = \int_\gamma d\bar{z}\, a(z, \bar{z})b(\bar{z}, z') \tag{5.42}$$

between two functions $a(z, z')$ and $b(z, z')$ in Keldysh space. It is easy to check that $c(z, z')$ belongs to the Keldysh space as well. The question we ask is how to express the Keldysh components of c in terms of the Keldysh components of a and b. We start by evaluating, e.g., the greater component. From the definition (5.37) and with the help of Fig. 5.2 we can write

$$c^>(t, t') = c(t_+, t'_-) = a(t_+, t'_-)b^\delta(t'_-) + a^\delta(t_+)b(t_+, t'_-) + \int_{t_0-}^{t'_-} d\bar{z}\, a^>(t_+, \bar{z})b^<(\bar{z}, t'_-)$$

$$+ \int_{t'_-}^{t_+} d\bar{z}\, a^>(t_+, \bar{z})b^>(\bar{z}, t'_-) + \int_{t_+}^{t_0-i\beta} d\bar{z}\, a^<(t_+, \bar{z})b^>(\bar{z}, t'_-)$$

$$= a^>(t, t')b^\delta(t') + a^\delta(t)b^>(t, t') + \int_{t_0}^{t'} d\bar{t}\, a^>(t, \bar{t})b^<(\bar{t}, t')$$

$$+ \int_{t'}^{t} d\bar{t}\, a^>(t, \bar{t})b^>(\bar{t}, t') + \int_{t}^{t_0} d\bar{t}\, a^<(t, \bar{t})b^>(\bar{t}, t') - i\int_0^\beta d\bar{\tau}\, a^\rceil(t, \bar{\tau})b^\lceil(\bar{\tau}, t').$$

The first integral in the last line is an ordinary integral on the real axis and can be rewritten as

$$\int_{t'}^{t} d\bar{t}\, a^>(t, \bar{t})b^>(\bar{t}, t') = \int_{t'}^{t_0} d\bar{t}\, a^>(t, \bar{t})b^>(\bar{t}, t') + \int_{t_0}^{t} d\bar{t}\, a^>(t, \bar{t})b^>(\bar{t}, t'),$$

which inserted in the previous expression leads to

$$c^>(t,t') = a^>(t,t')b^\delta(t') + a^\delta(t)b^>(t,t') - \int_{t_0}^{t'} d\bar{t}\, a^>(t,\bar{t})[b^>(\bar{t},t') - b^<(\bar{t},t')]$$

$$+ \int_{t_0}^{t} d\bar{t}\, [a^>(t,\bar{t}) - a^<(t,\bar{t})]b^>(\bar{t},t') - i\int_0^\beta d\bar{\tau}\, a^\rceil(t,\bar{\tau})b^\lceil(\bar{\tau},t'). \qquad (5.43)$$

We see that it is convenient to define two more Keldysh components with real-time arguments

$$\boxed{k^R(t,t') \equiv k^\delta(t)\delta(t-t') + \theta(t-t')[k^>(t,t') - k^<(t,t')]} \qquad (5.44)$$

$$\boxed{k^A(t,t') \equiv k^\delta(t)\delta(t-t') - \theta(t'-t)[k^>(t,t') - k^<(t,t')]} \qquad (5.45)$$

with the real-time Heaviside function $\theta(t) = 1$ for $t > 0$ and 0 otherwise. The *retarded* component $k^R(t,t')$ vanishes for $t < t'$, while the *advanced* component $k^A(t,t')$ vanishes for $t > t'$. The retarded and advanced functions can be used to rewrite (5.43) in a very elegant form:

$$c^>(t,t') = \int_{t_0}^\infty d\bar{t}\, [a^>(t,\bar{t})b^A(\bar{t},t') + a^R(t,\bar{t})b^>(\bar{t},t')] - i\int_0^\beta d\bar{\tau}\, a^\rceil(t,\bar{\tau})b^\lceil(\bar{\tau},t').$$

This formula can be made even more compact if we introduce a short-hand notation for convolutions between t_0 and ∞, and for convolutions between 0 and β. For two arbitrary functions f and g we define

$$f \cdot g \equiv \int_{t_0}^\infty d\bar{t}\, f(\bar{t})g(\bar{t}),$$

$$f \star g \equiv -i\int_0^\beta d\bar{\tau}\, f(\bar{\tau})g(\bar{\tau}).$$

Then the formula for $c^>$ becomes

$$\boxed{c^> = a^> \cdot b^A + a^R \cdot b^> + a^\rceil \star b^\lceil} \qquad (5.46)$$

In a similar way we can extract the lesser component and find that

$$\boxed{c^< = a^< \cdot b^A + a^R \cdot b^< + a^\rceil \star b^\lceil} \qquad (5.47)$$

Equations (5.46) and (5.47) can be used to extract the retarded and advanced components of c. Taking into account that the singular part c^δ is simply the product $a^\delta b^\delta$, and using the definition (5.44) we find

$$c^R(t,t') = c^\delta(t)\delta(t-t') + \theta(t-t')[c^>(t,t') - c^<(t,t')]$$

$$= a^\delta(t)b^\delta(t)\delta(t-t') + \theta(t-t')\int_{t_0}^\infty d\bar{t}\, a^R(t,\bar{t})[b^>(\bar{t},t') - b^<(\bar{t},t')]$$

$$+ \theta(t-t')\int_{t_0}^\infty d\bar{t}\, [a^>(t,\bar{t}) - a^<(t,\bar{t})]b^A(\bar{t},t').$$

Let us manipulate this expression a little. Similarly to the Matsubara component, we separate out the regular contribution from the retarded/advanced components: $k^{R/A}(t, t') = k^\delta(t)\delta(t - t') + k_r^{R/A}(t, t')$. Then we observe that the last term on the r.h.s. vanishes unless $t \geq t'$ and that $b^A(\bar{t}, t')$ vanishes unless $t' \geq \bar{t}$. We can, therefore, replace $[a^>(t, \bar{t}) - a^<(t, \bar{t})]$ with $\theta(t - \bar{t})[a^>(t, \bar{t}) - a^<(t, \bar{t})] = a_r^R(t, \bar{t})$ without changing the result. Next we consider the second term on the r.h.s.. Writing a^R as the sum of the singular and regular contributions and taking into account that $b^> - b^< = b_r^R - b_r^A$ we find

$$\theta(t - t') \int_{t_0}^\infty d\bar{t}\, a^R(t, \bar{t})[b^>(\bar{t}, t') - b^<(\bar{t}, t')]$$

$$= a^\delta(t)b_r^R(t, t') + \theta(t - t') \int_{t_0}^\infty d\bar{t}\, a_r^R(t, \bar{t})[b_r^R(\bar{t}, t') - b_r^A(\bar{t}, t')].$$

Collecting all these results, it is a matter of simple algebra to show that

$$\boxed{c^R = a^R \cdot b^R} \tag{5.48}$$

Using similar manipulations it is possible to show that

$$\boxed{c^A = a^A \cdot b^A} \tag{5.49}$$

It is worth noting that neither c^R nor c^A contains an integral along the vertical track of the contour.

Another important component that can be constructed from the greater and lesser functions is the *time-ordered* component. This is obtained by choosing both contour arguments in $k(z, z')$ on the forward branch

$$\boxed{k^T(t, t') \equiv k(t_-, t'_-) = k^\delta(t)\delta(t - t') + \theta(t - t')k^>(t, t') + \theta(t' - t)k^<(t, t')} \tag{5.50}$$

Similarly the *anti-time-ordered* component is obtained by choosing both z and z' on the backward branch

$$\boxed{k^{\bar{T}}(t, t') \equiv k(t_+, t'_+) = -k^\delta(t)\delta(t - t') + \theta(t' - t)k^>(t, t') + \theta(t - t')k^<(t, t')} \tag{5.51}$$

The reader can easily verify that

$$k^T = k^< + k^R = k^> + k^A,$$
$$k^{\bar{T}} = k^> - k^R = k^< - k^A.$$

Consequently, the time-ordered component of the convolution of two functions in Keldysh space can be extracted using the results for $c^{\lessgtr}$ and $c^{R,A}$:

$$c^T = c^> + c^A = a^R \cdot b^> + a^> \cdot b^A + a^A \cdot b^A + a^\rceil \star b^\lceil$$
$$= a^R \cdot b^> + a^T \cdot b^A + a^\rceil \star b^\lceil$$
$$= [a^T - a^<] \cdot b^> + a^T \cdot [b^T - b^>] + a^\rceil \star b^\lceil$$
$$= a^T \cdot b^T - a^< \cdot b^> + a^\rceil \star b^\lceil.$$

Similar manipulations lead to

$$c^{\bar{T}} = a^> \cdot b^< - a^{\bar{T}} \cdot b^{\bar{T}} + a^\rceil \star b^\lceil. \tag{5.52}$$

Next we show how to express the right and left components $c^\rceil$ and $c^\lceil$. Let us start with $c^\rceil$. In the definition (5.39) we choose, e.g., t_- as the first argument of c. Then we find

$$
\begin{aligned}
c^\rceil(t,\tau) &= a^\delta(t_-)b(t_-, t_0 - i\tau) + a(t_-, t_0 - i\tau)b^\delta(t_0 - i\tau) \\
&\quad + \int_{t_0-}^{t_-} d\bar{z}\, a^>(t_-, \bar{z})b^<(\bar{z}, t_0 - i\tau) + \int_{t_-}^{t_0+} d\bar{z}\, a^<(t_-, \bar{z})b^<(\bar{z}, t_0 - i\tau) \\
&\quad + \int_{t_0}^{t_0-i\tau} d\bar{z}\, a^<(t_-, \bar{z})b^<(\bar{z}, t_0 - i\tau) + \int_{t_0-i\tau}^{t_0-i\beta} d\bar{z}\, a^<(t_-, \bar{z})b^>(\bar{z}, t_0 - i\tau) \\
&= a^\delta(t)b^\rceil(t,\tau) + a^\rceil(t,\tau)b^\delta(\tau) + \int_{t_0}^{t} d\bar{t}\, [a^>(t, \bar{t}) - a^<(t, \bar{t})]b^\rceil(\bar{t}, \tau) \\
&\quad - i \int_0^\beta d\bar{\tau}\, a^\rceil(t, \bar{\tau})b_r^{\mathrm{M}}(\bar{\tau}, \tau).
\end{aligned}
$$

Recalling the definition (5.44) of the retarded component and taking into account (5.41) we arrive at the following compact formula

$$\boxed{c^\rceil = a^{\mathrm{R}} \cdot b^\rceil + a^\rceil \star b^{\mathrm{M}}} \tag{5.53}$$

The formula for $c^\lceil$ can be derived similarly and reads

$$\boxed{c^\lceil = a^\lceil \cdot b^{\mathrm{A}} + a^{\mathrm{M}} \star b^\lceil} \tag{5.54}$$

Finally, it is straightforward to prove that the Matsubara component of c is simply given by

$$\boxed{c^{\mathrm{M}} = a^{\mathrm{M}} \star b^{\mathrm{M}}} \tag{5.55}$$

since the integral along the forward branch cancels exactly the integral along the backward branch. This last identity exhausts the Langreth rules for the convolutions of two functions in Keldysh space. The results are summarized in the second column of Table 5.1.

There is another class of important identities which regards the product of two functions in Keldysh space

$$c(z, z') = a(z, z')b(z', z).$$

Unlike the convolution, the product does not always belong to the Keldysh space. If, for instance, the singular part a^δ is nonvanishing then the product c contains the term $\delta(z, z')a^\delta(z)b(z', z)$. The problems with this term are: (1) if b^δ is also nonvanishing then c contains the square of a δ-function and hence it does not belong to the Keldysh space; and (2) the function $b(z', z)$ is defined for $z' \to z$ but exactly in $z' = z$ it can be discontinuous and hence $\delta(z, z')b(z', z)$ is ill-defined. The product c is well defined only provided that the singular parts a^δ and b^δ are identically zero, in which case c belongs to the Keldysh space with $c^\delta = 0$. Below we consider only this sub-class of functions in Keldysh space. In practice we never have to deal with a product of functions with a singular part.

Definition	$c(z, z') = \int_\gamma d\bar{z}\, a(z, \bar{z}) b(\bar{z}, z')$	$c(z, z') = a(z, z') b(z', z)$ $[a^\delta = b^\delta = 0]$
$k^>(t, t') = k(t_+, t'_-)$	$c^> = a^> \cdot b^A + a^R \cdot b^> + a^\rceil \star b^\lceil$	$c^> = a^> b^<$
$k^<(t, t') = k(t_-, t'_+)$	$c^< = a^< \cdot b^A + a^R \cdot b^< + a^\rceil \star b^\lceil$	$c^< = a^< b^>$
$k^R(t, t') = k^\delta(t) \delta(t - t')$ $+\theta(t - t')[k^>(t, t') - k^<(t, t')]$	$c^R = a^R \cdot b^R$	$c^R = \begin{cases} a^R b^< + a^< b^A \\ a^R b^> + a^> b^A \end{cases}$
$k^A(t, t') = k^\delta(t) \delta(t - t')$ $-\theta(t' - t)[k^>(t, t') - k^<(t, t')]$	$c^A = a^A \cdot b^A$	$c^A = \begin{cases} a^A b^< + a^< b^R \\ a^A b^> + a^> b^R \end{cases}$
$k^T(t, t') = k(t_-, t'_-)$	$c^T = a^T \cdot b^T - a^< \cdot b^> + a^\rceil \star b^\lceil$	$c^T = \begin{cases} a^< b^T + a^R b^< \\ a^T b^> + a^> b^A \end{cases}$
$k^{\bar T}(t, t') = k(t_+, t'_+)$	$c^{\bar T} = a^> \cdot b^< - a^{\bar T} \cdot b^{\bar T} + a^\rceil \star b^\lceil$	$c^{\bar T} = \begin{cases} a^{\bar T} b^< - a^< b^A \\ a^> b^{\bar T} - a^R b^> \end{cases}$
$k^\rceil(t, \tau) = k(t_\pm, t_0 - i\tau)$	$c^\rceil = a^R \cdot b^\rceil + a^\rceil \star b^M$	$c^\rceil = a^\rceil b^\lceil$
$k^\lceil(\tau, t) = k(t_0 - i\tau, t_\pm)$	$c^\lceil = a^\lceil \cdot b^A + a^M \star b^\lceil$	$c^\lceil = a^\lceil b^\rceil$
$k^M(\tau, \tau') = k(t_0 - i\tau, t_0 - i\tau')$	$c^M = a^M \star b^M$	$c^M = a^M b^M$

Table 5.1 Definitions of Keldysh components (first column) and identities for the convolution (second column) and the product (third column) of two functions in Keldysh space.

For nonsingular functions the Keldysh components of c can easily be extracted in terms of those of a and b. The reader can check that

$$c^>(t, t') = a^>(t, t') b^<(t', t), \quad c^<(t, t') = a^<(t, t') b^>(t', t),$$
$$c^\rceil(t, \tau) = a^\rceil(t, \tau) b^\lceil(\tau, t), \quad c^\lceil(\tau, t) = a^\lceil(\tau, t) b^\rceil(t, \tau),$$
$$c^M(\tau, \tau') = a^M(\tau, \tau') b^M(\tau', \tau).$$

The retarded/advanced component is then obtained using the above identities. Taking into account that $c^\delta = 0$ we have

$$c^R(t, t') = \theta(t - t')[a^>(t, t') b^<(t', t) - a^<(t, t') b^>(t', t)].$$

We could eliminate the θ-function by adding and subtracting either $a^< b^<$ or $a^> b^>$ and rearranging the terms. The final result is

$$c^R(t, t') = a^R(t, t') b^<(t', t) + a^<(t, t') b^A(t', t)$$
$$= a^R(t, t') b^>(t', t) + a^>(t, t') b^A(t', t).$$

Similarly one finds

$$c^A(t, t') = a^A(t, t') b^<(t', t) + a^<(t, t') b^R(t', t)$$
$$= a^A(t, t') b^>(t', t) + a^>(t, t') b^R(t', t).$$

The time-ordered and anti-time-ordered functions can be derived in a similar way, and the reader can consult the third column of Table 5.1 for the complete list of identities.

Exercise 5.2. Let $a(z, z') = a^\delta(z)\delta(z, z')$ be a function in Keldysh space with only a singular part. Show that the convolution $c(z, z') = \int_\gamma d\bar{z}\, a(z, \bar{z})b(\bar{z}, z')$ has the following components:

$$c^{\lessgtr}(t, t') = a^\delta(t)b^{\lessgtr}(t, t'), \qquad c^{\rceil}(t, \tau) = a^\delta(t)b^{\rceil}(t, \tau),$$
$$c^{\lceil}(\tau, t) = a^\delta(\tau)b^{\lceil}(\tau, t), \qquad c^{M}(\tau, \tau') = a^\delta(\tau)b^{M}(\tau, \tau').$$

Find also the components for the convolution $c(z, z') = \int_\gamma d\bar{z}\, b(z, \bar{z})a(\bar{z}, z')$.

Exercise 5.3. Let a, b, c, be three functions in Keldysh space with $b(z, z') = b^\delta(z)\delta(z, z')$. Denoting by

$$d(z, z') = \int_\gamma d\bar{z}d\bar{z}'\, a(z, \bar{z})b(\bar{z}, \bar{z}')c(\bar{z}', z')$$

the convolution between the three functions, show that

$$d^{\lessgtr}(t, t') = \int_{t_0}^\infty d\bar{t}\, \left[a^{\lessgtr}(t, \bar{t})b^\delta(\bar{t})c^{A}(\bar{t}, t') + a^{R}(t, \bar{t})b^\delta(\bar{t})c^{\lessgtr}(\bar{t}, t') \right]$$
$$- i \int_0^\beta d\bar{\tau}a^{\rceil}(t, \bar{\tau})b^\delta(\bar{\tau})c^{\lceil}(\bar{\tau}, t').$$

Exercise 5.4. Let c be the convolution between two functions a and b in Keldysh space. Show that

$$\int_\gamma dz\, c(z, z^\pm) = (-i)^2 \int_0^\beta d\tau d\tau'\, a^{M}(\tau, \tau')b^{M}(\tau', \tau^\pm),$$

and hence the result involves only integrals along the imaginary track. Hint: use the fact that in accordance with (5.38) $c(z, z^\pm) = c^{\lessgtr}(t, t)$ for both $z = t_-$ and $z = t_+$.

Exercise 5.5. Let $a(z, z')$ be a function in Keldysh space and $f(z)$ a function on the contour with $f(t_+) = f(t_-)$ and $f(t_0 - i\tau) = 0$. Show that

$$\int_\gamma dz'\, a(t_\pm, z')f(z') = \int_{t_0}^\infty dt'\, a^{R}(t, t')f(t'),$$

and

$$\int_\gamma dz'\, f(z')a(z', t_\pm) = \int_{t_0}^\infty dt'\, f(t')a^{A}(t', t).$$

6

One-particle Green's function

In this chapter we get acquainted with the one-particle Green's function G, or simply the Green's function. The chapter is divided in three parts. In the first part (Section 6.1) we illustrate what kind of physical information can be extracted from the different Keldysh components of G. The aim of this first part is to introduce some general concepts without being too formal. In the second part (Section 6.2) we calculate the noninteracting Green's function. Finally in the third part (Sections 6.3 and 6.4) we consider the interacting Green's function and derive several exact properties. We also discuss other physical (and measurable) quantities that can be calculated from G and that are relevant to the analysis of the following chapters.

6.1 What can we learn from G?

We start our overview with a preliminary discussion on the different character of the space-spin and time dependence in $G(1; 2)$. In the Dirac formalism the time-dependent wavefunction $\Psi(\mathbf{x}, t)$ of a single particle is the inner product between the position–spin ket $|\mathbf{x}\rangle$ and the time evolved ket $|\Psi(t)\rangle$. In other words, the wavefunction $\Psi(\mathbf{x}, t)$ is the representation of the ket $|\Psi(t)\rangle$ in the position–spin basis. Likewise, the Green's function $G(1; 2)$ can be thought of as the representation in the position–spin basis of a (single-particle) operator in first quantization

$$\hat{\mathcal{G}}(z_1, z_2) = \int d\mathbf{x}_1 d\mathbf{x}_2 \, |\mathbf{x}_1\rangle G(1; 2)\langle\mathbf{x}_2|,$$

with matrix elements

$$\langle\mathbf{x}_1|\hat{\mathcal{G}}(z_1, z_2)|\mathbf{x}_2\rangle = G(1; 2). \tag{6.1}$$

It is important to appreciate the difference between $\hat{\mathcal{G}}$ and the operator correlator $\hat{G}_1$ of (4.58). The former is an operator in first quantization (and according to our notation is denoted by a calligraphic letter) while the latter is an operator in Fock space. The Green's function operator $\hat{\mathcal{G}}$ is a very useful quantity. Consider, for instance, the expansion (1.60) of the field operators

$$\hat{\psi}(\mathbf{x}) = \sum_i \varphi_i(\mathbf{x})\hat{d}_i, \qquad \hat{\psi}^\dagger(\mathbf{x}) = \sum_i \varphi_i^*(\mathbf{x})\hat{d}_i^\dagger,$$

over some more convenient basis for the problem at hand (if the new basis is not complete the expansion is an approximated expansion). Then, it is also more convenient to work with the Green's function

$$G_{ji}(z_1, z_2) = \frac{1}{i} \frac{\text{Tr}\left[e^{-\beta \hat{H}^{\text{M}}} \mathcal{T}\left\{\hat{d}_{j,H}(z_1)\hat{d}^\dagger_{i,H}(z_2)\right\}\right]}{\text{Tr}\left[e^{-\beta \hat{H}^{\text{M}}}\right]},$$

rather than with $G(1; 2)$. Now the point is that $G_{ji}(z_1, z_2)$ and $G(1; 2)$ are just different matrix elements of the same Green's function operator. Inserting the expansion of the field operators in $G(1; 2)$ we find

$$G(1; 2) = \sum_{ji} \varphi_j(\mathbf{x}_1) G_{ji}(z_1, z_2) \varphi_i^*(\mathbf{x}_2) = \sum_{ji} \langle \mathbf{x}_1 | j \rangle G_{ji}(z_1, z_2) \langle i | \mathbf{x}_2 \rangle,$$

and comparing this equation with (6.1) we conclude that

$$\hat{\mathcal{G}}(z_1, z_2) = \sum_{ji} |j\rangle G_{ji}(z_1, z_2) \langle i|,$$

from which the result

$$\langle j | \hat{\mathcal{G}}(z_1, z_2) | i \rangle = G_{ji}(z_1, z_2)$$

follows directly. Another advantage of working with the Green's function operator rather than with its matrix elements is that one can cast the equations of motion in an invariant form, i.e., in a form that is independent of the basis. The analogue in quantum mechanics would be to work with kets rather than with wavefunctions. For simplicity, let us consider the equations of motion of G for a system of noninteracting particles, $\hat{H}_{\text{int}} = 0$,

$$i\frac{d}{dz_1} G(1; 2) - \int d3\, h(1; 3) G(3; 2) = \delta(1; 2), \tag{6.2}$$

$$-i\frac{d}{dz_2} G(1; 2) - \int d3\, G(1; 3) h(3; 2) = \delta(1; 2), \tag{6.3}$$

where we define

$$h(1; 2) \equiv \delta(z_1, z_2) \langle \mathbf{x}_1 | \hat{h}(z_1) | \mathbf{x}_2 \rangle.$$

Equations (6.2) and (6.3) are the generalization of (5.17) and (5.18) to single-particle Hamiltonians $\hat{h}$ which are not diagonal in position–spin space. We see by inspection that (6.2) and (6.3) are obtained by sandwiching with $\langle \mathbf{x}_1 |$ and $| \mathbf{x}_2 \rangle$ the following equations for operators in first quantization

$$\left[i\frac{d}{dz_1} - \hat{h}(z_1)\right] \hat{\mathcal{G}}(z_1, z_2) = \delta(z_1, z_2), \tag{6.4}$$

$$\hat{\mathcal{G}}(z_1, z_2) \left[-i\frac{\overleftarrow{d}}{dz_2} - \hat{h}(z_2)\right] = \delta(z_1, z_2). \tag{6.5}$$

As usual the arrow over d/dz_2 specifies that the derivative acts on the left. The operator formulation helps us to visualize the structure of the equations of motion. In particular we see that (6.5) resembles the adjoint of (6.4). From now on we refer to (6.5) as the adjoint equation of motion.[1] In a similar way one can construct the (n-particle) operator in first quantization for the n-particle Green's function and cast the whole Martin–Schwinger hierarchy in an operator form. This, however, goes beyond the scope of the section; let us go back to the Green's function.

6.1.1 The inevitable emergence of memory

All the fundamental equations encountered so far are linear *differential equations* in time. An example is the time-dependent Schrödinger equation according to which the state of the system $|\Psi(t + \Delta_t)\rangle$ infinitesimally after the time t can be calculated from the state of the system $|\Psi(t)\rangle$ at time t, see Section 3.2. This means that we do not need to "remember" $|\Psi(t')\rangle$ for all $t' < t$ in order to evolve the state from t to $t + \Delta_t$; the time-dependent Schrödinger equation has no memory. Likewise, the Martin–Schwinger hierarchy for the many-particle Green's functions is a set of coupled *differential equations* in (contour) time that can be used to calculate all the G_ns from their values at an infinitesimally earlier (contour) time. Again, there is no memory involved. The *brute force* solution of the Schrödinger equation or of the Martin–Schwinger hierarchy is, however, not a viable route to make progress. In the presence of many particles the huge number of degrees of freedom of the state of the system (for the Schrödinger equation) or of the many-particle Green's functions G, G_2, G_3, ... (for the Martin–Schwinger hierarchy) renders these equations practically unsolvable. Luckily, we are typically not interested in the full knowledge of these quantities. For instance, to calculate the density $n(\mathbf{x}, t)$ we only need to know the Green's function G, see (5.4). Then the question arises whether it is possible to construct an exact effective equation for the quantities we are interested in by "embedding" all the other degrees of freedom into such an effective equation. As we see below and later on in this book the answer is affirmative, but the embedding procedure inevitably leads to the appearance of memory. This is a very profound concept and deserves a careful explanation.

Let us consider a system consisting of two subsystems coupled to each other. A possible realization is the Fano model where the first subsystem is the atom and the second subsystem is the metal, and the Hamiltonian for a single particle in first quantization is [see Section 2.3.2]

$$\hat{h} = \underbrace{\sum_k \epsilon_k |k\rangle\langle k|}_{\text{metal}} + \underbrace{\epsilon_0 |\epsilon_0\rangle\langle\epsilon_0|}_{\text{atom}} + \underbrace{\sum_k \left(T_k |k\rangle\langle\epsilon_0| + T_k^* |\epsilon_0\rangle\langle k| \right)}_{\text{coupling}} . \tag{6.6}$$

The evolution of a single-particle ket $|\Psi\rangle$ is determined by the time-dependent Schrödinger equation $i\frac{d}{dt}|\Psi(t)\rangle = \hat{h}|\Psi(t)\rangle$. Taking the inner product of the Schrödinger equation with $\langle\epsilon_0|$ and $\langle k|$ we find a coupled system of equations for the amplitudes $\varphi_0(t) = \langle\epsilon_0|\Psi(t)\rangle$

---

[1] In fact (6.5) is the adjoint of (6.4) for any z_1 and z_2 on the horizontal branches of the contour. This is proved in Chapter 9.

and $\varphi_k(t) = \langle k | \Psi(t) \rangle$:

$$i \frac{d}{dt} \varphi_0(t) = \epsilon_0 \varphi_0(t) + \sum_k T_k^* \varphi_k(t),$$

$$i \frac{d}{dt} \varphi_k(t) = \epsilon_k \varphi_k(t) + T_k \varphi_0(t).$$

The second equation can easily be solved for $\varphi_k(t)$. The function

$$g_k^R(t, t_0) = -i\theta(t - t_0) e^{-i\epsilon_k(t - t_0)} \tag{6.7}$$

obeys[2]

$$\left[i \frac{d}{dt} - \epsilon_k \right] g_k^R(t, t_0) = \delta(t - t_0),$$

and therefore for any $t > t_0$

$$\varphi_k(t) = i g_k^R(t, t_0) \varphi_k(t_0) + \int_{t_0}^{\infty} dt' \, g_k^R(t, t') T_k \varphi_0(t'). \tag{6.8}$$

The first term on the r.h.s. is the solution of the homogeneous equation and correctly depends on the boundary condition. Taking into account that for $t \to t_0^+$ the integral vanishes ($g_k^R(t, t')$ is zero for $t' > t$), whereas $g_k^R(t, t_0) \to -i$ we see that (6.8) is an equality in this limit. For times $t > t_0$ the reader can verify that by acting with $\left[i \frac{d}{dt} - \epsilon_k \right]$ on both sides of (6.8) we recover the differential equation for φ_k. Substitution of (6.8) into the equation for $\varphi_0(t)$ gives

$$\left[i \frac{d}{dt} - \epsilon_0 \right] \varphi_0(t) = i \sum_k T_k^* g_k^R(t, t_0) \varphi_k(t_0) + \int_{t_0}^{\infty} dt' \, \Sigma_{\text{em}}^R(t, t') \varphi_0(t'), \tag{6.9}$$

where we define the so-called (retarded) *embedding self-energy*

$$\Sigma_{\text{em}}^R(t, t') = \sum_k T_k^* g_k^R(t, t') T_k. \tag{6.10}$$

The embedding self-energy takes into account that the particle can escape from the atom at time t', wander in the metal and then come back to the atom at time t. For the isolated atom all the T_ks vanish and the solution of (6.9) reduces to $\varphi_0(t) = e^{-i\epsilon_0(t - t_0)} \varphi_0(t_0)$. The important message carried by (6.9) is that we can propagate in time the atomic amplitude φ_0 without knowing the metallic amplitudes φ_k (except for their value at the initial time). The price to pay, however, is that (6.9) is an *integro-differential equation* with a *memory kernel* given by the embedding self-energy: to calculate $\varphi_0(t)$ we must know $\varphi_0(t')$ at all previous times. In other words, memory appears because we embedded the degrees of freedom of one subsystem so as to have an exact effective equation for the degrees of freedom of the

---

[2]As we shall see, $g_k^R(t, t')$ is the retarded component of the Green's function of a noninteracting system with single-particle Hamiltonian $\hat{h}_k = \epsilon_k |k\rangle\langle k|$.

other subsystem. This result is completely general: the consequence of the embedding is the appearence of a memory kernel.

It is instructive (and mandatory for a book on Green's functions) to show how memory appears in the Green's function language. For this purpose we consider a noninteracting system described by the Fano Hamiltonian $\hat{h}$ in (6.6) and initially in equilibrium at a given temperature and chemical potential. The Green's function can be calculated from (6.4) and (6.5) with

$$\hat{h}(z) = \begin{cases} \hat{h}^{\mathrm{M}} = \hat{h} - \mu & \text{for } z = t_0 - i\tau \\ \hat{h} & \text{for } z = t_\pm \end{cases},$$

and by imposing the KMS boundary conditions. By sandwiching (6.4) with $\langle \epsilon_0 |$ and $| \epsilon_0 \rangle$ and with $\langle k |$ and $| \epsilon_0 \rangle$ we find the following coupled system of equations:

$$\left[i\frac{d}{dz_1} - h_{00}(z_1) \right] G_{00}(z_1, z_2) - \sum_k h_{0k}(z_1) G_{k0}(z_1, z_2) = \delta(z_1, z_2),$$

$$\left[i\frac{d}{dz_1} - h_{kk}(z_1) \right] G_{k0}(z_1, z_2) - h_{k0}(z_1) G_{00}(z_1, z_2) = 0,$$

with the obvious notation $h_{00}(z) = \langle \epsilon_0 | \hat{h}(z) | \epsilon_0 \rangle$, $h_{0k}(z) = \langle \epsilon_0 | \hat{h}(z) | k \rangle$, etc., and similarly for the Green's function. As in the example on the single-particle wavefunction, we solve the second of these equations by introducing the Green's function g of the isolated metal whose matrix elements $g_{kk'} = \delta_{kk'} g_k$ obey

$$\left[i\frac{d}{dz_1} - h_{kk}(z_1) \right] g_k(z_1, z_2) = \delta(z_1, z_2),$$

with KMS boundary condition $g_k(t_{0-}, z_2) = \pm g_k(t_0 - i\beta, z_2)$ (upper/lower sign for bosons/fermions). As we see in the next section, the retarded component of g_k is exactly the function in (6.7). We then have

$$G_{k0}(z_1, z_2) = \int_\gamma d\bar{z}\, g_k(z_1, \bar{z}) T_k G_{00}(\bar{z}, z_2),$$

where we take into account that $h_{k0}(z) = T_k$ for all z. From this result the reason for imposing the KMS boundary condition on g_k should be clear: any other boundary condition would have generated a G_{k0} which would not fulfill the KMS relations. Substitution of G_{k0} into the equation for G_{00} leads to an exact effective equation for the atomic Green's function

$$\left[i\frac{d}{dz_1} - h_{00}(z_1) \right] G_{00}(z_1, z_2) - \int_\gamma d\bar{z}\, \Sigma_{\mathrm{em}}(z_1, \bar{z}) G_{00}(\bar{z}, z_2) = \delta(z_1, z_2), \qquad (6.11)$$

where we define the embedding self-energy on the contour according to

$$\Sigma_{\mathrm{em}}(z_1, z_2) = \sum_k T_k^* g_k(z_1, z_2) T_k.$$

The adjoint equation of (6.11) can be derived similarly. Once again, the embedding of the metallic degrees of freedom leads to an equation for the atomic Green's function that contains a memory kernel. Equation (6.11) can be solved without knowing G_{k0} or $G_{kk'}$; we may say that memory emerges when leaving out some information. The equations of motion (6.4) and (6.5) clearly show that memory cannot emerge if *all* the matrix elements of the Green's function are taken into account.

Interacting systems are more complicated since the equations of motion for G involve the two-particle Green's function G_2. Therefore, even considering all the matrix elements of G, the equations of motion do not form a close set of equations for G. We can use the equations of motion for G_2 to express G_2 in terms of G but these equations also involve the three-particle Green's function G_3, so we should first determine G_3. More generally, the equations of motion couple G_n to $G_{n\pm 1}$ and the problem of finding an exact effective equation for G is rather complicated. Nevertheless, a formal solution exists and it has the same mathematical structure as (6.11):

$$\left[\mathrm{i}\frac{d}{dz_1} - \hat{h}(z_1) \right] \hat{\mathcal{G}}(z_1, z_2) - \int_\gamma d\bar{z}\, \hat{\Sigma}(z_1, \bar{z})\hat{\mathcal{G}}(\bar{z}, z_2) = \delta(z_1, z_2). \tag{6.12}$$

In this equation the memory kernel $\hat{\Sigma}$ (a single-particle operator like $\hat{\mathcal{G}}$) depends only on the Green's function and is known as the *many-body self-energy*. We prove (6.12) in Chapter 9. For the moment it is important to appreciate that one more time, and in a completely different context, the embedding of degrees of freedom (all the G_n with $n \geq 2$) leads to an exact effective *integro-differential equation* for the Green's function that contains a memory kernel. Due to the same mathematical structure of (6.11) and (6.12), the solution of the Fano model provides us with interesting physical information on the behavior of an interacting many-body system.

6.1.2   Matsubara Green's function and initial preparations

The Matsubara component of the Green's function follows from $\hat{\mathcal{G}}(z_1, z_2)$ by setting $z_1 = t_0 - \mathrm{i}\tau_1$ and $z_2 = t_0 - \mathrm{i}\tau_2$. The generic matrix element then reads

$$G_{ji}^{\mathrm{M}}(\tau_1, \tau_2) = \frac{1}{\mathrm{i}} \left\{ \theta(\tau_1 - \tau_2) \frac{\mathrm{Tr}\left[e^{(\tau_1 - \tau_2 - \beta)\hat{H}^{\mathrm{M}}} \hat{d}_j e^{(\tau_2 - \tau_1)\hat{H}^{\mathrm{M}}} \hat{d}_i^\dagger \right]}{\mathrm{Tr}\left[e^{-\beta\hat{H}^{\mathrm{M}}} \right]} \right.$$
$$\left. \pm \theta(\tau_2 - \tau_1) \frac{\mathrm{Tr}\left[e^{(\tau_2 - \tau_1 - \beta)\hat{H}^{\mathrm{M}}} \hat{d}_i^\dagger e^{(\tau_1 - \tau_2)\hat{H}^{\mathrm{M}}} \hat{d}_j \right]}{\mathrm{Tr}\left[e^{-\beta\hat{H}^{\mathrm{M}}} \right]} \right\}, \tag{6.13}$$

and does not contain any information on how the system evolves in time. Instead, it contains information on how the system is initially prepared. As already pointed out in Chapter 4, the initial state of the system can be the thermodynamic equilibrium state (in which case $\hat{H}^{\mathrm{M}} = \hat{H}(t_0) - \mu\hat{N}$) or any other state. It is easy to show that *from the equal-time Matsubara Green's function we can calculate the initial ensemble average of any one-body operator*

$$\hat{O} = \sum_{ij} O_{ij} \hat{d}_i^\dagger \hat{d}_j, \tag{6.14}$$

since

$$O = \frac{\text{Tr}\left[e^{-\beta \hat{H}^{\text{M}}} \hat{O}\right]}{\text{Tr}\left[e^{-\beta \hat{H}^{\text{M}}}\right]} = \sum_{ij} O_{ij} \frac{\text{Tr}\left[e^{-\beta \hat{H}^{\text{M}}} \hat{d}_i^\dagger \hat{d}_j\right]}{\text{Tr}\left[e^{-\beta \hat{H}^{\text{M}}}\right]} = \pm i \sum_{ij} O_{ij} G_{ji}^{\text{M}}(\tau, \tau^+), \qquad (6.15)$$

with τ^+ a time infinitesimally larger than τ. The r.h.s. is independent of τ since all matrix elements of $\hat{\mathcal{G}}^{\text{M}}(\tau_1, \tau_2)$ depend on the time difference $\tau_1 - \tau_2$ only. We further observe that the KMS relations imply

$$\hat{\mathcal{G}}^{\text{M}}(0, \tau) = \pm \hat{\mathcal{G}}^{\text{M}}(\beta, \tau), \qquad \hat{\mathcal{G}}^{\text{M}}(\tau, 0) = \pm \hat{\mathcal{G}}^{\text{M}}(\tau, \beta), \qquad (6.16)$$

i.e., the Matsubara Green's function is a periodic function for bosons and an antiperiodic function for fermions and the period is given by the inverse temperature β. We can therefore expand the Matsubara Green's function in a Fourier series according to

$$\boxed{\hat{\mathcal{G}}^{\text{M}}(\tau_1, \tau_2) = \frac{1}{-i\beta} \sum_{m=-\infty}^{\infty} e^{-\omega_m(\tau_1 - \tau_2)} \hat{\mathcal{G}}^{\text{M}}(\omega_m)} \qquad (6.17)$$

with *Matsubara frequencies*

$$\omega_m = \begin{cases} \dfrac{2m\pi}{-i\beta} & \text{for bosons} \\[2mm] \dfrac{(2m+1)\pi}{-i\beta} & \text{for fermions} \end{cases}.$$

Let us calculate the coefficients of the expansion for a noninteracting density matrix, i.e., for a density matrix with a one-body operator

$$\hat{H}^{\text{M}} = \sum_{ij} h_{ij}^{\text{M}} \hat{d}_i^\dagger \hat{d}_j.$$

Setting $z_1 = t_0 - i\tau_1$ and $z_2 = t_0 - i\tau_2$ in the equation of motion (6.4) we find

$$\left[-\frac{d}{d\tau_1} - \hat{h}^{\text{M}}\right] \hat{\mathcal{G}}^{\text{M}}(\tau_1, \tau_2) = \delta(-i\tau_1 + i\tau_2) = i\delta(\tau_1 - \tau_2),$$

where in the last step we use $\delta(-i\tau) = i\delta(\tau)$, see the discussion before (5.41). Inserting (6.17) into the above equation and exploiting the identities (see Appendix A),

$$\delta(\tau) = \frac{1}{\beta} \sum_{m=-\infty}^{\infty} \begin{cases} e^{-i\frac{2m\pi}{\beta}\tau} \\ e^{-i\frac{(2m+1)\pi}{\beta}\tau} \end{cases} = \frac{1}{\beta} \sum_{m=-\infty}^{\infty} e^{-\omega_m \tau},$$

we can easily extract the coefficients

$$\hat{\mathcal{G}}^{\text{M}}(\omega_m) = \frac{1}{\omega_m - \hat{h}^{\text{M}}}. \qquad (6.18)$$

To familiarize ourselves with these new formulas we now calculate the occupation n_0 of the Fano model and show that the result agrees with (2.27).

The occupation operator for the atomic site in the Fano model is $\hat{n}_0 = \hat{d}_0^\dagger \hat{d}_0$, and from (6.15) (lower sign for fermions) the ensemble average of $\hat{n}_0$ is

$$n_0 = -iG_{00}^{\mathrm{M}}(\tau, \tau^+) = \frac{-i}{-i\beta} \sum_m e^{\eta \omega_m} G_{00}^{\mathrm{M}}(\omega_m).$$

The second equality follows from (6.17) in which we set $\tau_1 = \tau$ and $\tau_2 = \tau^+ = \tau + \eta$, where η is an infinitesimal positive constant. We consider the system in thermodynamic equilibrium at a given temperature and chemical potential so that $\hat{h}^{\mathrm{M}} = \hat{h} - \mu$. The expansion coefficients can be calculated from (6.18) following the same steps leading to (2.25) and read

$$G_{00}^{\mathrm{M}}(\omega_m) = \langle \epsilon_0 | \frac{1}{\omega_m - \hat{h}^{\mathrm{M}}} | \epsilon_0 \rangle = \frac{1}{\omega_m + \mu - \epsilon_0 - \Sigma_{\mathrm{em}}(\omega_m + \mu)}.$$

To evaluate their sum we use the following trick. We first observe that the function $Q(\zeta) \equiv G_{00}^{\mathrm{M}}(\zeta)$ of the complex variable ζ is analytic everywhere except along the real axis[3] where it has poles (and/or branch-cuts when the spectrum ϵ_k becomes a continuum). For any such function we can write

$$\frac{1}{-i\beta} \sum_{m=-\infty}^{\infty} e^{\eta \omega_m} Q(\omega_m) = \int_{\Gamma_a} \frac{d\zeta}{2\pi} f(\zeta) e^{\eta \zeta} Q(\zeta), \tag{6.19}$$

where $f(\zeta) = 1/(e^{\beta \zeta} + 1)$ is the Fermi function with simple poles in $\zeta = \omega_m$ and residues $-1/\beta$, while Γ_a is the contour of Fig. 6.1(a) that encircles all Matsubara frequencies clockwisely.[4] Taking into account that $\eta > 0$ we can deform the contour Γ_a into the contour Γ_b of Fig. 6.1(b) since $\lim_{\zeta \to \pm\infty} e^{\eta \zeta} f(\zeta) = 0$, thus obtaining

$$\frac{1}{-i\beta} \sum_{m=-\infty}^{\infty} e^{\eta \omega_m} Q(\omega_m) = \lim_{\delta \to 0^+} \left[\int_{-\infty}^{\infty} \frac{d\omega}{2\pi} f(\omega) Q(\omega - i\delta) + \int_{\infty}^{-\infty} \frac{d\omega}{2\pi} f(\omega) Q(\omega + i\delta) \right]$$

$$= \lim_{\delta \to 0^+} \int_{-\infty}^{\infty} \frac{d\omega}{2\pi} f(\omega) \left[Q(\omega - i\delta) - Q(\omega + i\delta) \right].$$

---

[3]Suppose that the denominator of $G_{00}^{\mathrm{M}}(\zeta)$ vanishes for some complex $\zeta = x + iy$. Then, taking the imaginary part of the denominator we find

$$y \left(1 + \sum_k \frac{|T_k|^2}{(x + \mu - \epsilon_0)^2 + y^2} \right) = 0,$$

which can be satisfied only for $y = 0$.

[4]Recall that the Cauchy residue theorem for a meromorphic function $f(z)$ in a domain D states that

$$\oint_\gamma dz f(z) = 2\pi i \sum_j \lim_{z \to z_j} (z - z_j) f(z),$$

where γ is an anticlockwisely oriented contour in D and the sum runs over the simple poles z_j of $f(z)$ contained in γ (a minus sign in front of the formula appears for clockwisely oriented contours like Γ_a).

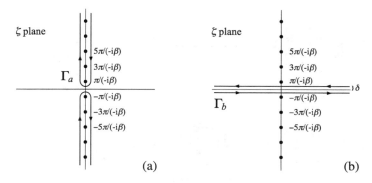

Figure 6.1 Contours for evaluation of the sum $\sum_m e^{\eta \omega_m} G_{00}^{\mathrm{M}}(\omega_m)$. The points displayed are the fermionic Matsubara frequencies $\omega_m = (2m+1)\pi/(-\mathrm{i}\beta)$.

Using the same notation as in (2.26) we have

$$\lim_{\delta \to 0^+} [G_{00}^{\mathrm{M}}(\omega - \mathrm{i}\delta) - G_{00}^{\mathrm{M}}(\omega + \mathrm{i}\delta)] = -2\mathrm{i} \operatorname{Im} \frac{1}{\omega + \mu - \epsilon_0 - \Lambda(\omega + \mu) + \frac{\mathrm{i}}{2}\Gamma(\omega + \mu) + \mathrm{i}\delta},$$

and hence

$$n_0 = -2 \int_{-\infty}^{\infty} \frac{d\omega}{2\pi} f(\omega - \mu) \operatorname{Im} \frac{1}{\omega - \epsilon_0 - \Lambda(\omega) + \frac{\mathrm{i}}{2}\Gamma(\omega) + \mathrm{i}\delta}. \tag{6.20}$$

This result generalizes (2.27) to finite temperatures and correctly reduces to (2.27) at zero temperature.

6.1.3 Lesser/greater Green's function: relaxation and quasi-particles

To access the dynamical properties of a system it is necessary to know: (1) how the system is initially prepared; and (2) how the system evolves in time. If we restrict ourselves to the time-dependent ensemble average of one-body operators as in (6.14), both information (1) and (2) are encoded in the lesser and greater Green's functions which, by definition, read

$$G_{ji}^{<}(t, t') = \mp \mathrm{i} \frac{\operatorname{Tr} \left[e^{-\beta \hat{H}^{\mathrm{M}}} \hat{d}_{i,H}^{\dagger}(t') \hat{d}_{j,H}(t) \right]}{\operatorname{Tr} \left[e^{-\beta \hat{H}^{\mathrm{M}}} \right]} = \mp \mathrm{i} \sum_k \rho_k \langle \Psi_k | \hat{d}_{i,H}^{\dagger}(t') \hat{d}_{j,H}(t) | \Psi_k \rangle, \tag{6.21}$$

$$G_{ji}^{>}(t, t') = -\mathrm{i} \frac{\operatorname{Tr} \left[e^{-\beta \hat{H}^{\mathrm{M}}} \hat{d}_{j,H}(t) \hat{d}_{i,H}^{\dagger}(t') \right]}{\operatorname{Tr} \left[e^{-\beta \hat{H}^{\mathrm{M}}} \right]} = -\mathrm{i} \sum_k \rho_k \langle \Psi_k | \hat{d}_{j,H}(t) \hat{d}_{i,H}^{\dagger}(t') | \Psi_k \rangle. \tag{6.22}$$

The $\hat{d}$-operators in these formulas are in the standard (as opposed to contour) Heisenberg picture (3.16). In the second equalities we simply introduce a complete set of eigenkets $|\Psi_k\rangle$

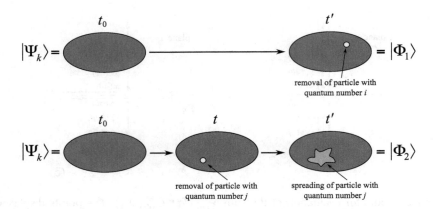

Figure 6.2 Representation of the states $|\Phi_1\rangle$ and $|\Phi_2\rangle$ appearing in (6.25).

of $\hat{H}^{\mathrm{M}}$ with eigenvalue E_k^{M} so that $\rho_k = e^{-\beta E_k^{\mathrm{M}}}/Z$. It is straightforward to verify that the greater/lesser Green's function has the property

$$\boxed{[G_{ji}^>(t,t')]^* = -G_{ij}^>(t',t)} \qquad \boxed{[G_{ji}^<(t,t')]^* = -G_{ij}^<(t',t)} \qquad (6.23)$$

and consequently the retarded and advanced Green's functions are related by

$$\boxed{\hat{\mathcal{G}}^{\mathrm{R}}(t,t') = \theta(t-t')\left[\hat{\mathcal{G}}^>(t,t') - \hat{\mathcal{G}}^<(t,t')\right] = \left[\hat{\mathcal{G}}^{\mathrm{A}}(t',t)\right]^\dagger} \qquad (6.24)$$

Below we discuss the lesser Green's function and leave it as an exercise for the reader to go through the same logical and mathematical steps in the case of the greater Green's function.

A generic term of the sum in (6.21) contains the quantity

$$\langle \Psi_k | \hat{d}_{i,H}^\dagger(t')\hat{d}_{j,H}(t) | \Psi_k \rangle = \underbrace{\langle \Psi_k | \hat{U}(t_0,t')\hat{d}_i^\dagger}_{\langle \Phi_1 |} \underbrace{\hat{U}(t',t)\,\hat{d}_j\,\hat{U}(t,t_0)|\Psi_k\rangle}_{|\Phi_2\rangle}, \qquad (6.25)$$

with $\hat{U}$ the evolution operator (3.15). This quantity is proportional to the probability amplitude that evolving $|\Psi_k\rangle$ from t_0 to t, then removing a particle with quantum number j and letting the new state evolve from t to t' (this state is $|\Phi_2\rangle$) we find the same state as evolving $|\Psi_k\rangle$ from t_0 to t', at which time a particle with quantum number i is removed (this state is $|\Phi_1\rangle$), see Fig. 6.2. As suggested by the figure, when time passes the disturbance (removal of a particle) "spreads" if j is not a good quantum number and, therefore, it is reasonable to expect that the probability amplitude vanishes for $|t - t'| \to \infty$ independently of the quantum number i. Of course, this expectation makes sense provided that the system has infinitely many degrees of freedom coupled to each other. If the system has only a finite number of degrees of freedom (as, e.g., in the PPP model for benzene or in the Hubbard dimer discussed in Chapter 2) then the probability amplitude exhibits an oscillatory behavior.

Let us go back to (6.25). We note that for $t = t'$ the disturbance has no time to spread. In this special (but relevant) case the probability amplitude reduces to the overlap between $\hat{d}_j|\Psi_k(t)\rangle$ and $\hat{d}_i|\Psi_k(t)\rangle$, where $|\Psi_k(t)\rangle = \hat{U}(t, t_0)|\Psi_k\rangle$. For $i = j$ the overlap becomes the probability of finding a particle with quantum number i when the system is in $|\Psi_k(t)\rangle$, i.e., the overlap becomes the quantum average of the occupation operator $\hat{n}_i = \hat{d}_i^\dagger \hat{d}_i$.[5] This is not surprising since the time-dependent ensemble average of $\hat{n}_i$ is $n_i(t) = \pm i\, G_{ii}(z, z^+) = \pm i\, G_{ii}^<(t, t)$. More generally, *the lesser Green's function allows us to calculate the time-dependent ensemble average of any one-body operator*, since given an operator $\hat{O}$ as in (6.14) we have

$$O(t) = \pm i \sum_{ij} O_{ij} G_{ji}^<(t, t). \tag{6.26}$$

For $t = t_0$ the average $O(t_0)$ must be equal to the initial average (6.15) and therefore

$$\hat{\mathcal{G}}^<(t_0, t_0) = \hat{\mathcal{G}}^M(\tau, \tau^+) \quad \text{for all } \tau. \tag{6.27}$$

This identity can also be directly deduced from the definition of the Matsubara and lesser components of the Green's function.

A special case that is worth discussing in more detail is a system with Hamiltonian $\hat{H}(t) = \hat{H}(t_0) = \hat{H}$ constant in time. The corresponding Green's function can describe either: (1) a system with Hamiltonian $\hat{H}' \equiv \hat{H}^M + \mu \hat{N}$ initially in thermodynamic equilibrium[6] and then driven out of equilibrium by the sudden switch-on of a perturbation $\Delta \hat{H} = \hat{H} - \hat{H}'$, so that $\hat{H}(t > t_0) = \hat{H}' + \Delta \hat{H} = \hat{H}$; or (2) a system with Hamiltonian $\hat{H}$ initially prepared in an excited (not necessarily stationary) configuration described by a density matrix $\hat{\rho} = e^{-\beta \hat{H}^M}/Z$. No matter what the physical situation, the evolution operator is a simple exponential and (6.21) becomes

$$G_{ji}^<(t, t') = \mp i \sum_k \rho_k \langle \Psi_k | e^{i\hat{H}(t'-t_0)} \hat{d}_i^\dagger e^{-i\hat{H}(t'-t)} \hat{d}_j e^{-i\hat{H}(t-t_0)} | \Psi_k \rangle. \tag{6.28}$$

In general the lesser Green's function is *not* a function of the time difference $t - t'$ only. This is a consequence of the fact that if the $|\Psi_k\rangle$ are not eigenstates of $\hat{H}$ then their evolution is not given by a multiplicative (time-dependent) phase factor.

Relaxation

From (6.28) we can discuss the "spreading" illustrated in Fig. 6.2 more quantitatively. Let us consider again the Hamiltonian $\hat{H}$ of the Fano model. As in Section 2.3.2, we define $\hat{c}_\lambda$ ($\hat{c}_\lambda^\dagger$) as the annihilation (creation) operators that bring the Hamiltonian into a diagonal form,

---

[5]It may be worth commenting on how a probability amplitude can become a probability. We show it with an example. Given an unnormalized state $|\Psi\rangle$ [like $|\Phi_1\rangle$ or $|\Phi_2\rangle$ in (6.25)], the quantity $\langle \Psi | \Psi \rangle$ is proportional to the probability amplitude of finding $|\Psi\rangle$ when the system is in $|\Psi\rangle$. Now if we write $|\Psi\rangle = |\Psi_1\rangle \langle \Psi_2 | \Phi \rangle$, with $|\Psi_1\rangle$, $|\Psi_2\rangle$ and $|\Phi\rangle$ normalized to 1 then $\langle \Psi | \Psi \rangle = |\langle \Psi_2 | \Phi \rangle|^2$ equals the probability of finding $|\Psi_2\rangle$ when the system is in $|\Phi\rangle$. The reader can generalize this argument to the case in which $|\Psi_1\rangle \langle \Psi_2 |$ is replaced by the operator $\hat{d}_i$.

[6]The equilibrium density matrix of a system with Hamiltonian $\hat{H}'$ is $\hat{\rho} = e^{-\beta \hat{H}^M}/Z$ with $\hat{H}^M = \hat{H}' - \mu \hat{N}$.

$\hat{H} = \sum_\lambda \epsilon_\lambda \hat{c}^\dagger_\lambda \hat{c}_\lambda$. To calculate the lesser Green's function $G^<_{00}$ on the atomic site we expand the operators $\hat{d}_0$ and $\hat{d}^\dagger_0$ in terms of the $\hat{c}$-operators (see Section 1.5),

$$\hat{d}_0 = \sum_\lambda \langle \epsilon_0 | \lambda \rangle \, \hat{c}_\lambda, \quad \hat{d}^\dagger_0 = \sum_\lambda \langle \lambda | \epsilon_0 \rangle \, \hat{c}^\dagger_\lambda,$$

and use the identities

$$\hat{c}_{\lambda,H}(t) = e^{i\hat{H}(t-t_0)} \hat{c}_\lambda e^{-i\hat{H}(t-t_0)} = e^{-i\epsilon_\lambda(t-t_0)} \hat{c}_\lambda,$$
$$\hat{c}^\dagger_{\lambda,H}(t) = e^{i\hat{H}(t-t_0)} \hat{c}^\dagger_\lambda e^{-i\hat{H}(t-t_0)} = e^{i\epsilon_\lambda(t-t_0)} \hat{c}^\dagger_\lambda,$$

that follow directly from the equation of motion $i\frac{d}{dt}\hat{c}_{\lambda,H}(t) = [\hat{c}_{\lambda,H}(t), \hat{H}]_- = \epsilon_\lambda \hat{c}_{\lambda,H}(t)$ and the like for $\hat{c}^\dagger_\lambda$. Equation (6.28) then gives

$$G^<_{00}(t,t') = \sum_{\lambda\lambda'} \langle \lambda | \epsilon_0 \rangle \langle \epsilon_0 | \lambda' \rangle \, e^{i[\epsilon_\lambda(t'-t_0) - \epsilon_{\lambda'}(t-t_0)]} G^<_{\lambda'\lambda}(t_0, t_0)$$

$$= \langle \epsilon_0 | e^{-i\hat{h}(t-t_0)} \hat{G}^<(t_0, t_0) e^{i\hat{h}(t'-t_0)} | \epsilon_0 \rangle, \tag{6.29}$$

where $\hat{h}$ is the single-particle Hamiltonian with eigenkets $|\lambda\rangle$ and eigenvalues ϵ_λ. We could have gone through the same steps to calculate any matrix element of $\hat{G}^<$. Thus, more generally, given an arbitrary density matrix (or equivalently an arbitrary $\hat{H}^M$) the time dependence of the lesser Green's function of a system with Hamiltonian $\hat{H}(t > t_0) = \sum_{ij} h_{ij} \hat{d}^\dagger_i \hat{d}_j$ is given by

$$\hat{G}^<(t,t') = e^{-i\hat{h}(t-t_0)} \hat{G}^<(t_0, t_0) e^{i\hat{h}(t'-t_0)}. \tag{6.30}$$

As anticipated, *the r.h.s. of (6.29) is an oscillatory function for systems with a finite number of degrees of freedom* (in the present case a finite number of single-particle eigenvalues ϵ_λ). If $\hat{G}^<(t_0, t_0)$ is the lesser Green's function of the Fano model in thermodynamic equilibrium, then (6.29) solves the equation of motion (6.11) with $z_1 = t_-$ and $z_2 = t'_+$, i.e.,

$$\left[i\frac{d}{dt} - \epsilon_0 \right] G^<_{00}(t,t') - \int_{t_0}^\infty d\bar{t} \, \left[\Sigma^R_{em}(t,\bar{t}) G^<_{00}(\bar{t},t') + \Sigma^<_{em}(t,\bar{t}) G^A_{00}(\bar{t},t') \right]$$

$$- (-i) \int_0^\beta d\bar{\tau} \, \Sigma^\rceil_{em}(t,\bar{\tau}) G^\lceil_{00}(\bar{\tau},t') = 0, \tag{6.31}$$

where we use the Langreth rules of Table 5.1 to evaluate the integral along the contour. It is then clear that also the embedding self-energy must be an oscillatory function in finite systems. Due to the formal similarity between (6.11) and (6.12) we can argue that in an interacting system with a finite number of degrees of freedom the many-body self-energy is an oscillatory function as well.[7]

When the spectrum ϵ_λ becomes a continuum (in the language of Section 2.3.2 this occurs in the limit $\Delta\epsilon\Delta Q \to 0$) the sum over λ and λ' in (6.29) becomes a double integral

---

[7]This is indeed the case but the reader must wait for the development of a perturbative many-body scheme in order to make the above intuitive argument a rigorous statement.

and the lesser Green's function takes the following mathematical form:

$$G_{00}^<(t,t') = \int d\lambda \, d\lambda' \, e^{i[\epsilon_\lambda t' - \epsilon_{\lambda'} t]} f(\lambda, \lambda'),$$ (6.32)

with $f(\lambda, \lambda')$ a function of the continuous quantum numbers λ and λ'. There is a standard manipulation to transform these kinds of integral into energy integrals. Given a function $g(\lambda)$ of the quantum number λ we have

$$\int d\lambda \, g(\lambda) = \int \frac{d\omega}{2\pi} \underbrace{\int d\lambda \, 2\pi \delta(\omega - \epsilon_\lambda) g(\lambda)}_{\tilde{g}(\omega)} = \int \frac{d\omega}{2\pi} \, \tilde{g}(\omega).$$ (6.33)

In the special case that $g(\lambda) = g(\epsilon_\lambda)$ depends on λ only through the eigenvalue ϵ_λ the function $\tilde{g}(\omega) = D(\omega)g(\omega)$ where

$$D(\omega) = 2\pi \int d\lambda \, \delta(\omega - \epsilon_\lambda)$$ (6.34)

is the *density of single particle states* with energy ω. Using this trick the double integral in (6.32) can be rewritten as a double energy integral

$$G_{00}^<(t,t') = \int \frac{d\omega}{2\pi} \frac{d\omega'}{2\pi} \, e^{i[\omega t' - \omega' t]} \tilde{f}(\omega, \omega').$$

At any fixed time t the r.h.s. of this equation approaches zero when $t' \to \infty$, provided that $\int \frac{d\omega'}{2\pi} e^{-i\omega' t} \tilde{f}(\omega, \omega')$ is an integrable function of ω. In mathematics this result is known as the Riemann–Lebesgue theorem.[8] We thus find the anticipated result on the "spreading": *a quantum number that is coupled to infinitely many other quantum numbers* (and hence it is not a good quantum number) *propagates in time and distributes smoothly over all of them.* This is the case of the quantum number 0 which is coupled by the Hamiltonian to all the quantum numbers λ, i.e., $\langle \lambda | \hat{h} | 0 \rangle \neq 0$ for all λ. The same argument can be used to demonstrate that the r.h.s. of (6.32) vanishes at any fixed t' when $t \to \infty$. The "spreading" of the removed particle resembles the relaxation process that takes place when a drop of ink falls into a bucket of water, and for this reason we refer to it as the *relaxation*. We deepen this aspect further in Section 6.3.4.

---

[8]According to the Riemann–Lebesgue theorem

$$\lim_{t \to \infty} \int d\omega \, e^{i\omega t} f(\omega) = 0,$$

provided that f is an integrable function. An intuitive proof of this result consists in splitting the integral over ω into the sum of integrals over a window $\Delta\omega$ around $\omega_n = n\Delta\omega$. If $f(\omega)$ is finite and $\Delta\omega$ is small enough we can approximate $f(\omega) \sim f(\omega_n)$ in each window and write

$$\lim_{t \to \infty} \int d\omega \, e^{i\omega t} f(\omega) \sim \sum_{n=-\infty}^{\infty} f(\omega_n) \lim_{t \to \infty} \int_{\omega_n - \Delta\omega/2}^{\omega_n + \Delta\omega/2} d\omega \, e^{i\omega t}$$

$$= \sum_{n=-\infty}^{\infty} f(\omega_n) \lim_{t \to \infty} \frac{2e^{i\omega_n t} \sin(\Delta\omega t/2)}{t} = 0.$$

Relaxation and memory

The equation of motion (6.31) tells us that for $G_{00}^<(t, t')$ to approach zero when $|t - t'| \to \infty$ the self-energy $\Sigma_{em}(t, t')$ must approach zero in the same limit. Vice versa we could state that for relaxation to occur, the equation of motion for the "sub-system" Green's function (the atomic Green's function of the Fano model or the full Green's function of an interacting system) must have memory and this memory must decay in time. *A sub-system with no memory or with an infinitely long memory does not, in general, relax.* It is instructive to re-consider the behavior of $G_{00}^<$ from this new perspective. Below we discuss an elementary example of both kinds of memory. Let us consider the equation of motion (6.31) with $t' = t_0$. The advanced Green's function vanishes for $\bar{t} > t'$ and hence the second term in the first integral does not contribute, and (6.31) simplifies to

$$\left[i \frac{d}{dt} - \epsilon_0 \right] G^<(t) - \int_{t_0}^\infty d\bar{t} \, \Sigma_{em}^R(t, \bar{t}) G^<(\bar{t}) = f(t),$$

where we introduce the short-hand notation $G^<(t) \equiv G_{00}^<(t, t_0)$ and

$$f(t) = -i \int_0^\beta d\bar{\tau} \, \Sigma_{em}^\rceil(t, \bar{\tau}) G_{00}^\lceil(\bar{\tau}, t_0).$$

Since $G_{00}^\lceil(\bar{\tau}, t_0) = G_{00}^M(\bar{\tau}, 0)$ the function $f(t)$ is completely determined by the initial density matrix and can be considered as an external driving force. The general solution of the differential equation is the sum of the homogeneous solution $G_{hom}^<$ and of the particular solution. The homogeneous solution satisfies

$$\left[i \frac{d}{dt} - \epsilon_0 \right] G_{hom}^<(t) - \int_{t_0}^\infty d\bar{t} \, \Sigma_{em}^R(t, \bar{t}) G_{hom}^<(\bar{t}) = 0. \tag{6.35}$$

The explicit form of Σ_{em}^R is obtained by inserting (6.7) into (6.10) and reads

$$\Sigma_{em}^R(t, t') = -i\theta(t - t') \sum_k |T_k|^2 e^{-i\epsilon_k(t - t')} = \int \frac{d\omega}{2\pi} e^{-i\omega(t - t')} \sum_k \frac{|T_k|^2}{\omega - \epsilon_k + i\eta}, \tag{6.36}$$

with η an infinitesimal positive constant. Note that the Fourier transform $\Sigma_{em}^R(\omega)$ is equal to the quantity $\Sigma_{em}(\omega + i\eta)$ already encountered in (2.26).

- *Infinitely long memory.* As an example of long memory kernels we consider the case for which

$$\Sigma_{em}^R(\omega) = \sum_k \frac{|T_k|^2}{\omega - \epsilon_k + i\eta} = \frac{L}{\omega + i\eta} \quad \Rightarrow \quad \Sigma_{em}^R(t, t') = -i\theta(t - t')L.$$

The constant L must be a real positive constant for the imaginary part of $L/(\omega + i\eta)$ to be negative. Inserting this self-energy into (6.35) and differentiating once with respect to t we find

$$\left[i \frac{d^2}{dt^2} - \epsilon_0 \frac{d}{dt} + iL \right] G_{hom}^<(t) = 0,$$

the solution of which is $G_{\text{hom}}^<(t) = Ae^{\alpha_+ t} + Be^{\alpha_- t}$ with

$$\alpha_\pm = \frac{\epsilon_0 \pm \sqrt{\epsilon_0^2 + 4L}}{2i}.$$

Both roots are purely imaginary and therefore $G_{\text{hom}}^<(t)$ does not vanish for large t, i.e., there is no relaxation. Long memory kernels are typical of systems with a finite number of degrees of freedom.

- *No memory.* An example of kernel with no memory is a $\Sigma_{\text{em}}^R(\omega)$ independent of ω:

$$\Sigma_{\text{em}}^R(\omega) = \sum_k \frac{|T_k|^2}{\omega - \epsilon_k + i\eta} = \Sigma_{\text{em}}^R \quad \Rightarrow \quad \Sigma_{\text{em}}^R(t, t') = \delta(t - t')\Sigma_{\text{em}}^R.$$

A purely imaginary constant $\Sigma_{\text{em}}^R = -i\Gamma/2$ corresponds to the WBLA encountered in Section 2.3.2. Even though the WBLA generates a retarded kernel with no memory the lesser component of the kernel decays smoothly in time.[9] This remains true as long as the constant Σ_{em}^R has an imaginary part. On the contrary, for real Σ_{em}^R the lesser component of the kernel vanishes.[10] Inserting the no-memory self-energy into (6.35) we find for the homogeneous solution

$$G_{\text{hom}}^<(t) = e^{-i(\epsilon_0 + \Sigma_{\text{em}}^R)(t - t_0)} G_{\text{hom}}^<(t_0).$$

We thus see that for real Σ_{em}^R (no memory) there is no relaxation whereas in the WBLA ($\Sigma_{\text{em}}^R = -i\Gamma/2$) or for complex Σ_{em}^R there is memory and, as expected, the Green's function decays (exponentially) in time.

Quasi-particles

Exploiting one more time the formal analogy between the Fano model and an interacting many-particle system we could say that if the operator $\hat{h} + \hat{\Sigma}^R(\omega)$ has an eigenvalue ϵ_λ with vanishing imaginary part, like the $\epsilon_0 + \Sigma_{\text{em}}^R$ of the previous example, then the creation or annihilation of a particle with energy ϵ_λ is not followed by relaxation. In other words the particle with energy ϵ_λ has an infinitely long life-time. In a truly noninteracting system these particles always exist since the many-body self-energy $\hat{\Sigma}^R(\omega) = 0$. In an interacting system

---

[9]In the next section we calculate all Keldysh components of the Green's function of a noninteracting system. The reader can then verify that $g_k^<(t, t') = \mp i f(\epsilon_k - \mu)e^{-i\epsilon_k(t-t')}$ and therefore

$$\Sigma_{\text{em}}^<(t, t') = \mp i \sum_k |T_k|^2 f(\epsilon_k - \mu)e^{-i\epsilon_k(t-t')} = \mp i \int \frac{d\omega}{2\pi} f(\omega - \mu)e^{-i\omega(t-t')} 2\pi \sum_k |T_k|^2 \delta(\omega - \epsilon_k).$$

The sum inside the integral is proportional to the imaginary part of $\Sigma_{\text{em}}^R(\omega)$. Thus, for a purely imaginary and frequency-independent kernel (WBLA) $\Sigma_{\text{em}}^<(t, t') \propto \int d\omega f(\omega - \mu)e^{-i\omega(t-t')}$ which vanishes only in the limit $|t - t'| \to \infty$.

[10]For the specific form (6.36) of the embedding self-energy it follows that if the constant Σ_{em}^R is real then it must be zero. This is due to the fact that $\text{Im}[\Sigma_{\text{em}}^R(\omega)] = -\pi \sum_k |T_k|^2 \delta(\omega - \epsilon_k)$ consists of a sum of terms with the same sign. Thus, a real Σ_{em}^R implies that all T_k vanish and hence $\Sigma_{\text{em}}^R = 0$.

almost all eigenvalues of $\hat{h} + \hat{\Sigma}^{\mathrm{R}}(\omega)$ have a finite imaginary part and hence the concept of "particles" with well-defined energy must be abandoned. However, if the imaginary part is small (with respect to some energy scale) these particles have a long life-time and behave almost like normal particles. For this reason they are called *quasi-particles*. As we see in Chapter 13, the concept of quasi-particles allows us to develop many useful and insightful approximations.

6.2 Noninteracting Green's function

In this section we calculate the Green's function for a system of noninteracting particles that is initially described by a noninteracting density matrix. This means that the Hamiltonian along the contour is

$$\hat{H}(z) = \sum_{ij} h_{ij}(z)\hat{d}_i^\dagger \hat{d}_j = \sum_{ij} \langle i|\hat{h}(z)|j\rangle \, \hat{d}_i^\dagger \hat{d}_j,$$

with $\hat{h}(z = t_0 - i\tau) = \hat{h}^{\mathrm{M}}$ the constant single-particle Hamiltonian along the imaginary track and $\hat{h}(z = t_\pm) = \hat{h}(t)$ the single-particle Hamiltonian along the horizontal branches. The noninteracting Green's function operator satisfies the equations of motion (6.4) and (6.5). To solve them, we write $\hat{\mathcal{G}}$ as

$$\hat{\mathcal{G}}(z_1, z_2) = \hat{\mathcal{U}}_L(z_1)\hat{\mathcal{F}}(z_1, z_2)\hat{\mathcal{U}}_R(z_2),$$

where the (first quantization) operators $\hat{\mathcal{U}}_{L/R}(z)$ fulfill

$$\mathrm{i}\frac{d}{dz}\hat{\mathcal{U}}_L(z) = \hat{h}(z)\hat{\mathcal{U}}_L(z), \qquad -\mathrm{i}\frac{d}{dz}\hat{\mathcal{U}}_R(z) = \hat{\mathcal{U}}_R(z)\hat{h}(z),$$

with boundary conditions $\hat{\mathcal{U}}_L(t_{0-}) = \hat{\mathcal{U}}_R(t_{0-}) = \hat{1}$. The structure of these equations is the same as that of the evolution operator on the contour, compare with (4.33) and (4.34). The explicit form of $\hat{\mathcal{U}}_{L/R}(z)$ is therefore the same as that in (4.32), i.e.,

$$\hat{\mathcal{U}}_L(z) = \mathcal{T}\left\{e^{-\mathrm{i}\int_{t_{0-}}^z d\bar{z}\,\hat{h}(\bar{z})}\right\}, \qquad \hat{\mathcal{U}}_R(z) = \bar{\mathcal{T}}\left\{e^{+\mathrm{i}\int_{t_{0-}}^z d\bar{z}\,\hat{h}(\bar{z})}\right\}. \tag{6.37}$$

The operator $\hat{\mathcal{U}}_L(z)$ can be seen as the single-particle forward evolution operator on the contour (from t_{0-} to z), whereas $\hat{\mathcal{U}}_R(z)$ can be seen as the single-particle backward evolution operator on the contour (from z to t_{0-}). This is also in agreement with the fact that $\hat{\mathcal{U}}_L(z)\hat{\mathcal{U}}_R(z) = \hat{\mathcal{U}}_R(z)\hat{\mathcal{U}}_L(z) = \hat{1}$, which follows directly from their definition. Substituting $\hat{\mathcal{G}}$ into the equations of motion (6.4) and (6.5) we obtain the following differential equations for $\hat{\mathcal{F}}$

$$\mathrm{i}\frac{d}{dz_1}\hat{\mathcal{F}}(z_1, z_2) = \delta(z_1, z_2), \qquad -\mathrm{i}\frac{d}{dz_2}\hat{\mathcal{F}}(z_1, z_2) = \delta(z_1, z_2).$$

The most general solution of these differential equations is

$$\hat{\mathcal{F}}(z_1, z_2) = \theta(z_1, z_2)\hat{\mathcal{F}}^> + \theta(z_2, z_1)\hat{\mathcal{F}}^<,$$

where the constant operators $\hat{\mathcal{F}}^>$ and $\hat{\mathcal{F}}^<$ are constrained by

$$\hat{\mathcal{F}}^> - \hat{\mathcal{F}}^< = -\mathrm{i}\,\hat{1}. \tag{6.38}$$

To determine $\hat{\mathcal{F}}^>$ (or $\hat{\mathcal{F}}^<$) we can use one of the two KMS relations. Choosing, e.g.,

$$\hat{\mathcal{G}}(t_{0-}, z') = \pm\hat{\mathcal{G}}(t_0 - \mathrm{i}\beta, z') \qquad \left\{ \begin{array}{l} + \text{ for bosons} \\ - \text{ for fermions} \end{array} \right.,$$

it is straightforward to find

$$\hat{\mathcal{F}}^< = \pm\hat{\mathcal{U}}_L(t_0 - \mathrm{i}\beta)\hat{\mathcal{F}}^> = \pm e^{-\beta\hat{h}^M}\hat{\mathcal{F}}^> \qquad \left\{ \begin{array}{l} + \text{ for bosons} \\ - \text{ for fermions} \end{array} \right..$$

Solving this equation for $\hat{\mathcal{F}}^>$ and inserting the result into (6.38) we get

$$\hat{\mathcal{F}}^< = \mp\mathrm{i}\frac{1}{e^{\beta\hat{h}^M} \mp \hat{1}} = \mp\mathrm{i}f(\hat{h}^M),$$

where $f(\omega) = 1/[e^{\beta\omega} \mp 1]$ is the Bose/Fermi function. Consequently, the operator $\hat{\mathcal{F}}^>$ reads

$$\hat{\mathcal{F}}^> = \pm\mathrm{i}\frac{1}{e^{-\beta\hat{h}^M} \mp 1} = -\mathrm{i}\bar{f}(\hat{h}^M),$$

with $\bar{f}(\omega) = 1 \pm f(\omega) = e^{\beta\omega}f(\omega)$. It is left as an exercise for the reader to show that we would have got the same results using the other KMS relation, i.e., $\hat{\mathcal{G}}(z, t_{0-}) = \pm\hat{\mathcal{G}}(z, t_0 - \mathrm{i}\beta)$. To summarize, the noninteracting Green's function can be written as

$$\boxed{\hat{\mathcal{G}}(z_1, z_2) = -\mathrm{i}\hat{\mathcal{U}}_L(z_1)\left[\theta(z_1, z_2)\bar{f}(\hat{h}^M) \pm \theta(z_2, z_1)f(\hat{h}^M)\right]\hat{\mathcal{U}}_R(z_2)} \tag{6.39}$$

with $\hat{\mathcal{U}}_{L/R}$ given in (6.37). Having the Green's function on the contour we can now extract all its Keldysh components.

6.2.1 Matsubara component

The Matsubara Green's function follows from (6.39) by setting $z_1 = t_0 - \mathrm{i}\tau_1$ and $z_2 = t_0 - \mathrm{i}\tau_2$ and reads

$$\boxed{\hat{\mathcal{G}}^M(\tau_1, \tau_2) = -\mathrm{i}\left[\theta(\tau_1 - \tau_2)\bar{f}(\hat{h}^M) \pm \theta(\tau_2 - \tau_1)f(\hat{h}^M)\right]e^{-(\tau_1-\tau_2)\hat{h}^M}} \tag{6.40}$$

where we make use of the fact that

$$\hat{\mathcal{U}}_L(t_0 - \mathrm{i}\tau) = e^{-\tau\hat{h}^M}, \qquad\qquad \hat{\mathcal{U}}_R(t_0 - \mathrm{i}\tau) = e^{\tau\hat{h}^M},$$

which also implies that $\hat{\mathcal{U}}_{L/R}(t_0 - \mathrm{i}\tau)$ commutes with $\hat{h}^M$. Inserting a complete set of eigenkets $|\lambda^M\rangle$ of $\hat{h}^M$ with eigenvalues ϵ_λ^M the Matsubara Green's function can also be written as

$$\hat{\mathcal{G}}^M(\tau_1, \tau_2) = -\mathrm{i}\sum_\lambda \left[\theta(\tau_1 - \tau_2)\bar{f}(\epsilon_\lambda^M) \pm \theta(\tau_2 - \tau_1)f(\epsilon_\lambda^M)\right]e^{-(\tau_1-\tau_2)\epsilon_\lambda^M}|\lambda^M\rangle\langle\lambda^M|.$$

This result should be compared with (6.17). For noninteracting density matrices the coefficients of the expansion are given by (6.18) and hence

$$\hat{\mathcal{G}}^{\mathrm{M}}(\tau_1, \tau_2) = \frac{1}{-i\beta} \sum_{m=-\infty}^{\infty} \frac{e^{-\omega_m(\tau_1 - \tau_2)}}{\omega_m - \hat{h}^{\mathrm{M}}} \tag{6.41}$$

We thus have two different ways, (6.40) and (6.41), of writing the same quantity. We close this section by proving the equivalence between them.

The strategy is to perform the sum over the Matsubara frequencies using a generalization of (6.19). The Bose/Fermi function $f(\zeta) = 1/(e^{\beta\zeta} \mp 1)$ of the complex variable ζ has simple poles in $\zeta = \omega_m$ with residues $\pm 1/\beta$ (as usual upper/lower sign for bosons/fermions). Therefore, given a function $Q(\zeta)$ analytic around all Matsubara frequencies we can write the following two equivalent identities

$$\frac{1}{-i\beta} \sum_{m=-\infty}^{\infty} Q(\omega_m) e^{-\omega_m \tau} = \int_{\Gamma_a} \frac{d\zeta}{2\pi} Q(\zeta) e^{-\zeta\tau} \times \begin{cases} \mp f(\zeta) \\ -e^{\beta\zeta} f(\zeta) \end{cases}. \tag{6.42}$$

The functions in the first and second row of the curly bracket yield the same value in $\zeta = \omega_m$ since $e^{\beta\omega_m} = \pm 1$. The contour Γ_a must encircle all the Matsubara frequencies clockwisely without including any singularity (pole or branch point) of Q, i.e., Q must be analytic inside Γ_a. In particular we consider functions Q with singularities only in $\zeta = \epsilon_\lambda^{\mathrm{M}}$. Then, for fermionic systems we can choose the contour Γ_a as shown in Fig. 6.1(a). The bosonic case is a bit more subtle since the Matsubara frequency $\omega_0 = 0$ lies on the real axis and hence the contour Γ_a does not exist if Q has a singularity in $\zeta = 0$. At any finite temperature, however, the lowest eigenvalue of $\hat{h}^{\mathrm{M}}$ must be strictly positive for otherwise both (6.40) and (6.41) are ill-defined.[11] Thus $Q(\zeta)$ has no poles in $\zeta = 0$ and a possible choice of contour Γ_a is illustrated in Fig. 6.3(a). Next, we observe that if $\beta > \tau > 0$ we could use the bottom identity in (6.42) and deform the contour as shown in Fig. 6.1(b) (for fermions) and Fig. 6.3(b) for bosons since $\lim_{\zeta \to \pm\infty} e^{(\beta-\tau)\zeta} f(\zeta) = 0$. On the contrary, if $0 > \tau > -\beta$ we could use the top identity and still deform the contour as shown in Fig. 6.1(b) (for fermions) and Fig. 6.3(b) (for bosons) since $\lim_{\zeta \to \pm\infty} e^{-\tau\zeta} f(\zeta) = 0$. We then conclude that

$$\frac{1}{-i\beta} \sum_{m=-\infty}^{\infty} Q(\omega_m) e^{-\omega_m \tau} = \int_{\Gamma_b} \frac{d\zeta}{2\pi} Q(\zeta) e^{-\zeta\tau} \left[-\theta(\tau) e^{\beta\zeta} f(\zeta) \mp \theta(-\tau) f(\zeta) \right] \tag{6.43}$$

[11]In bosonic systems the lowest eigenvalue $\epsilon_0^{\mathrm{M}} = \epsilon_0 - \mu$ of $\hat{h}^{\mathrm{M}}$ approaches zero when $\beta \to \infty$ as $\epsilon_0^{\mathrm{M}} \sim 1/\beta$ so that the product $\beta\epsilon_0^{\mathrm{M}}$ remains finite. If the energy spectrum of $\hat{h}^{\mathrm{M}}$ is continuous and if the density of single-particle states $D(\omega)$, see (6.34), vanishes when $\omega \to 0$ then all bosons have the same quantum number ϵ_0^{M} in the zero-temperature limit. This phenomenon is known as the *Bose condensation*. Thus, in physical bosonic systems the density matrix $\hat{\rho}$ can be of the form $|\Phi_0\rangle\langle\Phi_0|$ (pure state) only at zero temperature and only in the presence of Bose condensation. Without Bose condensation $\hat{\rho}$ is a mixture also at zero temperature, see Exercise 6.3.

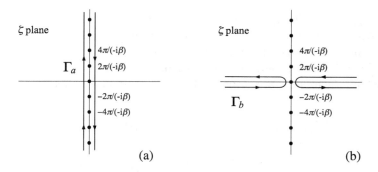

Figure 6.3 Contours for evaluation of the bosonic sum in (6.42). The points displayed are the bosonic Matsubara frequencies $\omega_m = 2m\pi/(-i\beta)$. The contour Γ_b contains all the poles of the function $Q(\zeta)$.

Let us specialize this formula to the function $Q(\zeta) = 1/(\zeta - \epsilon_\lambda^M)$. Since the simple pole in $\zeta = \epsilon_\lambda^M$ is inside Γ_b, the Cauchy residue theorem gives

$$\frac{1}{-i\beta} \sum_{m=-\infty}^{\infty} \frac{e^{-\omega_m \tau}}{\omega_m - \epsilon_\lambda^M} = -i \left[\theta(\tau)\bar{f}(\epsilon_\lambda^M) \pm \theta(-\tau)f(\epsilon_\lambda^M) \right] e^{-\tau \epsilon_\lambda^M}.$$

Multiplying both sides by $|\lambda^M\rangle\langle\lambda^M|$ and summing over λ we find the equivalence we were looking for.

6.2.2 Lesser and greater components

The operators $\hat{\mathcal{U}}_L(z)$ and $\hat{\mathcal{U}}_R(z)$ evaluated for z on the forward/backward branch of the contour reduce to the single-particle real-time evolution operators

$$\hat{\mathcal{U}}_L(t_\pm) \equiv \hat{\mathcal{U}}(t) = T\left\{ e^{-i\int_{t_0}^t d\bar{t}\,\hat{h}(\bar{t})} \right\},$$

$$\hat{\mathcal{U}}_R(t_\pm) = \hat{\mathcal{U}}^\dagger(t),$$

with T the chronological ordering operator introduced in (3.7). The action of $\hat{\mathcal{U}}(t)$ on a generic single-particle ket $|\Psi\rangle$ yields the time-evolved ket $|\Psi(t)\rangle$ which obeys the Schrödinger equation $i\frac{d}{dt}|\Psi(t)\rangle = \hat{h}(t)|\Psi(t)\rangle$. The lesser component of the noninteracting Green's function follows from (6.39) when setting $z_1 = t_{1-}$ and $z_2 = t_{2+}$. We find

$$\boxed{\hat{g}^<(t_1, t_2) = \mp i\hat{\mathcal{U}}(t_1)\,f(\hat{h}^M)\,\hat{\mathcal{U}}^\dagger(t_2)} \qquad (6.44)$$

Similarly, the greater component follows from (6.39) when setting $z_1 = t_{1+}$ and $z_2 = t_{2-}$ and reads

$$\boxed{\hat{g}^>(t_1, t_2) = -i\hat{\mathcal{U}}(t_1)\,\bar{f}(\hat{h}^M)\,\hat{\mathcal{U}}^\dagger(t_2)} \qquad (6.45)$$

Both $\hat{\mathcal{G}}^{>}(t_1, t_2)$ and $\hat{\mathcal{G}}^{<}(t_1, t_2)$ depend on the initial configuration through $f(\hat{h}^{\mathrm{M}})$. This should not come as a surprise since, e.g., the diagonal element $\pm \mathrm{i}\, G_{ii}^{<}(t, t)$ is the time-dependent ensemble average of the occupation operator $\hat{n}_i = \hat{d}_i^\dagger \hat{d}_i$. The physical content of the lesser/greater Green's function can be visualized more easily by inserting a complete set of eigenstates of $\hat{h}^{\mathrm{M}}$ between the Bose/Fermi function and the evolution operators:

$$\hat{\mathcal{G}}^{<}(t_1, t_2) = \mp \mathrm{i} \sum_{\lambda} f(\epsilon_\lambda^{\mathrm{M}})\, \hat{\mathcal{U}}(t_1)\, |\lambda^{\mathrm{M}}\rangle \langle \lambda^{\mathrm{M}}|\, \hat{\mathcal{U}}^\dagger(t_2)$$

$$= \mp \mathrm{i} \sum_{\lambda} f(\epsilon_\lambda^{\mathrm{M}})\, |\lambda^{\mathrm{M}}(t_1)\rangle\, \langle \lambda^{\mathrm{M}}(t_2)|.$$

In particular, the generic matrix element in the position–spin basis reads

$$G^{<}(1; 2) = \mp \mathrm{i} \sum_{\lambda} f(\epsilon_\lambda^{\mathrm{M}})\, \varphi_\lambda^{\mathrm{M}}(\mathbf{x}_1, t_1)\, \varphi_\lambda^{\mathrm{M}*}(\mathbf{x}_2, t_2),$$

with $\varphi_\lambda^{\mathrm{M}}(\mathbf{x}, t) = \langle \mathbf{x} | \lambda^{\mathrm{M}}(t) \rangle$ the time-evolved eigenfunction of $\hat{h}^{\mathrm{M}}$ (a similar formula can be derived for the greater component). Thus, the lesser (greater) Green's function of a noninteracting system can be constructed by populating the single-particle eigenfunctions of $\hat{h}^{\mathrm{M}}$ according to the Bose/Fermi function f ($\bar{f}$) and then evolving them according to the time-dependent Schrödinger equation with Hamiltonian $\hat{h}(t)$. The familiar result for the time-dependent density is readily recovered

$$n(\mathbf{x}, t) = \pm \mathrm{i}\, G^{<}(\mathbf{x}, t; \mathbf{x}, t) = \sum_{\lambda} f(\epsilon_\lambda^{\mathrm{M}})\, |\varphi_\lambda^{\mathrm{M}}(\mathbf{x}, t)|^2. \qquad (6.46)$$

The general dependence of $\hat{h}(t)$ on time prevents us from doing more analytic manipulations. Below we discuss the case in which $\hat{h}(t) = \hat{h}$ is time independent. Then, the evolution operator is simply $\hat{\mathcal{U}}(t) = \exp[-\mathrm{i}\hat{h}(t - t_0)]$ and (6.44) and (6.45) simplify to

$$\hat{\mathcal{G}}^{<}(t_1, t_2) = \mp \mathrm{i}\, e^{-\mathrm{i}\hat{h}(t_1 - t_0)}\, f(\hat{h}^{\mathrm{M}})\, e^{\mathrm{i}\hat{h}(t_2 - t_0)},$$

$$\hat{\mathcal{G}}^{>}(t_1, t_2) = -\mathrm{i}\, e^{-\mathrm{i}\hat{h}(t_1 - t_0)}\, \bar{f}(\hat{h}^{\mathrm{M}})\, e^{\mathrm{i}\hat{h}(t_2 - t_0)}.$$

These results are a special case of (6.30) in which the Matsubara Hamiltonian $\hat{H}^{\mathrm{M}}$ is noninteracting and described by $\hat{h}^{\mathrm{M}}$. We also observe that the lesser and greater Green's functions are not functions of the time difference $t_1 - t_2$, in agreement with the discussion below (6.28). The invariance under time translations requires that the system is prepared in an eigenstate (or in a mixture of eigenstates) of $\hat{h}$, i.e., that $\hat{h}^{\mathrm{M}}$ commutes with $\hat{h}$. In this case

$$\hat{\mathcal{G}}^{<}(t_1, t_2) = \mp \mathrm{i}\, f(\hat{h}^{\mathrm{M}})\, e^{-\mathrm{i}\hat{h}(t_1 - t_2)},$$

$$\hat{\mathcal{G}}^{>}(t_1, t_2) = -\mathrm{i}\, \bar{f}(\hat{h}^{\mathrm{M}})\, e^{-\mathrm{i}\hat{h}(t_1 - t_2)},$$

and the dependence on t_0 disappears. The time translational invariance allows us to define the Fourier transform

$$\hat{\mathcal{G}}^{\lessgtr}(t_1, t_2) = \int \frac{d\omega}{2\pi} e^{-i\omega(t_1 - t_2)} \hat{\mathcal{G}}^{\lessgtr}(\omega), \tag{6.47}$$

and we see by inspection that

$$\boxed{\hat{\mathcal{G}}^<(\omega) = \mp 2\pi i \, f(\hat{h}^{\mathrm{M}}) \, \delta(\omega - \hat{h})} \tag{6.48}$$

$$\boxed{\hat{\mathcal{G}}^>(\omega) = -2\pi i \, \bar{f}(\hat{h}^{\mathrm{M}}) \, \delta(\omega - \hat{h})} = \pm e^{\beta \hat{h}^{\mathrm{M}}} \, \hat{\mathcal{G}}^<(\omega). \tag{6.49}$$

6.2.3 All other components and a useful exercise

From a knowledge of the greater and lesser Green's functions we can extract all the remaining Keldysh components, see Table 5.1. By definition the retarded Green's function is

$$\hat{\mathcal{G}}^{\mathrm{R}}(t_1, t_2) = \theta(t_1 - t_2)[\hat{\mathcal{G}}^>(t_1, t_2) - \hat{\mathcal{G}}^<(t_1, t_2)] = -i\,\theta(t_1 - t_2)\hat{\mathcal{U}}(t_1)\hat{\mathcal{U}}^\dagger(t_2)$$

$$= -i\,\theta(t_1 - t_2)T\left\{ e^{-i \int_{t_2}^{t_1} d\bar{t}\, \hat{h}(\bar{t})} \right\}, \tag{6.50}$$

whereas the advanced Green's function reads

$$\hat{\mathcal{G}}^{\mathrm{A}}(t_1, t_2) = i\,\theta(t_2 - t_1)\bar{T}\left\{ e^{i \int_{t_1}^{t_2} d\bar{t}\, \hat{h}(\bar{t})} \right\} = [\hat{\mathcal{G}}^{\mathrm{R}}(t_2, t_1)]^\dagger. \tag{6.51}$$

It is interesting to observe that the retarded/advanced noninteracting Green's function does not depend on the initial density matrix. This means that $\hat{\mathcal{G}}^{\mathrm{R/A}}$ does not change by varying the initial number of particles or the distribution of the particles among the different energy levels. The information carried by $\hat{\mathcal{G}}^{\mathrm{R,A}}$ is the same as the information carried by the single-particle evolution operator $\hat{\mathcal{U}}$. We use this observation to rewrite $\hat{\mathcal{G}}^{\lessgtr}$ in (6.44) and (6.45) in terms of the retarded/advanced Green's function

$$\boxed{\hat{\mathcal{G}}^{\lessgtr}(t_1, t_2) = \hat{\mathcal{G}}^{\mathrm{R}}(t_1, t_0)\hat{\mathcal{G}}^{\lessgtr}(t_0, t_0)\hat{\mathcal{G}}^{\mathrm{A}}(t_0, t_2)} \tag{6.52}$$

Analogous relations can be derived for the left/right Green's functions,

$$\hat{\mathcal{G}}^{\rceil}(t, \tau) = \mp i\hat{\mathcal{U}}(t)f(\hat{h}^{\mathrm{M}})e^{\tau \hat{h}^{\mathrm{M}}} = i\,\hat{\mathcal{G}}^{\mathrm{R}}(t, t_0)\hat{\mathcal{G}}^{\mathrm{M}}(0, \tau), \tag{6.53}$$

$$\hat{\mathcal{G}}^{\lceil}(\tau, t) = -i\,e^{-\tau \hat{h}^{\mathrm{M}}}\bar{f}(\hat{h}^{\mathrm{M}})\hat{\mathcal{U}}^\dagger(t) = -i\,\hat{\mathcal{G}}^{\mathrm{M}}(\tau, 0)\hat{\mathcal{G}}^{\mathrm{A}}(t_0, t). \tag{6.54}$$

Finally, the time-ordered Green's function is

$$\hat{\mathcal{G}}^{\mathrm{T}}(t_1, t_2) = -i\hat{\mathcal{U}}(t_1)\left[\theta(t_1 - t_2)\bar{f}(\hat{h}^{\mathrm{M}}) \pm \theta(t_2 - t_1)f(\hat{h}^{\mathrm{M}})\right]\hat{\mathcal{U}}^\dagger(t_2).$$

A useful exercise: As an application of all these new formulas and of the Langreth rules of Table 5.1 it is particularly instructive to discuss a case that is often encountered in physics. Suppose that the single-particle Hamiltonian $\hat{h}(z)$ is the sum of an operator $\hat{h}_0(z)$ which is easy to handle and another operator $\hat{h}_1(z)$ which is problematic to deal with. To fix the ideas, we can take the Fano model and say that $\hat{h}_0$ is the metal+atom Hamiltonian whereas $\hat{h}_1$ is the contact Hamiltonian between the two subsystems. In these situations it is easier to work with the Keldysh Green's function $\hat{\mathcal{G}}_0$ of the $\hat{h}_0$ system than with the full Green's function $\hat{\mathcal{G}}$ and, therefore, it makes sense to look for an expansion of $\hat{\mathcal{G}}$ in terms of $\hat{\mathcal{G}}_0$. By definition $\hat{\mathcal{G}}_0$ satisfies the equations of motion (6.4) and (6.5) with $\hat{h}(z) = \hat{h}_0(z)$ and with KMS boundary conditions. Consequently, we can use $\hat{\mathcal{G}}_0$ to convert the equations of motion (6.4) and (6.5) for $\hat{\mathcal{G}}$ into an integral equation that embodies the KMS relations

$$\hat{\mathcal{G}}(z, z') = \hat{\mathcal{G}}_0(z, z') + \int_\gamma d\bar{z}\, \hat{\mathcal{G}}_0(z, \bar{z}) \hat{h}_1(\bar{z}) \hat{\mathcal{G}}(\bar{z}, z').$$

The reader can verify the correctness of this equation by applying $[\mathrm{i}\frac{d}{dz} - \hat{h}_0(z)]$ to both sides. These kinds of integral equation are known as the *Dyson equations* and in the following chapters we encounter them many times.[12] The Dyson equation is an iterative equation that allows us to expand $\hat{\mathcal{G}}$ in "powers" of $\hat{\mathcal{G}}_0$ by replacing $\hat{\mathcal{G}}$ on the r.h.s. with the whole r.h.s. *ad infinitum*:

$$\hat{\mathcal{G}}(z, z') = \hat{\mathcal{G}}_0(z, z') + \int_\gamma d\bar{z}\, \hat{\mathcal{G}}_0(z, \bar{z}) \hat{h}_1(\bar{z}) \hat{\mathcal{G}}_0(\bar{z}, z')$$

$$+ \int_\gamma d\bar{z}\, d\bar{z}'\, \hat{\mathcal{G}}_0(z, \bar{z}) \hat{h}_1(\bar{z}) \hat{\mathcal{G}}_0(\bar{z}, \bar{z}') \hat{h}_1(\bar{z}') \hat{\mathcal{G}}_0(\bar{z}', z') + \dots$$

$$= \hat{\mathcal{G}}_0(z, z') + \int_\gamma d\bar{z}\, \hat{\mathcal{G}}(z, \bar{z}) \hat{h}_1(\bar{z}) \hat{\mathcal{G}}_0(\bar{z}, z'),$$

where in the last step we resummed everything in an equivalent way. We refer to this equation as the adjoint Dyson equation and to the one before as the left Dyson equation. The exercise consists in using the Dyson equations to calculate, say, $\hat{\mathcal{G}}^<$ and to verify that the result agrees with (6.52). Let us define the singular operator in Keldysh space $\hat{h}_1(z, z') = \delta(z, z') \hat{h}_1(z)$. The Dyson equations contain a convolution between three functions in Keldysh space. The Langreth rules of Table 5.1 applied to the adjoint Dyson equation give

$$\hat{\mathcal{G}}^< = [\delta + \hat{\mathcal{G}}^R \cdot \hat{h}_1^R] \cdot \hat{\mathcal{G}}_0^< + \hat{\mathcal{G}}^< \cdot \hat{h}_1^A \cdot \hat{\mathcal{G}}_0^A + \hat{\mathcal{G}}^\rceil \star \hat{h}_1^M \star \hat{\mathcal{G}}_0^\lceil,$$

while the Dyson equation gives

$$\hat{\mathcal{G}}^{R/A} = \hat{\mathcal{G}}_0^{R/A} + \hat{\mathcal{G}}_0^{R/A} \cdot \hat{h}_1^{R/A} \cdot \hat{\mathcal{G}}^{R/A}. \tag{6.55}$$

The equation for $\hat{\mathcal{G}}^<$ contains $\hat{\mathcal{G}}^\rceil$ which has one argument on the vertical track of the contour. Due to this coupling the equations for the lesser, retarded and advanced Green's functions do not form a closed set unless $\hat{h}_1$ vanishes on the vertical track. Let us bring all terms containing $\hat{\mathcal{G}}^<$ onto the l.h.s.

$$\hat{\mathcal{G}}^< \cdot [\delta - \hat{h}_1^A \cdot \hat{\mathcal{G}}_0^A] = [\delta + \hat{\mathcal{G}}^R \cdot \hat{h}_1^R] \cdot \hat{\mathcal{G}}_0^< + \hat{\mathcal{G}}^\rceil \star \hat{h}_1^M \star \hat{\mathcal{G}}_0^\lceil.$$

To isolate $\hat{\mathcal{G}}^<$ we observe that

$$[\delta - \hat{h}_1^A \cdot \hat{\mathcal{G}}_0^A] \cdot [\delta + \hat{h}_1^A \cdot \hat{\mathcal{G}}^A] = \delta - \hat{h}_1^A \cdot [\hat{\mathcal{G}}_0^A + \hat{\mathcal{G}}_0^A \cdot \hat{h}_1^A \cdot \hat{\mathcal{G}}^A] + \hat{h}_1^A \cdot \hat{\mathcal{G}}^A = \delta,$$

---

[12]Our first encounter with a Dyson equation was in Section 5.2.

where in the last step we use the advanced Dyson equation. Therefore

$$\hat{\mathcal{G}}^< = [\delta + \hat{\mathcal{G}}^R \cdot \hat{h}_1^R] \cdot \hat{\mathcal{G}}_0^< \cdot [\delta + \hat{h}_1^A \cdot \hat{\mathcal{G}}^A] + \hat{\mathcal{G}}^{\rceil} \star \hat{h}_1^M \star \hat{\mathcal{G}}_0^{\lceil} \cdot [\delta + \hat{h}_1^A \cdot \hat{\mathcal{G}}^A].$$

From (6.52) and (6.54) we know that

$$\hat{\mathcal{G}}_0^< (t_1, t_2) = \hat{\mathcal{G}}_0^R (t_1, t_0)\hat{\mathcal{G}}_0^< (t_0, t_0)\hat{\mathcal{G}}_0^A (t_0, t_2), \qquad \hat{\mathcal{G}}_0^{\lceil}(\tau, t) = -i\hat{\mathcal{G}}_0^M (\tau, 0)\hat{\mathcal{G}}_0^A (t_0, t);$$

substituting these results into the above formula for $\hat{\mathcal{G}}^<$ and using the retarded/advanced Dyson equation we find

$$\hat{\mathcal{G}}^< (t_1, t_2) = \hat{\mathcal{G}}^R (t_1, t_0)\hat{\mathcal{G}}_0^< (t_0, t_0)\hat{\mathcal{G}}^A (t_0, t_2) - i[\hat{\mathcal{G}}^{\rceil} \star \hat{h}_1^M \star \hat{\mathcal{G}}_0^M](t, 0)\hat{\mathcal{G}}^A (t_0, t_2). \qquad (6.56)$$

It is interesting to observe that if $\hat{h}_1^M = 0$ then $\hat{\mathcal{G}}_0^< (t_0, t_0) = \hat{\mathcal{G}}^< (t_0, t_0)$ and (6.52) would be recovered. To proceed further we need an equation for the right Green's function. Taking the right component of the adjoint Dyson equation and using the identity (6.53) for $\hat{\mathcal{G}}_0^{\rceil}$ we find

$$\hat{\mathcal{G}}^{\rceil} (t, \tau) = i\hat{\mathcal{G}}^R (t, t_0)\hat{\mathcal{G}}_0^M (0, \tau) + [\hat{\mathcal{G}}^{\rceil} \star \hat{h}_1^M \star \hat{\mathcal{G}}_0^M](t, \tau)$$

and hence

$$\hat{\mathcal{G}}^{\rceil} \star [\delta - \hat{h}_1^M \star \hat{\mathcal{G}}_0^M](t, \tau) = i\hat{\mathcal{G}}^R (t, t_0)\hat{\mathcal{G}}_0^M (0, \tau).$$

This equation can be solved for $\hat{\mathcal{G}}^{\rceil}$. We have

$$[\delta - \hat{h}_1^M \star \hat{\mathcal{G}}_0^M] \star [\delta + \hat{h}_1^M \star \hat{\mathcal{G}}^M] = \delta - \hat{h}_1^M \star [\hat{\mathcal{G}}_0^M + \hat{\mathcal{G}}_0^M \star \hat{h}_1^M \star \hat{\mathcal{G}}^M] + \hat{h}_1^M \star \hat{\mathcal{G}}^M = \delta,$$

where in the last step we use the fact that the Matsubara component of the Dyson equation is

$$\hat{\mathcal{G}}^M = \hat{\mathcal{G}}_0^M + \hat{\mathcal{G}}_0^M \star \hat{h}_1^M \star \hat{\mathcal{G}}^M.$$

Therefore, the right Green's function reads $\hat{\mathcal{G}}^{\rceil} (t, \tau) = i\hat{\mathcal{G}}^R (t, t_0)\hat{\mathcal{G}}^M (0, \tau)$, which agrees with (6.53). Substituting this result into (6.56) and taking into account that $\hat{\mathcal{G}}_0^< (t_0, t_0) = \hat{\mathcal{G}}_0^M (0, 0^+)$ [see (6.27)], we find

$$\hat{\mathcal{G}}^< (t_1, t_2) = \hat{\mathcal{G}}^R (t_1, t_0)\hat{\mathcal{G}}_0^M (0, 0^+)\hat{\mathcal{G}}^A (t_0, t_2) + \hat{\mathcal{G}}^R (t, t_0)[\hat{\mathcal{G}}^M \star \hat{h}_1^M \star \hat{\mathcal{G}}_0^M](0, 0^+)\hat{\mathcal{G}}^A (t_0, t_2)$$
$$= \hat{\mathcal{G}}^R (t_1, t_0)\hat{\mathcal{G}}^M (0, 0^+)\hat{\mathcal{G}}^A (t_0, t_2),$$

which correctly agrees with (6.52).

We conclude this section by considering again a system with Hamiltonian $\hat{h}(t) = \hat{h}$ constant in time. Then, from (6.50) and (6.51) the retarded/advanced Green's functions become

$$\hat{\mathcal{G}}^R (t_1, t_2) = -i\,\theta(t_1 - t_2)e^{-i\hat{h}(t_1 - t_2)},$$
$$\hat{\mathcal{G}}^A (t_1, t_2) = +i\,\theta(t_2 - t_1)e^{+i\hat{h}(t_2 - t_1)},$$

which, as expected, depend only on the time difference $t_1 - t_2$. This allows us to define the Fourier transform of $\hat{\mathcal{G}}^{R/A}$ according to

$$\hat{\mathcal{G}}^{R,A} (t_1, t_2) = \int \frac{d\omega}{2\pi} e^{-i\omega(t_1 - t_2)} \hat{\mathcal{G}}^{R,A} (\omega).$$

The calculation of the Fourier transform $\hat{\mathcal{G}}^{\mathrm{R,A}}(\omega)$ is most easily done by using the representation of the Heaviside function

$$\theta(t_1 - t_2) = \mathrm{i} \int \frac{d\omega}{2\pi} \frac{e^{-\mathrm{i}\omega(t_1 - t_2)}}{\omega + \mathrm{i}\eta}, \tag{6.57}$$

with η an infinitesimal positive constant. We find

$$\hat{\mathcal{G}}^{\mathrm{R}}(t_1, t_2) = \int \frac{d\omega}{2\pi} \frac{e^{-\mathrm{i}(\omega + \hat{h})(t_1 - t_2)}}{\omega + \mathrm{i}\eta},$$

and changing the variable of integration $\omega \to \omega - \hat{h}$ we can identify the Fourier transform with

$$\boxed{\hat{\mathcal{G}}^{\mathrm{R}}(\omega) = \frac{1}{\omega - \hat{h} + \mathrm{i}\eta} = \sum_\lambda \frac{|\lambda\rangle\langle\lambda|}{\omega - \epsilon_\lambda + \mathrm{i}\eta}} \tag{6.58}$$

where the sum runs over a complete set of eigenkets $|\lambda\rangle$ of $\hat{h}$ with eigenvalues ϵ_λ. To calculate the Fourier transform of the advanced Green's function we note that the property (6.51) implies

$$\hat{\mathcal{G}}^{\mathrm{A}}(\omega) = [\hat{\mathcal{G}}^{\mathrm{R}}(\omega)]^\dagger$$

and hence

$$\boxed{\hat{\mathcal{G}}^{\mathrm{A}}(\omega) = \frac{1}{\omega - \hat{h} - \mathrm{i}\eta} = \sum_\lambda \frac{|\lambda\rangle\langle\lambda|}{\omega - \epsilon_\lambda - \mathrm{i}\eta}} \tag{6.59}$$

The retarded Green's function is analytic in the upper half of the complex ω plane, whereas the advanced Green's function is analytic in its lower half. As we see in the next section, this analytic structure is completely general since it is a direct consequence of *causality*: $\hat{\mathcal{G}}^{\mathrm{R}}(t_1, t_2)$ vanishes for $t_1 < t_2$ whereas $\hat{\mathcal{G}}^{\mathrm{A}}(t_1, t_2)$ vanishes for $t_1 > t_2$.

The results (6.58) and (6.59) are interesting also for another reason. From the definition of the retarded/advanced component of a Keldysh function we have the general relation (omitting the time arguments)

$$\hat{\mathcal{G}}^{\mathrm{R}} - \hat{\mathcal{G}}^{\mathrm{A}} = \hat{\mathcal{G}}^> - \hat{\mathcal{G}}^<,$$

which tells us that the *difference* between the lesser and greater Green's function can be expressed in terms of the difference between the retarded and advanced Green's function. We now show that a system prepared in a stationary excited state has lesser and greater Green's functions that can *separately* be written in terms of $\hat{\mathcal{G}}^{\mathrm{R}} - \hat{\mathcal{G}}^{\mathrm{A}}$. For the system to be in a stationary excited state the Hamiltonian $\hat{h}^{\mathrm{M}}$ must commute with $\hat{h}$. Then $\hat{\mathcal{G}}^{\lessgtr}$ is given by (6.48) and (6.49) which, taking into account (6.58) and (6.59) together with the Cauchy relation (2.23), can also be written as

$$\boxed{\hat{\mathcal{G}}^<(\omega) = \pm f(\hat{h}^{\mathrm{M}})[\hat{\mathcal{G}}^{\mathrm{R}}(\omega) - \hat{\mathcal{G}}^{\mathrm{A}}(\omega)]}$$

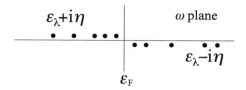

Figure 6.4 Location of the poles of the time-ordered Green's function for a system of fermions at zero temperature.

$$\hat{\mathcal{G}}^{>}(\omega) = \bar{f}(\hat{h}^{\mathrm{M}})[\hat{\mathcal{G}}^{\mathrm{R}}(\omega) - \hat{\mathcal{G}}^{\mathrm{A}}(\omega)]$$

In the special case when $\hat{h}^{\mathrm{M}} = \hat{h} - \mu$ describes the system in thermodynamic equilibrium, the above relations reduce to the so called *fluctuation–dissipation theorem*

$$\hat{\mathcal{G}}^{<}(\omega) = \pm f(\omega - \mu)[\hat{\mathcal{G}}^{\mathrm{R}}(\omega) - \hat{\mathcal{G}}^{\mathrm{A}}(\omega)],$$
$$\hat{\mathcal{G}}^{>}(\omega) = \bar{f}(\omega - \mu)[\hat{\mathcal{G}}^{\mathrm{R}}(\omega) - \hat{\mathcal{G}}^{\mathrm{A}}(\omega)].$$

As we see in the next section, the fluctuation–dissipation theorem is valid in interacting systems as well.

Let us also calculate the time-ordered Green's function for the case in which $\hat{h}^{\mathrm{M}}$ commutes with $\hat{h}$. Using the results (6.48) and (6.49) as well as the representation (6.57) of the Heaviside function we find

$$\hat{\mathcal{G}}^{\mathrm{T}}(t_1, t_2) = \int \frac{d\omega}{2\pi} e^{-i\omega(t_1 - t_2)} \underbrace{\left[\frac{\bar{f}(\hat{h}^{\mathrm{M}})}{\omega - \hat{h} + i\eta} \mp \frac{f(\hat{h}^{\mathrm{M}})}{\omega - \hat{h} - i\eta} \right]}_{\hat{\mathcal{G}}^{\mathrm{T}}(\omega)}.$$

The Fourier transform of the time-ordered Green's function has poles on both sides of the complex plane. For fermions in equilibrium at zero temperature, $\hat{h}^{\mathrm{M}} = \hat{h} - \epsilon_{\mathrm{F}}$, with $\epsilon_{\mathrm{F}} = \lim_{\beta \to \infty} \mu$ the Fermi energy, and $f(\epsilon_\lambda - \mu) = \theta(\epsilon_{\mathrm{F}} - \epsilon_\lambda)$. The equilibrium configuration corresponds to populating all eigenkets $|\lambda\rangle$ of $\hat{h}$ with energy $\epsilon_\lambda < \epsilon_{\mathrm{F}}$. Then

$$\hat{\mathcal{G}}^{\mathrm{T}}(\omega) = \sum_{\epsilon_\lambda > \epsilon_{\mathrm{F}}} \frac{|\lambda\rangle\langle\lambda|}{\omega - \epsilon_\lambda + i\eta} + \sum_{\epsilon_\lambda < \epsilon_{\mathrm{F}}} \frac{|\lambda\rangle\langle\lambda|}{\omega - \epsilon_\lambda - i\eta}. \qquad \begin{array}{l}\text{(fermions at}\\\text{zero temperature)}\end{array}$$

As illustrated in Fig. 6.4, the poles of $\hat{\mathcal{G}}^{\mathrm{T}}(\omega)$ with real part smaller than ϵ_{F} lie on the upper half of the complex ω-plane, whereas those with real part larger than ϵ_{F} lie on the lower half of the complex ω-plane. This property is mantained in interacting systems as well.

Exercise 6.1. Consider a system of particles in one dimension with single-particle Hamiltonian $\hat{h} = \hat{p}^2/2$. Taking into account that the single-particle eigenkets are the momentum-spin kets $|p\sigma\rangle$ with eigenvalue $p^2/2$, use (6.58) to show that

$$\langle x\sigma | \hat{\mathcal{G}}^{\mathrm{R}}(\omega) | x'\sigma' \rangle = \delta_{\sigma\sigma'} G^{\mathrm{R}}(x, x'; \omega),$$

with

$$G^{\mathrm{R}}(x, x'; \omega) = -\frac{1}{\sqrt{2|\omega|}} \begin{cases} \mathrm{i}\, e^{\mathrm{i}\sqrt{2|\omega|}\,|x-x'|} & \omega > 0 \\ e^{-\sqrt{2|\omega|}\,|x-x'|} & \omega < 0 \end{cases}.$$

Exercise 6.2. Consider the same system of Exercise 6.1 in which a δ-like potential $\hat{V} = \lambda\delta(\hat{x})$, $\lambda > 0$, is added to the free Hamiltonian. Use the retarded Dyson equation (6.55) with $\hat{h}_1^{\mathrm{R}}(t, t') = \delta(t - t')\hat{V}$ to show that the new Green's function is given by

$$G^{\mathrm{R}}(x, x'; \omega) = -\frac{1}{\sqrt{2|\omega|}} \begin{cases} \mathrm{i}\, e^{\mathrm{i}\sqrt{2|\omega|}\,|x-x'|} + \dfrac{\lambda e^{\mathrm{i}\sqrt{2|\omega|}\,(|x|+|x'|)}}{\sqrt{2|\omega|} + \mathrm{i}\lambda} & \omega > 0 \\[4mm] e^{-\sqrt{2|\omega|}\,|x-x'|} - \dfrac{\lambda e^{-\sqrt{2|\omega|}\,(|x|+|x'|)}}{\sqrt{2|\omega|} + \lambda} & \omega < 0 \end{cases}.$$

How does this result change if $\lambda < 0$? (Think about the formation of bound states.)

Exercise 6.3. Consider the single-level noninteracting Hamiltonian $\hat{H} = \epsilon\,\hat{d}^\dagger\hat{d}$ with eigenkets $|\Psi_k\rangle = \frac{(d^\dagger)^k}{\sqrt{k!}}|0\rangle$ and eigenvalues $E_k = k\epsilon$. Show that if the average occupation $\mathrm{Tr}[\hat{\rho}\,\hat{d}^\dagger\hat{d}] = n$ then the density matrix in thermodynamic equilibrium reads

$$\hat{\rho} = \frac{e^{-\beta(\hat{H}-\mu\hat{N})}}{\mathrm{Tr}\left[e^{-\beta(\hat{H}-\mu\hat{N})}\right]} = \frac{\sum_k \left(\frac{n}{n\pm1}\right)^k |\Psi_k\rangle\langle\Psi_k|}{\sum_k \left(\frac{n}{n\pm1}\right)^k},$$

where the upper/lower sign applies to bosons/fermions and the sum over k runs between 0 and ∞ in the case of bosons and between 0 and 1 in the case of fermions. We then see that in fermionic systems $\hat{\rho}$ is a pure state for $n \to 0$ or $n \to 1$ whereas in bosonic systems $\hat{\rho}$ is a mixture of states for all values of $n \neq 0$.

6.3 Interacting Green's function and Lehmann representation

In this section we discuss the Green's function of an interacting system with Hamiltonian $\hat{H}(t) = \hat{H}$ constant in time. The *Lehmann representation* of the n-particle Green's function G_n is simply a rewriting of the definition of G_n in which every evolution operator is expanded over a complete set of eigenstates of $\hat{H}$. The resulting expression is, in general, rather cumbersome already for the one-particle Green's function G; for instance in (6.28) one should expand three evolution operators. Below we consider two special cases. The first case is the equal-time Green's function relevant to calculating the time-dependent ensemble average of any one-body operator. The second case is the one treated in most textbooks and pertains to systems initially in thermodynamic equilibrium, $\hat{H}^{\mathrm{M}} = \hat{H} - \mu\hat{N}$.

6.3.1 Steady-states, persistent oscillations, initial-state dependence

The most general form of the equal-time $G^<$ of a system with Hamiltonian $\hat{H}(t) = \hat{H}$ is given in (6.28) with $t' = t$. Let us expand the evolution operators over a complete set of eigenstates of $\hat{H}$. The Hamiltonian $\hat{H}$ can have both discrete eigenkets $|\Psi_l\rangle$ with energy E_l and continuous eigenkets $|\Psi_\alpha\rangle$ with energy E_α. We normalize the former according to $\langle\Psi_l|\Psi_{l'}\rangle = \delta_{ll'}$ and the latter according to $\langle\Psi_\alpha|\Psi_{\alpha'}\rangle = \delta(\alpha-\alpha')$, so that the completeness relation reads

$$\hat{1} = \sum_l |\Psi_l\rangle\langle\Psi_l| + \int d\alpha\, |\Psi_\alpha\rangle\langle\Psi_\alpha|.$$

It is convenient to define the label a that runs over both discrete, $a = l$, and continuous, $a = \alpha$, eigenstates and use the short-hand notation $\hat{1} = \int da\, |\Psi_a\rangle\langle\Psi_a|$. Inserting in (6.28) the completeness relation twice we obtain the Lehmann representation of the equal-time $G^<$,

$$G^<_{ji}(t,t) = \int da\,da'\, e^{i(E_a - E_{a'})t} f_{ji}(a, a'), \tag{6.60}$$

with

$$f_{ji}(a, a') = \mp i\langle\Psi_{a'}|\hat{\rho}|\Psi_a\rangle\langle\Psi_a|\hat{d}_i^\dagger \hat{d}_j|\Psi_{a'}\rangle e^{-i(E_a - E_{a'})t_0} = -f_{ij}^*(a', a)$$

and $\hat{\rho} = \sum_k \rho_k|\Psi_k\rangle\langle\Psi_k|$ the initial density matrix. No further analytic manipulations can be performed for finite times t; this is why we stated before that the Lehmann representation is a simple rewriting. Simplifications, however, do occur if we consider the limit of long times, a limit which is relevant in several contexts of physics and especially of modern physics like, e.g., quantum transport or ultracold gases. In these contexts one is interested in studying the relaxation of a system which is either prepared in an excited configuration or is perturbed by a time-independent driving field. This kind of relaxation has nothing to do with the relaxation discussed in Section 6.1.3 and illustrated in Fig. 6.2. There we were adding or removing a particle, here we are disturbing the system with external fields without altering the number of particles. There the relaxation pertained to the behavior of $G^<(t, t')$ for large $|t - t'|$, here it pertains to the behavior of $G^<(t, t)$ for t much larger than the time at which the external field is switched on. In order not to mix these two different concepts we refer to this new relaxation as the *thermalization*.

Splitting the integrals over a and a' into a sum over discrete states and an integral over continuum states, (6.60) yields three different kinds of contribution: discrete–discrete, discrete–continuum, and continuum–continuum. The generic term of the discrete–continuum contribution reads

$$\int d\alpha \left[e^{i(E_l - E_\alpha)t} f_{ji}(l, \alpha) - e^{-i(E_l - E_\alpha)t} f_{ij}^*(l, \alpha) \right]$$

$$= \int \frac{d\omega}{2\pi} \left[e^{i(E_l - \omega)t} \tilde{f}_{ji}(l, \omega) - e^{-i(E_l - \omega)t} \tilde{f}_{ij}^*(l, \omega) \right],$$

where on the r.h.s. we have converted the integral over α into an integral over energy using the same trick as in (6.33): $\tilde{f}_{ji}(l, \omega) = \int d\alpha\, 2\pi\delta(\omega - E_\alpha) f_{ji}(l, \alpha)$. In most physical situations the quantity $\tilde{f}_{ji}(l, \omega)$ is an integrable function of ω and we can invoke the Riemann–Lebesgue theorem to assert that the discrete–continuum contribution vanishes when $t \to \infty$ [see discussion below (6.32)]. Then we have

$$\lim_{t\to\infty} G_{ji}^{<}(t,t) = \lim_{t\to\infty} \sum_{ll'} e^{i(E_l - E_{l'})t} f_{ji}(l, l')$$

$$+ \lim_{t\to\infty} \int d\alpha d\alpha' \, e^{i(E_\alpha - E_{\alpha'})t} f_{ji}(\alpha, \alpha'). \tag{6.61}$$

The discrete–discrete contribution gives rise to persistent oscillations whose amplitude depends on how the system is initially prepared. The analysis of the continuum–continuum contribution requires a preliminary discussion on the possible mathematical structure of $f_{ji}(\alpha, \alpha')$. By definition

$$f_{ji}(\alpha, \alpha') = \mp i \, \langle \Psi_{\alpha'} | \hat{\rho} | \Psi_\alpha \rangle \langle \Psi_\alpha | \hat{d}_i^\dagger \hat{d}_j | \Psi_{\alpha'} \rangle e^{-i(E_\alpha - E_{\alpha'})t_0}.$$

If $\hat{H}^{\mathrm{M}}$ commutes with $\hat{H}$ (as in the case of thermodynamic equilibrium) then we can choose the $|\Psi_\alpha\rangle$ to be also eigenkets of $\hat{H}^{\mathrm{M}}$ and hence of $\hat{\rho}$ with eigenvalues ρ_α, and the above formula simplifies to

$$f_{ji}(\alpha, \alpha') = \mp i \, \delta(\alpha - \alpha') \rho_\alpha \langle \Psi_\alpha | \hat{d}_i^\dagger \hat{d}_j | \Psi_\alpha \rangle.$$

This example tells us that the quantity $f_{ji}(\alpha, \alpha')$ can be singular for $\alpha = \alpha'$. In general, $f_{ji}(\alpha, \alpha')$ can have different kinds of singularity (not necessarily a δ-function). Here we only discuss δ-like singularities; the interested reader can work out more general cases. We then assume the following structure:

$$f_{ji}(\alpha, \alpha') = f_{ji}^\delta(\alpha)\delta(\alpha - \alpha') + f_{ji}^{ns}(\alpha, \alpha'),$$

with f^{ns} a nonsingular function. Inserting this f_{ji} into (6.61) and invoking one more time the Riemann–Lebesgue theorem to prove that the term with f^{ns} vanishes, we arrive at the very elegant result

$$\lim_{t\to\infty} G_{ji}^{<}(t,t) = \lim_{t\to\infty} \sum_{ll'} e^{i(E_l - E_{l'})t} f_{ji}(l, l') + \int d\alpha f_{ji}^\delta(\alpha). \tag{6.62}$$

The physics behind (6.62) is that a perfect destructive interference takes place for states with $|E_\alpha - E_{\alpha'}| \gtrsim 1/t$ (Riemann–Lebesgue theorem), and the continuum–continuum contribution becomes time-independent for large t. We refer to this mechanism as the *dephasing* and we call the continuum–continuum contribution the *steady-state* value. Mathematically it is the dephasing that leads to thermalization.

   At this point a natural question arises: how sensitive is the steady-state value to variations of the initial preparation? The answer is relevant to understanding to what extent the system remembers its initial state after the thermalization. The dependence of f^δ on the initial state is all contained in the matrix elements $\langle \Psi_\alpha | \hat{\rho} | \Psi_\alpha \rangle$. We could speculate that if $\hat{H}^{\mathrm{M}} \to \hat{H}^{\mathrm{M}} + \hat{L}$ with $\hat{L}$ a "local" perturbation, then the singular part f^δ does not change

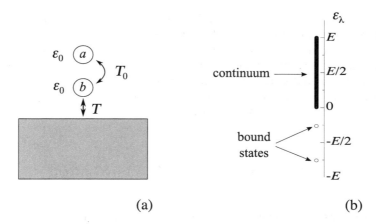

Figure 6.5 (a) The model described by the Hamiltonian (6.64). (b) Distribution of the single-particle eigenvalues of $\hat{H}$.

since $\hat{L}$ can change the average

$$\langle \Psi_\alpha | \hat{\rho} | \Psi_\alpha \rangle = \frac{\langle \Psi_\alpha | e^{-\beta(\hat{H}^{\mathrm{M}} + \hat{L})} | \Psi_\alpha \rangle}{\mathrm{Tr} \left[e^{-\beta(\hat{H}^{\mathrm{M}} + \hat{L})} \right]} = \frac{\langle \Psi_\alpha | e^{-\beta \hat{H}^{\mathrm{M}}} | \Psi_\alpha \rangle}{\mathrm{Tr} \left[e^{-\beta \hat{H}^{\mathrm{M}}} \right]} + \mathcal{O}(1/\mathrm{V})$$

only up to a term that scales with the inverse of the system volume V. Vice versa, we would expect that a "global" perturbation $\hat{G}$, $\hat{H}^{\mathrm{M}} \to \hat{H}^{\mathrm{M}} + \hat{G}$, affects f^δ and hence the steady-state value. Below we discuss two examples with the aim of rendering these intuitive arguments more quantitative. In both examples we take $\hat{H} = \sum_\lambda \epsilon_\lambda \hat{c}_\lambda^\dagger \hat{c}_\lambda$ as noninteracting so that $G_{ji}^<(t,t)$ simplifies to [see (6.30)]

$$G_{ji}^<(t,t) = \sum_{\lambda\lambda'} e^{i(\epsilon_\lambda - \epsilon_{\lambda'})t} \underbrace{\left(\langle j | \lambda' \rangle \, G_{\lambda'\lambda}^<(t_0,t_0) \langle \lambda | i \rangle \, e^{-i(\epsilon_\lambda - \epsilon_{\lambda'})t_0} \right)}_{f_{ji}(\lambda,\lambda')}. \qquad (6.63)$$

Local perturbations: Let us introduce an extension of the Fano model which will be useful also for later purposes. The model consists of a two-level system (molecule) coupled to a metal and is described by the Hamiltonian

$$\hat{H} = \underbrace{\sum_{k=0}^{N} \epsilon_k \hat{d}_k^\dagger \hat{d}_k}_{\text{metal}} + \underbrace{\epsilon_0 \sum_{i=a,b} \hat{d}_i^\dagger \hat{d}_i + T_0(\hat{d}_a^\dagger \hat{d}_b + \hat{d}_b^\dagger \hat{d}_a)}_{\text{molecule}} + \underbrace{\sum_{k=0}^{N} T_k(\hat{d}_k^\dagger \hat{d}_b + \hat{d}_b^\dagger \hat{d}_k)}_{\text{coupling}}, \qquad (6.64)$$

where the $\hat{d}$-operators are fermionic operators. A schematic illustration of the model is shown in Fig. 6.5(a). We take the energies ϵ_k of the metal to be equally spaced between 0

and E, $\epsilon_k = (k/N)E$, $k = 1, \ldots, N$, and the couplings $T_k = T/\sqrt{N}$ independent of k.[13]
In Fig. 6.5(b) we show the distribution of the single-particle eigenvalues ϵ_λ for $\epsilon_0 = -E$,
$T_0 = -E/4$, $T = -E/8$ and $N = 200$; there are two discrete (bound-state) energies below
zero and, as expected, a very dense distribution in the region of the metallic spectrum.
When $N \to \infty$ the bound-state energies converge to a value close to the value displayed
in the figure and the remaining energies form a continuum between 0 and E. Labeling the
bound-state energies with $\lambda = 1, 2$ we can isolate the discrete–discrete contribution from
the sum in (6.63),

$$G_{ji}^<(t, t) = \sum_{\lambda\lambda'=1}^{2} e^{i(\epsilon_\lambda - \epsilon_{\lambda'})t} f_{ji}(\lambda, \lambda') + \delta G_{ji}^<(t, t), \qquad (6.65)$$

where $\delta G^<$ contains the discrete–continuum part as well as the continuum–continuum
part. In the first term of this expression the sum over $\lambda = \lambda'$ yields a time-independent
contribution that carries information on the initial state, whereas the sum over $\lambda \neq \lambda'$ is
responsible for persistent oscillations whose amplitude depends on the initial state but whose
frequency depends only on the parameters of $\hat{H}$; the frequency is therefore insensitive to
a change of the initial configuration. The second term in (6.65) should instead approach a
constant in the limit $t \to \infty$. Below we solve the model numerically, check the correctness
of our predictions and address the issue whether $\lim_{t\to\infty} \delta G^<(t, t)$ depends on the initial
state or not.

We prepare the system in the ground state ($\beta \to \infty$) of the operator $\hat{H}^M$. The quantities
β and $\hat{H}^M$ specify the initial preparation of the system through the relation with the density
matrix $\hat{\rho} = e^{-\beta\hat{H}^M}/Z$. Below we consider $\hat{H}^M = \hat{H} - \mu\hat{N}$ with $\hat{H}$ identical to (6.64) but ϵ_0
is replaced by ϵ_0^M.[14] We choose ϵ_0^M larger than ϵ_0 and set the Fermi energy $\mu = \epsilon_F = E/4$.
Figure 6.6(a) shows the time-dependent occupation $n_b(t) = -iG_{bb}^<(t, t)$ of the b molecular
site for $\epsilon_0^M = -(1/4)E$ and $\epsilon_0^M = -(4/10)E$ (we recall that $\epsilon_0 = -E$). The calculations
have been performed with a large but finite N. The results agree with the $N \to \infty$ limit
up to a time $t_{\max} \sim N/E$; for $t > t_{\max}$ the dephasing is no longer effective. As expected,
the initial occupation $n_b(0)$ is larger for the more attractive ϵ_0^M. After some transient time
we observe the development of persistent oscillations whose frequency is the same for both
initial states and is given by $|\epsilon_1 - \epsilon_2|$, with ϵ_1, ϵ_2 the bound-state energies appearing in
(6.65). On the contrary, the amplitude of the oscillations as well as the average value of the
occupation depends on the initial state [55]. In accordance with (6.65), the average value of
the occupation when $t \to \infty$ is given by

$$n_b^\infty = \lim_{t\to\infty} \frac{1}{t} \int_0^t dt'\, n_b(t') = -i \sum_{\lambda=1}^{2} f_{bb}(\lambda, \lambda) + \underbrace{\lim_{t\to\infty} -i\delta G_{bb}^<(t, t)}_{\delta n_b(t)}.$$

---

[13]To have a sensible limit when $N \to \infty$ the couplings must scale as $1/\sqrt{N}$, see (2.19).

[14]This physical situation is identical to the situation in which the system is in thermodynamic equilibrium at zero
temperature with an initial potential on the molecule given by ϵ_0^M, and then it is driven out of equilibrium by the
sudden switch-on of an external potential $\delta\epsilon_0 = \epsilon_0 - \epsilon_0^M$ on the molecule, see also discussion above (6.28).

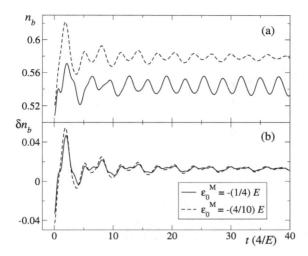

Figure 6.6 (a) Time-dependent occupation of the b molecular site, and (b) the same quantity from which the discrete–discrete contribution has been subtracted. The system is initially in the ground state of the Hamiltonian (6.64) with $\epsilon_0 \to \epsilon_0^M$. The curves refer to $\epsilon_0^M = -(1/4)E$ (solid) and $\epsilon_0^M = -(4/10)E$ (dashed). The rest of the parameters are $\epsilon_0 = -E$, $T_0 = -E/4$, $T = -E/8$, Fermi energy $\mu = \epsilon_F = E/4$, and $N = 200$.

In order to disentangle the initial-state dependence in n_b^∞ we display $\delta n_b(t)$ in Fig. 6.6(b). In both cases $\delta n_b(t)$ approaches the *same* steady-state value, i.e., the continuum–continuum contribution is insensitive to how the system is prepared. This fact agrees with the intuitive picture discussed before: the two Hamiltonians $\hat{H}^M$ differ by a *local* perturbation of the form $\hat{L} = \delta\epsilon_0^M \sum_{i=a,b} \hat{d}_i^\dagger \hat{d}_i$ and hence the bulk (metallic) averages of the initial density matrix are independent of the value of $\delta\epsilon_0^M$. Having an explicit example we can also provide a formal proof. The steady-state value of $\delta n_b(t)$ is given by the singular part of $f_{bb}(\lambda, \lambda')$ with λ, λ' the labels of two continuum eigenstates. The quantity $f_{bb}(\lambda, \lambda')$ is defined in (6.63) and the dependence on the initial state is all contained in $G^<_{\lambda\lambda'}(t_0, t_0) = G^M_{\lambda\lambda'}(\tau, \tau^+)$, see (6.27). We can calculate $G^M_{\lambda\lambda'}$ by taking the zero-temperature limit, $\beta \to \infty$, of (6.41) where

$$\hat{h}^M = \hat{h} - \mu\hat{\mathbb{1}} + \hat{L}, \qquad \hat{L} = (\epsilon_0^M - \epsilon_0) \sum_{i=a,b} |i\rangle\langle i|,$$

and $\hat{h}$ is the single-particle Hamiltonian at positive times. Using the, by now, familiar identity

$$\frac{1}{\omega_m - \hat{h}^M} = \frac{1}{(\omega_m + \mu) - \hat{h}} \left[\hat{\mathbb{1}} + \hat{L}\frac{1}{\omega_m - \hat{h}^M} \right],$$

we find

$$G^M_{\lambda\lambda'}(\tau, \tau^+) = \lim_{\beta\to\infty} \frac{1}{-i\beta} \sum_{m=-\infty}^{\infty} \frac{e^{\eta\omega_m}}{(\omega_m + \mu) - \epsilon_\lambda} \left[\delta(\lambda - \lambda') + \langle\lambda|\hat{L}\frac{1}{\omega_m - \hat{h}^M}|\lambda'\rangle \right].$$

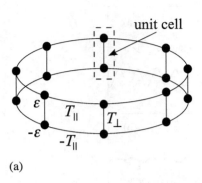

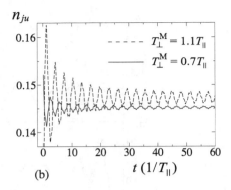

(a)   (b)

Figure 6.7 (a) Representation of a system consisting of two coupled rings. (b) Time-dependent occupation for two different initial values of the transverse coupling $T_\perp^{\rm M}/T_\parallel = 0.7, 1.1$. The rest of the parameters are (in units of $T_\parallel$) $\epsilon = 2$, $T_\perp = 1$, Fermi energy $\epsilon_{\rm F} = 0$ and $N = 1000$.

The first term in the square bracket is singular and independent of $\epsilon_0^{\rm M}$. The second term is instead a smooth function of λ and λ' since the overlap $\langle \lambda | i = a, b \rangle$ between a continuum state and a localized state is a smooth function of λ.

Global perturbations: The system of Fig. 6.7(a) belongs to the class of systems that can be diagonalized exactly using the Bloch theorem of Section 2.3.1. The unit cell consists of an up, u, and a down, d, site and the matrices h and T in (2.6) and (2.7) are 2×2 matrices with the following structure

$$h = \begin{pmatrix} \epsilon & T_\perp \\ T_\perp & -\epsilon \end{pmatrix}, \qquad T = \begin{pmatrix} T_\parallel & 0 \\ 0 & -T_\parallel \end{pmatrix}.$$

It is a simple exercise to show that the eigenvalues of $h_k = h + Te^{-ik} + T^\dagger e^{ik}$ are $\epsilon_{k\pm} = \pm\epsilon_k$ with $\epsilon_k = \sqrt{(\epsilon + 2T_\parallel \cos k)^2 + T_\perp^2} > 0$. For rings of N sites k takes the values $k = 2\pi m/N$ where, for odd N, $m = -\frac{(N-1)}{2}, \ldots, \frac{(N-1)}{2}$. In the limit $N \to \infty$ the eigenvalues form two bands separated by a finite energy gap and no discrete states are present. We denote by $|k\pm\rangle$ the single-particle eigenkets with energy $\epsilon_{k\pm}$. In accordance with the Bloch theorem, the amplitudes $\langle ju|k\pm\rangle$ and $\langle jd|k\pm\rangle$ on the jth site of the up and down ring are given by the eigenvectors of h_k multiplied by the normalized plane wave $e^{ikj}/\sqrt{N}$. Assuming $\epsilon \geq 2|T_\parallel|$ and $T_\perp > 0$ the reader can verify that these amplitudes are

$$\begin{pmatrix} \langle ju|k\pm\rangle \\ \langle jd|k\pm\rangle \end{pmatrix} = \frac{e^{ikj}}{\sqrt{N}} \frac{1}{\sqrt{2}} \begin{pmatrix} \sqrt{1 \pm \cos\theta_k} \\ \pm\sqrt{1 \mp \cos\theta_k} \end{pmatrix}, \qquad \cos\theta_k = \frac{\epsilon + 2T_\parallel \cos k}{\epsilon_k} \geq 0, \qquad (6.66)$$

and that the eigenkets are correctly normalized, $\langle k\nu|k'\nu'\rangle = \delta_{kk'}\delta_{\nu\nu'}$.

We intend to study the evolution of the system starting from the ground state of the same Hamiltonian but with a different $T_\perp$, i.e., $T_\perp \to T_\perp^{\mathrm{M}}$. Let us take the Fermi energy $\epsilon_{\mathrm{F}} = 0$ (bottom band fully occupied) and calculate the occupation $n_{ju}(t)$ on the jth site of the upper ring (due to rotational invariance the density does not depend on j). We use the natural convention that quantities calculated with $T_\perp^{\mathrm{M}}$ instead of $T_\perp$ carry the superscript "M". We then have

$$n_{ju}(t) = \sum_k \underbrace{\left| \langle ju | e^{-i\hat{h}t} | k^{\mathrm{M}} - \rangle \right|^2}_{|\varphi_{k-}^{\mathrm{M}}(ju,t)|^2} = \sum_k \left| \sum_{k'\nu} e^{-i\epsilon_{k'\nu}t} \langle ju | k'\nu \rangle \langle k'\nu | k^{\mathrm{M}} - \rangle \right|^2, \qquad (6.67)$$

since only the eigenkets $|k^{\mathrm{M}}-\rangle$ with negative energies contribute to the density. The overlap between $|k^{\mathrm{M}}-\rangle$ and the eigenkets $|k\nu\rangle$ of $\hat{H}$ can be deduced from (6.66),

$$\langle k' \pm | k^{\mathrm{M}} - \rangle = \frac{\delta_{kk'}}{2} \left[\sqrt{(1 \pm \cos\theta_k)(1 - \cos\theta_k^{\mathrm{M}})} \mp \sqrt{(1 \mp \cos\theta_k)(1 + \cos\theta_k^{\mathrm{M}})} \right].$$

Inserting this result into (6.67), using (6.66) and recalling that $\epsilon_{k\pm} = \pm\epsilon_k$, after some elementary algebra we arrive at the result below

$$n_{ju}(t) = \frac{1}{2N} \sum_k \left[(1 - \cos\theta_k^{\mathrm{M}}) + 2\sin\theta_k \sin(\theta_k - \theta_k^{\mathrm{M}}) \sin^2(\epsilon_k t) \right],$$

with $\sin\theta_k = T_\perp/\epsilon_k > 0$ and the like for $\sin\theta_k^{\mathrm{M}}$. We see that for $T_\perp^{\mathrm{M}} = T_\perp$, and hence $\theta_k = \theta_k^{\mathrm{M}}$, the second term in the square bracket vanishes and the occupation equals the ground-state occupation at all times, as it should. In Fig. 6.7(b) we plot $n_{ju}(t)$ for two different initial configurations corresponding to $T_\perp^{\mathrm{M}}/T_\parallel = 0.7$ and 1.1, whereas at positive times $T_\perp/T_\parallel = 1$. As in the previous example, we use a finite $N = 1000$ so as to reproduce the $N \to \infty$ results up to times $t_{\max} \sim N/T_\parallel$. The occupation exhibits damped oscillations and eventually attains a steady-state value, in agreement with the fact that there are no discrete states. The limit $n_{ju}^\infty = \lim_{t\to\infty} n_{ju}(t)$ does, however, depend on $T_\perp^{\mathrm{M}}$, i.e., on the initial state. This behavior should be compared with the behavior of $\delta n_b(t)$ in Fig. 6.6(b): in that case the value δn_b^∞ was independent of the initial state. This example confirms our intuitive expectation that if $\hat{H}^{\mathrm{M}} \to \hat{H}^{\mathrm{M}} + \hat{G}$ with $\hat{G}$ a *global* perturbation, then the memory of the initial state is not washed out in the long-time limit.

To conclude, we observe that many interesting physical problems fall into the class of global perturbations. Later in this book we investigate the quantum transport problem: two bulk metals connected by a nanoscale junction are initially in equilibrium and then driven out of equilibrium by the sudden switch-on of a bias. Denoting by $\hat{H}_{\mathrm{eq}}$ the Hamiltonian of the unperturbed system and by $\hat{H}_V$ the bias perturbation we have $\hat{H}^{\mathrm{M}} = \hat{H}_{\mathrm{eq}} - \mu\hat{N}$ and $\hat{H} = \hat{H}_{\mathrm{eq}} + \hat{H}_V$. We could, however, look at the problem from a different (but equivalent) perspective: the system is initially prepared in an excited configuration of $\hat{H}$ which is described by $\hat{H}^{\mathrm{M}} = \hat{H} - \hat{H}_V - \mu\hat{N}$. Since the bias is a global perturbation we expect that the steady-state depends on the initial state, i.e., on the bias.

6.3.2 Fluctuation-dissipation theorem and other exact properties

Systems prepared in a stationary excited configuration of $\hat{H}$ are described by a Matsubara operator $\hat{H}^{\mathrm{M}}$ that commutes with $\hat{H}$, i.e., $[\hat{H}^{\mathrm{M}}, \hat{H}]_{-} = 0$. To this class of initial states belongs the equilibrium state for which $\hat{H}^{\mathrm{M}} = \hat{H} - \mu\hat{N}$. The eigenkets $|\Psi_k\rangle$ of $\hat{H}^{\mathrm{M}}$ in (6.21) and (6.22) can then be chosen to be also eigenkets of $\hat{H}$ with eigenvalue E_k, and the general formula for the lesser and greater Green's function simplifies to

$$G_{ji}^{<}(t, t') = \mp i \sum_{k} \rho_k \langle \Psi_k | \hat{d}_i^{\dagger} e^{-i(\hat{H} - E_k)(t' - t)} \hat{d}_j | \Psi_k \rangle, \tag{6.68}$$

$$G_{ji}^{>}(t, t') = -i \sum_{k} \rho_k \langle \Psi_k | \hat{d}_j e^{-i(\hat{H} - E_k)(t - t')} \hat{d}_i^{\dagger} | \Psi_k \rangle. \tag{6.69}$$

As expected, all matrix elements of $\hat{\mathcal{G}}^{\lessgtr}(t, t')$ depend only on the difference $t - t'$ and, therefore, can be Fourier transformed as in (6.47). In Fourier space the relation between $\hat{\mathcal{G}}^{\lessgtr}(\omega)$ and $\hat{\mathcal{G}}^{\mathrm{R/A}}(\omega)$ is particularly elegant. Using the representation (6.57) of the Heaviside function we have

$$\hat{\mathcal{G}}^{\mathrm{R}}(t, t') = i \int \frac{d\omega}{2\pi} \frac{e^{-i\omega(t - t')}}{\omega + i\eta} \int \frac{d\omega'}{2\pi} e^{-i\omega'(t - t')} \left[\hat{\mathcal{G}}^{>}(\omega') - \hat{\mathcal{G}}^{<}(\omega') \right],$$

from which it follows that

$$\hat{\mathcal{G}}^{\mathrm{R}}(\omega) = i \int \frac{d\omega'}{2\pi} \frac{\hat{\mathcal{G}}^{>}(\omega') - \hat{\mathcal{G}}^{<}(\omega')}{\omega - \omega' + i\eta}. \tag{6.70}$$

For the Fourier transform of the advanced Green's function we observe that (6.24) implies $\hat{\mathcal{G}}^{\mathrm{A}}(\omega) = [\hat{\mathcal{G}}^{\mathrm{R}}(\omega)]^{\dagger}$ and hence

$$\hat{\mathcal{G}}^{\mathrm{A}}(\omega) = i \int \frac{d\omega'}{2\pi} \frac{\hat{\mathcal{G}}^{>}(\omega') - \hat{\mathcal{G}}^{<}(\omega')}{\omega - \omega' - i\eta}, \tag{6.71}$$

where we take into account that the operator $i[\hat{\mathcal{G}}^{>}(\omega') - \hat{\mathcal{G}}^{<}(\omega')]$ is self-adjoint, see (6.23).

Going back to (6.68) and (6.69) we expand the evolution operator over a complete set of eigenstates of $\hat{H}$ and obtain the Lehmann representation

$$\boxed{G_{ji}^{<}(t, t') = \mp i \sum_{pk} \rho_k \, \Phi_{pk}^{*}(i) \Phi_{pk}(j) e^{-i(E_p - E_k)(t' - t)}} \tag{6.72}$$

$$\boxed{G_{ji}^{>}(t, t') = -i \sum_{pk} \rho_k \, \Phi_{kp}(j) \Phi_{kp}^{*}(i) e^{-i(E_p - E_k)(t - t')}} \tag{6.73}$$

where we define the amplitudes

$$\Phi_{kp}(i) = \langle \Psi_k | \hat{d}_i | \Psi_p \rangle. \tag{6.74}$$

For these amplitudes to be different from zero, $|\Psi_p\rangle$ must contain one particle more than $|\Psi_k\rangle$. To gain some insight into the physical meaning of the Φ_{kp} let us consider a noninteracting Hamiltonian $\hat{H} = \sum_\lambda \hat{c}_\lambda^\dagger \hat{c}_\lambda$. Then, the generic eigenket with N particles is

$$|\Psi_k\rangle = \hat{c}_{\lambda_1}^\dagger \dots \hat{c}_{\lambda_N}^\dagger |0\rangle,$$

and the only nonvanishing amplitudes are those with $|\Psi_p\rangle = \hat{c}_\lambda^\dagger |\Psi_k\rangle$, i.e., $|\Psi_p\rangle$ must be an eigenket with $N + 1$ particles, N of which are in the energy levels $\lambda_1, \dots, \lambda_N$. In this case

$$\Phi_{kp}(i) = \langle \Psi_k | \hat{d}_i \hat{c}_\lambda^\dagger | \Psi_k \rangle = C \varphi_\lambda(i)$$

is proportional to the wavefunction of the λth single-particle eigenstate;[15] in particular, if $\hat{d}_i$ is the field operator $\hat{\psi}(\mathbf{x})$ then $\Phi_{kp}(\mathbf{x}) = C \varphi_\lambda(\mathbf{x})$. In interacting systems the Φ_{kp} cannot be identified with a single-particle eigenfunction. Nevertheless, they satisfy a single-particle Schrödinger-like equation that we now derive. Let $\hat{H} = \sum_{ij} h_{ij} \hat{d}_i^\dagger \hat{d}_j + \hat{H}_{\text{int}}$ be the Hamiltonian of the interacting system and consider the sandwich of the commutator

$$\left[\hat{d}_i, \hat{H}\right]_- = \sum_j h_{ij} \hat{d}_j + \left[\hat{d}_i, \hat{H}_{\text{int}}\right]_-$$

with the states $\langle \Psi_k |$ and $|\Psi_p\rangle$. Taking into account the definition (6.74) we find

$$\sum_j h_{ij} \Phi_{kp}(j) + \sum_q \left[\Phi_{kq}(i) I_{qp} - I_{kq} \Phi_{qp}(i)\right] = (E_p - E_k) \Phi_{kp}(i),$$

where

$$I_{qp} = \langle \Psi_q | \hat{H}_{\text{int}} | \Psi_p \rangle.$$

For $\hat{H}_{\text{int}} = 0$ the single-particle Schrödinger-like equation is solved by $E_p - E_k = \epsilon_\lambda$ and $\Phi_{kp}(i) = \varphi_\lambda(i)$, in agreement with the discussion above. In the general case we refer to the $\Phi_{kp}(i)$ as the *quasi-particle wavefunctions*.

From (6.72) and (6.73) we can immediately read the Fourier transform of the lesser and greater Green's functions

$$\boxed{G_{ji}^<(\omega) = \mp 2\pi i \sum_{pk} \rho_k \, \Phi_{pk}^*(i) \Phi_{pk}(j) \delta(\omega - E_k + E_p)} \tag{6.75}$$

$$\boxed{G_{ji}^>(\omega) = -2\pi i \sum_{pk} \rho_k \, \Phi_{kp}(j) \Phi_{kp}^*(i) \delta(\omega - E_p + E_k)} \tag{6.76}$$

The diagonal elements of $iG^\lessgtr$ have a well defined sign for all frequencies ω:

$$iG_{jj}^>(\omega) \geq 0, \qquad iG_{jj}^<(\omega) \begin{array}{l} \geq 0 \quad \text{for bosons} \\ \leq 0 \quad \text{for fermions} \end{array}. \tag{6.77}$$

---

[15]The constant of proportionality C is 1 for fermions and $n_\lambda!$ for bosons, where n_λ is the number of bosons in the λth energy level, see Chapter 1.

Substituting these results into (6.70) and (6.71) we find

$$G_{ji}^{R/A}(\omega) = \sum_{pk} \frac{\Phi_{kp}(j)\Phi_{kp}^*(i)}{\omega - E_p + E_k \pm i\eta}\,[\,\rho_k \mp \rho_p\,] \qquad (6.78)$$

where in $G^<$ we renamed the summation indices $k \leftrightarrow p$. We see that the Fourier transform of the retarded Green's function is analytic in the upper half of the complex ω plane whereas the Fourier transform of the advanced Green's function is analytic in the lower half of the complex ω plane, in agreement with the results of Section 6.2.3.

The Fourier transforms can be used to derive an important result for systems in thermodynamic equilibrium, $\hat{H}^{M} = \hat{H} - \mu\hat{N}$. In this case

$$\rho_k = \frac{e^{-\beta(E_k - \mu N_k)}}{\mathrm{Tr}\left[e^{-\beta(\hat{H} - \mu\hat{N})}\right]} = e^{-\beta(E_k - E_p) + \beta\mu(N_k - N_p)}\rho_p,$$

with N_k the number of particles in the state $|\Psi_k\rangle$. Substituting this result into $G^>$, renaming the summation indices $k \leftrightarrow p$ and taking into account that the only nonvanishing Φ_{kp} are those for which $N_k - N_p = -1$, we find

$$\hat{\mathcal{G}}^>(\omega) = \pm e^{\beta(\omega - \mu)}\hat{\mathcal{G}}^<(\omega) \qquad (6.79)$$

Next we recall that by the very same definition of retarded/advanced functions

$$\hat{\mathcal{G}}^>(\omega) - \hat{\mathcal{G}}^<(\omega) = \hat{\mathcal{G}}^R(\omega) - \hat{\mathcal{G}}^A(\omega).$$

The combination of these last two identities leads to the *fluctuation–dissipation theorem*

$$\hat{\mathcal{G}}^<(\omega) = \pm f(\omega - \mu)[\hat{\mathcal{G}}^R(\omega) - \hat{\mathcal{G}}^A(\omega)] \qquad (6.80)$$

$$\hat{\mathcal{G}}^>(\omega) = \bar{f}(\omega - \mu)[\hat{\mathcal{G}}^R(\omega) - \hat{\mathcal{G}}^A(\omega)] \qquad (6.81)$$

which was previously demonstrated only for noninteracting particles. There also exists another way of deriving (6.79) and hence the fluctuation–dissipation theorem. By definition the left Green's function is

$$G_{ji}^{\lceil}(\tau, t') = -i \sum_k \rho_k \langle \Psi_k | \underbrace{e^{i(\hat{H} - \mu\hat{N})(-i\tau)}\hat{d}_j e^{-i(\hat{H} - \mu\hat{N})(-i\tau)}}_{\hat{d}_{j,H}(t_0 - i\tau)}\,\underbrace{e^{i\hat{H}(t' - t_0)}\hat{d}_i^\dagger e^{-i\hat{H}(t' - t_0)}}_{\hat{d}_{i,H}^\dagger(t')} |\Psi_k\rangle,$$

$$= -i e^{\mu\tau} \sum_k \rho_k \langle \Psi_k | \hat{d}_j e^{-i(\hat{H} - E_k)(t_0 - i\tau - t')}\hat{d}_i^\dagger |\Psi_k\rangle,$$

and comparing this result with (6.69) we conclude that

$$\hat{\mathcal{G}}^{\lceil}(\tau, t') = e^{\mu\tau}\hat{\mathcal{G}}^>(t_0 - i\tau, t').$$

A similar relation can be derived for the lesser and right Green's function

$$\hat{\mathcal{G}}^{\rceil}(t, \tau) = e^{-\mu\tau} \hat{\mathcal{G}}^{<}(t, t_0 - \mathrm{i}\tau).$$

Combining these results with the KMS relations we find

$$
\begin{aligned}
\hat{\mathcal{G}}^{<}(t_0, t') &= \hat{\mathcal{G}}(t_{0-}, t'_+) \\
&= \pm \hat{\mathcal{G}}(t_0 - \mathrm{i}\beta, t'_+) \\
&= \pm \hat{\mathcal{G}}^{\lceil}(\beta, t') \\
&= \pm e^{\mu\beta} \hat{\mathcal{G}}^{>}(t_0 - \mathrm{i}\beta, t').
\end{aligned}
$$

Taking the Fourier transform of both sides, we recover (6.79).

For systems in thermodynamic equilibrium we can also derive an important relation between the Matsubara Green's function and the retarded/advanced Green's function. The starting point is the formula (6.13). Expanding the exponentials in a complete set of eigenstates of $\hat{H}^{\mathrm{M}} = \hat{H} - \mu\hat{N}$ we find

$$G_{ji}^{\mathrm{M}}(\tau_1, \tau_2) = \frac{1}{\mathrm{i}} \sum_{kp} \rho_p \left[\theta(\tau_1 - \tau_2) e^{\beta(E_p^{\mathrm{M}} - E_k^{\mathrm{M}})} \pm \theta(\tau_2 - \tau_1) \right] e^{-(\tau_1 - \tau_2)(E_p^{\mathrm{M}} - E_k^{\mathrm{M}})} \Phi_{kp}(j) \Phi_{kp}^*(i),$$

where $E_p^{\mathrm{M}} = E_p - \mu N_p$. To obtain this result we have simply renamed the summation indices $k \leftrightarrow p$ in the first term and used the fact that $\rho_k = e^{\beta(E_p^{\mathrm{M}} - E_k^{\mathrm{M}})} \rho_p$. In Section 6.2.1 we proved the identity (recall that $\bar{f}(E) = e^{\beta E} f(E)$)

$$\frac{1}{-\mathrm{i}\beta} \sum_{m=-\infty}^{\infty} \frac{e^{-\omega_m \tau}}{\omega_m - E} = \frac{1}{\mathrm{i}} \left[\theta(\tau) e^{\beta E} \pm \theta(-\tau) \right] f(E) e^{-\tau E}.$$

The r.h.s. of this equation has precisely the same structure appearing in G^{M}. Therefore we can rewrite the Matsubara Green's function as

$$G_{ji}^{\mathrm{M}}(\tau_1, \tau_2) = \frac{1}{-\mathrm{i}\beta} \sum_{m=-\infty}^{\infty} e^{-\omega_m(\tau_1 - \tau_2)} \underbrace{\sum_{kp} \frac{\rho_p}{f(E_p^{\mathrm{M}} - E_k^{\mathrm{M}})} \frac{\Phi_{kp}(j) \Phi_{kp}^*(i)}{\omega_m - E_p^{\mathrm{M}} + E_k^{\mathrm{M}}}}_{G_{ji}^{\mathrm{M}}(\omega_m)}.$$

Let us manipulate this formula. We have

$$\frac{\rho_p}{f(E_p^{\mathrm{M}} - E_k^{\mathrm{M}})} = \rho_p \left(e^{\beta(E_p^{\mathrm{M}} - E_k^{\mathrm{M}})} \mp 1 \right) = \rho_k \mp \rho_p.$$

Furthermore, the only nonvanishing terms in the sum over p and k are those for which $N_p - N_k = 1$ and hence

$$E_p^{\mathrm{M}} - E_k^{\mathrm{M}} = E_p - E_k - \mu(N_p - N_k) = E_p - E_k - \mu.$$

Inserting these results into the formula for G^{M} we find

$$G_{ji}^{\mathrm{M}}(\omega_m) = \sum_{kp} \frac{\Phi_{kp}(j)\Phi_{kp}^*(i)}{\omega_m + \mu - E_p + E_k} \left[\, \rho_k \mp \rho_p \,\right]$$

and a comparison with (6.78) leads to the important relation

$$\hat{\mathcal{G}}^{\mathrm{M}}(\zeta) = \begin{cases} \hat{\mathcal{G}}^{\mathrm{R}}(\zeta + \mu) & \text{for } \mathrm{Im}[\zeta] > 0 \\ \hat{\mathcal{G}}^{\mathrm{A}}(\zeta + \mu) & \text{for } \mathrm{Im}[\zeta] < 0 \end{cases}. \tag{6.82}$$

Thus $\hat{\mathcal{G}}^{\mathrm{M}}(\zeta)$ is analytic everywhere except along the real axis where it can have poles or branch points. The reader can easily check that (6.82) agrees with the formulas (6.18) and (6.58), (6.59) for noninteracting Green's functions. In particular (6.82) implies that for $\zeta = \omega \pm \mathrm{i}\eta$

$$\hat{\mathcal{G}}^{\mathrm{M}}(\omega \pm \mathrm{i}\eta) = \hat{\mathcal{G}}^{\mathrm{R/A}}(\omega + \mu) \tag{6.83}$$

according to which $\hat{\mathcal{G}}^{\mathrm{M}}$ has a discontinuity given by the difference $\hat{\mathcal{G}}^{\mathrm{R}} - \hat{\mathcal{G}}^{\mathrm{A}}$ when the complex frequency crosses the real axis.

6.3.3 Spectral function and probability interpretation

The Lehmann representation (6.72) and (6.73) simplifies further for systems that are initially in a pure state and hence $\hat{\rho} = |\Psi_{N,0}\rangle\langle\Psi_{N,0}|$. In the discussion that follows the eigenstate $|\Psi_{N,0}\rangle$ with eigenenergy $E_{N,0}$ can be either the ground state or an excited state of $\hat{H}$ with N particles. We denote by $|\Psi_{N\pm1,m}\rangle$ the eigenstates of $\hat{H}$ with $N \pm 1$ particles and define the *quasi-particle wavefunctions* P_m and the *quasi-hole wavefunctions* Q_m according to

$$P_m(i) = \langle\Psi_{N,0}|\hat{d}_i|\Psi_{N+1,m}\rangle, \qquad Q_m(i) = \langle\Psi_{N-1,m}|\hat{d}_i|\Psi_{N,0}\rangle. \tag{6.84}$$

Then, the lesser and greater Green's functions become

$$G_{ji}^<(t,t') = \mp\mathrm{i}\sum_m Q_m(j)Q_m^*(i)e^{-\mathrm{i}(E_{N-1,m}-E_{N,0})(t'-t)}, \tag{6.85}$$

$$G_{ji}^>(t,t') = -\mathrm{i}\sum_m P_m(j)P_m^*(i)e^{-\mathrm{i}(E_{N+1,m}-E_{N,0})(t-t')}, \tag{6.86}$$

where $E_{N\pm1,m}$ is the energy eigenvalue of $|\Psi_{N\pm1,m}\rangle$. From (6.85) we see that the probability amplitude analyzed in Fig. 6.2 can be written as the sum of oscillatory functions whose frequency $(E_{N,0}-E_{N-1,m})$ corresponds to a possible *ionization energy* (also called *removal energy*) of the system. Similarly, the probability amplitude (described by $G^>$) that by adding a particle with quantum number i at time t' and then evolving until t we find the same state as evolving until t and then adding a particle with quantum number j, can be written as the sum of oscillatory functions whose frequency $(E_{N+1,m} - E_{N,0})$ corresponds to a possible *affinity* (also called *addition energy*) of the system.

It is especially interesting to look at the Fourier transforms of these functions:

$$G_{ji}^{<}(\omega) = \mp 2\pi\mathrm{i} \sum_m Q_m(j) Q_m^*(i)\, \delta(\omega - [E_{N,0} - E_{N-1,m}])$$

(6.87)

$$G_{ji}^{>}(\omega) = -2\pi\mathrm{i} \sum_m P_m(j) P_m^*(i)\, \delta(\omega - [E_{N+1,m} - E_{N,0}])$$

(6.88)

The Fourier transform of $G^{<}$ is peaked at the removal energies whereas the Fourier transform of $G^{>}$ is peaked at the addition energies. We can say that by removing (adding) a particle the system gets excited in a combination of eigenstates with one particle less (one particle more). In noninteracting systems the matrix elements $G_{\lambda\lambda'}^{\lessgtr}(\omega)$ in the basis that diagonalizes $\hat{h}$ can be extracted from (6.48) and (6.49) and read

$$G_{\lambda\lambda'}^{<}(\omega) = \mp\delta_{\lambda\lambda'} 2\pi\mathrm{i} f(\epsilon_\lambda^{\mathrm{M}})\delta(\omega - \epsilon_\lambda), \qquad G_{\lambda\lambda'}^{>}(\omega) = -\delta_{\lambda\lambda'} 2\pi\mathrm{i} \bar{f}(\epsilon_\lambda^{\mathrm{M}})\delta(\omega - \epsilon_\lambda).$$

Thus, the removal (addition) of a particle with quantum number λ excites the system in only one way since the lesser (greater) Green's function is peaked at the removal (addition) energy ϵ_λ and is zero otherwise. This property reflects the fact that in a noninteracting system there exist *stationary* single-particle states, i.e., states in which the particle is not scattered by any other particle and hence its energy is well defined or, equivalently, its life-time is infinitely long. Instead the removal (addition) of a particle with quantum numbers other than λ does not generate an excitation with a well-defined energy, and the matrix elements $G_{ji}^{<}$ $(G_{ji}^{>})$ exhibit peaks at all possible energies ϵ_λ with weights proportional to the product of overlaps $\langle j|\lambda\rangle\langle\lambda|i\rangle$. Interacting systems behave in a similar way since there exist no good single-particle quantum numbers; a particle scatters with all the other particles and its energy cannot be sharply defined. It would be useful to construct a frequency-dependent operator $\hat{A}(\omega)$ whose average $A_{jj}(\omega) = \langle j|\hat{A}(\omega)|j\rangle$ contains information about the probability for an added/removed particle with quantum number j to have energy ω. From the above discussion a natural proposal for this operator is

$$\hat{A}(\omega) = \mathrm{i}\,[\hat{\mathcal{G}}^{>}(\omega) - \hat{\mathcal{G}}^{<}(\omega)] = \mathrm{i}\,[\hat{\mathcal{G}}^{\mathrm{R}}(\omega) - \hat{\mathcal{G}}^{\mathrm{A}}(\omega)]$$

(6.89)

In noninteracting systems[16] we have $\hat{A}(\omega) = 2\pi\delta(\omega - \hat{h})$ and hence

$$A_{jj}(\omega) = 2\pi \sum_\lambda |\langle j|\lambda\rangle|^2 \delta(\omega - \epsilon_\lambda) \geq 0.$$

(6.90)

The standard interpretation of this result is that the probability for a particle with quantum number j to have energy ω is zero unless ω is one of the single-particle energies, in which case the probability is proportional to $|\langle j|\lambda\rangle|^2$. This interpretation is sound also for another reason. The probability that the particle has energy ω in the range $(-\infty, \infty)$ should be 1 for any j, and indeed from (6.90) we have

$$\int \frac{d\omega}{2\pi} A_{jj}(\omega) = \sum_\lambda |\langle j|\lambda\rangle|^2 = 1.$$

---

[16]See also the discussion in Section 2.3.2.

Furthermore, the sum over the complete set j of $A_{jj}(\omega)$ should correspond to the density of states with energy ω and indeed

$$D(\omega) = \sum_j A_{jj}(\omega) = 2\pi \sum_\lambda \delta(\omega - \epsilon_\lambda),$$

which agrees with (6.34). A suitable name for the operator $\hat{A}(\omega)$ is *spectral function* operator, since it contains information on the energy spectrum of a single particle. The spectral function is a very useful mathematical quantity but the above probabilistic interpretation is questionable. Is $A_{jj}(\omega)$ the probability that a removed particle or an added particle with quantum number j has energy ω? Are we sure that $A_{jj}(\omega) \geq 0$ also in the interacting case?

In the interacting case the matrix element $A_{jj}(\omega)$ can be derived from (6.87) and (6.88) and reads

$$A_{jj}(\omega) = 2\pi \left[\sum_m |P_m(j)|^2 \delta(\omega - [E_{N+1,m} - E_{N,0}]) \right.$$

$$\left. \mp \sum_m |Q_m(j)|^2 \delta(\omega - [E_{N,0} - E_{N-1,m}]) \right]. \tag{6.91}$$

A comparison with (6.90) shows how natural it is to interpret the functions P_m and Q_m as quasi-particle and quasi-hole wavefunctions. In fermionic systems $A_{jj}(\omega) \geq 0$ but in bosonic systems $A_{jj}(\omega)$ can be negative since the second term on the r.h.s. of (6.91) is nonpositive. For instance one can show that the spectral function of the bosonic Hubbard model is negative for some ωs [56, 57], see also Exercise 6.8. As we see in the next section, the particle and hole contributions to $A(\omega)$ can be measured separately and therefore the quantities $iG^>(\omega) > 0$ and $\pm iG^<(\omega) > 0$, see (6.77), are more fundamental than $A(\omega)$. Even though these quantities do not integrate to unity we can always normalize them and interpret $iG^>(\omega)$ as the probability that an added particle has energy ω and $\pm iG^<(\omega)$ as the probability that a removed particle has energy ω.

Despite the nonpositivity of the bosonic spectral function the normalization condition is fulfilled for both fermions and bosons. Upon integration of (6.91) over ω we end up with the sums $\sum_m |P_m(j)|^2$ and $\sum_m |Q_m(j)|^2$. From the definitions (6.84) it is easy to see that

$$\sum_m |P_m(j)|^2 = \langle \Phi_{N,0} | \hat{d}_j \hat{d}_j^\dagger | \Phi_{N,0} \rangle, \quad \sum_m |Q_m(j)|^2 = \langle \Phi_{N,0} | \hat{d}_j^\dagger \hat{d}_j | \Phi_{N,0} \rangle,$$

and hence

$$\int \frac{d\omega}{2\pi} A_{jj}(\omega) = \langle \Phi_{N,0} | \left[\hat{d}_j, \hat{d}_j^\dagger \right]_\mp | \Phi_{N,0} \rangle = 1.$$

More generally, the integral of the matrix elements of the spectral function operator satisfies the sum rule

$$\boxed{\int \frac{d\omega}{2\pi} A_{ji}(\omega) = \delta_{ji}} \tag{6.92}$$

which can be verified similarly. The reader can also derive this result in the position–spin basis; using the field operators instead of the $\hat{d}$-operators the matrix elements of $\hat{A}$ are functions $A(\mathbf{x}, \mathbf{x}'; \omega)$ of the position–spin coordinates, and the r.h.s. of the sum rule (6.92) is replaced by $\delta(\mathbf{x} - \mathbf{x}')$.

We have discussed the physical interpretation of $\hat{A}$ only for systems in a pure state. However the definition (6.89) makes sense for any initial configuration such that $[\hat{H}^{M}, \hat{H}]_{-} = 0$, like for instance the equilibrium configuration at finite temperature. In this case $G^{<}$ and $G^{>}$ are given by (6.75) and (6.76), and the curious reader can easily generalize the previous discussion as well as check that (6.92) is still satisfied. It is important to stress that for systems in thermodynamic equilibrium the knowledge of $\hat{A}$ is enough to calculate all Keldysh components of $\hat{\mathcal{G}}$ with real-time arguments. From (6.70) and (6.71) we have

$$\hat{\mathcal{G}}^{R}(\omega) = \int \frac{d\omega'}{2\pi} \frac{\hat{A}(\omega')}{\omega - \omega' + i\eta} \qquad \hat{\mathcal{G}}^{A}(\omega) = \int \frac{d\omega'}{2\pi} \frac{\hat{A}(\omega')}{\omega - \omega' - i\eta} \qquad (6.93)$$

and from the fluctuation–dissipation theorem (6.80) and (6.81)

$$\hat{\mathcal{G}}^{<}(\omega) = \mp i f(\omega - \mu)\hat{A}(\omega) \qquad \hat{\mathcal{G}}^{>}(\omega) = -i\bar{f}(\omega - \mu)\hat{A}(\omega) \qquad (6.94)$$

In a *zero temperature* fermionic system the function $f(\omega - \mu) = \theta(\mu - \omega)$ and hence the spectral function has peaks only at the addition energies for $\omega > \mu$ and peaks only at the removal energies for $\omega < \mu$. This nice separation is not possible at finite temperature. We further observe that if we write the retarded/advanced Green's functions as the sum of a Hermitian and anti-Hermitean quantity:

$$\hat{\mathcal{G}}^{R/A}(\omega) = \hat{C}(\omega) \mp \frac{i}{2}\hat{A}(\omega),$$

then (6.93) implies that $\hat{C}$ and $\hat{A}$ are connected by a Hilbert transformation, similarly to the real and imaginary part of the embedding self-energy in (2.26). As we see, this is a general property of several many-body quantities.

Finally, we point out that the shape of the spectral function can be used to predict the time-dependent behavior of the Green's function and vice versa. If a matrix element of the spectral function has δ-like peaks, like the $A_{\lambda\lambda}(\omega) = 2\pi\delta(\omega - \epsilon_{\lambda})$ of a noninteracting system or like the $A_{ji}(\omega)$ of a finite system, then the corresponding matrix element of the Green's function oscillates in time. As already discussed in Section 6.1.3, for the Green's function $G_{ji}^{\lessgtr}(t, t')$ to decay with $|t - t'|$ the quantum numbers j and i must be coupled to infinitely many degrees of freedom. This occurs both in *macroscopic noninteracting* systems, if j and i are not good quantum numbers (e.g., $|j\rangle = |\epsilon_0\rangle$ in the Fano model), or in *macroscopic interacting* systems. The effect of the coupling is to broaden the δ-like peaks, thus giving a finite lifetime to (almost) every single-particle excitation. This can be seen in (6.90) (for noninteracting systems) and (6.91) (for interacting systems); $A_{jj}(\omega)$ becomes a continuous function of ω since the sum over discrete states becomes an integral over a *continuum* of states. The transition from sharp δ-like peaks to a continuous function often generates some consternation. In Appendix I we show that there is nothing mysterious in this transition. Our first example of a continuous, broadened spectral function has been the Lorentzian $A_{00}(\omega)$ of the Fano model, see Fig. 2.8. In the next section we discuss a second instructive example in which the broadening is induced by the interaction.

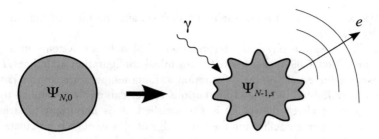

Figure 6.8 A photoemission experiment. After absorption of the photon γ of energy ω_0 the system is left in an excited state formed by the photoelectron and $N-1$ electrons in the sth excited state.

6.3.4 Photoemission experiments and interaction effects

At the moment the spectral function is a pure mathematical object providing information on the underlying physical system (we have seen that it has δ-like peaks at the removal and addition energies). Can we also relate it to some measurable quantity? In this section we show that the *spectral function can be measured with a photoemission experiment.*

Let us start by considering a system of electrons (fermionic system). A photoemission experiment consists in irradiating the system with light of frequency $\omega_0 > 0$ and then measuring the number of ejected electrons (or photoelectrons) with energy ϵ, see Fig. 6.8. Without loss of generality we can set the threshold of the continuum (scattering) states to zero so that $\epsilon > 0$. Due to energy conservation, $E_{N,0} + \omega_0 = E_{N-1,s} + \epsilon$ where $E_{N,0}$ is the energy of the initially unperturbed system with N particles and $E_{N-1,s}$ is the energy of the sth excited state in which the system is left after the photoelectron has been kicked out.[17] Clearly, if the frequency ω_0 is smaller than $E_{N-1,0} - E_{N,0}$, with $E_{N-1,0}$ the ground-state energy of the system with $N-1$ particles, no electron will be ejected. Let us develop a simple theory to calculate the result of a photoemission experiment. The electromagnetic field is described by a monochromatic vector potential $\mathbf{A}(\mathbf{r}, t) = \mathbf{A}(\mathbf{r})e^{i\omega_0 t} + \mathbf{A}^*(\mathbf{r})e^{-i\omega_0 t}$. Experimentally one observes that the photocurrent (number of ejected electrons per unit time) is proportional to the intensity of the electromagnetic field, and hence we can discard the A^2 term appearing in the single-particle Hamiltonian (3.1). The time-dependent perturbation which couples the light to the electrons is then proportional to $\hat{\mathbf{p}} \cdot \mathbf{A}(\hat{\mathbf{r}}, t) + \mathbf{A}(\hat{\mathbf{r}}, t) \cdot \hat{\mathbf{p}}$. Expanding the field operators over some convenient basis the most general form of this perturbation in second quantization reads

$$\hat{H}_{l-e}(t) = \sum_{ij} (h_{ij} e^{i\omega_0 t} + h_{ij}^* e^{-i\omega_0 t}) \hat{d}_i^\dagger \hat{d}_j.$$

According to the *Fermi Golden rule*[18] the probability per unit time of the transition from the initial state $|\Psi_{N,0}\rangle$ with energy $E_{N,0}$ to an excited state $|\Psi_{N,m}\rangle$ with energy $E_{N,m}$ is given

---

[17]Even though in most experiments the system is initially in its ground state our treatment is applicable also to situations in which $E_{N,0}$ refers to some excited state, read again the beginning of Section 6.3.3.

[18]In Chapter 14 we present a pedagogical derivation of the Fermi Golden rule and clarify several subtle points.

by

$$P_{0 \to m} = 2\pi \left| \langle \Psi_{N,m} | \sum_{ij} h_{ij}^* \hat{d}_i^\dagger \hat{d}_j | \Psi_{N,0} \rangle \right|^2 \delta(\omega_0 - [E_{N,m} - E_{N,0}]).$$

The photocurrent $I_{\mathrm{ph}}(\epsilon)$ with electrons of energy ϵ is proportional to the sum of all the transition probabilities with excited states $|\Psi_{N,m}\rangle = \hat{d}_\epsilon^\dagger |\Psi_{N-1,s}\rangle$. These states describe an electron outside the system with energy ϵ and $N-1$ electrons inside the system in the sth excited state, see again Fig. 6.8. For energies ϵ sufficiently large the photoelectron does not feel the presence of the $N-1$ electrons left behind and therefore $|\Psi_{N,m}\rangle$ is simply obtained by creating the photoelectron over the interacting $(N-1)$-particle excited state. If we restrict ourselves to these kinds of excited state and rename the corresponding transition probability $P_{0 \to m} = P_s(\epsilon)$ we see that the index i of the perturbation must be equal to ϵ for $P_s(\epsilon)$ not to vanish.[19] We then have

$$P_s(\epsilon) = 2\pi \left| \langle \Psi_{N-1,s} | \sum_j h_{\epsilon j}^* \hat{d}_j | \Psi_{N,0} \rangle \right|^2 \delta(\omega_0 - \epsilon - [E_{N-1,s} - E_{N,0}]).$$

Summing over all the excited states of the $N-1$ particle system and comparing the result with (6.87) we obtain

$$I_{\mathrm{ph}}(\epsilon) \propto \sum_s P_s(\epsilon) = -i \sum_{jj'} h_{\epsilon j}^* h_{\epsilon j'} G_{jj'}^<(\epsilon - \omega_0). \tag{6.95}$$

Thus the photocurrent is proportional to $-iG^<(\epsilon - \omega_0)$. If the system is initially in equilibrium at zero temperature then $E_{N,0}$ is the energy of the ground-state, and the photocurrent is proportional to the spectral function at energies $\epsilon - \omega_0 < \mu$, see the observation below (6.94).[20]

In a noninteracting system, $I_{\mathrm{ph}}(\epsilon) \neq 0$ provided that $\epsilon - \omega_0$ is one of the eigenenergies ϵ_λ of the occupied single-particle states *and* the matrix element $h_{\epsilon\lambda} \neq 0$. Consider for example a crystal. Here the quantum number $\lambda = \mathbf{k}\nu$, see Section 2.3.1 and also Appendix E, and due to momentum conservation $h_{\epsilon\lambda} \neq 0$ only if $\mathbf{k}$ is the difference between the momentum of the incoming photon and the momentum $\mathbf{q}$ of the ejected electron. Photoemission experiments can therefore be used to measure the energy of electrons with a given crystal momentum, or equivalently the band structure of the crystal, by resolving the photocurrent in the momentum $\mathbf{q}$ of the photoelectron [58]: $I_{\mathrm{ph}}(\epsilon) = \int d\mathbf{q}\, I_{\mathrm{ph}}(\epsilon, \mathbf{q})$. In the real world electrons interact with each other but the concept of band structure remains a very useful concept since the removal of electrons with a given crystal momentum generates one main excitation with a long lifetime.[21] Consequently, the photocurrent $I_{\mathrm{ph}}(\epsilon, \mathbf{q})$ remains a peaked function of ϵ for a given $\mathbf{q}$, although the position of the peak is in general different from that of the "noninteracting crystal." The development of perturbative methods for including interaction effects in the Green's function is crucial to improving the results of

---

[19]The ground state $|\Psi_{N,0}\rangle$ does not contain photoelectrons.

[20]At zero temperature $\mu = E_{N,0} - E_{N-1,0} \equiv I$ is the ionization potential and therefore $I_{\mathrm{ph}}(\epsilon) = 0$ for $\omega_0 < \epsilon - I$, as expected.

[21]In particular, the excitation generated by removal of an electron with momentum on the Fermi surface has an infinitely long lifetime and hence the corresponding matrix element of the spectral function has a sharp δ-peak, see also Section 15.5.4.

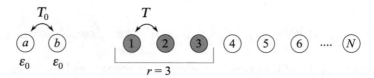

Figure 6.9 A model system in which a two-level molecule is weakly coupled to a metallic chain and the electrons in the molecule interact with electrons in the first r atomic sites of the chain.

a noninteracting treatment, and hence to deepen our understanding of interacting many-particle systems.

If we had started from a system of bosons we could have gone through the same logical and mathematical steps as before, to then get the same expression for the photocurrent in terms of a sum of transition probabilities $P_s(\epsilon)$. In the bosonic case, however, the sum over all s of $P_s(\epsilon)$ gives a linear combination of $+iG^<_{jj'}(\epsilon - \omega_0)$. Thus the sign of the bosonic photocurrent is opposite to the sign of the hole part of $A_{jj'}(\epsilon-\omega_0)$. This argument provides a physical explanation of the possible nonpositivity of the bosonic spectral function.

The outcome of a photoemission experiment relates the photocurrent to $G^<$ independently of the statistics of the particles. To have access to $G^>$ we must perform an *inverse photoemission experiment*. This experiment consists in directing a beam of particles of well-defined energy at the sample. The incident particles penetrate the surface and decay in low-energy unoccupied states by emitting photons. With considerations similar to those of a photoemission experiment we can develop a simple theory to determine the number of emitted photons of energy ω_0. Not surprisingly the result is proportional to $G^>$. Thus $G^>$ and $G^<$ can be separately measured and interpreted. Below we calculate the spectral function of a molecule adsorbed on a surface, and show that the interparticle interaction can modify substantially the noninteracting spectral properties.

Let us consider the Hamiltonian (6.64) in which a repulsive interaction between the electrons is included. For simplicity we treat the electrons as spinless particles since the qualitative results are independent of spin; the reader can easily extend the analysis by including the spin degrees of freedom. We model the metal as a one-dimensional chain with N atomic sites and label with 1 the atom of the metal that is closest to the molecule, see Fig. 6.9. The repulsion between an electron in the molecule and an electron in the metal is taken to be a constant U if the electron in the metal lies in one of the first r atomic sites and zero otherwise

$$\hat{H}_{\text{int}} = U \Big(\underbrace{\sum_{i=a,b} \hat{n}_i - 1}_{\hat{N}_{\text{mol}}} \Big) \Big(\underbrace{\sum_{j \leq r} \hat{n}_j - \bar{N}_r}_{\hat{N}_r} \Big). \tag{6.96}$$

In this formula 1 and $\bar{N}_r$ are the ground-state values of the number of electrons on the molecule and on the first r atomic sites.[22] This form of the interaction guarantees that

---

[22]We are therefore implicitly assigning a Fermi energy for which there is only one electron on the molecule. Since in nature every physical system is charge neutral, our model describes a molecule with an effective nuclear charge equal to 1.

if the two subsystems are charge neutral then there is no interaction between them. To avoid unnecessary complications we neglect the interaction between two electrons in the molecule and between two electrons in the metal; we further assume that the coupling between the molecule and the metal is very small so that $T_k \sim 0$ for all k. The Hamiltonian that describes the system is then

$$\hat{H} = T \underbrace{\sum_{j=1}^{N-1} (\hat{d}_j^\dagger \hat{d}_{j+1} + \hat{d}_{j+1}^\dagger \hat{d}_j)}_{\hat{H}_{\text{met}}} + \underbrace{\epsilon_0 \sum_{i=a,b} \hat{d}_i^\dagger \hat{d}_i + T_0(\hat{d}_a^\dagger \hat{d}_b + \hat{d}_b^\dagger \hat{d}_a)}_{\hat{H}_{\text{mol}}} + \hat{H}_{\text{int}}. \qquad (6.97)$$

A similar Hamiltonian but with $T_k \neq 0$ was originally introduced by Nozières and De Dominicis to study X-ray absorption/emission of metals [59]. In this context the discrete levels do not describe a molecule but the deep core hole left behind by the emitted electron, and the model is known as the *interacting resonant level model*. More recently the Hamiltonian (6.97) has been proposed [60] to illustrate the importance of polarization effects in the renormalization of the quasi-particle levels of molecules on metal surfaces, see also Refs. [61, 62]. This is the context that we should have in mind in the following analysis.

Our first observation is that the operator $\hat{N}_{\text{mol}} = \sum_{i=a,b} \hat{d}_i^\dagger \hat{d}_i$ commutes with $\hat{H}$ and therefore the many-body eigenstates of $\hat{H}$ have a well defined number of electrons on the molecule. We denote by $\hat{c}_{H,L} = \frac{1}{\sqrt{2}} (\hat{d}_a \pm \hat{d}_b)$ the operators that bring $\hat{H}_{\text{mol}} = (\epsilon_H \hat{c}_H^\dagger \hat{c}_H + \epsilon_L \hat{c}_L^\dagger \hat{c}_L)$ into a diagonal form, with the single-particle molecular energies $\epsilon_H = \epsilon_0 - |T_0|$, $\epsilon_L = \epsilon_0 + |T_0|$. Recalling the philosophy behind the idea of working with a finite basis set (see Section 2.1), these energies should correspond to the most relevant physical states of the molecule, which are the *Highest Occupied Molecular Orbital* (HOMO) and the *Lowest Unoccupied Molecular Orbital* (LUMO). Hence, the ground state of the system has one electron on the HOMO level and zero electrons on the LUMO level. The eigenstates of $\hat{H}$ with one electron on the HOMO level can be written as $|\Psi\rangle = \hat{c}_H^\dagger |\Psi_{\text{met}}\rangle$, where $|\Psi_{\text{met}}\rangle$ is a many-body state with no electrons in the molecule, i.e., $\hat{c}_H |\Psi_{\text{met}}\rangle = \hat{c}_L |\Psi_{\text{met}}\rangle = |\emptyset\rangle$. From the eigenvalue equation $\hat{H}|\Psi\rangle = E|\Psi\rangle$ it is easy to see that $|\Psi_{\text{met}}\rangle$ must satisfy the equation

$$\hat{H}_{\text{met}}|\Psi_{\text{met}}\rangle = (E - \epsilon_H)|\Psi_{\text{met}}\rangle. \qquad (6.98)$$

The operator $\hat{H}_{\text{met}} = \sum_{ij} h_{\text{met},ij} \hat{d}_i^\dagger \hat{d}_j$ is a one-body operator that can be diagonalized in the usual manner: we find the eigenkets $|\lambda\rangle$ of $\hat{h}_{\text{met}}$ with eigenvalues ϵ_λ, construct the operators $\hat{c}_\lambda = \sum_j \langle \lambda|j\rangle \hat{d}_j$ and rewrite $\hat{H}_{\text{met}} = \sum_\lambda \epsilon_\lambda \hat{c}_\lambda^\dagger \hat{c}_\lambda$. Then, the eigenstates of (6.98) with M electrons have the form $|\Psi_{\text{met}}\rangle = \hat{c}_{\lambda_M}^\dagger \dots \hat{c}_{\lambda_1}^\dagger |0\rangle = |\lambda_1 \dots \lambda_M\rangle$ with eigenvalues $E - \epsilon_H = \epsilon_{\lambda_1} + \dots + \epsilon_{\lambda_M}$. In particular, the ground state $|\Psi_{\text{met},0}\rangle$ is obtained by populating the lowest M energy levels.

Having described the model and its eigenstates we now study the probability $A_{HH}(\omega)$ that an electron on the HOMO level has energy ω. Let us order the single-particle energies ϵ_λ of $\hat{H}_{\text{met}}$ as $\epsilon_1 \leq \epsilon_2 \leq \dots \leq \epsilon_N$. Then, the ground state $|\Psi_0\rangle = \hat{c}_H^\dagger |\Psi_{\text{met},0}\rangle$ of $\hat{H}$ with M electrons in the metal has energy $E_0 = \epsilon_H + \epsilon_1 + \dots + \epsilon_M$. The lesser and greater components of the Green's function with $\hat{\rho} = |\Psi_0\rangle\langle\Psi_0|$ can be calculated from (6.68) and

(6.69) and read

$$G^<_{HH}(t,t') = \mathrm{i}\,\langle\Psi_0|\hat{c}^\dagger_H e^{-\mathrm{i}(\hat{H}-E_0)(t'-t)}\hat{c}_H|\Psi_0\rangle = \mathrm{i}\,\langle\Psi_{\mathrm{met},0}|e^{-\mathrm{i}(\hat{H}^-_{\mathrm{met}}-E_0)(t'-t)}|\Psi_{\mathrm{met},0}\rangle,$$

$$(6.99)$$

with $\hat{H}^-_{\mathrm{met}} = \hat{H}_{\mathrm{met}} - U(\hat{N}_r - \bar{N}_r)$, and

$$G^>_{HH}(t,t') = -\mathrm{i}\,\langle\Psi_0|\hat{c}_H e^{-\mathrm{i}(\hat{H}-E_0)(t-t')}\hat{c}^\dagger_H|\Psi_0\rangle = 0, \qquad (6.100)$$

where in this last equation we take into account that there cannot be two electrons in the HOMO level due to the Pauli exclusion principle ($\hat{c}^\dagger_H\hat{c}^\dagger_H = 0$) and hence $\hat{c}^\dagger_H|\Psi_0\rangle = 0$.[23] Taking into account that $G^>_{HH} = 0$ the spectral function $A_{HH}(\omega) = -\mathrm{i}G^<_{HH}(\omega)$, and from (6.99) it follows that

$$A_{HH}(\omega) = 2\pi\langle\Psi_{\mathrm{met},0}|\delta(\omega - E_0 + \hat{H}^-_{\mathrm{met}})|\Psi_{\mathrm{met},0}\rangle.$$

For $U = 0$ (no interaction) $|\Psi_{\mathrm{met},0}\rangle$ is an eigenstate of $\hat{H}^-_{\mathrm{met}}$ with eigenvalue $E_0 - \epsilon_H$ and hence $A_{HH}(\omega) = 2\pi\delta(\omega - \epsilon_H)$, i.e., an electron on the HOMO level has a well-defined energy and hence an infinitely long lifetime. For $U \neq 0$ an operative way to calculate $A_{HH}(\omega)$ consists in using the single particle eigenkets $|\lambda^-\rangle$ of $\hat{h}^-_{\mathrm{met}}$ with eigenvalues ϵ^-_λ in order to expand $|\Psi_{\mathrm{met},0}\rangle$ as [see (1.59)]

$$|\Psi_{\mathrm{met},0}\rangle = \frac{1}{M!}\sum_{\lambda_1\ldots\lambda_M}|\lambda^-_1\ldots\lambda^-_M\rangle\langle\lambda^-_1\ldots\lambda^-_M|\Psi_{\mathrm{met},0}\rangle.$$

By construction the M-electron kets $|\lambda^-_1\ldots\lambda^-_M\rangle$ are eigenstates of $\hat{H}^-_{\mathrm{met}}$ with eigenvalues $\epsilon^-_{\lambda_1} + \ldots + \epsilon^-_{\lambda_M} + U\bar{N}_r$ and therefore

$$A_{HH}(\omega) = -\frac{2}{M!}\sum_{\lambda_1\ldots\lambda_M}\mathrm{Im}\frac{|\langle\lambda^-_1\ldots\lambda^-_M|1\ldots M\rangle|^2}{\omega - E_0 + \epsilon^-_{\lambda_1} + \ldots + \epsilon^-_{\lambda_M} + U\bar{N}_r + \mathrm{i}\eta}, \qquad (6.101)$$

where we use the identity $\delta(\omega) = -\frac{1}{\pi}\mathrm{Im}\frac{1}{\omega+\mathrm{i}\eta}$ and, as usual, the limit $\eta \to 0$ is understood. In order to plot $A_{HH}(\omega)$ we take a finite but small η so that the sharp δ-peaks become Lorentzians of width η.

---

[23]It is interesting to observe that $G^<_{HH}$ is obtained by evolving in time $|\Psi_{\mathrm{met},0}\rangle = |1\ldots M\rangle$ with the Hamiltonian $\hat{H}^-_{\mathrm{met}}$ and then by taking the overlap with $|\Psi_{\mathrm{met},0}\rangle$. Due to the one-body nature of $\hat{H}^-_{\mathrm{met}} = \sum_{ij}h^-_{\mathrm{met},ij}\hat{d}^\dagger_i\hat{d}_j + U\bar{N}_r$ we have $e^{-\mathrm{i}\hat{H}^-_{\mathrm{met}}t}|1\ldots M\rangle = e^{-\mathrm{i}U\bar{N}_r t}|1(t)\ldots M(t)\rangle$, where $|\lambda(t)\rangle = e^{-\mathrm{i}\hat{h}^-_{\mathrm{met}}t}|\lambda\rangle$ is the time-evolved single-particle ket (since $e^{-\mathrm{i}\hat{h}^-_{\mathrm{met}}t}$ is unitary then the states $|\lambda(t)\rangle$ form an orthonormal basis at any time). Then, according to (1.63), $G^<_{HH}$ can also be written as a Slater determinant since

$$\langle 1\ldots M|1(t)\ldots M(t)\rangle = \begin{vmatrix} \langle 1|1(t)\rangle & \cdots & \langle 1|M(t)\rangle \\ \vdots & \cdots & \vdots \\ \langle M|1(t)\rangle & \cdots & \langle M|M(t)\rangle \end{vmatrix}_-.$$

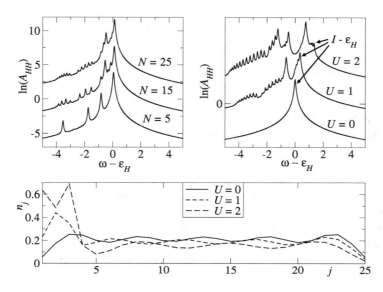

Figure 6.10 Top-left: Spectral function at filling factor $M/N = 1/5$, interaction $U = 0.5$ for different chain lengths $N = 5, 15, 25$. The curves for $N = 15, 25$ are shifted upward. Top-right: Spectral function with filling factor $M/N = 1/5$, chain length $N = 25$ for different interactions $U = 0, 1, 2$. The curves for $U = 1, 2$ are shifted upward. Bottom: Ground states density of $\hat{H}_{\mathrm{met}} - U\hat{N}_r$ for filling factor $M/N = 1/5$ and $U = 0, 1, 2$. All energies are in units of $|T|$, the range of the interaction $r = 3$, and the parameter $\eta = 0.04$.

Let us start our analysis by fixing the interaction parameter $U = 0.5$, the range $r = 3$ and the filling factor $M/N = 1/5$ (average electron density in the metal). In the top-left panel of Fig. 6.10 we show the spectral function (in logarithmic scale) for different length chains N. For $N = 5$ there is only one electron in the metal ($M = 1$) and we expect five peaks, in agreement with the bottom curve (even though one peak is hardly visible). With increasing N the number of peaks grow and already for $N = 25$ we can anticipate the $N \to \infty$ behavior: the peaks become denser and denser and eventually merge to form a continuum. Thus, the probability that an electron on the HOMO level has energy ω becomes a continuous function of ω. The transition from sharp δ-like peaks at finite N to a continuous function for $N \to \infty$ is addressed in Appendix I. There we show that it is just a matter of taking the limit $N \to \infty$ and $\eta \to 0$ in the proper order.

The top-right panel of Fig. 6.10 illustrates how the spectral function varies by increasing the interaction. We are here in the large N limit ($N = 25$) with a range $r = 3$ and a filling factor $M/N = 1/5$. For $U = 0$ the spectral function exhibits one single peak, in agreement with the fact that the HOMO state is, in this case, an eigenstate of $\hat{H}$. With increasing U the HOMO peak broadens and new structures appear in different spectral regions. This figure clearly shows the effect of the interactions: due to the fact that an electron in the molecule can scatter with an electron in the metal the energy of a single particle state cannot be sharply defined. In the figure we also see that the position of the ionization energy[24] $I = E_0 - \min_{\lambda_1,...,\lambda_M}(\epsilon^-_{\lambda_1} + ... + \epsilon^-_{\lambda_M}) - U\bar{N}_r$ moves to the right as U increases, meaning that the electron repulsion lowers the ground-state energy of the system without the HOMO electron. To understand this increase in the ionization energy we show in the bottom panel of Fig. 6.10 the ground-state electronic density n_j of $\hat{H}^-_{\text{met}}$ as a function of the atomic site j for different values of U. The $U = 0$ curve can also be seen as the ground-state density of the chain *with* the HOMO electron. In the proximity of the molecule (first three sites of the chain) the ground-state density without the HOMO electron is always larger than that with the HOMO electron (image charge effect). This is due to the fact that by removing an electron from the HOMO the molecule becomes positively charged and hence electrons in the metal are attracted by the molecule. Equivalently, we can say that the positive charge on the molecule induces an attractive potential $-U\hat{N}_r$ on the first r sites of the metal. The surplus of metallic charge binds with the missing HOMO electron (HOMO hole), thereby lowering the energy of the system with one particle less.

To estimate I we may use the Hellmann–Feynman theorem. Let us define $\hat{H}_{\text{met}}(u) = \hat{H}_{\text{met}} + u\hat{N}_r$ with ground state $|\Psi_{\text{met},0}(u)\rangle$ and ground energy $E_{\text{met},0}(u)$. Then the ionization energy can be written as

$$I(U) = \underbrace{\epsilon_H + E_{\text{met},0}(0)}_{E_0} - E_{\text{met},0}(-U) - U\bar{N}_r$$

$$= \epsilon_H - U\bar{N}_r + \int_{-U}^0 du \, \langle\Psi_{\text{met},0}(u)|\frac{\partial\hat{H}_{\text{met}}(u)}{\partial u}|\Psi_{\text{met},0}(u)\rangle$$

$$= \epsilon_H - U\bar{N}_r + \int_{-U}^0 du \, N_r(u),$$

where $N_r(u) = \langle\Psi_{\text{met},0}(u)|\hat{N}_r|\Psi_{\text{met},0}(u)\rangle$ is the quantum average of $\hat{N}_r$ over the ground state with interaction u. Since $N_r(-|u|) > N_r(0) = \bar{N}_r$, the ionization energy increases with the strength of the repulsion

We could follow the same steps to calculate the spectral function $A_{LL}(\omega)$ of the LUMO level. In this case $G^<_{LL}(t,t') = 0$ while

$$G^>_{LL}(t,t') = -i\langle\Psi_{\text{met},0}|e^{-i(\epsilon_L+\epsilon_H+\hat{H}^+_{\text{met}}-E_0)(t-t')}|\Psi_{\text{met},0}\rangle,$$

where $\hat{H}^+_{\text{met}} = \hat{H}_{\text{met}} + U(\hat{N}_r - \bar{N}_r)$ is the metal Hamiltonian with one electron more on the molecule. Then

$$A_{LL}(\omega) = 2\pi\langle\Psi_{\text{met},0}|\delta(\omega - \epsilon_L - \epsilon_H - \hat{H}^+_{\text{met}} + E_0)|\Psi_{\text{met},0}\rangle. \tag{6.102}$$

As expected $A_{LL}(\omega) = 2\pi\delta(\omega - \epsilon_L)$ in the noninteracting case ($U = 0$). In the interacting case we can calculate the affinity[25] (or removal energy) $A(U)$ using again the

---

[24]In our case the ionization energy is the difference between the ground-state energy of the system with $M + 1$ electrons, which is E_0, and the ground-state energy of the system with M electrons *all* in the metal.

[25]In our case the affinity is the difference between the ground state energy with $M + 2$ electrons (M of which are in the metal) and the ground state energy E_0 with $M + 1$ electrons.

Hellmann–Feynman theorem. Since the ground-state energy of the system with two electrons on the molecule is $\epsilon_L + \epsilon_H + E_{\mathrm{met},0}(U) - U\bar{N}_r$, we have

$$A(U) = \epsilon_L + \epsilon_H + E_{\mathrm{met},0}(U) - U\bar{N}_r - E_0 = \epsilon_L - U\bar{N}_r + E_{\mathrm{met},0}(U) - E_{\mathrm{met},0}(0)$$

$$= \epsilon_L - U\bar{N}_r + \int_0^U du\, N_r(u).$$

For positive u, $N_r(u) < \bar{N}_r$ and the affinity decreases with the strength of the interation. We can interpret the difference $A - I$ as the renormalized gap between the LUMO and HOMO levels. This difference corresponds to the distance between the peaks of $A_{LL} + A_{HH}$ and equals $\epsilon_L - \epsilon_H = 2|T_0|$ for $U = 0$. Our analysis shows that electron correlations are responsible for closing the gap between the HOMO and LUMO levels. We come back to the physical mechanisms behind the reduction of the gap in Section 13.3.1.

Exercise 6.4. Using the results of Exercise 6.2 show that the spectral function $A(x, x'; \omega) = \mathrm{i}\left[G^R(x, x'; \omega) - G^A(x, x'; \omega)\right]$ of a noninteracting system of particles with single-particle Hamiltonian $\hat{h} = \hat{p}^2/2 + \lambda\delta(\hat{x})$ is given by

$$A(x, x'; \omega) = \frac{2}{\sqrt{2\omega}}\left\{\cos(\sqrt{2\omega}|x - x'|) + \frac{\lambda}{2\omega + \lambda^2}\mathrm{Im}\left[(\sqrt{2\omega} - \mathrm{i}\lambda)e^{\mathrm{i}\sqrt{2\omega}\,(|x|+|x'|)}\right]\right\}$$

for $\omega > 0$ and $A(x, x'; \omega) = 0$ for $\omega < 0$. Show further that $A(x, x; \omega) \geq 0$.

Exercise 6.5. Consider the spectral function of Exercise 6.4. Use the fluctuation–dissipation theorem to show that the density $n(x\sigma) = \pm\mathrm{i}\int\frac{d\omega}{2\pi}G^<(x, x; \omega)$ for particles in position x with spin σ is given by

$$n(x\sigma) = 2\int_0^\infty \frac{dq}{2\pi}\frac{1}{e^{\beta(\frac{q^2}{2} - \mu)}\mp 1}\left\{1 + \lambda\frac{q\sin(2q|x|) - \lambda\cos(2q|x|)}{q^2 + \lambda^2}\right\}.$$

Further show that

$$\lim_{x \to \pm\infty} n(x\sigma) = \int_{-\infty}^\infty \frac{dq}{2\pi}\frac{1}{e^{\beta(\frac{q^2}{2} - \mu)}\mp 1} = n_0,$$

where n_0 is the density per spin of the homogeneous system, i.e., with $\lambda = 0$.

Exercise 6.6. Consider the density of Exercise 6.5 for fermions at zero temperature. In this case the Fermi function $\lim_{\beta \to \infty}(e^{\beta(\frac{q^2}{2} - \mu)} + 1)^{-1}$ is unity for $|q| < p_F$ and zero otherwise, where the Fermi momentum $p_F = \sqrt{2\mu}$. Show that $p_F = \pi n_0$ and that the density at the origin is given by

$$n(0\sigma) = n_0\left[1 - \frac{\lambda}{p_F}\arctan(\frac{p_F}{\lambda})\right].$$

Show further that for $\lambda \to \infty$ the density profile becomes

$$n(x\sigma) = n_0\left[1 - \frac{\sin(2p_F|x|)}{2p_F|x|}\right].$$

Therefore the density exhibits spatial oscillations with wavevector $2p_F$. These are known as the *Friedel oscillations*. How does this result change for $\lambda \to -\infty$?

Exercise 6.7. Show that the current of photons of energy ω_0 in an inverse photoemission experiment can be expressed solely in terms of $G^>$.

Exercise 6.8. Consider a single-level bosonic Hamiltonian $\hat{H} = E(\hat{n})$ where $\hat{n} = \hat{d}^\dagger \hat{d}$ is the occupation operator and $E(x)$ is an arbitrary real function of x. Show that in thermodynamic equilibrium the spectral function is

$$A(\omega) = 2\pi \sum_{k=0}^{\infty} (k+1) \left[\rho_k - \rho_{k+1} \right] \delta(\omega - E(k+1) + E(k)),$$

with $\rho_k = e^{-\beta(E(k)-\mu k)}/Z$ and Z the partition function. From this result we see that if $E(k+1) - \mu < E(k)$ for some k then $[\rho_k - \rho_{k+1}] < 0$ and hence the spectral function can be negative for some frequency.

Exercise 6.9. Show that for a system of fermions the diagonal matrix element $G_{jj}^{\mathrm{R}}(\zeta)$ is nonvanishing in the upper half of the complex ζ plane. Similarly $G_{jj}^{\mathrm{A}}(\zeta)$ is nonvanishing in the lower half of the complex ζ plane. Hint: calculate the imaginary part from (6.93) and use the fact that $A_{jj}(\omega) \geq 0$.

Exercise 6.10. Show that the eigenvalues of the Hermitian matrices $iG_{ij}^>(\omega)$ and $\pm iG_{ij}^<(\omega)$ are non-negative for all ω. This is a generalization of the property (6.77). Show also that for fermionic systems the eigenvalues of the spectral function are non-negative. Hint: use the fact that if the expectation value of a Hermitian operator is non-negative for all states then its eigenvalues are non-negative.

6.4 Total energy from the Galitskii–Migdal formula

In our introductory discussion on the lesser and greater Green's function we showed that the time-dependent ensemble average of any one-body operator can be computed from the equal-time lesser Green's function $\hat{\mathcal{G}}^<(t,t)$, see (6.26). In this section we show that from a knowledge of the full $\hat{\mathcal{G}}^<(t,t')$ we can also calculate the time-dependent energy of the system. This result is highly nontrivial since the Hamiltonian contains an interaction part that is a two-body operator.

Let us start by clarifying a point which is often the source of some confusion: how is the total energy operator defined in the presence of a time-dependent external field? This is a very general question and, as such, the answer cannot depend on the details of the system. We, therefore, consider the simple case of a quantum mechanical particle in free space with Hamiltonian $\hat{h} = \hat{p}^2/2m$. At time t_0 we switch on an external electromagnetic field and ask the question: how does the energy of the particle change in time? Let $|\Psi(t)\rangle$ be the ket of the particle at time t. The time evolution is governed by the Schrödinger equation $i\frac{d}{dt}|\Psi(t)\rangle = \hat{h}(t)|\Psi(t)\rangle$ with

$$\hat{h}(t) = \frac{1}{2m} \left(\hat{\boldsymbol{p}} - \frac{q}{c}\mathbf{A}(\hat{\boldsymbol{r}}, t) \right)^2 + qV(\hat{\boldsymbol{r}}, t).$$

Is the energy of the particle given by the average over $|\Psi(t)\rangle$ of $\hat{p}^2/2m$ (which describes the original system with no external fields) or of the full Hamiltonian $\hat{h}(t)$? The correct answer is neither of the two. The average of $\hat{h}(t)$ must be ruled out since it contains the coupling energy between the particle and the *external* field $qV(\hat{r}, t)$. Our system is a free particle and therefore its energy is simply the average of the kinetic energy operator, i.e., the velocity operator squared over $2m$. In the presence of a vector potential, however, the velocity operator is $[\hat{p} - \frac{q}{c}\mathbf{A}(\hat{r}, t)]$ and not just $\hat{p}$. We conclude that the energy of the system at time t must be calculated by averaging the operator $\frac{1}{2m}\left(\hat{p} - \frac{q}{c}\mathbf{A}(\hat{r}, t)\right)^2 = \hat{h}(t) - qV(\hat{r}, t)$ over $|\Psi(t)\rangle$. Further evidence in favour of the above choice comes from the fact that all physical quantities must be invariant under a gauge transformation. For the case of a single particle the gauge transformation $\mathbf{A} \rightarrow \mathbf{A} + \nabla\Lambda$, $V \rightarrow V - \frac{1}{c}\frac{\partial}{\partial t}\Lambda$ implies that the ket changes according to $|\Psi(t)\rangle \rightarrow \exp\left[i\frac{q}{c}\Lambda(\hat{r}, t)\right]|\Psi(t)\rangle$, see (3.24). The reader can easily verify that neither $\langle\Psi(t)|\hat{p}^2/2m|\Psi(t)\rangle$ nor $\langle\Psi(t)|\hat{h}(t)|\Psi(t)\rangle$ is gauge invariant, while $\langle\Psi(t)|\hat{h}(t) - qV(\hat{r}, t)|\Psi(t)\rangle$ is.

Let us now apply the same reasoning to a system of interacting identical particles with Hamiltonian $\hat{H}(t) = \hat{H}_0(t) + \hat{H}_{\text{int}}$. According to the previous discussion the energy of the system at a generic time t_1, $E_S(t_1)$, is the time-dependent ensemble average of the operator

$$\hat{H}_S(t_1) \equiv \hat{H}(t_1) - q\int d\mathbf{x}_1\,\hat{n}(\mathbf{x}_1)\delta V(1),$$

where $\delta V(1) = V(\mathbf{r}_1, t_1) - V(\mathbf{r}_1)$ is the difference between the total potential at time t_1 and the initial potential, i.e., $\delta V(1)$ is the *external* potential. Then we have

$$E_S(z_1) = \sum_k \rho_k \langle\Psi_k|\hat{U}(t_{0-}, z_1)\left[\int d\mathbf{x}_1 d\mathbf{x}_2\,\hat{\psi}^\dagger(\mathbf{x}_1)\langle\mathbf{x}_1|\hat{h}_S(z_1)|\mathbf{x}_2\rangle\hat{\psi}(\mathbf{x}_2)\right.$$
$$\left.+ \frac{1}{2}\int d\mathbf{x}_1 d\mathbf{x}_2\,v(\mathbf{x}_1, \mathbf{x}_2)\,\hat{\psi}^\dagger(\mathbf{x}_1)\hat{\psi}^\dagger(\mathbf{x}_2)\hat{\psi}(\mathbf{x}_2)\hat{\psi}(\mathbf{x}_1)\right]\hat{U}(z_1, t_{0-})|\Psi_k\rangle,$$

where $\hat{h}_S(z_1) = \hat{h}(z_1) - q\delta V(\hat{r}_1, z_1)$ is the single-particle Hamiltonian of the system. We remind the reader that the energy $E_S(t_1)$ as a function of the physical time t_1 is given by $E_S(t_{1\pm})$, see again (4.10). The reason for introducing the contour time z_1 is that we now recognize on the r.h.s. the one- and two-particle Green's functions with a precise order of the contour-time variables. It is easy to show that

$$E_S(z_1) = \pm i\int d\mathbf{x}_1 d2\,h_S(1; 2)G(2; 1^+) - \frac{1}{2}\int d\mathbf{x}_1 d2\,v(1; 2)G_2(1, 2; 1^+, 2^+). \qquad (6.103)$$

The particular form of the second integral allows us to express E_S in terms of G only. Adding the equation of motion for G to its adjoint [see (5.7) and (5.8)] and then setting $2 = 1^+$ we find

$$\left[\left(i\frac{d}{dz_1} - i\frac{d}{dz_2}\right)G(1; 2)\right]_{2=1^+} - \int d3\left[h(1; 3)G(3; 1^+) + G(1; 3^+)h(3; 1)\right]$$
$$= \pm 2i\int d3\,v(1; 3)G_2(1, 3; 1^+, 3^+).$$

Inserting this result into $E_S(z_1)$ we arrive at the very interesting formula

$$E_S(z_1) = \pm i \int dx_1 \langle \mathbf{x}_1| \left[\hat{h}_S(z_1) - \frac{1}{2}\hat{h}(z_1) \right] \hat{G}(z_1, z_1^+)|\mathbf{x}_1\rangle$$

$$\pm \frac{i}{4} \int dx_1 \left[\left(i\frac{d}{dz_1} - i\frac{d}{dz_2} \right) \langle \mathbf{x}_1|\hat{G}(z_1, z_2)|\mathbf{x}_1\rangle \right]_{z_2 = z_1^+} . \quad (6.104)$$

This formula yields the initial energy of the system E_S^M when $z_1 = t_0 - i\tau_1$ and the time-dependent energy $E_S(t_1)$ when $z_1 = t_{1\pm}$. In the former case $\hat{h}_S(z_1) = \hat{h}(z_1) = \hat{h}^M$ and expanding $\hat{G}$ in the Matsubara series (6.17) we get

$$E_S^M = \pm\frac{i}{2}\frac{1}{-i\beta}\sum_m e^{\eta\omega_m}\int dx\, \langle \mathbf{x}|(\omega_m + \hat{h}^M)\hat{G}^M(\omega_m)|\mathbf{x}\rangle \qquad (6.105)$$

For the time-dependent energy $E_S(t)$ we recall that $\hat{h}_S(t) = \hat{h}(t) - q\delta V(\hat{\mathbf{r}}, t)$ and hence from (6.104) with $z_1 = t_\pm$

$$E_S(t) = \pm\frac{i}{4}\int dx\, \langle \mathbf{x}| \left(i\frac{d}{dt} - i\frac{d}{dt'} + 2\hat{h}(t) \right) \hat{G}^<(t, t')|\mathbf{x}\rangle \Big|_{t'=t} - q\int dx\, n(\mathbf{x}, t)\delta V(\mathbf{r}, t).$$

In the special case that the Hamiltonian $\hat{H}(t) = \hat{H}$ does not depend on time (hence $\delta V = 0$) and the system is initially in a stationary excited configuration of $\hat{H}$, the lesser Green's function depends on the time difference $t - t'$ only. Then, Fourier transforming $\hat{G}^<$ as in (6.47) the energy simplifies to

$$E_S = \pm\frac{i}{2}\int\frac{d\omega}{2\pi}\int dx\, \langle \mathbf{x}|(\omega + \hat{h})\hat{G}^<(\omega)|\mathbf{x}\rangle \qquad (6.106)$$

This result represents a generalization of the so-called *Galitskii–Migdal formula* since it is valid not only for systems initially in thermodynamic equilibrium but also for systems in a stationary excited configuration. In the case of thermodynamic equilibrium $E_S^M = E_S - \mu N$, and hence (6.106) provides an alternative formula to (6.105) for calculation of the initial energy.

As an example of stationary excited configuration we consider the case of noninteracting systems. Then the lesser Green's function is given by (6.48), and (6.106) yields

$$E_S = \sum_\lambda f(\epsilon_\lambda^M)\epsilon_\lambda; \qquad (6.107)$$

the energy of the system is the weighted sum of the single-particle energies. As expected, the equilibrium energy is recovered for $\hat{h}^M = \hat{h} - \mu$. In general, however, we can calculate the energy of an arbitrary excited configuration by a proper choice of $\hat{h}^M$. We use (6.107) in Section 7.3.2 to study the spin-polarized ground-state of an electron gas.

7

Mean field approximations

7.1 Introduction

The effect of interparticle interactions is to correlate the motion of a particle to the motion of all the other particles. For the two-body interactions considered so far, the action of $\hat{H}_{\mathrm{int}}$ on a many-body ket $|\Phi\rangle = \hat{d}_{n_1}^{\dagger} \hat{d}_{n_2}^{\dagger} \dots |0\rangle$ (one particle in φ_{n_1}, another one in φ_{n_2} and so on) yields a linear combination of kets in which at most two particles have changed their state. This implies that the matrix element $\langle \Phi' | \hat{H}_{\mathrm{int}} | \Phi \rangle$ is zero for all those $|\Phi'\rangle$ differing from $|\Phi\rangle$ by more than two $\hat{d}^{\dagger}$ operators. We may say that a particle can scatter at most with another particle, and after the scattering the two particles end up in new states. Therefore if we know how two particles propagate in the system, i.e., if we know G_2, then we can deduce how a single particle propagates in the system, i.e., we can determine G. This is another way to understand the appearance of G_2 in the equation of motion for G. We emphasize that this is true only for two-body interactions. For an interaction Hamiltonian that is an n-body operator, i.e., an interaction Hamiltonian that is a linear combination of products of $2n$ field operators, the scattering involves n particles and the equations of motion for G contain the n-particle Green's function G_n.

Since from any approximate G_2 we can extract an approximate G, let us use some physical intuition to generate reasonable approximations to G_2. For the time being we ignore the fact that the particles are identical and suppose that among them there exist two "special" ones which cannot interact directly. These two special particles feel the presence of all other particles but they are insensitive to their mutual position. Then, the probability amplitude for the first particle to go from $1'$ to 1 *and* the second particle to go from $2'$ to 2 is simply the product of the probability amplitudes of the two separate events, or in the Green's function language

$$G_2(1, 2; 1', 2') \sim G_{2,\mathrm{H}}(1, 2; 1', 2') \equiv e^{\mathrm{i}\alpha} G(1; 1') G(2; 2'). \tag{7.1}$$

If we represent the Green's function $G(1; 1')$ with a line going from $1'$ to 1 and the two-particle Green's function $G_2(1, 2; 1', 2')$ with two lines that start in $1'$ and $2'$, enter a square where processes of arbitrary complexity can occur and go out in 1 and 2, then (7.1) can be represented diagrammatically as in Fig. 7.1. The phase factor $e^{\mathrm{i}\alpha}$ in (7.1) can be determined by observing that $G_2(1, 2; 1^+, 2^+)$ is the ensemble average of $-\hat{n}_H(1)\hat{n}_H(2)$ and hence

$$\lim_{2 \to 1} G_2(1, 2; 1^+, 2^+) = \text{real negative number}.$$

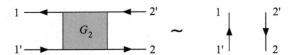

Figure 7.1 Hartree approximation to the two-particle Green's function.

Evaluating the r.h.s. of (7.1) in the same point and taking into account that $G(1;1^+) = \mp in(1)$ (upper/lower sign for bosons/fermions) is a negative/positive imaginary number we conclude that $e^{i\alpha} = 1$. As we see in Chapter 10, the approximation (7.1) can also be derived from the perturbative formula (5.34) which naturally fixes the phase factor $e^{i\alpha}$ to be 1. The approximation (7.1) with $e^{i\alpha} = 1$ is known as the *Hartree approximation* and neglects the direct interaction between two particles.

The Hartree approximation also ignores that the particles are identical. In the exact case, for two identical particles initially in $(1',2')$ it is not possible to distinguish the event in which they will be detected in $(1,2)$ from the event in which they will be detected in $(2,1)$. Quantum mechanics teaches us that if an event can occur through two different paths then the total probability amplitude is the sum of the probability amplitudes of each path. In the Green's function language this leads to the so called *Hartree–Fock approximation* for the two-particle Green's function,

$$G_2(1,2;1',2') \sim G_{2,\mathrm{HF}}(1,2;1',2') \equiv G(1;1')G(2;2') + e^{i\beta}G(1;2')G(2;1'), \qquad (7.2)$$

whose diagrammatic representation is illustrated in Fig. 7.2. The phase factor $e^{i\beta}$ can be determined using the symmetry properties of G_2. From the general definition (5.1) we have $G_2(1,2;1',2') = \pm G_2(1,2;2',1')$ where the upper/lower sign applies to bosons/fermions, and hence $e^{i\beta} = \pm 1$. We have recovered the approximation (5.14). The Hartree–Fock approximation (like the Hartree approximation) can also be deduced from the perturbative formula (5.34).

Both the Hartree and Hartree–Fock approximations neglect the direct interaction between two particles, see also the discussion in Section 5.2. In these approximations a particle moves like a free particle under the influence of an effective potential which depends on the position of all the other particles. This is precisely the idea of a *mean field approximation*: to include the effects of the interaction through an effective potential. In bulk systems this idea makes sense if there are so many particles, or equivalently if the density is so high, that we can treat the interaction of a particle with all the other particles as an effective average interaction, i.e., as an effective field. In the next sections we calculate the Green's function G at the Hartree and Hartree–Fock level and apply the general results to some concrete examples. We further give a mathematical form to the effective potential and explore the physical content as well as the limitations of these approximations.

Figure 7.2 Hartree–Fock approximation to the two-particle Green's function.

7.2 Hartree approximation

Let us consider a system of identical particles with mass m and charge q. Substituting the Hartree approximation (7.1) into the equation of motion (5.7) for G we find

$$i\frac{d}{dz_1}G(1;2) - \int d3\,[h(1;3) + qV_{\mathrm{H}}(1)\delta(1;3)]\,G(3;2) = \delta(1;2), \tag{7.3}$$

where we consider the general case of a single-particle Hamiltonian $\hat{h}$ which is nondiagonal in space and spin, $h(1;3) = \delta(z_1,z_3)\langle\mathbf{x}_1|\hat{h}|\mathbf{x}_3\rangle$, and where

$$V_{\mathrm{H}}(1) = V_{\mathrm{H}}(\mathbf{x}_1,z_1) = \pm\frac{i}{q}\int d3\,v(1;3)G(3;3^+) = \frac{1}{q}\int d3\,v(1;3)n(3), \tag{7.4}$$

is the effective potential of the Hartree approximation, also known as the *Hartree potential*. For Coulombic interactions $v(1;2) = q^2\delta(z_1,z_2)/|\mathbf{r}_1 - \mathbf{r}_2|$, and the Hartree potential coincides with the classical potential generated by a density distribution $n(1) = n(\mathbf{x}_1,z_1)$:

$$V_{\mathrm{H}}(1) = q\int d\mathbf{x}\,\frac{n(\mathbf{x},z_1)}{|\mathbf{r}_1 - \mathbf{r}|} \qquad \text{(for Coulombic interactions).} \tag{7.5}$$

Equations (7.3) and (7.4) form a system of coupled equations for the Green's function. In practical calculations they are solved by making an initial (and reasonable) guess for the Green's function, $G^{(0)}$, which is then used to determine an initial guess for the Hartree potential, $V_{\mathrm{H}}^{(0)}$, using (7.4); $V_{\mathrm{H}}^{(0)}$ is then inserted in (7.3) to obtain a new Green's function $G^{(1)}$ which is then used in (7.4) to determine a new Hartree potential $V_{\mathrm{H}}^{(1)}$ and so on and so forth until convergence is achieved. In other words the Green's function must be calculated *self-consistently* since the operator acting on G depends on G itself.

For evaluation of both the Green's function and the Hartree potential we describe a method based on solving a system of coupled equations for single-particle wavefunctions. Except for the fact that V_{H} depends on G, the equation of motion (7.3) is formally identical to the equation of motion of noninteracting particles. The general solution is therefore obtained as described in Section 6.2 by replacing $\hat{h}(z)$ with

$$\hat{h}_{\mathrm{H}}(z) \equiv \hat{h}(z) + q\hat{\mathcal{V}}_{\mathrm{H}}(z),$$

where

$$\hat{\mathcal{V}}_{\mathrm{H}}(z) = \int d\mathbf{x}_1 d\mathbf{x}_2\,|\mathbf{x}_1\rangle\delta(\mathbf{x}_1 - \mathbf{x}_2)V_{\mathrm{H}}(\mathbf{x}_1,z)\langle\mathbf{x}_2|$$

$$= \int d\mathbf{x}\,|\mathbf{x}\rangle V_{\mathrm{H}}(\mathbf{x},z)\langle\mathbf{x}|, \tag{7.6}$$

is the Hartree potential operator in first quantization. The reader can easily check that (7.3) can be written as (6.4) with $\hat{h}_{\mathrm{H}}$ in place of $\hat{h}$. By definition $\hat{\mathcal{V}}_{\mathrm{H}}(z)$ is a constant operator along the imaginary track ($z = t_0 - i\tau$) which we denote by $\hat{\mathcal{V}}_{\mathrm{H}}^{\mathrm{M}}$, consistently with our notation.

7.2.1 Hartree equations

Let us consider an interacting system initially in equilibrium in some external potential $V(\mathbf{x} = \mathbf{r}\sigma)$ which depends on both the position and the spin projection along the z axis (the inclusion of a vector potential is straightforward). The single-particle Hamiltonian $\hat{h}^{\mathrm{M}}$ is then

$$\hat{h}^{\mathrm{M}} = \hat{h} - \mu = \frac{\hat{p}^2}{2m} + qV(\hat{\mathbf{r}}, \hat{S}_z) - \mu,$$

and the interaction $v^{\mathrm{M}} = v$. To calculate the Green's function we first need to solve the eigenvalue problem

$$\left(\hat{h}^{\mathrm{M}} + q\hat{\mathcal{V}}_{\mathrm{H}}^{\mathrm{M}}\right)|\lambda\rangle = (\epsilon_\lambda - \mu)|\lambda\rangle,$$

where $(\epsilon_\lambda - \mu) = \epsilon_\lambda^{\mathrm{M}}$ are the single-particle eigenvalues of $\left(\hat{h}^{\mathrm{M}} + q\hat{\mathcal{V}}_{\mathrm{H}}^{\mathrm{M}}\right) = \hat{h}_{\mathrm{H}}^{\mathrm{M}}$ and $|\lambda\rangle = |\lambda^{\mathrm{M}}\rangle$ are the corresponding eigenkets. In order to lighten the notation we here omit the superscript "M" in the eigenkets. By sandwiching the eigenvalue equation with the bra $\langle\mathbf{x}|$ we find

$$\boxed{\left[-\frac{\nabla^2}{2m} + qV(\mathbf{x}) + qV_{\mathrm{H}}^{\mathrm{M}}(\mathbf{x})\right]\varphi_\lambda(\mathbf{x}) = \epsilon_\lambda\varphi_\lambda(\mathbf{x})} \tag{7.7}$$

where according to (7.4) and to the formula (6.46) for the density of noninteracting particles,

$$\boxed{V_{\mathrm{H}}^{\mathrm{M}}(\mathbf{x}) = \frac{1}{q}\int d\mathbf{x}'\, v(\mathbf{x}, \mathbf{x}') \overbrace{\sum_\nu f(\epsilon_\nu - \mu)|\varphi_\nu(\mathbf{x}')|^2}^{n(\mathbf{x}')}} \tag{7.8}$$

The coupled equations (7.7) and (7.8) for the single-particle wavefunctions $\{\varphi_\lambda\}$ are known as the *Hartree equations* [63]. Note that the Hartree wavefunctions $\{\varphi_\lambda\}$ form an orthonormal basis since they are the eigenfunctions of the Hermitian operator $\hat{h}^{\mathrm{M}} + q\hat{\mathcal{V}}_{\mathrm{H}}^{\mathrm{M}}$.

Zero temperature fermions

In a system of fermions at zero temperature, only states with energy $\epsilon_\lambda < \mu$ contribute to the sum over λ in (7.8). The number of these states, say N, is the number of fermions in the system. For $N = 1$ the *exact* solution of the problem is the same as the noninteracting solution since one single fermion cannot interact with itself. Does the Hartree approximation give the exact result for $N = 1$? The Hartree equation for the occupied wave function of a single fermion reads

$$\left[-\frac{\nabla^2}{2m} + qV(\mathbf{x}) + \int d\mathbf{x}'\, v(\mathbf{x}, \mathbf{x}')|\varphi(\mathbf{x}')|^2\right]\varphi(\mathbf{x}) = \epsilon\varphi(\mathbf{x}), \tag{7.9}$$

which differs from the noninteracting eigenvalue equation. Thus the Hartree approximation is not exact even for $N = 1$ since the Hartree potential is equal to the classical potential generated by the fermion itself. This is an intrinsic feature of the Hartree approximation known as the *self-interaction error*: each fermion feels the potential generated by itself. The self-interaction error varies as $1/N$ and it is therefore vanishingly small in bulk systems but it can be rather large in finite systems. As we shall see, the Hartree–Fock approximation cures this problem for the occupied states.

Zero temperature bosons

In bosonic systems at zero temperature bosons can condense in the lowest energy level φ, meaning that all occupied φ_λ in (7.7) collapse in the same wavefunction φ. Then, as for one single fermion at zero temperature, the Hartree equations reduce to a single nonlinear equation

$$\left[-\frac{\nabla^2}{2m} + qV(\mathbf{x}) + N \int d\mathbf{x}'\, v(\mathbf{x}, \mathbf{x}') |\varphi(\mathbf{x}')|^2 \right] \varphi(\mathbf{x}) = \epsilon \varphi(\mathbf{x}), \qquad (7.10)$$

where N is the number of bosons in the system. Also the bosonic case suffers from the self-interaction error. In particular for $N = 1$ the above equation does not reduce to the eigenvalue equation for one single boson. As we shall see the Hartree–Fock approximation does *not* cure this problem in bosonic systems.

If we multiply (7.10) by $\sqrt{N}$ and define $\tilde{\varphi} = \sqrt{N}\varphi$ we obtain exactly (7.9). These kinds of equation are called *nonlinear Schrödinger equations*. A renowned physical example of a nonlinear Schrödinger equation is that of a system of hard-core bosons with interparticle interaction $v(\mathbf{x}, \mathbf{x}') = v_0 \delta(\mathbf{x} - \mathbf{x}')$. In this case (7.10) becomes

$$\left[-\frac{\nabla^2}{2m} + qV(\mathbf{x}) + v_0 |\tilde{\varphi}(\mathbf{x})|^2 \right] \tilde{\varphi}(\mathbf{x}) = \epsilon \tilde{\varphi}(\mathbf{x}),$$

which is called the *Gross–Pitaevskii equation* [64,65]. The Gross–Pitaevskii equation is today enjoying increasing popularity due to recent experimental advances in trapping and cooling weakly interacting atoms using lasers. The first ever observation of a Bose condensate dates back to 1995, and since then many groups have been able to reproduce this remarkable phenomenon. For a review on Bose condensation in these systems see Ref. [66].

Total energy in the Hartree approximation

We have already pointed out that in the Hartree approximation the particles behave as free particles. Then we would (naively) expect that the total energy of the system in thermodynamic equilibrium is the noninteracting energy (6.107) with single-particle Hartree energies ϵ_λ and occupations $f_\lambda \equiv f(\epsilon_\lambda^M) = f(\epsilon_\lambda - \mu)$. This is, however, not so! Let us clarify this subtle point. If we multiply (7.7) by $\varphi_\lambda^*(\mathbf{x})$ and integrate over $\mathbf{x}$ we can express the generic eigenvalue in terms of the matrix elements of $\hat{h} + q\hat{\mathcal{V}}_{\mathrm{H}}^{\mathrm{M}}$,

$$\epsilon_\lambda = h_{\lambda\lambda} + \int d\mathbf{x}d\mathbf{x}' \, v(\mathbf{x},\mathbf{x}')n(\mathbf{x}')|\varphi_\lambda(\mathbf{x})|^2$$

$$= h_{\lambda\lambda} + \sum_\nu f_\nu \int d\mathbf{x}d\mathbf{x}' \, \varphi_\lambda^*(\mathbf{x})\varphi_\nu^*(\mathbf{x}')v(\mathbf{x},\mathbf{x}')\varphi_\nu(\mathbf{x}')\varphi_\lambda(\mathbf{x})$$

$$= h_{\lambda\lambda} + \sum_\nu f_\nu v_{\lambda\nu\nu\lambda},$$

where we use the definitions (1.84) and (1.85). Let us now evaluate the total energy. In (6.106) $\hat{h}$ is the time-independent single-particle Hamiltonian and *not* the single-particle Hartree Hamiltonian. The latter is $\hat{h}_{\mathrm{H}} \equiv \hat{h} + q\hat{\mathcal{V}}_{\mathrm{H}} = \hat{h} + q\hat{\mathcal{V}}_{\mathrm{H}}^{\mathrm{M}}$, where we have taken into account that in equilibrium $\hat{\mathcal{V}}_{\mathrm{H}}(t) = \hat{\mathcal{V}}_{\mathrm{H}}^{\mathrm{M}}$. In the Hartree approximation

$$\hat{\mathcal{G}}^<(\omega) = \mp 2\pi i \, f(\hat{h}_{\mathrm{H}}^{\mathrm{M}}) \, \delta(\omega - \hat{h}_{\mathrm{H}}),$$

and using $\int d\mathbf{x}\langle \mathbf{x}| \ldots |\mathbf{x}\rangle = \sum_\lambda \langle\lambda| \ldots |\lambda\rangle$ (invariance of the trace under unitary transformations) we find

$$E = \frac{1}{2}\int \frac{d\omega}{2\pi} \sum_\lambda \langle\lambda|(\hat{h}+\omega)f(\underbrace{\hat{h}+q\hat{\mathcal{V}}_{\mathrm{H}}^{\mathrm{M}}}_{\hat{h}_{\mathrm{H}}^{\mathrm{M}}} - \mu) \, 2\pi\,\delta(\omega - [\underbrace{\hat{h}+q\hat{\mathcal{V}}_{\mathrm{H}}^{\mathrm{M}}}_{\hat{h}_{\mathrm{H}}}])|\lambda\rangle$$

$$= \frac{1}{2}\sum_\lambda f_\lambda \langle\lambda|\hat{h} + \epsilon_\lambda|\lambda\rangle = \sum_\lambda f_\lambda \epsilon_\lambda - \frac{1}{2}\sum_{\lambda\nu} f_\lambda f_\nu v_{\lambda\nu\nu\lambda}, \qquad (7.11)$$

where in the last equality we express $h_{\lambda\lambda}$ in terms of ϵ_λ and the Coulomb integrals. Why does this energy differ from the energy of a truly noninteracting system? The explanation of this apparent paradox is simple. The eigenvalue ϵ_λ contains the interaction energy $v_{\lambda\nu\nu\lambda}$ between a particle in φ_λ and a particle in φ_ν. If we sum over all λ this interaction energy is counted twice. In (7.11) the double counting is correctly removed by subtracting the last term.

Time-dependent Hartree equations

Once the equilibrium problem is solved we can construct the Matsubara Green's function as, e.g., in (6.40). For all other components, however, we need to propagate the eigenstates of $\hat{h}_{\mathrm{H}}^{\mathrm{M}}$ in time. If the interacting system is exposed to a time-dependent electromagnetic field then the single particle Hamiltonian $\hat{h}(t)$ is time-dependent and so will be the Hartree Hamiltonian $\hat{h}_{\mathrm{H}}(t) = \hat{h}(t) + q\hat{\mathcal{V}}_{\mathrm{H}}(t)$. The evolution is governed by the Schrödinger equation $i\frac{d}{dt}|\lambda(t)\rangle = \hat{h}_{\mathrm{H}}(t)|\lambda(t)\rangle$ with $|\lambda(t_0)\rangle = |\lambda\rangle$ the eigenkets of $\hat{h}_{\mathrm{H}}^{\mathrm{M}}$. By sandwiching the time-dependent Schrödinger equation with the bra $\langle\mathbf{x}|$ we get

$$i\frac{d}{dt}\varphi_\lambda(\mathbf{x},t) = \left[\frac{1}{2m}\left(-i\boldsymbol{\nabla} - \frac{q}{c}\mathbf{A}(\mathbf{r},t)\right)^2 + qV(\mathbf{x},t) + qV_{\mathrm{H}}(\mathbf{x},t)\right]\varphi_\lambda(\mathbf{x},t), \qquad (7.12)$$

where $\mathbf{A}$ and V are the external time-dependent vector and scalar potentials.[1] The time-dependent Hartree potential is given in (7.4); expressing the density in terms of the time-evolved eigenfunctions we have

$$V_{\mathrm{H}}(\mathbf{x},t) = \frac{1}{q} \int d\mathbf{x}' \, v(\mathbf{x},\mathbf{x}') \overbrace{\sum_{\lambda} f(\epsilon_\lambda - \mu) |\varphi_\lambda(\mathbf{x}',t)|^2}^{n(\mathbf{x}',t)}. \tag{7.13}$$

The coupled equations (7.12) and (7.13) for the wavefunctions $\varphi_\lambda(\mathbf{x},t)$ are known as the *time-dependent Hartree equations*. For the unperturbed system the Hartree equations are solved by $\varphi_\lambda(\mathbf{x},t) = e^{-i\epsilon_\lambda t}\varphi_\lambda(\mathbf{x})$, as they should be.

7.2.2 Electron gas

The *electron gas* is a system of interacting electrons (spin 1/2 fermions) with single-particle Hamiltonian $\hat{h}(t) = \hat{h} = \hat{p}^2/2$ and with a spin-independent interparticle interaction $v(\mathbf{x}_1, \mathbf{x}_2) = v(\mathbf{r}_1 - \mathbf{r}_2)$. The reader can consult Refs. [45, 67] for a detailed and thorough presentation of the physics of the electron gas. Below we consider the electron gas in thermodynamic equilibrium so that $\hat{h}^{\mathrm{M}} = \hat{h} - \mu$ and $v^{\mathrm{M}} = v$. The full Hamiltonian $\hat{H}_0 + \hat{H}_{\mathrm{int}}$ is invariant under space and time translations and consequently all physical quantities have the same property. In particular the electron density per spin $n(\mathbf{x},z) = n/2$ is independent of $\mathbf{x}$ and z and so is the Hartree potential[2]

$$V_{\mathrm{H}}(1) = -\int d\mathbf{x} \, v(\mathbf{x}_1, \mathbf{x})\frac{n}{2} = -\sum_\sigma \int d\mathbf{r} \, v(\mathbf{r}_1 - \mathbf{r})\frac{n}{2} \equiv -n\tilde{v}_0, \tag{7.14}$$

where $\tilde{v}_0$ is the Fourier transform $\tilde{v}_{\mathbf{p}} = \int d\mathbf{r} \, e^{-i\mathbf{p}\cdot\mathbf{r}} v(\mathbf{r})$ with $\mathbf{p} = 0$. The eigenkets of the Hartree Hamiltonian $\hat{h}_{\mathrm{H}}^{\mathrm{M}} = \hat{h} - \hat{V}_{\mathrm{H}}^{\mathrm{M}} - \mu$ are the momentum-spin kets $|\mathbf{p}\sigma\rangle$ with eigenenergy $p^2/2 + n\tilde{v}_0 - \mu$. The lesser Green's function can then be calculated using (6.48) and reads

$$\hat{\mathcal{G}}^<(\omega) = 2\pi i \sum_\sigma \int \frac{d\mathbf{p}}{(2\pi)^3} \, |\mathbf{p}\sigma\rangle\langle\mathbf{p}\sigma| \frac{\delta(\omega - \frac{p^2}{2} - n\tilde{v}_0)}{e^{\beta(\frac{p^2}{2} + n\tilde{v}_0 - \mu)} + 1},$$

from which we can extract the value n of the density

$$\begin{aligned}
\frac{n}{2} &= -i \, \langle\mathbf{x}|\hat{\mathcal{G}}^<(t,t)|\mathbf{x}\rangle = -i \int \frac{d\omega}{2\pi} \langle \mathbf{r}\sigma|\hat{\mathcal{G}}^<(\omega)|\mathbf{r}\sigma\rangle \\
&= \int \frac{d\mathbf{p}}{(2\pi)^3} \frac{1}{e^{\beta(\frac{p^2}{2} + n\tilde{v}_0 - \mu)} + 1},
\end{aligned} \tag{7.15}$$

where we use $\langle\mathbf{r}|\mathbf{p}\rangle = e^{i\mathbf{p}\cdot\mathbf{r}}$, see (1.10). For any given initial temperature $T = 1/(K_{\mathrm{B}}\beta)$ and chemical potential μ, (7.15) provides a self-consistent equation for the density. Thus, for an

---

[1] For simplicity in (7.12) the coupling between the time-dependent magnetic field and the spin degrees of freedom has not been included.

[2] We recall that in our units the electron charge $q = e = -1$ while the electron mass $m = m_e = 1$.

electron gas the solution of the Hartree equations reduces to the solution of (7.15): once n is known the Hartree problem is solved. The solution of (7.15) when $\tilde{v}_0 > 0$ (repulsive interaction) deserves comment. The r.h.s. is a decreasing function of n, as can easily be checked by taking the derivative with respect to n. Consequently the self-consistent density is always smaller than the noninteracting density. For the interacting system to have the same density as the noninteracting system we have to increase μ, i.e., the free-energy per electron. This agrees with our intuitive picture that the energy needed to bring together interacting particles increases with the strength of the repulsion. Of course the opposite is true for attractive ($\tilde{v}_0 < 0$) interactions.

From the above result we also expect that the pressure of the interacting system is larger or smaller than that of the noninteracting system depending on whether the interaction is repulsive or attractive [68]. We recall that the pressure P is obtained from the density n by performing the integral (see Appendix D)

$$P(\beta, \mu) = \int_{-\infty}^{\mu} d\mu' \, n(\beta, \mu') \quad \Rightarrow \quad dP = nd\mu, \tag{7.16}$$

where $n(\beta, \mu)$ is the density given by the solution of (7.15). In the low density limit $\beta\mu \to -\infty$ and the self-consistent equation (7.15) simplifies to

$$n(\beta, \mu) = 2 \int \frac{d\mathbf{p}}{(2\pi)^3} e^{-\beta(\frac{p^2}{2} + n\tilde{v}_0 - \mu)} = 2 \frac{e^{-\beta(n\tilde{v}_0 - \mu)}}{\sqrt{(2\pi\beta)^3}}.$$

Differentiating the above equation at fixed β we get $dn = -\beta\tilde{v}_0 n dn + \beta n d\mu$ from which we find $d\mu = \tilde{v}_0 dn + \frac{1}{\beta}\frac{dn}{n}$. Substituting this result into (7.16) and integrating over n between 0 and n we deduce the equation of state for an electron gas in the Hartree approximation

$$P = nK_BT + \frac{1}{2}\tilde{v}_0 n^2, \tag{7.17}$$

where we use the fact that the pressure at zero density vanishes. The correction to the equation of state $P = nK_BT$ (noninteracting system) is positive in the repulsive case and negative otherwise, as expected. However, in the attractive case the result does not make sense for too low temperatures. The derivative of P with respect to n should always be positive since

$$\left(\frac{\partial P}{\partial n}\right)_T = \left(\frac{\partial P}{\partial \mu}\right)_T \left(\frac{\partial \mu}{\partial n}\right)_T = n\left(\frac{\partial \mu}{\partial n}\right)_T > 0.$$

The inequality follows from the positivity of the density n and of the derivative

$$\left(\frac{\partial \mu}{\partial n}\right)_T^{-1} = \left(\frac{\partial n}{\partial \mu}\right)_T = \frac{1}{V}\frac{\partial}{\partial \mu} \frac{\mathrm{Tr}\left[e^{-\beta(\hat{H}-\mu\hat{N})}\hat{N}\right]}{\mathrm{Tr}\left[e^{-\beta(\hat{H}-\mu\hat{N})}\right]} = \frac{\beta}{V}\langle\left(\hat{N} - N\right)^2\rangle,$$

where $\langle\ldots\rangle$ denotes the ensemble average. If we take the derivative of (7.17) we find

$$\left(\frac{\partial P}{\partial n}\right)_T = K_BT + \tilde{v}_0 n,$$

which becomes negative for temperatures $T < T_c \equiv -\tilde{v}_0 n/K_B$. This meaningless result does actually contain some physical information. As the temperature decreases from values above T_c, the derivative $(\partial P/\partial n)_T$ approaches zero from positive values and hence the fluctuation in the number of particles, $\langle (\hat{N} - N)^2 \rangle$, diverges in this limit. Typically the occurrence of large fluctuations in physical quantities signals the occurrence of an instability. In fact, the electron gas with attractive interparticle interaction undergoes a phase transition at sufficiently low temperatures, turning into a superconductor.

It is also interesting to observe that (7.17) has the form of the van der Waals equation $(P - \alpha n^2)(V - V_{exc}) = n V K_B T$ with α a constant, V the volume of the system and V_{exc} the exclusion volume, i.e., the hard-core impenetrable volume of the particles. In the Hartree approximation $\alpha = -\frac{1}{2}\tilde{v}_0$ while $V_{exc} = 0$. The fact that $V_{exc} = 0$ is somehow expected. As already pointed out the Hartree approximation treats the particles as effectively independent, meaning that two particles can near each other without expending energy. To have a nonvanishing exclusion volume we must go beyond the mean field approximation and introduce correlations in the two-particle Green's function.[3]

The free nature of the electrons in the Hartree approximation is most evident from the spectral function. From the definition (6.89) we find

$$\langle \mathbf{p}\sigma | \hat{\mathcal{A}}(\omega) | \mathbf{p}'\sigma' \rangle = 2\pi \delta_{\sigma\sigma'} \delta(\mathbf{p} - \mathbf{p}') \delta(\omega - \frac{p^2}{2} - n\tilde{v}_0).$$

According to our interpretation of the spectral function, the above result tells us that an electron with definite momentum and spin has a well-defined energy, $\epsilon_\mathbf{p} = \frac{p^2}{2} + n\tilde{v}_0$, and hence an infinitely long life-time. This is possible only provided that there is no scattering between the electrons.

7.2.3   Quantum discharge of a capacitor

During recent decades, the size of electronic circuits has been continuously reduced. Today, systems like quantum wires and quantum dots are routinely produced on the nanometer scale. The seemingly ultimate limit of miniaturization has been achieved by several experimental groups who have been able to place single molecules between two macroscopic electrodes [69,70]. In this section we start discussing the quantum properties of the electric current flowing through nanoscale structures. The motivation for discussing this system is two-fold. First, we can apply the Hartree approximation in a modern context, and second we introduce a paradigmatic model Hamiltonian which is used later to illustrate the physical content of more sophisticated approximations.

Let us consider a classical capacitor of capacitance C whose plates are connected by a wire of resistance R, see Fig. 7.3(a). At time t_0 we change the potential of one plate by V_0 and a current $I(t)$ will *instantaneously* start to flow. According to the Kirchhoff laws $R\dot{I}(t) = -I(t)/C$ and hence $I(t) = I_0 e^{-t/RC}$ with $I_0 = V_0/R$. How does this classical picture change when the plates are connected by a molecule or by some other kind of nanoscale junction? To answer this question quantum modeling of the problem is needed.

Let $\{\varphi_{k\alpha}\}$ be the basis functions for an electron, or more generally a fermion of charge q, in the $\alpha = L$ (left) and $\alpha = R$ (right) plates, and $\{\varphi_m\}$ be the basis functions for

---

[3]This is nicely discussed in Ref. [68].

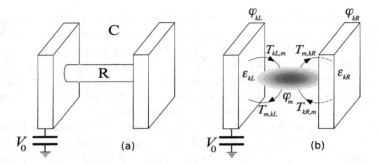

Figure 7.3 Capacitor whose plates are connected by a wire of resistance R (a), and by a nanojunction like, e.g., a molecule or a quantum dot (b).

a fermion in the nanojunction (molecule). Without loss of generality we can choose the $\{\varphi_{k\alpha}\}$ as the eigenstates of the one-body Hamiltonian of the α plate with energy $\epsilon_{k\alpha}$. The most general one-body Hamiltonian of the whole system can then be written as

$$\hat{H}_0 = \sum_{k\alpha} \epsilon_{k\alpha}\hat{n}_{k\alpha} + \sum_{mn} T_{mn}\hat{d}_m^\dagger \hat{d}_n + \sum_{m,k\alpha} (T_{m\,k\alpha}\hat{d}_m^\dagger \hat{d}_{k\alpha} + T_{k\alpha\,m}\hat{d}_{k\alpha}^\dagger \hat{d}_m), \qquad (7.18)$$

where $\hat{n}_{k\alpha} \equiv \hat{d}_{k\alpha}^\dagger \hat{d}_{k\alpha}$ is the occupation operator for fermions in $\varphi_{k\alpha}$. In (7.18) the first term refers to the plates (or electrodes), the second term to the nanojunction, while the last term describes the contact between the different parts of the system, see Fig. 7.3(b). This Hamiltonian can also be regarded as a generalization of the Fano model of Section 2.3.2 to finite systems (atoms or molecules) with many localized orbitals and in contact with more than one continuum of states.

To model the interaction we assume that the plates are symmetric and hence that the electrostatic forces vanish when the number of fermions in the left plate, N_L, and in the right plate, N_R, are equal. The simplest interaction Hamiltonian with such a physical ingredient is

$$\hat{H}_{\text{int}} = \frac{v_0}{2}(\hat{N}_L - \hat{N}_R)^2$$

$$= \frac{v_0}{2}(\hat{N}_L + \hat{N}_R) + \frac{v_0}{2}\sum_{kk'}\left[\sum_\alpha \hat{d}_{k\alpha}^\dagger \hat{d}_{k'\alpha}^\dagger \hat{d}_{k'\alpha}\hat{d}_{k\alpha} - 2\hat{d}_{kL}^\dagger \hat{d}_{k'R}^\dagger \hat{d}_{k'R}\hat{d}_{kL}\right], \qquad (7.19)$$

with $\hat{N}_\alpha = \sum_k \hat{n}_{k\alpha}$ the total number of particle operators in electrode α. We can relate the interaction energy $v_0 > 0$ to the capacitance C by calculating the energy cost to transfer one fermion from L to R. Starting from the equilibrium configuration $N_L = N_R = N$ (zero interaction energy), the energy cost to reach the configuration with $N_L = N - 1$ and $N_R = N + 1$ is $2v_0$. In classical electrostatics the same process costs an energy of $q^2/2C$ and therefore $v_0 = q^2/4C$. Can the resistance R also be related to the parameters of the Hamiltonian? The resistance of a bulk conductor originates from the inelastic scattering between the electrons and the lattice vibrations (phonons). In our model Hamiltonian we did not include the electron–phonon interaction and hence we should (naively) expect that $R = 0$. As we shall see, however, in the quantum world there is a minimum value of the resistance below which it is not possible to go.

The total Hamiltonian $\hat{H} = \hat{H}_0 + \hat{H}_{\text{int}}$ describes a system of interacting particles that can be treated within the Hartree approximation. The first step is to express the Hartree potential operator in the $\{\varphi_i\}$ basis with $i \in \{k\alpha\}$ (for the α electrode) or $i \in \{m\}$ (for the molecule). From (7.6) and (7.4) we easily obtain the following general formula

$$
\begin{aligned}
\langle i | q\hat{\mathcal{V}}_{\text{H}}^{\text{M}} | j \rangle &= -i \int d\mathbf{x}_1 d\mathbf{x}_2 \langle i | \mathbf{x}_1 \rangle v(\mathbf{x}_1, \mathbf{x}_2) \langle \mathbf{x}_2 | \hat{\mathcal{G}}^<(t_0, t_0) | \mathbf{x}_2 \rangle \langle \mathbf{x}_1 | j \rangle \\
&= -i \sum_{pq} v_{ipqj} \langle q | \hat{\mathcal{G}}^<(t_0, t_0) | p \rangle,
\end{aligned}
\tag{7.20}
$$

with the Coulomb integrals v_{ipqj} defined as in (1.85). For the present problem the v_{ipqj} can be extracted from the second row of (7.19). Indeed, the second term has exactly the same structure as the second term in (1.83) and therefore

$$
v_{ipqj} = v_0 \delta_{ij} \delta_{pq} \times \begin{cases} 1 & i = k\alpha, \ p = k'\alpha' \text{ and } \alpha = \alpha' \\ -1 & i = k\alpha, \ p = k'\alpha' \text{ and } \alpha \neq \alpha' \\ 0 & \text{otherwise} \end{cases}.
\tag{7.21}
$$

The first term in the second row of (7.19) is a one-body operator that can be absorbed in $\hat{H}_0$ by shifting the single-particle energies $\epsilon_{k\alpha} \to \epsilon_{k\alpha} + v_0/2$. If we now insert the v_{ipqj} of (7.21) into (7.20) we can readily derive that

$$
q\hat{\mathcal{V}}_{\text{H}}^{\text{M}} = v_0(N_L - N_R) \sum_k (|kL\rangle\langle kL| - |kR\rangle\langle kR|).
\tag{7.22}
$$

Thus the Hartree potential depends only on the average number of particles N_α in electrode $\alpha = L, R$ and not on how the particles distribute over the basis functions $\{\varphi_{k\alpha}\}$.

We are interested in calculating the current flowing between the plates when the system is initially in the equilibrium configuration of $\hat{H}_0 + \hat{H}_{\text{int}}$ and then is driven out of equilibrium by a bias. Then the single-particle Hartree Hamiltonian along the imaginary track is $\hat{h}_{\text{H}}^{\text{M}} = \hat{h} + q\hat{\mathcal{V}}_{\text{H}}^{\text{M}} - \mu$ while at positive times $\hat{h}_{\text{H}}(t) = \hat{h} + q\hat{\mathcal{V}}_{\text{H}}(t) + q\hat{\mathcal{V}}(t)$, with $\hat{\mathcal{V}}(t)$ the external potential (bias). To keep the mathematical complications at a minimum without altering the qualitative picture we use only one basis ket $|\epsilon_0\rangle$ for the molecule and introduce the shorthand notation $\epsilon_0 \equiv T_{00}$ for the onsite energy and $T_{k\alpha} \equiv T_{0k\alpha} = T_{k\alpha 0}^*$ for the hopping parameters. The single-particle Hamiltonian $\hat{h}$ for this simplified version of the model is

$$
\hat{h} = \hat{\mathcal{E}} + \hat{\mathcal{T}}, \qquad \begin{cases} \hat{\mathcal{E}} = \sum_{k\alpha} \epsilon_{k\alpha} |k\alpha\rangle\langle k\alpha| + \epsilon_0 |\epsilon_0\rangle\langle\epsilon_0| \\ \hat{\mathcal{T}} = \sum_{k\alpha} (T_{k\alpha} |\epsilon_0\rangle\langle k\alpha| + T_{k\alpha}^* |k\alpha\rangle\langle\epsilon_0|) \end{cases},
$$

where $\hat{\mathcal{E}}$ describes the different parts of the system while $\hat{\mathcal{T}}$ accounts for the contacts between them.

Equilibrium problem

In equilibrium the bias $\hat{\mathcal{V}}(t) = 0$ and all real-time Green's functions depend only on the time difference. The self-consistent equation for the number of particles N_α is

$$N_\alpha = -i \sum_k \langle k\alpha | \hat{\mathcal{G}}^<(t,t) | k\alpha \rangle = -i \int \frac{d\omega}{2\pi} \sum_k \langle k\alpha | \hat{\mathcal{G}}^<(\omega) | k\alpha \rangle. \tag{7.23}$$

The basis functions $\varphi_{k\alpha}$ are not eigenfunctions of the Hartree Hamiltonian and therefore the evaluation of the matrix elements $\langle k\alpha | \hat{\mathcal{G}}^<(\omega) | k\alpha \rangle$ is less trivial than in the example of the electron gas. Our strategy is to use the fluctuation–dissipation theorem (6.80) and to exploit the identity [see also (2.24)]

$$\underbrace{\frac{1}{\omega - \hat{h}_{\mathrm{H}} \pm i\eta}}_{\hat{\mathcal{G}}^{\mathrm{R,A}}(\omega)} = \frac{1}{\omega - (\hat{\mathcal{E}} + q\hat{\mathcal{V}}_{\mathrm{H}}^{\mathrm{M}}) \pm i\eta} + \frac{1}{\omega - (\hat{\mathcal{E}} + q\hat{\mathcal{V}}_{\mathrm{H}}^{\mathrm{M}}) \pm i\eta} \hat{T} \frac{1}{\omega - \hat{h}_{\mathrm{H}} \pm i\eta}, \tag{7.24}$$

to determine the matrix elements of the retarded/advanced Green's function. An alternative strategy would be to work with the Matsubara Green's function. Indeed, N_α can also be written as

$$N_\alpha = -i \sum_k \langle k\alpha | \hat{\mathcal{G}}^{\mathrm{M}}(\tau, \tau^+) | k\alpha \rangle = \frac{1}{\beta} \sum_{m=-\infty}^{\infty} \sum_k e^{\omega_m \eta} \langle k\alpha | \frac{1}{\omega_m - \hat{h}_{\mathrm{H}}^{\mathrm{M}}} | k\alpha \rangle,$$

and then the identity (7.24) with $\omega \pm i\eta \to \omega_m - \mu$ could be used to determine all matrix elements. We follow the first route here, since the formulas that we derive are relevant to the time-dependent case as well. The reader can, however, verify that the second route leads to the very same results.

As usual we denote by G_{ij} the matrix element $\langle i | \hat{\mathcal{G}} | j \rangle$ of the Green's function. The sandwich of (7.24) with $\langle k\alpha |$ and $| \epsilon_0 \rangle$ and with $\langle \epsilon_0 |$ and $| \epsilon_0 \rangle$ yields the following linear system of coupled equations for the matrix elements of the retarded Green's function (similar relations can be obtained for the advanced Green's function):

$$G_{k\alpha 0}^{\mathrm{R}}(\omega) = \frac{T_{k\alpha}^*}{\omega - \tilde{\epsilon}_{k\alpha} + i\eta} G_{00}^{\mathrm{R}}(\omega), \tag{7.25}$$

$$G_{00}^{\mathrm{R}}(\omega) = \frac{1}{\omega - \epsilon_0 + i\eta} + \frac{1}{\omega - \epsilon_0 + i\eta} \sum_{k\alpha} T_{k\alpha} G_{k\alpha 0}^{\mathrm{R}}(\omega), \tag{7.26}$$

where we define the eigenvalues of $\hat{\mathcal{E}} + q\hat{\mathcal{V}}_{\mathrm{H}}^{\mathrm{M}}$ as [see (7.22)]

$$\tilde{\epsilon}_{kL} = \epsilon_{kL} + v_0(N_L - N_R), \qquad \tilde{\epsilon}_{kR} = \epsilon_{kR} + v_0(N_R - N_L).$$

Substituting (7.25) into (7.26) we arrive at the solution

$$G_{00}^{\mathrm{R}}(\omega) = \frac{1}{\omega - \epsilon_0 - \Sigma_{\mathrm{em}}^{\mathrm{R}}(\omega) + i\eta},$$

where

$$\Sigma_{\mathrm{em}}^{\mathrm{R}}(\omega) = \sum_{k\alpha} \frac{|T_{k\alpha}|^2}{\omega - \tilde{\epsilon}_{k\alpha} + i\eta} = \underbrace{P \sum_{k\alpha} \frac{|T_{k\alpha}|^2}{\omega - \tilde{\epsilon}_{k\alpha}}}_{\Lambda(\omega)} - \frac{i}{2} \underbrace{2\pi \sum_{k\alpha} |T_{k\alpha}|^2 \delta(\omega - \tilde{\epsilon}_{k\alpha})}_{\Gamma(\omega)} \qquad (7.27)$$

is the retarded *embedding self-energy*. In (7.27) the embedding self-energy is separated into a real $[\Lambda(\omega)]$ and imaginary $[-\frac{1}{2}\Gamma(\omega)]$ part using the Cauchy relation (2.23). The embedding self-energy accounts for processes in which a particle in φ_0 hops to $\varphi_{k\alpha}$ (with amplitude $T_{k\alpha}^*$), stays in the α electrode for some time (the denominator in (7.27) is the Fourier transform of the propagator $\theta(t)e^{-i\tilde{\epsilon}_{k\alpha}t}$) and then hops back in φ_0 (with amplitude $T_{k\alpha}$). In other words, the embedding self-energy appears because the nanojunction is *open*, i.e., it can exchange particles and energy with the electrodes. Notice that G_{00}^{R} reduces to the Green's function of the isolated molecule for $\Sigma_{\mathrm{em}}^{\mathrm{R}} = 0$.

At this point we simplify the problem further by making the *Wide Band Limit Approximation* (WBLA) already discussed in the context of the Fano model. In the WBLA $\Gamma(\omega) \sim \Gamma$ is a constant independent of frequency and hence the real part $\Lambda(\omega)$ vanishes. Consequently the embedding self-energy $\Sigma_{\mathrm{em}}^{\mathrm{R}}(\omega) = -\frac{i}{2}\Gamma$ is a pure imaginary number and the retarded Green's function of the molecule simplifies to

$$G_{00}^{\mathrm{R}}(\omega) = \frac{1}{\omega - \epsilon_0 + \frac{i}{2}\Gamma}.$$

The WBLA is a good approximation provided that the applied bias is much smaller than the energy scale over which $\Gamma(\omega)$ varies.

Let us now evaluate the sum $\sum_k G_{k\alpha\,k\alpha}^{\mathrm{R}}(\omega)$ [needed to calculate the r.h.s. of the self-consistent equation (7.23)] in the WBLA. We denote by Γ_α the α contribution to Γ resulting from summing over k at fixed α in (7.27). From the identity (7.24) and its adjoint it is a matter of simple algebra to show that

$$G_{k\alpha\,p\beta}^{\mathrm{R}}(\omega) = \frac{\delta_{\alpha\beta}\delta_{kp}}{\omega - \tilde{\epsilon}_{k\alpha} + i\eta} + \frac{T_{k\alpha}^*}{\omega - \tilde{\epsilon}_{k\alpha} + i\eta} G_{00}^{\mathrm{R}}(\omega) \frac{T_{p\beta}}{\omega - \tilde{\epsilon}_{p\beta} + i\eta}.$$

Setting $p\beta = k\alpha$ and summing over k we find

$$\sum_k G_{k\alpha\,k\alpha}^{\mathrm{R}}(\omega) = \sum_k \frac{1}{\omega - \tilde{\epsilon}_{k\alpha} + i\eta} + G_{00}^{\mathrm{R}}(\omega) \underbrace{\sum_k \frac{|T_{k\alpha}|^2}{(\omega - \tilde{\epsilon}_{k\alpha} + i\eta)^2}}_{=0\ \text{in WBLA}}.$$

The second term is proportional to the ω-derivative of the α contribution to the embedding self-energy which, in our case, is simply $-\frac{i}{2}\Gamma_\alpha$. Thus the second term vanishes in the WBLA and the self-consistent equations (7.23) reduce to

$$N_L = \sum_k f(\epsilon_{kL} + v_0(N_L - N_R) - \mu), \qquad N_R = \sum_k f(\epsilon_{kR} + v_0(N_R - N_L) - \mu). \qquad (7.28)$$

These equations can be seen as a generalization of the self-consistent equation (7.15) for the electron gas. We also notice that in the WBLA the nanojunction plays no role in determining the self-consistent densities: the particles distribute as if the plates of the capacitor were perfectly isolated (all hopping parameters equal zero). This is due to the macroscopic size of the plates as compared to the size of the molecule.

Solving (7.28) for N_L and N_R amounts to determining the Hartree Hamiltonian and hence all equilibrium Green's functions in the Hartree approximation. Before moving to the time-dependent problem we wish to observe that the matrix element of the spectral function operator

$$\langle \epsilon_0 | \hat{\mathcal{A}}(\omega) | \epsilon_0 \rangle = -2\text{Im} \langle \epsilon_0 | \hat{\mathcal{G}}^{\text{R}}(\omega) | \epsilon_0 \rangle = \frac{\Gamma}{[(\omega - \epsilon_0)^2 + \Gamma^2/4]}$$

is a Lorentzian centered around ϵ_0. How do we reconcile this result with the fact that in the Hartree approximation the particles behave as free particles and hence have well-defined energies? The answer is the same as in the case of noninteracting systems. The state φ_0 is not an eigenstate of the Hartree Hamiltonian and hence it does not have a well-defined energy: a particle in φ_0 has a finite probability of having any energy, the most probable energy being ϵ_0. In the limit $\Gamma \to 0$ (no contacts between the plates and the molecule) the Lorentzian approaches a δ-function in agreement with the fact that φ_0 becomes an eigenstate with eigenvalue ϵ_0. As we saw in Section 6.3.4 the effect of the interparticle interaction is to broaden *all* matrix elements of the spectral function and hence to destroy the single-particle picture. In order to capture this effect, however, we must go beyond the mean-field description.

Time-dependent problem

In analogy with the classical example we switch on a bias $\hat{V}(t)$ in, e.g., the left plate, and drive the system out of equilibrium. The time-dependent perturbation in second quantization reads

$$\hat{H}_V(t) = qV_0(t)\hat{N}_L, \tag{7.29}$$

and the total Hamiltonian at times $t > t_0$ is $\hat{H}(t) = \hat{H}_0 + \hat{H}_{\text{int}} + \hat{H}_V(t)$. A current will start flowing through the molecule till the potential energy difference between the L and R plates is leveled out. The electric current I_α from the plate α is given by the change of N_α per unit time multiplied by the charge q of the particles, $I_\alpha(t) = q\frac{d}{dt}N_\alpha(t)$. Since $N_\alpha(t)$ is the ensemble average of the number operator in the Heisenberg picture, $\hat{N}_{\alpha,H}(t)$, its time derivative is the ensemble average of the commutator $[\hat{N}_{\alpha,H}(t), \hat{H}_H(t)]_-$, see (3.19). The operator $\hat{N}_\alpha$ commutes with $\hat{H}_{\text{int}}$, $\hat{H}_V(t)$ and with all the terms of $\hat{H}_0$ except the one describing the contacts between the molecule and the plates [third term in (7.18)]. It is a matter of simple algebra to show that

$$\left[\hat{N}_\alpha, \hat{H}(t)\right]_- = \sum_k \left[\hat{n}_{k\alpha}, \hat{H}(t)\right]_- = \sum_k \left(-T_{k\alpha}\hat{d}_0^\dagger \hat{d}_{k\alpha} + T_{k\alpha}^* \hat{d}_{k\alpha}^\dagger \hat{d}_0\right),$$

and therefore

$$I_\alpha(t) = 2q \sum_k \text{Re}\left[T_{k\alpha}G_{k\alpha\,0}^<(t,t)\right], \tag{7.30}$$

where we take into account that the lesser Green's function is anti-Hermitian, see (6.23). The above formula is an exact result that relates the current to the lesser Green's function. It is natural to ask whether or not (7.30) preserves its form when $N_\alpha(t)$ is *not* exact but it rather comes from the Hartree approximation. We can answer this question by subtracting the Hartree equation of motion from its adjoint

$$\left[i\frac{d}{dz} + i\frac{d}{dz'}\right]\hat{\mathcal{G}}(z,z') = \left(\hat{h}(z) + q\hat{\mathcal{V}}_H(z)\right)\hat{\mathcal{G}}(z,z') - \hat{\mathcal{G}}(z,z')\left(\hat{h}(z') + q\hat{\mathcal{V}}_H(z')\right),$$

where $\hat{h}(z) = \hat{h} + q\hat{\mathcal{V}}(z)$ is the biased Hamiltonian. Setting $z = t_-$ and $z' = t_+$ the l.h.s. becomes the total derivative $i\frac{d}{dt}\hat{\mathcal{G}}^<(t,t)$. Therefore, if we sandwich with $\langle k\alpha|$ and $|k\alpha\rangle$ and sum over k we obtain the equation for $\frac{d}{dt}N_\alpha(t)$ in the Hartree approximation. In this equation the terms containing the Hartree potential $\hat{\mathcal{V}}_H$ cancel out and the current formula (7.30) is recovered. More generally, for an approximate two-particle Green's function G_2 to reproduce the current formula (7.30) it is crucial that the terms originating from G_2 cancel out when calculating $\frac{d}{dt}N_\alpha(t)$ as above. The approximations that preserve the continuity equation as well as other basic conservation laws are known as the *conserving approximations*, and the Hartree approximation is one of them. In Chapters 8 and 9 we develop a simple and elegant formalism to recognize whether an approximation is conserving or not. This formalism also provides us with a systematic way to generate conserving approximations of increasing accuracy.

We continue our analysis by evaluating the current $I_\alpha(t)$ in (7.30) to *linear order* in the bias $\hat{\mathcal{V}}(t)$. This can be done in several different ways.[4] Here we present an instructive derivation which requires use of the Langreth rules. Let us write the time-dependent Hartree Hamiltonian as the sum of the equilibrium Hamiltonian $\hat{h}_H(t_0)$ and the time-dependent operator $q\delta\hat{\mathcal{V}}_{\text{eff}}(t) \equiv q\hat{\mathcal{V}}(t) + q\delta\hat{\mathcal{V}}_H(t)$, where $\hat{\mathcal{V}}(t) = V_0(t)\sum_k |kL\rangle\langle kL|$ is the external bias while $\delta\hat{\mathcal{V}}_H(t) = \hat{\mathcal{V}}_H(t) - \hat{\mathcal{V}}_H(t_0)$ is the change of the Hartree potential induced by the bias. The equation of motion for the Green's function can be integrated using the equilibrium Green's function $\hat{\mathcal{G}}_{\text{eq}}$. The result is the following integral equation on the contour:

$$\hat{\mathcal{G}}(z,z') = \hat{\mathcal{G}}_{\text{eq}}(z,z') + \int_\gamma d\bar{z}\,\hat{\mathcal{G}}_{\text{eq}}(z,\bar{z})\,q\delta\hat{\mathcal{V}}_{\text{eff}}(\bar{z})\,\hat{\mathcal{G}}(\bar{z},z'). \tag{7.31}$$

We can easily verify that the above $\hat{\mathcal{G}}$ satisfies the Hartree equation of motion by acting with $[i\frac{d}{dz} - \hat{h}_H(t_0)]$ on both sides and by taking into account that, by definition,

$$\left[i\frac{d}{dz} - \hat{h}_H(t_0)\right]\hat{\mathcal{G}}_{\text{eq}}(z,z') = \delta(z,z').$$

Equation (7.31) is a special case of the *Dyson equation* whose main distinctive feature is the possibility of expanding the full Green's function in "powers" of some other Green's function ($\hat{\mathcal{G}}_{\text{eq}}$ in this case) by iterations, i.e., by replacing the $\hat{\mathcal{G}}$ in the last term of the r.h.s. with the whole r.h.s.[5]

---

[4]A thorough discussion on linear response theory is presented in Chapters 14 and 15.

[5]Another example of the Dyson equation is equation (7.24) for the Fourier transform of the retarded/advanced Green's function. A Dyson equation similar to (7.31) was also derived in the useful exercise of Section 6.2.3.

We observe that for $V_0(t) = 0$ (zero bias) the Hartree potential does not change and hence $\delta\hat{\mathcal{V}}_{\text{eff}}(z) = 0$, which implies $\hat{\mathcal{G}} = \hat{\mathcal{G}}_{\text{eq}}$, as expected. It is also important to observe that in (7.31) the information on the boundary conditions is all contained in $\hat{\mathcal{G}}_{\text{eq}}$: the Green's function $\hat{\mathcal{G}}$ fulfills the same KMS boundary conditions as the equilibrium Green's function. To first order in $\delta\hat{\mathcal{V}}_{\text{eff}}$ the change $\delta\hat{\mathcal{G}} = \hat{\mathcal{G}} - \hat{\mathcal{G}}_{\text{eq}}$ follows from the first iteration of (7.31):

$$\delta\hat{\mathcal{G}}(z, z') = \int_\gamma d\bar{z}\, \hat{\mathcal{G}}_{\text{eq}}(z, \bar{z})\, q\delta\hat{\mathcal{V}}_{\text{eff}}(\bar{z})\, \hat{\mathcal{G}}_{\text{eq}}(\bar{z}, z'). \tag{7.32}$$

To calculate the components of $\delta\hat{\mathcal{G}}$ relevant to the current we sandwich (7.32) with $\langle k\alpha|$ and $|\epsilon_0\rangle$, set $z = t_-$, $z' = t_+$ and use the Langreth rules of Table 5.1. The r.h.s. of (7.32) can be seen as the convolution of three functions in Keldysh space, one of which, $\delta(z, z')\delta\hat{\mathcal{V}}_{\text{eff}}(z)$, has only a singular part. Taking into account that $\delta\hat{\mathcal{V}}_{\text{eff}}(t_0 - i\tau)$ (vertical track of the contour) vanishes we find [see Exercise 5.3]

$$\delta G^<_{k\alpha 0}(t, t) = \sum_{p\beta} \int_{t_0}^\infty d\bar{t}\, \left[G^{\text{R}}_{k\alpha p\beta}(t, \bar{t}) G^<_{p\beta 0}(\bar{t}, t) + G^<_{k\alpha p\beta}(t, \bar{t}) G^{\text{A}}_{p\beta 0}(\bar{t}, t) \right] q\delta V_\beta(\bar{t}),$$
$$\tag{7.33}$$

with the effective potential [see (7.22)]

$$q\delta V_L(t) = qV_0(t) + v_0\left(\delta N_L(t) - \delta N_R(t)\right),$$
$$q\delta V_R(t) = v_0\left(\delta N_R(t) - \delta N_L(t)\right).$$

On the r.h.s. of (7.33) the Green's functions are equilibrium Green's functions and depend only on the time difference $\Delta_t = t - \bar{t}$. Inserting (7.33) in the current formula (7.30) we find

$$I_\alpha(t) = 2q^2 \sum_\beta \int_{t_0}^t d\bar{t}\, \text{Re}\left[C^{(1)}_{\alpha\beta}(\Delta_t) + C^{(2)}_{\alpha\beta}(\Delta_t) \right] \delta V_\beta(\bar{t}), \tag{7.34}$$

with

$$C^{(1)}_{\alpha\beta}(\Delta_t) = \int \frac{d\omega\, d\omega'}{2\pi\, 2\pi} e^{-i(\omega-\omega')\Delta_t} \sum_{kp} T_{k\alpha} G^{\text{R}}_{k\alpha p\beta}(\omega) G^<_{p\beta 0}(\omega'),$$

$$C^{(2)}_{\alpha\beta}(\Delta_t) = \int \frac{d\omega\, d\omega'}{2\pi\, 2\pi} e^{-i(\omega-\omega')\Delta_t} \sum_{kp} T_{k\alpha} G^<_{k\alpha p\beta}(\omega) G^{\text{A}}_{p\beta 0}(\omega').$$

The calculation of the kernels $C^{(1)}$ and $C^{(2)}$ can easily be carried out by inserting the explicit expression of the matrix elements of the retarded/advanced Green's function previously derived, and by using the fluctuation–dissipation theorem for the lesser Green's function. In doing so the following structures appear

$$\sum_p \frac{|T_{p\beta}|^2}{(\omega - \tilde{\epsilon}_{p\beta} \pm i\eta)(\omega' - \tilde{\epsilon}_{p\beta} \pm i\eta)} = \frac{\sum_p \frac{|T_{p\beta}|^2}{\omega - \tilde{\epsilon}_{p\beta} \pm i\eta} - \sum_p \frac{|T_{p\beta}|^2}{\omega' - \tilde{\epsilon}_{p\beta} \pm i\eta}}{\omega' - \omega} = 0,$$

$$\sum_p \frac{|T_{p\beta}|^2}{(\omega - \tilde{\epsilon}_{p\beta} + i\eta)(\omega' - \tilde{\epsilon}_{p\beta} - i\eta)} = \frac{\sum_p \frac{|T_{p\beta}|^2}{\omega - \tilde{\epsilon}_{p\beta} + i\eta} - \sum_p \frac{|T_{p\beta}|^2}{\omega' - \tilde{\epsilon}_{p\beta} - i\eta}}{\omega' - \omega - 2i\eta} = \frac{-i\Gamma_\beta}{\omega' - \omega - 2i\eta},$$

where we take into account that in the WBLA the β contribution to the embedding self-energy is a purely imaginary constant. Performing the sum over k and then integrating over one of the frequencies,[6] we find

$$C_{\alpha\beta}^{(1)}(\Delta_t) = \Gamma_\alpha \int \frac{d\omega}{2\pi} \frac{f(\omega-\mu)}{\omega-\zeta_0} \left(\delta_{\alpha\beta} - \frac{i}{2}\Gamma_\beta \frac{1-e^{i(\omega-\zeta_0^*)\Delta_t}}{\omega-\zeta_0^*} \right),$$

$$C_{\alpha\beta}^{(2)}(\Delta_t) = -\Gamma_\alpha \int \frac{d\omega}{2\pi} \frac{f(\omega-\mu)}{\omega-\zeta_0} \left(\delta_{\alpha\beta} - \frac{i}{2}\Gamma_\beta \frac{1}{\omega-\zeta_0^*} \right) \left(1-e^{-i(\omega-\zeta_0)\Delta_t} \right),$$

where $\zeta_0 = \epsilon_0 + \frac{i}{2}\Gamma$. Inserting these results into (7.34) we obtain the following integral equation:

$$I_\alpha(t) = 2q^2\Gamma_\alpha \sum_\beta \int_{t_0}^t d\bar{t}\, \mathrm{Re} \left[\int \frac{d\omega}{2\pi} f(\omega-\mu) \frac{e^{i(\omega-\zeta_0^*)\Delta_t}}{\omega-\zeta_0^*} \left(\delta_{\alpha\beta} + i\frac{\Gamma_\beta}{\omega-\zeta_0} \right) \right] \delta V_\beta(\bar{t}).$$

(7.35)

The integral of $I_\alpha(t)$ between t_0 and t yields the variation of the number of particles in electrode α at time t. The effective potential δV_β can therefore be expressed in terms of the currents as

$$q\delta V_L(t) = qV_0(t) + \frac{v_0}{q} \int_{t_0}^t d\bar{t}\, (I_L(\bar{t}) - I_R(\bar{t})),$$

$$q\delta V_R(t) = \frac{v_0}{q} \int_{t_0}^t d\bar{t}\, (I_R(\bar{t}) - I_L(\bar{t})).$$

(7.36)

Equations (7.35) and (7.36) form a system of coupled integral equations for the time dependent current in the Hartree approximation. Let us discuss their physical content.

We focus on a sudden switch-on process $V_0(t) = \theta(t)V_0$ in order to compare the results with those of the classical case. For noninteracting fermions $v_0 = 0$ and the equations decouple. In this case the current can be calculated by solving the integral in (7.35) with $\delta V_R = 0$ and $\delta V_L = V_0$. The results are shown in Fig. 7.4. Three main features capture our attention. First, in contrast to the classical case the current does not start flowing instantaneously. Second, the current reaches a constant value after some damping time $\tau_d \sim 1/\Gamma$. During the transient we observe oscillations of frequency $|\epsilon_0 - \mu|$ due to virtual transitions between the Fermi level μ and the molecular level ϵ_0. In the classical case $v_0 = 0$ corresponds to $C = \infty$ and hence $I(t) = V_0/R$ at all times, i.e., the steady-state value is reached immediately after the switch-on of the bias. Third, the long-time limit of the current is finite despite the absence of dissipative mechanisms leading to a finite resistance. This is the anticipated quantum behavior: the microscopic cross-section of the nanojunction prevents a macroscopic number of fermions from passing through it [54, 71]. The steady-state value of the current, $I_\alpha^{(S)}$, can be obtained from (7.35) in the limit $t \to \infty$. Since the kernel in the square bracket vanishes when $\Delta_t \to \infty$ we can safely extend the integral to

---

[6]The integral over the frequency is done by closing the contour in the upper or lower complex ω-plane.

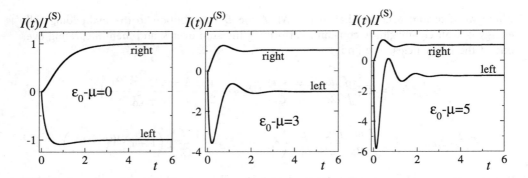

Figure 7.4 Time-dependent current in units of the steady current (7.37) for noninteracting particles. The model parameters are $T = 0$ (zero temperature), $\mu = 0$, $\Gamma_L = \Gamma_R = 1$ and bias $V_0 = 1$. The transient oscillations have frequency $|\epsilon_0 - \mu|$ and are due to virtual transitions between the Fermi level μ and the molecular level ϵ_0. At the steady-state $I_L(t) = -I_R(t)$ as it should.

$t_0 = -\infty$, thus finding

$$I_R^{(S)} = 2q^2\Gamma_L\Gamma_R \int \frac{d\omega}{2\pi}\text{Re}\left[\frac{f(\omega - \mu)}{(\omega - \zeta_0^*)^2(\omega - \zeta_0)}\right]V_0 \xrightarrow{T \to 0} \frac{q^2}{2\pi}\frac{\Gamma_L\Gamma_R}{(\epsilon_0 - \mu)^2 + (\Gamma/2)^2}V_0, \tag{7.37}$$

where in the last step the zero temperature limit has been taken. The resistance $\text{R} = V_0/I_R^{(S)}$ is minimum when the on-site energy ϵ_0 is aligned to the Fermi energy μ and the molecule is symmetrically coupled to the plates, $\Gamma_L = \Gamma_R = \Gamma/2$. The minimum of R at $\epsilon_0 = \mu$ is physically sound. Indeed for small biases only fermions close to the Fermi energy contribute to the current and the probability of being reflected is minimum when their energy matches the height of the molecular barrier, i.e., ϵ_0. The inverse of the minimum resistance for particles with unit charge is denoted by σ_0 and is called the *quantum of conductance*. From (7.37) we have $\sigma_0 = \frac{1}{2\pi}$, or reinserting the fundamental constants $\sigma_0 = e^2/h = 1/(25 \text{ kOhm})$, with $-|e|$ the electric charge and h the Planck constant.

The interacting results obtained by solving numerically (7.35) and (7.36) are shown in Fig. 7.5 where the left/right currents (a), and effective potentials (b), are calculated for different values of the interaction strength v_0. As in the classical case, the currents are exponentially damped and approach zero when the potential energy difference vanishes. The damping is in remarkably good agreement with that of a classical circuit having resistance $\text{R} = V_0/I_R^{(S)} = 2\pi$ [from (7.37) with the parameters of the figure] and capacitance $\text{C} = q^2/4v_0$. The classical current $I(t)/I_R^{(S)} = e^{-t/\text{RC}}$ is shown in panel (a) with a thin line. Finally we note that the present microscopic approach allows for appreciating the delay of $I_R(t)$ with respect to $I_L(t)$ near $t = 0$ (the left current starts much faster than the right current). This is due to the finite speed v_F (Fermi velocity) of the particles: a perturbation in a point P produces a change in P' after a time $\sim$ (distance between P and P')/v_F.

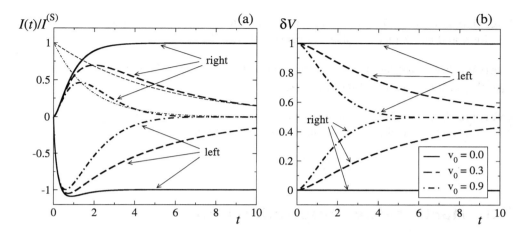

Figure 7.5 Time-dependent Hartree current in units of the steady current (7.37) (a), and effective potential (b), for $v_0 = 0.0, 0.3, 0.9$. The model parameters are $T = 0$ (zero temperature), $\mu = 0$, $\Gamma_L = \Gamma_R = 1$, $\epsilon_0 = 0$, $q = -1$ and bias $V_0 = 1$. In panel (a) the current of a classical circuit is shown with thin lines.

Exercise 7.1. Consider the time-dependent Gross–Pitaevskii equation in one dimension with $V = \mathbf{A} = 0$:

$$i\frac{d}{dt}\varphi(x,t) = -\frac{1}{2m}\frac{d^2}{dx^2}\varphi(x,t) + v_0|\varphi(x,t)|^2\varphi(x,t),$$

where for simplicity we omit the spin index. Show that for $v_0 > 0$ a possible solution of the Gross–Pitaevskii equation is

$$\varphi(x,t) = \tanh\left[\sqrt{mv_0}(x - \mathrm{v}t)\right] e^{i\left[m\mathrm{v}x - \left(m\frac{\mathrm{v}^2}{2} + v_0\right)t\right]},$$

whereas for $v_0 < 0$ a possible solution is

$$\varphi(x,t) = \mathrm{sech}\left[\sqrt{-mv_0}(x - \mathrm{v}t)\right] e^{i\left[m\mathrm{v}x - \left(m\frac{\mathrm{v}^2}{2} + \frac{v_0}{2}\right)t\right]},$$

where v is an arbitrary velocity. These solutions are solitary waves, or *solitons*. Indeed the function $|\varphi(x,t)|^2$ propagates in the medium with velocity v without changing its profile. Readers interested in the theory of solitons should consult Ref. [72].

Exercise 7.2. Using (6.106) show that the energy of the electron gas in the Hartree approximation is given by

$$E = \mathrm{V}\left[\frac{1}{2}\tilde{v}_0 n^2 + 2\int \frac{d\mathbf{p}}{(2\pi)^3}\frac{p^2/2}{e^{\beta\left(\frac{p^2}{2} + \tilde{v}_0 n - \mu\right)} + 1}\right],$$

where V is the volume of the system.

Exercise 7.3. Using (7.15) plot μ as a function of temperature T for different values of the density n and check that the plots are consistent with the result $\beta\mu \to -\infty$ for $n \to 0$.

Exercise 7.4. Calculate the steady-state current $I_L^{(S)}$ from (7.35) with $v_0 = 0$ and show that $I_L^{(S)} = -I_R^{(S)}$ with $I_R^{(S)}$ given in (7.37).

Exercise 7.5. Calculate the time-dependent occupation $n_0(t) = -\mathrm{i}\delta G_{00}^<(t,t)$ from (7.32) and show that the total number of particles is conserved, i.e.,

$$q\frac{dn_0(t)}{dt} = -I_L(t) - I_R(t).$$

7.3 Hartree–Fock approximation

The main advance of the Hartree–Fock approximation over the Hartree approximation is the incorporation of exchange effects: the two-particle Green's function $G_{2,\mathrm{HF}}$ has the *same* symmetry properties of the exact G_2 under the interchange of the space–spin–time arguments. If we approximate G_2 with (7.2) in the equation of motion for the Green's function we find

$$\mathrm{i}\frac{d}{dz_1}G(1;2) - \int d3\left[h(1;3) + qV_{\mathrm{H}}(1)\delta(1;3) + \mathrm{i}v(1;3)G(1;3^+)\right]G(3;2) = \delta(1;2).$$
(7.38)

We notice at once an important difference with respect to the Hartree approximation. The quantity $qV_{\mathrm{H}}(1)\delta(1;2)$ is proportional to $\delta(\mathbf{r}_1 - \mathbf{r}_2)\delta(z_1,z_2)$, i.e., it is local both in time and space, while the quantity

$$v(1;2)G(1;2^+) = \delta(z_1,z_2)v(\mathbf{x}_1,\mathbf{x}_2)G(\mathbf{x}_1,z_1;\mathbf{x}_2;z_1^+)$$

is local in time but is not local in space. Does this difference constitute a conceptual complication? We show below that it does not: what we know about noninteracting Green's functions is still enough. We define the *Hartree–Fock potential* as

$$V_{\mathrm{HF}}(\mathbf{x}_1,\mathbf{x}_2,z) = \left[\delta(\mathbf{x}_1 - \mathbf{x}_2)V_{\mathrm{H}}(\mathbf{x}_1,z) + \frac{\mathrm{i}}{q}v(\mathbf{x}_1,\mathbf{x}_2)G(\mathbf{x}_1,z;\mathbf{x}_2,z^+)\right]$$

$$= \frac{1}{q}\,\delta(\mathbf{x}_1 - \mathbf{x}_2)\int d\mathbf{x}\,v(\mathbf{x}_1,\mathbf{x})n(\mathbf{x},z) \pm \frac{1}{q}\,v(\mathbf{x}_1,\mathbf{x}_2)n(\mathbf{x}_1,\mathbf{x}_2,z),\quad (7.39)$$

where

$$n(\mathbf{x}_1,\mathbf{x}_2,z) = \pm\mathrm{i}\,G(\mathbf{x}_1,z;\mathbf{x}_2,z^+)$$

is the time-dependent *one-particle density matrix*. We also introduce the Hartree–Fock potential operator in first quantization in the usual manner

$$\hat{\mathcal{V}}_{\mathrm{HF}}(z) = \int d\mathbf{x}_1 d\mathbf{x}_2\,|\mathbf{x}_1\rangle V_{\mathrm{HF}}(\mathbf{x}_1,\mathbf{x}_2,z)\langle\mathbf{x}_2|.$$
(7.40)

We then see that (7.38) is the sandwich with $\langle \mathbf{x}_1 |$ and $| \mathbf{x}_2 \rangle$ of the following equation for operators in first quantization

$$\left[i \frac{d}{dz_1} - \hat{h}(z_1) - q\hat{\mathcal{V}}_{\mathrm{HF}}(z_1) \right] \hat{\mathcal{G}}(z_1, z_2) = \delta(z_1, z_2), \qquad (7.41)$$

which has exactly the same structure as the equation of motion of a noninteracting Green's function! This is a completely general result: given an equation of the form

$$i \frac{d}{dz_1} G(1; 2) - \int d3 \left[h(1; 3) + \Sigma(1; 3) \right] G(3; 2) = \delta(1; 2), \qquad (7.42)$$

with $\Sigma(1; 2) = \delta(z_1, z_2) E(\mathbf{x}_1, \mathbf{x}_2, z_1)$, we can define the operator

$$\hat{\mathcal{E}}(z) = \int d\mathbf{x}_1 d\mathbf{x}_2 \, |\mathbf{x}_1\rangle E(\mathbf{x}_1, \mathbf{x}_2, z) \langle \mathbf{x}_2|,$$

and recognize that (7.42) is the sandwich with $\langle \mathbf{x}_1 |$ and $| \mathbf{x}_2 \rangle$ of

$$\left[i \frac{d}{dz_1} - \hat{h}(z_1) - \hat{\mathcal{E}}(z_1) \right] \hat{\mathcal{G}}(z_1, z_2) = \delta(z_1, z_2). \qquad (7.43)$$

More generally, any approximation to G_2 leading to an equation like (7.43) is a mean-field approximation, which discards the direct interaction between the particles. The corresponding mean-field G has the same structure as that of a noninteracting G. In the next chapters we develop a perturbative approach to approximate G_2, and we show that any approximation beyond Hartree–Fock leads to an equation of motion of the form (7.42) but with a Σ that is nonlocal in time.

The Green's function in (7.41) must be solved self-consistently and, as in the Hartree approximation, this can be done by solving a set of coupled equations for single-particle wavefunctions. These equations are known as the *Hartree–Fock equations* and are derived below.

7.3.1 Hartree–Fock equations

The Hartree–Fock G can be calculated as in Section 6.2 by replacing $\hat{h}(z)$ with

$$\hat{h}_{\mathrm{HF}}(z) = \hat{h}(z) + q\hat{\mathcal{V}}_{\mathrm{HF}}(z). \qquad (7.44)$$

Along the imaginary track the Hartree–Fock potential operator is a constant operator that is denoted by $\hat{\mathcal{V}}_{\mathrm{HF}}^{\mathrm{M}}$. We again specialize the discussion to particles of mass m and charge q initially in equilibrium in some external potential $V(\mathbf{x})$.[7] The single-particle Hamiltonian which describes this situation is therefore $\hat{h}^{\mathrm{M}} = \hat{p}^2/(2m) + qV(\hat{\mathbf{r}}, \hat{S}_z) - \mu$ and the interaction is $v^{\mathrm{M}} = v$. The first step consists in finding the kets $|\lambda\rangle$ which solve the eigenvalue problem

$$\left[\hat{h}^{\mathrm{M}} + q\hat{\mathcal{V}}_{\mathrm{HF}}^{\mathrm{M}} \right] |\lambda\rangle = (\epsilon_\lambda - \mu)|\lambda\rangle.$$

---

[7]The inclusion of vector potentials and spin–flip interactions is straightforward.

By sandwiching with the bra $\langle \mathbf{x} |$ and using the explicit form of $\hat{h}^{\mathrm{M}}$, the eigenvalue equation reads

$$\left[-\frac{\nabla^2}{2m} + qV(\mathbf{x}) \right] \varphi_\lambda(\mathbf{x}) + \int d\mathbf{x}' \, q V_{\mathrm{HF}}^{\mathrm{M}}(\mathbf{x}, \mathbf{x}') \varphi_\lambda(\mathbf{x}') = \epsilon_\lambda \varphi_\lambda(\mathbf{x}) \qquad (7.45)$$

with

$$V_{\mathrm{HF}}^{\mathrm{M}}(\mathbf{x}, \mathbf{x}') = \delta(\mathbf{x} - \mathbf{x}') V_{\mathrm{H}}^{\mathrm{M}}(\mathbf{x}) \pm \frac{1}{q} v(\mathbf{x}, \mathbf{x}') \sum_\nu f(\epsilon_\nu - \mu) \varphi_\nu(\mathbf{x}) \varphi_\nu^*(\mathbf{x}') \qquad (7.46)$$

and the Hartree potential $V_{\mathrm{H}}^{\mathrm{M}}$ given in (7.8). In (7.46) we use the result (6.40) according to which

$$G(\mathbf{x}, t_0 - \mathrm{i}\tau; \mathbf{x}', t_0 - \mathrm{i}\tau^+) = G^{\mathrm{M}}(\mathbf{x}, \tau; \mathbf{x}', \tau^+) = \mp\mathrm{i} \sum_\nu f(\epsilon_\nu - \mu) \langle \mathbf{x} | \varphi_\nu \rangle \langle \varphi_\nu | \mathbf{x}' \rangle.$$

We thus obtain a coupled system of nonlinear equations for the eigenfunctions φ_λ. These equations are known as the *Hartree–Fock equations* [73, 74]. As usual the upper/lower sign in (7.46) refers to bosons/fermions.

We have already mentioned that the Hartree–Fock approximation cures the self-interaction problem in fermionic systems. Using (7.46) the second term on the l.h.s. of (7.45) becomes

$$\int d\mathbf{x}' \, q V_{\mathrm{HF}}^{\mathrm{M}}(\mathbf{x}, \mathbf{x}') \varphi_\lambda(\mathbf{x}') = \sum_\nu f(\epsilon_\nu - \mu) \int d\mathbf{x}' \, v(\mathbf{x}, \mathbf{x}')$$

$$\times \left[|\varphi_\nu(\mathbf{x}')|^2 \varphi_\lambda(\mathbf{x}) \pm \varphi_\nu(\mathbf{x}) \varphi_\nu^*(\mathbf{x}') \varphi_\lambda(\mathbf{x}') \right]. \quad (7.47)$$

The term $\nu = \lambda$ in the above sum vanishes for fermions while it yields twice the Hartree contribution for bosons. Thus, for a system of bosons at zero temperature the Hartree–Fock equations are identical to the Hartree equations (7.10) with $N \to 2N$. It is also noteworthy that if the interaction is spin-independent, i.e., $v(\mathbf{x}_1, \mathbf{x}_2) = v(\mathbf{r}_1 - \mathbf{r}_2)$, the second term in the square bracket vanishes unless ν and λ have the same spin projection. In other words there are no exchange contributions coming from particles of different spin projection.

Once the equilibrium problem is solved we can construct the Matsubara Green's function. For all other Keldysh components we must propagate the $|\lambda\rangle$s in time according to the Schrödinger equation $\mathrm{i}\frac{d}{dt}|\lambda(t)\rangle = [\hat{h}(t) + q\hat{V}_{\mathrm{HF}}(t)]|\lambda(t)\rangle$ with initial conditions $|\lambda(t_0)\rangle = |\lambda\rangle$. By sandwiching again with $\langle \mathbf{x} |$ we find the so called *time-dependent Hartree–Fock equations*. They are simply obtained from the static equations (7.45) by replacing $[-\nabla^2/(2m) + qV(\mathbf{x})]$ with the time-dependent single-particle Hamiltonian $h(\mathbf{r}, -\mathrm{i}\nabla, \mathbf{S}, t)$, the static wavefunctions $\varphi_\lambda(\mathbf{x})$ with the time-evolved wavefunctions $\varphi_\lambda(\mathbf{x}, t)$, the static Hartree–Fock potential $V_{\mathrm{HF}}^{\mathrm{M}}(\mathbf{x}, \mathbf{x}')$ with the time-dependent Hartree–Fock potential $V_{\mathrm{HF}}(\mathbf{x}, \mathbf{x}', t)$, and ϵ_λ with $\mathrm{i}\frac{d}{dt}$.

Total energy in the Hartree-Fock approximation

In thermodynamic equilibrium ($\hat{h}^{\mathrm{M}} = \hat{h} - \mu$ and $v^{\mathrm{M}} = v$) the single-particle Hamiltonian $\hat{h}(t) = \hat{h}$ is independent of time. As a consequence the Hartree–Fock Hamiltonian $\hat{h}_{\mathrm{HF}} =$

$\hat{h} + q\hat{\mathcal{V}}_{\mathrm{HF}} = \hat{h} + q\hat{\mathcal{V}}_{\mathrm{HF}}^{\mathrm{M}}$ is also independent of time and the lesser Green's function reads

$$\hat{\mathcal{G}}^{<}(\omega) = \mp 2\pi\mathrm{i}\, f(\underbrace{\hat{h} + q\hat{\mathcal{V}}_{\mathrm{HF}}^{\mathrm{M}} - \mu}_{\hat{h}_{\mathrm{HF}}^{\mathrm{M}}})\, \delta(\omega - [\underbrace{\hat{h} + q\hat{\mathcal{V}}_{\mathrm{HF}}^{\mathrm{M}}}_{\hat{h}_{\mathrm{HF}}}]).$$

Substituting this $\hat{\mathcal{G}}^{<}$ in (6.106) and inserting a complete set of eigenstates of $\hat{h} + q\hat{\mathcal{V}}_{\mathrm{HF}}^{\mathrm{M}}$ we get

$$E = \sum_{\lambda} f_{\lambda} \left[\epsilon_{\lambda} - \frac{1}{2} \int d\mathbf{x} \langle \mathbf{x} | q\hat{\mathcal{V}}_{\mathrm{HF}}^{\mathrm{M}} | \lambda \rangle \langle \lambda | \mathbf{x} \rangle \right]$$

$$= \sum_{\lambda} f_{\lambda} \left[\epsilon_{\lambda} - \frac{1}{2} \sum_{\nu} f_{\nu} (v_{\lambda\nu\nu\lambda} \pm v_{\lambda\nu\lambda\nu}) \right], \qquad f_{\lambda} \equiv f(\epsilon_{\lambda} - \mu), \qquad (7.48)$$

where in the last equality we use (7.47). As expected, the total energy is not the weighted sum of the single-particle Hartree–Fock energies, see discussion below (7.11). It is instructive to express the ϵ_{λ}s in terms of $h_{\lambda\lambda}$ and Coulomb integrals. If we multiply (7.45) by $\varphi_{\lambda}^{*}(\mathbf{x})$ and integrate over $\mathbf{x}$ we obtain

$$\epsilon_{\lambda} = h_{\lambda\lambda} + \sum_{\nu} f_{\nu} (v_{\lambda\nu\nu\lambda} \pm v_{\lambda\nu\lambda\nu}), \qquad (7.49)$$

from which it is evident that the self-interaction energy, i.e., the contribution $\lambda = \nu$ in the sum, vanishes in the case of fermions.

Koopmans' theorem

Since the total energy $E \neq \sum_{\lambda} f_{\lambda}\epsilon_{\lambda}$ we cannot interpret the ϵ_{λ} as the energy of a particle. Can we still give a physical interpretation to the ϵ_{λ}? To answer this question we insert (7.49) into (7.48) and find

$$E = \sum_{\lambda} f_{\lambda} h_{\lambda\lambda} + \frac{1}{2} \sum_{\lambda\nu} f_{\lambda} f_{\nu} (v_{\lambda\nu\nu\lambda} \pm v_{\lambda\nu\lambda\nu}). \qquad (7.50)$$

We now consider an ultrafast ionization process in which the particle in the ρth level is suddenly taken infinitely far away from the system. If we measure the energy of the ionized system before it has time to relax to some lower energy state we find

$$E_{\rho} = \sum_{\lambda \neq \rho} f_{\lambda} h_{\lambda\lambda} + \frac{1}{2} \sum_{\lambda\nu \neq \rho} f_{\lambda} f_{\nu} (v_{\lambda\nu\nu\lambda} \pm v_{\lambda\nu\lambda\nu}).$$

The difference between the initial energy E and the energy E_{ρ} is

$$E - E_{\rho} = f_{\rho} h_{\rho\rho} + \frac{1}{2} f_{\rho} f_{\rho} (v_{\rho\rho\rho\rho} \pm v_{\rho\rho\rho\rho})$$

$$+ \frac{1}{2} \sum_{\nu \neq \rho} f_{\rho} f_{\nu} (v_{\rho\nu\nu\rho} \pm v_{\rho\nu\rho\nu}) + \frac{1}{2} \sum_{\lambda \neq \rho} f_{\lambda} f_{\rho} (v_{\lambda\rho\rho\lambda} \pm v_{\lambda\rho\lambda\rho}).$$

Using the symmetry $v_{ijkl} = v_{jilk}$ of the Coulomb integrals we can rewrite $E - E_\rho$ in the following compact form:

$$E - E_\rho = f_\rho \left[\epsilon_\rho - \frac{1}{2} f_\rho (v_{\rho\rho\rho\rho} \pm v_{\rho\rho\rho\rho}) \right].$$

Thus, for a system of fermions the eigenvalue ϵ_ρ multiplied by the occupation f_ρ is the difference between the energy of the initial system and the energy of the ionized and unrelaxed system. At zero temperature we expect that the removal of a particle from the highest occupied level (HOMO) does not cause a dramatic relaxation, and hence $\epsilon_{\rho_{\text{HOMO}}}$ should provide a good estimate of the ionization energy. This result is known as *Koopmans' theorem*. For a system of bosons a similar interpretation is not possible.

7.3.2   Coulombic electron gas and spin-polarized solutions

Let us consider again an electron gas in equilibrium with single-particle Hamiltonian $\hat{h}^{\text{M}} = \hat{p}^2/2 - V_0 - \mu$ where V_0 is a constant energy shift (for electrons $q = -1$). Due to translational invariance the eigenkets of $\hat{h}^{\text{M}} - \hat{\mathcal{V}}^{\text{M}}_{\text{HF}}$ are the momentum-spin kets $|\mathbf{p}\sigma\rangle$:

$$\left[\hat{h}^{\text{M}} - \hat{\mathcal{V}}^{\text{M}}_{\text{HF}} \right] |\mathbf{p}\sigma\rangle = (\epsilon_{\mathbf{p}} - \mu)|\mathbf{p}\sigma\rangle. \tag{7.51}$$

The Matsubara Green's function (needed to evaluate $\hat{\mathcal{V}}^{\text{M}}_{\text{HF}}$) is given in (6.40) and reads

$$G^{\text{M}}(\mathbf{x}_1, \tau; \mathbf{x}_2, \tau^+) = i \sum_\sigma \int \frac{d\mathbf{k}}{(2\pi)^3} f_{\mathbf{k}} \langle \mathbf{x}_1 | \mathbf{k}\sigma\rangle \langle \mathbf{k}\sigma | \mathbf{x}_2\rangle, \tag{7.52}$$

where we use the short-hand notation $f_{\mathbf{k}} \equiv f(\epsilon_{\mathbf{k}} - \mu)$ and insert the completeness relation (1.11). Fourier transforming the interparticle interaction

$$v(\mathbf{r}_1 - \mathbf{r}_2) = \int \frac{d\mathbf{q}}{(2\pi)^3} e^{i\mathbf{q}\cdot(\mathbf{r}_1 - \mathbf{r}_2)} \tilde{v}_{\mathbf{q}} = \int \frac{d\mathbf{q}}{(2\pi)^3} \tilde{v}_{\mathbf{q}} \langle \mathbf{r}_1 | \mathbf{q}\rangle \langle \mathbf{q} | \mathbf{r}_2\rangle,$$

and using the identity $\langle \mathbf{r}_1 | \mathbf{k}\rangle \langle \mathbf{r}_1 | \mathbf{q}\rangle = \langle \mathbf{r}_1 | \mathbf{k} + \mathbf{q}\rangle$ we can rewrite the Hartree–Fock operator (7.40) in a diagonal form

$$\hat{\mathcal{V}}^{\text{M}}_{\text{HF}} = \int d\mathbf{x}_1 d\mathbf{x}_2 \, |\mathbf{x}_1\rangle\langle\mathbf{x}_1| \left[-n\tilde{v}_0 + \sum_\sigma \int \frac{d\mathbf{k}d\mathbf{q}}{(2\pi)^6} |\mathbf{q}\sigma\rangle f_{\mathbf{k}} \tilde{v}_{\mathbf{q}-\mathbf{k}} \langle\mathbf{q}\sigma| \right] |\mathbf{x}_2\rangle\langle\mathbf{x}_2|$$

$$= \sum_\sigma \int \frac{d\mathbf{q}}{(2\pi)^3} |\mathbf{q}\sigma\rangle \left[-n\tilde{v}_0 + \int \frac{d\mathbf{k}}{(2\pi)^3} f_{\mathbf{k}} \tilde{v}_{\mathbf{q}-\mathbf{k}} \right] \langle\mathbf{q}\sigma|. \tag{7.53}$$

Substituting this result into the eigenvalue equation (7.51) we obtain the following expression for $\epsilon_{\mathbf{p}}$:

$$\epsilon_{\mathbf{p}} = \frac{p^2}{2} - V_0 + n\tilde{v}_0 - \int \frac{d\mathbf{k}}{(2\pi)^3} f_{\mathbf{k}} \tilde{v}_{\mathbf{p}-\mathbf{k}}.$$

The total energy is the integral over all momenta of $\epsilon_{\mathbf{p}}$ minus one half the interaction energy weighted with the Fermi function [see (7.48)]

$$E = 2V \int \frac{d\mathbf{p}}{(2\pi)^3} f_{\mathbf{p}} \left[\frac{p^2}{2} - V_0 + \frac{1}{2} n\tilde{v}_0 - \frac{1}{2} \int \frac{d\mathbf{k}}{(2\pi)^3} f_{\mathbf{k}} \tilde{v}_{\mathbf{p}-\mathbf{k}} \right], \quad (7.54)$$

where the factor of 2 comes from spin, and where V is the volume of the system.

Let us specialize these formulas to the Coulombic interaction $v(\mathbf{r}_1 - \mathbf{r}_2) = 1/|\mathbf{r}_1 - \mathbf{r}_2|$. The Fourier transform is in this case $\tilde{v}_{\mathbf{p}} = 4\pi/p^2$ and hence $\tilde{v}_0$ diverges! Is this physical? The answer is yes. In the absence of any positive charge that attracts and binds the electrons together, the energy that we must spend to put an extra electron in a box containing a macroscopic number of electrons is infinite. We therefore consider the physical situation in which the space is permeated by a uniform density $n_b = n$ of positive charge, in such a way that the whole system is charge neutral. Then the potential V_0 felt by an electron in $\mathbf{r}$ is $V_0 = n_b \int d\mathbf{r}_1 \, v(\mathbf{r} - \mathbf{r}_1) = n\tilde{v}_0$ which is also divergent, and the single-particle eigenvalues turn out to be finite

$$\epsilon_{\mathbf{p}} = \frac{p^2}{2} - 4\pi \int \frac{d\mathbf{k}}{(2\pi)^3} \frac{f_{\mathbf{k}}}{|\mathbf{p} - \mathbf{k}|^2}. \quad (7.55)$$

The energy E is, however, still divergent. This is due to the fact that E contains only the electronic energy. Adding to E the energy of the positive background $E_b = \frac{1}{2} \int d\mathbf{r}_1 d\mathbf{r}_2 n_b^2 / |\mathbf{r}_1 - \mathbf{r}_2| = \frac{1}{2} Vn^2\tilde{v}_0$, the total energy of the system electrons+background is finite and reads[8]

$$E_{\text{tot}} = E + E_b = V \int \frac{d\mathbf{p}}{(2\pi)^3} f_{\mathbf{p}} \left[2\frac{p^2}{2} - 4\pi \int \frac{d\mathbf{k}}{(2\pi)^3} \frac{f_{\mathbf{k}}}{|\mathbf{p} - \mathbf{k}|^2} \right]. \quad (7.56)$$

Let us evaluate (7.55) and (7.56) at zero temperature.

At zero temperature the chemical potential μ coincides with the Fermi energy ϵ_{F} and we can write $f_{\mathbf{p}} = \theta(\epsilon_{\text{F}} - \epsilon_{\mathbf{p}})$. The eigenstates with energy below ϵ_{F} are filled while the others are empty. The density (per spin) is therefore

$$\frac{n}{2} = \int \frac{d\mathbf{p}}{(2\pi)^3} \theta(\epsilon_{\text{F}} - \epsilon_{\mathbf{p}}) = p_{\text{F}}^3/(6\pi^2), \quad (7.57)$$

where p_{F} is the Fermi momentum, i.e., the value of the modulus of $\mathbf{p}$ for which $\epsilon_{\mathbf{p}} = \epsilon_{\text{F}}$. Note that the Fermi momentum is the same as that of a noninteracting gas with the same density. As we see in Section 11.6 this result is a general consequence of the conserving nature of the Hartree–Fock approximation. We also observe that the eigenvalue $\epsilon_{\mathbf{p}}$ in (7.55) depends only on $p = |\mathbf{p}|$. This is true for all interactions $v(\mathbf{r}_1 - \mathbf{r}_2)$ that are invariant under rotations. The calculation of the integral in (7.55) can be performed by expanding $|\mathbf{p} - \mathbf{k}|^2 = p^2 + k^2 - 2kp\cos\theta$ (with θ the angle between $\mathbf{p}$ and $\mathbf{k}$) and changing the measure from cartesian to polar coordinates $d\mathbf{k} = d\varphi d(\cos\theta) dk\, k^2$. The integral of the angular part yields

$$4\pi \int \frac{d\mathbf{k}}{(2\pi)^3} \frac{f_{\mathbf{k}}}{|\mathbf{p} - \mathbf{k}|^2} = \frac{1}{\pi p} \int_0^{p_{\text{F}}} dk\, k \ln \left| \frac{p+k}{p-k} \right| = \frac{p_{\text{F}}}{\pi x} \int_0^1 dy\, y \ln \left| \frac{x+y}{x-y} \right|,$$

---

[8]Take into account that the density of the electron gas is given by $n = 2 \int \frac{d\mathbf{p}}{(2\pi)^3} f_{\mathbf{p}}$.

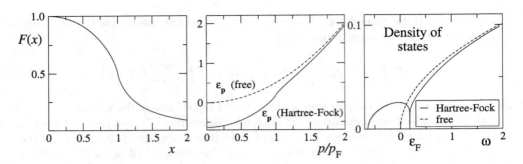

Figure 7.6 Left panel. The function $F(x)$ whose derivative diverges in $x = 1$. Middle panel. Comparison between the free-electron eigenvalues and the Hartree–Fock eigenvalues in units of p_F^2. Right panel. Density of states $D(\omega)/2\pi$ in units of p_F (energy is in units of p_F^2). In the middle and right panel $p_F = 1$ which corresponds to $r_s \sim 1.92$.

where in the last equality we introduce the dimensionless variables $x = p/p_F$ and $y = k/p_F$. The remaining integral is a standard logarithmic integral

$$I(a,b) \equiv \int_0^1 dy\, y\, \ln\left|\frac{b+ay}{b-ay}\right| = \frac{b}{a} + \frac{1}{2}\left(1 - \frac{b^2}{a^2}\right)\ln\left|\frac{b+a}{b-a}\right|, \tag{7.58}$$

and we thus arrive at the result

$$4\pi \int \frac{d\mathbf{k}}{(2\pi)^3} \frac{f_\mathbf{k}}{|\mathbf{p}-\mathbf{k}|^2} = \frac{2p_F}{\pi} F\left(\frac{p}{p_F}\right), \tag{7.59}$$

where

$$F(x) = \frac{1}{2} + \frac{1-x^2}{4x}\ln\left|\frac{1+x}{1-x}\right|. \tag{7.60}$$

The function $F(x)$ as well as a comparison between the free-electron eigenvalue $p^2/2$ and the Hartree–Fock eigenvalue $\epsilon_\mathbf{p}$ is shown in Fig. 7.6 for $p_F = 1$. The exchange contribution lowers the free-electron eigenvalues:

$$\underbrace{\epsilon_\mathbf{p} = \frac{p^2}{2}}_{\text{free}} \quad\to\quad \underbrace{\epsilon_\mathbf{p} = \frac{p^2}{2} - \frac{2p_F}{\pi} F\left(\frac{p}{p_F}\right)}_{\text{Hartree–Fock}}. \tag{7.61}$$

This is a very important result since it provides a (partial) explanation of the stability of matter. Without the exchange correction the electrons would behave as free particles and hence they could easily escape from a solid. We can calculate the binding energy due to the exchange term by substituting (7.59) into (7.56) with $f_\mathbf{p} = \theta(\epsilon_F - \epsilon_\mathbf{p})$. Changing the measure from cartesian to polar coordinates and exploiting the analytic result,

$$\int dx\, x(1-x^2)\ln\left|\frac{1+x}{1-x}\right| = \frac{1}{2}x - \frac{1}{6}x^3 - \frac{1}{4}(1-x^2)^2\ln\left|\frac{1+x}{1-x}\right|,$$

we find that the total energy at zero temperature is

$$E_{\text{tot}} = nV \left[\frac{3}{5} \frac{p_{\text{F}}^2}{2} - \frac{3}{4} \frac{p_{\text{F}}}{\pi} \right] \tag{7.62}$$

It is common to express this result in terms of the "average distance" between the electrons. This simply means that if there is one electron in a sphere of radius R then the density of the gas is $n^{-1} = \frac{4}{3}\pi R^3$. Writing the radius $R = r_s a_{\text{B}}$ in units of the Bohr radius $a_{\text{B}} = 1$ a.u. and taking into account that $p_{\text{F}} = (3\pi^2 n)^{1/3}$ [see (7.57)] we find $p_{\text{F}} = (\frac{9\pi}{4})^{\frac{1}{3}} \frac{1}{r_s}$ and hence

$$\frac{E_{\text{tot}}}{nV} = \frac{1}{2} \left[\frac{2.21}{r_s^2} - \frac{0.916}{r_s} \right] \quad \text{(in atomic units)} \tag{7.63}$$

Most metals have r_s in the range $2-6$ and therefore the exchange energy gives an important contribution to the binding energy.[9]

So far so good, but the Hartree–Fock approximation also suffers serious problems. For instance, the density of states $D(\omega)$ defined in (6.34) is

$$D(\omega) = 2\pi \int \frac{d\mathbf{p}}{(2\pi)^3} \delta(\omega - \epsilon_{\mathbf{p}}) = \frac{1}{\pi} \int dp\, p^2 \frac{\delta(p - p(\omega))}{|\partial \epsilon_{\mathbf{p}}/\partial p|_{p(\omega)}} = \frac{p^2(\omega)}{\pi} \left| \frac{d\epsilon_{\mathbf{p}}}{dp} \right|_{p(\omega)}^{-1},$$

with $p(\omega)$ the value of the momentum p for which $\epsilon_{\mathbf{p}} = \omega$. In the noninteracting gas $p(\omega) = \sqrt{2\omega}$ and the density of states is proportional to $\sqrt{\omega}$. In the Hartree–Fock approximation $D(\omega)$ is zero at the Fermi energy, see Fig. 7.6, due to the divergence of the derivative of $F(x)$ in $x = 1$. This is in neat contrast to experimentally observed density of states of metals which are smooth and finite across ϵ_{F}. Another problem of the Hartree–Fock approximation is the overestimation of the band-widths of metals and of the band-gaps of semiconductors and insulators. These deficiencies are due to the lack of screening effects. As the interparticle interaction is repulsive an electron pushes away electrons around it and remains surrounded by a positive charge (coming from the positive background charge). The consequence of the screening is that the effective repulsion between two electrons is reduced (and, as we shall see, retarded).

A further problem of the Hartree–Fock approximation that we wish to discuss is related to the stability of the solution. Consider preparing the system in a different initial configuration by changing the single-particle Hamiltonian $\hat{h}^{\text{M}}$ under the constraint that $\hat{h}_{\text{HF}}^{\text{M}}$ and $\hat{h}_{\text{HF}}(t_0)$ commute.[10] Then the eigenkets $|\lambda\rangle$ of $\hat{h}_{\text{HF}}^{\text{M}}$ are also eigenkets of $\hat{h}_{\text{HF}}(t_0)$ with eigenvalues ϵ_λ. The time-evolved kets $|\lambda(t)\rangle = e^{-i\epsilon_\lambda(t-t_0)}|\lambda\rangle$ are the self-consistent solution of the time-dependent Hartree–Fock equations since in this case $\hat{h}_{\text{HF}}(t) = \hat{h}_{\text{HF}}(t_0)$ is time-independent and, at the same time, $i\frac{d}{dt}|\lambda(t)\rangle = \hat{h}_{\text{HF}}(t)|\lambda(t)\rangle$. The corresponding real-time Green's function depends only on the time difference and hence it describes an excited stationary configuration. According to (6.48) the lesser Green's function is obtained by populating the kets $|\lambda\rangle$ with occupations $f(\epsilon_\lambda^{\text{M}})$ and, in principle, may give a total energy which is lower

---

[9] The dimensionless radius r_s is also called the *Wigner–Seitz radius*.

[10] Recall that $\hat{h}^{\text{M}}$ specifies the initial preparation and does not have to represent the physical Hamiltonian $\hat{h}$ of the system. In thermodynamic equilibrium $\hat{h}^{\text{M}} = \hat{h} - \mu$. Any other choice of $\hat{h}^{\text{M}}$ describes an excited configuration.

than (7.62). If so the Hartree–Fock approximation is unstable and this instability indicates either that the approximation is not a good approximation or that the system would like to rearrange the electrons in a different way. Changing the Hamiltonian $\hat{h}^{\mathrm{M}}$ and checking the stability of an approximation is a very common procedure to assess the quality of the approximation and/or to understand if the ground state of the system has the same symmetry as the noninteracting ground state. Let us illustrate how it works with an example. As a trial single-particle Hamiltonian we consider

$$\hat{h}^{\mathrm{M}} = \hat{h} + \sum_{\sigma} \int \frac{d\mathbf{p}}{(2\pi)^3} \, \epsilon_{\sigma} |\mathbf{p}\sigma\rangle\langle\mathbf{p}\sigma| - \mu, \qquad \hat{h} = \hat{p}^2/2 - V_0. \tag{7.64}$$

As we shall see shortly, this choice of $\hat{h}^{\mathrm{M}}$ corresponds to an initial configuration with different number of spin up and down electrons. We are therefore exploring the possibility that a *spin-polarized* electron gas has lower energy than the spin-unpolarized one.

Having broken the spin symmetry, the single particle eigenvalues of the Hartree–Fock Hamiltonian $\hat{h}^{\mathrm{M}}_{\mathrm{HF}} = \hat{h}^{\mathrm{M}} - \hat{\mathcal{V}}^{\mathrm{M}}_{\mathrm{HF}}$ are spin dependent. The extra term in (7.64), however, does not break the translational invariance and therefore the momentum–spin kets $|\mathbf{p}\sigma\rangle$ are still good eigenkets. The Matsubara Green's function needed to calculate $\hat{\mathcal{V}}^{\mathrm{M}}_{\mathrm{HF}}$ reads

$$G^{\mathrm{M}}(\mathbf{x}_1, \tau; \mathbf{x}_2, \tau^+) = i \sum_{\sigma} \int \frac{d\mathbf{k}}{(2\pi)^3} \, f_{\mathbf{k}\sigma} \langle\mathbf{x}_1|\mathbf{k}\sigma\rangle\langle\mathbf{k}\sigma|\mathbf{x}_2\rangle,$$

where $f_{\mathbf{k}\sigma} = \theta(\epsilon_{\mathrm{F}} - \epsilon_{\mathbf{k}\sigma} - \epsilon_{\sigma})$ and $\epsilon_{\mathbf{k}\sigma} + \epsilon_{\sigma} - \epsilon_{\mathrm{F}}$ are the eigenvalues of $\hat{h}^{\mathrm{M}}_{\mathrm{HF}}$. Inserting this result into the Hartree–Fock potential (7.39) we find the following Hartree–Fock potential operator [compare with (7.53)]

$$\hat{\mathcal{V}}^{\mathrm{M}}_{\mathrm{HF}} = \sum_{\sigma} \int \frac{d\mathbf{q}}{(2\pi)^3} \, |\mathbf{q}\sigma\rangle \left[-(n_{\uparrow} + n_{\downarrow})\tilde{v}_0 + \int \frac{d\mathbf{k}}{(2\pi)^3} \, f_{\mathbf{k}\sigma}\tilde{v}_{\mathbf{q}-\mathbf{k}} \right] \langle\mathbf{q}\sigma|,$$

with $n_{\sigma} = \int \frac{d\mathbf{p}}{(2\pi)^3} f_{\mathbf{p}\sigma}$ the uniform density for electrons of spin σ. Since we are at zero temperature, $n_{\sigma} = p_{\mathrm{F}\sigma}^3/(6\pi^2)$ where the spin-dependent Fermi momentum is defined as the solution of $\epsilon_{\mathbf{p}\sigma} = \epsilon_{\mathrm{F}} - \epsilon_{\sigma}$; thus we can treat either ϵ_{σ} or n_{σ} or $p_{\mathrm{F}\sigma}$ as the independent variable.

The Hartree–Fock Hamiltonian at positive times is time-independent and given by

$$\hat{h}_{\mathrm{HF}} = \frac{\hat{p}^2}{2} - V_0 - \hat{\mathcal{V}}^{\mathrm{M}}_{\mathrm{HF}},$$

with eigenvalues $\epsilon_{\mathbf{p}\sigma}$. We consider again Coulombic interactions and neutralize the system with a positive background charge $n_b = n_{\uparrow} + n_{\downarrow}$. Then the eigenvalues of $\hat{h}_{\mathrm{HF}}$ are given by

$$\epsilon_{\mathbf{p}\sigma} = \frac{p^2}{2} - \frac{2p_{\mathrm{F}\sigma}}{\pi} F\left(\frac{p}{p_{\mathrm{F}\sigma}}\right),$$

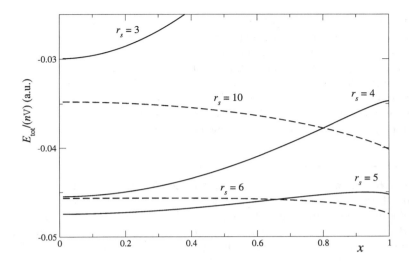

Figure 7.7 Energy density of the electron gas for the spin polarized Hartree–Fock approximation.

and we find

$$\hat{\mathcal{G}}^<(\omega) = 2\pi i f(\hat{h}_{\mathrm{HF}}^{\mathrm{M}})\delta(\omega - \hat{h}_{\mathrm{HF}})$$

$$= 2\pi i \sum_\sigma \int \frac{d\mathbf{p}}{(2\pi)^3} |\mathbf{p}\sigma\rangle f_{\mathbf{p}\sigma} \underbrace{\delta\left(\omega - \frac{p^2}{2} + \frac{2p_{\mathrm{F}\sigma}}{\pi}F(\frac{p}{p_{\mathrm{F}\sigma}})\right)}_{\delta(\omega - \epsilon_{\mathbf{p}\sigma})} \langle\mathbf{p}\sigma|.$$

The total energy of the spin-polarized electron gas can readily be obtained using (6.106) and reads

$$E_{\mathrm{tot}} = \mathbf{V}\sum_\sigma \int \frac{d\mathbf{p}}{(2\pi)^3} f_{\mathbf{p}\sigma}\left[\frac{p^2}{2} - \frac{p_{\mathrm{F}\sigma}}{\pi}F(\frac{p}{p_{\mathrm{F}\sigma}})\right] = \mathbf{V}\sum_\sigma n_\sigma \left[\frac{3}{5}\frac{p_{\mathrm{F}\sigma}^2}{2} - \frac{3}{4}\frac{p_{\mathrm{F}\sigma}}{\pi}\right].$$

Let us study what this formula predicts at fixed $n_\uparrow + n_\downarrow \equiv n$. If we define the spin polarization $x = (n_\uparrow - n_\downarrow)/n$ we can write $n_\uparrow = \frac{n}{2}(1+x)$ and $n_\downarrow = \frac{n}{2}(1-x)$, and hence we can express E_{tot} in terms of n and x only (recall that $p_{\mathrm{F}\sigma} = (6\pi^2 n_\sigma)^{1/3}$). In Fig. 7.7 we plot $E_{\mathrm{tot}}/(n\mathbf{V})$ as a function of the spin polarization x for values of $r_s = \left(\frac{3}{4\pi n}\right)^{1/3}$. For small r_s the state of minimum energy is spin-unpolarized since the energy has a minimum in $x = 0$. However, by increasing r_s, or equivalently by decreasing the density, the system prefers to break the spin symmetry and to relax in a spin-polarized state. The tendency toward spin-polarized solutions at small densities can be easily understood. The exchange interaction between two electrons is negative if the electrons have the same spin and it is zero otherwise [see discussion below (7.47)]. Consider a spin-unpolarized electron gas and flip the spin of one electron. Due to the Pauli exclusion principle, particles with the same spin cannot occupy the same level and therefore after the spin-flip we must put the electron into a state with momentum larger than p_F. If the density is small this increase in kinetic energy is lower than the exchange energy gain. Then we can lower the energy further by flipping the spin of another electron and then of another one and so on and so forth until we start to expend too much kinetic energy. If the fraction of flipped spin is finite then the more stable solution is spin-polarized. The results in Fig. 7.7 predict that the transition from spin-unpolarized to spin-polarized solutions occurs at a critical value of r_s between 5 and 6 (the precise value is $r_s = 5.45$). This critical value is very unphysical since it is in the range of the r_ss of ordinary metals (ferromagnetic materials have r_s an order of magnitude larger).

Exercise 7.6. The atoms of a simple metal have Z electrons (or holes) in the conduction band. Since 1 gram of protons corresponds to $N_A = 0.6 \times 10^{24}$ protons (N_A is Avogadro's number) the density of conduction electrons per cubic centimeter is $n = N_A \rho (Z/A)$ where ρ is the density of the metal measured in g/cm^3 and A is the mass number (number of protons plus neutrons) of the element. Calculate $r_s = (\frac{3}{4\pi n})^{1/3}$ for Cu, Ag, Au, Al, Ga using the following values for Z: $Z_{Cu} = 1$, $Z_{Ag} = 1$, $Z_{Au} = 1$, $Z_{Al} = 3$, $Z_{Ga} = 3$. Take the values of ρ and A from the periodic table. In all cases r_s is between 2 and 3. Looking at the number of outer-shell electrons in these elements try to justify the choice of the numbers Z.

Exercise 7.7. Consider a system of fermions. Show that in the Hartree–Fock approximation

$$\Gamma_2(\mathbf{x}, \mathbf{x}'; \mathbf{x}, \mathbf{x}') = n(\mathbf{x})n(\mathbf{x}') - |n(\mathbf{x}, \mathbf{x}')|^2,$$

where Γ_2 is the two-particle density matrix defined in Section 1.7. In Appendix C we show that Γ_2 is the probability of finding a particle at $\mathbf{x}$ and another at $\mathbf{x}'$. Therefore Γ_2 approaches the independent product of probabilities $n(\mathbf{x})n(\mathbf{x}')$ for $|\mathbf{r} - \mathbf{r}'| \to \infty$. A measure of the correlation between the particles is provided by the *pair correlation function* $g(\mathbf{x}, \mathbf{x}') = \Gamma_2(\mathbf{x}, \mathbf{x}'; \mathbf{x}, \mathbf{x}')/n(\mathbf{x})n(\mathbf{x}')$. In the Hartree–Fock approximation

$$g(\mathbf{x}, \mathbf{x}') = 1 - \frac{|n(\mathbf{x}, \mathbf{x}')|^2}{n(\mathbf{x})n(\mathbf{x}')}.$$

In an electron gas the Hartree–Fock pair correlation function is equal to $(1 - \delta_{\sigma\sigma'})$ for $\mathbf{r} = \mathbf{r}'$ and approaches 1 for $|\mathbf{r} - \mathbf{r}'| \to \infty$, see Appendix C. Thus electrons far away are uncorrelated while electrons that are close to each other are correlated only through the Pauli principle. The reader can find more on the pair correlation function in the electron gas in Refs. [67, 75, 76].

8

Conserving approximations: two-particle Green's function

8.1 Introduction

In this and the next two chapters we lay down the basis to go beyond the Hartree–Fock approximation. We develop a powerful approximation method with two main distinctive features. The first feature is that the approximations are *conserving*. A major difficulty in the theory of nonequilibrium processes consists in generating approximate Green's functions yielding a particle density $n(\mathbf{x}, t)$ and a current density $\mathbf{J}(\mathbf{x}, t)$ which satisfy the continuity equation, or a total momentum $\mathbf{P}(t)$ and a Lorentz force $\mathbf{F}(t)$ which is the time-derivative of $\mathbf{P}(t)$, or a total energy $E(t)$ and a total power which is the time-derivative of $E(t)$, etc. All these fundamental relations express the conservation of some physical quantity. In equilibrium the current is zero and hence the density is conserved, i.e., it is the same at all times; the force is zero and hence the momentum is conserved; the power fed into the system is zero and hence the energy is conserved, etc. The second important feature is that every approximation has a clear physical interpretation. In Chapter 7 we showed that the two-particle Green's function in the Hartree and Hartree–Fock approximation can be represented diagrammatically as in Fig. 7.1 and Fig. 7.2 respectively. In these diagrams two particles propagate from $(1', 2')$ to $(1, 2)$ as if there was no interaction. Accordingly, the Green's function in both approximations has the structure of a noninteracting G. The diagrammatic representation turns out to be extremely useful to visualize the scattering processes contained in an approximation and then to convert these processes into mathematical expressions. Without being too rigorous we could already give a taste of how the diagrammatic formalism works, just to show how intuitive and simple it is. Suppose that there is a region of space where the particle density is smaller than everywhere else. Then we do not expect a mean-field approximation to work in this region since the concept of effective potential is not justified for low densities. We therefore must construct an approximation that takes into account the direct interaction between the particles. This means that the propagation from $(1', 2')$ to $(1, 2)$ must take into account possible intermediate scatterings in $(3, 4)$ as illustrated in Fig. 8.1. In the first diagram two particles propagate from $(1', 2')$ to $(3, 4)$ and then scatter. The scattering is caused by the interaction (wiggly line) $v(3; 4)$, and after scattering the propagation continues until the particles arrive in $(1, 2)$. The second

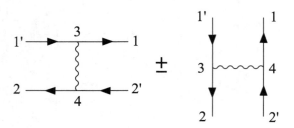

Figure 8.1 Direct scattering between two particles.

diagram contains the same physical information but the final state $(1, 2)$ is exchanged with $(2, 1)$. As in the Hartree–Fock case the second diagram guarantees that G_2 has the correct symmetry under the interchange of its arguments. If we now associate a Green's function with each straight line and an interparticle interaction with each wiggly line we may try to improve over the Hartree–Fock approximation by adding to $G_{2,\mathrm{HF}}$ a term proportional to the diagrams in Fig. 8.1, i.e.,

$$\int d3d4 \; v(3;4) \left[G(1;3)G(3;1')G(2;4)G(4;2') \pm G(1;4)G(4;2')G(2;3)G(3;1') \right],$$

whose dominant contribution comes from the integral over the low density region. This is, in essence, how new physical mechanisms can be incorporated in an approximation within the diagrammatic approach. There are, of course, plenty of other diagrams that we can dream up and that may be relevant to the problem at hand; Fig. 8.1 does not exhaust all possible scattering processes. For instance, the particles may interact one more time before reaching the final state, see Fig. 8.2(a). Another possibility is that the particle in $3'$ interacts "with itself" in $4'$ and during the propagation between $3'$ and $4'$ scatters with the particle from $2'$, see Fig. 8.2(b). This process seems a bit bizarre, but as we shall see it exists.

The diagrammatic approach is very appealing but, at this stage, not rigorous. There are several questions that we can ask: what are the exact rules to convert a diagram into a formula? How should we choose the diagrams so as to have an approximation which is conserving? And also: what is the prefactor in front of each diagram? Is there a systematic way to include all possible diagrams? Including all possible diagrams with an appropriate prefactor do we get the exact Green's function? To answer all these questions we proceed as follows. In this chapter we establish the conditions for an approximate G_2 to yield a conserving Green's function. Then, in Chapter 9 we show that G_2 is not the most natural object to construct a conserving approximation and introduce a much more suitable quantity, i.e., the *self-energy*. The self-energy also has a diagrammatic representation and the rules to convert a diagram into a formula are given. Finally, in Chapter 10 we describe a method to construct the exact self-energy in terms of the Green's function and the interparticle interaction.

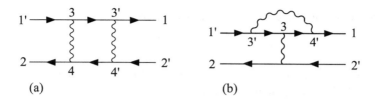

Figure 8.2 Examples of other scattering processes.

8.2 Conditions on the approximate G_2

We consider a system of interacting particles with mass m and charge q under the influence of an external electromagnetic field. The single-particle Hamiltonian is

$$h(1;2) = \langle \mathbf{x}_1 | \hat{h}(z_1) | \mathbf{x}_2 \rangle \delta(z_1, z_2) = \left[\frac{1}{2m}(-\mathrm{i}\,\mathbf{D}_1)^2 + qV(1) \right] \delta(1;2), \qquad (8.1)$$

with the gauge-invariant derivative

$$\mathbf{D}_1 \equiv \mathbf{\nabla}_1 - \mathrm{i}\frac{q}{c}\mathbf{A}(1). \qquad (8.2)$$

The fields $\mathbf{A}(1)$ and $V(1)$ are the vector and scalar potentials. The single-particle Hamiltonian $\hat{h}$ is the same on the forward and backward branch, and it is time-independent along the imaginary track of the contour. Therefore, the scalar and vector potentials with arguments on the contour are defined as

$$\mathbf{A}(\mathbf{x}, z = t_\pm) = \mathbf{A}(\mathbf{x}, t), \qquad \mathbf{A}(\mathbf{x}, z = t_0 - \mathrm{i}\tau) = \mathbf{A}(\mathbf{x}), \qquad (8.3)$$

and

$$V(\mathbf{x}, z = t_\pm) = V(\mathbf{x}, t), \qquad V(\mathbf{x}, z = t_0 - \mathrm{i}\tau) = V(\mathbf{x}). \qquad (8.4)$$

We observe that neither $\mathbf{A}$ nor V has an actual dependence on spin. In the following we use either $V(\mathbf{x}, z)$ or $V(\mathbf{r}, z)$ to represent the scalar potential and either $\mathbf{A}(\mathbf{x}, z)$ or $\mathbf{A}(\mathbf{r}, z)$ to represent the vector potential. The choice of $\mathbf{r}$ or $\mathbf{x}$ is dictated by the principle of having the notation as light as possible.

The equations of motion for the Green's function with $h(1;2)$ from (8.1) read

$$[\,\mathrm{i}\frac{d}{dz_1} + \overbrace{\frac{\nabla_1^2}{2m} - \frac{\mathrm{i}q}{2mc}[(\mathbf{\nabla}_1\cdot\mathbf{A}(1)) + 2\mathbf{A}(1)\cdot\mathbf{\nabla}_1] - w(1)}^{\frac{1}{2m}D_1^2 - qV(1)}\,]G(1;2)$$

$$= \delta(1;2) \pm \mathrm{i}\int d3\, v(1;3)G_2(1,3;2,3^+), \qquad (8.5)$$

$$[\,-\mathrm{i}\frac{d}{dz_2} + \overbrace{\frac{\nabla_2^2}{2m} + \frac{\mathrm{i}q}{2mc}[(\mathbf{\nabla}_2\cdot\mathbf{A}(2)) + 2\mathbf{A}(2)\cdot\mathbf{\nabla}_2] - w(2)}^{\frac{1}{2m}(D_2^2)^* - qV(2)}\,]G(1;2)$$

$$= \delta(1;2) \pm \mathrm{i}\int d3\, G_2(1,3^-;2,3)v(3;2), \qquad (8.6)$$

where we have introduced the short-hand notation [see also (3.34)]

$$w(1) = qV(1) + \frac{q^2}{2mc^2}A^2(1).$$

As always in this book the upper/lower sign refers to bosons/fermions. In what follows we shall prove that if the approximate two-particle Green's function fulfills:

(A1) the symmetry condition

$$G_2(1,2;1^+,2^+) = G_2(2,1;2^+,1^+),$$

(A2) the same boundary conditions as the exact G_2,

then the approximate G obtained from (8.5) and (8.6) satisfies all conservation laws. This important result was derived by Baym and Kadanoff in 1961 [77]. As we shall see, condition (A2) is needed only to prove the energy conservation law. For all other conservation laws condition (A1) is enough. Note that condition (A1) is a special case of the more general symmetry property of the two-particle Green's function,

$$G_2(1,2;3,4) = G_2(2,1;4,3), \tag{8.7}$$

which follows directly from the definition (5.1).

Exercise 8.1. Show that both the Hartree and Hartree–Fock approximations fulfill (A1) and (A2).

8.3 Continuity equation

Subtracting (8.6) from (8.5) and writing $\nabla_1^2 - \nabla_2^2 = (\nabla_1 + \nabla_2)\cdot(\nabla_1 - \nabla_2)$ we find

$$\left[i\frac{d}{dz_1} + i\frac{d}{dz_2}\right]G(1;2) + (\nabla_1 + \nabla_2)\cdot\frac{\nabla_1 - \nabla_2}{2m}G(1;2)$$

$$-\frac{iq}{2mc}\left[(\nabla_1\cdot A(1)) + (\nabla_2\cdot A(2)) + 2A(1)\cdot\nabla_1 + 2A(2)\cdot\nabla_2\right]G(1;2)$$

$$-[w(1) - w(2)]G(1;2) = \pm i\int d3\,[v(1;3)G_2(1,3;2,3^+) - v(2;3)G_2(1,3^-;2,3)]. \tag{8.8}$$

This is an important equation since it constitutes the starting point to prove all conservation laws. We can extract an equation for the density by setting $2 = 1^+ = (x_1, z_1^+)$. Then the first term on the l.h.s. becomes the derivative of the density

$$i\frac{d}{dz_1}G(1;1^+) = \pm\frac{d}{dz_1}n(1).$$

The second and the third term can be grouped to form a total divergence

$$\text{2nd + 3rd term} = \nabla_1\cdot\left[\left(\frac{\nabla_1 - \nabla_2}{2m}G(1;2)\right)_{2=1^+} - \frac{iq}{mc}A(1)G(1;1^+)\right]. \tag{8.9}$$

It is easy to see that the remaining terms cancel out. In particular the term containing G_2 vanishes *for all approximate two-particle Green's functions* since v is local in time and hence

$$[v(1;3)G_2(1,3;1^+,3^+) - G_2(1,3^-;1^+,3)v(1^+;3)] = 0.$$

In fact, the conservation of the number of particles is a direct consequence of the equations of motion alone. The time-dependent ensemble averages of the paramagnetic and diamagnetic current density operators are expressed in terms of the Green's function according to [see (5.4) and (5.5)]

$$\mathbf{j}(1) = \pm \left(\frac{\boldsymbol{\nabla}_1 - \boldsymbol{\nabla}_2}{2m} G(1;2) \right)_{2=1^+}, \tag{8.10}$$

and

$$\mathbf{j}_d(1) = -\frac{q}{mc} \mathbf{A}(1) n(1) = \mp \frac{iq}{mc} \mathbf{A}(1) G(1;1^+).$$

Hence the r.h.s. of (8.9) is the divergence of the total current density $\mathbf{J}(1) = \mathbf{j}(1) + \mathbf{j}_d(1)$ already defined in (3.31). We thus derive the continuity equation

$$\frac{d}{dz_1} n(1) + \boldsymbol{\nabla}_1 \cdot \mathbf{J}(1) = 0, \tag{8.11}$$

which tells us that accumulation of charge in a certain region of space is related to current flow. This is certainly an important relation that one wants to have satisfied in nonequilibrium systems such as, e.g., the quantum discharge of a capacitor discussed in Section 7.2.3.

The reader may not expect to find that the continuity equation (8.11) is an equation on the contour instead of an equation on the real time axis. Does (8.11) contain the same information as the continuity equation on the real time axis? Strictly speaking (8.11) contains some more information, even though the extra information is rather obvious. It is instructive to analyze (8.11) in detail since the same analysis is valid for other physical quantities. To extract physical information from (8.11) we let the contour variable z_1 lie on a different part of the contour. For $z_1 = t_{1-}$ on the forward branch we have

$$\frac{d}{dz_1} n(\mathbf{x}_1, z_1) = \lim_{\epsilon \to 0} \frac{n(\mathbf{x}_1, (t_1+\epsilon)_-) - n(\mathbf{x}_1, t_{1-})}{\epsilon} = \lim_{\epsilon \to 0} \frac{n(\mathbf{x}_1, t_1+\epsilon) - n(\mathbf{x}_1, t_1)}{\epsilon}$$

$$= \frac{d}{dt_1} n(\mathbf{x}_1, t_1) = -\boldsymbol{\nabla}_1 \cdot \mathbf{J}(\mathbf{x}_1, t_1),$$

and similarly for $z_1 = t_{1+}$ on the backward branch

$$\frac{d}{dz_1} n(\mathbf{x}_1, z_1) = \lim_{\epsilon \to 0} \frac{n(\mathbf{x}_1, (t_1-\epsilon)_+) - n(\mathbf{x}_1, t_{1+})}{-\epsilon} = \lim_{\epsilon \to 0} \frac{n(\mathbf{x}_1, t_1-\epsilon) - n(\mathbf{x}_1, t_1)}{-\epsilon}$$

$$= \frac{d}{dt_1} n(\mathbf{x}_1, t_1) = -\boldsymbol{\nabla}_1 \cdot \mathbf{J}(\mathbf{x}_1, t_1).$$

In these two equations we use the definition of the derivative on the contour, see Section 4.4, as well as that for all operators $\hat{O}(t)$ built from the field operators, like the density and the current density, the time-dependent ensemble average $O(t_\pm) = O(t)$. The continuity equation on the contour contains the same information as the continuity equation on the real time axis when z_1 lies on the horizontal branches. The extra information contained in (8.11) comes from z_1 on the vertical track. In this case both the density and the current density are independent of z_1 and the continuity equation reduces to $\boldsymbol{\nabla}_1 \cdot \mathbf{J}(1) = 0$. This result tells us that for a system in equilibrium the current density is divergenceless, and hence there can be no density fluctuations in any portion of the system.

The continuity equation expresses the *local* conservation of particles, since it is a differential equation for the density $n(\mathbf{x}_1, z_1)$ in the point $\mathbf{x}_1$. The time-derivative of the local momentum, angular momentum, or energy is not a total divergence. For instance the time-derivative of the local momentum $\mathbf{P}(1) = m\mathbf{J}(1)$ is a total divergence plus the local Lorentz force, see Section 3.5. Below we prove that momentum, angular momentum, and energy are conserved in a *global* sense, i.e., we prove the conservation laws for the integral of the corresponding local quantities.

8.4 Momentum conservation law

We saw in (3.32) that the total momentum of a system is the integral over all space and spin of the current density $\mathbf{J}$ multiplied by the mass of the particles

$$\mathbf{P}(t) \equiv m \int d\mathbf{x}\, \mathbf{J}(\mathbf{x}, t).$$

To calculate the time derivative of $\mathbf{P}$ we calculate separately the time derivative of $\mathbf{j}$ (paramagnetic current density) and $\mathbf{j}_d$ (diamagnetic current density). We introduce the shorthand notation $f_1 = f(1)$ for a scalar function and $f_{1,p} = f_p(1)$ for the pth component of a vector function $\mathbf{f}(1)$. Then, using the Einstein convention of summing of repeated indices we can rewrite (8.8) as

$$\left[\mathrm{i}\frac{d}{dz_1} + \mathrm{i}\frac{d}{dz_2} \right] G(1; 2) + (\partial_{1,p} + \partial_{2,p})\frac{\partial_{1,p} - \partial_{2,p}}{2m} G(1; 2)$$

$$- \frac{\mathrm{i}q}{2mc}\left[(\partial_{1,p}A_{1,p}) + (\partial_{2,p}A_{2,p}) + 2A_{1,p}\partial_{1,p} + 2A_{2,p}\partial_{2,p} \right] G(1; 2)$$

$$- [w_1 - w_2]G(1; 2) = \pm\mathrm{i}\int d3\, [v(1; 3)G_2(1, 3; 2, 3^+) - v(2; 3)G_2(1, 3^-; 2, 3)],$$

$$\text{(8.12)}$$

where $\partial_{1,p}$ is the pth component of $\boldsymbol{\nabla}_1$. To obtain an equation for the paramagnetic current density we act on (8.12) with $(\partial_{1,k} - \partial_{2,k})/2m$ and evaluate the result in $2 = 1^+$. The first term in (8.12) yields

$$\left[\left(\frac{\partial_{1,k} - \partial_{2,k}}{2m} \right)\left(\mathrm{i}\frac{d}{dz_1} + \mathrm{i}\frac{d}{dz_2} \right) G(1; 2) \right]_{2=1^+} = \mathrm{i}\frac{d}{dz_1}\left[\frac{\partial_{1,k} - \partial_{2,k}}{2m} G(1; 2) \right]_{2=1^+} = \pm\mathrm{i}\frac{d}{dz_1}j_{1,k},$$

where we use (8.10). The second term of (8.12) gives a total divergence since

$$\frac{\partial_{1,k} - \partial_{2,k}}{2m}(\partial_{1,p} + \partial_{2,p})\frac{\partial_{1,p} - \partial_{2,p}}{2m}G(1;2)\bigg|_{2=1^+}$$

$$= \partial_{1,p}\left[\frac{\partial_{1,k} - \partial_{2,k}}{2m}\frac{\partial_{1,p} - \partial_{2,p}}{2m}G(1;2)\right]_{2=1^+}.$$

It is easy to show that the quantity in the square bracket is proportional to the time-dependent ensemble average of the momentum stress-tensor operator defined in (3.36). Next we consider the second row of (8.12). For any continuous function f the quantity $(\partial_{1,k} - \partial_{2,k})(f_1 + f_2)$ vanishes in $2 = 1^+$. Therefore the first two terms give

$$-\frac{iq}{2mc}\left[[(\partial_{1,p}A_{1,p}) + (\partial_{2,p}A_{2,p})]\frac{\partial_{1,k} - \partial_{2,k}}{2m}G(1;2)\right]_{2=1^+} = \mp\frac{iq}{mc}(\partial_{1,p}A_{1,p})j_{1,k}. \quad (8.13)$$

For the last two terms we use the identity

$$(\partial_{1,k} - \partial_{2,k})(A_{1,p}\partial_{1,p} + A_{2,p}\partial_{2,p}) = (\partial_{1,k}A_{1,p})\partial_{1,p} - (\partial_{2,k}A_{2,p})\partial_{2,p}$$
$$+ (A_{1,p}\partial_{1,p} + A_{2,p}\partial_{2,p})(\partial_{1,k} - \partial_{2,k}),$$

and find

$$-\frac{iq}{mc}\left[\frac{\partial_{1,k} - \partial_{2,k}}{2m}(A_{1,p}\partial_{1,p} + A_{2,p}\partial_{2,p})G(1;2)\right]_{2=1^+}$$

$$= \mp\frac{iq}{mc}\left[j_{1,p}(\partial_{1,k}A_{1,p}) + A_{1,p}(\partial_{1,p}j_{1,k})\right].$$

Note that adding (8.13) to the second term on the r.h.s. of this equation we get a total divergence. Finally, the last two terms on the l.h.s of (8.12) yield

$$-\left[\frac{\partial_{1,k} - \partial_{2,k}}{2m}[w_1 - w_2]G(1;2)\right]_{2=1^+} = \pm\frac{i}{m}n_1\partial_{1,k}w_1$$

$$= \pm\frac{iq}{m}n_1\left(\partial_{1,k}V_1 + \frac{q}{mc^2}A_{1,p}\partial_{1,k}A_{1,p}\right).$$

Putting together all these results we obtain the equation for the paramagnetic current density

$$\frac{d}{dz_1}j_{1,k} = \frac{q}{mc}J_{1,p}\partial_{1,k}A_{1,p} - \frac{q}{m}n_1\partial_{1,k}V_1 + \text{total divergence}$$

$$+ \underbrace{\int d3\left[\frac{\partial_{1,k} - \partial_{2,k}}{2m}[v(1;3)G_2(1,3;2,3^+) - v(2;3)G_2(1,3^-;2,3)]\right]_{2=1^+}}_{\frac{1}{m}\int d3\,[\partial_{1,k}v(1;3)]G_2(1,3;1^+,3^+)}.$$

The derivative of the diamagnetic current density follows directly from the continuity equation and reads

$$\frac{d}{dz_1}(j_d)_{1,k} = -\frac{q}{mc}n_1\frac{d}{dz_1}A_{1,k} + \frac{q}{mc}A_{1,k}\partial_{1,p}J_{1,p}$$

$$= -\frac{q}{mc}n_1\frac{d}{dz_1}A_{1,k} - \frac{q}{mc}J_{1,p}\partial_{1,p}A_{1,k} + \frac{q}{mc}\underbrace{\partial_{1,p}(A_{1,k}J_{1,p})}_{\text{total divergence}}.$$

Summing the last two equations we find the equation for the current density:

$$\frac{d}{dz_1} J_{1,k} = \frac{q}{m} n_1 \left(-\partial_{1,k} V_1 - \frac{1}{c} \frac{d}{dz_1} A_{1,k} \right) + \frac{q}{mc} J_{1,p} \left(\partial_{1,k} A_{1,p} - \partial_{1,p} A_{1,k} \right)$$

$$+ \text{total divergence} + \frac{1}{m} \int d3 \, [\partial_{1,k} v(1;3)] G_2(1,3;1^+,3^+). \tag{8.14}$$

In the first row we recognize the kth component of the electric field and the kth component of the vector product between $\mathbf{J}$ and the magnetic field, see (3.37) and the discussion below. Upon integration over $\mathbf{x}_1$ the total divergence does not contribute and we end up with

$$\frac{d}{dz_1} \mathbf{P}(z_1) = \int d\mathbf{x}_1 \left(qn(1)\mathbf{E}(1) + \frac{q}{c} \mathbf{J}(1) \times \mathbf{B}(1) \right)$$

$$+ \frac{1}{m} \int d\mathbf{x}_1 d\mathbf{x}_3 \, [\boldsymbol{\nabla}_1 v(\mathbf{x}_1, \mathbf{x}_3)] \, G_2(1,3;1^+,3^+) \Big|_{z_3 = z_1},$$

where we use the fact that the interparticle interaction is local in time, $v(1;2) = \delta(z_1, z_2) v(\mathbf{x}_1, \mathbf{x}_2)$. We thus see that if the last term on the r.h.s. vanishes then the derivative of the total momentum coincides with the total applied force, and hence the approximation to the two-particle Green's function is conserving. The vanishing of the last term is a direct consequence of the symmetry condition (A1): under the interchange $1 \leftrightarrow 3$ the G_2 does not change while the gradient of v picks up a minus sign, see again the discussion below (3.37).

8.5 Angular momentum conservation law

The angular momentum of a system is conserved if the Hamiltonian is invariant under rotations. In the absence of external electromagnetic fields ($\mathbf{A} = V = 0$) this invariance requires that the interparticle interaction $v(\mathbf{x}_1, \mathbf{x}_2)$ depends only on the distance $|\mathbf{r}_1 - \mathbf{r}_2|$. Below we show that for such interparticle interactions the time-derivative of the total angular momentum

$$\mathbf{L}(t) \equiv m \int d\mathbf{x} \, \mathbf{r} \times \mathbf{J}(\mathbf{x}, t)$$

is the total applied torque provided that the approximate G_2 satisfies the symmetry condition (A1).

We start by rewriting (8.14) in terms of the local Lorentz force $\mathbf{F}(1) \equiv qn(1)\mathbf{E}(1) + \frac{q}{c} \mathbf{J}(1) \times \mathbf{B}(1)$ given by the first two terms on the r.h.s.:

$$m \frac{d}{dz_1} J_{1,k} = F_{1,k} \pm \frac{i}{m} \partial_{1,p} \left[\frac{\partial_{1,k} - \partial_{2,k}}{2} \frac{\partial_{1,p} - \partial_{2,p}}{2} G(1;2) \right]_{2=1^+}$$

$$+ \frac{q}{c} \partial_{1,p} [A_{1,p} j_{1,k} + A_{1,k} J_{1,p}] + \int d3 \, [\partial_{1,k} v(1;3)] G_2(1,3;1^+,3^+), \tag{8.15}$$

where we have explicitly written down the total divergence (2nd and 3rd term). The lth component of the angular momentum is obtained by integrating over space and spin the *l*th component of $[\mathbf{r}_1 \times \mathbf{J}(1)]_l = \varepsilon_{lik} r_{1,i} J_{1,k}$, where ε_{lik} is the Levi–Civita tensor and summation over repeated indices is understood. Let us consider the total divergence. The second term is the derivative of a symmetric function of k and p. The third term is also the derivative of a symmetric function of k and p since

$$A_{1,p} j_{1,k} + A_{1,k} J_{1,p} = A_{1,p} j_{1,k} + A_{1,k} j_{1,p} - \frac{q}{mc} n_1 A_{1,k} A_{1,p}.$$

We now show that the quantity $\varepsilon_{lik} r_{1,i} \partial_{1,p} S_{1,kp}$ is a total divergence for every symmetric function $S_{1,kp} = S_{1,pk}$, and therefore this quantity vanishes upon integration over space. We have

$$\varepsilon_{lik} r_{1,i} \partial_{1,p} S_{1,kp} = \partial_{1,p} \left[\varepsilon_{lik} r_{1,i} S_{1,kp} \right] - \varepsilon_{lik} S_{1,kp} \underbrace{\partial_{1,p} r_{1,i}}_{\delta_{pi}}$$

$$= \partial_{1,p} \left[\varepsilon_{lik} r_{1,i} S_{1,kp} \right] - \varepsilon_{lik} S_{1,ki},$$

and the last term is zero due to the symmetry of S. Therefore, multiplying (8.15) by $\varepsilon_{lik} r_{1,i}$, integrating over $\mathbf{x}_1$ and reintroducing the vector notation we find

$$\frac{d}{dz_1} \mathbf{L}(z_1) = \int d\mathbf{x}_1 (\mathbf{r}_1 \times \mathbf{F}(1)) + \int d\mathbf{x}_1 d\mathbf{x}_3 \left(\mathbf{r}_1 \times [\boldsymbol{\nabla}_1 v(\mathbf{x}_1, \mathbf{x}_3)] \right) G_2(1, 3; 1^+, 3^+) \Big|_{z_3 = z_1}.$$

The first term on the r.h.s. is the total torque applied to the system. Hence the angular momentum is conserved provided that the last term vanishes. Exploiting the symmetry condition (A1) and the antisymmetry of the gradient of v we have

$$\int d\mathbf{x}_1 d\mathbf{x}_3 (\mathbf{r}_1 \times [\boldsymbol{\nabla}_1 v(\mathbf{x}_1, \mathbf{x}_3)]) G_2(1, 3; 1^+, 3^+) \Big|_{z_3 = z_1}$$

$$= \frac{1}{2} \int d\mathbf{x}_1 d\mathbf{x}_3 ((\mathbf{r}_1 - \mathbf{r}_3) \times [\boldsymbol{\nabla}_1 v(\mathbf{x}_1, \mathbf{x}_3)]) G_2(1, 3; 1^+, 3^+) \Big|_{z_3 = z_1}.$$

Now recall that we are considering an interparticle interaction $v(\mathbf{x}_1, \mathbf{x}_3)$ which depends only on the distance $|\mathbf{r}_1 - \mathbf{r}_3|$. For any function $f(|\mathbf{r}|)$ the quantity $\mathbf{r} \times \boldsymbol{\nabla} f(|\mathbf{r}|)$ is zero since $\boldsymbol{\nabla} f(|\mathbf{r}|)$ is parallel to $\mathbf{r}$. In conclusion the term with G_2 vanishes and the approximation is conserving.

8.6 Energy conservation law

According to the discussion in Section 6.4 the energy of the system at a generic time t_1, $E_S(t_1)$, is the time-dependent ensemble average of the operator

$$\hat{H}_S(t_1) \equiv \hat{H}(t_1) - q \int d\mathbf{x}_1 \, \hat{n}(\mathbf{x}_1) \delta V(1),$$

where $\delta V(1) = V(\mathbf{r}_1, t_1) - V(\mathbf{r}_1)$ is the *external* potential. Before showing that any approximate G_2 fulfilling conditions (A1) and (A2) is energy conserving we derive the exact equation for $\frac{d}{dt_1}E_S(t_1)$, since this is the equation that we must reproduce using the Green's function approach.

We write the density matrix operator as $\hat{\rho} = \sum_k \rho_k |\Psi_k\rangle\langle\Psi_k|$. The time-dependent ensemble average of $\hat{H}_S(t_1)$ is then

$$E_S(t_1) = \sum_k \rho_k \langle\Psi_k(t_1)|\hat{H}(t_1) - q\int d\mathbf{x}_1\, \hat{n}(\mathbf{x}_1)\delta V(1)|\Psi_k(t_1)\rangle$$

$$= \sum_k \rho_k \langle\Psi_k(t_1)|\hat{H}(t_1)|\Psi_k(t_1)\rangle - q\int d\mathbf{x}_1\, n(1)\delta V(1).$$

To calculate the time derivative of the first term we must differentiate the bras $\langle\Psi_k(t_1)|$, the kets $|\Psi_k(t_1)\rangle$, as well as the Hamiltonian operator. Since all kets $|\Psi_k(t_1)\rangle$ evolve according to the same Schrödinger equation, $i\frac{d}{dt}|\Psi_k(t)\rangle = \hat{H}(t)|\Psi_k(t)\rangle$, the term coming from $\frac{d}{dt_1}\langle\Psi_k(t_1)|$ cancels the term coming from $\frac{d}{dt_1}|\Psi_k(t_1)\rangle$. The time derivative of $\hat{H}(t_1)$ is the same as the time derivative of the noninteracting Hamiltonian $\hat{H}_0(t_1)$ (the physical interaction is independent of time). Writing $\hat{H}_0(t_1)$ as in (3.29) we find

$$\frac{d}{dt_1}E_S(t_1) = \int d\mathbf{x}_1 \left[-\frac{q}{c}\mathbf{j}(1)\cdot\frac{d\mathbf{A}(1)}{dt_1} + qn(1)\frac{d}{dt_1}\left(\delta V(1) + \frac{q}{2mc^2}A^2(1)\right)\right]$$

$$\qquad - q\frac{d}{dt_1}\int d\mathbf{x}_1\, n(1)\delta V(1)$$

$$= \int d\mathbf{x}_1 \left[-\frac{q}{c}\mathbf{J}(1)\cdot\frac{d\mathbf{A}(1)}{dt_1} - q\delta V(1)\frac{d}{dt_1}n(1)\right].$$

The last term can be further manipulated using the continuity equation and integrating by parts; we thus arrive at the following important result

$$\frac{d}{dt_1}E_S(t_1) = \int d\mathbf{x}_1\, q\mathbf{J}(1)\cdot\mathbf{E}_{\text{ext}}(1), \qquad (8.16)$$

with $\mathbf{E}_{\text{ext}} = -\boldsymbol{\nabla}\delta V - \frac{1}{c}\frac{d}{dt}\mathbf{A}$ the *external* electric field. The r.h.s. of (8.16) is the scalar product between the current density and the Lorentz force, i.e., it is the power fed into the system (the magnetic field generates a force orthogonal to $\mathbf{J}$ and hence does not contribute to the power). Below we prove that (8.16) is satisfied by those Gs obtained from (8.5) and (8.6) with a G_2 fulfilling conditions (A1) and (A2).

The proof of energy conservation is a nice example of the importance of using the most convenient notation when carrying out lengthy calculations. We start from (6.103) which expresses the energy of the system in terms of G and G_2:

$$E_S(z_1) = \pm i\int d\mathbf{x}_1\langle\mathbf{x}_1|\hat{h}_S(z_1)\hat{\mathcal{G}}(z_1,z_1^+)|\mathbf{x}_1\rangle \underbrace{\qquad}_{E_{\text{one}}(z_1)} - \frac{1}{2}\int d\mathbf{x}_1 d3\, v(1;3)G_2(1,3;1^+,3^+), \quad (8.17)$$

$$\underbrace{\hspace{4.5cm}}_{E_{\text{one}}(z_1)}\underbrace{\hspace{5cm}}_{E_{\text{int}}(z_1)}$$

with

$$\hat{h}_S(z_1) \equiv \hat{h}(z_1) - q\delta V(\hat{\mathbf{r}}, z_1),$$

the single-particle Hamiltonian of the original system. In (8.17) the energy E_{one} is the one-body part of the total energy, i.e., the sum of the kinetic and potential energy whereas E_{int} is the interaction energy. Since E_{one} is a trace in the one-particle Hilbert space, below we often use the cyclic property to rearrange expressions in the most convenient way. The derivative of $E_{\text{one}}(z_1)$ is the sum of two contributions: one containing the derivative of $\hat{h}_S(z_1)$ and another containing the derivative of $\hat{\mathcal{G}}(z_1, z_1^+)$. To calculate the derivative of the latter we use the equations of motion (5.7) and (5.8). Subtracting the second equation from the first and setting $z_2 = z_1^+$ we find

$$\frac{d}{dz_1}\langle \mathbf{x}_2|\hat{\mathcal{G}}(z_1, z_1^+)|\mathbf{x}_1\rangle = \langle \mathbf{x}_2|\left(\frac{d}{dz_1} + \frac{d}{dz_2}\right)\hat{\mathcal{G}}(z_1, z_2)|\mathbf{x}_1\rangle\Big|_{z_2=z_1^+}$$

$$= -\mathrm{i}\,\langle \mathbf{x}_2|\left[\hat{h}(z_1), \hat{\mathcal{G}}(z_1, z_1^+)\right]_-|\mathbf{x}_1\rangle$$

$$\pm \int d3\left[v(1;3)G_2(1,3;2,3^+) - v(2;3)G_2(1,3^-;2,3)\right]\Big|_{z_2=z_1^+}.$$

The derivative of $E_{\text{one}}(z_1)$ can then be rewritten as

$$\frac{d}{dz_1}E_{\text{one}}(z_1) = P(z_1) + W_{\text{one}}(z_1),$$

with

$$P(z_1) = \pm\mathrm{i}\int d\mathbf{x}_1\langle \mathbf{x}_1|\left(\frac{d\hat{h}_S(z_1)}{dz_1} - \mathrm{i}\left[\hat{h}_S(z_1), \hat{h}(z_1)\right]_-\right)\hat{\mathcal{G}}(z_1, z_1^+)|\mathbf{x}_1\rangle, \qquad (8.18)$$

and

$$W_{\text{one}}(z_1) = \mathrm{i}\int d\mathbf{x}_1 d\mathbf{x}_2 d3\,\langle \mathbf{x}_1|\hat{h}_S(z_1)|\mathbf{x}_2\rangle$$

$$\times \left[v(1;3)G_2(1,3;2,3^+) - v(2;3)G_2(1,3^-;2,3)\right]_{z_2=z_1^+}.$$

In (8.18) the cyclic property of the trace has been used, $\mathrm{Tr}\left[\hat{A}[\hat{B}, \hat{C}]_-\right] = \mathrm{Tr}\left[[\hat{A}, \hat{B}]_-\hat{C}\right]$. We now show that $P(z_1)$ is exactly the power fed into the system. Let us manipulate the operator to the left of $\hat{\mathcal{G}}(z_1, z_1^+)$. We have

$$\frac{d\hat{h}_S(z_1)}{dz_1} = \frac{d}{dz_1}\left[\frac{1}{2m}\left(\hat{p}^2 - \frac{q}{c}\hat{p}\cdot\mathbf{A}(\hat{r}, z_1) - \frac{q}{c}\mathbf{A}(\hat{r}, z_1)\cdot\hat{p} + \frac{q^2}{c^2}A^2(\hat{r}, z_1)\right) + qV(\hat{r})\right]$$

$$= -\frac{q}{2mc}\hat{p}\cdot\frac{d\mathbf{A}(\hat{r}, z_1)}{dz_1} - \frac{q}{2mc}\frac{d\mathbf{A}(\hat{r}, z_1)}{dz_1}\cdot\hat{p} + \frac{q^2}{mc^2}\mathbf{A}(\hat{r}, z_1)\cdot\frac{d\mathbf{A}(\hat{r}, z_1)}{dz_1}.$$

Taking into account that the commutator between the momentum operator $\hat{p}$ and an arbitrary function $f(\hat{r})$ of the position operator is $[\hat{p}, f(\hat{r})] = -\mathrm{i}\boldsymbol{\nabla}f(\hat{r})$, we also have

$$-\mathrm{i}\left[\hat{h}_S(z_1), \hat{h}(z_1)\right]_- = -\mathrm{i}q\left[\hat{h}_S(z_1), \delta V(\hat{r}, z_1)\right]_-$$

$$= -\frac{q}{2m}\hat{p}\cdot\boldsymbol{\nabla}\delta V(\hat{r}, z_1) - \frac{q}{2m}\boldsymbol{\nabla}\delta V(\hat{r}, z_1)\cdot\hat{p} + \frac{q^2}{mc}\mathbf{A}(\hat{r}, z_1)\cdot\boldsymbol{\nabla}\delta V(\hat{r}, z_1).$$

Putting these results together we find

$$\frac{d\hat{h}_S(z_1)}{dz_1} - i\left[\hat{h}_S(z_1), \hat{h}(z_1)\right]_-$$
$$= q\left(\frac{\hat{p}\cdot\mathbf{E}_{\text{ext}}(\hat{r}, z_1) + \mathbf{E}_{\text{ext}}(\hat{r}, z_1)\cdot\hat{p}}{2m} - \frac{q}{mc}\mathbf{A}(\hat{r}, z_1)\cdot\mathbf{E}_{\text{ext}}(\hat{r}, z_1)\right), \qquad (8.19)$$

where $\mathbf{E}_{\text{ext}}(\hat{r}, z_1) = -\boldsymbol{\nabla}\delta V(\hat{r}, z_1) - \frac{1}{c}\frac{d}{dz_1}\mathbf{A}(\hat{r}, z_1)$ is the external electric field operator. Inserting (8.19) into (8.18) we get

$$P(z_1) = \pm iq\int dx_1 \langle x_1|\frac{\left[\hat{p}, \hat{\mathcal{G}}(z_1, z_1^+)\right]_+}{2m} - \frac{q}{mc}\mathbf{A}(\hat{r}, z_1)\hat{\mathcal{G}}(z_1, z_1^+)|x_1\rangle \cdot \mathbf{E}_{\text{ext}}(1)$$
$$= \int dx_1\, q\mathbf{J}(1)\cdot\mathbf{E}_{\text{ext}}(1),$$

where we first use the cyclic property of the trace and then, in the last equality, the identities (1.12) along with the formula (8.10) to express the anticommutator in terms of the paramagnetic current density[1]

$$\pm i\,\langle x_1|\frac{\hat{p}\hat{\mathcal{G}}(z_1, z_1^+) + \hat{\mathcal{G}}(z_1, z_1^+)\hat{p}}{2m}|x_1\rangle = \pm\left(\frac{\boldsymbol{\nabla}_1 - \boldsymbol{\nabla}_2}{2m}G(1;2)\right)_{2=1^+} = \mathbf{j}(1).$$

The proof of energy conservation is then reduced to show that the sum of $W_{\text{one}}(z_1)$ and

$$W_{\text{int}}(z_1) \equiv dE_{\text{int}}(z_1)/dz_1$$

vanishes. Using again the identities (1.12), the quantity $W_{\text{one}}(z_1)$ can be rewritten as follows:

$$W_{\text{one}}(z_1) = i\int dx_1 d3\left[v(1;3)G_2(1,3;2,3^+)h_S(\mathbf{r}_2, i\overleftarrow{\boldsymbol{\nabla}}_2, z_1)\right.$$
$$\left. - v(2;3)h_S(\mathbf{r}_1, -i\overrightarrow{\boldsymbol{\nabla}}_1, z_1)G_2(1,3^-;2,3)\right]_{2=1^+}$$
$$= i\int dx_1 d3\, v(1;3)\left[\frac{D_1^2 - (D_2^2)^*}{2m}G_2(1,3;2,3^+)\right]_{2=1^+}.$$

The quantity $W_{\text{int}}(z_1)$ can be separated into two contributions

$$W_{\text{int}}(z_1) = -\frac{1}{2}\int dx_1 d3\left[\frac{d}{dz_1}v(1;3)\right]G_2(1,3;1^+,3^+)$$
$$- \frac{1}{2}\int dx_1 d3\, v(1;3)\left[\left(\frac{d}{dz_1} + \frac{d}{dz_2}\right)G_2(1,3;2,3^+)\right]_{2=1^+}. \qquad (8.20)$$

---

[1]To prove this equation consider for instance the first term on the r.h.s. We have

$$\langle x_1|\hat{p}\hat{\mathcal{G}}(z_1, z_1^+)|x_1\rangle = \lim_{x_2\to x_1}\int dx_3\langle x_1|\hat{p}|x_3\rangle\langle x_3|\hat{\mathcal{G}}(z_1, z_1^+)|x_2\rangle$$
$$= -i\lim_{x_2\to x_1}\int dx_3\boldsymbol{\nabla}_1\delta(x_1 - x_3)\langle x_3|\hat{\mathcal{G}}(z_1, z_1^+)|x_2\rangle$$
$$= -i\lim_{x_2\to x_1}\boldsymbol{\nabla}_1\langle x_1|\hat{\mathcal{G}}(z_1, z_1^+)|x_2\rangle = -i\,\boldsymbol{\nabla}_1 G(1;2)|_{2=1^+}.$$

Let us consider the first term on the r.h.s.. Since $v(1;3) = \delta(z_1 - z_3)v(\mathbf{x}_1, \mathbf{x}_3)$ we have $\frac{d}{dz_1}v(1;3) = -\frac{d}{dz_3}v(1;3)$ and hence

$$-\frac{1}{2}\left[\frac{d}{dz_1}v(1;3)\right]G_2(1,3;1^+,3^+) = \frac{1}{2}\frac{d}{dz_3}\left[v(1;3)G_2(1,3;1^+,3^+)\right]$$

$$-\frac{1}{2}v(1;3)\frac{d}{dz_3}G_2(1,3;1^+,3^+). \qquad (8.21)$$

We have arrived at a crucial point in our derivation. The first term on the r.h.s. of (8.21) is a total derivative with respect to z_3. Therefore the integral over z_3 yields the difference of $v(1;3)G_2(1,3;1^+,3^+)$ evaluated in t_{0-} and in $t_0 - i\beta$. The interparticle interaction is the same in these two points since the contour δ-function is periodic: $\delta(-i\tau) = \delta(-i\tau - i\beta)$. The same is true for the two-particle Green's function due to condition (A2). Hence the total derivative does not contribute. It is worth noting that this is the only place where condition (A2) is used in the derivation. Let us now consider the second term in (8.21). Using the symmetry condition (A1) it is easy to see that this term is equal to the integrand in the second line of (8.20). In conclusion the sum $W(z_1) \equiv W_{\text{one}}(z_1) + W_{\text{int}}(z_1)$ takes the form

$$W(z_1) = i\int d\mathbf{x}_1 d3\, v(1;3)\left[\left(i\frac{d}{dz_1} + i\frac{d}{dz_2} + \frac{D_1^2 - (D_2^2)^*}{2m}\right)G_2(1,3;2,3^+)\right]_{2=1^+}.$$

To show that $W(z_1)$ is zero is now straightforward. We multiply the equation of motion (8.5) by $[-i\frac{d}{dz_2} + \frac{1}{2m}(D_2^2)^*]$, the adjoint equation (8.6) by $[i\frac{d}{dz_1} + \frac{1}{2m}D_1^2]$, subtract one from the other and set $2 = 1^+$. The result is

$$-qV(1)\left[\left(i\frac{d}{dz_1} + i\frac{d}{dz_2} + \frac{D_1^2 - (D_2^2)^*}{2m}\right)G(1;2)\right]_{2=1^+}$$

$$= \pm i\int d3\, v(1;3)\left[\left(i\frac{d}{dz_1} + i\frac{d}{dz_2} + \frac{D_1^2 - (D_2^2)^*}{2m}\right)G_2(1,3;2,3^+)\right]_{2=1^+}. \qquad (8.22)$$

The quantity to the right of $qV(1)$ is the difference between (8.5) and (8.6) in $2 = 1^+$. From Section 8.3 we know that this quantity is equal to $\pm[\frac{d}{dz_1}n(1) + \boldsymbol{\nabla}_1 \cdot \mathbf{J}(1)]$, which is zero due to satisfaction of the continuity equation. Thus $W(z_1)$ vanishes since the integral over 3 vanishes *for all* $\mathbf{x}_1$. This concludes our proof of energy conservation.

9

Conserving approximations: self-energy

9.1 Self-energy and Dyson equations I

From the equations of motion of the Green's function it emerges that the full knowledge of G_2 is redundant: it is indeed enough to know the two-particle Green's function with the second and fourth arguments in 3 and 3^+, see again (8.5) and (8.6). This gives us a strong reason to introduce a quantity that contains only the necessary information to determine G. This quantity is called the *self-energy* Σ and is defined by

$$\int d3\, \Sigma(1;3)G(3;2) = \pm i \int d3\, v(1;3)G_2(1,3;2,3^+). \qquad (9.1)$$

The definition of the self-energy stems naturally from the equations of motion for G. Note that (9.1) fixes Σ up to the addition of a function σ such that $\int d3\, \sigma(1;3)G(3;2) = 0$. As we soon see, the only σ satisfying this constraint is $\sigma = 0$. Equation (9.1) allows us to cast the equation of motion (8.5) in the form

$$i\frac{d}{dz_1}G(1;2) - \int d3\, [h(1;3) + \Sigma(1;3)]\, G(3;2) = \delta(1;2), \qquad (9.2)$$

with $h(1;3) = \delta(z_1,z_3)\langle \mathbf{x}_1|\hat{h}(z_1)|\mathbf{x}_3\rangle$. We have already encountered an equation of this kind when we were playing with the Martin–Schwinger hierarchy in Section 5.2, see (5.15) and (5.16), as well as in the context of the Hartree–Fock approximation in Section 7.3. In those cases, however, we did not give a definition of the exact self-energy. Here we are saying that (9.2) is an exact equation for G provided that the self-energy is defined as in (9.1). The self-energy is the most natural object to describe the propagation of a particle in an interacting medium. Moreover it has a clear physical interpretation. We can understand the origin of its name from (9.2): Σ is added to h and their sum defines a sort of *self*-consistent Hamiltonian (or *energy* operator) which accounts for the effects of the interparticle interaction. Since the exact self-energy is nonlocal in time the energy operator $h + \Sigma$ not only considers the instantaneous position of the particle but also where the particle was. These kinds of nonlocal (in time) effects are usually referred to as *retardation effects*. We discuss them at length in the following chapters. We also observe that Σ is a function belonging to the

Keldysh space. Thus the equation of motion (9.2) contains a convolution on the contour between two functions, G and Σ, in Keldysh space. To extract an equation for the lesser, greater, etc. components we have to use the Langreth rules. How to solve the equations of motion on the contour is discussed later in this chapter.

For the self-energy to be a useful quantity we must be able to express also the adjoint equation of motion (8.6) in terms of Σ. This requires some manipulations. The strategy is as follows. We first define the quantity $\tilde{\Sigma}$ from[1]

$$\int d3\, G(1;3)\tilde{\Sigma}(3;2) = \pm i \int d3\, G_2(1,3^-;2,3)v(3;2), \tag{9.3}$$

and then prove that $\Sigma = \tilde{\Sigma}$, so that (8.6) becomes

$$-i\frac{d}{dz_2}G(1;2) - \int d3\, G(1;3)\left[h(3;2) + \Sigma(3;2)\right] = \delta(1;2). \tag{9.4}$$

To prove that $\Sigma = \tilde{\Sigma}$ we evaluate the two-particle Green's function with the second and fourth arguments in 3 and 3^+

$$G_2(1,3;2,3^+) = \mp\langle\mathcal{T}\left\{\hat{\psi}_H(1)\hat{n}_H(3)\hat{\psi}_H^\dagger(2)\right\}\rangle,$$

where the shorthand notation $\langle\ldots\rangle \equiv \mathrm{Tr}[e^{-\beta\hat{H}^M}\ldots]/\mathrm{Tr}[e^{-\beta\hat{H}^M}]$ denotes the ensemble average. If we define the operators[2]

$$\hat{\gamma}_H(1) \equiv \int d3\, v(1;3)\hat{n}_H(3)\hat{\psi}_H(1),$$

$$\hat{\gamma}_H^\dagger(1) \equiv \int d3\, v(1;3)\hat{\psi}_H^\dagger(1)\hat{n}_H(3),$$

then the definitions of Σ and $\tilde{\Sigma}$ can be rewritten as

$$\int d3\, \Sigma(1;3)G(3;2) = -i\langle\mathcal{T}\left\{\hat{\gamma}_H(1)\hat{\psi}_H^\dagger(2)\right\}\rangle, \tag{9.5}$$

$$\int d3\, G(1;3)\tilde{\Sigma}(3;2) = -i\langle\mathcal{T}\left\{\hat{\psi}_H(1)\hat{\gamma}_H^\dagger(2)\right\}\rangle, \tag{9.6}$$

where we use the fact that in the $\mathcal{T}$ product the field operators (anti)commute. The operators $\hat{\gamma}_H$ and $\hat{\gamma}_H^\dagger$ should not be new to the reader; they also appear in the equations of motion of the field operators and, of course, this is not a coincidence. Let us cast (4.44) and (4.45) in terms of $\hat{\gamma}_H$ and $\hat{\gamma}_H^\dagger$:

$$\int d1\, \left[\delta(4;1)\,i\frac{d}{dz_1} - h(4;1)\right]\hat{\psi}_H(1) = \hat{\gamma}_H(4), \tag{9.7}$$

---

[1] Like Σ, $\tilde{\Sigma}$ is also defined up to the addition of a function $\tilde{\sigma}(3;2)$ such that $\int d3\, G(1;3)\tilde{\sigma}(3;2) = 0$. We see that the properties of G fix $\tilde{\sigma} = 0$.

[2] The operator $\hat{\gamma}_H^\dagger$ is the adjoint of $\hat{\gamma}_H$ only for contour arguments on the horizontal branches.

$$\int d2\,\hat{\psi}_H^\dagger(2)\left[-\mathrm{i}\frac{\overleftarrow{d}}{dz_2}\delta(2;4)-h(2;4)\right]=\hat{\gamma}_H^\dagger(4).\tag{9.8}$$

We now act on (9.5) from the right with $[-\mathrm{i}\frac{\overleftarrow{d}}{dz_2}\delta(2;4)-h(2;4)]$ and integrate over 2.[3] Using (9.4) in which $\Sigma\to\tilde{\Sigma}$ (since we have not yet proved that $\Sigma=\tilde{\Sigma}$) as well as (9.8) we get

$$\Sigma(1;4)+\int d2d3\,\Sigma(1;3)G(3;2)\tilde{\Sigma}(2;4)=\delta(z_1,z_4)\langle\left[\hat{\gamma}_H(\mathbf{x}_1,z_1),\hat{\psi}_H^\dagger(\mathbf{x}_4,z_1)\right]_\mp\rangle$$
$$-\mathrm{i}\,\langle\mathcal{T}\left\{\hat{\gamma}_H(1)\hat{\gamma}_H^\dagger(4)\right\}\rangle.\tag{9.9}$$

The (anti)commutator in the first term on the r.h.s. originates from the derivative of the contour θ-functions implicit in the $\mathcal{T}$ product (as always the upper/lower sign refers to bosons/fermions). We see that if Σ is a solution of (9.9) then $\Sigma+\sigma$ is a solution only for $\sigma=0$ (recall that $\int d3\,\sigma(1;3)G(3;2)=0$). There is therefore only one solution of (9.1). A similar equation can be derived by acting on (9.6) from the left with $[\delta(4;1)\mathrm{i}\frac{d}{dz_1}-h(4;1)]$ and integrating over 1. Taking into account (9.2) and (9.7) we get

$$\tilde{\Sigma}(4;2)+\int d1d3\,\Sigma(4;1)G(1;3)\tilde{\Sigma}(3;2)=\delta(z_2,z_4)\langle\left[\hat{\psi}_H(\mathbf{x}_4,z_2),\hat{\gamma}_H^\dagger(\mathbf{x}_2,z_2)\right]_\mp\rangle$$
$$-\mathrm{i}\,\langle\mathcal{T}\left\{\hat{\gamma}_H(4)\hat{\gamma}_H^\dagger(2)\right\}\rangle,\tag{9.10}$$

from which we also deduce that for $\tilde{\Sigma}$ and $\tilde{\Sigma}+\tilde{\sigma}$ both to be solutions of (9.10) $\tilde{\sigma}$ must be zero. Comparing (9.9) with (9.10) we conclude that the difference between Σ and $\tilde{\Sigma}$ is

$$\Sigma(1;2)-\tilde{\Sigma}(1;2)=\delta(z_1,z_2)\langle\left[\hat{\gamma}_H(\mathbf{x}_1,z_1),\hat{\psi}_H^\dagger(\mathbf{x}_2,z_1)\right]_\mp-\left[\hat{\psi}_H(\mathbf{x}_1,z_1),\hat{\gamma}_H^\dagger(\mathbf{x}_2,z_1)\right]_\mp\rangle.\tag{9.11}$$

Let us calculate the first (anti)commutator on the r.h.s.. From the definition (4.38) of operators in the contour Heisenberg picture we have

$$\langle\left[\hat{\gamma}_H(\mathbf{x}_1,z_1),\hat{\psi}_H^\dagger(\mathbf{x}_2,z_1)\right]_\mp\rangle=\langle\hat{U}(t_{0-},z_1)\left[\hat{\gamma}(\mathbf{x}_1),\hat{\psi}^\dagger(\mathbf{x}_2)\right]_\mp\hat{U}(z_1,t_{0-})\rangle.$$

Using the identity

$$\left[\hat{A}\hat{B},\hat{C}\right]_\mp=\hat{A}\left[\hat{B},\hat{C}\right]_\mp\pm\left[\hat{A},\hat{C}\right]_-\hat{B}$$

it is a matter of simple algebra to derive

$$\left[\hat{\gamma}(\mathbf{x}_1),\hat{\psi}^\dagger(\mathbf{x}_2)\right]_\mp=\delta(\mathbf{x}_1-\mathbf{x}_2)\int d\mathbf{x}_3\,v(\mathbf{x}_1,\mathbf{x}_3)\hat{n}(\mathbf{x}_3)\pm v(\mathbf{x}_1,\mathbf{x}_2)\hat{\psi}^\dagger(\mathbf{x}_2)\hat{\psi}(\mathbf{x}_1).$$

---

[3]We are assuming here that we can interchange integration and differentiation. In Appendix G we show how one can alternatively treat initial correlations by shrinking the vertical track γ^M to a point. In this approach Σ acquires a new singularity proportional to $\delta(z,t_{0-})-\delta(z,t_{0+})$ and the operations of integration and differentiation cannot be interchanged any longer. As a result $\Sigma\neq\tilde{\Sigma}$.

We leave it as an exercise for the reader to prove that the second (anti)comutator in (9.11) yields exactly the same result, and hence $\tilde{\Sigma} = \Sigma$.

Equation (9.9) is particularly interesting since it shows that the self-energy, as a function in Keldysh space, has the following structure [41]:

$$\Sigma(1;2) = \delta(z_1, z_2)\Sigma^\delta(\mathbf{x}_1, \mathbf{x}_2, z_1) + \theta(z_1, z_2)\Sigma^>(1;2) + \theta(z_2, z_1)\Sigma^<(1;2), \qquad (9.12)$$

where the singular part Σ^δ is given by the first term on the r.h.s. of (9.9). Taking into account the definition (7.39) of the Hartree–Fock potential we also see that $\Sigma^\delta = qV_{\text{HF}}$. This is the result anticipated at the end of Section 7.3: any approximation beyond the Hartree–Fock approximation leads to a nonlocal (in time) self-energy. We call $\Sigma_{\text{HF}}(1;2) = \delta(z_1, z_2)\Sigma^\delta(\mathbf{x}_1, \mathbf{x}_2, z_1)$ the *Hartree–Fock self-energy*

$$\boxed{\Sigma_{\text{HF}}(1;2) = \pm\mathrm{i}\,\delta(1;2)\int d3\, v(1;3)G(3;3^+) + \mathrm{i}\, v(1;2)G(1;2^+)} \qquad (9.13)$$

and the remaining part

$$\boxed{\Sigma_{\text{c}}(1;2) = \theta(z_1, z_2)\Sigma^>(1;2) + \theta(z_2, z_1)\Sigma^<(1;2)} \qquad (9.14)$$

the *correlation self-energy*. The correlation self-energy takes into account all effects beyond mean-field theory. It is also important to observe that the *exact* self-energy satisfies the same KMS relations as G, i.e.,

$$\Sigma(\mathbf{x}_1, z_1; \mathbf{x}_2, t_{0-}) = \pm\Sigma(\mathbf{x}_1, z_1; \mathbf{x}_2, t_0 - \mathrm{i}\beta), \qquad (9.15)$$

$$\Sigma(\mathbf{x}_1, t_{0-}; \mathbf{x}_2, z_2) = \pm\Sigma(\mathbf{x}_1, t_0 - \mathrm{i}\beta; \mathbf{x}_2, z_2). \qquad (9.16)$$

This result follows directly from the definition (9.1) and (9.3) and from the fact that both G and G_2 satisfy the KMS relations.

We conclude this section by stressing again that (9.2) and (9.4) are first-order integro-differential equations (in the contour arguments) to be solved with the KMS boundary conditions (5.6). Alternatively, we can define the noninteracting Green's function G_0 as the solution of the noninteracting equations of motion ($\Sigma = 0$) with KMS boundary conditions, and convert the integro-differential equations into two *equivalent* integral equations

$$\boxed{\begin{aligned} G(1;2) &= G_0(1;2) + \int d3d4\, G_0(1;3)\Sigma(3;4)G(4;2) \\ &= G_0(1;2) + \int d3d4\, G(1;3)\Sigma(3;4)G_0(4;2) \end{aligned}} \qquad (9.17)$$

also known as the *Dyson equations* for the Green's function. The reader can verify (9.17) by acting with $[\delta(1';1)\mathrm{i}\frac{d}{dz_1} - h(1';1)]$ and integrating over 1 or by acting with $[-\mathrm{i}\frac{\overleftarrow{d}}{dz_2}\delta(2;2') - h(2;2')]$ and integrating over 2. The full interacting G which solves the Dyson equation has the correct KMS boundary conditions built in automatically.

9.2 Conditions on the approximate Σ

The use of the self-energy instead of the two-particle Green's function raises a natural question: which conditions must an approximate Σ fulfill for the Green's function of (9.2) and (9.4) to be conserving? It is the purpose of this section to formulate the theory of conserving approximations in terms of Σ. As we shall see, the self-energy has to fulfill different conditions for different conservation laws, a fact that may discourage us from developing the new formulation any further. Remarkably, however, *all* these conditions follow from *one* single property: the theory of conserving approximations in terms of Σ turns out to be extremely elegant.

We start our discussion by rewriting (8.8) in terms of the self-energy. Subtracting (9.4) from (9.2) and using the explicit form (8.1) for $h(1;2)$ we find

$$
\left[i \frac{d}{dz_1} + i \frac{d}{dz_2} \right] G(1;2) + (\boldsymbol{\nabla}_1 + \boldsymbol{\nabla}_2) \cdot \frac{\boldsymbol{\nabla}_1 - \boldsymbol{\nabla}_2}{2m} G(1;2)
$$
$$
- \frac{iq}{2mc} \left[(\boldsymbol{\nabla}_1 \cdot \mathbf{A}(1)) + (\boldsymbol{\nabla}_2 \cdot \mathbf{A}(2)) + 2\mathbf{A}(1) \cdot \boldsymbol{\nabla}_1 + 2\mathbf{A}(2) \cdot \boldsymbol{\nabla}_2 \right] G(1;2)
$$
$$
- [w(1) - w(2)] G(1;2) = \int d3 \left[\Sigma(1;3) G(3;2) - G(1;3) \Sigma(3;2) \right]. \tag{9.18}
$$

Proceeding along the same lines as in Section 8.3 we get the following condition for the satisfaction of the continuity equation:

$$
(B1): \quad \int d3 \left[\Sigma(1;3) G(3;1^+) - G(1;3) \Sigma(3;1^+) \right] = 0.
$$

In a similar way we can easily derive the conditions for satisfaction of the momentum and angular momentum conservation laws:

$$
(B2): \quad \int d\mathbf{x}_1 d3 \left[\Sigma(1;3) \boldsymbol{\nabla}_1 G(3;1^+) - G(1;3) \boldsymbol{\nabla}_1 \Sigma(3;1^+) \right] = 0,
$$

$$
(B3): \quad \int d\mathbf{x}_1 d3 \, \mathbf{r}_1 \times \left[\Sigma(1;3) \boldsymbol{\nabla}_1 G(3;1^+) - G(1;3) \boldsymbol{\nabla}_1 \Sigma(3;1^+) \right] = 0.
$$

The condition for the energy conservation law is slightly more involved to derive but still simple. The definition of the self-energy in (9.1) and the equivalent definition in (9.3) lead to two equivalent formulas for the interaction energy $E_{\text{int}}(z_1)$ [second term on the r.h.s. of (8.17)]:

$$
\boxed{ E_{\text{int}}(z_1) = \pm \frac{i}{2} \int d\mathbf{x}_1 d3 \, \Sigma(1;3) G(3;1^+) = \pm \frac{i}{2} \int d\mathbf{x}_1 d3 \, G(1;3) \Sigma(3;1^+) } \tag{9.19}
$$

These two formulas for E_{int} certainly give the same result when evaluated with the *exact* self-energy (and hence with the exact G). For an approximate self-energy, however, the equivalence between the two formulas is not guaranteed. The set of approximate self-energies for which the two formulas are equivalent contains the set of self-energies which preserve the continuity equation, see condition (B1). For the energy conservation law we then assume that Σ fulfills (B1) and rewrite the interaction energy as the arithmetic average

$$E_{\text{int}}(z_1) = \pm \frac{\text{i}}{4} \int d\mathbf{x}_1 d3 \left[\Sigma(1; 3)G(3; 1^+) + G(1; 3)\Sigma(3; 1^+) \right].$$

The energy of the system is $E_{\text{S}}(z_1) = E_{\text{one}}(z_1) + E_{\text{int}}(z_1)$, with E_{one} given by the first term on the r.h.s. of (8.17). Proceeding along the same lines as in Section 8.6 we can calculate the derivative of $E_{\text{one}}(z_1)$ and find that it is the sum of the power fed into the system, $P(z_1)$, and a term $W_{\text{one}}(z_1)$ which reads

$$W_{\text{one}}(z_1) = \pm \int d\mathbf{x}_1 d\mathbf{x}_2 d3 \, \langle \mathbf{x}_1 | \hat{h}_{\text{S}}(z_1) | \mathbf{x}_2 \rangle \left[\Sigma(2; 3)G(3; 1^+) - G(2; 3)\Sigma(3; 1^+) \right] \Big|_{z_2 = z_1}$$

$$= \pm \int d\mathbf{x}_1 d3 \left[\left(-\frac{D_2^2}{2m} + qV(\mathbf{x}_2) \right) \left[\Sigma(2; 3)G(3; 1^+) - G(2; 3)\Sigma(3; 1^+) \right] \right]_{2=1}.$$

The energy conservation law is satisfied provided that the sum $W_{\text{one}}(z_1) + \frac{d}{dz_1}E_{\text{int}}(z_1)$ vanishes. Let us work out a simpler expression for $W_{\text{one}}(z)$ and derive a more transparent condition for the energy conservation law. In the definition of W_{one} the term with the *static* scalar potential, $qV(\mathbf{x}_2)$, vanishes due to condition (B1). The remaining term contains the integral of the difference between $[D_1^2 \Sigma(1; 3)]G(3; 1^+)$ and $[D_1^2 G(1; 3)]\Sigma(3; 1^+)$. The gauge invariant derivative $\mathbf{D}_1 = \boldsymbol{\nabla}_1 - \text{i}\frac{q}{c}\mathbf{A}(1)$ can be treated similarly to the normal gradient $\boldsymbol{\nabla}_1$ when integrating by parts. Indeed, given two functions $f(1)$ and $g(1)$ that vanish for $|\mathbf{r}_1| \to \infty$ we have

$$\int d\mathbf{r}_1 f(1)D_1^2 g(1) = -\int d\mathbf{r}_1 [\mathbf{D}_1^* f(1)] \cdot \mathbf{D}_1 g(1) = \int d\mathbf{r}_1 g(1)(D_1^2)^* f(1). \qquad (9.20)$$

Integrating by parts $[D_1^2 \Sigma(1; 3)]G(3; 1^+)$ with the help of (9.20), $W_{\text{one}}(z_1)$ can be rewritten as

$$W_{\text{one}}(z_1) = \pm \int d\mathbf{x}_1 d3 \left[\left(\frac{D_1^2}{2m}G(1; 3) \right) \Sigma(3; 1^+) - \Sigma(1; 3) \left(\frac{(D_1^2)^*}{2m}G(3; 1^+) \right) \right].$$

This result can be further manipulated using the analog of (8.22) in which $vG_2 \to \Sigma G$. We multiply the equation of motion (9.2) by $[-\text{i}\frac{d}{dz_2} + \frac{1}{2m}(D_2^2)^*]$, the adjoint equation (9.4) by $[\text{i}\frac{d}{dz_1} + \frac{1}{2m}(D_1^2)]$, subtract one from the other and set $2 = 1^+$. Then, taking into account that Σ satisfies (B1) we find

$$\int d3 \left\{ \left[\left(\text{i}\frac{d}{dz_1} + \frac{D_1^2}{2m} \right)G(1; 3) \right] \Sigma(3; 1^+) + \Sigma(1; 3) \left[\left(\text{i}\frac{d}{dz_1} - \frac{(D_1^2)^*}{2m} \right)G(3; 1^+) \right] \right\} = 0.$$

Thus we see that $W_{\text{one}}(z_1)$ can be expressed entirely in terms of the time derivatives of G. The condition for the energy conservation law is: the self-energy must fulfill (B1) and must give $W_{\text{one}}(z_1) + \frac{d}{dz_1}E_{\text{int}}(z_1) = 0$, i.e.,

$$(\text{B4}): \quad \int d\mathbf{x}_1 d3 \left\{ \frac{1}{4}\frac{d}{dz_1} \left[\Sigma(1; 3)G(3; 1^+) + G(1; 3)\Sigma(3; 1^+) \right] \right.$$

$$\left. - \left[\Sigma(1; 3) \left(\frac{d}{dz_1}G(3; 1^+) \right) + \left(\frac{d}{dz_1}G(1; 3) \right) \Sigma(3; 1^+) \right] \right\} = 0.$$

The conditions on the approximate Σ are very physical since 1, $\boldsymbol{\nabla}_1$, $\mathbf{r}_1 \times \boldsymbol{\nabla}_1$ and $\frac{d}{dz_1}$ are the generators of gauge transformations, spatial translations, rotations, and time translations, which are exactly the symmetries required for particle, momentum, angular momentum, and energy conservation. The formulation in terms of the self-energy replaces conditions (A1) and (A2) for G_2 with conditions (B1), (B2), (B3), and (B4). In the next section we show that all conditions of type (B) follow from one single property of the self-energy. Furthermore, if the self-energy has this property then it automatically satisfies the KMS relations (9.15) and (9.16).

9.3 Φ functional

In 1962 Baym proposed a very simple prescription to generate a conserving self-energy [78]. To understand the principle behind the general idea we first make a few introductory remarks. A close inspection of the (B) conditions reveals that they all contain quantities like $\Sigma(1;3)\eta_1 G(3;1^+)$ and $G(1;3)\eta_1 \Sigma(3;1^+)$ where η_1 is either a constant, or $\boldsymbol{\nabla}_1$, or $\mathbf{r}_1 \times \boldsymbol{\nabla}_1$, or a contour derivative. Let us consider one of these cases, say, $\eta_1 = \boldsymbol{\nabla}_1$. We know that given a continuous function $f(\mathbf{r}_1)$ and an infinitesimal vector $\boldsymbol{\varepsilon}$ the difference $\delta f(\mathbf{r}_1) = f(\mathbf{r}_1 + \boldsymbol{\varepsilon}) - f(\mathbf{r}_1)$ is, to first order in $\boldsymbol{\varepsilon}$, $\delta f(\mathbf{r}_1) = \boldsymbol{\varepsilon} \cdot \boldsymbol{\nabla}_1 f(\mathbf{r}_1)$. Now consider a functional $F[f]$ of the function f, i.e., an application that maps a function into a number. What is the first-order variation of the functional when $f \to f + \delta f$? Using the appropriate extension of the chain rule the answer is simply

$$\delta F = F[f + \delta f] - F[f] = \int d\mathbf{r}_1 \frac{\delta F[f]}{\delta f(\mathbf{r}_1)} \delta f(\mathbf{r}_1) = \boldsymbol{\varepsilon} \cdot \int d\mathbf{r}_1 \frac{\delta F[f]}{\delta f(\mathbf{r}_1)} \boldsymbol{\nabla}_1 f(\mathbf{r}_1).$$

More generally, for an infinitesimal variation of the function f that can be written as $\delta f(\mathbf{r}_1) = \varepsilon \, \eta_1 f(\mathbf{r}_1)$, the variation of the functional F is

$$\delta F = \varepsilon \int d\mathbf{r}_1 Q(\mathbf{r}_1)\eta_1 f(\mathbf{r}_1),$$

with $Q(\mathbf{r}_1) \equiv \delta F/\delta f(\mathbf{r}_1)$ the functional derivative of F. Thus, if the functional F is symmetric under some transformation, i.e., if it does not change when adding to f the variation δf, then $\int d\mathbf{r}_1 Q(\mathbf{r}_1)\eta_1 f(\mathbf{r}_1) = 0$.

These considerations prompt us to look for a suitable functional $\Phi[G]$ of the Green's function G from which to obtain the self-energy as

$$\boxed{\Sigma(1;2) = \frac{\delta \Phi[G]}{\delta G(2;1^+)}} \tag{9.21}$$

Indeed, if the functional is symmetric under some infinitesimal variation of the Green's function then

$$0 = \delta \Phi = \int d1 d2 \, \Sigma(1;2)\delta G(2;1^+), \tag{9.22}$$

which resembles very closely the structure of the (B) conditions. The functional Φ must, in our case, be invariant under gauge transformations, space and time translations, and rotations. These fundamental symmetries guarantee the conservation of particles, momentum, energy, and angular momentum. Since *every scattering process* preserves these quantities the functional Φ must be the amplitude, or the sum of amplitudes, of a scattering process. Let us convert this intuitive argument into mathematical formulas.

We construct a functional of the Green's function according to the following *diagrammatic rules* (we come back to these rules in Chapter 11):

- Draw a diagram with a certain number of disconnected and oriented loops and a certain number of wiggly lines starting from some point (vertex) of a loop and ending in another point (vertex) of the same or of another loop. The resulting diagram must be connected, i.e., there must be at least a wiggly line between two different loops.[4]

- Label all vertices of the diagram with integer numbers $1, \ldots, N$, where N is the total number of vertices.

- Associate with each wiggly line between vertex i and vertex j the interparticle interaction $v(i; j)$.

- Associate with each oriented line going from vertex i to vertex j with no vertices in between the Green's function $G(j; i^{+})$.

- The functional corresponding to this diagram is obtained by integrating over $1, \ldots, N$ the product of all interparticle interactions and Green's functions.

We will demonstrate that a functional $\Phi[G]$ obtained by taking an arbitrary linear combination of functionals constructed according to the above diagrammatic rules yields, through (9.21), a conserving self-energy. Let us first consider a few examples. Below we show the only two diagrams with one interaction (wiggly) line:

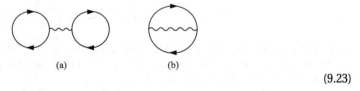

(a)   (b)

$$(9.23)$$

The functionals corresponding to diagrams (a) and (b) are

$$\Phi_a[G] = \int d1 d2 \, G(1; 1^{+}) v(1; 2) G(2; 2^{+}),$$

and

$$\Phi_b[G] = \int d1 d2 \, G(1; 2^{+}) v(1; 2) G(2; 1^{+}).$$

It is important to realize that if G belongs to the Keldysh space then the contour integrals reduce to integrals along the imaginary track since the forward branch is exactly cancelled by the backward branch (see for instance Exercise 5.4). The functional Φ, however, is defined

---

[4] For a diagram with n loops to be connected the minimum number of wiggly lines is $n - 1$.

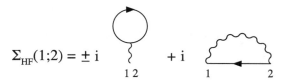

$$\Sigma_{\mathrm{HF}}(1;2) = \pm\, i \quad \begin{array}{c} \\ 1\ 2 \end{array} \quad + i \quad \begin{array}{c} \\ 1 \qquad\qquad 2 \end{array}$$

Figure 9.1 Representation of the Hartree–Fock self-energy.

for *any* G. This is a crucial point. When we take the functional derivative we must allow arbitrary variations of the Green's function including those variations that bring G away from the Keldysh space. For this reason we cannot reduce the contour integrals to the vertical track, and a variation of Φ induced by a variation of $G(1;2)$ with z_1 and/or z_2 on the horizontal branches is, in general, different from zero. In particular the variation $\delta\Phi/\delta G = \Sigma$ evaluated at a Green's function in Keldysh space is different from zero. That said, let us derive the form of the self-energy when $\Phi = \alpha\Phi_a + \beta\Phi_b$ is a linear combination of the above functionals. From (9.21) we have

$$\Sigma(1;2) = \alpha\frac{\delta\Phi_a[G]}{\delta G(2;1^+)} + \beta\frac{\delta\Phi_b[G]}{\delta G(2;1^+)}$$
$$= 2\alpha\,\delta(1;2)\int d3\, v(1;3)G(3;3^+) + 2\beta\, v(1;2)G(1;2^+). \qquad (9.24)$$

Comparing (9.24) with (9.13) we see that for $\alpha = \pm i/2$ and $\beta = i/2$ this self-energy coincides with the Hartree–Fock self-energy. The self-energies which are functional derivatives of some Φ are called Φ-*derivable*. The Hartree–Fock self-energy is therefore Φ-derivable. Like Φ, Σ also has a diagrammatic representation. The rules to convert a diagram into a formula are the same as for the Φ-diagrams, with the exception that we should not integrate over those vertices that are the end-points of a curve. A diagrammatic representation of the Hartree–Fock self-energy is given in Fig 9.1.

Examples of Φ-diagrams with two interaction lines are shown in Fig. 9.2, together with the corresponding self-energies. From the diagrammatic representation we see that the action of taking the functional derivative is the same as removing a G-line in all possible ways. For instance, in the top Φ-diagram of Fig. 9.2 we can remove four different G-lines. This leads to four identical self-energy diagrams and hence to a factor of four in Σ. The same is true for the bottom Φ-diagram in the same figure.

In Fig. 9.3 we show an example Φ-diagram with three interaction lines as well as the diagrams of the corresponding self-energy. This example is important since the functional derivative of Φ is not represented by only one diagram. A linear combination of the diagrams in Fig. 9.3, with coefficients that differ from those of the figure by more than an overall multiplicative constant, is *not* a Φ-derivable self-energy and hence it is *not* conserving. The Φ-derivability property allows only special linear combinations of self-energy diagrams.

Let us now come back to the conserving nature of the Φ-derivable self-energies. The first important property of a Φ-derivable Σ is that it satisfies the KMS relations (9.15) and (9.16), provided that the Green's function satisfies the KMS relations. This property follows directly from the diagrammatic representation of Σ and it is crucial to prove condition

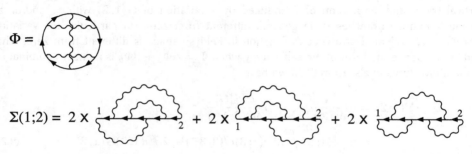

Figure 9.2 Φ-diagrams with two interaction lines and the corresponding Φ-derivable self-energy diagrams.

$$\Phi =$$

$$\Sigma(1;2) = 2 \times \quad + 2 \times \quad + 2 \times$$

Figure 9.3 Example of a Φ-diagram with three interaction lines (top). The corresponding self-energy is also displayed (bottom).

(B4). The second property is that all (B) conditions follow from the invariance of Φ under special variations of G. Consider the variation of G induced by the infinitesimal gauge transformation

$$G(1;2) \to e^{i\Lambda(1)}G(1;2)e^{-i\Lambda(2)} \quad \Rightarrow \quad \delta G(1;2) = i\,[\Lambda(1) - \Lambda(2)]G(1;2), \qquad (9.25)$$

with $\Lambda(\mathbf{x}, t_{0-}) = \Lambda(\mathbf{x}, t_0 - i\beta)$ to preserve the KMS relations. This transformation leaves Φ unchanged since with every vertex j of a Φ-diagram is associated an ingoing Green's function, $G(\ldots; j^+)$, and an outgoing Green's function, $G(j; \ldots)$. Thus, from (9.22) we find[5]

$$0 = \delta\Phi[G] = i\int d1d2\,\Sigma(1;2)[\Lambda(2) - \Lambda(1)]G(2;1^+)$$

$$= -i\int d1d2\,\left[\Sigma(1;2)G(2;1^+) - G(1;2)\Sigma(2;1^+)\right]\Lambda(1), \qquad (9.26)$$

---

[5]The arguments of G and Σ in the second term of the second line of (9.26) can be understood as follows. Renaming the integration variables $1 \leftrightarrow 2$ in the first line of (9.26) we obtain the term $G(1;2^+)\Sigma(2;1)$. The self-energy has, in general, a singular part $\delta(z_2, z_1)\Sigma^\delta(\mathbf{x}_2, \mathbf{x}_1, z_1)$ and a regular part (the correlation self-energy) $\Sigma_c(2;1)$. The product $G(1;2^+)\delta(z_2, z_1)\Sigma^\delta(\mathbf{x}_2, \mathbf{x}_1, z_1) = G(1;2)\delta(z_2, z_1^+)\Sigma^\delta(\mathbf{x}_2, \mathbf{x}_1, z_1)$, while the product of the Green's function with the correlation self-energy gives the same result with or without the superscript "+": $\int d2\,G(1;2^+)\Sigma_c(2;1) = \int d2\,G(1;2)\Sigma_c(2;1) = \int d2\,G(1;2)\Sigma_c(2;1^+)$ since the point $1 = 2$ has zero measure in the domain of integration. Therefore $\int d2\,G(1;2^+)\Sigma(2;1) = \int d2\,G(1;2)\Sigma(2;1^+)$.

and due to the arbitrariness of Λ

$$\int d2 \ [\Sigma(1;2)G(2;1^+) - G(1;2)\Sigma(2;1^+)] = 0,$$

which is exactly condition (B1). For interparticle interactions $v(\mathbf{x}_1, \mathbf{x}_2)$ which depend only on the difference $\mathbf{r}_1 - \mathbf{r}_2$ the Φ functional is also invariant under the transformation

$$G(1;2) \rightarrow G((\mathbf{r}_1 + \mathbf{R}(z_1))\sigma_1, z_1; (\mathbf{r}_2 + \mathbf{R}(z_2))\sigma_2, z_2),$$

with $\mathbf{R}(t_{0-}) = \mathbf{R}(t_0 - i\beta)$. A Φ-diagram contains an integral over all spatial coordinates and hence the shift $\mathbf{r}_i \rightarrow \mathbf{r}_i - \mathbf{R}(z_i)$ brings the transformed G back to the original G without changing $v(i;j)$, since $v(i;j) \rightarrow \delta(z_i, z_j)v(\mathbf{r}_i - \mathbf{R}(z_i) - \mathbf{r}_j + \mathbf{R}(z_j)) = v(i;j)$ due to the locality in time. To first order in $\mathbf{R}$ the variation in G is

$$\delta G(2;1^+) = [\mathbf{R}(z_2) \cdot \boldsymbol{\nabla}_2 + \mathbf{R}(z_1) \cdot \boldsymbol{\nabla}_1] G(2;1^+),$$

and (9.22) yields

$$0 = \delta \Phi = \int d1 d2 \ \Sigma(1;2) [\mathbf{R}(z_2) \cdot \boldsymbol{\nabla}_2 + \mathbf{R}(z_1) \cdot \boldsymbol{\nabla}_1] G(2;1^+)$$

$$= \int d1 d2 \ [\Sigma(1;2)\boldsymbol{\nabla}_1 G(2;1^+) - G(1;2)\boldsymbol{\nabla}_1\Sigma(2;1^+)] \mathbf{R}(z_1), \qquad (9.27)$$

where in the last equality we have performed an integration by part and renamed $1 \leftrightarrow 2$. Since (9.27) is true for all vector functions $\mathbf{R}(z)$, condition (B2) follows. Similarly, for rotationally invariant interparticle interactions we can exploit the invariance of Φ under the transformation

$$G(1;2) \rightarrow G((R[\boldsymbol{\alpha}(z_1)]\mathbf{r}_1)\sigma_1, z_1; (R[\boldsymbol{\alpha}(z_2)]\mathbf{r}_2)\sigma_2, z_2), \qquad (9.28)$$

where $R[\boldsymbol{\alpha}]$ is the 3×3 matrix which rotates a vector by an angle $\boldsymbol{\alpha}$, and $\boldsymbol{\alpha}(t_{0-}) = \boldsymbol{\alpha}(t_0 - i\beta)$. To first order in $\boldsymbol{\alpha}$ the variation of the spatial coordinates is $\delta \mathbf{r}_i = \boldsymbol{\alpha} \times \mathbf{r}_i$, and again using (9.22) one can easily prove condition (B3). Finally, condition (B4) follows from the invariance of Φ under the transformation

$$G(1;2) \rightarrow \left(\frac{dw(z_1)}{dz_1}\right)^{1/4} G(\mathbf{r}_1\sigma_1, w(z_1); \mathbf{r}_2\sigma_2, w(z_2)) \left(\frac{dw(z_2)}{dz_2}\right)^{1/4},$$

where $w(z)$ is an invertible function for z on the contour with $w(t_{0-}) = t_{0-}$ and $w(t_0 - i\beta) = t_0 - i\beta$. This is an invariance since for every interaction line $v(i;j)$ there are four Gs that have the integration variables $z_i = z_j$ in common.[6] These four Gs supply a net factor dw/dz that changes the measure from dz to dw. For the infinitesimal transformation $w(z) = z + \varepsilon(z)$ the variation in the Green's function is, to first order,

$$\delta G(1;2) = \left\{ \frac{1}{4} \left[\frac{d\varepsilon(z_1)}{dz_1} + \frac{d\varepsilon(z_2)}{dz_2} \right] + \left[\varepsilon(z_1)\frac{d}{dz_1} + \varepsilon(z_2)\frac{d}{dz_2} \right] \right\} G(1;2). \qquad (9.29)$$

---

[6]We recall that the interparticle interaction is local in time.

Inserting this variation into (9.22), integrating by parts and taking into account that the self-energy satisfies the KMS relations we recover condition (B4).

In Chapter 11 we construct the *exact* functional Φ, i.e., the Φ whose functional derivative is the exact self-energy. The exact Φ is the sum of an infinite number of diagrams each multiplied by a well defined prefactor.

Exercise 9.1. Draw all possible Φ-diagrams with three interaction lines and calculate the corresponding self-energies.

Exercise 9.2. Calculate to first order in α the variation δG from (9.28) and use (9.22) to prove condition (B3).

Exercise 9.3. Prove condition (B4) from (9.29) and (9.22).

9.4 Kadanoff–Baym equations

To solve the equations of motion (9.2) and (9.4) in practice we need to transform them into ordinary integro-differential equations for quantities with real-time arguments (as opposed to contour-time arguments). This can be done by taking the contour times of the Green's function and self-energy on different branches of the contour and then using the Langreth rules of Table 5.1. The Green's function has no singular contribution, i.e., $G^{\delta} = 0$ while the self-energy has a singular contribution given by the Hartree–Fock part.

Let us define the self-energy operator in first quantization, $\hat{\Sigma}$, in a similar way as we already have for the Green's function and the Hartree–Fock potential:

$$\hat{\Sigma}(z_1, z_2) = \int d\mathbf{x}_1 d\mathbf{x}_2 \, |\mathbf{x}_1\rangle \Sigma(1; 2) \langle \mathbf{x}_2|,$$

and rewrite the equations of motion (9.2) and (9.4) in an operator form:

$$\left[\mathrm{i}\frac{d}{dz_1} - \hat{h}(z_1) \right] \hat{\mathcal{G}}(z_1, z_2) = \delta(z_1, z_2) + \int_{\gamma} dz \, \hat{\Sigma}(z_1, z) \hat{\mathcal{G}}(z, z_2), \qquad (9.30)$$

$$\hat{\mathcal{G}}(z_1, z_2) \left[-\mathrm{i}\frac{\overleftarrow{d}}{dz_2} - \hat{h}(z_2) \right] = \delta(z_1, z_2) + \int_{\gamma} dz \, \hat{\mathcal{G}}(z_1, z) \hat{\Sigma}(z, z_2). \qquad (9.31)$$

These equations reduce to the equations of motion for the noninteracting Green's function operator (6.4) and (6.5) when $\hat{\Sigma} = 0$. In the interacting case (9.30) and (9.31) provide an approximate solution for the Green's function once an approximate functional form for the self-energy $\Sigma[G]$ is inserted. If the functional form $\Sigma[G]$ is exact then the Green's function will be exact also.

Taking both arguments of $\hat{\mathcal{G}}$ on the vertical track, $z_{1,2} = t_0 - \mathrm{i}\tau_{1,2}$, we obtain a pair of integro-differential equations for the Matsubara Green's function

$$\left[-\frac{d}{d\tau_1} - \hat{h}^{\mathrm{M}} \right] \hat{\mathcal{G}}^{\mathrm{M}}(\tau_1, \tau_2) = \mathrm{i}\delta(\tau_1 - \tau_2) + \left[\hat{\Sigma}^{\mathrm{M}} \star \hat{\mathcal{G}}^{\mathrm{M}} \right](\tau_1, \tau_2), \qquad (9.32)$$

$$\hat{\mathcal{G}}^{\mathrm{M}}(\tau_1, \tau_2) \left[\frac{\overleftarrow{d}}{d\tau_2} - \hat{h}^{\mathrm{M}} \right] = i\delta(\tau_1 - \tau_2) + \left[\hat{\mathcal{G}}^{\mathrm{M}} \star \hat{\Sigma}^{\mathrm{M}} \right](\tau_1, \tau_2). \tag{9.33}$$

At this point it is important to recall the observation we made in Section 5.3. If we are interested in calculating the Green's function with real-time arguments up to a maximum time T then we do not need a contour γ that goes all the way to ∞. It is enough that γ reaches T. This implies that *if the external vertices of a self-energy diagram are smaller than T then all integrals over the internal vertices that go from T to ∞ cancel with those from ∞ to T.* Consequently to calculate $\hat{\Sigma}(z_1, z_2)$ with z_1 and z_2 up to a maximum real-time T we only need the Green's function $\hat{\mathcal{G}}(z_1, z_2)$ with real-times up to T. In particular, to calculate $\hat{\Sigma}^{\mathrm{M}}$ we can set $T = t_0$ and therefore we only need $\hat{\mathcal{G}}(z_1, z_2)$ with arguments on the vertical track, i.e., we only need $\hat{\mathcal{G}}^{\mathrm{M}}$. Let us consider, for instance, the diagrams in Fig. 9.3; each diagram contains the convolution of five Green's functions, and for z_1 and z_2 on γ^{M} these diagrams can be written as the convolution along γ^{M} of five Matsubara Green's functions, see (5.55). Another example is the diagram in the first row of Fig. 9.2. In this case we have to integrate the vertices of the bubble over the entire contour; however, since the interaction (wiggly lines) is local in time this integral reduces to $G(\mathbf{x}_3, z_1; \mathbf{x}_4, z_2)G(\mathbf{x}_4, z_2; \mathbf{x}_3, z_1)$, which is the product of two Matsubara Green's functions when $z_{1,2} = t_0 - i\tau_{1,2}$. We then conclude that (9.32) and (9.33) constitute a close system of equations for $\hat{\mathcal{G}}^{\mathrm{M}}$, i.e., there is no mixing between $\hat{\mathcal{G}}^{\mathrm{M}}$ and the other components of the Green's function. We also infer that the two equations are not independent. Since the single particle Hamiltonian $\hat{h}(z)$ is constant along the vertical track, $\hat{\mathcal{G}}^{\mathrm{M}}(\tau_1, \tau_2)$ depends only on the difference $\tau_1 - \tau_2$ and, consequently, $\hat{\Sigma}^{\mathrm{M}}$ also depends only on the time difference. Therefore one of the two equations is redundant. Taking into account the KMS relations, we can expand $\hat{\mathcal{G}}^{\mathrm{M}}$ and $\hat{\Sigma}^{\mathrm{M}}$ as

$$\hat{\mathcal{G}}^{\mathrm{M}}(\tau_1, \tau_2) = \frac{1}{-i\beta} \sum_{m=-\infty}^{\infty} e^{-\omega_m(\tau_1 - \tau_2)} \hat{\mathcal{G}}^{\mathrm{M}}(\omega_m),$$

$$\hat{\Sigma}^{\mathrm{M}}(\tau_1, \tau_2) = \frac{1}{-i\beta} \sum_{m=-\infty}^{\infty} e^{-\omega_m(\tau_1 - \tau_2)} \hat{\Sigma}^{\mathrm{M}}(\omega_m),$$

with ω_m the Matsubara frequencies, and convert (9.32) and (9.33) into a system of algebraic equations for the $\hat{\mathcal{G}}^{\mathrm{M}}(\omega_m)$s:

$$\hat{\mathcal{G}}^{\mathrm{M}}(\omega_m) = \frac{1}{\omega_m - \hat{h}^{\mathrm{M}} - \hat{\Sigma}^{\mathrm{M}}(\omega_m)}. \tag{9.34}$$

In general $\hat{\Sigma}^{\mathrm{M}}(\omega_m)$ depends on all the Matsubara coefficients $\{\hat{\mathcal{G}}^{\mathrm{M}}(\omega_n)\}$. Thus (9.34) is a coupled system of equations for the unknown $\hat{\mathcal{G}}^{\mathrm{M}}(\omega_n)$. The solution of this system is the preliminary step to solving the equations of motion and it amounts to determining the initial preparation.

With the Matsubara component at our disposal we can calculate all other components by time-propagation. The equation for the right Green's function is obtained by taking $z_1 = t_-$

or t_+ and $z_2 = t_0 - \mathrm{i}\tau$ in (9.30) and reads

$$\boxed{\left[\mathrm{i}\frac{d}{dt} - \hat{h}(t)\right]\hat{\mathcal{G}}^\rceil(t,\tau) = \left[\hat{\Sigma}^\mathrm{R}\cdot\hat{\mathcal{G}}^\rceil + \hat{\Sigma}^\rceil\star\hat{\mathcal{G}}^\mathrm{M}\right](t,\tau)}$$ (9.35)

Similarly, for $z_1 = t_0 - \mathrm{i}\tau$ and $z_2 = t_-$ or t_+ the equation of motion (9.31) yields

$$\boxed{\hat{\mathcal{G}}^\lceil(\tau,t)\left[-\mathrm{i}\frac{\overleftarrow{d}}{dt} - \hat{h}(t)\right] = \left[\hat{\mathcal{G}}^\lceil\cdot\hat{\Sigma}^\mathrm{A} + \hat{\mathcal{G}}^\mathrm{M}\star\hat{\Sigma}^\lceil\right](\tau,t)}$$ (9.36)

At fixed τ (9.35) and (9.36) are first order integro-differential equations in t which must be solved with initial conditions

$$\hat{\mathcal{G}}^\rceil(0,\tau) = \hat{\mathcal{G}}^\mathrm{M}(0,\tau), \qquad \hat{\mathcal{G}}^\lceil(\tau,0) = \hat{\mathcal{G}}^\mathrm{M}(\tau,0).$$ (9.37)

The retarded/advanced as well as the left/right components of the self-energy in (9.35) and (9.36) depend not only on $G^\rceil$ and $G^\lceil$ but also on the lesser and greater Green's functions. Therefore, (9.35) and (9.36) do not form a close set of equations for $G^\rceil$ and $G^\lceil$. To close the set we need the equations of motion for $G^\lessgtr$. These can easily be obtained by setting $z_1 = t_{1\pm}$ and $z_2 = t_{2\mp}$ in (9.30) and (9.31)

$$\boxed{\left[\mathrm{i}\frac{d}{dt_1} - \hat{h}(t_1)\right]\hat{\mathcal{G}}^\lessgtr(t_1,t_2) = \left[\hat{\Sigma}^\lessgtr\cdot\hat{\mathcal{G}}^\mathrm{A} + \hat{\Sigma}^\mathrm{R}\cdot\hat{\mathcal{G}}^\lessgtr + \hat{\Sigma}^\rceil\star\hat{\mathcal{G}}^\lceil\right](t_1,t_2)}$$ (9.38)

$$\boxed{\hat{\mathcal{G}}^\lessgtr(t_1,t_2)\left[-\mathrm{i}\frac{\overleftarrow{d}}{dt_2} - \hat{h}(t_2)\right] = \left[\hat{\mathcal{G}}^\lessgtr\cdot\hat{\Sigma}^\mathrm{A} + \hat{\mathcal{G}}^\mathrm{R}\cdot\hat{\Sigma}^\lessgtr + \hat{\mathcal{G}}^\rceil\star\hat{\Sigma}^\lceil\right](t_1,t_2)}$$ (9.39)

which must be solved with initial conditions

$$\hat{\mathcal{G}}^<(t_0,t_0) = \hat{\mathcal{G}}^\mathrm{M}(0,0^+), \qquad \hat{\mathcal{G}}^>(t_0,t_0) = \hat{\mathcal{G}}^\mathrm{M}(0^+,0).$$ (9.40)

The set of equations (9.35), (9.36), (9.38), and (9.39) are known as the *Kadanoff–Baym equations* [41, 68]. The Kadanoff–Baym equations, together with the initial conditions (9.37) and (9.40), completely determine the Green's function with one and two real-times once a choice for the self-energy is made. It is worth noticing that the Kadanoff–Baym equations are invariant under a gauge transformation for any Φ-derivable self-energy. Indeed, if $G(1;2) \to e^{\mathrm{i}\Lambda(1)}G(1;2)e^{-\mathrm{i}\Lambda(2)}$ then the self-energy $\Sigma(1;2) \to e^{\mathrm{i}\Lambda(1)}\Sigma(1;2)e^{-\mathrm{i}\Lambda(2)}$ as follows directly from the structure of the self-energy diagrams.

Practical applications of the Kadanoff–Baym equations are discussed in Chapter 16 where we present results on the dynamics of interacting systems for different approximate self-energies. We still need to learn how to construct the self-energy before we can use these equations. Nevertheless, we can already derive some exact relations that are of great help in the actual implementation. From the definition of the Green's function, we have already derived the relation (6.23) which we rewrite below for convenience

$$\boxed{\hat{\mathcal{G}}^\lessgtr(t_1,t_2) = -\left[\hat{\mathcal{G}}^\lessgtr(t_2,t_1)\right]^\dagger}$$ (9.41)

In a similar way it is easy to show that (as always the upper/lower sign refers to bosons/fermions)

$$\hat{\mathcal{G}}^{\lceil}(\tau, t) = \mp \left[\hat{\mathcal{G}}^{\rceil}(t, \beta - \tau) \right]^{\dagger} \tag{9.42}$$

and

$$\hat{\mathcal{G}}^{M}(\tau_1, \tau_2) = - \left[\hat{\mathcal{G}}^{M}(\tau_1, \tau_2) \right]^{\dagger} \tag{9.43}$$

These properties of the Green's function can be transferred directly to the self-energy. Consider, for instance, the adjoint of (9.32). Taking into account that the derivative of $\hat{\mathcal{G}}^{M}(\tau_1, \tau_2)$ with respect to τ_1 is *minus* the derivative with respect to τ_2, and using (9.43), we find

$$-\hat{\mathcal{G}}^{M}(\tau_1, \tau_2) \left[\frac{\overleftarrow{d}}{d\tau_2} - \hat{h}^{M} \right] = -i\delta(\tau_1 - \tau_2) - i \int_0^{\beta} d\tau \left(-\hat{\mathcal{G}}^{M}(\tau, \tau_2) \right) \left[\hat{\Sigma}^{M}(\tau_1, \tau) \right]^{\dagger}.$$

We rename the integration variable $\tau \to \tau_1 + \tau_2 - \tau$. Then the argument of the Green's function becomes $\tau_1 - \tau$ while the argument of the self-energy becomes $\tau - \tau_2$ (recall that these quantities depend only on the time difference). Furthermore the domain of integration remains $(0, \beta)$ since G and Σ are (anti)periodic. Comparing the resulting equation with (9.33) we conclude that

$$\int_0^{\beta} d\tau \, \hat{\mathcal{G}}^{M}(\tau_1, \tau) \hat{\Sigma}^{M}(\tau, \tau_2) = - \int_0^{\beta} d\tau \, \hat{\mathcal{G}}^{M}(\tau_1, \tau) \left[\hat{\Sigma}^{M}(\tau, \tau_2) \right]^{\dagger}.$$

This equation must be true for all τ_1 and τ_2, and since $\hat{\mathcal{G}}^{M}$ is invertible, i.e., $\hat{\mathcal{G}}^{M} \star \hat{\Sigma}^{M} = 0$ only for $\hat{\Sigma}^{M} = 0$ [see Section 9.1], it follows that[7]

$$\hat{\Sigma}^{M}(\tau_1, \tau_2) = - \left[\hat{\Sigma}^{M}(\tau_1, \tau_2) \right]^{\dagger} \tag{9.44}$$

Vice versa, if the self-energy satisfies (9.44) then the Matsubara Green's function satisfies (9.43). In a similar way we can deduce the relations between the left, right, lesser, greater self-energy and the corresponding adjoint quantities. In this case, however, we must use the four coupled equations (9.35), (9.36), (9.38), and (9.39) at the same time. The derivation is a bit more lengthy but the final result is predictable. We therefore leave it as an exercise for the reader to prove that

$$\hat{\Sigma}^{\lessgtr}(t_1, t_2) = - \left[\hat{\Sigma}^{\lessgtr}(t_2, t_1) \right]^{\dagger} \tag{9.45}$$

$$\hat{\Sigma}^{\lceil}(\tau, t) = \mp \left[\hat{\Sigma}^{\rceil}(t, \beta - \tau) \right]^{\dagger} \tag{9.46}$$

---

[7]We could also have derived (9.44) directly from (9.34).

In particular (9.45) implies

$$\boxed{\hat{\Sigma}^{\mathrm{R}}(t_1, t_2) = \left[\hat{\Sigma}^{\mathrm{A}}(t_2, t_1)\right]^\dagger} \tag{9.47}$$

This last relation could have been deduced also from the retarded/advanced component of the Dyson equation (9.17) together with the property (6.24). To summarize, the self-energy has the same symmetry properties as the Green's function under complex conjugation.

The r.h.s. of the Kadanoff–Baym equations is a Keldysh component of either the convolution

$$\hat{\mathcal{I}}_L(z_1, z_2) \equiv \int_\gamma dz\, \hat{\Sigma}(z_1, z)\hat{\mathcal{G}}(z, z_2),$$

or the convolution

$$\hat{\mathcal{I}}_R(z_1, z_2) \equiv \int_\gamma dz\, \hat{\mathcal{G}}(z_1, z)\hat{\Sigma}(z, z_2).$$

The quantities $\hat{\mathcal{I}}_L$ and $\hat{\mathcal{I}}_R$ belong to the Keldysh space and are usually referred to as the *collision integrals* since they contain information on how the particles scatter. From the above relations for G and Σ it follows that

$$\hat{\mathcal{I}}_L^{\lceil}(\tau, t) = \mp \left[\hat{\mathcal{I}}_R^{\rceil}(t, \beta - \tau)\right]^\dagger, \tag{9.48}$$

$$\hat{\mathcal{I}}_L^{\lessgtr}(t_1, t_2) = - \left[\hat{\mathcal{I}}_R^{\lessgtr}(t_2, t_1)\right]^\dagger. \tag{9.49}$$

Therefore, it is sufficient to calculate, say, $\hat{\mathcal{I}}_L^<$, $\hat{\mathcal{I}}_R^>$, and $\hat{\mathcal{I}}_L^{\lceil}$ in order to solve the Kadanoff–Baym equations.

A final remark is concerned with the solution of the Kadanoff–Baym equations for systems in equilibrium. We know that in this case $\hat{\mathcal{G}}^{\lessgtr}$, and hence also $\hat{\Sigma}^{\lessgtr}$, depends on the time difference only. However, it is not obvious at all that $\hat{\mathcal{G}}^{\lessgtr}(t_1 - t_2)$ can be a solution of (9.38) and (9.39), since neither the two real-time convolutions between t_0 and ∞ nor the convolution along the vertical track depend on $t_1 - t_2$. Is there anything wrong with these equations? Before answering we need to derive some exact properties of the self-energy for systems in equilibrium.

Exercise 9.4. Prove (9.42) and (9.43).

Exercise 9.5. Prove (9.45) and (9.46).

9.5 Fluctuation-dissipation theorem for the self-energy

Let us consider a system in equilibrium and hence with $\hat{H}^{\mathrm{M}} = \hat{H} - \mu\hat{N}$. In Section 6.3.2 we saw that the equilibrium Green's function satisfies

$$\hat{\mathcal{G}}^{\lceil}(\tau, t') = e^{\mu\tau}\hat{\mathcal{G}}^>(t_0 - i\tau, t'), \qquad \hat{\mathcal{G}}^{\rceil}(t, \tau') = \hat{\mathcal{G}}^<(t, t_0 - i\tau')e^{-\mu\tau'}. \tag{9.50}$$

In a completely similar way one can also prove that

$$\hat{\mathcal{G}}^{\mathrm{M}}(\tau,\tau') = \begin{cases} e^{\mu\tau}\hat{\mathcal{G}}^{>}(t_0 - i\tau, t_0 - i\tau')e^{-\mu\tau'} & \tau > \tau' \\ e^{\mu\tau}\hat{\mathcal{G}}^{<}(t_0 - i\tau, t_0 - i\tau')e^{-\mu\tau'} & \tau < \tau' \end{cases}. \tag{9.51}$$

An alternative strategy to derive (9.50) and (9.51) consists in considering the Green's function $\hat{\mathcal{G}}_\mu$ of a system with constant Hamiltonian $\hat{H}(z) = \hat{H}^{\mathrm{M}} = \hat{H} - \mu\hat{N}$ along the entire contour. The Green's function $\hat{\mathcal{G}}_\mu$ differs from $\hat{\mathcal{G}}$ since in $\hat{\mathcal{G}}$ the Hamiltonian $\hat{H}(z) = \hat{H} - \mu\hat{N}$ for z on the vertical track and $\hat{H}(z) = \hat{H}$ for z on the horizontal branches. The relation between $\hat{\mathcal{G}}$ and $\hat{\mathcal{G}}_\mu$ is

$$\hat{\mathcal{G}}(z,z') = \hat{\mathcal{G}}_\mu(z,z') \times \begin{cases} e^{-i\mu(z-z')} & z, z' \text{ on the horizontal branches} \\ e^{-i\mu(z-t_0)} & z \text{ on the horizontal branches and } z' \text{ on } \gamma^{\mathrm{M}} \\ e^{i\mu(z'-t_0)} & z \text{ on } \gamma^{\mathrm{M}} \text{ and } z' \text{ on the horizontal branches} \\ 1 & z, z' \text{ on } \gamma^{\mathrm{M}} \end{cases}$$

as it follows directly from

$$e^{i(\hat{H}-\mu\hat{N})(t-t_0)}\,\hat{\psi}(\mathbf{x})\,e^{-i(\hat{H}-\mu\hat{N})(t-t_0)} = e^{i\mu(t-t_0)}\,e^{i\hat{H}(t-t_0)}\,\hat{\psi}(\mathbf{x})\,e^{-i\hat{H}(t-t_0)}$$

and the like for the creation operator. The crucial observation is that $\hat{\mathcal{G}}_\mu^{\lessgtr}(z,z')$ depends only on $z - z'$ for *all* z and z' since the Hamiltonian is independent of z. This implies that if we calculate $\hat{\mathcal{G}}_\mu^{>}(t_0, t')$ and then replace $t_0 \to t_0 - i\tau$ we get

$$\hat{\mathcal{G}}_\mu^{>}(t_0 - i\tau, t') = \hat{\mathcal{G}}_\mu^{>}(z = t_0 - i\tau, t'_\pm) = \hat{\mathcal{G}}_\mu^{\lceil}(\tau, t').$$

Using the relation between $\hat{\mathcal{G}}$ and $\hat{\mathcal{G}}_\mu$ the first identity in (9.50) follows. The second identity in (9.50) as well as the identity (9.51) can be deduced in a similar manner. This alternative derivation allows us to prove a fluctuation–dissipation theorem for several many-body quantities. We show here how it works for the self-energy. Like the Green's function $\hat{\mathcal{G}}_\mu^{\lessgtr}(z,z')$ also the self-energy $\hat{\Sigma}_\mu^{\lessgtr}(z,z')$ depends only on $z - z'$. Therefore if we calculate $\hat{\Sigma}_\mu^{>}(t_0, t')$ and then replace $t_0 \to t_0 - i\tau$ we get

$$\hat{\Sigma}_\mu^{>}(t_0 - i\tau, t') = \hat{\Sigma}_\mu^{>}(z = t_0 - i\tau, t'_\pm) = \hat{\Sigma}_\mu^{\lceil}(\tau, t').$$

Consider now a generic self-energy diagram for $\hat{\Sigma}_\mu$. Expressing every $\hat{\mathcal{G}}_\mu$ in terms of $\hat{\mathcal{G}}$ the phase factors cancel out in all internal vertices. Consequently the relation between $\hat{\Sigma}_\mu$ and $\hat{\Sigma}$ is the same as the relation between $\hat{\mathcal{G}}_\mu$ and $\hat{\mathcal{G}}$. Combining this relation with the above result we get $\hat{\Sigma}^{\lceil}(\tau, t') = e^{\mu\tau}\hat{\Sigma}^{>}(t_0 - i\tau, t')$. In a similar way we can work out the other combinations of contour arguments. In conclusion we have

$$\hat{\Sigma}^{\lceil}(\tau, t') = e^{\mu\tau}\hat{\Sigma}^{>}(t_0 - i\tau, t'), \qquad \hat{\Sigma}^{\rceil}(t, \tau') = \hat{\Sigma}^{<}(t, t_0 - i\tau')e^{-\mu\tau'}, \tag{9.52}$$

and

$$\hat{\Sigma}^{M}(\tau,\tau') = \begin{cases} e^{\mu\tau}\hat{\Sigma}^{>}(t_0 - i\tau, t_0 - i\tau')e^{-\mu\tau'} & \tau > \tau' \\ e^{\mu\tau}\hat{\Sigma}^{<}(t_0 - i\tau, t_0 - i\tau')e^{-\mu\tau'} & \tau < \tau' \end{cases}. \qquad (9.53)$$

We can combine these relations with the KMS boundary conditions (9.15) and (9.16) to get

$$\begin{aligned} \hat{\Sigma}^{<}(t_0, t') &= \hat{\Sigma}(t_{0-}, t'_{+}) \\ &= \pm\hat{\Sigma}(t_0 - i\beta, t'_{+}) \\ &= \pm\hat{\Sigma}^{\lceil}(\beta, t') \\ &= \pm e^{\mu\beta}\hat{\Sigma}^{>}(t_0 - i\beta, t'). \end{aligned}$$

Fourier transforming both sides of this equation we find

$$\hat{\Sigma}^{>}(\omega) = \pm e^{\beta(\omega - \mu)}\hat{\Sigma}^{<}(\omega), \qquad (9.54)$$

which allows us to prove a fluctuation–dissipation theorem for the self-energy. If we define the *rate operator*[8]

$$\boxed{\hat{\Gamma}(\omega) \equiv i[\hat{\Sigma}^{>}(\omega) - \hat{\Sigma}^{<}(\omega)] = i[\hat{\Sigma}^{R}(\omega) - \hat{\Sigma}^{A}(\omega)]} \qquad (9.55)$$

then we can express the lesser and greater self-energy in terms of $\hat{\Gamma}$ as follows

$$\boxed{\begin{aligned} \hat{\Sigma}^{<}(\omega) &= \mp i f(\omega - \mu)\hat{\Gamma}(\omega) \\ \hat{\Sigma}^{>}(\omega) &= -i\bar{f}(\omega - \mu)\hat{\Gamma}(\omega) \end{aligned}} \qquad (9.56)$$

The rate operator is self-adjoint since the retarded self-energy is the adjoint of the advanced self-energy, see (9.47). We further observe that in (9.55) we could replace $\hat{\Sigma} \to \hat{\Sigma}_c$ since $\hat{\Sigma}_{HF}^{\lessgtr} = 0$ (or, equivalently $\hat{\Sigma}_{HF}^{R} - \hat{\Sigma}_{HF}^{A} = 0$). The rate operator for the self-energy is the analogue of the spectral function operator for the Green's function. From a knowledge of $\hat{\Gamma}$ we can determine all Keldysh components of $\hat{\Sigma}$ with real-time arguments. The lesser and greater self-energies are obtained from (9.56). The retarded and advanced correlation self-energies follow from the Fourier transform of $\hat{\Sigma}_c^{R}(t,t') = \theta(t - t')[\hat{\Sigma}^{>}(t,t') - \hat{\Sigma}^{<}(t,t')]$ and the like for $\hat{\Sigma}_c^{A}$, and read

$$\boxed{\hat{\Sigma}_c^{R/A}(\omega) = \int \frac{d\omega'}{2\pi} \frac{\hat{\Gamma}(\omega')}{\omega - \omega' \pm i\eta}} \qquad (9.57)$$

Another important relation that we can derive in equilibrium pertains to the connection between $\hat{\Sigma}^{M}$ and $\hat{\Sigma}^{R/A}$. Taking the retarded/advanced component of the Dyson equation (9.17) and Fourier transforming we find

$$\hat{G}^{R/A}(\omega) = \hat{G}_0^{R/A}(\omega) + \hat{G}_0^{R/A}(\omega)\hat{\Sigma}^{R/A}(\omega)\hat{G}^{R/A}(\omega),$$

---

[8]The reason for the name "rate operator" becomes clear in Chapter 13.

where the noninteracting Green's function is $\hat{\mathcal{G}}_0^{R/A}(\omega) = 1/(\omega - \hat{h} \pm i\eta)$, see (6.58) and (6.59). Solving for $\hat{\mathcal{G}}^{R/A}(\omega)$ we get

$$\hat{\mathcal{G}}^{R/A}(\omega) = \frac{1}{\omega - \hat{h} - \hat{\Sigma}^{R/A}(\omega) \pm i\eta}.$$

Comparing this result with (9.34) and using (6.82) we conclude that

$$\hat{\Sigma}^M(\zeta) = \begin{cases} \hat{\Sigma}^R(\zeta + \mu) & \text{for Im}[\zeta] > 0 \\ \hat{\Sigma}^A(\zeta + \mu) & \text{for Im}[\zeta] < 0 \end{cases}, \tag{9.58}$$

which is the same relation satisfied by the Green's function. In particular, for $\zeta = \omega \pm i\eta$ we find

$$\boxed{\hat{\Sigma}^M(\omega \pm i\eta) = \hat{\Sigma}^{R/A}(\omega + \mu)} \tag{9.59}$$

We observe that for a system of fermions at zero temperature the Fermi function $f(\omega) = \theta(-\omega)$. Assuming that $\hat{\Sigma}^{\lessgtr}(\omega)$ is a continuous function of ω, (9.56) implies $\hat{\Sigma}^{\lessgtr}(\mu) = 0$ and hence $\hat{\Gamma}(\mu) = 0$. Consequently the retarded self-energy is equal to the advanced self-energy for $\omega = \mu$. Then (9.59) tells us that the Matsubara Green's function is continuous when the complex frequency crosses the real axis in $\omega = 0$.

9.6 Recovering equilibrium from the Kadanoff-Baym equations

The fluctuation–dissipation theorem for the Green's function and the self-energy is all that we need to show that in equilibrium the r.h.s. of (9.38) and (9.39) depends only on $t_1 - t_2$ *and* is independent of t_0. In particular we can prove that

$$\left[\hat{\Sigma}_c^{\lessgtr} \cdot \hat{\mathcal{G}}^A + \hat{\Sigma}_c^R \cdot \hat{\mathcal{G}}^{\lessgtr} + \hat{\Sigma}_c^\rceil \star \hat{\mathcal{G}}^\lceil\right](t_1, t_2) = \int \frac{d\omega}{2\pi} e^{-i\omega(t_1 - t_2)} \left[\hat{\Sigma}_c^{\lessgtr}(\omega)\hat{\mathcal{G}}^A(\omega) + \hat{\Sigma}_c^R(\omega)\hat{\mathcal{G}}^{\lessgtr}(\omega)\right] \tag{9.60}$$

$$\left[\hat{\mathcal{G}}^{\lessgtr} \cdot \hat{\Sigma}_c^A + \hat{\mathcal{G}}^R \cdot \hat{\Sigma}_c^{\lessgtr} + \hat{\mathcal{G}}^\rceil \star \hat{\Sigma}_c^\lceil\right](t_1, t_2) = \int \frac{d\omega}{2\pi} e^{-i\omega(t_1 - t_2)} \left[\hat{\mathcal{G}}^{\lessgtr}(\omega)\hat{\Sigma}_c^A(\omega) + \hat{\mathcal{G}}^R(\omega)\hat{\Sigma}_c^{\lessgtr}(\omega)\right] \tag{9.61}$$

for all t_0. In these equations we have replaced $\hat{\Sigma} \to \hat{\Sigma}_c$ since

$$\hat{\Sigma}^{\lessgtr, \lceil, \rceil} = \hat{\Sigma}_c^{\lessgtr, \lceil, \rceil}, \qquad \hat{\Sigma}^{R/A} = \hat{\Sigma}_{HF}^{R/A} + \hat{\Sigma}_c^{R/A},$$

and the products $\hat{\Sigma}_{HF}^R \cdot \hat{\mathcal{G}}^{\lessgtr}$ and $\hat{\mathcal{G}}^{\lessgtr} \cdot \hat{\Sigma}_{HF}^A$ depend only on the time difference when $\hat{\mathcal{G}}^{\lessgtr}$ depends only on the time difference ($\hat{\Sigma}_{HF}^{R/A}$ is local in time). Thus, for our purposes it is enough to prove (9.60) and (9.61). We should point out that standard derivations of (9.60) and (9.61) typically require a few extra (but superfluous) assumptions. Either one uses the adiabatic assumption, in which case $\hat{\Sigma}_c^\rceil = \hat{\Sigma}_c^\lceil = 0$ (since the self-energy vanishes along the imaginary track) while the convolutions along the real-time axis become the r.h.s. of (9.60) and (9.61) (since $t_0 \to -\infty$). Or, alternatively, one assumes that

$$\lim_{t_0 \to -\infty} \hat{\Sigma}_c^{\rceil}(t, \tau) = \lim_{t_0 \to -\infty} \hat{\Sigma}_c^{\lceil}(\tau, t) = 0, \tag{9.62}$$

which is often satisfied in systems with infinitely many degrees of freedom. However, in systems with an arbitrary large but finite single-particle basis the Green's function is an oscillatory function and so is the self-energy, see again the discussion in Section 6.1.3. Therefore (9.62) is not satisfied in real or model systems with a discrete spectrum. It is, however, more important to realize that the standard derivations "prove" (9.60) and (9.61) only for $t_0 \to -\infty$. In the following we prove (9.60) and (9.61) for all t_0 and without any extra assumptions. The derivation below nicely illustrates how the apparent dependence on t_0 disappears. Equations (9.60) and (9.61) are therefore much more fundamental than they are commonly thought to be.

Let us consider the lesser version of (9.60). The l.h.s. is the lesser component of the collision integral $\hat{\mathcal{I}}_L^<(t_1, t_2)$. We use the identity (9.50) for $\hat{\mathcal{G}}^{\lceil}$ and the identity (9.52) for $\hat{\Sigma}^{\rceil} = \hat{\Sigma}_c^{\rceil}$. Expanding all Green's functions and self-energies in Fourier integrals we get

$$\hat{\mathcal{I}}_L^<(t_1, t_2) = \int \frac{d\omega_1}{2\pi} \frac{d\omega_2}{2\pi} e^{-i\omega_1 t_1 + i\omega_2 t_2} \left[\int_{t_0}^{t_2} dt\, e^{i(\omega_1 - \omega_2)t} \hat{\Sigma}_c^<(\omega_1) \hat{\mathcal{G}}^A(\omega_2) \right.$$
$$\left. + \int_{t_0}^{t_1} dt\, e^{i(\omega_1 - \omega_2)t} \hat{\Sigma}_c^R(\omega_1) \hat{\mathcal{G}}^<(\omega_2) - i \int_0^{\beta} d\tau\, e^{(\omega_1 - \omega_2)(it_0 + \tau)} \hat{\Sigma}_c^<(\omega_1) \hat{\mathcal{G}}^>(\omega_2) \right].$$

For all these integrals to be well behaved when $t_0 \to -\infty$ we give to ω_1 a small negative imaginary part, $\omega_1 \to \omega_1 - i\eta/2$, and to ω_2 a small positive imaginary part, $\omega_2 \to \omega_2 + i\eta/2$. This is just a regularization and it has nothing to do with the adiabatic assumption.[9] With this regularization (9.60) is easily recovered in the limit $t_0 \to -\infty$. In the following we keep t_0 finite and show that (9.60) is still true. Performing the integrals over time, and using the fluctuation–dissipation theorem to express everything in terms of retarded and advanced

---

[9]Suppose that we want to recover the result

$$\int_{-\infty}^{\infty} dt\, e^{i\omega t} = 2\pi\delta(\omega)$$

from the integral $I(T) = \int_{-T}^{T} dt\, e^{i\omega t}$ when $T \to \infty$. Then we can write

$$I(T) = \int_{-T}^{0} dt\, e^{i\omega t} + \int_0^T dt\, e^{i\omega t} = \int_{-T}^{0} dt\, e^{i(\omega - i\eta)t} + \int_0^T dt\, e^{i(\omega + i\eta)t},$$

where in the last step we have simply regularized the integral so that it is well behaved for $T \to \infty$. At the end of the calculation we send $\eta \to 0$. Performing the integral we find

$$I(T) = \frac{1 - e^{-i(\omega - i\eta)T}}{i(\omega - i\eta)} + \frac{e^{i(\omega + i\eta)T} - 1}{i(\omega + i\eta)} \xrightarrow{T \to \infty} \frac{1}{i(\omega - i\eta)} - \frac{1}{i(\omega + i\eta)}.$$

Using the Cauchy relation $1/(\omega \pm i\eta) = P(1/\omega) \mp i\pi\delta(\omega)$, with P the principal part, we recover the δ-function result as the regularized limit of $I(T)$.

quantities, we find

$$
\hat{\mathcal{I}}_L^<(t_1, t_2) = \int \frac{d\omega_1}{2\pi} \frac{d\omega_2}{2\pi} \frac{e^{-i\omega_1 t_1 + i\omega_2 t_2}}{i(\omega_1 - \omega_2 - i\eta)}
$$
$$
\times \left[\pm f_1 \left(e^{i(\omega_1 - \omega_2)t_2} - e^{i(\omega_1 - \omega_2)t_0} \right) \left(\hat{\Sigma}_{c,1}^R - \hat{\Sigma}_{c,1}^A \right) \hat{\mathcal{G}}_2^A \right.
$$
$$
\pm f_2 \left(e^{i(\omega_1 - \omega_2)t_1} - e^{i(\omega_1 - \omega_2)t_0} \right) \hat{\Sigma}_{c,1}^R \left(\hat{\mathcal{G}}_2^R - \hat{\mathcal{G}}_2^A \right)
$$
$$
\left. \pm f_1 \bar{f}_2 \left(e^{(\omega_1 - \omega_2)\beta} - 1 \right) e^{i(\omega_1 - \omega_2)t_0} \left(\hat{\Sigma}_{c,1}^R - \hat{\Sigma}_{c,1}^A \right) \left(\hat{\mathcal{G}}_2^R - \hat{\mathcal{G}}_2^A \right) \right], \quad (9.63)
$$

where we have introduced the short-hand notation f_1 to denote the Bose/Fermi function $f(\omega_1)$ and similarly $f_2 = f(\omega_2)$ and *mutatis mutandis* the self-energy and the Green's function. Next we observe that

$$
f_1 \bar{f}_2 \left(e^{(\omega_1 - \omega_2)\beta} - 1 \right) = f_2 - f_1.
$$

Using this relation in (9.63) we achieve a considerable simplification since many terms cancel out and we remain with

$$
\hat{\mathcal{I}}_L^<(t_1, t_2) = \int \frac{d\omega_1}{2\pi} \frac{d\omega_2}{2\pi} \frac{e^{-i\omega_1 t_1 + i\omega_2 t_2}}{i(\omega_1 - \omega_2 - i\eta)} \left[e^{i(\omega_1 - \omega_2)t_2} \hat{\Sigma}_{c,1}^< \hat{\mathcal{G}}_2^A + e^{i(\omega_1 - \omega_2)t_1} \hat{\Sigma}_{c,1}^R \hat{\mathcal{G}}_2^< \right.
$$
$$
\left. - e^{i(\omega_1 - \omega_2)t_0} \hat{\Sigma}_{c,1}^< \hat{\mathcal{G}}_2^R - e^{i(\omega_1 - \omega_2)t_0} \hat{\Sigma}_{c,1}^A \hat{\mathcal{G}}_2^< \right]. \quad (9.64)
$$

To get rid of the t_0-dependence we exploit the identity

$$
0 = \int_{t_0}^{t_2} dt\, \hat{\Sigma}_c^<(t_1, t) \hat{\mathcal{G}}^R(t, t_2),
$$

which follows from the fact that the retarded Green's function vanishes whenever its first argument is smaller than the second. In Fourier space this identity looks much more interesting since

$$
0 = \int \frac{d\omega_1}{2\pi} \frac{d\omega_2}{2\pi} \frac{e^{-i\omega_1 t_1 + i\omega_2 t_2}}{i(\omega_1 - \omega_2 - i\eta)} \left(e^{i(\omega_1 - \omega_2)t_2} - e^{i(\omega_1 - \omega_2)t_0} \right) \hat{\Sigma}_{c,1}^< \hat{\mathcal{G}}_2^R.
$$

Thus, we see that we can replace t_0 with t_2 in the first term of the second line of (9.64). In a similar way we can show that the second t_0 in (9.64) can be replaced with t_1 and hence the collision integral can be rewritten as

$$
\hat{\mathcal{I}}_L^<(t_1, t_2) = \int \frac{d\omega_1}{2\pi} e^{-i\omega_1(t_1 - t_2)} \hat{\Sigma}_c^<(\omega_1) \underbrace{\int \frac{d\omega_2}{2\pi} \frac{i[\hat{\mathcal{G}}^R(\omega_2) - \hat{\mathcal{G}}^A(\omega_2)]}{\omega_1 - \omega_2 - i\eta}}_{\hat{\mathcal{G}}^A(\omega_1)}
$$
$$
+ \int \frac{d\omega_2}{2\pi} e^{-i\omega_2(t_1 - t_2)} \underbrace{\int \frac{d\omega_1}{2\pi} \frac{i[\hat{\Sigma}_c^R(\omega_1) - \hat{\Sigma}_c^A(\omega_1)]}{\omega_2 - \omega_1 + i\eta}}_{\hat{\Sigma}_c^R(\omega_2)} \hat{\mathcal{G}}^<(\omega_2).
$$

The quantities below the underbraces are the result of the frequency integral, see (6.93) and (9.57). We have thus proved the lesser version of (9.60) for all t_0. The reader can verify that the greater version of (9.60) as well as (9.61) can be derived in a similar manner. We make use of these relations in Chapter 13 to derive an exact expression for the interaction energy in terms of a real-frequency integral of self-energies and Green's functions.

9.7 Formal solution of the Kadanoff–Baym equations

The aim of solving the Kadanoff–Baym equations is to obtain the lesser and greater Green's functions from which to calculate, e.g., the time-dependent ensemble average of any one-body operator, the time-dependent total energy, the addition and removal energies, etc.. In this section we generalize the formal solution (6.52) to interacting systems using the Dyson equation (i.e., the integral form of the Kadanoff–Baym equations). From a practical point of view it is much more advantageous to solve the Kadanoff–Baym equations than the integral Dyson equation. However, there exist situations for which the formal solution simplifies considerably. As we shall see, the simplified solution is extremely useful to derive analytic results and/or to set up numerical algorithms for calculating $\hat{\mathcal{G}}^{\lessgtr}$ without explicitly propagating the Green's function in time.

We follow the derivation of Ref. [54]. The starting point is the Dyson equation (9.17) for the Green's function

$$\hat{\mathcal{G}}(z, z') = \hat{\mathcal{G}}_0(z, z') + \int_\gamma d\bar{z} d\bar{z}' \, \hat{\mathcal{G}}(z, \bar{z}) \hat{\Sigma}(\bar{z}, \bar{z}') \hat{\mathcal{G}}_0(\bar{z}', z'), \tag{9.65}$$

where $\hat{\mathcal{G}}_0$ is the noninteracting Green's function. We separate the self-energy $\hat{\Sigma} = \hat{\Sigma}_{\mathrm{HF}} + \hat{\Sigma}_{\mathrm{c}}$ into the Hartree–Fock self-energy and the correlation self-energy (9.14). Then the Dyson equation can be rewritten as

$$\hat{\mathcal{G}}(z, z') = \hat{\mathcal{G}}_{\mathrm{HF}}(z, z') + \int_\gamma d\bar{z} d\bar{z}' \, \hat{\mathcal{G}}(z, \bar{z}) \hat{\Sigma}_{\mathrm{c}}(\bar{z}, \bar{z}') \hat{\mathcal{G}}_{\mathrm{HF}}(\bar{z}', z'), \tag{9.66}$$

with $\hat{\mathcal{G}}_{\mathrm{HF}}$ the Hartree–Fock Green's function. The equivalence between these two forms of the Dyson equation can easily be verified by acting on (9.65) from the right with $[-i\frac{\overleftarrow{d}}{dz'} - \hat{h}(z')]$ and on (9.66) from the right with $[-i\frac{\overleftarrow{d}}{dz'} - \hat{h}_{\mathrm{HF}}(z')]$. In both cases the result is the equation of motion (9.31). The advantage of using (9.66) is that the correlation self-energy is nonlocal in time and, in macroscopic interacting systems, often decays to zero when the separation between its time arguments approaches infinity. As we shall see this fact is at the basis of an important simplification in the long-time limit. Using the Langreth rules the lesser Green's function reads

$$\hat{\mathcal{G}}^< = \left[\delta + \hat{\mathcal{G}}^{\mathrm{R}} \cdot \hat{\Sigma}_{\mathrm{c}}^{\mathrm{R}} \right] \cdot \hat{\mathcal{G}}_{\mathrm{HF}}^< + \hat{\mathcal{G}}^< \cdot \hat{\Sigma}_{\mathrm{c}}^{\mathrm{A}} \cdot \hat{\mathcal{G}}_{\mathrm{HF}}^{\mathrm{A}} + \left[\hat{\mathcal{G}}^{\mathrm{R}} \cdot \hat{\Sigma}_{\mathrm{c}}^< + \hat{\mathcal{G}}^\rceil \star \hat{\Sigma}_{\mathrm{c}}^\lceil \right] \cdot \hat{\mathcal{G}}_{\mathrm{HF}}^{\mathrm{A}}$$

$$+ \hat{\mathcal{G}}^{\mathrm{R}} \cdot \hat{\Sigma}_{\mathrm{c}}^\rceil \star \hat{\mathcal{G}}_{\mathrm{HF}}^\lceil + \hat{\mathcal{G}}^\rceil \star \hat{\Sigma}_{\mathrm{c}}^{\mathrm{M}} \star \hat{\mathcal{G}}_{\mathrm{HF}}^\lceil,$$

and solving for $\hat{\mathcal{G}}^<$

$$\hat{\mathcal{G}}^< = \left[\delta + \hat{\mathcal{G}}^R \cdot \hat{\Sigma}_c^R\right] \cdot \hat{\mathcal{G}}_{HF}^< \cdot \left[\delta + \hat{\Sigma}_c^A \cdot \hat{\mathcal{G}}^A\right] + \left[\hat{\mathcal{G}}^R \cdot \hat{\Sigma}_c^< + \hat{\mathcal{G}}^\rceil \star \hat{\Sigma}_c^\lceil\right] \cdot \hat{\mathcal{G}}^A$$
$$+ \left[\hat{\mathcal{G}}^R \cdot \hat{\Sigma}_c^\rceil \star \hat{\mathcal{G}}_{HF}^\lceil + \hat{\mathcal{G}}^\rceil \star \hat{\Sigma}_c^M \star \hat{\mathcal{G}}_{HF}^\lceil\right] \cdot \left[\delta + \hat{\Sigma}_c^A \cdot \hat{\mathcal{G}}^A\right].$$

In obtaining this result we use the obvious identity (see also the useful exercise in Section 6.2.3)

$$\left[\delta - \hat{\Sigma}_c^A \cdot \hat{\mathcal{G}}^A\right] \cdot \left[\delta + \hat{\Sigma}_c^A \cdot \hat{\mathcal{G}}_{HF}^A\right] = \delta,$$

which is a direct consequence of the advanced Dyson equation. Next we observe that (6.52) is valid for both noninteracting and mean-field Green's functions and hence

$$\hat{\mathcal{G}}_{HF}^<(t, t') = \hat{\mathcal{G}}_{HF}^R(t, t_0)\hat{\mathcal{G}}_{HF}^<(t_0, t_0)\hat{\mathcal{G}}_{HF}^A(t_0, t'), \tag{9.67}$$

and similarly from (6.54)

$$\hat{\mathcal{G}}_{HF}^\lceil(\tau, t') = -i\hat{\mathcal{G}}_{HF}^M(\tau, 0)\hat{\mathcal{G}}_{HF}^A(t_0, t').$$

Therefore

$$\hat{\mathcal{G}}^<(t, t') = \hat{\mathcal{G}}^R(t, t_0)\hat{\mathcal{G}}_{HF}^<(t_0, t_0)\hat{\mathcal{G}}^A(t_0, t') + \left[\hat{\mathcal{G}}^R \cdot \hat{\Sigma}_c^< \cdot \hat{\mathcal{G}}^A\right](t, t')$$
$$+ \left[\hat{\mathcal{G}}^\rceil \star \hat{\Sigma}_c^\lceil \cdot \hat{\mathcal{G}}^A\right](t, t') - i\left[\hat{\mathcal{G}}^R \cdot \hat{\Sigma}_c^\rceil \star \hat{\mathcal{G}}_{HF}^M + \hat{\mathcal{G}}^\rceil \star \hat{\Sigma}_c^M \star \hat{\mathcal{G}}_{HF}^M\right](t, t_0)\hat{\mathcal{G}}^A(t_0, t').$$

In order to eliminate the right Green's function from this equation we again use the Dyson equation. The right component of (9.66) yields

$$\hat{\mathcal{G}}^\rceil \star \left[\delta - \hat{\Sigma}_c^M \star \hat{\mathcal{G}}_{HF}^M\right] = \left[\delta + \hat{\mathcal{G}}^R \cdot \hat{\Sigma}_c^R\right] \cdot \hat{\mathcal{G}}_{HF}^\rceil + \hat{\mathcal{G}}^R \cdot \hat{\Sigma}_c^\rceil \star \hat{\mathcal{G}}_{HF}^M.$$

Using the identity

$$\left[\delta - \hat{\Sigma}_c^M \star \hat{\mathcal{G}}^M\right] \cdot \left[\delta + \hat{\Sigma}_c^M \star \hat{\mathcal{G}}_{HF}^M\right] = \delta,$$

which follows from the Matsubara–Dyson equation, as well as (6.53) for Hartree–Fock Green's functions,

$$\hat{\mathcal{G}}_{HF}^\rceil(t, \tau) = i\hat{\mathcal{G}}_{HF}^R(t, t_0)\hat{\mathcal{G}}_{HF}^M(0, \tau),$$

we find

$$\hat{\mathcal{G}}^\rceil(t, \tau) = i\hat{\mathcal{G}}^R(t, t_0)\hat{\mathcal{G}}^M(0, \tau) + \left[\hat{\mathcal{G}}^R \cdot \hat{\Sigma}_c^\rceil \star \hat{\mathcal{G}}^M\right](t, \tau). \tag{9.68}$$

Substituting this result in the equation for $\hat{\mathcal{G}}^<$ we obtain the generalization of (6.52)

$$\boxed{\begin{aligned}\hat{\mathcal{G}}^<(t, t') &= \hat{\mathcal{G}}^R(t, t_0)\hat{\mathcal{G}}^<(t_0, t_0)\hat{\mathcal{G}}^A(t_0, t') + \left[\hat{\mathcal{G}}^R \cdot \left(\hat{\Sigma}_c^< + \hat{\Sigma}_c^\rceil \star \hat{\mathcal{G}}^M \star \hat{\Sigma}_c^\lceil\right) \cdot \hat{\mathcal{G}}^A\right] \\ &+ i\hat{\mathcal{G}}^R(t, t_0)\left[\hat{\mathcal{G}}^M \star \hat{\Sigma}_c^\lceil \cdot \hat{\mathcal{G}}^A\right](t_0, t') - i\left[\hat{\mathcal{G}}^R \cdot \hat{\Sigma}_c^\rceil \star \hat{\mathcal{G}}^M\right](t, t_0)\hat{\mathcal{G}}^A(t_0, t')\end{aligned}} \tag{9.69}$$

The formula for the greater Green's function is identical to (9.69) but with all the superscripts "$<$" replaced by "$>$". We see that for $\hat{\Sigma}_{\mathrm{c}} = 0$ only the first term survives, in agreement with the fact that in this case the Green's function is a mean-field Green's function and hence (9.67) holds. For nonvanishing correlation self-energies all terms must be retained and use of (9.69) for calculation of $\hat{\mathcal{G}}^{<}$ is computationally very demanding due to the large number of time convolutions involved.

It is important to clarify a point about the formal solution (9.69). Suppose that we have solved the equilibrium problem and hence that we know $\hat{\mathcal{G}}^{\mathrm{M}}$. Taking into account that $\hat{\mathcal{G}}^{\mathrm{R/A}}$ is defined in terms of $\hat{\mathcal{G}}^{\lessgtr}$, couldn't we use (9.68), (9.69) and the analogous equations for $\hat{\mathcal{G}}^{\lceil}$ and $\hat{\mathcal{G}}^{>}$ to determine $\hat{\mathcal{G}}^{\lessgtr}$, $\hat{\mathcal{G}}^{\rceil}$ and $\hat{\mathcal{G}}^{\lceil}$? If the answer were positive we would bump into a serious conundrum since these equations do not know anything about the Hamiltonian at positive times! In other words we would have the same Green's function independently of the external fields. Thus the answer to the above question must be negative and the equations (9.68), (9.69) and the like for $\hat{\mathcal{G}}^{\lceil}$ and $\hat{\mathcal{G}}^{>}$ cannot be independent from one another. We can easily show that this is the case in a system of noninteracting particles. Here we have

$$\hat{\mathcal{G}}^{\lessgtr}(t,t') = -[\hat{\mathcal{G}}^{>}(t,t_0) - \hat{\mathcal{G}}^{<}(t,t_0)]\,\hat{\mathcal{G}}^{\lessgtr}(t_0,t_0)\,[\hat{\mathcal{G}}^{>}(t_0,t') - \hat{\mathcal{G}}^{<}(t_0,t')],$$

where we have used $\hat{\mathcal{G}}^{\mathrm{R}}(t,t_0) = \hat{\mathcal{G}}^{>}(t,t_0) - \hat{\mathcal{G}}^{<}(t,t_0)$ and $\hat{\mathcal{G}}^{\mathrm{A}}(t_0,t') = -[\hat{\mathcal{G}}^{>}(t_0,t') - \hat{\mathcal{G}}^{<}(t_0,t')]$. Now consider the above equation with, e.g., $t' = t_0$,

$$\hat{\mathcal{G}}^{<}(t,t_0) = \mathrm{i}\,[\hat{\mathcal{G}}^{>}(t,t_0) - \hat{\mathcal{G}}^{<}(t,t_0)]\,\hat{\mathcal{G}}^{<}(t_0,t_0),$$

$$\hat{\mathcal{G}}^{>}(t,t_0) = \mathrm{i}\,[\hat{\mathcal{G}}^{>}(t,t_0) - \hat{\mathcal{G}}^{<}(t,t_0)]\,\hat{\mathcal{G}}^{>}(t_0,t_0),$$

where we take into account that $\hat{\mathcal{G}}^{>}(t_0,t_0) - \hat{\mathcal{G}}^{<}(t_0,t_0) = -\mathrm{i}$. This is a system of coupled equations for $\hat{\mathcal{G}}^{<}(t,t_0)$ and $\hat{\mathcal{G}}^{>}(t,t_0)$. We see, however, that the system admits infinitely many solutions since subtracting the first equation from the second we get a trivial identity. To conclude the set of equations (9.68), (9.69) and the like for $\hat{\mathcal{G}}^{\lceil}$ and $\hat{\mathcal{G}}^{>}$ it is not sufficient to calculate the Green's function. To form a complete system of equations we must include the retarded or advanced Dyson equation, i.e., $\hat{\mathcal{G}}^{\mathrm{R/A}} = \hat{\mathcal{G}}^{\mathrm{R/A}}_{\mathrm{HF}} + \hat{\mathcal{G}}^{\mathrm{R/A}}_{\mathrm{HF}} \cdot \hat{\Sigma}^{\mathrm{R/A}}_{\mathrm{c}} \cdot \hat{\mathcal{G}}^{\mathrm{R/A}}$. This equation depends on the external fields through the Hartree–Fock Green's function and, therefore, it is certainly independent of the other equations.

There exist special circumstances in which a considerable simplification occurs. Suppose that we are interested in the behavior of $\hat{\mathcal{G}}^{<}$ for times t, t' much larger than t_0. In most macroscopic interacting systems the memory carried by $\hat{\Sigma}_{\mathrm{c}}$ vanishes when the separation between the time arguments increases. If so then the Green's function vanishes also in the same limit. This is the relaxation phenomenon discussed in Section 6.1.3 and amounts to saying that initial correlations and initial-state dependences are washed out in the long-time limit. In these cases only one term remains in (9.69):

$$\lim_{t,t' \to \infty} \hat{\mathcal{G}}^{\lessgtr}(t,t') = \left[\hat{\mathcal{G}}^{\mathrm{R}} \cdot \hat{\Sigma}^{\lessgtr}_{\mathrm{c}} \cdot \hat{\mathcal{G}}^{\mathrm{A}}\right](t,t'). \tag{9.70}$$

As we see in Chapter 13, this is an *exact* result *for all* times t, t' and for both finite and macroscopic systems provided that we are in thermodynamic equilibrium or, equivalently, provided that the fluctuation–dissipation theorem is valid. In other words, for systems in equilibrium all terms in (9.69) vanish except for the convolution $\hat{\mathcal{G}}^{\mathrm{R}} \cdot \hat{\Sigma}^{\lessgtr}_{\mathrm{c}} \cdot \hat{\mathcal{G}}^{\mathrm{A}}$. For systems out of equilibrium, instead, this simplification occurs only for long times and only for systems with decaying memory.[10] We further observe that if the Hamiltonian $\hat{H}(t)$ becomes independent of time when $t \to \infty$ then it is reasonable to expect that the Green's functions and self-energies depend only on the time difference (for large times). In this case the nonequilibrium $\hat{\mathcal{G}}$ and $\hat{\Sigma}$ can be Fourier transformed and the convolution in (9.70) becomes a simple product in frequency space

$$\hat{\mathcal{G}}^{\lessgtr}(\omega) = \hat{\mathcal{G}}^{R}(\omega)\hat{\Sigma}_{c}^{\lessgtr}(\omega)\hat{\mathcal{G}}^{A}(\omega). \tag{9.71}$$

This result allows us to calculate steady-state quantities without solving the Kadanoff–Baym equations. Indeed, under the above simplifying assumptions the correlation self-energy with real times can be written solely in terms of the Green's function with real times. Consider a generic self-energy diagram with external vertices on the forward or backward branches at a distance t and t' from the origin. If both t and t' tend to infinity then all internal convolutions along the imaginary track tend to zero. In this limit $\hat{\Sigma}_{c}^{\lessgtr}$ depends only on $\hat{\mathcal{G}}^{<}$ and $\hat{\mathcal{G}}^{>}$.[11] Then (9.71) and the retarded or advanced Dyson equation,

$$\hat{\mathcal{G}}^{R/A}(\omega) = \hat{\mathcal{G}}_{HF}^{R/A}(\omega) + \hat{\mathcal{G}}_{HF}^{R/A}(\omega)\hat{\Sigma}_{c}^{R/A}(\omega)\hat{\mathcal{G}}^{R/A}(\omega), \tag{9.72}$$

constitute a close set of coupled equations for $\hat{\mathcal{G}}^{<}(\omega)$ and $\hat{\mathcal{G}}^{>}(\omega)$ to be solved self-consistently.

---

[10]The same reasoning can also be applied to special matrix elements of the $\hat{\mathcal{G}}^{\lessgtr}$ of macroscopic noninteracting systems like the $G_{00}^{<}$ of the Fano model where the embedding self-energy plays the role of the correlation self-energy, see Section 6.1.1.

[11]The dependence is, of course, determined by the approximation we make for the functional Φ.

10

MBPT for the Green's function

The Hartree and Hartree–Fock approximations introduced in Chapter 7 were based on mere physical intuition. How to go beyond these approximations in a systematic way is the topic of this chapter. We present an efficient perturbative method to expand the Green's function and the self-energy in powers of the interparticle interaction. The starting point is the formula (5.32) for the Green's function, which we rewrite again for convenience:

$$
G(a;b) = \frac{\displaystyle\sum_{k=0}^{\infty} \frac{1}{k!} \left(\frac{i}{2}\right)^k \int v(1;1')..v(k;k') \begin{vmatrix} G_0(a;b) & G_0(a;1^+) & \dots & G_0(a;k'^+) \\ G_0(1;b) & G_0(1;1^+) & \dots & G_0(1;k'^+) \\ \vdots & \vdots & \ddots & \vdots \\ G_0(k';b) & G_0(k';1^+) & \dots & G_0(k';k'^+) \end{vmatrix}_\pm}{\displaystyle\sum_{k=0}^{\infty} \frac{1}{k!} \left(\frac{i}{2}\right)^k \int v(1;1')..v(k;k') \begin{vmatrix} G_0(1;1^+) & G_0(1;1'^+) & \dots & G_0(1;k'^+) \\ G_0(1';1^+) & G_0(1';1'^+) & \dots & G_0(1';k'^+) \\ \vdots & \vdots & \ddots & \vdots \\ G_0(k';1^+) & G_0(k';1'^+) & \dots & G_0(k';k'^+) \end{vmatrix}_\pm} .
$$

$$(10.1)$$

This equation gives explicitly all the terms needed to calculate G to all orders in the interaction strength. What we need to do is to find an efficient way to collect them. We introduce a graphical method which consists in representing every term of the MBPT expansion with a diagram. This method was invented by Feynman in 1948 and has two main appealing features. On the one hand, it is much easier to manipulate diagrams than lengthy and intricate mathematical expressions. On the other hand, the Feynman diagrams explicitly unravel the underlying physical content of the various terms of (10.1).

10.1 Getting started with Feynman diagrams

To get some experience with (10.1) let us work out some low order terms explicitly. We start with the denominator which is the ratio Z/Z_0 between the partition function of the interacting and noninteracting system, see (5.35). To first order we have

$$\left(\frac{Z}{Z_0}\right)^{(1)} = \frac{i}{2}\int d1d1'\, v(1;1') \begin{vmatrix} G_0(1;1^+) & G_0(1;1'^+) \\ G_0(1';1^+) & G_0(1';1'^+) \end{vmatrix}_\pm$$

$$= \frac{i}{2}\int d1d1'\, v(1;1')\left[G_0(1;1^+)G_0(1';1'^+) \pm G_0(1;1'^+)G_0(1';1^+)\right].$$

$$(10.2)$$

The basic idea of the Feynman diagrams is to provide a simple set of rules to convert a drawing into a well defined mathematical quantity, like (10.2). Since (10.2) contains only Green's functions and interparticle interactions we must assign to G_0 and v a graphical object. We use the graphical notation already introduced in Section 9.3, according to which a Green's function $G_0(1; 2^+)$ is represented by an oriented line going from 2 to 1:

$$G_0(1;2^+)\ =\ 1\ \longleftarrow\ 2$$

The Green's function line is oriented to distinguish $G_0(1; 2^+)$ from $G_0(2; 1^+)$. The orientation is, of course, a pure convention. We could have chosen the opposite orientation as long as we consistently use the same orientation for all Green's functions. The convention above is the standard one. It stems from the intuitive picture that in $G_0(1; 2^+)$ we create a particle in 2 and destroy it back in 1. Thus the particle "moves" from 2 to 1. The interaction $v(1; 2)$ is represented by a wiggly line:

$$v(1;2)\ =\ 1\ \sim\!\sim\!\sim\ 2$$

which has no direction since $v(1; 2) = v(2; 1)$. Then, the two terms in (10.2) have the graphical form

where integration over all internal vertices (in this case 1 and $1'$) is understood. The diagrams above are exactly the same as those of the Hartree–Fock approximation (9.23) with the difference that here the lines represent G_0 instead of G. It is important to observe that the infinitesimal contour-time shift in $G_0(1; 2^+)$ plays a role only when the starting and end points of G are the same (as in the first diagram above), or when points 1 and 2 are joined by an iteration line $v(1; 2) = \delta(z_1, z_2)v(\mathbf{x}_1, \mathbf{x}_2)$ (as in the second diagram above). In all other cases we can safely discard the shift since $z_1 = z_2$ is a set of zero measure in the integration domain.

> *In the remainder of the book we sometimes omit the infinitesimal shift. If a diagram contains a Green's function with the same contour-time arguments then the second argument is understood to be infinitesimally later than the first.*

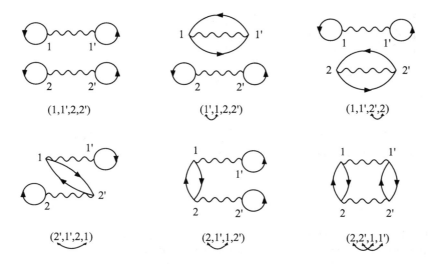

Figure 10.1 Some of the second-order diagrams of the MBPT expansion of Z/Z_0.

It is also worth noting that the prefactor of the diagrams is determined by (10.2); it is $(i/2)$ for the first diagram and $(\pm i/2)$ for the second diagram. More generally, to any order in the interaction strength the prefactors follow directly from (10.1). We do not need to come up with ad hoc solutions to fix them. From now on the prefactor of each diagram is incorporated in the diagram itself, i.e., to each diagram corresponds an integral of Green's functions and interactions with the appropriate prefactor.

To evaluate the second order contribution to Z/Z_0 we must expand the permanent/determinant of a 4×4 matrix which yields $4! = 24$ terms,

$$\left(\frac{Z}{Z_0}\right)^{(2)} = \frac{1}{2!}\frac{i^2}{2^2} \int d1d1'd2d2' \, v(1;1')v(2;2')$$
$$\times \sum_P (\pm)^P G_0(1;P(1))G_0(1';P(1'))G_0(2;P(2))G_0(2';P(2')). \quad (10.3)$$

In Fig. 10.1 we show some of the diagrams originating from this expansion. Below each diagram we indicate the permutation that generates it. The prefactor is simply $\frac{1}{2!}\left(\frac{i}{2}\right)^2$ times the sign of the permutation. Going to higher order in v the number of diagrams grows and their topology becomes increasingly more complicated. However, they all have a common feature: their mathematical expression contains an integral over all vertices. We refer to these diagrams as the *vacuum diagrams*.[1] Thus, given a vacuum diagram the rules to convert it into a mathematical expression are:

- Number all vertices and assign an interaction line $v(i;j)$ to a wiggly line between j and i and a Green's function $G_0(i;j^+)$ to an oriented line from j to i.

---

[1] The Φ-diagrams of Section 9.3 are examples of vacuum diagrams.

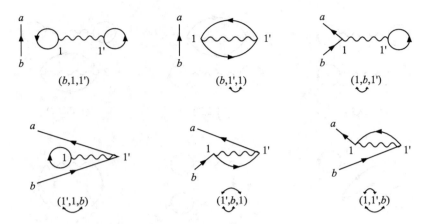

Figure 10.2 First order diagrams for the numerator $N(a;b)$. Below each diagram we report the permutation that generates it. The arrows help to visualize the number of interchanges (the first interchange is the arrow under the triple of labels).

- Integrate over all vertices and multiply by $[(\pm)^P \frac{1}{k!}(\frac{i}{2})^k]$, where $(\pm)^P$ is the sign of the permutation and k is the number of interaction lines.

Let us now turn our attention to the numerator, $N(a;b)$, of (10.1). To first-order in the interaction strength we must evaluate the permanent/determinant of a 3×3 matrix. The expansion along the first column leads to

$$N^{(1)}(a;b) = \frac{i}{2}G_0(a;b)\int d1d1'v(1;1')\begin{vmatrix} G_0(1;1^+) & G_0(1;1'^+) \\ G_0(1';1^+) & G_0(1';1'^+) \end{vmatrix}_{\pm}$$

$$\pm \frac{i}{2}\int d1d1'v(1;1')G_0(1;b)\begin{vmatrix} G_0(a;1^+) & G_0(a;1'^+) \\ G_0(1';1^+) & G_0(1';1'^+) \end{vmatrix}_{\pm}$$

$$+ \frac{i}{2}\int d1d1'v(1;1')G_0(1';b)\begin{vmatrix} G_0(a;1^+) & G_0(a;1'^+) \\ G_0(1;1^+) & G_0(1;1'^+) \end{vmatrix}_{\pm}. \qquad (10.4)$$

To each term we can easily give a diagrammatic representation. In the first line of (10.4) we recognize the ratio $(Z/Z_0)^{(1)}$ of (10.2) multiplied by $G_0(a;b)$. The corresponding diagrams are simply those of $(Z/Z_0)^{(1)}$ with an extra line going from b to a. The remaining terms can be drawn in a similar manner and the full set of diagrams (together with the corresponding permutations) is shown in Fig. 10.2. The prefactor is simply $(i/2)$ times the sign of the permutation. The diagrams for $N(a;b)$ are different from the vacuum diagrams since there are two *external* vertices (a and b) over which we do not integrate. We refer to these diagrams as the *Green's function diagrams*. The rules to convert a Green's function diagram into a mathematical expression are the same as those for the vacuum diagrams with the exception that there is no integration over the external vertices. In contrast, a vacuum diagram contains only *internal* vertices.

At this point the only quantity which is a bit awkward to determine is the sign of the permutation. It would be useful to have a simple rule to fix the sign by giving the diagram a cursory glance. This is the topic of the next section.

Exercise 10.1. Calculate all the second-order terms of Z/Z_0 and draw the corresponding diagrams.

Exercise 10.2. Show that the diagrammatic representation of the last two lines of (10.4) is the last four diagrams of Fig. 10.2.

10.2 Loop rule

From (10.1) or (10.3) we see that the sign of a diagram is determined by the sign of the permutation that changes the second argument of the Green's functions. In graphical terms this amounts to a permutation of the starting points of the Green's function lines of a diagram. Since every permutation can be obtained by successive interchanges of pairs of labels (i, j) we only need to investigate how such interchanges modify a diagram. A vacuum diagram consists of a certain number of loops and, therefore, an interchange can occur either between two starting points of the same loop or between two starting points of different loops. In the former case we have the generic situation:

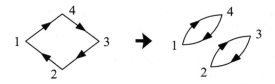

in which we have interchanged the starting points 2 and 4 so that $G_0(1;2)G_0(2;3)G_0(3;4)$ $G_0(4;1) \to G_0(1;4)G_0(2;3)G_0(3;2)G_0(4;1)$ and hence the number of loops increases by one. In the latter case we have the generic situation:

in which we have interchanged the starting points 3 and 6 so that $G_0(1;2)G_0(2;3)G_0(3;1)$ $G_0(5;4)G_0(4;6)G_0(6;5) \to G_0(1;2)G_0(2;6)G_0(3;1)G_0(5;4)G_0(4;3)G_0(6;5)$ and hence the number of loops decreases by one. It is not difficult to convince ourselves that this is a general rule: an interchange of starting points changes the number of loops by one. For a Green's function diagram, in addition to the interchanges just considered, we have two more possibilities: either the interchange occurs between two starting points on the path connecting b to a or between a starting point on the path and a starting point on a loop. In the first case we have the generic situation:

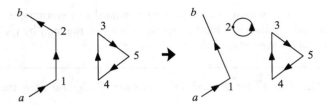

in which we have interchanged the starting points 1 and 2, and the number of loops increases by one. The reader can easily verify that the number of loops would have increased by one also by an interchange of a and 2 or of a and 1. In the second case an interchange of, e.g., the starting points a and 3 leads to:

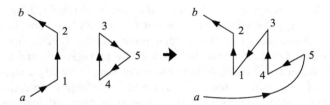

and the number of loops decreases by one. Again, this is completely general: an interchange of starting points changes the number of loops by one. Taking into account that for the identity permutation the sign of a diagram, be it a vacuum or a Green's function diagram, is $+$ and the number of loops is even, we can state the so called *loop rule*: $(\pm)^P = (\pm)^l$ where l is the number of loops! As an example we show in Fig. 10.3 some of the $5! = 120$ second-order Green's function diagrams together with the corresponding permutation; in all cases the loop rule is fulfilled. This example also shows that there are several diagrams [(a) to (e)] which are products of a *connected diagram* (connected to a and b) and a vacuum diagram. It turns out that the vacuum diagrams are cancelled out by the denominator of the Green's function in (10.1), thus leading to a large reduction of the number of diagrams to be considered. Furthermore, there are diagrams that have the same numerical value [e.g., (a)-(b)-(e), (c)-(d) and (f)-(g)]. This is due to a permutation and mirror symmetry of the interaction lines $v(j; j')$. We can achieve a large reduction in the number of diagrams by taking these symmetries into account. In the next two sections we discuss how to do it.

Exercise 10.3. Draw the G-diagram and the vacuum diagram of order n, i.e., with n interaction lines, corresponding to the identity permutation.

10.3 Cancellation of disconnected diagrams

We have already observed that the number of terms generated by (10.1) grows very rapidly when going to higher order in the interaction strength. In this section we show that the

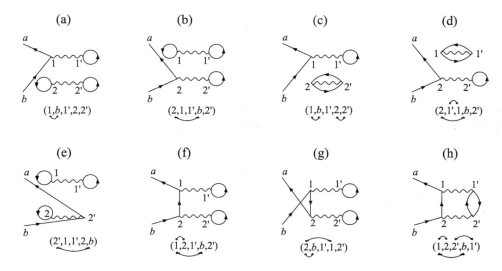

Figure 10.3 Some of the second-order Green's function diagrams and the corresponding permutations.

disconnected vacuum diagrams of the numerator are exactly cancelled by the vacuum diagrams of the denominator. Let us start with an example. Consider the following three Green's function diagrams that are part of the expansion of the numerator of (10.1) to third order:

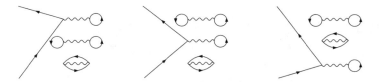

They are all related by a permutation of the interaction lines $v(i; i') \leftrightarrow v(j; j')$ that preserves the structure of two disjoint pieces which are: (piece 1) the connected Green's function diagram corresponding to the third diagram of Fig. 10.2, and (piece 2) the vacuum diagram corresponding to the third diagram of Fig. 10.1. Furthermore, they all have the same prefactor $(\pm)^4 \frac{1}{3!} (\frac{i}{2})^3$ so that their total contribution is

$$3 \times \frac{1}{3!} \left(\frac{i}{2}\right)^3 \int [G_0 G_0 G_0 v] \int [G_0 G_0 G_0 G_0 vv], \tag{10.5}$$

with the obvious notation that the first factor $\int [G_0 G_0 G_0 v]$ refers to the Green's function diagram and the second factor $\int [G_0 G_0 G_0 G_0 vv]$ refers to the vacuum diagram. Let us now consider the product of diagrams

As the first diagram is $-\frac{i}{2} \int [G_0 G_0 G_0 v]$ and the second diagram is $-\frac{1}{2!} \left(\frac{i}{2}\right)^2 \int [G_0 G_0 G_0 G_0 v v]$, their product equals the sum of the three diagrams above, i.e.,

This result is readily seen to be generally valid. To kth order there are $\begin{pmatrix} k \\ n \end{pmatrix}$ ways to construct a given connected nth order Green's function diagram out of the interaction lines $v(1; 1'), \ldots, v(k; k')$ which all give the same contribution $G_{c,i}^{(n)}(a; b) = \int [G_0 \ldots G_0 v \ldots v]_i$, where i labels the given diagram. In our example $\begin{pmatrix} 3 \\ 1 \end{pmatrix} = 3$. The remaining part $V_j^{(k-n)} = \int [G_0 \ldots G_0 v \ldots v]_j$ is proportional to the jth vacuum diagram (consisting of one or more disjoint pieces) of order $k - n$. Thus, the total contribution of the kth order term of the numerator of (10.1) is

$$N^{(k)}(a; b) = \frac{1}{k!} \left(\frac{i}{2}\right)^k \sum_{n=0}^{k} \begin{pmatrix} k \\ n \end{pmatrix} \sum_{\substack{i= \text{ G-connected} \\ \text{diagrams}}} \sum_{\substack{j= \text{ vacuum} \\ \text{diagrams}}} (\pm)^{l_i + l_j} G_{c,i}^{(n)}(a; b) V_j^{(k-n)}$$

$$= \sum_{n=0}^{k} \frac{1}{n!} \left(\frac{i}{2}\right)^n \sum_{\substack{i= \text{ G-connected} \\ \text{diagrams}}} (\pm)^{l_i} G_{c,i}^{(n)}(a; b)$$

$$\times \frac{1}{(k-n)!} \left(\frac{i}{2}\right)^{k-n} \sum_{\substack{j= \text{ vacuum} \\ \text{diagrams}}} (\pm)^{l_j} V_j^{(k-n)},$$

where l_i and l_j are the number of loops in the diagrams i and j. The third line in this equation is exactly $(Z/Z_0)^{(k-n)}$. Therefore, if we denote by

$$G_c^{(n)}(a; b) = \frac{1}{n!} \left(\frac{i}{2}\right)^n \sum_{\substack{i= \text{ G-connected} \\ \text{diagrams}}} (\pm)^{l_i} G_{c,i}^{(n)}(a; b)$$

the sum of all nth order connected diagrams of $N(a; b)$, we have

$$
N(a; b) = \sum_{k=0}^{\infty} \sum_{n=0}^{k} G_c^{(n)}(a; b) \left(\frac{Z}{Z_0} \right)^{(k-n)} = \sum_{n=0}^{\infty} \sum_{k=n}^{\infty} G_c^{(n)}(a; b) \left(\frac{Z}{Z_0} \right)^{(k-n)}
$$

$$
= \left(\frac{Z}{Z_0} \right) \sum_{n=0}^{\infty} G_c^{(n)}(a; b).
$$

We have just found the important and beautiful result that all vacuum diagrams of the denominator of (10.1) are cancelled out by the disconnected part of the numerator! The MBPT formula (10.1) simplifies to

$$
\boxed{
G(a; b) = \sum_{n=0}^{\infty} \frac{1}{n!} \left(\frac{\mathrm{i}}{2} \right)^n \int v(1; 1') \dots v(n; n')
\begin{vmatrix}
G_0(a; b) & G_0(a; 1^+) & \dots & G_0(a; n'^+) \\
G_0(1; b) & G_0(1; 1^+) & \dots & G_0(1; n'^+) \\
\vdots & \vdots & \ddots & \vdots \\
G_0(n'; b) & G_0(n'; 1^+) & \dots & G_0(n'; n'^+)
\end{vmatrix}_{c}^{\pm}
}
$$

$$
(10.6)
$$

where the symbol $| \dots |_{c}^{\pm}$ signifies that in the expansion of the permanent/determinant only the terms represented by connected diagrams are retained.

10.4 Summing only the topologically inequivalent diagrams

The cancellation of disconnected diagrams reduces drastically the number of terms in the MBPT expansion of G, but we can do even better. If we write down the diagrams for G we realize that there are still many connected diagrams with the same value. In first order, for instance, the 3rd and 4th diagram as well as the 5th and 6th diagram of Fig. 10.2 clearly lead to the same integrals, for only the labels 1 and $1'$ are interchanged. In second order each connected diagram comes in eight variants all with the same value. An example is the diagrams of Fig. 10.4, in which (b) is obtained from (a) by mirroring the interaction line $v(1; 1')$, (c) is obtained from (a) by mirroring the interaction line $v(2; 2')$, and (d) is obtained from (a) by mirroring both interaction lines. For an nth order diagram we thus have 2^n such mirroring operations ($2^2 = 4$ in Fig. 10.2). The second row of the figure is obtained by interchanging the interaction lines $v(1; 1') \leftrightarrow v(2; 2')$ in the diagrams of the first row. If we have n interaction lines then there are $n!$ possible permutations ($2! = 2$ in Fig. 10.2). We conclude that there exist $2^n n!$ diagrams with the same value to order n. Since these diagrams are obtained by mirroring and permutations of interaction lines they are also topologically equivalent, i.e., they are obtained from one another by a continuous deformation. Therefore we only need to consider diagrams with different topology and multiply by $2^n n!$ where n is the number of interaction lines. For these diagrams the new value of the prefactor becomes

$$
2^n n! \frac{1}{n!} \left(\frac{\mathrm{i}}{2} \right)^n (\pm)^l = \mathrm{i}^n (\pm)^l,
$$

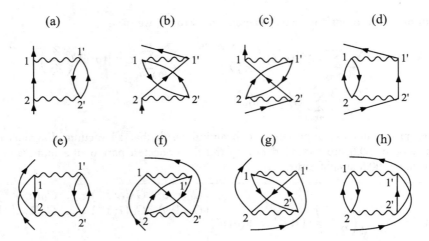

Figure 10.4 A class of eight equivalent second-order connected diagrams for the Green's function.

and (10.6) can be rewritten as

$$
G(a;b) = \sum_{n=0}^{\infty} \mathrm{i}^n \int v(1;1') \dots v(n;n') \begin{vmatrix} G_0(a;b) & G_0(a;1^+) & \dots & G_0(a;n'^+) \\ G_0(1;b) & G_0(1;1^+) & \dots & G_0(1;n'^+) \\ \vdots & \vdots & \ddots & \vdots \\ G_0(n';b) & G_0(n';1^+) & \dots & G_0(n';n'^+) \end{vmatrix}_{\substack{c \\ t.i.}}^{\pm}
$$

(10.7)

where the symbol $\left| \dots \right|_{\substack{c \\ t.i.}}^{\pm}$ signifies that in the expansion of the permanent/determinant only the terms represented by connected and topologically inequivalent diagrams are retained. From now on we work only with these diagrams. Thus, the new rules to convert a Green's function diagram into a mathematical expression are:

- Number all vertices and assign an interaction line $v(i;j)$ to a wiggly line between j and i and a Green's function $G_0(i;j^+)$ to an oriented line from j to i;

- Integrate over all internal vertices and multiply by $\mathrm{i}^n(\pm)^l$ where l is the number of loops and n is the number of interaction lines.

Using (10.7) we can easily expand the Green's function to second order; the result is shown in Fig. 10.5. We find two first-order diagrams and ten second-order diagrams. This is an enormous simplification! Without using the cancellation of disconnected diagrams and resummation of topologically equivalent diagrams the Green's function to second-order $G^{(2)} = (N^{(0)} + N^{(1)} + N^{(2)}) - (Z/Z_0)^{(1)}(N^{(0)} + N^{(1)}) + \{[(Z/Z_0)^{(1)}]^2 + (Z/Z_0)^{(2)}\}N^{(0)}$ would consist of $(1 + 3! + 5!) + 2!(1 + 3!) + (2!)^2 + 4! = 169$ diagrams.

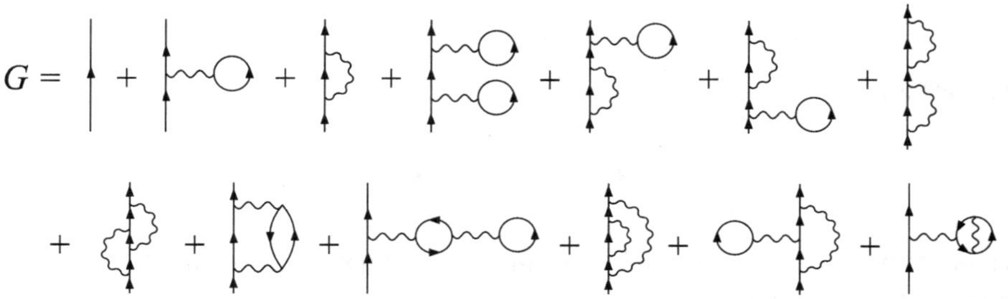

$$G = \quad + \quad + \quad + \quad + \quad + \quad + \quad + \quad$$

$$+ \quad + \quad + \quad + \quad + \quad + \quad$$

Figure 10.5 MBPT expansion of the Green's function to second order using the new Feynman rules.

Exercise 10.4. Prove that G to second order in v is given by Fig. 10.5.

10.5 Self-energy and Dyson equations II

To reduce the number of diagrams further it is necessary to introduce a new quantity which, as we shall see, turns out to be the *self-energy* Σ. It is clear from the diagrammatic structure that the Green's function has the general form

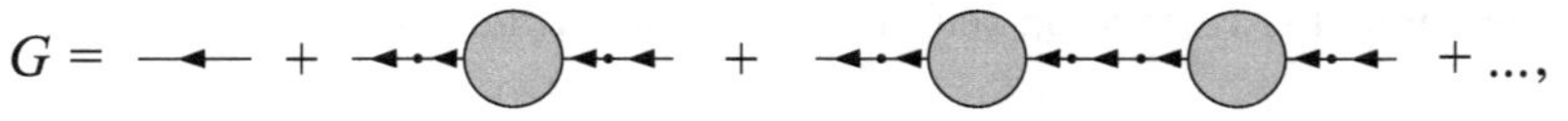

$$G = \quad + \quad + \quad + ...,$$

(10.8)

where the self-energy

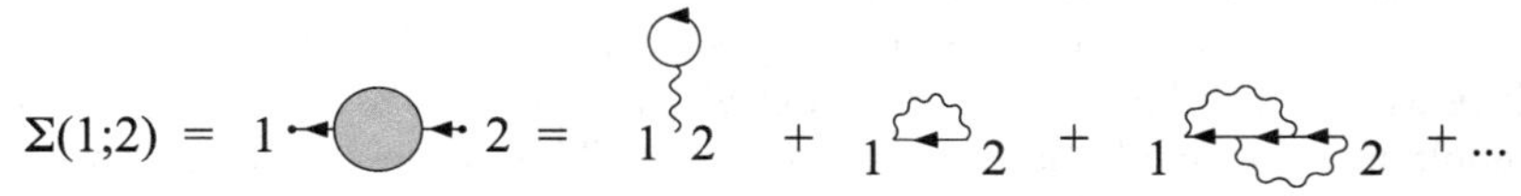

$$\Sigma(1;2) = 1 \quad 2 = 1 \quad 2 \quad + \quad 1 \quad 2 \quad + \quad 1 \quad 2 \quad + ...$$

The self-energy consists of all diagrams that do not break up into two disjoint pieces by cutting a single G_0-line. So, for instance, the 4th, 5th, 6th and 7th diagram of Fig. 10.5 belong to the third term on the r.h.s. of (10.8) while all diagrams in the second row of the same figure belong to the second term. The self-energy diagrams are called *one-particle irreducible diagrams* or simply *irreducible diagrams*. By construction $\Sigma = \Sigma[G_0, v]$ depends on the noninteracting Green's function G_0 and on the interaction v. If we represent the interacting Green's function by an oriented double line

$$G(1;2) = \quad 1 \quad 2$$

then we can rewrite (10.8) as

$$\hspace{9cm}(10.9)$$

or, equivalently,

$$\hspace{9cm}(10.10)$$

Equations (10.9) and (10.10) are readily seen to generate (10.8) by iteration. The mathematical expression of these equations is

$$G(1;2) = G_0(1;2) + \int d3d4\, G_0(1;3)\Sigma(3;4)G(4;2)$$

$$= G_0(1;2) + \int d3d4\, G(1;3)\Sigma(3;4)G_0(4;2), \hspace{2cm}(10.11)$$

which is identical to (9.17): we have again found the Dyson equation! However, whereas in Chapter 9 we did not know how to construct approximations for the self-energy, in this chapter we have learned how to use the Wick's theorem to expand the self-energy to any order in the interaction strength.

It is interesting to observe that the self-energy (in contrast to the Green's function) does not have a mathematical expression in terms of a contour-ordered average of operators. The mathematical expression for Σ which is closest to a contour-ordered average can be deduced from (9.9). Here the structure $\Sigma G\Sigma$ is reducible and hence does not belong to Σ. The first term on the r.h.s. of (9.9) is the singular (Hartree–Fock) part of the self-energy, see (9.13). Consequently the correlation self-energy can be written as

$$\boxed{\Sigma_{\rm c}(1;2) = -{\rm i}\,\langle \mathcal{T}\left\{\hat{\gamma}_H(1)\hat{\gamma}_H^\dagger(2)\right\}\rangle_{\rm irr}} \hspace{2cm}(10.12)$$

where $\langle\ldots\rangle_{\rm irr}$ signifies that in the Wick's expansion of the average we only retain those terms whose diagrammatic representation is an irreducible diagram [41].

Thanks to the introduction of the self-energy we can reduce the number of diagrams even further since we only need to consider topologically inequivalent Σ-diagrams (these diagrams are, by definition, connected and one-particle irreducible). Then the Dyson equation (10.11) allows us to sum a large number of diagrams to *infinite order*. Indeed, a finite number of Σ-diagrams implies an infinite number of Green's function diagrams through (10.8).

For the diagrammatic construction of the self-energy the rules are the same as those for the Green's function. For instance, to first order we have

or in formulas

$$\Sigma^{(1)}(1;2) = \pm{\rm i}\,\delta(1;2)\int d3\, v(1;3)G_0(3;3^+) + {\rm i}\,v(1;2)G_0(1;2^+). \hspace{2cm}(10.13)$$

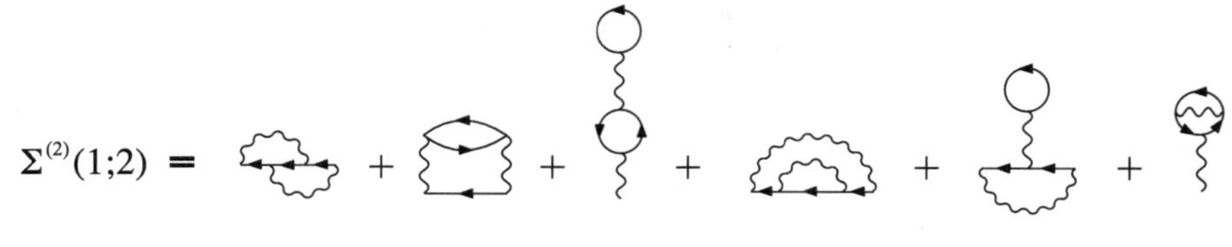

$$\Sigma^{(2)}(1;2) =$$

Figure 10.6 Self-energy diagrams to second order in interaction strength.

This is exactly the Hartree–Fock self-energy of Fig. 9.1 [see also (9.13)] in which G has been replaced by G_0. According to our new convention the prefactor is now included in the diagram and coincides with the prefactor derived heuristically in Section 7.1. As anticipated, the perturbative expansion fixes the prefactors uniquely. In the Hartree term of (10.13) we have also added a δ-function since in the corresponding Green's function diagram (second diagram of Fig. 10.5) we have $1 = 2$. This applies to all self-energy diagrams that start and end with the *same* interaction vertex (see, e.g., the 3rd and 6th diagram in the second row of Fig. 10.5). The second-order self-energy diagrams are shown in Fig. 10.6. There are only six diagrams to be considered against the ten second-order Green's function diagrams. In the next two sections we achieve another reduction in the number of diagrams by introducing a very useful topological concept.

Exercise 10.5. Evaluate the r.h.s. of (10.12) to second order in the interaction v and show that it is given by the sum of the 1st and 2nd diagrams in Fig. 10.6 (all other diagrams in the same figure contain a $\delta(z_1, z_2)$ and hence they are part of the singular (Hartree–Fock) self-energy).

10.6 G-skeleton diagrams

A G-skeleton diagram for the self-energy is obtained by removing all self-energy insertions from a given diagram. A self-energy insertion is a piece that can be cut away from a diagram by cutting two Green's function lines. For example, the diagram below:

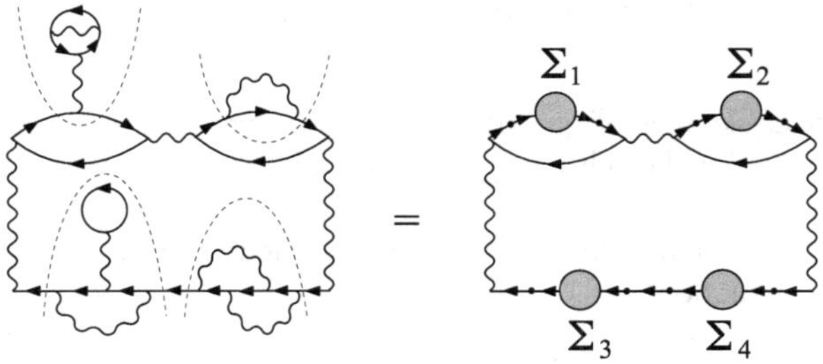

$$(10.14)$$

has four self-energy insertions residing inside the thin-dashed parabolic lines. The r.h.s. highlights the structure of the diagram and implicitly defines the self-energy insertions Σ_i, $i = 1, 2, 3, 4$. The G-skeleton diagram corresponding to (10.14) is therefore

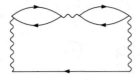

$$(10.15)$$

The G-skeleton diagrams allow us to express the self-energy in terms of the interacting (dressed) Green's function G rather than the noninteracting Green's function G_0. Consider again the example (10.15). If we sum over all possible self-energy insertions we find

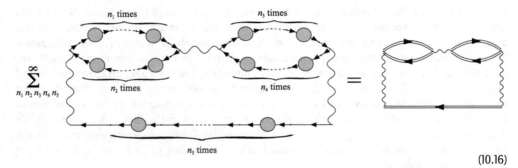

$$(10.16)$$

where each grey circle represents the exact self-energy. Thus, the sum over n_i on the l.h.s. gives the dressed G and the result is the G-skeleton diagram (10.15) in which G_0 is replaced by G. This procedure yields the self-energy $\Sigma = \Sigma_s[G, v]$ as a functional of the interaction v and of the dressed Green's function G. The subscript "s" specifies that the functional is constructed by taking only the G-skeleton diagram from the functional $\Sigma = \Sigma[G_0, v]$ and then replacing G_0 with G: $\Sigma[G_0, v] = \Sigma_s[G, v]$. In the next chapter we show how to use this result to construct the functional $\Phi[G] = \Phi[G, v]$ whose functional derivative with respect to G is the exact self-energy. Using G-skeleton diagrams we can write the self-energy up to second order in the interaction as

$$\Sigma_s[G,v] \;=\; \text{[diagrams]} \;+\; \text{[diagram]} \;+\; \text{[diagram]} \;+\; \text{[diagram]} \;+\; \dots$$

$$(10.17)$$

There are only two G-skeleton diagrams of second order against the six diagrams in Fig. 10.6. The approximation for the self-energy corresponding to these four G-skeleton diagrams is called the *second-Born approximation*.

For any given approximation to $\Sigma_s[G, v]$ we can calculate an approximate Green's function from the Dyson equation

$$G(1;2) = G_0(1;2) + \int d3d4\, G_0(1;3)\Sigma_s[G,v](3;4)G(4;2)$$

$$= G_0(1;2) + \int d3d4\, G(1;3)\Sigma_s[G,v](3;4)G_0(4;2),$$

$$(10.18)$$

which is a *nonlinear* integral equation for G. Alternatively, we could apply $[\delta(1';1)\mathrm{i}\frac{d}{dz_1} - h(1';1)]$ to the first row and integrate over 1, $[-\mathrm{i}\frac{\overleftarrow{d}}{dz_2}\delta(2;2') - h(2;2')]$ to the second row and integrate over 2, and convert (10.18) into a coupled system of nonlinear integro-differential equations,

$$\int d1 \left[\delta(1';1)\,\mathrm{i}\frac{d}{dz_1} - h(1';1)\right] G(1;2) = \delta(1';2) + \int d4\,\Sigma_s[G,v](1';4)G(4;2), \quad (10.19)$$

$$\int d2\, G(1;2) \left[-\mathrm{i}\frac{\overleftarrow{d}}{dz_2}\delta(2;2') - h(2;2')\right] = \delta(1;2') + \int d3\, G(1;3)\Sigma_s[G,v](3;2'), \quad (10.20)$$

to be solved with the KMS boundary conditions. These equations are exactly the Kadanoff–Baym equations discussed in Section 9.4. From the results of the previous chapter we also know that if the approximate self-energy is Φ-derivable then the Green's function obtained from (10.19) and (10.20) preserves all basic conservation laws. In particular, the solution of the Kadanoff–Baym equations with the exact self-energy (obtained by summing all the G-skeleton self-energy diagrams) yields the exact Green's function.

10.7 *W*-skeleton diagrams

The topological concept of G-skeleton diagrams can also be applied to the interaction lines and leads to a further reduction of the number of diagrams. Let us call a piece of diagram a *polarization* insertion if it can be cut away by cutting two interaction lines. For example

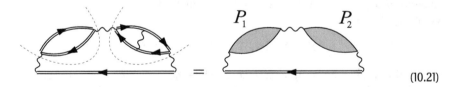

(10.21)

is a G-skeleton diagram with two polarization insertions

$$P_1 = \quad\overleftrightarrow{\bigcirc}\qquad\qquad P_2 = \quad\overleftrightarrow{\bigcirc}$$

(10.22)

The polarization diagrams, like the ones above, must be one-interaction-line irreducible, i.e., they cannot break into two disjoint pieces by cutting an interaction line. Thus, even though we could cut away the piece $P_1 v P_2$ by cutting two interaction lines on the l.h.s. of (10.21), the diagram $P_1 v P_2$ is not a polarization diagram since it breaks into the disjoint pieces P_1 and P_2 by cutting the interaction line in the middle.

The polarization diagrams can be used to define the *dressed* or *screened interaction* W according to

$$W(1;2) = \quad = \quad + \quad + \quad + \dots$$

$$= \quad + $$

$$(10.23)$$

In this diagrammatic equation P is the sum of all possible polarization diagrams. We refer to P as the *polarizability*. Similarly to the self-energy, the polarizability can be considered either as a functional of the noninteracting Green's function G_0 and the interaction v, $P = P[G_0, v]$, or as a functional of the dressed G and v, $P = P_s[G, v]$. Equation (10.23) has the form of the Dyson equation (10.8) for the Green's function. In formulas it reads

$$W(1;2) = v(1;2) + \int v(1;3)P(3;4)v(4;2) + \int v(1;3)P(3;4)v(4;5)P(5;6)v(6;2) + \dots$$

$$= v(1;2) + \int v(1;3)P(3;4)W(4;2), \tag{10.24}$$

where the integral is over all repeated variables.

Let us express $\Sigma_s[G, v]$ in terms of the screened interaction W. We say that a diagram is a W-*skeleton diagram* if it does not contain polarization insertions. Then the desired expression for Σ is obtained by discarding all those diagrams which are not W-skeletonic in the expansion (10.17), and then replacing v with W. The only diagram for which we should not proceed with the replacement is the Hartree diagram [first diagram in (10.17)] since here every polarization insertion is equivalent to a self-energy insertion, see for instance the third diagram of Fig. 10.6. In the Hartree diagram the replacement $v \to W$ would lead to double counting. Therefore

$$\Sigma = \Sigma_{ss}[G, W] = \Sigma_{\mathrm{H}}[G, v] + \Sigma_{ss,\mathrm{xc}}[G, W], \tag{10.25}$$

where $\Sigma_{\mathrm{H}}[G, v]$ is the Hartree self-energy while the remaining part is the so called *exchange-correlation* (XC) self-energy which includes the Fock (exchange) diagram and all other diagrams accounting for nonlocal in time effects (correlations), see again the discussion in Section 6.1.1 and Section 7.3. The subscript "ss" specifies that to construct the functional Σ_{ss} or $\Sigma_{ss,\mathrm{xc}}$ we must take into account only those self-energy diagrams which are skeletonic with respect to both Green's function and interaction lines. The skeletonic expansion of the self-energy in terms of G and W (up to third order in W) is shown in Fig. 10.7(a). Similarly, the MBPT expansion of the polarizability in terms of G and W is obtained by taking only G- and W-skeleton diagrams and then replacing $G_0 \to G$ and $v \to W$. This operation leads to a polarizability $P = P_{ss}[G, W]$ which can be regarded as a functional of G and W and whose expansion (up to second order in W) is shown in Fig. 10.7(b).

The diagrammatic expansion in terms of W instead of v does not only have the *mathematical* advantage of reducing the number of diagrams. There is also a *physical* advantage [79]. In bulk systems and for long-range interactions, like the Coulomb interaction, many self-energy diagrams are *divergent*. The effect of the polarization insertions is to cut off

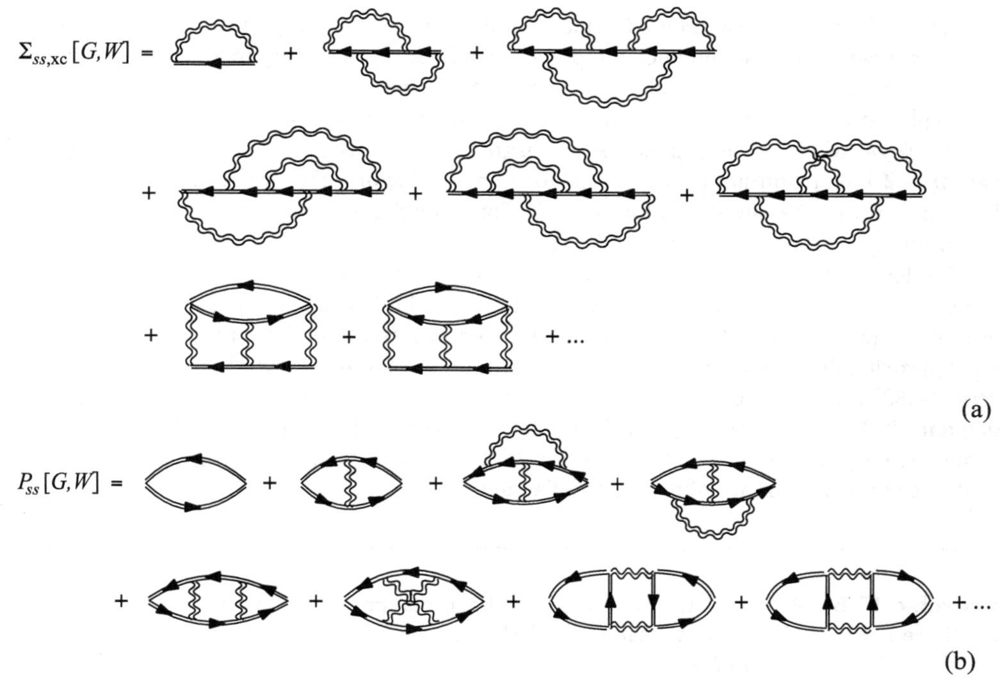

$$\Sigma_{ss,\mathrm{xc}}[G,W] =$$

(a)

$$P_{ss}[G,W] =$$

(b)

Figure 10.7 MBPT expansion in terms of G and W of (a) the self-energy up to third order, and (b) the polarization up to second order.

the long-range nature of v. Replacing v with W makes these diagrams finite and physically interpretable. We see an example of divergent self-energy diagrams in Section 13.5.

From the diagrammatic structure of the polarization diagrams we see that

$$\boxed{P(1;2) = P(2;1)} \tag{10.26}$$

and, as a consequence, also

$$\boxed{W(1;2) = W(2;1)} \tag{10.27}$$

which tells us that the screened interaction between the particles is symmetric, as one would expect. It is worth noting, however, that this symmetry property is not fulfilled by every single diagram. For instance, the 3rd and 4th diagrams of Fig. 10.7(b) are, separately, not symmetric. The symmetry is recovered only when they are summed. This is a general property of the polarization diagrams: either they are symmetric or they come in pairs of mutually *reversed* diagrams. By a reversed polarization diagram we mean the same diagram in which the end points are interchanged. If we label the left and right vertices of the 3rd and 4th diagrams of Fig. 10.7(b) with 1 and 2, then the 4th diagram with relabeled vertices 2 and 1 (reversed diagram) becomes identical to the 3rd diagram.

Let us derive the diagrammatic rules for the polarization diagrams. In an nth order Σ-diagram with n interaction lines the prefactor is $i^n(\pm)^l$ since every interaction line comes with a factor i. We want to maintain the same diagrammatic rules when the bare interaction v is replaced by W. Let us consider a screened interaction diagram W_k of the form, e.g., $W_k = vP_kv$, where P_k is a polarization diagram. If P_k has m interaction lines then W_k has $m+2$ interaction lines. For W_k to have the prefactor i the prefactor of P_k must be $i^{m+1}(\pm)^l$, with l the number of loops in P_k. For example, the prefactors of the diagrams in (10.22) are $i(\pm)^1$ for P_1 and $i^2(\pm)^1$ for P_2.

The lowest order approximation one can make in Fig. 10.7 is $\Sigma_{ss,\text{xc}}(1;2) = iG(1;2^+)$ $W(1;2)$ (first diagram) and $P(1;2) = \pm iG(1;2)G(2;1)$ (first diagram). This approximation was introduced by Hedin in 1965 and is today known as the GW approximation [79]. The GW approximation has been rather successful in describing the spectral properties of many crystals [80] and is further discussed in the next chapters. In general, for any given choice of diagrams in $\Sigma_{ss}[G,W]$ and $P_{ss}[G,W]$ we have to solve the Dyson equation (10.18) and (10.24) simultaneously to find an approximation to G and W which, we stress again, corresponds to the resummation of an infinite set of diagrams.

Exercise 10.6. Show that the G- and W-skeleton diagrams for Σ up to third-order in W are those of Fig. 10.7(a) and that the G- and W-skeleton diagrams for P up to second-order in W are those of Fig. 10.7(b).

10.8 Summary and Feynman rules

We finally summarize the achievements of this chapter. Rather than working out all diagrams for the Green's function it is more advantageous to work out the diagrams for the self-energy. In this way many diagrams are already summed to infinite order by means of the Dyson equation. The Feynman rules to construct the self-energy are:

(I) Undressed case: $\Sigma = \Sigma[G_0, v]$

- Draw all topologically inequivalent self-energy diagrams using G_0 and v. By definition a self-energy diagram is connected and one-particle irreducible.

- If the diagram has n interaction lines and l loops then the prefactor is $i^n(\pm)^l$.

- Integrate over all internal vertices of the diagram.

(II) Partially dressed case: $\Sigma = \Sigma_s[G, v]$

- Same as in (I) but in the first point we should consider only the G-skeleton diagrams (no self-energy insertions) and replace G_0 with G.

(III) Fully dressed case: $\Sigma = \Sigma_{ss}[G, W]$, $P = P_{ss}[G, W]$

- Same as in (I) but in the first point we should consider only diagrams which are both G-skeletonic (no self-energy insertions) and W-skeletonic (no polarization insertion), and replace G_0 with G and v with W.

- In this case we also need to construct the polarization diagrams. For m interaction lines and l loops the prefactor of these diagrams is $i^{m+1}(\pm)^l$.

The presentation of the diagrammatic expansion has been done with (1.82) as the reference Hamiltonian. For this Hamiltonian the building blocks of a diagram are the Green's function and the interaction, and they combine in the following manner:

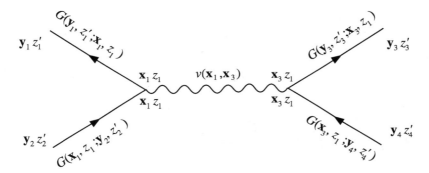

With the exception of special systems like, e.g., the electron gas, the evaluation of a generic diagram with pencil and paper is essentially impossible and we must resort to computer programs. In order to evaluate a diagram numerically, however, we must expand the field operators over some discrete (and incomplete) basis. In this case the Green's function $G(\mathbf{x}, z; \mathbf{x}', z')$ becomes a matrix $G_{ii'}(z, z')$, the self-energy $\Sigma(\mathbf{x}, z; \mathbf{x}', z')$ becomes a matrix $\Sigma_{ii'}(z, z')$, the interaction $v(\mathbf{x}, \mathbf{x}')$ becomes a four-index tensor (Coulomb integrals) v_{ijmn}, the screened interaction $W(\mathbf{x}, z; \mathbf{x}', z')$ becomes a four-index tensor $W_{ijmn}(z, z')$, and also the polarizability $P(\mathbf{x}, z; \mathbf{x}', z')$ becomes a four-index tensor $P_{ijmn}(z, z')$. The diagrammatic rules to construct these quantities are exactly the same as those already derived, and the entire discussion of this and the following chapters remains valid provided that the building blocks (Green's function and interaction) are combined according to

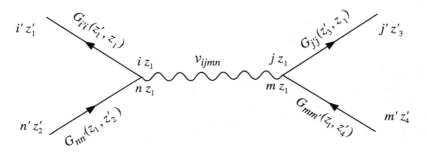

We see that there is no difference in the time-label of a vertex since the interaction remains local in time. On the other hand, the orbital–spin-label of every vertex duplicates, since the orbital–spin-label of the G that enters a vertex is, in general, different from the orbital–spin-label of the G that exits from the same vertex. The matrix elements v_{ijmn} join the outgoing Greens functions with entries i and j to the ingoing Green's functions with entries m and n. The labels (i,n) and (j,m) in v_{ijmn} correspond to basis functions calculated in the same position–spin coordinate, see (1.85), and therefore must be placed next to each other. For instance the self-energy up to second order in the interaction is (omitting the time coordinates):

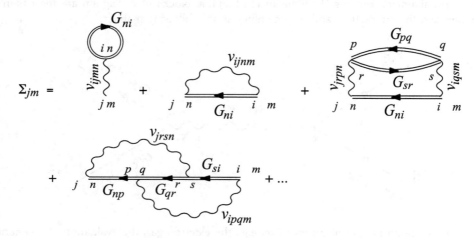

while the diagrammatic expansion of the screened interaction looks like:

$$
\begin{array}{ccccc}
n \quad \overset{W_{ijmn}}{} \quad m \\[-4pt]
\underset{i}{}\underset{j}{}
\end{array}
=
\begin{array}{c}
n \quad v_{ijmn} \quad m \\
i j
\end{array}
+
\begin{array}{c}
n \ v_{iqpn} \ p \quad \overset{P_{qrsp}}{} \quad r \ v_{sjmr} \ m \\
i q s j
\end{array}
+ \dots
$$

where the diagrammatic expansion of the polarizability has the structure:

$$
P_{ijmn} =
\quad
\begin{array}{c}
\overset{G_{im}}{} \\
i m \\
n j \\
\underset{G_{jn}}{}
\end{array}
\quad + \quad
\begin{array}{c}
\overset{G_{ip} \ p \ q \ G_{qm}}{} \\
i m \\
n \ G_{rn} \ r \ s \ G_{js} \ j
\end{array}
\quad + \dots
$$

As in the continuum case, once a diagram has been drawn we must integrate the internal times over the contour and sum the internal orbital–spin-labels over the discrete basis. Remember that for spin-independent interactions the matrix elements v_{ijmn} are zero unless the spin of i is equal to the spin of n and/or the spin of j is equal to the spin of m. This implies that in the discrete case the spin is conserved at every vertex also, as it should be.

11

MBPT and variational principles for the grand potential

11.1 Linked cluster theorem

In the previous chapter we briefly discussed the diagrammatic expansion of the partition function Z/Z_0 in terms of G_0 and v. We saw that this expansion involves the sum of vacuum diagrams and we also enunciated the diagrammatic rules in Section 10.1. The vacuum diagrams are either connected or consist of disjoint pieces which are, therefore, proportional to the product of connected vacuum diagrams. It would be nice to get rid of the disjoint pieces and to derive a formula for Z/Z_0 in terms of connected diagrams only. To show how this can be done let us start with an example. Consider the diagram

$$= \frac{1}{5!} \left(\frac{i}{2}\right)^5 (\pm)^6 \int [\underbrace{v \ldots \ldots v}_{\text{5 interactions}} \times \underbrace{G_0 \ldots \ldots G_0}_{\text{10 Green's functions}}], \qquad (11.1)$$

where we have used the Feynman rules of Section 10.1 to convert the diagram into the mathematical expression on the r.h.s. Corresponding to this diagram there are several others that yield the same value. For instance.

295

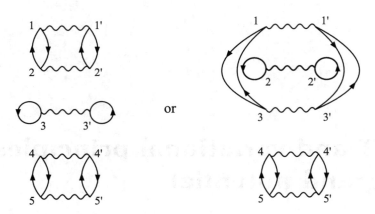

These diagrams simply correspond to different choices of the interaction lines $v(k; k')$ used to draw the connected vacuum diagrams. Let us calculate how many of these diagrams there are. To draw the top diagram in (11.1) we have to choose two interaction lines out of five and this can be done in $\binom{5}{2} = 10$ ways. The middle diagram in (11.1) can then be drawn by choosing two interaction lines out of the remaining three and hence we have $\binom{3}{2} = 3$ possibilities. Finally, the bottom diagram requires the choice of one interaction line out of the only one remaining which can be done in only one way. Since the top and middle diagrams have the same form, and the order in which we construct them does not matter, we still have to divide by 2!. Indeed, in our construction the step

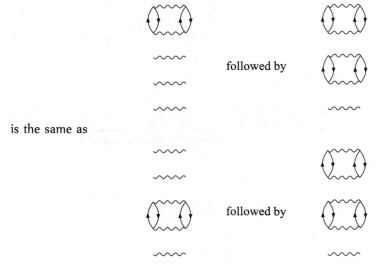

We thus find $\frac{1}{2!}\binom{5}{2}\binom{3}{2} = 15$ diagrams that yield the same value (11.1). It is now easy to make this argument general.

Let us label all the connected vacuum diagrams with an index $i = 1, \ldots, \infty$ and denote by $V_{c,i} = \int [v \ldots v G_0 \ldots G_0]_i$ the integral corresponding to the ith diagram. The number n_i of interaction lines in $V_{c,i}$ is the order of the diagram. A general (and hence not necessarily connected) vacuum diagram contains k_1 times the diagram $V_{c,1}$, k_2 times the diagram $V_{c,2}$ and so on, and therefore its order is

$$n = n_1 k_1 + n_2 k_2 + \ldots,$$

with k_1, $k_2, \ldots$ integers between 0 and ∞. There are $\begin{pmatrix} n \\ n_1 \end{pmatrix}$ ways to construct the first diagram $V_{c,1}$, $\begin{pmatrix} n - n_1 \\ n_1 \end{pmatrix}$ ways to construct the second diagram $V_{c,1}$, etc. When we exhaust the diagrams of type $V_{c,1}$ we can start drawing the diagrams $V_{c,2}$. The first diagram $V_{c,2}$ can be constructed in $\begin{pmatrix} n - n_1 k_1 \\ n_2 \end{pmatrix}$ ways, etc. Finally, we have to divide by $k_1! k_2! \ldots$ in order to compensate for the different order in which the same diagrams can be drawn. We thus find that the vacuum diagram considered appears

$$\frac{1}{k_1! k_2! \ldots} \begin{pmatrix} n \\ n_1 \end{pmatrix} \begin{pmatrix} n - n_1 \\ n_1 \end{pmatrix} \ldots \begin{pmatrix} n - n_1 k_1 \\ n_2 \end{pmatrix} \ldots = \frac{1}{k_1! k_2! \ldots} \frac{n!}{(n_1!)^{k_1} (n_2!)^{k_2} \ldots}$$

times in the perturbative expansion of Z/Z_0. The total contribution of this type of diagram to Z/Z_0 is therefore

$$\frac{1}{k_1! k_2! \ldots} \frac{n!}{(n_1!)^{k_1} (n_2!)^{k_2} \ldots} \frac{1}{n!} \left(\frac{i}{2} \right)^n (\pm)^l \int [\underbrace{v \ldots v}_{n \text{ times}} \times \underbrace{G_0 \ldots G_0}_{2n \text{ times}}]$$

$$= \frac{1}{k_1!} \left[\frac{1}{n_1!} \left(\frac{i}{2} \right)^{n_1} (\pm)^{l_1} V_{c,1} \right]^{k_1} \frac{1}{k_2!} \left[\frac{1}{n_2!} \left(\frac{i}{2} \right)^{n_2} (\pm)^{l_2} V_{c,2} \right]^{k_2} \ldots,$$

where l_i is the number of loops in diagram $V_{c,i}$ and $l = k_1 l_1 + k_2 l_2 + \ldots$ is the total number of loops. We have obtained a product of separate connected vacuum diagrams

$$D_{c,i} \equiv \frac{1}{n_i!} \left(\frac{i}{2} \right)^{n_i} (\pm)^{l_i} V_{c,i}$$

with the right prefactors. To calculate Z/Z_0 we simply have to sum the above expression over all k_i between 0 and ∞:

$$\frac{Z}{Z_0} = \sum_{k_1 = 0}^{\infty} \frac{1}{k_1!} (D_{c,1})^{k_1} \sum_{k_2 = 0}^{\infty} \frac{1}{k_2!} (D_{c,2})^{k_2} \ldots = e^{D_{c,1} + D_{c,2} + \cdots}. \tag{11.2}$$

This elegant result is known as the *linked cluster theorem*, since it shows that for calculating the partition function it is enough to consider connected vacuum diagrams. Another way to

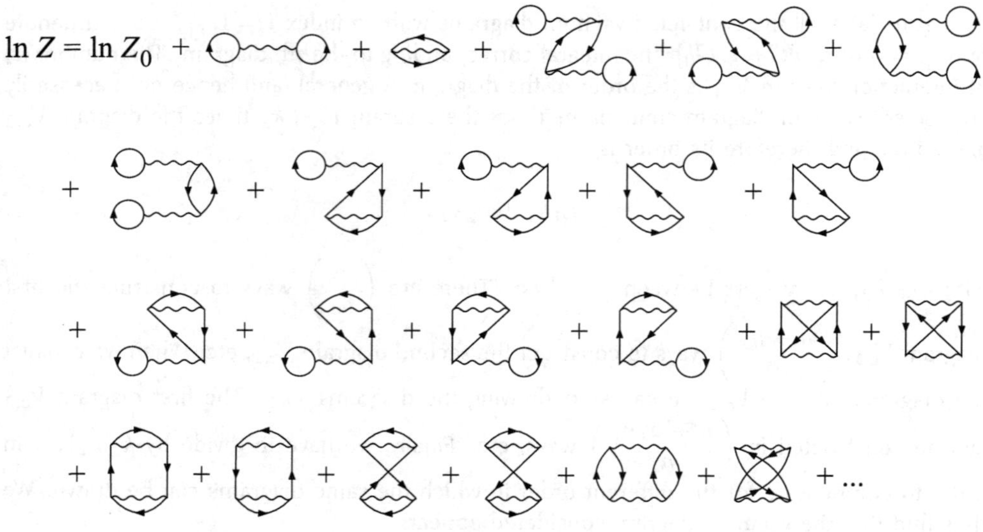

$$\ln Z = \ln Z_0 + \cdots$$

Figure 11.1 Expansion of $\ln Z$ up to second order in the interaction.

write (11.2) is

$$\ln \frac{Z}{Z_0} = \sum_i D_{c,i}$$

$$= \sum_{k=0}^{\infty} \frac{1}{k!} \left(\frac{i}{2}\right)^k \int v(1;1') \ldots v(k;k') \begin{vmatrix} G_0(1;1^+) & G_0(1;1'^+) & \ldots & G_0(1;k'^+) \\ G_0(1';1^+) & G_0(1';1'^+) & \ldots & G_0(1';k'^+) \\ \vdots & \vdots & \ddots & \vdots \\ G_0(k';1^+) & G_0(k';1'^+) & \ldots & G_0(k';k'^+) \end{vmatrix}_c^{\pm}$$

(11.3)

where the integral is over all variables and the symbol $|\ldots|_c^{\pm}$ signifies that in the expansion of the permanent/determinant only the terms represented by connected diagrams are retained. As the logarithm of the partition function is related to the grand potential Ω through the relation $\Omega = -\frac{1}{\beta} \ln Z$, see Appendix D, the linked cluster theorem is a MBPT expansion of Ω in terms of connected vacuum diagrams. In Fig. 11.1 we show the diagrammatic expansion of $\ln Z$ up to second order in the interaction. There are two first-order diagrams and 20 second-order diagrams against the $4! = 24$ second-order diagrams resulting from the expansion of the permanent/determinant in (5.35). Thus, the achieved reduction is rather modest.[1] Nevertheless, there is still much symmetry in these diagrams. For instance the first four second-order diagrams have the same numerical value; the same is true for the next eight diagrams, for the last two diagrams of the third row, for the first four diagrams of the last row and for the last two diagrams. We could therefore reduce the number of diagrams

---

[1]It should be noted, however, that 20 diagrams for $\ln Z$ correspond to infinitely many diagrams for $Z = e^{\ln Z}$.

further if we understand how many vacuum diagrams have the same topology. As we see in the next section, this is less straightforward than for the Green's function diagrams (see Section 10.4) due to the absence of external vertices.

11.2 Summing only the topologically inequivalent diagrams

In Section 10.4 we showed that each nth order connected diagram for the Green's function comes in $2^n n!$ variants. An example was given in Fig. 10.4 where we drew eight topologically equivalent second-order diagrams. Unfortunately, counting the number of variants in which a vacuum diagram can appear is not as easy. For example, the last two diagrams of Fig. 11.1 are topologically equivalent; since there are no other diagrams with the same topology the number of variants is, in this case, only two. For the first diagram in the last row of Fig. 11.1 the number of variants is four, for the first diagram in the third row of Fig. 11.1 the number of variants is eight, etc. The smaller number of variants for some of the vacuum diagrams is due to the fact that not all permutations and/or mirrorings of the interaction lines lead to a different vacuum diagram, i.e., to a different term in the expansion (11.3). If we label the internal vertices of the last two diagrams of Fig. 11.1:

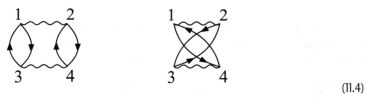

$$\tag{11.4}$$

we see that the simultaneous mirroring $v(1;2) \to v(2;1)$ and $v(3;4) \to v(4;3)$ maps each diagram into itself while the single mirroring $v(1;2) \to v(2;1)$ or $v(3;4) \to v(4;3)$ maps each diagram into the other. It is then clear that to solve the counting problem we must count the symmetries of a given diagram.

Consider a generic vacuum diagram D of order k and label all its internal vertices from 1 to $2k$. We define $\mathcal{G}$ to be the set of oriented Green's function lines and $\mathcal{V}$ to be the set of interaction lines. For the left diagram in (11.4) we have $\mathcal{G} = \{(1;3),(3;1),(2;4),(4;2)\}$ and $\mathcal{V} = \{(1;2),(3;4)\}$. Note that for a v-line $(i;j) = (j;i)$ while for a G_0-line we must consider $(i;j)$ and $(j;i)$ as two different elements. The mirrorings and permutations of the v-lines give rise to $2^k k!$ relabelings of the diagram D which map $\mathcal{V}$ to itself. If such a relabeling also maps $\mathcal{G}$ into $\mathcal{G}$ we call it a symmetry and the total number of symmetries is denoted by N_S. For the left diagram in (11.4) we have the following symmetries:

$$
\begin{aligned}
s_1(1,2,3,4) &= (1,2,3,4) &&\text{identity} \\
s_2(1,2,3,4) &= (2,1,4,3) &&\text{mirroring both interaction lines} \\
s_3(1,2,3,4) &= (3,4,1,2) &&\text{permutation of interaction lines} \\
s_4(1,2,3,4) &= (4,3,2,1) &&\text{s_3 after s_2}
\end{aligned}
\tag{11.5}
$$

and hence $N_S = 4$. It is easy to see that the set of relabeling forms a group of order $2^k k!$ and the set of symmetries form a subgroup of order N_S. Let s_i, $i = 1, \ldots, N_S$ be the symmetries of D. If we now take a relabeling g different from these s_i we obtain a new vacuum diagram D'. For instance, $g(1,2,3,4) = (2,1,3,4)$ maps the left diagram of (11.4) into the right diagram and vice versa. The diagram D' also has N_S symmetries s_i' given by

$$s_i' = g \circ s_i \circ g^{-1},$$

which map D' into itself. Clearly $s_i' \neq s_j'$ if $s_i \neq s_j$. By taking another relabeling h different from s_i, s_i', $i = 1, \ldots, N_S$ we obtain another vacuum diagram D'' which also has N_S symmetries $s_i'' = h \circ s_i \circ h^{-1}$. Continuing in this way we finally obtain

$$N = \frac{2^k k!}{N_S}$$

different vacuum diagrams with the same topology. Thus, to know the number N of variants we must determine the number N_S of symmetries. Once N_S is known the diagrammatic expansion of $\ln Z$ can be performed by including only connected and topologically inequivalent vacuum diagrams $V_c = \int [v \ldots v G_0 \ldots G_0]$ with prefactor

$$\frac{2^k k!}{N_S} \frac{1}{k!} \left(\frac{\mathrm{i}}{2}\right)^k (\pm)^l = \frac{\mathrm{i}^k}{N_S} (\pm)^l.$$

From now on we work only with connected and topologically inequivalent vacuum diagrams. It is therefore convenient to change the Feynman rules for the vacuum diagrams similarly to the way we did in Section 10.4 for the Green's function:

- Number all vertices and assign an interaction line $v(i;j)$ to a wiggly line between j and i and a Green's function $G_0(i;j^+)$ to an oriented line from j to i;

- Integrate over all vertices and multiply by $\mathrm{i}^k(\pm)^l$ where l is the number of loops and k is the number of interaction lines.

These are the same Feynman rules as for the Green's function diagrams. Since the symmetry factor $1/N_S$ is not included in the new rules each diagram will be explicitly multiplied by $1/N_S$. For example, the diagrammatic expansion of $\ln Z$ up to second-order in the interaction is represented as in Fig. 11.2. Note the drastic reduction of second-order diagrams: from 20 in Fig. 11.1 to five in Fig. 11.2. The diagrammatic expansion of Fig. 11.2 is still an expansion in noninteracting Green's functions G_0. By analogy to what we did in Chapter 10 we may try to expand $\ln Z$ in terms of the dressed Green's function G. As we shall see, however, the dressed expansion of $\ln Z$ is a bit more complicated than the dressed expansion of Σ. In the next section we illustrate where the problem lies while in Section 11.4 we show how to overcome it.

11.3   How to construct the Φ functional

Let us consider the vacuum diagram

$$\ln Z = \ln Z_0 + \frac{1}{2} \; \text{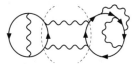}$$

Figure 11.2 Expansion of $\ln Z$ up to second order in the interaction with the new Feynman rules.

where the thin-dashed lines help to visualize the self-energy insertions. This diagram can be regarded either as

$$\Sigma_1 = \qquad \text{inserted into} \qquad \Sigma_1$$

or as

$$\Sigma_2 = \qquad \text{inserted into} \qquad \Sigma_2$$

so that we could choose the G-skeleton diagram to be one of the following two diagrams:

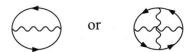

Remember that a diagram is G-skeletonic if it cannot be broken into two disjoint pieces by cutting two Green's function lines, see Section 10.6. In the case of the Green's function diagrams this problem does not arise since there is a unique G-skeleton diagram with one ingoing and one outgoing line. Due to this difficulty it is more convenient to proceed along a new line of argument. We derive the dressed expansion of $\ln Z$ from the self-energy. For this purpose we must establish the relation between a vacuum diagram and a self-energy diagram. This relation at the same time provides us with a different way of determining the number of symmetries N_S of a vacuum diagram.

Let us start with some examples. The nonskeletonic vacuum diagram

consists of the set of G_0-lines $\mathcal{G} = \{(1;1),(3;2),(2;3),(4;4)\}$ and the set of v-lines $\mathcal{V} = \{(1;2),(3;4)\}$. Except for the identity the only other symmetry of this diagram is

$$s(1,2,3,4) = (4,3,2,1),$$

which corresponds to a permutation followed by a simultaneous mirroring of the two v-lines. We now study what happens when we remove a G_0-line from the diagram above. If we remove the G_0-lines $(2;3)$ or $(3;2)$ we produce two topologically equivalent diagrams with the structure below:

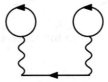

This is a *reducible* self-energy diagram since it can be broken into two disjoint pieces by cutting a G_0-line. The reducible self-energy Σ_r is simply the series (10.8) for the Green's function in which the external G_0-lines are removed, i.e.,

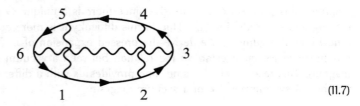

Thus Σ_r contains all Σ-diagrams plus all diagrams that we can make by joining an arbitrary number of Σ-diagrams with Green's function lines. In terms of Σ_r we can rewrite the Dyson equations (10.11) as

$$G(1;2) = G_0(1;2) + \int d3d4\, G_0(1;3)\Sigma_r(3;4)G_0(4;2), \tag{11.6}$$

which implicitly defines the reducible self-energy. If we instead cut the G_0-lines $(1;1)$ or $(4;4)$ we obtain the third diagram of Fig. 10.6. This is a nonskeletonic self-energy diagram since it is one-particle irreducible but contains a self-energy insertion.

A second example is the G-skeletonic vacuum diagram

$$\tag{11.7}$$

which consists of the set of G_0-lines $\mathcal{G} = \{(2;1), (3;2), (4;3), (5;4), (6;5), (1;6)\}$ and the set of v-lines $\mathcal{V} = \{(1;5), (2;4), (3;6)\}$. The symmetries of this diagram are the identity and

$$s(1,2,3,4,5,6) = (4,5,6,1,2,3), \tag{11.8}$$

which corresponds to a permutation of the vertical v-lines followed by a mirroring of all the v-lines.[2] If we remove $G_0(5;4)$ we obtain the diagram

$$\tag{11.9}$$

which is a (G-skeletonic) self-energy diagram. A topologically equivalent diagram is obtained if we remove the Green's function $G_0(2;1)$

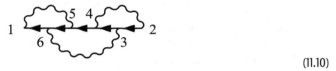

$$\tag{11.10}$$

On the other hand, if we remove the Green's function $G_0(4;3)$ we obtain the diagram

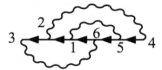

which is again a (G-skeletonic) self-energy diagram but with a different topology.

From these examples we see that by removing G_0-lines from a vacuum diagram we generate reducible self-energy diagrams of the same or different topology. We call two G_0-lines *equivalent* if their removal leads to topologically equivalent Σ_r-diagrams. Since the diagrams (11.9) and (11.10) have the same topology the G_0-lines $(2;1)$ and $(5;4)$ are equivalent. It is easy to verify that the diagram (11.7) splits into three classes of equivalent lines

$$C_1 = \{(2;1), (5;4)\}, \quad C_2 = \{(3;2), (6;5)\}, \quad C_3 = \{(4;3), (1;6)\},$$

which correspond to the self-energy diagrams

---

[2]Looking at (11.7), the simultaneous mirroring of the vertical v-lines may look like a symmetry since it corresponds to rotating the diagram by 180 degrees along the axis passing through 3 and 6. However, this rotation also changes the orientation of the G-lines. One can check that $\mathcal{G}$ is not a map to itself under the mirrorings $v(1;5) \rightarrow v(5;1)$ and $v(2;4) \rightarrow v(4;2)$.

already encountered in Fig. 9.3. We observe that all classes C_i contain a number of elements equal to the number of symmetry operations. Furthermore, an element of class C_i can be generated from one in the same class by applying the symmetry operation (11.8). We now show that this fact is generally true: every vacuum diagram with N_S symmetry operations splits into classes of N_S equivalent lines. This statement stems from the following three properties:

- If s is a symmetry of a vacuum diagram D, then the G_0-lines $(i;j)$ and $(s(i);s(j))$ are equivalent.

  Proof. By the symmetry s the diagram D is mapped onto a topologically equivalent diagram D' with relabeled vertices. Superimposing D and D' it is evident that the removal of line $(i;j)$ from D gives the same reducible self-energy diagram as the removal of line $(s(i);s(j))$ from D'.

- If two G_0-lines $(i;j)$ and $(k;l)$ in a vacuum diagram D are equivalent then there exists a symmetry operation s of D with the property $(s(i);s(j)) = (k;l)$.

  Proof. Let $\Sigma_r^{(i;j)}$ and $\Sigma_r^{(k;l)}$ be the reducible self-energy diagrams obtained by removing from D the G_0-lines $(i;j)$ and $(k;l)$ respectively. By hypothesis the diagrams $\Sigma_r^{(i;j)}$ and $\Sigma_r^{(k;l)}$ are topologically equivalent. Therefore, superimposing the two self-energies we find a one-to-one mapping between the vertex labels that preserves the topological structure. Hence, this mapping is a symmetry operation. For example, by superimposing the diagrams (11.9) and (11.10) and comparing the labels we find the one-to-one relation $(4, 3, 2, 1, 6, 5) \leftrightarrow (1, 6, 5, 4, 3, 2)$ which is exactly the symmetry operation (11.8).

- If a symmetry operation s of a vacuum diagram maps a G_0-line to itself, i.e., $(s(i);s(j)) = (i;j)$, then s must be the identity operation.

  Proof. A symmetry must preserve the connectivity of the diagram and therefore it is completely determined by the mapping of one vertex. For example the symmetry s_2 in (11.5) maps the vertex 1 into 2, i.e., $s(1) = 2$. Since in 1 a G_0-line arrives from 3 and 1 is mapped to 2 in which a G_0-line arrives from 4 then $s(3) = 4$. Furthermore, 1 is connected to 2 by a v-line and hence $s(2) = 1$. Finally, in 2 a G_0-line arrives from 4 and 2 is mapped to 1 in which a G_0-line arrives from 3 and hence $s(4) = 3$. Similarly, for the diagram in (11.7) the symmetry (11.8) maps the vertex 1 into 4, i.e., $s(1) = 4$. Since in 1 a G_0-line arrives from 6 and 1 is mapped to 4 in which a G_0-line arrives from 3 then $s(6) = 3$. In a similar way one can reconstruct the mapping of all the remaining vertices. Thus, if a G_0-line is mapped onto itself the symmetry operation must be the identity.

From these properties it follows that a symmetry maps the classes into themselves and that the elements of the same class are related by a symmetry operation. Since for every symmetry $(s(i);s(j)) \neq (i;j)$ unless s is the identity operation we must have $(s(i);s(j)) \neq (s'(i);s'(j))$ for $s \neq s'$, for otherwise $s^{-1} \circ s' = 1$ and hence $s = s'$. Thus, the application of two different symmetry operations to a G_0-line leads to different G_0-lines in the same class. Consequently, all classes must contain the same number of elements and this number equals the number of symmetries N_S. Taking into account that in an nth order diagram the number of G_0-lines is $2n$, the number of classes is given by

$$N_C = \frac{2n}{N_S}. \tag{11.11}$$

For example, in the diagram (11.7) we have $n = 3$ and $N_S = 2$, so there are $N_C = (2 \times 3)/2 = 3$ classes.

　　Our analysis has led to a different way of calculating the symmetry number N_S. Rather than finding the permutations and mirrorings of interaction lines that do not change the vacuum diagram we can count how many G_0-lines yield the same Σ_r-diagram. Mathematically the act of removing a G_0-line corresponds to taking the functional derivative of the vacuum diagram with respect to G_0. This was clearly illustrated in Section 9.3: the removal of a G_0-line from the vacuum diagrams of Fig. 9.2 generates the self-energy diagrams to the right of the same figure multiplied by the symmetry factor $N_S = 4$. Similarly, the removal of a G_0-line from the vacuum diagram of Fig. 9.3 generates the self-energy diagrams at the bottom of the same figure multiplied by the symmetry factor $N_S = 2$. Since the removal of a G_0-line reduces the number of loops by one we conclude that if D_c is a connected vacuum diagram then

$$\frac{\delta}{\delta G_0(2; 1^+)} \frac{1}{N_S} D_c = \pm \Sigma_r^{(D_c)}(1; 2),$$

where $\Sigma_r^{(D_c)}$ is a contribution to the reducible self-energy consisting of N_C diagrams. For instance, if D_c is the left (or right) diagram of (11.4) which has $N_S = 4$ then we have

$$\frac{1}{4} D_c = \frac{1}{4} \times i^2 (\pm)^2 \int d1d2d3d4 \; v(1; 2)v(3; 4)G_0(1; 3)G_0(3; 1)G_0(2; 4)G_0(4; 2),$$

and its functional derivative with respect to $G_0(6; 5^+)$ gives

$$\frac{1}{4} \frac{\delta D_c}{\delta G_0(6; 5^+)} = i^2 (\pm)^2 \int d2d4 \; v(5; 2)v(6; 4)G_0(5; 6)G_0(2; 4)G_0(4; 2) = \pm \Sigma_r^{(D_c)}(5; 6),$$

where the self-energy $\Sigma_r^{(D_c)}$ is the second diagram of Fig. 10.6 ($N_C = 1$ in this case). Thus, if we write

$$\ln Z = \ln Z_0 \pm \Phi_r[G_0, v], \tag{11.12}$$

where Φ_r is plus/minus (for bosons/fermions), the sum of all connected and topologically inequivalent vacuum diagrams each multiplied by the symmetry factor $1/N_S$, then we have

$$\Sigma_r(1; 2) = \Sigma_r[G_0, v](1; 2) = \frac{\delta \Phi_r}{\delta G_0(2; 1^+)}. \tag{11.13}$$

In the second equality we emphasize that Σ_r is a functional of G_0 and v. It is worth stressing that (11.13) is an identity because the functional derivative of Φ_r contains *all possible reducible self-energy diagrams only once*. If one diagram were missing then the vacuum diagram obtained by closing the missing Σ_r-diagram with a G_0-line would be connected and topologically different from all other diagrams in Φ_r, in contradiction of the hypothesis

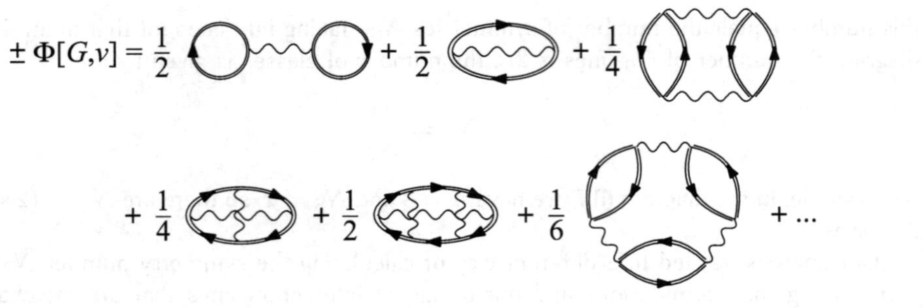

Figure 11.3 Expansion of the Φ functional.

that Φ_r contains *all* connected vacuum diagrams. Furthermore a Σ_r-diagram cannot be generated by two different vacuum diagrams, and hence there is no multiple counting in (11.13).

Equation (11.13) establishes that the reducible self-energy can be obtained as the functional derivative of a functional Φ_r with respect to G_0. We can also go in the opposite direction and "integrate" (11.13) to express Φ_r in terms of reducible self-energy diagrams. We know that there are always N_C topologically inequivalent Σ_r-diagrams originating from the same vacuum diagram. Therefore, if we close these diagrams with a G_0-line we obtain the same vacuum diagram N_C times multiplied by $(\pm)$ due to the creation of a loop. From (11.11) we have $1/(N_C N_S) = 1/2n$ and hence

$$\Phi_r[G_0, v] = \sum_{n=1}^{\infty} \frac{1}{2n} \int d1d2 \; \Sigma_r^{(n)}[G_0, v](1;2)G_0(2;1^+), \tag{11.14}$$

where $\Sigma_r^{(n)}$ denotes the sum of all reducible and topologically inequivalent self-energy diagrams of order n.

An important consequence of this result is that it tells us how to construct the functional $\Phi = \Phi[G, v]$ whose functional derivative with respect to the *dressed* Green's function G is the self-energy $\Sigma = \Sigma_s[G, v]$ introduced in Section 10.6:

$$\Sigma(1;2) = \Sigma_s[G, v](1;2) = \frac{\delta\Phi}{\delta G(2;1^+)}. \tag{11.15}$$

By the very same reasoning $\Phi[G, v]$ is plus/minus the sum of all connected, topologically inequivalent and G-skeletonic vacuum diagrams each multiplied by the corresponding symmetry factor $1/N_S$ and in which $G_0 \to G$, see Fig. 11.3. This is so because a nonskeletonic vacuum diagram can only generate nonskeletonic and/or one-particle reducible Σ_r-diagrams while a G-skeleton vacuum diagram generates only G-skeleton Σ-diagrams. The functional Φ can be written in the same fashion as Φ_r, i.e.,

$$\Phi[G, v] = \sum_{n=1}^{\infty} \frac{1}{2n} \int d1d2 \; \Sigma_s^{(n)}[G, v](1;2)G(2;1^+), \tag{11.16}$$

where $\Sigma_s^{(n)}$ is the sum of all topologically inequivalent and G-skeleton self-energy diagrams of order n in which $G_0 \to G$. It is convenient to introduce the shorthand notation

$$\text{tr}_\gamma[fg] \equiv \int d1d2 \; f(1;2)g(2;1) = \text{tr}_\gamma[gf] \tag{11.17}$$

for the trace over space and spin and for the convolution along the contour. Then the expansion of Φ takes the compact form[3]

$$\Phi[G,v] = \sum_{n=1}^{\infty} \frac{1}{2n} \text{tr}_\gamma \left[\Sigma_s^{(n)}[G,v]G \right] \tag{11.18}$$

In order to obtain an expression for $\ln Z$ in terms of the dressed Green's function G rather than G_0 it is tempting to replace $\Phi_r[G_0, v]$ with $\Phi[G, v]$ in (11.12). Unfortunately the value of $\Phi[G, v]$ is not equal to $\Phi_r[G_0, v]$ due to the double counting problem mentioned at the beginning of the section. Nevertheless, it is possible to derive an exact and simple formula for the correction $\Phi_r[G_0, v] - \Phi[G, v]$; this is the topic of the next section.

11.4 Dressed expansion of the grand potential

Let us consider a system described by the Hamiltonian $\hat{H}_\lambda(z)$ with rescaled interaction $v \to \lambda v$: $\hat{H}_\lambda(z) = \hat{H}_0(z) + \lambda \hat{H}_{\text{int}}(z)$. The grand potential for this Hamiltonian is

$$\Omega_\lambda = -\frac{1}{\beta} \ln Z_\lambda = -\frac{1}{\beta} \ln \text{Tr} \left[e^{-\beta \hat{H}_\lambda^M} \right] = -\frac{1}{\beta} \ln \text{Tr} \left[\mathcal{T} \left\{ e^{-i \int_\gamma dz \hat{H}_\lambda(z)} \right\} \right]. \tag{11.19}$$

The derivative of (11.19) with respect to λ is

$$\frac{d\Omega_\lambda}{d\lambda} = \frac{i}{\beta} \int_\gamma dz_1 \frac{\text{Tr} \left[\mathcal{T} \left\{ e^{-i \int_\gamma dz \hat{H}_\lambda(z)} \hat{H}_{\text{int}}(z_1) \right\} \right]}{\text{Tr} \left[\mathcal{T} \left\{ e^{-i \int_\gamma dz \hat{H}_\lambda(z)} \right\} \right]} = \frac{i}{\beta} \int_\gamma dz_1 \langle \hat{H}_{\text{int}}(z_1) \rangle_\lambda, \tag{11.20}$$

where we have introduced the short-hand notation

$$\langle \ldots \rangle_\lambda = \text{Tr} \left[\mathcal{T} \left\{ e^{-i \int_\gamma dz \hat{H}_\lambda(z)} \ldots \right\} \right] / Z_\lambda,$$

in which any string of operators can be inserted in the dots. The integrand on the r.h.s. of (11.20) multiplied by λ is the interaction energy $E_{\text{int},\lambda}(z_1)$ with rescaled interaction. This energy can be expressed in terms of the rescaled Green's function G_λ and self-energy $\Sigma_\lambda = \Sigma_s[G_\lambda, \lambda v]$ as

$$E_{\text{int},\lambda}(z_1) = \lambda \langle \hat{H}_{\text{int}}(z_1) \rangle_\lambda = \pm \frac{i}{2} \int d\mathbf{x}_1 d2 \; \Sigma_\lambda(1;2)G_\lambda(2;1^+), \tag{11.21}$$

---

[3]We observe that the definition (11.17) with $f = \Sigma$ and $g = G$ is ambiguous only for the singular Hartree–Fock self-energy since the Green's function should then be evaluated at equal times. MBPT tells us how to interpret (11.17) in this case: the second time argument of the Green's function must be infinitesimally later than the first time argument.

in accordance with (9.19). Inserting this result into (11.20) and integrating over λ between 0 and 1 we find

$$\Omega = \Omega_0 + \frac{i}{\beta} \int_0^1 \frac{d\lambda}{\lambda} \int_\gamma dz_1 \left(\pm \frac{i}{2} \right) \int d\mathbf{x}_1 d2 \, \Sigma_\lambda(1;2) G_\lambda(2;1^+)$$

$$= \Omega_0 \mp \frac{1}{2\beta} \int_0^1 \frac{d\lambda}{\lambda} \, \mathrm{tr}_\gamma \left[\Sigma_\lambda G_\lambda \right]. \tag{11.22}$$

Equation (11.22) can be used to derive (11.14) in an alternative way. Since (omitting arguments and integrals)

$$\Sigma G = \Sigma[G_0 + G_0 \Sigma G_0 + G_0 \Sigma G_0 \Sigma G_0 + \ldots]$$
$$= [\Sigma + \Sigma G_0 \Sigma + \Sigma G_0 \Sigma G_0 \Sigma + \ldots] G_0 = \Sigma_r G_0,$$

we can rewrite (11.22) as

$$\Omega = \Omega_0 \mp \frac{1}{2\beta} \int_0^1 \frac{d\lambda}{\lambda} \, \mathrm{tr}_\gamma \left[\Sigma_r[G_0, \lambda v] G_0 \right]$$

$$= \Omega_0 \mp \frac{1}{2\beta} \sum_{n=1}^\infty \int_0^1 \frac{d\lambda}{\lambda} \lambda^n \mathrm{tr}_\gamma \left[\Sigma_r^{(n)}[G_0, v] G_0 \right]$$

$$= \Omega_0 \mp \frac{1}{\beta} \sum_{n=1}^\infty \frac{1}{2n} \, \mathrm{tr}_\gamma \left[\Sigma_r^{(n)}[G_0, v] G_0 \right].$$

Multiplying both sides by $-\beta$ and comparing with the definition of Φ_r in (11.12) we reobtain (11.14).

Let us now consider the functional Φ in (11.16) for a system with rescaled interaction λv:

$$\Phi[G_\lambda, \lambda v] = \sum_{n=1}^\infty \frac{1}{2n} \, \mathrm{tr}_\gamma \left[\Sigma_s^{(n)}[G_\lambda, \lambda v] G_\lambda \right]$$

$$= \sum_{n=1}^\infty \frac{\lambda^n}{2n} \, \mathrm{tr}_\gamma \left[\Sigma_s^{(n)}[G_\lambda, v] G_\lambda \right].$$

The derivative of this expression with respect to λ gives

$$\frac{d}{d\lambda} \Phi[G_\lambda, \lambda v] = \sum_{n=1}^\infty \frac{\lambda^{n-1}}{2} \, \mathrm{tr}_\gamma \left[\Sigma_s^{(n)}[G_\lambda, v] G_\lambda \right] + \mathrm{tr}_\gamma \left[\frac{\delta\Phi[G_\lambda, \lambda v]}{\delta G_\lambda} \frac{dG_\lambda}{d\lambda} \right]$$

$$= \mathrm{tr}_\gamma \left[\frac{1}{2\lambda} \Sigma_\lambda G_\lambda + \Sigma_\lambda \frac{dG_\lambda}{d\lambda} \right], \tag{11.23}$$

where the first term originates from the variation of the interaction lines and the second term from the variation of the Green's function lines. The first term in (11.23) appears also in (11.22), which can therefore be rewritten as

$$\Omega = \Omega_0 \mp \frac{1}{\beta} \int_0^1 d\lambda \left(\frac{d}{d\lambda} \Phi[G_\lambda, \lambda v] - \mathrm{tr}_\gamma \left[\Sigma_\lambda \frac{dG_\lambda}{d\lambda} \right] \right)$$

$$= \Omega_0 \mp \frac{1}{\beta} \left(\Phi - \mathrm{tr}_\gamma [\Sigma G] \right) \mp \frac{1}{\beta} \int_0^1 d\lambda \, \mathrm{tr}_\gamma \left[\frac{d\Sigma_\lambda}{d\lambda} G_\lambda \right], \tag{11.24}$$

where in the last equality we have integrated by parts and taken into account that $\Phi[G_\lambda, \lambda v]$ and Σ_λ vanish for $\lambda = 0$ while they equal $\Phi = \Phi[G, v]$ and $\Sigma = \Sigma_s[G, v]$ for $\lambda = 1$. To calculate the last term we observe that[4]

$$
\begin{aligned}
\frac{d}{d\lambda} \mathrm{tr}_\gamma \left[\ln(1 - G_0 \Sigma_\lambda) \right] &= -\frac{d}{d\lambda} \mathrm{tr}_\gamma \left[G_0 \Sigma_\lambda + \frac{1}{2} G_0 \Sigma_\lambda G_0 \Sigma_\lambda + \frac{1}{3} G_0 \Sigma_\lambda G_0 \Sigma_\lambda G_0 \Sigma_\lambda + \ldots \right] \\
&= -\mathrm{tr}_\gamma \left[G_0 \frac{d\Sigma_\lambda}{d\lambda} + G_0 \Sigma_\lambda G_0 \frac{d\Sigma_\lambda}{d\lambda} + G_0 \Sigma_\lambda G_0 \Sigma_\lambda G_0 \frac{d\Sigma_\lambda}{d\lambda} + \ldots \right] \\
&= -\mathrm{tr}_\gamma \left[G_\lambda \frac{d\Sigma_\lambda}{d\lambda} \right].
\end{aligned}
\tag{11.25}
$$

Inserting this result in (11.24) we obtain an elegant formula for the grand potential,

$$
\beta(\Omega - \Omega_0) = \mp \left\{ \Phi - \mathrm{tr}_\gamma \left[\Sigma G + \ln(1 - G_0 \Sigma) \right] \right\}.
\tag{11.26}
$$

This formula provides the MBPT expansion of Ω (and hence of $\ln Z$) in terms of the dressed Green's function G since we know how to expand the functional $\Phi = \Phi[G, v]$ and the self-energy $\Sigma = \Sigma_s[G, v]$ in G-skeleton diagrams.

11.5 Luttinger–Ward and Klein functionals

Equation (11.26) can also be regarded as the definition of a functional $\Omega[G, v]$ which takes the value of the grand potential when G is the Green's function of the underlying physical system, i.e., $G = G_0 + G_0 \Sigma G$. Any physical Green's function belongs to the Keldysh space and hence the contour integrals in (11.26) reduce to integrals along the imaginary track γ^{M}. If, on the other hand, we evaluate $\Omega[G, v]$ at a G which does not belong to the Keldysh space then the contour integrals cannot be reduced to integrals along γ^{M} and the full contour must be considered. The functional $\Omega[G, v]$ was first introduced by Luttinger and Ward [81] and we therefore refer to it as the *Luttinger–Ward functional*. To distinguish the functional from the grand potential (which is a number) we denote the former by $\Omega_{\mathrm{LW}}[G, v]$:

$$
\boxed{\Omega_{\mathrm{LW}}[G, v] = \Omega_0 \mp \frac{1}{\beta} \left\{ \Phi[G, v] - \mathrm{tr}_\gamma \left[\Sigma_s[G, v] G + \ln(1 - G_0 \Sigma_s[G, v]) \right] \right\}}
\tag{11.27}
$$

A remarkable feature of the Luttinger–Ward (LW) functional is its variational property. If we change the Green's function from G to $G + \delta G$ then the change of Ω_{LW} reads

$$
\delta\Omega_{\mathrm{LW}} = \mp \frac{1}{\beta} \left\{ \delta\Phi - \mathrm{tr}_\gamma \left[\delta\Sigma G + \Sigma \delta G - (G_0 + G_0 \Sigma G_0 + \ldots) \delta\Sigma \right] \right\}.
$$

Since $\delta\Phi = \mathrm{tr}_\gamma[\Sigma \delta G]$ we conclude that the variation $\delta\Omega_{\mathrm{LW}}$ vanishes when (omitting arguments and integrals)

$$
G = G_0 + G_0 \Sigma G_0 + G_0 \Sigma G_0 \Sigma G_0 + \ldots,
\tag{11.28}
$$

---

[4] By definition the function of an operator $\hat{A}$ is defined by its Taylor expansion and therefore $-\ln(1 - \hat{A}) = \sum_n \frac{1}{n} \hat{A}^n$.

i.e., when G is the self-consistent solution of $G = G_0 + G_0 \Sigma G$ with $\Sigma = \Sigma_s[G, v]$. Therefore Ω_{LW} equals the grand potential at the stationary point. We further observe that the LW functional preserves the variational property for any approximate Φ provided that the self-energy $\Sigma_s[G, v]$ is calculated as the functional derivative of such Φ. In other words, Ω_{LW} is stationary at the approximate G which satisfies the Dyson equation (11.28) with $\Sigma_s[G, v] = \delta\Phi[G, v]/\delta G$.[5] It is also worth stressing that in the functional $\Omega_{\mathrm{LW}}[G, v]$ there is an explicit reference to the underlying physical system through the noninteracting Green's function G_0, see last term in (11.27). This is not the case for the Φ functional (and hence neither for the self-energy $\Sigma_s[G, v]$) for which a stationary principle would not make sense.[6]

The variational property of the LW functional naturally introduces a new concept to tackle the equilibrium many-body problem. This concept is based on variational principles instead of diagrammatic MBPT expansions. Since Ω_{LW} is stationary with respect to changes in G, the value of Ω_{LW} at a Green's function that deviates from the stationary G by δG leads to an error in Ω_{LW} which is only of second order in δG. Therefore the quality of the results depend primarily on the chosen approximation for Φ. Note that although Ω_{LW} is stationary at the self-consistent G the stationary point does not have to be a minimum.[7] Finally we observe that the advantage of calculating Ω from variational expressions such as the LW functional is that in this way we avoid solving the Dyson equation. If the self-consistent solution gives accurate Ω then the LW functional will produce approximations to these grand potentials with much less computational effort [82, 83].

At this point we draw the attention of the reader to an interesting fact. The variational schemes are by no means unique [84]. By adding to Ω_{LW} any functional $F[D]$, where

$$D[G, v](1; 2) = G(1; 2) - G_0(1; 2) - \int d3 d4 \, G(1; 3)\Sigma_s[G, v](3; 4)G_0(4; 2), \qquad (11.29)$$

obeying

$$F[D = 0] = \left(\frac{\delta F}{\delta D}\right)_{D=0} = 0, \qquad (11.30)$$

one obtains a new variational functional having the same stationary point and the same value at the stationary point. It might, however, be designed to give a second derivative which also vanishes at the stationary point (something that would be of utmost practical value).

Choosing to add inside the curly brackets of (11.27) the functional

$$F[D] = \mathrm{tr}_\gamma \left[\ln(1 + D G_0^{-1}) - D \overleftarrow{G}_0^{-1} \right], \qquad (11.31)$$

with $\overleftarrow{G}_0^{-1}(1; 2) = -i\frac{\overleftarrow{d}}{dz_1}\delta(1; 2) - h(1; 2)$ the differential operator (acting on quantities to its left) for which $\int d3 \, G_0(1; 3) \, \overleftarrow{G}_0^{-1}(3; 2) = \delta(1; 2)$, leads to the functional

$$\boxed{\Omega_{\mathrm{K}}[G, v] = \Omega_0 \mp \frac{1}{\beta}\left\{ \Phi[G, v] + \mathrm{tr}_\gamma \left[\ln(G\overleftarrow{G}_0^{-1}) + 1 - G\overleftarrow{G}_0^{-1} \right] \right\}} \qquad (11.32)$$

---

[5]From Chapter 9 we know that this Green's function preserves all basic conservation laws.

[6]There is no reason for Φ to be stationary at the Green's function of the physical system since Φ does not know anything about it!

[7]In Ref. [82] it was shown that Ω_{LW} is minimum at the stationary point when Φ is approximated by the infinite sum of ring diagrams. This is the so called RPA approximation and it is discussed in detail in Section 15.5.

As the functional F in (11.31) has the desired properties (11.30), this new functional is stationary at the Dyson equation and equals the grand potential at the stationary point. The functional (11.32) was first proposed by Klein [85] and we refer to it as the *Klein functional*. The Klein functional is much easier to evaluate and manipulate as compared to the LW functional but, unfortunately, it is less stable (large second derivative) at the stationary point [83].

We conclude this section by showing that Ω_0 can be written as

$$\Omega_0 = \mp \frac{1}{\beta} \mathrm{tr}_\gamma \left[\ln(-G_0) \right] \tag{11.33}$$

according to which the Klein functional can also be written as

$$\Omega_{\mathrm{K}}[G, v] = \mp \frac{1}{\beta} \left\{ \Phi[G, v] + \mathrm{tr}_\gamma \left[\ln(-G) + 1 - G \overleftarrow{G}_0^{-1} \right] \right\}. \tag{11.34}$$

We start by calculating the tr_γ of an arbitrary power of G_0

$$\mathrm{tr}_\gamma \left[G_0^m \right] = \int d\mathbf{x} \langle \mathbf{x} | \int_\gamma dz_1 \ldots dz_m \, \hat{\mathcal{G}}_0(z_1, z_2) \ldots \hat{\mathcal{G}}_0(z_m, z_1^+) | \mathbf{x} \rangle$$

$$= \int d\mathbf{x} \langle \mathbf{x} | (-\mathrm{i})^m \int_0^\beta d\tau_1 \ldots d\tau_m \, \hat{\mathcal{G}}_0^{\mathrm{M}}(\tau_1, \tau_2) \ldots \hat{\mathcal{G}}_0^{\mathrm{M}}(\tau_m, \tau_1^+) | \mathbf{x} \rangle,$$

where in the last step we use the fact that for any G_0 in Keldysh space the integral along the forward branch cancels the integral along the backward branch, see also Exercise 5.4. Expanding the Matsubara Green's function as in (6.17) and using the identity [see (A.5)]

$$\int_0^\beta d\tau \, e^{-(\omega_p - \omega_q)\tau} = \beta \delta_{pq},$$

we can rewrite the trace as

$$\mathrm{tr}_\gamma \left[G_0^m \right] = \sum_{p=-\infty}^\infty e^{\eta \omega_p} \int d\mathbf{x} \langle \mathbf{x} | \frac{1}{(\omega_p - \hat{h}^{\mathrm{M}})^m} | \mathbf{x} \rangle = \sum_\lambda \sum_{p=-\infty}^\infty e^{\eta \omega_p} \frac{1}{(\omega_p - \epsilon_\lambda^{\mathrm{M}})^m},$$

with $\epsilon_\lambda^{\mathrm{M}}$ the eigenvalues of $\hat{h}^{\mathrm{M}}$. Therefore we define

$$\mathrm{tr}_\gamma \left[\ln(-G_0) \right] = \sum_\lambda \sum_{p=-\infty}^\infty e^{\eta \omega_p} \ln \frac{1}{\epsilon_\lambda^{\mathrm{M}} - \omega_p}.$$

To evaluate the sum over the Matsubara frequencies we use the trick (6.43) and get

$$\mathrm{tr}_\gamma \left[\ln(-G_0) \right] = \mp (-\mathrm{i}\beta) \sum_\lambda \int_{\Gamma_b} \frac{d\zeta}{2\pi} \, e^{\eta \zeta} f(\zeta) \ln \frac{1}{\epsilon_\lambda^{\mathrm{M}} - \zeta}$$

$$= \mathrm{i} \sum_\lambda \int_{\Gamma_b} \frac{d\zeta}{2\pi} \frac{e^{\eta \zeta} \ln(1 \mp e^{-\beta \zeta})}{\zeta - \epsilon_\lambda^{\mathrm{M}}} = -\sum_\lambda \ln(1 \mp e^{-\beta \epsilon_\lambda^{\mathrm{M}}}), \tag{11.35}$$

where we first perform an integration by parts and then uses the Cauchy residue theorem. To relate this result to Ω_0 we observe that

$$Z_0 = \text{Tr}\left[e^{-\beta \hat{H}_0^{\text{M}}}\right] = \prod_\lambda \sum_n e^{-\beta \epsilon_\lambda^{\text{M}} n} = \prod_\lambda \left(1 \mp e^{-\beta \epsilon_\lambda^{\text{M}}}\right)^{\mp 1},$$

where in the last equality we have taken into account that the sum over n runs between 0 and ∞ in the case of bosons and between 0 and 1 in the case of fermions. Therefore

$$\Omega_0 = -\frac{1}{\beta} \ln Z_0 = \pm \frac{1}{\beta} \sum_\lambda \ln(1 \mp e^{-\beta \epsilon_\lambda^{\text{M}}}).$$

Comparing this result with (11.35) we find (11.33).

11.6   Luttinger–Ward theorem

In this section we discuss a few nice consequences of the variational idea in combination with conserving approximations. Using (11.33) the Luttinger–Ward functional can be rewritten as

$$\Omega_{\text{LW}} = \mp \frac{1}{\beta} \left\{ \Phi - \text{tr}_\gamma \left[\Sigma G + \ln(\Sigma - G_0^{-1}) \right] \right\},$$

where we omit the explicit dependence of Φ and Σ on G and v. If we evaluate this functional for a physical Green's function (hence belonging to the Keldysh space) then the contour integrals reduce to integrals along the imaginary track. Expanding G and Σ in Matsubara frequencies we find

$$\Omega_{\text{LW}} = \mp \frac{1}{\beta} \Phi \pm \frac{1}{\beta} \sum_{p=-\infty}^\infty e^{\eta \omega_p} \int d\mathbf{x}\, \langle \mathbf{x} | \hat{\Sigma}^{\text{M}}(\omega_p) \hat{\mathcal{G}}^{\text{M}}(\omega_p) + \ln\left(\hat{\Sigma}^{\text{M}}(\omega_p) - \omega_p + \hat{h}^{\text{M}}\right) | \mathbf{x} \rangle,$$

where in the last term we use (6.18). From quantum statistical mechanics we know that the total number of particles in the system is given by minus the derivative of the grand potential Ω with respect to the chemical potential μ, see Appendix D. Is it true that for an approximate self-energy (and hence for an approximate G)

$$N = -\frac{\partial \Omega_{\text{LW}}}{\partial \mu}? \tag{11.36}$$

The answer is positive provided that Σ is Φ-derivable and that G is self-consistently calculated from the Dyson equation. In this case Ω_{LW} is stationary with respect to changes in G and therefore in (11.36) we only need to differentiate with respect to the explicit dependence on μ. This dependence is contained in $\hat{h}^{\text{M}} = \hat{h} - \mu$ and therefore

$$-\frac{\partial \Omega_{\text{LW}}}{\partial \mu} = \mp \frac{1}{\beta} \sum_{p=-\infty}^\infty e^{\eta \omega_p} \int d\mathbf{x}\, \langle \mathbf{x} | \frac{-1}{\hat{\Sigma}^{\text{M}}(\omega_p) - \omega_p + \hat{h}^{\text{M}}} | \mathbf{x} \rangle$$

$$= \mp \frac{1}{\beta} \sum_{p=-\infty}^\infty e^{\eta \omega_p} \int d\mathbf{x}\, \langle \mathbf{x} | \hat{\mathcal{G}}^{\text{M}}(\omega_p) | \mathbf{x} \rangle$$

$$= \pm i \int d\mathbf{x}\, G^{\text{M}}(\mathbf{x}, \tau; \mathbf{x}, \tau^+). \tag{11.37}$$

The r.h.s. of (11.37) is exactly the total number of particles N calculated using the G for which Ω_{LW} is stationary. This result was obtained by Baym [78] and shows that conserving approximations preserve (11.36).

Equation (11.36) has an interesting consequence in systems of fermions at zero temperature. Let us rewrite the first line of (11.37) in a slightly different form (lower sign for fermions)

$$N = -\frac{\partial \Omega_{\mathrm{LW}}}{\partial \mu} = \frac{1}{\beta} \sum_{p=-\infty}^{\infty} e^{\eta \omega_p} \left[\frac{\partial}{\partial \omega_p} \int d\mathbf{x}\, \langle \mathbf{x}| \ln\left(\hat{\Sigma}^{\mathrm{M}}(\omega_p) - \omega_p + \hat{h}^{\mathrm{M}} \right) |\mathbf{x}\rangle \right.$$

$$\left. + \int d\mathbf{x}\, \langle \mathbf{x}| \hat{\mathcal{G}}^{\mathrm{M}}(\omega_p) \frac{\partial \hat{\Sigma}^{\mathrm{M}}(\omega_p)}{\partial \omega_p} |\mathbf{x}\rangle \right]. \tag{11.38}$$

We now prove that at zero temperature the last term of this equation vanishes. For $\beta \to \infty$ the Matsubara frequency $\omega_p = (2p+1)\pi/(-\mathrm{i}\beta)$ becomes a continuous variable ζ and the sum over p becomes an integral over ζ. Since $d\zeta \equiv \omega_{p+1} - \omega_p = 2\pi/(-\mathrm{i}\beta)$ we have

$$\lim_{\beta \to \infty} \frac{1}{\beta} \sum_{p=-\infty}^{\infty} \int d\mathbf{x}\, \langle \mathbf{x}| \hat{\mathcal{G}}^{\mathrm{M}}(\omega_p) \frac{\partial \hat{\Sigma}^{\mathrm{M}}(\omega_p)}{\partial \omega_p} |\mathbf{x}\rangle = \frac{1}{2\pi\mathrm{i}} \int_{-\mathrm{i}\infty}^{\mathrm{i}\infty} d\zeta \int d\mathbf{x}\, \langle \mathbf{x}| \hat{\mathcal{G}}^{\mathrm{M}}(\zeta) \frac{\partial \hat{\Sigma}^{\mathrm{M}}(\zeta)}{\partial \zeta} |\mathbf{x}\rangle$$

$$= \frac{1}{2\pi\mathrm{i}} \int d\mathbf{x} \int_{-\mathrm{i}\infty}^{\mathrm{i}\infty} d\zeta\, \langle \mathbf{x}| \underbrace{\frac{\partial}{\partial \zeta} \left(\hat{\mathcal{G}}^{\mathrm{M}}(\zeta) \hat{\Sigma}^{\mathrm{M}}(\zeta) \right)}_{\text{total derivative}} - \hat{\Sigma}^{\mathrm{M}}(\zeta) \frac{\partial \hat{\mathcal{G}}^{\mathrm{M}}(\zeta)}{\partial \zeta} |\mathbf{x}\rangle$$

$$= -\frac{1}{2\pi\mathrm{i}} \int d\mathbf{x} \int_{-\mathrm{i}\infty}^{\mathrm{i}\infty} d\zeta\, \langle \mathbf{x}| \frac{\delta \Phi}{\delta \hat{\mathcal{G}}^{\mathrm{M}}(\zeta)} \frac{\partial \hat{\mathcal{G}}^{\mathrm{M}}(\zeta)}{\partial \zeta} |\mathbf{x}\rangle, \tag{11.39}$$

where we have taken into account that the integral over ζ of the total derivative yields zero and that the self-energy is Φ-derivable. We also set to unity the exponential factor $e^{\eta \zeta}$ since the integral along the imaginary axis is convergent. Due to energy conservation every Φ-diagram is invariant under the change $\hat{\mathcal{G}}^{\mathrm{M}}(\zeta) \to \hat{\mathcal{G}}^{\mathrm{M}}(\zeta + \delta\zeta)$. Thus the variation $\delta\Phi = 0$ under this transformation. However, the variation $\delta\Phi$ is formally given by the last line of (11.39) which, therefore, must vanish.

To evaluate N we have to perform a sum over Matsubara frequencies. The argument of the logarithm in (11.38) is $-[\hat{\mathcal{G}}^{\mathrm{M}}(\omega_p)]^{-1}$. Then we can again use the trick (6.43), where Γ_b is the contour in Fig. 6.1, since $\hat{\mathcal{G}}^{\mathrm{M}}(\zeta)$ is analytic in the complex ζ plane except along the real axis, see (6.82), and the eigenvalues of $\hat{\mathcal{G}}^{\mathrm{M}}(\zeta)$ are nonzero, see Exercise 6.9. We find

$$N = -\int_{-\infty}^{\infty} \frac{d\omega}{2\pi\mathrm{i}}\, f(\omega) e^{\eta \omega} \frac{\partial}{\partial \omega} \int d\mathbf{x}\, \langle \mathbf{x}| \ln\left(-\hat{\mathcal{G}}^{\mathrm{M}}(\omega - \mathrm{i}\delta) \right) - \ln\left(-\hat{\mathcal{G}}^{\mathrm{M}}(\omega + \mathrm{i}\delta) \right) |\mathbf{x}\rangle.$$

Integrating by parts and taking into account that in the zero temperature limit the derivative of the Fermi function $\partial f(\omega)/\partial \omega = -\delta(\omega)$, we get

$$N = \frac{1}{2\pi\mathrm{i}} \int d\mathbf{x}\, \langle \mathbf{x}| \ln\left(\hat{\Sigma}^{\mathrm{M}}(-\mathrm{i}\delta) + \mathrm{i}\delta + \hat{h} - \mu \right) - \ln\left(\hat{\Sigma}^{\mathrm{M}}(\mathrm{i}\delta) - \mathrm{i}\delta + \hat{h} - \mu \right) |\mathbf{x}\rangle, \tag{11.40}$$

where we insert back the explicit expression for $\hat{\mathcal{G}}^{\mathrm{M}}$ in terms of $\hat{\Sigma}^{\mathrm{M}}$ and use the fact that $\hat{h}^{\mathrm{M}} = \hat{h} - \mu$. According to (9.59) and the discussion below it we have $\hat{\Sigma}^{\mathrm{M}}(\pm \mathrm{i}\delta) = \hat{\Sigma}^{\mathrm{R}}(\mu) = \hat{\Sigma}^{\mathrm{A}}(\mu)$, since the rate operator vanishes at $\omega = \mu$. This also implies that $\hat{\Sigma}^{\mathrm{R}}(\mu)$ is self-adjoint. Let us calculate (11.40) for an electron gas.

In the electron gas $\hat{h}$ and $\hat{\Sigma}^{\mathrm{R}}$ are diagonal in the momentum-spin basis, i.e., $\hat{h}|\mathbf{p}\sigma\rangle = \frac{p^2}{2}|\mathbf{p}\sigma\rangle$ and $\hat{\Sigma}^{\mathrm{R}}(\mu)|\mathbf{p}\sigma\rangle = \Sigma^{\mathrm{R}}(\mathbf{p},\mu)|\mathbf{p}\sigma\rangle$. Inserting the completeness relation (1.11) in the bracket $\langle \mathbf{x}|\ldots|\mathbf{x}\rangle$ we find

$$N = 2\frac{\mathrm{V}}{2\pi\mathrm{i}} \int \frac{d\mathbf{p}}{(2\pi)^3} \left[\ln\left(\Sigma^{\mathrm{R}}(\mathbf{p},\mu) + \mathrm{i}\delta + \frac{p^2}{2} - \mu \right) - \ln\left(\Sigma^{\mathrm{R}}(\mathbf{p},\mu) - \mathrm{i}\delta + \frac{p^2}{2} - \mu \right) \right],$$

where the factor of 2 comes from spin and $\mathrm{V} = \int d\mathbf{r}$ is the volume of the system. Since

$$\ln(a \pm \mathrm{i}\delta) = \begin{cases} \ln a & a > 0 \\ \ln|a| \pm \mathrm{i}\pi & a < 0 \end{cases}$$

we can rewrite the total number of particles at zero temperature as

$$\boxed{N = 2\mathrm{V} \int \frac{d\mathbf{p}}{(2\pi)^3}\, \theta\left(\mu - \frac{p^2}{2} - \Sigma^{\mathrm{R}}(\mathbf{p},\mu)\right)} \tag{11.41}$$

This result is known as the *Luttinger–Ward theorem* and is valid *for any conserving approximation*. Equation (7.57) is an example of the Luttinger–Ward theorem in the Hartree–Fock approximation (remember that the Fermi energy ϵ_{F} is the zero-temperature limit of the chemical potential). Due to the rotational invariance, $\epsilon_{\mathbf{p}}$ and $\Sigma^{\mathrm{R}}(\mathbf{p},\mu)$ depend only on the modulus $p = |\mathbf{p}|$ of the momentum. If we define the Fermi momentum p_{F} as the solution of

$$\mu - \frac{p_{\mathrm{F}}^2}{2} - \Sigma^{\mathrm{R}}(p_{\mathrm{F}}, \mu) = 0, \tag{11.42}$$

then the equation for the total number of particles reduces to

$$\frac{N}{\mathrm{V}} = n = 2 \int_{p < p_{\mathrm{F}}} \frac{d\mathbf{p}}{(2\pi)^3} = \frac{p_{\mathrm{F}}^3}{3\pi^2}. \tag{11.43}$$

In Appendix K we show that the momentum distribution $n_{\mathbf{p}}$ (average number of electrons with momentum $\mathbf{p}$) is discontinuous at $|\mathbf{p}| = p_{\mathrm{F}}$. Therefore, in an electron gas with N particles the discontinuity in $n_{\mathbf{p}}$ occurs at the same Fermi momentum independently of the interaction strength. However, the value of the chemical potential (or Fermi energy) yielding the same number of particles in an interacting and noninteracting system is different. This fact has already been emphasized in the context of the Hartree approximation, see discussion below (7.15).

11.7 Relation between the reducible polarizability and the Φ functional

The LW functional is based on the MBPT expansion in terms of the dressed Green's function G and interaction v. For long-range interactions like the Coulomb interaction this is not necessarily the best choice of variables. For instance in bulk systems the most natural and meaningful variable is the screened interaction W [79]. The aim of the next sections is to construct a variational many body scheme in terms of G and W. To achieve this goal we need a preliminary discussion on the relation between the so called *reducible polarizability* or *density response function* χ and the Φ functional.

The reducible polarizability χ is defined diagrammatically in terms of the polarizability P by the equation below:

$$\chi(1;2) = \underset{1 \quad \quad 2}{\overset{P}{\bigcirc}} + \underset{1 \quad \quad \quad \quad 2}{\overset{P \quad \quad P}{\bigcirc\!\!\sim\!\!\bigcirc}} + \ldots$$

$$= P(1;2) + \int d3d4\, P(1;3)v(3;4)\chi(4;2). \tag{11.44}$$

In contrast to P-diagrams, χ-diagrams can be broken into two disjoint pieces by cutting an interaction line. This is the reason for the adjective "reducible" in the name of χ (if it was not for the widespread convention of using the letter χ we could have used the symbol P_r). The Feynman rules to convert a χ-diagram into a mathematical expression are the same as those for the polarizability, i.e., a prefactor $i^{m+1}(\pm)^l$ for a diagram with m interaction lines and l loops. Using (11.44) in the Dyson equation for the screened interaction, see (10.24), we obtain

$$W(1;2) = v(1;2) + \int d3d4\, v(1;3)\chi(3;4)v(4;2). \tag{11.45}$$

The reader should appreciate the analogy between (11.45) and the Dyson equation (11.6); due to the presence of the reducible self-energy we replaced $G \to G_0$ under the integral sign.

It is evident that if we close a χ-diagram with a v-line we obtain a vacuum diagram. Vice versa, cutting off a v-line in a vacuum diagram we obtain a χ-diagram, with only one exception. The diagram which generates the Hartree self-energy (first diagram of Fig. 11.3) is the only vacuum diagram to be one-interaction-line reducible. Therefore, cutting off the v-line from the Hartree diagram does not generate a χ-diagram and, conversely, no χ-diagram can generate the Hartree diagram if closed with a v-line. We point out that the "exceptional" vacuum diagrams would be infinitely many in a formulation in terms of G_0 rather than G since there would be infinitely many one-interaction-line reducible diagrams (consider, e.g., the first diagram in the second row of Fig. 11.2). This is the main reason for us to present a formulation in terms of dressed vacuum diagrams.

As the operation of cutting off a v-line corresponds to taking a functional derivative with respect to v we expect that there is a simple relation between χ and $\delta\Phi/\delta v$. This is indeed the case even though we must be careful in defining the functional derivative with respect to v. The difficulty that arises here, and that does not arise for the functional derivatives with respect to G, has to do with the symmetry $v(i;j) = v(j;i)$. When we take a functional derivative we must allow for arbitrary variations of the interaction, including *nonsymmetric variations*. To clarify this point consider the vacuum diagram (11.7). The integral associated with it is

$$V_1 = \int v(1;5)v(2;4)v(3;6)G(1;6)G(6;5)G(5;4)G(4;3)G(3;2)G(2;1),$$

where the integral is over all variables. Since v is symmetric we could also write the same integral as

$$V_2 = \int v(5;1)v(2;4)v(3;6)G(1;6)G(6;5)G(5;4)G(4;3)G(3;2)G(2;1),$$

where in V_2 we simply replace $v(1;5) \to v(5;1)$. For a symmetric interaction $V_1 = V_2$ but otherwise V_1 is different from V_2. If we now take the functional derivative of V_1 with respect to $v(a;b)$ and then evaluate the result for a symmetric interaction we get

$$\frac{\delta V_1}{\delta v(a;b)} = \int v(2;4)v(3;6)G(a;6)G(6;b)G(b;4)G(4;3)G(3;2)G(2;a)$$

$$+ \int v(1;5)v(3;6)G(1;6)G(6;5)G(5;b)G(b;3)G(3;a)G(a;1)$$

$$+ \int v(1;5)v(2;4)G(1;b)G(b;5)G(5;4)G(4;a)G(a;2)G(2;1)$$

$$= I_{\chi,1}(a;b) + I_{\chi,2}(a;b) + I_{\chi,3}(a;b), \tag{11.46}$$

which is proportional to the following sum of χ-diagrams:

$$\tag{11.47}$$

On the other hand, if we take the functional derivative of V_2 with respect to $v(a;b)$ and then evaluate the result for a symmetric interaction we get

$$\frac{\delta V_2}{\delta v(a;b)} = I_{\chi,1}(b;a) + I_{\chi,2}(a;b) + I_{\chi,3}(a;b) = 2I_{\chi,2}(a;b) + I_{\chi,3}(a;b).$$

In the last equality we have taken into account that under the interchange $a \leftrightarrow b$ the first diagram in (11.47) [representing $I_{\chi,1}(a;b)$] becomes identical to the second diagram. We thus conclude that $\delta V_1/\delta v(a;b) \neq \delta V_2/\delta v(a;b)$. To remove this ambiguity we define the symmetric derivative

$$\frac{\delta}{\delta v(a;b)} \cdots \bigg|_{\text{S}} = \frac{1}{2}\left[\frac{\delta}{\delta v(a;b)} \cdots + \frac{\delta}{\delta v(b;a)} \cdots \right],$$

where any diagram can be inserted in the dots. The reader can easily check that $\frac{\delta V_1}{\delta v(a;b)}\big|_{\text{S}} = \frac{\delta V_2}{\delta v(a;b)}\big|_{\text{S}}$. Clearly, there is not such a problem for $\delta/\delta G$ since G is not symmetric and hence there is only one way to choose the arguments of G in a diagram.[8]

---

[8]Remember that even though the operation $\delta/\delta G$ is unambiguous, in this case also we must allow for arbitrary variations of G including those variations that bring G away from the Keldysh space, see discussion before (9.24).

We say that two χ-diagrams are related by a *reversal* symmetry if one diagram can be obtained from the other by interchanging the external vertices. The first two diagrams in (11.47) are clearly each the reversal of the other. Not all χ-diagrams become another diagram after a reversal operation. For instance the third diagram in (11.47) is the reversal of itself.

In graphical terms the symmetric derivative corresponds to cutting a v-line, adding the reversed diagram and dividing the result by 2. The operation of cutting and symmetrizing allows us to define two v-lines as *equivalent* if they lead to the same χ-diagram(s). For instance, cutting a v-line from the diagram (11.7) and then symmetrizing yields (up to a prefactor):

$$\frac{1}{2}\left[\text{(diagram)} + \text{(diagram)}\right] \qquad \text{by cutting } (1;5)$$

$$\frac{1}{2}\left[\text{(diagram)} + \text{(diagram)}\right] \qquad \text{by cutting } (2;4)$$

$$\text{(diagram)} \qquad \text{by cutting } (3;6)$$

Thus, the v-lines $(1;5)$ and $(2;4)$ are equivalent while $(3;6)$ is not equivalent to any other line.

Similarly to the G-lines, the v-lines have the following properties:

- If s is a symmetry of a vacuum diagram then the v-lines $(i;j)$ and $(s(i);s(j))$ are equivalent;

- If two v-lines $(i;j)$ and $(k;l)$ in a vacuum diagram are equivalent then there exists a symmetry such that $(s(i),s(j)) = (k;l)$ [remember that for a v-line $(i;j) = (j;i)$].

The proof of both statements is identical to the proof for the Green's function lines (see Section 11.3) and it is left as an exercise for the reader. Let us explore the implications of the above two properties. Consider a vacuum diagram with N_S symmetry operations. Given a v-line $(i;j)$ there are two possibilities: (1) there exists a symmetry s_R that reverses the interaction line, i.e., $s_R(i) = j$ and $s_R(j) = i$, or (2) the symmetry s_R does not exist. In case (1) by cutting $(i;j)$ and symmetrizing we get a single χ-diagram which is the reversal of itself. Then, by applying all N_S symmetry operations to $(i;j)$ we obtain only $N_S/2$ topologically equivalent χ-diagrams since the symmetry s and the symmetry $s \circ s_R$ generate the same χ-diagram and, obviously, $s_1 \circ s_R \neq s_2 \circ s_R$ for $s_1 \neq s_2$. For instance, the only symmetry of the diagram (11.7) is a reversal symmetry of the v-line $(3;6)$ and consequently cutting $(3;6)$

and symmetrizing leads to only $N_S/2 = 2/2 = 1$ diagram. In case (2) by cutting $(i;j)$ and symmetrizing we get two χ-diagrams which are the reversal of each other, and hence there must be a symmetry s_R that maps a diagram into the other. By applying the N_S symmetry operations to $(i;j)$ we then obtain $N_S/2$ couples of topologically equivalent χ-diagrams.

From this analysis it follows that if D_c is a connected vacuum diagram of order n with l loops and V_c is the corresponding integral, $D_c = i^n(\pm)^l V_c$, we have

$$\frac{1}{N_S}\frac{\delta D_c}{\delta v(1;2)}\bigg|_S = i^n(\pm)^l \frac{1}{N_S}\frac{\delta V_c}{\delta v(1;2)}\bigg|_S = \frac{1}{2}\chi^{(D_c)}(1;2),$$

where $\chi^{(D_c)}$ is a contribution of order $n-1$ to the reducible polarizability consisting of $n/(N_S/2) = 2n/N_S$ diagrams. Therefore, summing over all connected, topologically inequivalent and G-skeleton vacuum diagrams each multiplied by the symmetry factor $1/N_S$, with the exclusion of the Hartree diagram, we generate a functional from which to calculate the full χ by a functional derivative. This functional is exactly $\pm(\Phi - \Phi_H)$ where Φ_H is the Hartree functional:

$$\Phi_H[G,v] = \pm\frac{1}{2}i\,(\pm)^2 \int d1d2\, G(1;1^+)v(1;2)G(2;2^+).$$

The difference

$$\Phi_{xc}[G,v] = \Phi[G,v] - \Phi_H[G,v],$$

is the exchange-correlation (XC) part of the Φ functional (whose functional derivative yields the XC self-energy). We can then write the important and elegant formula

$$\boxed{\frac{\delta\Phi_{xc}}{\delta v(1;2)}\bigg|_S = \pm\frac{1}{2}\chi(1;2)} \tag{11.48}$$

11.8 Ψ functional

The result (11.48) constitutes the starting point to switch from the variables (G,v) to the variables (G,W). There is only one ingredient missing and that is the Ψ functional. Let us construct the functional $\Psi[G,W]$ by removing from Φ_{xc} all diagrams which contain polarization insertions and then replacing $v \to W$. By the very same reasoning that led to (11.15) and (11.48), the functional derivatives of Ψ with respect to G and W give the self-energy and the polarizability

$$\boxed{\Sigma_{xc}(1;2) = \Sigma_{ss,xc}[G,W](1;2) = \frac{\delta\Psi}{\delta G(2;1^+)}} \tag{11.49}$$

$$\boxed{P(1;2) = P_{ss}[G,W](1;2) = \pm 2\,\frac{\delta\Psi}{\delta W(1;2)}\bigg|_S} \tag{11.50}$$

Similarly to (11.16), the Ψ functional can be expanded in G- and W-skeleton *self-energy* diagrams as

$$\Psi[G,W] = \sum_{n=1}^{\infty} \frac{1}{2n} \int d1d2 \, \Sigma_{ss,\mathrm{xc}}^{(n)}[G,W](1;2)G(2,1^+)$$

$$= \sum_{n=1}^{\infty} \frac{1}{2n} \mathrm{tr}_\gamma \left[\Sigma_{\mathrm{xc}}^{(n)} G \right]. \tag{11.51}$$

An alternative expression in terms of G- and W-skeleton *polarization* diagrams can be deduced from the obvious invariance $\Psi[G,W] = \Psi[\alpha^{-1/2}G, \alpha W]$. Taking the derivative with respect to α and evaluating the result in $\alpha = 1$ we find

$$0 = \left. \frac{\partial \Psi}{\partial \alpha} \right|_{\alpha=1} = -\frac{1}{2}\mathrm{tr}_\gamma \left[\Sigma_{\mathrm{xc}} G \right] \pm \frac{1}{2}\mathrm{tr}_\gamma \left[PW \right].$$

By equating orders in W we then obtain

$$\mathrm{tr}_\gamma \left[\Sigma_{\mathrm{xc}}^{(n)} G \right] = \pm \mathrm{tr}_\gamma \left[P^{(n-1)} W \right], \tag{11.52}$$

and substitution into (11.51) gives

$$\boxed{\Psi[G,W] = \pm \sum_{n=1}^{\infty} \frac{1}{2n} \mathrm{tr}_\gamma \left[P^{(n-1)} W \right]} \tag{11.53}$$

A word of caution about the equality (11.52) with $n = 1$ is required. The l.h.s. is simply the Fock (or exchange) diagram and reads

$$\mathrm{i} \int d1d2 \, W(1;2)G(1;2^+)G(2;1^+). \tag{11.54}$$

Since the zeroth order polarizability is $P^{(0)}(1;2) = \pm \mathrm{i} G(1;2)G(2;1)$, the r.h.s. of (11.52) reads

$$\mathrm{i} \int d1d2 \, W(1;2)G(1;2)G(2;1). \tag{11.55}$$

These two expressions are the same only provided that W is not proportional to $\delta(z_1, z_2)$. If W is approximated with, e.g., $W(1;2) \sim \delta(z_1, z_2)W(\mathbf{x}_1, \mathbf{x}_2)$ then (11.55) is not well defined as it contains two Green's functions with the same time arguments. The ambiguity is resolved by (11.54) which tells us to evaluate the Green's function with the second time argument infinitesimally later than the first, i.e., $G(1;2)G(2;1) \rightarrow G(1;2^+)G(2;1^+)$. This agrees with the general remark on notation in Section 10.1.

There is a simple relation between the Ψ functional and the Φ functional. Let $W = W[G,v]$ be the solution of $W = v + vPW$ (Dyson equation (10.24) for the screened interaction) with $P = P_{ss}[G,W]$. At the self-consistent W we have $P_{ss}[G,W] = P_s[G,v]$ and hence $\Sigma_{ss}[G,W] = \Sigma_s[G,v]$.[9] If we rescale the interaction $v \rightarrow \lambda v$ *without changing*

---

[9]This means that a Ψ-derivable self-energy is conserving provided that W is the self-consistent solution of the Dyson equation.

G, the self-consistent W changes to $W[G, \lambda v] \equiv \lambda W_\lambda$. Then, the derivative of $\Psi[G, \lambda W_\lambda]$ with respect to λ is the sum of two contributions: one comes from W_λ and the other comes from the explicit dependence on λ. The latter can be calculated in the same way as (11.23) and therefore

$$\frac{d}{d\lambda} \Psi[G, \lambda W_\lambda] = \text{tr}_\gamma \left[\frac{1}{2\lambda} \Sigma_{s,\text{xc}}[G, \lambda v] G \pm \frac{1}{2} \lambda P_s[G, \lambda v] \frac{dW_\lambda}{d\lambda} \right], \qquad (11.56)$$

where we use (11.50) for the last term. Next we separate out the Hartree contribution from Φ and calculate the derivative of $\Phi[G, \lambda v]$ with respect to λ

$$\frac{d}{d\lambda} \Phi[G, \lambda v] = \frac{d}{d\lambda} \Phi_\text{H}[G, \lambda v] + \text{tr}_\gamma \left[\frac{1}{2\lambda} \Sigma_{s,\text{xc}}[G, \lambda v] G \right].$$

Comparing with (11.56) we get

$$\frac{d}{d\lambda} \Phi[G, \lambda v] = \frac{d}{d\lambda} \Phi_\text{H}[G, \lambda v] + \frac{d}{d\lambda} \Psi[G, \lambda W_\lambda] \mp \frac{1}{2} \text{tr}_\gamma \left[\lambda P_s[G, \lambda v] \frac{dW_\lambda}{d\lambda} \right]. \qquad (11.57)$$

The last term in this equation can be written as a total derivative with respect to λ using the same trick leading to (11.25). For convenience we define $P_\lambda \equiv P_s[G, \lambda v]$. Then we have

$$\frac{d}{d\lambda} \text{tr}_\gamma \left[\ln(1 - v\lambda P_\lambda) \right] = -\frac{d}{d\lambda} \text{tr}_\gamma \left[v(\lambda P_\lambda) + \frac{1}{2} v(\lambda P_\lambda) v(\lambda P_\lambda) + \dots \right]$$

$$= -\text{tr}_\gamma \left[v \frac{d(\lambda P_\lambda)}{d\lambda} + v(\lambda P_\lambda) v \frac{d(\lambda P_\lambda)}{d\lambda} + \dots \right]$$

$$= -\text{tr}_\gamma \left[W_\lambda \frac{d(\lambda P_\lambda)}{d\lambda} \right] = -\frac{d}{d\lambda} \text{tr}_\gamma \left[W_\lambda \lambda P_\lambda \right] + \text{tr}_\gamma \left[\lambda P_\lambda \frac{dW_\lambda}{d\lambda} \right].$$

Inserting this result into (11.57) and integrating over λ between 0 and 1 we eventually arrive at the result

$$\boxed{\Phi[G, v] = \Phi_\text{H}[G, v] + \Psi[G, W] \mp \frac{1}{2} \text{tr}_\gamma \left[WP + \ln(1 - vP) \right]} \qquad (11.58)$$

with $W = W[G, v]$ and $P = P_s[G, v]$.

11.9   Screened functionals

We are now ready to formulate a variational many-body theory of the grand potential in terms of the independent quantities G and W. First we observe that by inserting (11.58) into (11.26) we obtain the doubly-dressed MBPT expansion of the grand potential

$$\beta(\Omega - \Omega_0) = \mp \left\{ \Phi_\text{H} + \Psi - \text{tr}_\gamma [\Sigma G + \ln(1 - G_0 \Sigma)] \mp \frac{1}{2} \text{tr}_\gamma [PW + \ln(1 - vP)] \right\}.$$

$$(11.59)$$

This formula provides the diagrammatic expansion of Ω in the variables G and W, since we know how to expand the quantities Ψ, Σ, and P using skeletonic diagrams in G and W.

We can regard (11.59) as the definition of a functional $\Omega[G, W]$ which equals the grand potential when G and W are the Green's function and the screened interaction of the underlying physical system. This functional was first proposed in Ref. [86] and is denoted by $\Omega_{\text{sLW}}[G, W]$ to distinguish it from the grand potential Ω which is a number (sLW stands for screened Luttinger–Ward). We then have

$$\Omega_{\text{sLW}}[G, W] = \Omega_0 \mp \frac{1}{\beta} \left\{ \Phi_{\text{H}}[G, v] + \Psi[G, W] \right.$$

$$- \text{tr}_\gamma \left[\Sigma_{ss}[G, W]G + \ln(1 - G_0\Sigma_{ss}[G, W]) \right]$$

$$\left. \mp \frac{1}{2}\text{tr}_\gamma \left[P_{ss}[G, W]W + \ln(1 - vP_{ss}[G, W]) \right] \right\}. \quad (11.60)$$

A precursor of this functional can be found in Appendix B of Hedin's work [79]. The possibility of constructing W-based variational schemes is also implicit in earlier works by De Dominicis and Martin [87,88]. The variation of the sLW functional induced by a variation δG and δW is

$$\delta\Omega_{\text{sLW}} = \mp \frac{1}{\beta} \left\{ \delta\Phi_{\text{H}} + \delta\Psi - \text{tr}_\gamma \left[\Sigma\delta G + G\delta\Sigma - (G_0 + G_0\Sigma G_0 + \ldots)\delta\Sigma \right] \right.$$

$$\left. \mp \frac{1}{2}\text{tr}_\gamma \left[P\delta W + W\delta P - (v + vPv + \ldots)\delta P \right] \right\}.$$

Taking into account that $\delta\Phi_{\text{H}} = \text{tr}_\gamma[\Sigma_{\text{H}}\delta G]$ and that $\delta\Psi = \text{tr}_\gamma[\Sigma_{\text{xc}}\delta G \pm \frac{1}{2}P\delta W]$ we see that the sLW functional is stationary when

$$G = G_0 + G_0\Sigma G_0 + G_0\Sigma G_0\Sigma G_0 + \ldots$$

and

$$W = v + vPv + vPvPv + \ldots,$$

i.e., when G is the self-consistent Green's function $G = G_0 + G_0\Sigma G$ and W is the self-consistent screened interaction $W = v + vPW$. At the stationary point $\Omega_{\text{sLW}} = \Omega$.

Just as in the case of the LW functional we can add to the sLW functional any $F[D]$ which fulfills (11.30). The resulting functional is stationary in the same point and it takes the same value at the stationary point. If we add the functional (11.31), with $D = D[G, W]$ defined in (11.29) in which $\Sigma_s[G, v] \rightarrow \Sigma_{ss}[G, W]$, we obtain the Klein version of the sLW functional

$$\Omega_{\text{sK}} = \Omega_0 \mp \frac{1}{\beta} \left\{ \Phi_{\text{H}} + \Psi + \text{tr}_\gamma \left[\ln(G\overset{\leftarrow}{G}_0^{-1}) + 1 - G\overset{\leftarrow}{G}_0^{-1} \right] \mp \frac{1}{2}\text{tr}_\gamma \left[PW + \ln(1 - vP) \right] \right\},$$

where the functional dependence of the various quantities is understood. There is, however, an additional freedom which we can use to design more stable functionals [84]. We can add an arbitrary functional $K[Q]$ of a functional $Q[G, W]$ defined by

$$Q[G, W](1; 2) = W(1; 2) - v(1; 2) - \int d3d4\, W(1; 2)P_{ss}[G, W](3; 4)v(4; 2),$$

with the properties

$$K[Q=0] = \left(\frac{\delta K}{\delta Q}\right)_{Q=0} = 0.$$

We then obtain a new functional for the grand potential with the same stationary point and the same value at the stationary point. Choosing to add inside the curly bracket of (11.60) the functional

$$K[Q] = \pm\frac{1}{2}\mathrm{tr}_\gamma\left[\,\ln(1 + Q\overleftarrow{v}^{-1}) - Q\overleftarrow{v}^{-1}\,\right],$$

with $\overleftarrow{v}^{-1}(1;2)$ the differential operator (acting on quantities to its left) for which $\int d3\, v(1;3)\,\overleftarrow{v}^{-1}(3;2) = \delta(1;2)$,[10] we find

$$\Omega_{\mathrm{ssK}} = \Omega_0 \mp \frac{1}{\beta}\Big\{\,\Phi_{\mathrm{H}} + \Psi - \mathrm{tr}_\gamma\big[\Sigma G + \ln(1 - G_0\Sigma)\big]\Big\} \mp \frac{1}{2}\mathrm{tr}_\gamma\big[\,\ln(Wv^{-1}) + 1 - Wv^{-1}\,\big]\Big\}.$$

Here, ssK stands for the simple version of the screened Klein functional based on the construction of Ref. [86], see also Refs. [84, 89].

We conclude this section by pointing out that conserving approximations to Σ not only have the merit of preserving basic conservation laws. At zero temperature the grand potential equals the ground-state energy $E_{\mathrm{S}}^{\mathrm{M}}$ of a system with Hamiltonian $\hat{H}^{\mathrm{M}}$. A further advantage of conserving approximations is that the energy calculated from $E_{\mathrm{S}}^{\mathrm{M}} = \lim_{\beta\to\infty}\Omega$, with Ω *any* of the grand potential functionals previously discussed, or calculated from the Galitskii–Migdal formula (6.105), is the same provided that Σ is conserving. Indeed, the derivation of the Galitskii–Migdal formula is based on the sole assumption that G satisfies the Dyson equation for some Σ. Therefore, if $\Sigma = \delta\Phi/\delta G$ and if G is calculated self-consistently from $G = G_0 + G_0\Sigma G$ then the two methods yield the same result. Other interesting properties of conserving approximations are, e.g., the Luttinger–Ward theorem, the satisfaction of the virial theorem, see Appendix J, the discontinuity of the momentum distribution in a zero-temperature electron gas, see Appendix K, and the Ward identities, see Chapter 15.

---

[10]For instance for the Coulomb interaction $v(1,2) = \delta(z_1, z_2)/|\mathbf{r}_1 - \mathbf{r}_2|$ and for particles of spin 0 we have $\overleftarrow{v}^{-1}(1;2) = \frac{1}{4\pi}\delta(z_1, z_2)\delta(\mathbf{r}_1 - \mathbf{r}_2)\overleftarrow{\nabla}_2^2$.

12

MBPT for the two-particle Green's function

We complete our presentation of MBPT by discussing the two-particle Green's function G_2. At the end of this chapter the reader will have the necessary tools to extend the diagrammatic techniques to the general n-particle Green's function. In most cases a knowledge of G and G_2 is sufficient to determine the physical quantities in which one is usually interested. There exist situations, however, which necessitate the calculation of higher order Green's functions. For instance the G_3 is needed to study the nucleon–nucleon interaction in deuterium [90] or the Coster–Kronig preceded Auger processes in solids [91]. Note that in a three-particle system with two-body interactions the expansion of G_3 can be cast in the form of a recursive relation known as the *Faddeev equation* [92].

12.1 Diagrams for G_2 and loop rule

The expansion of G_2 in terms of the noninteracting Green's function G_0 and interaction v is given in (5.34) which we write again below:

$$G_2(a, b; c, d)$$

$$= \frac{\sum_{k=0}^{\infty} \frac{1}{k!} \left(\frac{i}{2}\right)^k \int v(1; 1') \dots v(k; k') \begin{vmatrix} G_0(a;c) & G_0(a;d) & \dots & G_0(a;k'^+) \\ G_0(b;c) & G_0(b;d) & \dots & G_0(b;k'^+) \\ \vdots & \vdots & \ddots & \vdots \\ G_0(k';c) & G_0(k';d) & \dots & G_0(k';k'^+) \end{vmatrix}_{\pm}}{\sum_{k=0}^{\infty} \frac{1}{k!} \left(\frac{i}{2}\right)^k \int v(1; 1') \dots v(k; k') \begin{vmatrix} G_0(1;1^+) & G_0(1;1'^+) & \dots & G_0(1;k'^+) \\ G_0(1';1^+) & G_0(1';1'^+) & \dots & G_0(1';k'^+) \\ \vdots & \vdots & \ddots & \vdots \\ G_0(k';1^+) & G_0(k';1'^+) & \dots & G_0(k';k'^+) \end{vmatrix}_{\pm}} .$$

$$(12.1)$$

The MBPT for the two-particle Green's function is completely defined by (12.1). To see what kind of simplifications we can make let us familiarize ourselves with the diagrammatic

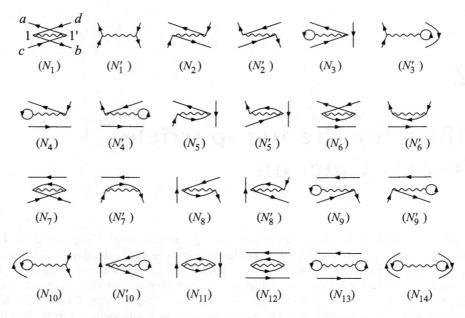

Figure 12.1 The 24 first-order diagrams of the numerator of G_2. The diagrams N_j and N'_j are obtained one from the other by a mirroring of the interaction lines. The last four diagrams can be written as the product of a G_2-diagram and a vacuum diagram.

representation of the various terms. Let $N(a, b; c, d)$ be the numerator of (12.1). To zeroth order in the interaction, N is

$$N^{(0)}(a, b; c, d) = \begin{vmatrix} G_0(a; c) & G_0(a; d) \\ G_0(b; c) & G_0(b; d) \end{vmatrix} = G_0(a; c)G_0(b; d) \pm G_0(a; d)G_0(b; c). \quad (12.2)$$

This is precisely the Hartree–Fock approximation to G_2 derived heuristically in Section 7.1 where G is replaced by G_0. The two terms in the r.h.s. of (12.2) are represented by the diagrams below:

$$N^{(0)}(a,b;c,d) \;=\; \begin{matrix} a \\ \uparrow \\ c \end{matrix} \quad \begin{matrix} d \\ \downarrow \\ b \end{matrix} \quad + \quad \begin{matrix} a \longleftarrow d \\ \\ c \longrightarrow b \end{matrix}$$

$$(12.3)$$

where the prefactor is incorporated in the diagrams and is given by the sign of the permutation. The calculation of $N(a, b; c, d)$ to first order requires expansion of the permanent/determinant of a 4×4 matrix. This yields $4! = 24$ terms whose diagrammatic representation is displayed in Fig. 12.1. The prefactor of each diagram is $(i/2)$ times the sign of the permutation and $N^{(1)}(a, b; c, d)$ is simply the sum of all of them. Going to higher order we generate diagrams with four external vertices a, b, c, d and a certain number of interaction lines. The prefactor of an nth order diagram is $\frac{1}{n!}(i/2)^n(\pm)^P$ with $(\pm)^P$ the sign of the permutation. It would be useful to have a graphical method (similar to the loop rule for the Green's function) to determine the sign of a diagram. If we have n interaction lines then the identity permutation gives the diagram

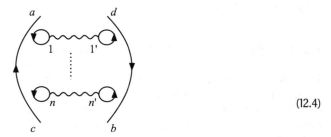

$$(12.4)$$

whose sign is $+$. A subsequent permutation that interchanges c and d gives the diagram

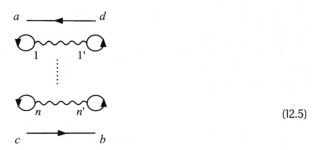

$$(12.5)$$

whose sign is $(\pm)$. Any subsequent interchange starting from (12.4) or (12.5) always changes the number of loops by one but keeps the vertices (a,c) and (b,d) in (12.4) or (a,d) and (b,c) in (12.5) connected. We then consider a G_2-diagram for $G_2(a,b;c,d)$ with l' loops. By closing this diagram with the Green's function lines $G_0(c;a)$ and $G_0(d;b)$

we obtain a diagram with a number of loops $l = l' + 2$ if the G_2-diagram originates from (12.4) [and hence (a,c) and (b,d) are connected] and $l = l' + 1$ if the G_2-diagram originates from (12.5) [and hence (a,d) and (b,c) are connected].[1] Therefore, the sign of a diagram can easily be fixed according to the loop rule for the two-particle Green's function

$$(\pm)^P = (\pm)^l,$$

---

[1]Note that these diagrams are not vacuum diagrams. The closing G-lines are used only for graphical purposes to count the number of loops.

where l is the number of loops of the closed G_2-diagram. In the next section we show how to reduce the number of diagrams by exploiting their symmetry properties as well as the topological notion of skeleton diagrams.

Exercise 12.1. Show that the diagrams in Fig. 12.1 correspond to the 24 terms of the expansion of $N^{(1)}(a, b; c, d)$.

12.2 Bethe–Salpeter equation

Similarly to the one-particle Green's function, the vacuum diagrams appearing in the numerator $N(a, b; c, d)$ are cancelled out by the denominator of (12.1). For instance the last four diagrams in Fig. 12.1 can be discarded. Then we can rewrite (12.1) without the denominator provided that we retain only connected G_2-diagrams in the expansion of N (a G_2-diagram is connected if it does not contain vacuum diagrams). To first order G_2 is the sum of the two diagrams in (12.3) and the first 20 diagrams in Fig. 12.1. The number of diagrams can be further reduced by taking into account that any permutation or mirroring of the interaction lines leads to a topologically equivalent diagram. Thus, to order n each diagram appears in $2^n n!$ variants. For $n = 1$ the number of variants is 2 as it is clearly illustrated in Fig. 12.1. Here the diagrams N_j and N'_j with $j = 1, \ldots, 10$ are obtained from one another by mirroring the interaction line $v(1; 1') \to v(1'; 1)$. In conclusion, the MBPT formula (12.1) for G_2 simplifies to

$$G_2(a, b; c, d) = \sum_{k=0}^{\infty} \mathrm{i}^n \int v(1; 1') \ldots v(k; k') \begin{vmatrix} G_0(a; c) & G_0(a; d) & \ldots & G_0(a; k'^+) \\ G_0(b; c) & G_0(b; d) & \ldots & G_0(b; k'^+) \\ \vdots & \vdots & \ddots & \vdots \\ G_0(k'; c) & G_0(k'; d) & \ldots & G_0(k'; k'^+) \end{vmatrix}_{\substack{\pm \\ c \\ t.i.}},$$

$$\tag{12.6}$$

where the symbol $\left| \ldots \right|_{\substack{\pm \\ c \\ t.i.}}$ signifies that in the expansion of the permanent/determinant only the terms represented by connected and topologically inequivalent diagrams are retained. In this way the number of first order diagrams reduces from 20 to 10.

From (12.6) we see that it is convenient to change the Feynman rules for calculating the prefactors. From now on we use the following Feynman rules for the G_2-diagrams:

- Number all vertices and assign an interaction line $v(i; j)$ to a wiggly line between j and i and a Green's function $G_0(i; j^+)$ to an oriented line from j to i;

- Integrate over all internal vertices and multiply by $\mathrm{i}^n (\pm)^l$ where n is the number of interaction lines and l is the number of loops in the closed G_2-diagram.

The expansion (12.6) contains also diagrams with self-energy insertions, see Fig. 12.1. A further reduction in the number of diagrams can be achieved by expanding G_2 in terms of the dressed Green's function G. This means to remove from (12.6) all diagrams with self-energy insertions and then replace G_0 with G. In this way the expansion of G_2 to first order in the interaction contains only four diagrams and reads

$$(12.7)$$

This first order approximation to G_2 generates the second-Born approximation (10.17) for Σ through the relation (9.1), which we rewrite below for convenience

$$\int d3\, \Sigma(1;3)G(3;2) = \pm i \int d3\, v(1;3)G_2(1,3;2,3^+).$$
$$(12.8)$$

Finally, we could remove from (12.6) all diagrams with polarization insertions and then replace v with W. Keeping only G- and W-skeleton diagrams, (12.6) provides us with an expansion of G_2 in terms of the dressed Green's function G and interaction W. Below we investigate the structure of the expansion of G_2 in G-skeleton diagrams and postpone to Section 12.5 the expansion in G- and W-skeleton diagrams.

We represent $G_2(1,2;3,4)$ as a grey square with two outgoing lines starting from opposite vertices of the square and ending in 1 and 2, and two ingoing lines starting from 3 and 4 and ending in the remaining vertices of the square:

$$G_2(1,2;3,4) \;=\;$$

$$(12.9)$$

This diagrammatic representation is unambiguous if we specify that the third variable (which is 3 in this case) is located to the right of an imaginary oriented line connecting diagonally the first to the second variable (1 and 2 in this case). Indeed, the only other possible source of ambiguity is which variable is first; looking at (12.9) we could say that it represents either $G_2(1,2;3,4)$ or $G_2(2,1;4,3)$. However, the exact as well as any conserving approximation to G_2 is such that $G_2(1,2;3,4) = G_2(2,1;4,3)$, and therefore we do not need to bother about which variable is first. From the G-skeletonic expansion it follows that the general structure of G_2 is

$$(12.10)$$

or, in formula,

$$G_2(1,2;3,4) = G(1;3)G(2;4) \pm G(1;4)G(2;3)$$
$$+ \int G(1;1')G(3';3)K_r(1',2';3',4')G(4';4)G(2;2'),$$
$$(12.11)$$

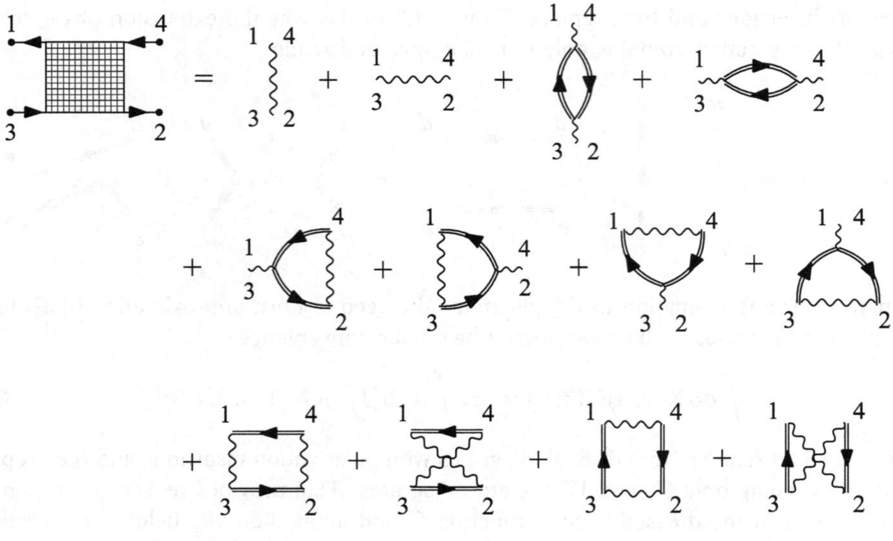

Figure 12.2 Diagrams for the kernel K_r up to second order in the interaction.

where the kernel K_r is represented by the square grid and accounts for all G_2-diagrams of order larger than zero. These diagrams are all connected since if the piece with external vertices, say, 1 and 3 were disconnected from the piece with external vertices 2 and 4 [like the first diagram in (12.10)] then there would certainly be a self-energy insertion. The rules to assign the variables to the square grid $K_r(1, 2; 3, 4)$ are the same as those for G_2: the first two variables 1 and 2 label the opposite vertices with an outgoing line and the third variable 3 labels the vertex located to the right of an imaginary oriented line going from 1 to 2. The diagrams for K_r up to second order in the interaction are shown in Fig. 12.2. The Feynman rules to convert them into mathematical expressions are the same as those for G_2 (see beginning of the section). The only extra rule is that if two external vertices i and j coincide, as in the first eight diagrams of Fig. 12.2, then we must multiply the diagram by $\delta(i; j)$. This is the same rule that we introduced for the Hartree self-energy diagram, see discussion below (10.13).

From Fig. 12.2 we see that in the 4th, 5th, 6th, and 9th diagram the vertices 1 and 3 can be disconnected from the vertices 2 and 4 by cutting two G-lines. The remaining diagrams are instead *two-particle irreducible* since $(1, 3)$ and $(2, 4)$ cannot be disjoint by cutting two G-lines. This is the reason for the subscript r in the kernel K_r which contains also two-particle *reducible* diagrams and can therefore be called a reducible kernel. Let us denote by K the *irreducible kernel*.

> K is obtained from K_r by removing all two-particle reducible diagrams. The Feynman rules for K are identical to the Feynman rules for K_r.

A generic K_r-diagram has the structure $K^{(1)}GGK^{(2)}GG\ldots K^{(n)}$ with the $K^{(i)}$ two-particle irreducible diagrams. Let us check how the sign of a K_r-diagram is related to the sign of the constituent $K^{(i)}$ irreducible diagrams. Let l be the number of loops of a closed K_r-diagram with the structure $K^{(1)}GGK^{(2)}$

where the irreducible kernels are represented by a square with vertical stripes. We see that the closed $K^{(1)}$-diagram and $K^{(2)}$-diagram are obtained from the above diagram by an interchange of the starting points of the connecting GG-double-lines

Since a single interchange adds or removes a loop we conclude that $(\pm)^l = (\pm)^{l_1+l_2+1}$ where l_1 and l_2 are the number of loops in the closed diagrams $K^{(1)}$ and $K^{(2)}$. This property is readily seen to be general. If $K_r = K^{(1)}GGK^{(2)}GG\ldots K^{(n)}$ contains n irreducible kernels with signs $(\pm)^{l_i}$, then the sign of the K_r-diagram is $(\pm)^{l_1+\ldots+l_n+n-1}$. Thus we can alternatively formulate the Feynman rule for the prefactor of a K_r-diagram as:

- If $K_r = K^{(1)}GGK^{(2)}GG\ldots K^{(n)}$ with $K^{(i)}$ irreducible diagrams of order k_i and sign $(\pm)^{l_i}$, then the prefactor of K_r is

$$\mathrm{i}^{k_1+\ldots+k_n}(\pm)^{l_1+\ldots+l_n+n-1}.$$

In other words every GG-double-line contributes with a $(\pm)$ to the overall sign.

Let us consider few examples. For the 4th K_r-diagram of Fig. 12.2 the prefactor is

$$\mathrm{i}^2(\pm)^l = \mathrm{i}^2(\pm)^3 = \pm\mathrm{i}^2. \tag{12.12}$$

The same diagram can be written as $K^{(1)}GGK^{(2)}$ where $K^{(1)}(1,2;3,4) = \text{prefactor} \times \delta(1;3)\delta(2;4)\,v(1;2)$ and $K^{(2)} = K^{(1)}$. Since the closed $K^{(1)}$-diagram has two loops the prefactor of $K^{(1)}$ is

$$\mathrm{i}(\pm)^{l_1} = \mathrm{i}(\pm)^2 = \mathrm{i}.$$

The $K^{(2)}$-diagram has the same prefactor and since we have only one GG-double-line the overall prefactor will be $\pm\mathrm{i}^2$ in agreement with (12.12). A second example is the 5th K_r-diagram of Fig. 12.2 whose prefactor is

$$\mathrm{i}^2(\pm)^l = \mathrm{i}^2(\pm)^2 = \mathrm{i}^2. \tag{12.13}$$

This diagram also has the structure $K^{(1)}GGK^{(2)}$ where $K^{(1)}$ is the same as before while $K^{(2)}(1,2;3,4) = \text{prefactor} \times \delta(1;4)\delta(2;3)v(1;3)$. Since the closed $K^{(2)}$-diagram has one loop the prefactor of $K^{(2)}$ is

$$\mathrm{i}(\pm)^{l_2} = \mathrm{i}(\pm)^1 = \pm\mathrm{i}.$$

Again the product of the prefactors of $K^{(1)}$ and $K^{(2)}$ times the ($\pm$) sign coming from the GG-double-line agrees with the prefactor (12.13). Therefore we can write the following Dyson-like equation for the reducible kernel (integral over primed variables is understood)

$$K_r(1,2;3,4) = K(1,2;3,4) \pm \int K(1,2';3,4')G(4';1')G(3';2')K_r(1',2;3',4), \quad (12.14)$$

which is represented by the diagrammatic equation (remember that every GG-double-line contributes with a $\pm$)

$$(12.15)$$

Equation (12.14) is known as the *Bethe–Salpeter equation* for the reducible kernel. Inserting (12.15) into (12.10) we find

with the grey blob

$$L(1,2;3,4) \equiv \pm\big[G_2(1,2;3,4) - G(1;3)G(2;4)\big]. \quad (12.16)$$

The function L fulfills the diagrammatic equation

or, in formulas (integral over primed variables is understood),

$$\boxed{L(1,2;3,4) = G(1;4)G(2;3) \pm \int G(1;1')G(3';3)K(1',2';3',4')L(4',2;2',4)} \quad (12.17)$$

The reader can easily check the correctness of these equations by iterating them. We refer to L as the *two-particle XC function* since it is obtained from G_2 by subtracting the Hartree diagram.

There are several reasons for introducing yet another quantity like L. On one hand we see that by setting $4 = 2$ and $3 = 1$ the L-diagrams are the same as the χ-diagrams, with

χ the reducible polarizability defined in Section 11.7. In fact, the two quantities differ by just a prefactor that can be deduced by comparing the Feynman rules for the polarizability with the Feynman rules for G_2. A polarization diagram of order n has prefactor $i^{n+1}(\pm)^l$ while the same diagram seen as an L-diagram in the limit $4 \to 2$ and $3 \to 1$ has prefactor $i^n(\pm)^l$ times the sign $(\pm)$ which comes from the definition of L, see (12.16). Thus the relation between χ and L is

$$\boxed{\chi(1;2) = \pm i\, L(1,2;1,2)}$$

(12.18)

This equation is not well defined if $L(1,2;1,2)$ is calculated as the difference between G_2 and GG. The reason is that in GG both Green's functions have the same time arguments. As already emphasized several times, these kinds of ambiguity are always removed by shifting the arguments of the starting points. The precise way of writing the relation between χ and L is $\chi(1;2) = \pm i\, L(1,2;1^+,2^+)$. However, if L is defined as the solution of the Bethe–Salpeter equation or as the sum of all G_2-diagrams with the exception of the Hartree diagram then (12.18) is well defined. Another reason for defining L is that it contains information on how the Green's function changes $G \to G + \delta G$ after a change in the single-particle Hamiltonian $h \to h + \delta h$. In Chapter 15 we prove the important relation

$$\delta G(1;3) = \int d2d4\, L(1,2;3,4)\, \delta h(4;2).$$

(12.19)

This identity allows us to study all possible linear response properties of a system.

Equation (12.17) is known as the *Bethe–Salpeter equation* for L. One of the most successful applications of the Bethe–Salpeter equation is the description of the optical spectra of semiconductors and insulators. Experimentally the optical spectrum of a system is obtained by irradiating a sample with photons of energy ω and then measuring the intensity of the absorbed light as a function of ω. In semiconductors or insulators a photon of high enough energy can excite an electron from the valence band to the conduction band. The lack of an electron in the valence band can be seen as a hole, i.e., a particle of positive charge, which is then attracted by the excited electron. This particle–hole pair can form a stable excited state known as the *exciton* [93], and in correspondence with its energy the optical spectrum exhibits a peak. In the next section we use a simplified model to show how to locate the position of the excitonic peaks using the Bethe–Salpeter equation.

Exercise 12.2. Show that the vacuum diagrams in N are cancelled out by the denominator of (12.1).

Exercise 12.3. Show that (12.8) is an identity with G_2 given by (12.7) and Σ given by the second-Born approximation.

Exercise 12.4. Show that the diagrams of K_r to second-order in v are all those illustrated in Fig. 12.2.

12.3 Excitons

Let us consider an insulator in the ground state with the valence band fully occupied and the conduction band completely empty, see Fig. 12.3(a). If the insulator is transparent to light of frequency ω then quantities like density, current, etc. remain unperturbed. Conversely, the absorption of light induces a change in these observable quantities. The time-dependent ensemble average of any one-body operator can be calculated from $G^<(\mathbf{x}_1, t, \mathbf{x}_2, t) = G(\mathbf{x}_1, z, \mathbf{x}_2, z^+)$ through (6.26) and hence its variation can be calculated from the variation $\delta G(\mathbf{x}_1, z, \mathbf{x}_2, z^+)$. The coupling between the electrons and the external electromagnetic field is contained in $\delta h(1;2) = \delta(z_1, z_2)\langle\mathbf{x}_1|\delta\hat{h}(z_1)|\mathbf{x}_2\rangle$ and is local in time. Therefore, from (12.19) it follows that the knowledge of

$$L(\mathbf{x}_1, \mathbf{x}_2, \mathbf{x}_3, \mathbf{x}_4; z, z') \equiv L(\mathbf{x}_1, z, \mathbf{x}_2, z'; \mathbf{x}_3, z, \mathbf{x}_4, z') \tag{12.20}$$

is enough to calculate the variations of density, current, etc. We have

$$\delta G(\mathbf{x}_1, z, \mathbf{x}_3, z^+) = \int d\mathbf{x}_2 d\mathbf{x}_4 \int_\gamma d\bar{z}\, L(\mathbf{x}_1, \mathbf{x}_2, \mathbf{x}_3, \mathbf{x}_4; z, \bar{z})\langle\mathbf{x}_4|\delta\hat{h}(\bar{z})|\mathbf{x}_2\rangle,$$

and setting $z = t_-$ or $z = t_+$ we find (see Exercise 5.5)

$$\delta G^<(\mathbf{x}_1, t, \mathbf{x}_3, t) = \int d\mathbf{x}_2 d\mathbf{x}_4 \int_{t_0}^\infty d\bar{t}\, L^R(\mathbf{x}_1, \mathbf{x}_2, \mathbf{x}_3, \mathbf{x}_4; t, \bar{t})\langle\mathbf{x}_4|\delta\hat{h}(\bar{t})|\mathbf{x}_2\rangle. \tag{12.21}$$

The choice of the time arguments in (12.20) corresponds to $z_1 = z_3 = z$ and $z_2 = z_4 = z'$ in the diagrams above (12.17). The lowest order diagram describes the propagation of a free particle–hole pair since either z is earlier than z' or it is the other way around. In higher order diagrams the particle interacts with the hole in many different ways depending on the approximation to the kernel K. Thus L can be interpreted as the interacting particle–hole propagator in a similar way as, e.g., $G^<$ has been interpreted as a hole propagator, see Section 6.1.3. An exciton is a bound (or quasi-bound) particle–hole excitation and, if it exists, the Fourier transform of L exhibits a peak at the exciton binding energy. Below we derive an equation for L based on a physically sound approximation for K.

Due to the presence of a gap between the bands, low energy excitations are strongly suppressed and the ground state is "rigid." We then approximate G with the noninteracting Green's function G_0, i.e., we ignore the effects of multiple self-energy insertions. For the kernel K we assume that the interaction is weak and consider the lowest order approximation in v given by the first two diagrams of Fig. 12.2 (lower sign for fermions):

$$K(1,2;3,4) = \vcenter{\hbox{[diagram]}} \;+\; \vcenter{\hbox{[diagram]}} = -\,i\delta(1;4)\delta(2;3)v(1;3) + i\delta(1;3)\delta(2;4)v(1;2)\cdot \tag{12.22}$$

Plugging this K in the Bethe–Salpeter equation (12.17) and replacing $G \to G_0$ we find the following approximate equation for L (integral over primed variables is understood)

$$L(1,2;3,4) = L_0(1,2;3,4) + i\int L_0(1,2';3,1')v(1';2')L(1',2;2',4)$$

$$-\,i\int L_0(1,1';3,1')v(1';2')L(2',2;2',4), \tag{12.23}$$

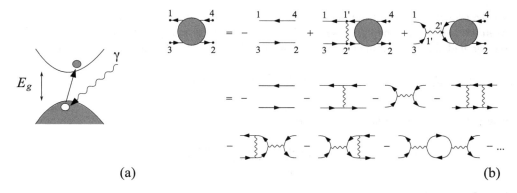

(a)   (b)

Figure 12.3 (a) Representation of the light absorption in an insulator with an electron which is promoted from the valence band to the conduction band and the hole left behind. (b) Diagrams for L when the kernel is approximated as in (12.22).

with $L_0(1,2;3,4) = G_0(1;4)G_0(2;3)$. The diagrammatic content of (12.23) is illustrated in Fig. 12.3(b). Setting in this equation $z_1 = z_3 = z$ and $z_2 = z_4 = z'$ and using the definition (12.20) we get

$$L(\mathbf{x}_1,\mathbf{x}_2,\mathbf{x}_3,\mathbf{x}_4;z,z') = L_0(\mathbf{x}_1,\mathbf{x}_2,\mathbf{x}_3,\mathbf{x}_4;z,z')$$
$$+ i\int d\mathbf{x}_1' d\mathbf{x}_2' \int d\bar{z}\, L_0(\mathbf{x}_1,\mathbf{x}_2';\mathbf{x}_3,\mathbf{x}_1';z,\bar{z})v(\mathbf{x}_1',\mathbf{x}_2')L(\mathbf{x}_1',\mathbf{x}_2,\mathbf{x}_2',\mathbf{x}_4;\bar{z},z')$$
$$- i\int d\mathbf{x}_1' d\mathbf{x}_2' \int d\bar{z}\, L_0(\mathbf{x}_1,\mathbf{x}_1';\mathbf{x}_3,\mathbf{x}_1';z,\bar{z})v(\mathbf{x}_1',\mathbf{x}_2')L(\mathbf{x}_2',\mathbf{x}_2,\mathbf{x}_2',\mathbf{x}_4;\bar{z},z').$$

This is an identity between functions in Keldysh space and contains a convolution along the contour. We can easily extract the retarded component from the Langreth rules of Table 5.1 and subsequently Fourier transform. Indeed both L and L_0 depend only on the time difference since the insulator is in equilibrium. The result is

$$L^{\mathrm{R}}(\mathbf{x}_1,\mathbf{x}_2,\mathbf{x}_3,\mathbf{x}_4;\omega) = L_0^{\mathrm{R}}(\mathbf{x}_1,\mathbf{x}_2,\mathbf{x}_3,\mathbf{x}_4;\omega)$$
$$+ i\int d\mathbf{x}_1' d\mathbf{x}_2'\, L_0^{\mathrm{R}}(\mathbf{x}_1,\mathbf{x}_2';\mathbf{x}_3,\mathbf{x}_1';\omega)v(\mathbf{x}_1',\mathbf{x}_2')L^{\mathrm{R}}(\mathbf{x}_1',\mathbf{x}_2,\mathbf{x}_2',\mathbf{x}_4;\omega)$$
$$- i\int d\mathbf{x}_1' d\mathbf{x}_2'\, L_0^{\mathrm{R}}(\mathbf{x}_1,\mathbf{x}_1';\mathbf{x}_3,\mathbf{x}_1';\omega)v(\mathbf{x}_1',\mathbf{x}_2')L^{\mathrm{R}}(\mathbf{x}_2',\mathbf{x}_2,\mathbf{x}_2',\mathbf{x}_4;\omega). \quad (12.24)$$

To calculate L_0^{R} we use the identities in the last column of Table 5.1. By definition

$$L_0(\mathbf{x}_1,\mathbf{x}_2,\mathbf{x}_3,\mathbf{x}_4;z,z') = G_0(\mathbf{x}_1,z;\mathbf{x}_4,z')G_0(\mathbf{x}_2,z';\mathbf{x}_3,z)$$

and hence

$$L_0^{\mathrm{R}}(\mathbf{x}_1,\mathbf{x}_2,\mathbf{x}_3,\mathbf{x}_4;\omega) = \int \frac{d\omega'}{2\pi} \big[G_0^{\mathrm{R}}(\mathbf{x}_1,\mathbf{x}_4;\omega+\omega')G_0^{<}(\mathbf{x}_2,\mathbf{x}_3;\omega')$$
$$+ G_0^{<}(\mathbf{x}_1,\mathbf{x}_4;\omega+\omega')G_0^{\mathrm{A}}(\mathbf{x}_2,\mathbf{x}_3;\omega')\big]. \quad (12.25)$$

The Fourier transforms of the Keldysh components of a noninteracting Green's function have been worked out in Chapter 6, see (6.48) for the lesser component and (6.58), (6.59) for the retarded/advanced components. If we sandwich these equations with $\langle \mathbf{x}|$ and $|\mathbf{x}'\rangle$ we get

$$G_0^<(\mathbf{x}, \mathbf{x}'; \omega) = \langle \mathbf{x}|\hat{\mathcal{G}}_0^<(\omega)|\mathbf{x}'\rangle = 2\pi i \sum_n f_n \delta(\omega - \epsilon_n)\varphi_n(\mathbf{x})\varphi_n^*(\mathbf{x}'), \qquad (12.26)$$

where $f_n = f(\epsilon_n - \mu)$ is the Fermi function with chemical potential μ, and

$$G_0^{R/A}(\mathbf{x}, \mathbf{x}'; \omega) = \langle \mathbf{x}|\hat{\mathcal{G}}_0^{R/A}(\omega)|\mathbf{x}'\rangle = \sum_m \frac{\varphi_m(\mathbf{x})\varphi_m^*(\mathbf{x}')}{\omega - \epsilon_m \pm i\eta}. \qquad (12.27)$$

In (12.26) and (12.27) the sum runs over *all* single particle eigenstates. The substitution of these expressions in (12.25) and the subsequent integration over ω' leads to

$$L_0^R(\mathbf{x}_1, \mathbf{x}_2, \mathbf{x}_3, \mathbf{x}_4; \omega)$$
$$= i \sum_{nm} f_n \bar{f}_m \left[\frac{\varphi_m(\mathbf{x}_1)\varphi_m^*(\mathbf{x}_4)\varphi_n(\mathbf{x}_2)\varphi_n^*(\mathbf{x}_3)}{\omega - (\epsilon_m - \epsilon_n) + i\eta} - \frac{\varphi_m(\mathbf{x}_2)\varphi_m^*(\mathbf{x}_3)\varphi_n(\mathbf{x}_1)\varphi_n^*(\mathbf{x}_4)}{\omega + (\epsilon_m - \epsilon_n) + i\eta} \right].$$
$$(12.28)$$

In this equation we have multiplied f_n by $\bar{f}_m = 1 - f_m$. This can be done since the quantity in the square bracket is antisymmetric under the interchange $n \leftrightarrow m$. At zero temperature $f_n = 0$ if $n = C$ is a conduction state while $\bar{f}_m = 0$ if $m = V$ is a valence state.

Suppose now that L^R has simple poles in $\omega = \omega_s$ inside the gap: $0 < \omega_s < E_g$. Then our observable quantities change when we shine light of frequency ω_s on the insulator, see again (12.21). Physically this implies that photons of energy ω_s are absorbed by the insulator, and since $\omega_s < E_g$ this absorption must occur via the excitation of a *bound* electron–hole pair, i.e., an exciton. Thus the poles of L^R correspond to the energy of a bound exciton. Since the poles of L_0^R are in $(\epsilon_C - \epsilon_V) \geq E_g$ and in $-(\epsilon_C - \epsilon_V) \leq -E_g$ we can discard L_0^R in the first term of (12.24) for $\omega \sim \omega_s$. We then multiply both sides of (12.24) by $\varphi_C^*(\mathbf{x}_1)\varphi_V(\mathbf{x}_3)$ and integrate over $\mathbf{x}_1$ and $\mathbf{x}_3$. Using the orthonormality of the wavefunctions we find

$$\left[\omega_s - \epsilon_C + \epsilon_V \right] L_{CV}(\mathbf{x}_2, \mathbf{x}_4; \omega_s)$$
$$= -\int d\mathbf{x}_1' d\mathbf{x}_2' \varphi_C^*(\mathbf{x}_1')\varphi_V(\mathbf{x}_2') v(\mathbf{x}_1', \mathbf{x}_2') L^R(\mathbf{x}_1', \mathbf{x}_2, \mathbf{x}_2', \mathbf{x}_4; \omega_s)$$
$$+ \int d\mathbf{x}_1' d\mathbf{x}_2' \varphi_C^*(\mathbf{x}_1')\varphi_V(\mathbf{x}_1') v(\mathbf{x}_1', \mathbf{x}_2') L^R(\mathbf{x}_2', \mathbf{x}_2, \mathbf{x}_2', \mathbf{x}_4; \omega_s), \qquad (12.29)$$

where we introduce the coefficients L_{CV} of the expansion of L^R according to

$$L^R(\mathbf{x}_1, \mathbf{x}_2, \mathbf{x}_3, \mathbf{x}_4; \omega) = \sum_{CV} \varphi_C(\mathbf{x}_1)\varphi_V^*(\mathbf{x}_3) L_{CV}(\mathbf{x}_2, \mathbf{x}_4; \omega)$$
$$+ \sum_{CV} \varphi_V(\mathbf{x}_1)\varphi_C^*(\mathbf{x}_3) L_{VC}(\mathbf{x}_2, \mathbf{x}_4; \omega). \qquad (12.30)$$

The expansion of L^R has the same structure as the expansion of L_0^R in (12.28), since in the Bethe–Salpeter equation the first and third arguments of L_0 are the same as those of L. From (12.28) we see that the coefficients $L_{0,VC}$ of the expansion of L_0^R have poles below $-E_g$. We therefore expect that for $\omega \sim \omega_s > 0$ the dominant contribution to L comes from the first sum in (12.30), since for weak interactions $L_{VC} \sim L_{0,VC}$. Approximating $L_{VC}(\mathbf{x}_2, \mathbf{x}_4; \omega_s) \sim 0$ and inserting (12.30) into (12.29) we find a *bound-state equation* for the coefficients L_{CV}

$$\big[\omega_s - \epsilon_C + \epsilon_V\big] L_{CV} = - \sum_{C'V'} \big[v_{CV'VC'} - v_{CV'C'V}\big] L_{C'V'}, \tag{12.31}$$

where we use the definition (1.85) of the Coulomb integrals.

In order to extract some physics from (12.31) we must specify eigenvalues and eigenfunctions of the noninteracting insulator as well as the nature of the interaction. Each eigenstate is characterized by a band index $\nu = c, v$ that specifies the conduction or valence band, a quasi-momentum $\mathbf{k}$ and a spin index σ, and hence the collective quantum numbers C and V stand for $C = cp\sigma_c$ and $V = vk\sigma_v$. Since the interaction is spin independent we have

$$v_{CV'VC'} = \delta_{\sigma_c\sigma_c'}\delta_{\sigma_v\sigma_v'} \int d\mathbf{r}_1 d\mathbf{r}_2\, \varphi_{cp}^*(\mathbf{r}_1)\varphi_{vk'}^*(\mathbf{r}_2)v(\mathbf{r}_1 - \mathbf{r}_2)\varphi_{vk}(\mathbf{r}_2)\varphi_{cp'}(\mathbf{r}_1),$$

$$v_{CV'C'V} = \delta_{\sigma_c\sigma_v}\delta_{\sigma_c'\sigma_v'} \int d\mathbf{r}_1 d\mathbf{r}_2\, \varphi_{cp}^*(\mathbf{r}_1)\varphi_{vk'}^*(\mathbf{r}_2)v(\mathbf{r}_1 - \mathbf{r}_2)\varphi_{cp'}(\mathbf{r}_2)\varphi_{vk}(\mathbf{r}_1),$$

where $\varphi_{\nu k}(\mathbf{r})$ is the orbital part of the eigenfunction: $\varphi_{\nu k\sigma'}(\mathbf{r}\sigma) = \delta_{\sigma\sigma'}\varphi_{\nu k}(\mathbf{r})$. To push the analytic calculations a bit further we approximate the eigenfunctions to be plane waves with a simple parabolic dispersion for both the valence states, $\epsilon_V = -k^2/(2m_v)$, and the conduction states, $\epsilon_C = E_g + p^2/(2m_c)$. However, we have to remember that states with different band index are orthogonal. Therefore we make the following approximation for the product of two eigenfunctions *under the integral sign*

$$\varphi_{\nu\mathbf{p}}^*(\mathbf{r})\varphi_{\nu\mathbf{p}'}(\mathbf{r}) = e^{-i(\mathbf{p}-\mathbf{p}')\cdot\mathbf{r}}, \quad \nu = c, v,$$

and

$$\varphi_{c\mathbf{p}}^*(\mathbf{r})\varphi_{v\mathbf{k}}(\mathbf{r}) \sim 0.$$

This last approximation would be exact for a constant interaction due to the orthogonality of the c and v states. With these approximations we find that $v_{CV'C'V} \sim 0$ while

$$v_{CV'VC'} = \delta_{\sigma_c\sigma_c'}\delta_{\sigma_v\sigma_v'}\tilde{v}_{\mathbf{p}-\mathbf{p}'} \times (2\pi)^3\delta(\mathbf{p} - \mathbf{k} - \mathbf{p}' + \mathbf{k}'),$$

with $\tilde{v}_\mathbf{p} = \int d\mathbf{r}\, e^{-i\mathbf{p}\cdot\mathbf{r}}v(\mathbf{r})$ the Fourier transform of the interaction. Inserting these results into (12.31) with

$$\sum_{C'V'} \rightarrow \sum_{\sigma_c'\sigma_v'} \int \frac{d\mathbf{p}'}{(2\pi)^3}\frac{d\mathbf{k}'}{(2\pi)^3},$$

it is a matter of simple algebra to arrive at the following equation:

$$\Big[\omega_s - E_g - \frac{p^2}{2m_c} - \frac{k^2}{2m_v}\Big] L_{\mathbf{p}\sigma_c\mathbf{k}\sigma_v} = -\int \frac{d\mathbf{p}'}{(2\pi)^3} \tilde{v}_{\mathbf{p}-\mathbf{p}'} L_{\mathbf{p}'\sigma_c\mathbf{p}'-(\mathbf{p}-\mathbf{k})\sigma_v}. \tag{12.32}$$

Let us study the case of small momentum transfer $\mathbf{P} = \mathbf{p} - \mathbf{k}$ which is the most relevant in optical transitions since it corresponds to excitons of low energy, and hence to the most stable excitons. We define the exciton amplitudes

$$A_{\mathbf{P}}(\mathbf{k}) = L_{\mathbf{P}+\mathbf{k}\sigma_c\mathbf{k}\sigma_v},$$

and for $|\mathbf{P}| \ll |\mathbf{k}|$ we approximate $p^2 \sim k^2$ in the eigenenergy of the conduction state. Then, we can rewrite (12.32) as

$$\frac{k^2}{2\mu_{\text{ex}}}A_{\mathbf{P}}(\mathbf{k}) - \int \frac{d\mathbf{k}'}{(2\pi)^3}\,\tilde{v}_{\mathbf{k}-\mathbf{k}'}A_{\mathbf{P}}(\mathbf{k}') = (\omega_s - E_g)A_{\mathbf{P}}(\mathbf{k}),$$

with the effective exciton mass $\mu_{\text{ex}}^{-1} = m_c^{-1} + m_v^{-1}$. This equation has the same structure as an eigenvalue equation for a particle of mass μ_{ex} in an external potential $-v$. If we multiply both sides by $e^{i\mathbf{k}\cdot\mathbf{r}}$ and integrate over $\mathbf{k}$ we find

$$\left[-\frac{\nabla^2}{2\mu_{\text{ex}}} - v(\mathbf{r})\right]A_{\mathbf{P}}(\mathbf{r}) = (\omega_s - E_g)A_{\mathbf{P}}(\mathbf{r}), \qquad (12.33)$$

with $A_{\mathbf{P}}(\mathbf{r}) = \int \frac{d\mathbf{k}}{(2\pi)^3}e^{i\mathbf{k}\cdot\mathbf{r}}A_{\mathbf{P}}(\mathbf{k})$ the Fourier transform of the exciton amplitudes. Since v is the repulsive interaction between the electrons $(-v)$ is an attractive potential. For Coulombic interactions $(-v)$ is the same potential as the hydrogen atom and therefore (12.33) admits solutions at $\omega_{s,n} = E_g - \mu_{\text{ex}}/(2n^2)$ with n integers. These eigenvalues correspond to the possible exciton binding energies. Excitonic features in optical spectra are manifest as peaks at $\omega = \omega_{s,n}$; at these frequencies L^R has simple poles and, according to (12.21), light with these frequencies causes a large change in $\delta G^<$.

Up until the time this book was published, the Bethe–Salpeter equation has been solved exclusively with the kernel (12.22), or variations in which the second interaction in (12.22) is replaced by a static, i.e., frequency-independent, screened interaction W. The first solutions date back to the late seventies when the Bethe–Salpeter equation was applied to calculate the optical spectrum of diamond and silicon [94]. The accuracy of the solution has been improved over the years and the calculated optical spectra of several systems with strong excitonic features have been found in good agreement with experiments. For a review see Ref. [95] and references therein. Going beyond the static approximation for W represents a major computational challenge [96]. An alternative route to extract L with more sophisticated kernels consists in solving the Kadanoff–Baym equations [97–99]. Let us explain how it works. For a given approximation to $\Sigma = \Sigma_s[G, v]$ we can use the Kadanoff–Baym equations to calculate the Green's function of a system with single-particle Hamiltonian h. In Chapter 15 we show that the first-order change δG induced by a change δh in the single-particle Hamiltonian is given by (12.19) where L fulfills the Bethe–Salpeter equation with kernel $K(1, 2; 3, 4) = \pm\delta\Sigma(1; 3)/\delta G(4; 2)$. Since the kernel (12.22) is just $\pm\delta\Sigma_{\text{HF}}/\delta G$ (see next section) it is clear that any approximation to the self-energy that goes beyond the Hartree–Fock approximation gives us access to Ls of a higher degree of sophistication. Applications of this method are presented in Section 16.8.

12.4 Diagrammatic proof of $K = \pm\delta\Sigma/\delta G$

In this section we prove the important relation

$$\boxed{K(1,2;3,4) = \pm\frac{\delta\Sigma(1;3)}{\delta G(4;2)}} \tag{12.34}$$

between the self-energy $\Sigma = \Sigma_s[G,v]$ and the kernel $K = K_s[G,v]$ of the Bethe–Salpeter equation. In (12.34) Σ_s and K_s are functionals of G and v (built using G-skeleton diagrams). The proof consists in showing that by cutting a G-line in all possible ways from every Σ-diagram[2] we get the full set of K-diagrams, each with the right prefactor. Let us first consider some examples. For the Hartree–Fock self-energy

$$\Sigma_{\mathrm{HF}}(1;3) = \quad \underset{1 \quad 3}{\bigcirc} \quad + \quad \overset{\frown}{\underset{1 \qquad 3}{\longrightarrow}}$$

the prefactor is $(\pm\mathrm{i})$ for the first diagram and $(+\mathrm{i})$ for the second diagram. The functional derivative with respect to $G(4;2)$ yields the K-diagrams

$$\pm\frac{\delta\Sigma_{\mathrm{HF}}(1;3)}{\delta G(4;2)} = \quad \underset{3 \quad\quad 2}{\overset{1 \quad\quad 4}{\sim\!\sim\!\sim}} \quad + \quad \underset{3 \quad 2}{\overset{1 \quad 4}{\}}} \tag{12.35}$$

whose prefactors are correctly given by plus/minus the prefactors of the Σ-diagrams from which they originate. Another example is the first bubble diagram of the second-Born approximation

$$\Sigma_{\mathrm{2B,bubble}}(1;3) = \quad \underset{1 \qquad\qquad 3}{\text{⬭}}$$

with prefactor $\pm\mathrm{i}^2$. Its functional derivative generates the following three diagrams for the kernel:

$$\pm\frac{\delta\Sigma_{\mathrm{2B,bubble}}(1;3)}{\delta G(4;2)} = \quad \underset{3 \quad 2}{\overset{1 \quad 4}{⬦}} \quad + \quad \underset{3 \qquad 2}{\overset{1 \qquad 4}{\square}} \quad + \quad \underset{3 \qquad 2}{\overset{1 \qquad 4}{\bowtie}}$$

The prefactor of the first diagram is $\mathrm{i}^2(\pm)^l = \mathrm{i}^2(\pm)^2 = \mathrm{i}^2$. Similarly one can calculate the prefactor of the other two diagrams and check that it is again given by i^2.

The general proof of (12.34) follows from a few statements:

---

[2]As already observed several times, to cut a G-line in all possible ways is equivalent to taking the functional derivative with respect to G.

- Closing a K-diagram $K^{(i)}(1, 2; 3, 4)$ with a Green's function $G(4; 2)$ leads to a skeletonic Σ-diagram up to a prefactor (we prove below that the prefactor is $\pm$).

Proof. Suppose that the self-energy diagram obtained from $K^{(i)}$ is not skeletonic. Then it must be of the form

$$\Sigma(1; 3) = \quad$$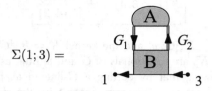

where A is a self-energy insertion. If the $G(4; 2)$ that we used to close the K-diagram is in A then the original K-diagram would be two-particle *reducible* since it could be divided into two disjoint pieces containing $(1, 3)$ and $(2, 4)$ by cutting two G-lines. This is in contradiction to the definition of the irreducible kernel K. The added line $G(4; 2)$ cannot be G_1 or G_2 for otherwise the original K-diagram would contain a self-energy insertion and hence it would not be skeletonic. For the very same reason the added line $G(4; 2)$ cannot be in B either. We conclude that the self-energy diagram obtained by closing a K-diagram with a G-line is skeletonic.

- Every K-diagram is obtained from a unique Σ-diagram. Moreover, every Σ-diagram with n G-lines gives n topologically inequivalent K-diagrams.

Proof. From the previous statement we know that by closing a K-diagram with a G-line we generate a Σ-diagram and hence there must be at least one Σ-diagram from which this K-diagram can be obtained. Furthermore, it is clear that the cutting of a G-line in two topologically inequivalent Σ-diagrams $\Sigma^{(1)}$ and $\Sigma^{(2)}$ cannot lead to topologically equivalent K-diagrams $K^{(1)}$ and $K^{(2)}$ since if we added back the G-line to $K^{(1)}$ and $K^{(2)}$ we would find that $\Sigma^{(1)}$ and $\Sigma^{(2)}$ have the same topology, contrary to the assumption. We thus conclude that every K-diagram comes from a unique Σ-diagram. The remaining question is then whether a given Σ-diagram can lead to two (or more) topologically equivalent K-diagrams $K^{(1)}$ and $K^{(2)}$ by cutting two different G-lines. In Section 11.3 we saw that by cutting a G-line from a vacuum diagram we obtain N_S topological equivalent Σ-diagrams, where N_S is the number of symmetries of the vacuum diagram. Similarly if $K^{(1)}$ and $K^{(2)}$ are topologically equivalent then there must be a symmetry s of the generating Σ-diagram that maps the set $\mathcal{G}$ of G-lines into itself. However, from the third statement of Section 11.3 it follows that the only possible symmetry is the identity permutation since the external vertices of Σ are fixed. We conclude that by removing two different G-lines we always obtain two topologically inequivalent K-diagrams.

- The prefactor of a K-diagram calculated from (12.34) agrees with the Feynman rules for the K-diagrams.

Proof. The prefactor of a K-diagram calculated from (12.34) is plus/minus the prefactor of the generating Σ-diagram, i.e., $i^k(\pm)^{l+1}$ where k is the number of interaction lines and l is the number of loops in the Σ-diagram. Since the Σ- and K-diagrams have the same number of interaction lines the factor i^k is certainly correct. We only need to check the sign. The sign of a K-diagram is, by definition, given by the number of loops in the diagram $K(1, 2; 3, 4)$ closed with $G(4; 2)$ and $G(3; 1)$. When closing the K-diagram with $G(4; 2)$ we get back $\Sigma(1; 3)$. If we now close $\Sigma(1; 3)$ with $G(3; 1)$ we increase the number of loops in Σ by one. This proves the statement.

These three statements constitute the diagrammatic proof of (12.34). For Φ-derivable self-energies we can also write

$$K(1, 2; 3, 4) = \pm \frac{\delta^2 \Phi}{\delta G(3; 1) \delta G(4; 2)}. \tag{12.36}$$

Thus K is obtained from Φ by differentiating twice with respect to G. Since the order of the functional derivatives does not matter, we have

$$K(1, 2; 3, 4) = \pm \frac{\delta^2 \Phi}{\delta G(3; 1) \delta G(4; 2)} = \pm \frac{\delta^2 \Phi}{\delta G(4; 2) \delta G(3; 1)} = K(2, 1; 4, 3), \tag{12.37}$$

which is indeed a symmetry of the two-particle Green's function, see (8.7). We can also regard (12.37) as a vanishing "curl" condition for $\Sigma = \Sigma_s[G, v]$,

$$\frac{\delta \Sigma(1; 3)}{\delta G(4; 2)} - \frac{\delta \Sigma(2; 4)}{\delta G(3; 1)} = 0, \tag{12.38}$$

which is a necessary condition for the existence of a functional Φ such that Σ is Φ-derivable. Thus, another way to establish whether a given Σ is Φ-derivable consists in checking the validity of (12.38).

12.5 Vertex function and Hedin equations

In this section we derive a closed system of equations for the various many-body quantities introduced so far. To this end we observe that the general structure of the self-energy $\Sigma = \Sigma_s[G, v]$ is

$$\Sigma(1; 2) = \quad \text{(diagram)} \quad + i \quad \text{(diagram)}$$

$$= \pm i\, \delta(1; 2) \int d3\, v(1; 3) G(3; 3^+) + i \int d3 d4\, v(1; 3) G(1; 4) \Lambda(4, 2; 3). \tag{12.39}$$

The so called *vertex function* (or just vertex)

$$\Lambda(1, 2; 3) = \quad \text{(diagram)}$$

contains all possible diagrams that we can enter in the second diagram of (12.39) to form a self-energy diagram. The zeroth order vertex function is

$$\Lambda_0(1,2;3) = \delta(1;2^+)\delta(3;2) = \quad \bullet\, ,$$

which is represented by a dot and yields the self-energy Fock diagram, since

$$i\int d3d4\, v(1;3)G(1;4)\delta(4;2^+)\delta(3;2) = i\,v(1;2)G(1;2^+).$$

We can deduce an equation for the vertex by inserting in (12.8) the expression (12.39) for Σ and the expression (12.11) for G_2. It is a simple and instructive exercise for the reader to show that

$$\Lambda(1,2;3) = \delta(1;2^+)\delta(3;2) \pm \int d4d5\, K_r(1,4;2,5)G(5;3)G(3;4).$$

We can give to this formula the following diagrammatic representation

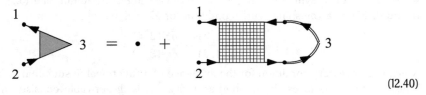

$$(12.40)$$

where we take into account that the GG-double-line gives a factor $(\pm)$ when the diagram is converted into a mathematical expression, see Section 12.2. Expanding the reducible kernel as in (12.14) we then obtain a Dyson equation for the vertex

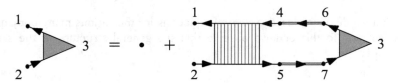

which in formulas reads (remember that $K = \pm\delta\Sigma/\delta G$)

$$\Lambda(1,2;3) = \delta(1;2^+)\delta(3;2) + \int d4d5d6d7\, \frac{\delta\Sigma(1;2)}{\delta G(4;5)}\, G(4;6)G(7;5)\Lambda(6,7;3). \qquad (12.41)$$

This is the Bethe–Salpeter equation for the vertex function. Equations (12.39) and (12.41) provide a coupled system of equations from which to obtain the self-energy iteratively. If we start with the zeroth order vertex function $\Lambda_0(1,2;3) = \delta(1;2^+)\delta(3;2)$, then (12.39) yields the Hartree–Fock self-energy, whose functional derivative with respect to G yields the kernel (12.35). The insertion of this kernel into (12.41) generates the vertex function

$$(12.42)$$

The solution of (12.42) gives an infinite series of bubble and ladder diagrams similar to those of Fig. 12.3(b). The new vertex can now be inserted back into (12.39) to obtain a new self-energy. A subsequent differentiation with respect to G then gives yet another vertex and we can continue to iterate *ad infinitum*. We are not aware of any proof that such iterative scheme converges nor that it generates all possible skeleton diagrams for the self-energy.

From the solution of (12.39) and (12.41) we get the self-energy as a functional of G and v. The natural question then arises whether we can derive a similar system of equations from which to get the self-energy as a functional of G and W. To have Σ in terms of G and W we must first replace $v \to W$ in the second diagram of (12.39), and then remove from Λ all diagrams containing a polarization insertion and again replace $v \to W$. In order to remove polarization insertions from Λ we need to introduce another kernel $\tilde{K}_r$:

> $\tilde{K}_r(1,2;3,4)$ *is obtained from* $K_r(1,2;3,4)$ *by discarding all those diagrams that are one-interaction line reducible, i.e., that can be broken into two disjoint pieces, one containing* $(1,3)$ *and the other containing* $(2,4)$*, by cutting an interaction line.*

In Fig. 12.2 the 2nd, 4th, 5th, and 6th diagrams are one-interaction line reducible while the remaining diagrams are one-interaction line irreducible. Alternatively we can say that we are removing from K_r all those diagrams that give rise to a polarization insertion when we close the right vertices with two G-lines (so as to form a Λ-diagram, see (12.40)). The reader can easily check that with the exception of the 2nd, 4th, 5th, and 6th diagrams of Fig. 12.2 no other diagram in the same figure gives rise to a polarization insertion when it is closed with two G-lines. Since the kernel $\tilde{K}_r$ is not two-particle irreducible we can introduce an irreducible kernel $\tilde{K}$ in the same fashion as we did for K_r.

> *The kernel* $\tilde{K}$ *is two-particle irreducible and one-interaction line irreducible.*

The only diagram in K which is one-interaction line reducible is the Hartree diagram [first diagram in (12.35)] and hence

$$\tilde{K}(1,2;3,4) = K(1,2;3,4) - i\,\delta(1;3)\delta(2;4)v(1;2). \tag{12.43}$$

This equation is represented by the diagrammatic relation

where the striped square with a tilde is $\tilde{K}$. The reducible kernel $\tilde{K}_r$ can then be expanded in terms of the irreducible kernel $\tilde{K}$ in a Bethe–Salpeter-like equation

So far we have regarded $\tilde{K} = \tilde{K}_s[G, v]$ as a functional of G and v. If we remove from it all diagrams with a polarization insertion and subsequently replace $v \to W$ we obtain a new functional $\tilde{K}_{ss}[G, W]$ which yields $\tilde{K}$ as a functional of G and W. Then, writing the self-energy as in (10.25), i.e.,

$$\Sigma = \Sigma_{ss}[G, W] = \Sigma_{\mathrm{H}}[G, v] + \Sigma_{ss,\mathrm{xc}}[G, W],$$

we can establish the following relation between $\tilde{K} = \tilde{K}_{ss}$ and $\Sigma_{\mathrm{xc}} = \Sigma_{ss,\mathrm{xc}}$

$$\boxed{\tilde{K}(1, 2; 3, 4) = \pm \frac{\delta \Sigma_{\mathrm{xc}}(1, 3)}{\delta G(4; 2)}} \tag{12.44}$$

The proof of (12.44) is identical to the proof of (12.34). We stress again that for (12.44) to be valid we must exclude from Σ_{xc} all self-energy diagrams with a polarization insertion and use W as the independent variable. We now have all ingredients to construct the vertex function $\Gamma = \Gamma_{ss}[G, W]$ without polarization insertions.[3] The new vertex function is obtained from $\tilde{K} = \tilde{K}_{ss}[G, W]$ in a similar way to that by which $\Lambda = \Lambda_s[G, v]$ is obtained from $K = K_s[G, v]$, see (12.41),

where Γ is represented by the black triangle. In the first equality $\tilde{K}_r = \tilde{K}_{ss,r}[G, W]$ is the sum of all the $\tilde{K}_r$-diagrams which do not contain polarization insertions and in which $v \to W$; similarly in the second equality $\tilde{K} = \tilde{K}_{ss}[G, W]$. The mathematical expression of this diagrammatic equation is

$$\boxed{\Gamma(1, 2; 3) = \delta(1; 2^+)\delta(3; 2) + \int d4d5d6d7 \, \frac{\delta \Sigma_{\mathrm{xc}}(1; 2)}{\delta G(4; 5)} \, G(4; 6)G(7; 5)\Gamma(6, 7; 3)} \tag{12.45}$$

Equation (12.45) is the Bethe–Salpeter equation for the G- and W-skeleton vertex function $\Gamma_{ss}[G, W]$. In the remainder of this section we regard all quantities as functionals of the independent variables G and W; in other words we do not specify that a quantity

---

[3] The reader may wonder why we did not use the symbol Λ_{ss} for this vertex. The vertex Γ is obtained from Λ by removing all diagrams containing a polarization insertion and then replacing $v \to W$. Now the point is that $\Lambda(1, 2; 3)$ can have a polarization insertion which is either internal or external. A Λ-diagram with an internal polarization insertion is, e.g., the diagram resulting from the 3rd term of Fig. 12.2, while a Λ-diagram with an external polarization insertion is, e.g., the diagram resulting from the 4th, 5th, or 6th term of Fig. 12.2. The typical structure of a Λ-diagram with an external polarization insertion is $\Lambda(1, 2; 3) \propto \int d4d5 \, \Lambda_i(1, 2; 4)v(4; 5)P_j(5; 3)$ with Λ_i some vertex diagram and P_j some polarization diagram. To construct Γ we have to remove from Λ the diagrams with internal and/or external polarization insertions. It is then clear that by expanding $W = v + vPv + \ldots$ in a Γ-diagram we get back only Λ-diagrams with *internal* polarization insertions, i.e., we cannot recover the full Λ. In other words $\Gamma_{ss}[G, W] \neq \Lambda_s[G, v]$. This is the motivation for using a different symbol. The situation is different for the self-energy and polarizability since the expansion of W in $\Sigma_{ss}[G, W]$ and $P_{ss}[G, W]$ gives back $\Sigma_s[G, v]$ and $P_s[G, v]$.

$Q = \Sigma$, Γ, $\tilde{K}$, P, etc. has to be understood as $Q_{ss}[G, W]$. The vertex Γ allows us to write the self-energy as a functional of G and W according to

$$\Sigma(1;2) =$$

or in formula

$$\Sigma(1;2) = \pm i\,\delta(1;2)\int d3\,v(1;3)G(3;3^+) + i\int d3d4\,W(1;3)G(1;4)\Gamma(4,2;3) \qquad (12.46)$$

We see that if we take a Γ-diagram and expand $W = v + vPv + \ldots$ in powers of v we get Λ-diagrams with no external polarization insertions and hence the product $W\Gamma$ does not lead to double counting.

The kernel $\tilde{K}_r$ can also be used to write the polarizability according to

$$P(1;2) =$$

where the prefactor i in the second term on the r.h.s. is due to the different Feynman rules for P (m interactions give a prefactor i^{m+1}, see Section 10.8) and for $\tilde{K}_r$ (m interactions give a prefactor i^m, see Section 12.2). The zeroth order polarization diagram can also be seen as a closed Γ-diagram in which Γ is approximated by $\delta\delta = \bullet$, i.e.,

It is easy to check that this is an equality: the diagram on the l.h.s. is a polarization diagram and hence the prefactor is $\pm i$ while the diagram on the r.h.s. is a closed Γ-diagram and hence the prefactor is $\pm$ due to the GG-double-line. This explains the presence of the factor i in front of it. We thus see that

or in formulas

$$P(1;2) = \pm i\int d3d4\,G(1;3)G(4;1)\Gamma(3,4;2) \qquad (12.47)$$

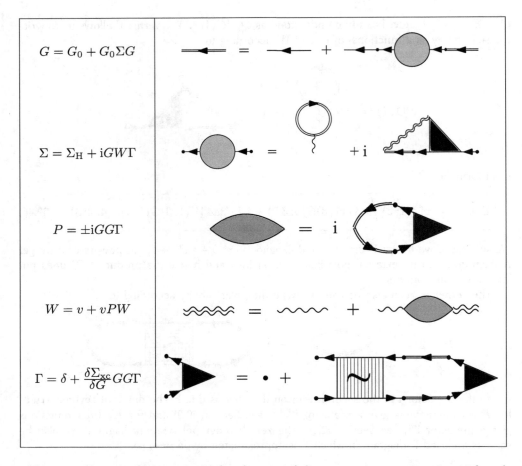

Table 12.1 Mathematical expression (left column) and diagrammatic representation (right column) of the Hedin equations.

Equations (12.45), (12.46), and (12.47) together with the Dyson equations $G = G_0 + G_0\Sigma G$ and $W = v + vPW$ are the famous *Hedin equations*. They form a set of coupled equations for G, Σ, P, W, and Γ whose solution yields the same G which solves the Martin–Schwinger hierarchy. We have thus achieved a major reduction of the equations needed to determine the Green's function: from the infinite Martin–Schwinger hierarchy to the five Hedin equations. In Table 12.1 we list the Hedin equations and their diagrammatic representation. Like the coupled equations (12.39) and (12.41) the Hedin equations can be iterated to obtain an expansion of Σ in terms of G and W. If we start with the zeroth order vertex $\Gamma(1, 2; 3) = \delta(1; 2^+)\delta(3; 2)$, we find

$$\Sigma_{\text{xc}}(1; 2) = i\, W(1; 2)G(1; 2) = \qquad\qquad\qquad \tag{12.48}$$

and

$$P(1;2) = \pm i\, G(1;2)G(2;1) = \quad\text{(diagram)}\qquad\text{(12.49)}$$

This is the GW approximation already encountered in Section 10.7. Let us iterate further by calculating the kernel from the GW self-energy and then a new vertex Γ from (12.45). We have

$$\pm\frac{\delta\Sigma_{\text{xc}}(1;3)}{\delta G(4;2)} = \pm i\,\delta(1;4)\delta(2;3)\,W(1;3) \pm i\,G(1;3)\frac{\delta W(1;3)}{\delta G(4;2)}.$$

The last term can be evaluated from the 4th Hedin equation $W = v + vPW$. We have

$$\frac{\delta W}{\delta G} = v\frac{\delta P}{\delta G}W + vP\frac{\delta W}{\delta G}.$$

This is a Dyson-like equation for $\delta W/\delta G$. We can expand $\delta W/\delta G$ in "powers" of v by iterations, i.e., by replacing the $\delta W/\delta G$ in the last term of the r.h.s. with the whole r.h.s.. In doing so we find the series

$$\frac{\delta W}{\delta G} = v\frac{\delta P}{\delta G}W + vPv\frac{\delta P}{\delta G}W + vPvPv\frac{\delta P}{\delta G}W + \ldots = (v + vPv + vPvPv + \ldots)\frac{\delta P}{\delta G}W$$
$$= W\frac{\delta P}{\delta G}W.$$

From (12.49) it follows that

$$\frac{\delta P(5;6)}{\delta G(4;2)} = \pm i\delta(4;5)\delta(2;6)G(6;5) \pm i\delta(4;6)\delta(2;5)G(5;6),$$

and therefore

$$\pm\frac{\delta\Sigma_{\text{xc}}(1;3)}{\delta G(4;2)} = \pm i\,\delta(1;4)\delta(2;3)\,W(1;3)$$
$$+ i^2\,G(1;3)W(1;4)G(2;4)W(2;3) + i^2 G(1;3)W(1;2)G(2;4)W(4;3)$$

Inserting this kernel into the 5th Hedin equation we obtain the following diagrammatic equation for the vertex:

From this vertex we can then calculate a new self-energy, etc. This has actually been done in the context of the Hubbard model in Ref. [100]. As for the coupled equations (12.39) and (12.41), we are not aware of any proof that such an iterative scheme converges, nor that it generates all possible skeleton diagrams for the self-energy.

In Appendix L we derive the Hedin equations using the so called *source field* method introduced by Martin and Schwinger [47]. The source field method does not use any diagrammatic concepts and allows us to arrive at the Hedin equations in few steps. We should emphasize, however, that the source field method does not tell us how to expand the various quantities Σ, P, Γ, etc. and hence does not have the same physical appeal as the diagrammatic method discussed before. Furthermore, every time the source field method generates an equation in which the Green's function has to be calculated at equal times, the most convenient way to resolve the ambiguity is to go back to the diagrammatic method.

Exercise 12.5. Prove (12.40).

13

Applications of MBPT to equilibrium problems

13.1 Lifetimes and quasi-particles

After the formal developments of the previous chapters it is time to use MBPT to solve some concrete problems. In this chapter we study systems in equilibrium. Applications of MBPT to time-dependent (or out-of-equilibrium) systems can be found in Chapter 16.

We start by deepening an aspect which was only mentioned in Section 6.1.3, i.e., the existence of quasi-particles in interacting systems. What we saw there was that the matrix elements $G_{ij}^<(t, t')$ vanish when $|t - t'| \to \infty$ provided that the self-energy carries memory. We explained this result with the absence of good single-particle quantum numbers: the probability that after removing a particle in state i at time t and putting it back in state j at time t' we find the system unchanged should approach zero when $|t - t'| \to \infty$. This is the *relaxation* phenomenon illustrated in Fig. 6.2. The Hartree–Fock approximation cannot account for relaxation effects since

$$\hat{\Sigma}_{\mathrm{HF}}(z_1, z_2) = \delta(z_1, z_2) q \hat{\mathcal{V}}_{\mathrm{HF}}(z_1)$$

is local in time and hence does not carry memory. Accordingly, there exist single-particle states [the Hartree–Fock eigenstates in (7.45)] with an infinitely long lifetime. An equivalent way to express the same concept is through the spectral function. In the Hartree–Fock approximation

$$\hat{A}(\omega) = 2\pi \delta(\omega - \hat{h}_{\mathrm{HF}}), \tag{13.1}$$

and using the fluctuation–dissipation theorem (6.94)

$$
\begin{aligned}
G_{ij}^<(t, t') &= \mp \mathrm{i} \int \frac{d\omega}{2\pi} e^{-\mathrm{i}\omega(t-t')} f(\omega - \mu)\, 2\pi \, \langle i | \delta(\omega - \hat{h}_{\mathrm{HF}}) | j \rangle \\
&= \mp \mathrm{i} \, \langle i | f(\hat{h}_{\mathrm{HF}} - \mu) e^{-\mathrm{i}\hat{h}_{\mathrm{HF}}(t-t')} | j \rangle.
\end{aligned}
$$

If $i = j$ is an eigenstate of $\hat{h}_{\mathrm{HF}}$ with eigenvalue ϵ_i then

$$G_{ii}^<(t, t') = \mp \mathrm{i} \, f(\epsilon_i - \mu) e^{-\mathrm{i}\epsilon_i(t-t')},$$

which does not decay when $|t - t'| \to \infty$. *The δ-like structure of the spectral function is the signature of the absence of relaxation mechanisms.* Therefore, going beyond the Hartree-Fock approximation we expect that the spectral function become a smooth integrable function (of course for systems with infinitely many degrees of freedom). Indeed in this case

$$G_{ii}^<(t, t') = \mp i \int \frac{d\omega}{2\pi} e^{-i\omega(t-t')} f(\omega - \mu) \langle i | \hat{A}(\omega) | i \rangle \xrightarrow[|t-t'| \to \infty]{} 0, \qquad (13.2)$$

due to the Riemann–Lebesgue theorem.

These considerations prompt us to look for a transparent relation between the spectral function and the self-energy. Let us write the Dyson equation for the retarded/advanced Green's function

$$\hat{\mathcal{G}}^{R/A}(\omega) = \hat{\mathcal{G}}_0^{R/A}(\omega) + \hat{\mathcal{G}}_0^{R/A}(\omega) \hat{\Sigma}^{R/A}(\omega) \hat{\mathcal{G}}^{R/A}(\omega),$$

where [see (6.58) and (6.59)]

$$\hat{\mathcal{G}}_0^{R/A}(\omega) = \frac{1}{\omega - \hat{h} \pm i\eta}.$$

We split the self-energy $\hat{\Sigma} = \hat{\Sigma}_{HF} + \hat{\Sigma}_c$ into the Hartree-Fock self-energy and the correlation self-energy. From (9.12) we know that $\hat{\Sigma}_c$ belongs to the Keldysh space and has no singular part. Since $\hat{\Sigma}_{HF}(z, z') = \delta(z, z') q \hat{V}_{HF}$ we have $\hat{\Sigma}^{R/A}(\omega) = q\hat{V}_{HF} + \hat{\Sigma}_c^{R/A}(\omega)$.[1] Recalling that the Hartree-Fock Hamiltonian is $\hat{h}_{HF} = \hat{h} + q\hat{V}_{HF}$ we then find

$$\hat{\mathcal{G}}^{R/A}(\omega) = \frac{1}{\omega - \hat{h}_{HF} - \hat{\Sigma}_c^{R/A}(\omega) \pm i\eta}.$$

A remark on the infinitesimal constant η appearing in the denominator is in order. If the imaginary part of $\hat{\Sigma}_c^R(\omega) = [\hat{\Sigma}_c^A(\omega)]^\dagger$ is nonzero we can safely discard η. However, if $\mathrm{Im}[\hat{\Sigma}_c^R(\omega)] = 0$ for some ω then η must absolutely be present for otherwise the Green's function does not have the correct analytic properties. Having said that, from now on we incorporate this infinitesimal constant into the retarded/advanced correlation self-energy. From the definition (6.89) of the spectral function we have

$$\hat{A}(\omega) = i[\hat{\mathcal{G}}^R(\omega) - \hat{\mathcal{G}}^A(\omega)] = i\hat{\mathcal{G}}^R(\omega) \left[\frac{1}{\hat{\mathcal{G}}^A(\omega)} - \frac{1}{\hat{\mathcal{G}}^R(\omega)} \right] \hat{\mathcal{G}}^A(\omega)$$

$$= i\hat{\mathcal{G}}^R(\omega) \left[\hat{\Sigma}_c^R(\omega) - \hat{\Sigma}_c^A(\omega) \right] \hat{\mathcal{G}}^A(\omega).$$

The difference between the retarded and advanced correlation self-energy is the same as the difference $\hat{\Sigma}^R(\omega) - \hat{\Sigma}^A(\omega)$ since the Hartree-Fock part cancels. Then, taking into account the definition (9.55) of the rate operator we can write the following exact relation for systems in equilibrium

$$\boxed{\hat{A}(\omega) = \hat{\mathcal{G}}^R(\omega) \hat{\Gamma}(\omega) \hat{\mathcal{G}}^A(\omega) = \frac{1}{\omega - \hat{h}_{HF} - \hat{\Sigma}_c^R(\omega)} \hat{\Gamma}(\omega) \frac{1}{\omega - \hat{h}_{HF} - \hat{\Sigma}_c^A(\omega)}} \qquad (13.3)$$

---

[1] In equilibrium the Hartree-Fock potential does not depend on time and is equal to $\hat{V}_{HF}^M$.

This is just the relation between the spectral function and the self-energy that we were looking for. In fermionic systems the spectral function operator is positive semidefinite (all eigenvalues are non-negative, see Exercise 6.10) and therefore also the rate operator is positive semidefinite, see Exercise 13.1.

Let us see how to recover the Hartree–Fock result. In this case $\hat{\Sigma}_c^{R/A}(\omega) = \mp i\eta$ and hence the rate operator is $\hat{\Gamma}(\omega) = 2\eta$. Inserting these values into (13.3)

$$\hat{A}(\omega) = 2\frac{\eta}{(\omega - \hat{h}_{HF})^2 + \eta^2} \xrightarrow{\eta \to 0} 2\pi\,\delta(\omega - \hat{h}_{HF}),$$

which correctly agrees with (13.1). From the fluctuation–dissipation theorem for G and Σ we also see that (13.3) implies

$$\hat{\mathcal{G}}^{\lessgtr}(\omega) = \hat{\mathcal{G}}^R(\omega)\hat{\Sigma}^{\lessgtr}(\omega)\hat{\mathcal{G}}^A(\omega).$$

As anticipated in Section 9.7, in equilibrium systems (9.70) is an exact relation valid for all times t and t'.

From (13.3) it is clear that for $\hat{A}(\omega)$ to be nonsingular, $\hat{\Gamma}(\omega)$ has to be finite. We can provide a justification of the name "rate operator" for $\hat{\Gamma}$ by estimating how fast $G_{ij}^<(t,t')$ decays with $|t - t'|$. We write the self-energy $\hat{\Sigma}_c^{R/A}$ as the sum of a Hermitian operator $\hat{\Lambda}$ and an anti-Hermitian operator $\mp\frac{i}{2}\hat{\Gamma}$ [compare with the embedding self-energy in (7.27)],

$$\hat{\Sigma}_c^{R/A}(\omega) = \hat{\Lambda}(\omega) \mp \frac{i}{2}\hat{\Gamma}(\omega).$$

The relation (9.57) implies that $\hat{\Lambda}$ is the Hilbert transform of $\hat{\Gamma}$

$$\hat{\Lambda}(\omega) = P\int\frac{d\omega'}{2\pi}\frac{\hat{\Gamma}(\omega')}{\omega - \omega'}. \tag{13.4}$$

Now consider a system invariant under translations, like the electron gas. In the absence of magnetic fields the momentum–spin kets are eigenkets of $\hat{h}_{HF}$ with the same eigenvalue for spin up and down:

$$\hat{h}_{HF}|\mathbf{p}\sigma\rangle = \epsilon_{\mathbf{p}}|\mathbf{p}\sigma\rangle \qquad \left(\text{in the electron gas: } \epsilon_{\mathbf{p}} = \frac{p^2}{2} - V_0 + \Sigma_{HF}(\mathbf{p})\right).$$

Furthermore, the matrix elements of $\hat{\Gamma}$ in the momentum–spin basis are nonvanishing only along the diagonal

$$\langle\mathbf{p}\sigma|\hat{\Gamma}(\omega)|\mathbf{p}'\sigma'\rangle = (2\pi)^3\,\delta_{\sigma\sigma'}\delta(\mathbf{p} - \mathbf{p}')\Gamma(\mathbf{p},\omega),$$

and, of course, the same holds true for $\hat{\Lambda}$ and $\hat{A}$. Denoting by $\Lambda(\mathbf{p},\omega)$ and $A(\mathbf{p},\omega)$ the corresponding diagonal matrix elements (13.3) gives

$$A(\mathbf{p},\omega) = \frac{\Gamma(\mathbf{p},\omega)}{(\omega - \epsilon_{\mathbf{p}} - \Lambda(\mathbf{p},\omega))^2 + \left(\frac{\Gamma(\mathbf{p},\omega)}{2}\right)^2}. \tag{13.5}$$

We use this result to calculate the matrix elements of $G^<$ in the momentum–spin basis

$$G^<_{\mathbf{p}\sigma\,\mathbf{p}'\sigma'}(t,t') = (2\pi)^3 \delta_{\sigma\sigma'}\delta(\mathbf{p}-\mathbf{p}')\,G^<(\mathbf{p},t-t').$$

From (13.2) we have

$$G^<(\mathbf{p},t-t') = \mp i \int \frac{d\omega}{2\pi} e^{-i\omega(t-t')} f(\omega-\mu)A(\mathbf{p},\omega). \tag{13.6}$$

To estimate this integral we define the so called *quasi-particle energy* $E_{\mathbf{p}}$ to be the solution of $\omega - \epsilon_{\mathbf{p}} - \Lambda(\mathbf{p},\omega) = 0$. To first order in $(\omega - E_{\mathbf{p}})$ we can write

$$\omega - \epsilon_{\mathbf{p}} - \Lambda(\mathbf{p},\omega) = \left(1 - \frac{\partial\Lambda}{\partial\omega}\right)_{\omega=E_{\mathbf{p}}} (\omega - E_{\mathbf{p}}).$$

If the function $\Gamma(\mathbf{p},\omega)$ is small for $\omega \sim E_{\mathbf{p}}$ and slowly varying in ω then the main contribution to the integral (13.6) comes from a region around $E_{\mathbf{p}}$. In this case we can approximate $f(\omega-\mu)$ with $f(E_{\mathbf{p}}-\mu)$ and

$$A(\mathbf{p},\omega) \sim Z_{\mathbf{p}} \frac{1/\tau_{\mathbf{p}}}{(\omega - E_{\mathbf{p}})^2 + (1/2\tau_{\mathbf{p}})^2}, \tag{13.7}$$

where

$$Z_{\mathbf{p}} = \frac{1}{1 - \frac{\partial\Lambda}{\partial\omega}\big|_{\omega=E_{\mathbf{p}}}}, \tag{13.8}$$

and

$$\frac{1}{\tau_{\mathbf{p}}} = Z_{\mathbf{p}}\,\Gamma(\mathbf{p},E_{\mathbf{p}}).$$

With these approximations (13.6) for $t > t'$ yields

$$G^<(\mathbf{p},t-t') = \mp i Z_{\mathbf{p}} f(E_{\mathbf{p}} - \mu) e^{-iE_{\mathbf{p}}(t-t')} e^{-(t-t')/(2\tau_{\mathbf{p}})}.$$

The probability of finding the system unchanged is the modulus square of the above quantity and decays as $e^{-(t-t')/\tau_{\mathbf{p}}}$. Thus $\tau_{\mathbf{p}}^{-1}$ is the decay rate of a removed particle with momentum $\mathbf{p}$. In a similar way one can prove that $\tau_{\mathbf{p}}^{-1}$ is also the decay rate of an added particle with momentum $\mathbf{p}$. This provides a justification of the name "rate operator" for $\hat{\Gamma}$. Interestingly $\tau_{\mathbf{p}}^{-1}$ is also the width of the peak of the spectral function (13.7). We then conclude that *the lifetime of a single-particle excitation can be estimated from the width of the corresponding peak in the spectral function.* The smaller is the width $\tau_{\mathbf{p}}^{-1}$, the longer is the life-time and the more particle-like is the behavior of the excitation. This is the reason for calling *quasiparticles* those excitations with a long lifetime. We stress that this interpretation makes sense only if $\Gamma(\mathbf{p},\omega)$ is small for $\omega \sim E_{\mathbf{p}}$.

In a zero-temperature fermionic system, $\hat{\Sigma}^>(\omega)$ vanishes below the chemical potential while $\hat{\Sigma}^<(\omega)$ vanishes above the chemical potential, see (9.56). Assuming that $\hat{\Sigma}^>(\omega)$ and

$\hat{\Sigma}^{<}(\omega)$ are continuous functions,[2] we infer that $\hat{\Gamma}(\mu) = 0$. We know from the Luttinger-Ward theorem that $E_{\mathbf{p}} \to \mu$ for $|\mathbf{p}| \to p_F$. Thus (13.7) becomes a δ-function with strength $Z = \lim_{|\mathbf{p}| \to p_F} Z_{\mathbf{p}}$ centered at $\omega = \mu$ for $|\mathbf{p}| = p_F$. The strength Z varies in the range

$$0 < Z \le 1,$$

since from the Hilbert transform relation (13.4)

$$\left. \frac{\partial \Lambda(\mathbf{p}, \omega)}{\partial \omega} \right|_{\omega = \mu} = - \int \frac{d\omega'}{2\pi} \frac{\Gamma(\mathbf{p}, \omega')}{(\mu - \omega')^2} < 0, \tag{13.9}$$

where we take into account that the rate function $\Gamma(\mathbf{p}, \omega')$ is non-negative, see observation below (13.3), and vanishes quadratically as $\omega' \to \mu$ (this will be proved in Section 13.3). As the exact spectral function integrates to unity, see (6.92), this δ-function is superimposed on a continuous function which integrates to $1 - Z$. Physically this means that a particle injected or removed with momentum $|\mathbf{p}| = p_F$ generates (with probability Z) an infinitely long-living excitation with energy μ plus other excitations with different energies. In the noninteracting case $Z = 1$ and therefore we can interpret Z as the renormalization of the square of the single-particle wavefunction due to interactions. More generally we refer to $Z_{\mathbf{p}}$ as the *quasi-particle renormalization factor*. From the viewpoint of a photoemission experiment (see Section 6.3.4) the spectral function $A(\mathbf{p}, \omega)$ gives the probability that when a momentum $\mathbf{p}$ is transferred to the electron gas the system will have changed its energy by ω. For $|\mathbf{p}|$ values close to p_F this energy transfer is most likely equal to $E_{\mathbf{p}}$ and the energy of the ejected electron is then most likely to be $\epsilon = \omega_0 - E_{\mathbf{p}}$ where ω_0 is the energy of the photon. In this way the quasi-particle energies can be measured experimentally in metals and experiments have indeed confirmed the quasi-particle picture. As we see in Chapter 15, the addition/removal of a particle with momentum $\mathbf{p}$ *away from the Fermi momentum* p_F has large probability of exciting particle–hole pairs as well as collective modes known as plasmons. These excitations introduce an uncertainty in the energy of the particle: the peak of the spectral function in $\omega = E_{\mathbf{p}}$ is broadened by an amount $\tau_{\mathbf{p}}^{-1}$ (mainly due to particle–hole excitations), and other prominent features appear in different spectral regions (mainly due to plasmons).

One more observation is about $\hat{\Lambda}$. While the imaginary part of $\hat{\Sigma}_c^{R}$ gives the width of the spectral peak the real part of $\hat{\Sigma}_c^{R}$ gives the energy shift in the position of this peak. Therefore the quantity $\hat{\Lambda}$ can be interpreted as the correction to the Hartree–Fock single-particle energy due to collisions with other particles. The real and imaginary parts of $\hat{\Sigma}_c^{R}$ are not independent but related through a Hilbert transformation, see (13.4). This is a general aspect of all many-body quantities of equilibrium systems and follows directly from the definition of retarded/advanced functions: real and imaginary parts of the Fourier transform of the retarded Green's function, self-energy, polarization, screened interaction, etc. are all related through a Hilbert transformation.[3]

---

[2]This is true provided that MBPT does not break down, see Section 13.3. An example of a system for which MBPT breaks down is a superconductor. In this case the MBPT expansion has to be done using a Green's function different from the noninteracting Green's function G_0. Typically the break-down of MBPT using G_0 as a starting point signals that the interacting system is very different from the noninteracting one, e.g, the symmetry of the ground state is different.

[3]And, of course, the same is true for the advanced quantities.

Exercise 13.1. The spectral function operator (13.3) has the structure $\hat{A} = \hat{G}\,\hat{\Gamma}\,\hat{G}^{\dagger}$ with $\hat{G} = \hat{G}^{\mathrm{R}}$. Therefore $\hat{\Gamma} = (1/\hat{G})\,\hat{A}\,(1/\hat{G}^{\dagger})$. Show that if $\hat{A}$ is positive semidefinite then $\langle i|\hat{\Gamma}|i\rangle \geq 0$ for all quantum numbers i and therefore also $\hat{\Gamma}$ is positive semidefinite.

13.2   Fluctuation-dissipation theorem for P and W

Like the self-energy, the polarizability and the screened interaction also fulfill a fluctuation-dissipation theorem. Let us first consider the polarizability. Every diagram for $P(1;2)$ starts with a couple of Green's functions $G(\mathbf{x}_1, z_1; \ldots)G(\ldots; \mathbf{x}_1, z_1)$ and ends with a couple of Green's functions $G(\ldots; \mathbf{x}_2, z_2)G(\mathbf{x}_2, z_2; \ldots)$. Introducing the operator (in first quantization) $\hat{P}(z_1, z_2)$ with matrix elements

$$\langle \mathbf{x}_1|\hat{P}(z_1, z_2)|\mathbf{x}_2\rangle = P(1;2),$$

the identities (9.50) and (9.51) imply that

$$\hat{P}^{\lceil}(\tau, t') = \hat{P}^{>}(t_0 - i\tau, t'), \qquad \hat{P}^{\rceil}(t, \tau') = \hat{P}^{<}(t, t_0 - i\tau'),$$

and

$$\hat{P}^{\mathrm{M}}(\tau, \tau') = \begin{cases} \hat{P}^{>}(t_0 - i\tau, t_0 - i\tau') & \tau > \tau' \\ \hat{P}^{<}(t_0 - i\tau, t_0 - i\tau') & \tau < \tau' \end{cases}.$$

To derive a fluctuation-dissipation theorem for P we need the boundary conditions. The polarizability $P(1;2)$ contains two Green's functions with argument 1 and two Green's functions with argument 2. Since the Green's function satisfies the KMS relations (5.6) the polarizability satisfies the KMS relations below

$$\hat{P}(z_1, t_{0-}) = \hat{P}(z_1, t_0 - i\beta), \qquad \hat{P}(t_{0-}, z_2) = \hat{P}(t_0 - i\beta, z_2).$$

Therefore we can write

$$\begin{aligned} \hat{P}^{<}(t_0, t') &= \hat{P}(t_{0-}, t'_+) \\ &= \hat{P}(t_0 - i\beta, t'_+) \\ &= \hat{P}^{\lceil}(\beta, t') \\ &= \hat{P}^{>}(t_0 - i\beta, t'). \end{aligned}$$

Fourier transforming both sides of this equation we find the important relation

$$\hat{P}^{>}(\omega) = e^{\beta\omega}\hat{P}^{<}(\omega).$$

Then the fluctuation-dissipation theorem for the polarizability reads

$$\boxed{\hat{\Pi}(\omega) \equiv i[\hat{P}^{>}(\omega) - \hat{P}^{<}(\omega)] \quad \Rightarrow \quad \begin{cases} \hat{P}^{<}(\omega) = -if(\omega)\hat{\Pi}(\omega) \\[2mm] \hat{P}^{>}(\omega) = -i\bar{f}(\omega)\hat{\Pi}(\omega) \end{cases}} \qquad (13.10)$$

where $f(\omega) = 1/(e^{\beta\omega} - 1)$ is the Bose function and $\bar{f}(\omega) = 1 + f(\omega)$. Clearly the density response function χ fullfils the same fluctuation–dissipation theorem as P, since the topology of a χ-diagram is the same as the topology of a P-diagram. In Section 15.2 we give an alternative proof of the fluctuation–dissipation theorem for χ based on the Lehmann representation of this quantity.

The derivation of the fluctuation–dissipation theorem for W goes along the same lines. Let us write W in terms of χ according to $W = v + v\chi v$, see (11.45). Since the interaction has only a singular part we have $v^{R} = v^{A} = v$ and $v^{\lessgtr} = 0$. Therefore the lesser and greater screened interaction is simply $W^{\lessgtr} = v\chi^{\lessgtr}v$ and similarly $W^{\lceil} = v\chi^{\lceil}v$ and $W^{\rceil} = v\chi^{\rceil}v$. The interaction v acts like a simple multiplicative factor for the time variable[4] and W fullfils the same relations as χ, which are identical to the relations for P. Therefore, introducing the operator (in first quantization) $\hat{\mathcal{W}}(z_1, z_2)$ with matrix elements

$$\langle \mathbf{x}_1 | \hat{\mathcal{W}}(z_1, z_2) | \mathbf{x}_2 \rangle = W(1; 2)$$

we can write

$$\hat{\mathcal{W}}^{>}(\omega) = e^{\beta\omega}\hat{\mathcal{W}}^{<}(\omega).$$

The fluctuation–dissipation theorem for W reads

$$\hat{\Omega}(\omega) \equiv \mathrm{i}[\hat{\mathcal{W}}^{>}(\omega) - \hat{\mathcal{W}}^{<}(\omega)] \quad \Rightarrow \quad \begin{cases} \hat{\mathcal{W}}^{<}(\omega) = -\mathrm{i}f(\omega)\hat{\Omega}(\omega) \\[2mm] \hat{\mathcal{W}}^{>}(\omega) = -\mathrm{i}\bar{f}(\omega)\hat{\Omega}(\omega) \end{cases} \tag{13.11}$$

with f the Bose function.

The fluctuation–dissipation theorem for P and W can be used to recover the equilibrium solution from the Dyson equation $W = v + vPW$, similarly to how we recovered the equilibrium solution from the Kadanoff–Baym equations in Section 9.6. Let us be more precise. We define the operator (in first quantization) of the bare interaction $\hat{v}(z_1, z_2) = \delta(z_1, z_2)\hat{v}$ with matrix elements

$$\langle \mathbf{x}_1 | \hat{v}(z_1, z_2) | \mathbf{x}_2 \rangle = \delta(z_1, z_2)v(\mathbf{x}_1, \mathbf{x}_2).$$

Then the lesser/greater component of the Dyson equation $W = v + vPW$ can be written as

$$\hat{\mathcal{W}}^{\lessgtr}(t, t') = \hat{v}\left[\hat{\mathcal{P}}^{\lessgtr} \cdot \hat{\mathcal{W}}^{A} + \hat{\mathcal{P}}^{R} \cdot \hat{\mathcal{W}}^{\lessgtr} + \hat{\mathcal{P}}^{\rceil} \cdot \hat{\mathcal{W}}^{\lceil}\right](t, t'). \tag{13.12}$$

For systems in equilibrium the l.h.s. depends only on the time difference. However, it is not obvious that this is true also for the r.h.s. since the time convolutions are either from t_0 to ∞ or from 0 to β. In order to prove that the r.h.s. depends only on $t - t'$ and is independent of the initial time t_0 we must use the fluctuation–dissipation theorem as well as the fact that real and imaginary parts of the retarded/advanced P and $\delta W = W - v$ are related by a Hilbert transformation (here δW is the regular part of the screened interaction).

---

[4]This is not true for the space variable since $v\chi v$ involves two space convolutions.

The mathematical steps are identical to those of Section 9.6 and here we write only the (expected) final result

$$\left[\hat{\mathcal{P}}^{\lessgtr} \cdot \delta \hat{\mathcal{W}}^{\mathrm{A}} + \hat{\mathcal{P}}^{\mathrm{R}} \cdot \delta \hat{\mathcal{W}}^{\lessgtr} + \hat{\mathcal{P}}^{\rceil} \cdot \delta \hat{\mathcal{W}}^{\lceil} \right] (t, t')$$
$$= \int \frac{d\omega}{2\pi} e^{-i\omega(t-t')} \left[\hat{\mathcal{P}}^{\lessgtr}(\omega) \delta \hat{\mathcal{W}}^{\mathrm{A}}(\omega) + \hat{\mathcal{P}}^{\mathrm{R}}(\omega) \delta \hat{\mathcal{W}}^{\lessgtr}(\omega) \right]. \qquad (13.13)$$

More generally, it is always true that:

> *If the system is in thermodynamic equilibrium then we can simplify the Langreth rules of Table 5.1 by taking $t_0 \to -\infty$ and discarding the vertical track. In this way real-time convolutions become simple products in frequency space.*

We have explicitly proved this result for the convolution of the self-energy with the Green's function, see Section 9.6. With the same kinds of manipulation one can show that the above statement is true for all convolutions appearing in MBPT.

Exercise 13.2. Prove (13.13).

13.3 Correlations in the second-Born approximation

The simplest approximation that includes the effects of particle collisions is the second-Born approximation (12.7) which we rewrite below

$$G_2(1,2;1',2') = G_{2,\mathrm{HF}}(1,2;1',2') \quad + \quad \cdots \quad + \quad \cdots \qquad (13.14)$$

The first term on the r.h.s. is the two-particle Green's function in the Hartree–Fock approximation. The second term describes the scattering process of two particles added to the system in $1'$ and $2'$, propagating to $\bar{1}$ and $\bar{2}$, interacting, and then continuing their propagation to 1 and 2, where they are removed. The third term contains the same physical information; it simply accounts for the correct (anti)symmetry property (exchange diagram) of G_2. The inclusion of these scattering processes in (12.8) leads to the self-energy in (10.17), i.e.,

$$\Sigma(1; \bar{1}) = \Sigma_{\mathrm{HF}}(1; \bar{1}) \quad + \quad \cdots \quad + \quad \cdots \qquad (13.15)$$

In the second term on the r.h.s. we recognize the process previously described. A particle coming from somewhere reaches $\bar{1}$ and interacts with a particle coming from $2'$ in $\bar{2}$. After the interaction the first particle propagates to 1 and the second particle propagates to $2 = 2'$. The third term in (13.15) stems from the exchange diagram in (13.14). Later we give to these self-energy diagrams an equivalent but more appealing physical interpretation in terms of polarization effects. Before, however, it is instructive to work out the explicit form of the second-Born self-energy and calculate the rate operator [68].

Using the Feynman rules of Section 10.8 the self-energy (13.15) for particles of spin S reads

$$\Sigma_c(1; \bar{1}) = \pm \, i^2 \int d2 d\bar{2} \; v(1; 2) v(\bar{1}; \bar{2}) \left[G(1; \bar{1}) G(2; \bar{2}) G(\bar{2}; 2) \pm G(1; \bar{2}) G(\bar{2}; 2) G(2; \bar{1}) \right]$$

$$= \pm \, i^2 \delta_{\sigma_1 \bar{\sigma}_1} \int d\mathbf{r}_2 d\bar{\mathbf{r}}_2 \; v(\mathbf{r}_1, \mathbf{r}_2) v(\bar{\mathbf{r}}_1, \bar{\mathbf{r}}_2)$$

$$\times \left[(2S + 1) G(\mathbf{r}_1, z_1; \bar{\mathbf{r}}_1, \bar{z}_1) G(\mathbf{r}_2, z_1; \bar{\mathbf{r}}_2, \bar{z}_1) G(\bar{\mathbf{r}}_2, \bar{z}_1; \mathbf{r}_2, z_1) \right.$$

$$\left. \pm \, G(\mathbf{r}_1, z_1; \bar{\mathbf{r}}_2, \bar{z}_1) G(\bar{\mathbf{r}}_2, \bar{z}_1; \mathbf{r}_2, z_1) G(\mathbf{r}_2, z_1; \bar{\mathbf{r}}_1, \bar{z}_1) \right], \tag{13.16}$$

where in the last equality we have integrated over times and summed over spin assuming that $v(1; 2) = \delta(z_1, z_2) v(\mathbf{r}_1, \mathbf{r}_2)$ is spin-independent and that $G(1; 2) = \delta_{\sigma_1 \sigma_2} G(\mathbf{r}_1, z_1; \mathbf{r}_2, z_2)$ is spin-diagonal. The interpretation of (13.16) is particularly transparent in a system invariant under translations, like the electron gas. In this case all quantities depend only on the relative coordinate and therefore can be Fourier transformed according to

$$G(1; 2) = \delta_{\sigma_1 \sigma_2} \int \frac{d\mathbf{p}}{(2\pi)^3} \, e^{i\mathbf{p} \cdot (\mathbf{r}_1 - \mathbf{r}_2)} G(\mathbf{p}; z_1, z_2), \tag{13.17}$$

$$\Sigma_c(1; 2) = \delta_{\sigma_1 \sigma_2} \int \frac{d\mathbf{p}}{(2\pi)^3} \, e^{i\mathbf{p} \cdot (\mathbf{r}_1 - \mathbf{r}_2)} \Sigma_c(\mathbf{p}; z_1, z_2). \tag{13.18}$$

It is a matter of simple algebra to show that in Fourier space (13.16) reads

$$\Sigma_c(\mathbf{p}; z_1, \bar{z}_1) = \pm i^2 \int \frac{d\mathbf{p}'}{(2\pi)^3} \frac{d\bar{\mathbf{p}}}{(2\pi)^3} \frac{d\bar{\mathbf{p}}'}{(2\pi)^3} \, (2\pi)^3 \delta(\mathbf{p} + \mathbf{p}' - \bar{\mathbf{p}} - \bar{\mathbf{p}}') B(\mathbf{p}, \bar{\mathbf{p}}, \bar{\mathbf{p}}')$$

$$\times G(\mathbf{p}'; \bar{z}_1, z_1) G(\bar{\mathbf{p}}; z_1, \bar{z}_1) G(\bar{\mathbf{p}}'; z_1, \bar{z}_1), \tag{13.19}$$

with B a quantity proportional to the differential cross-section for the scattering $\mathbf{p}, \mathbf{p}' \to \bar{\mathbf{p}}, \bar{\mathbf{p}}'$ in the Born approximation (see Appendix M):

$$B(\mathbf{p}, \bar{\mathbf{p}}, \bar{\mathbf{p}}') = B(\mathbf{p}, \bar{\mathbf{p}}', \bar{\mathbf{p}}) = \frac{2S + 1}{2} \left[\tilde{v}_{\mathbf{p} - \bar{\mathbf{p}}}^2 + \tilde{v}_{\mathbf{p} - \bar{\mathbf{p}}'}^2 \right] \pm \tilde{v}_{\mathbf{p} - \bar{\mathbf{p}}} \tilde{v}_{\mathbf{p} - \bar{\mathbf{p}}'}. \tag{13.20}$$

As usual, in this formula $\tilde{v}_{\mathbf{p}}$ is the Fourier transform of the interaction. From (13.19) we can easily extract the greater and lesser components of Σ_c. Going to frequency space we get

$$\Sigma_c^{\gtrless}(\mathbf{p}, \omega) = \pm i^2 \int \frac{d\mathbf{p}' d\omega'}{(2\pi)^4} \frac{d\bar{\mathbf{p}} d\bar{\omega}}{(2\pi)^4} \frac{d\bar{\mathbf{p}}' d\bar{\omega}'}{(2\pi)^4} \, (2\pi)^4 \delta(\mathbf{p} + \mathbf{p}' - \bar{\mathbf{p}} - \bar{\mathbf{p}}') \delta(\omega + \omega' - \bar{\omega} - \bar{\omega}')$$

$$\times B(\mathbf{p}, \bar{\mathbf{p}}, \bar{\mathbf{p}}') G^{\lessgtr}(\mathbf{p}', \omega') G^{\gtrless}(\bar{\mathbf{p}}, \bar{\omega}) G^{\gtrless}(\bar{\mathbf{p}}', \bar{\omega}'). \tag{13.21}$$

This formula has a transparent physical interpretation in terms of the scattering processes illustrated in (13.14). In Section 13.1 we learned that the rate $\Gamma(\mathbf{p}, \omega) = i[\Sigma_c^>(\mathbf{p}, \omega) - \Sigma_c^<(\mathbf{p}, \omega)]$ is a measure of the lifetime of quasi-particles with momentum $\mathbf{p}$ and energy ω. We could interpret $\Sigma_c^>$ as the decay rate of an added particle and $\Sigma_c^<$ as the decay rate of a removed particle (hole). In fact $\Sigma_c^>$ describes a process in which a particle with momentum-energy $\mathbf{p}, \omega$ hits a particle with momentum-energy $\mathbf{p}', \omega'$ and after the scattering one particle goes into the state $\bar{\mathbf{p}}, \bar{\omega}$ and the other goes into the state $\bar{\mathbf{p}}', \bar{\omega}'$. The two δ-functions guarantee the conservation of momentum and energy. In Appendix M we show that the differential cross-section for this process is proportional to B to lowest order in the interaction, i.e., in the Born approximation. It is also intuitive to understand the appearance of the product of three Gs. The probability of scattering off a particle with momentum-energy $\mathbf{p}', \omega'$ is given by the density of particles with this momentum-energy, i.e., $f(\omega' - \mu)A(\mathbf{p}', \omega') = \pm iG^<(\mathbf{p}', \omega')$. The probability that after the scattering the particles end up in the states $\bar{\mathbf{p}}, \bar{\omega}$ and $\bar{\mathbf{p}}', \bar{\omega}'$ is given by the density of holes (available states) $\bar{f}(\bar{\omega} - \mu)A(\bar{\mathbf{p}}, \bar{\omega}) = iG^>(\bar{\mathbf{p}}, \bar{\omega})$ and $\bar{f}(\bar{\omega}' - \mu)A(\bar{\mathbf{p}}', \bar{\omega}') = iG^>(\bar{\mathbf{p}}', \bar{\omega}')$. In a similar way we can discuss $\Sigma_c^<$. In the low-density limit $\beta\mu \to -\infty$ and from (9.54) we see that the rate $\Sigma_c^<$ to scatter *into* the state $\mathbf{p}, \omega$ is negligible compared to the rate $\Sigma_c^>$ to scatter *out* of the same state. On the contrary, at low temperatures the two rates are equally important.

Let us estimate $\Sigma_c^{\lessgtr}(\mathbf{p}, \omega)$ at zero temperature for frequencies $\omega \sim \mu$ in a system of fermions. We assume that the integral over all momenta of the product of $\delta(\mathbf{p} + \mathbf{p}' - \bar{\mathbf{p}} - \bar{\mathbf{p}}')$, the differential cross-section B and the three spectral functions $A(\mathbf{p}', \omega')$, $A(\bar{\mathbf{p}}, \bar{\omega})$, $A(\bar{\mathbf{p}}', \bar{\omega}')$ is a smooth function $F_{\mathbf{p}}(\omega', \bar{\omega}, \bar{\omega}')$ of the frequencies ω', $\bar{\omega}$, and $\bar{\omega}'$. This can rigorously be proven when $A(\mathbf{p}, \omega) = 2\pi\delta(\omega - p^2/2)$ is the noninteracting spectral function and the number of spatial dimensions is larger than one. At zero temperature the Fermi function $f(\omega - \mu) = \theta(\mu - \omega)$ and $\bar{f}(\omega - \mu) = 1 - \theta(\mu - \omega) = \theta(\omega - \mu)$ and therefore from (13.21) we have

$$\Sigma_c^>(\mathbf{p}, \omega) = \int \frac{d\omega'}{2\pi} \frac{d\bar{\omega}}{2\pi} \frac{d\bar{\omega}'}{2\pi} F_{\mathbf{p}}(\omega', \bar{\omega}, \bar{\omega}')\theta(\mu - \omega')\theta(\bar{\omega} - \mu)\theta(\bar{\omega}' - \mu)\delta(\omega + \omega' - \bar{\omega} - \bar{\omega}').$$
(13.22)

Because of the θs the frequency $\omega' < \mu$ and $\bar{\omega} > \mu$, $\bar{\omega}' > \mu$. Since $\bar{\omega} + \bar{\omega}' > \omega' + \mu$ the frequency ω must be larger than μ for otherwise the argument of the δ-function cannot vanish. This agrees with the fluctuation–dissipation theorem for the self-energy according to which $\Sigma_c^>(\mathbf{p}, \omega) \propto 1 - \theta(\mu - \omega) = \theta(\omega - \mu)$. For $\omega > \mu$ but very close to it the argument of the δ-function vanishes only in a region where ω', $\bar{\omega}$, and $\bar{\omega}'$ are very close to μ. Denoting by $u = \omega - \mu > 0$ the distance of the frequency ω from the chemical potential we can rewrite (13.22) as

$$\Sigma_c^>(\mathbf{p}, \omega) = \int_{\mu-u}^{\mu} \frac{d\omega'}{2\pi} \int_{\mu}^{\mu+u} \frac{d\bar{\omega}}{2\pi} \int_{\mu}^{\mu+u} \frac{d\bar{\omega}'}{2\pi} F_{\mathbf{p}}(\omega', \bar{\omega}, \bar{\omega}')\delta(\omega + \omega' - \bar{\omega} - \bar{\omega}')$$

$$\sim \frac{F_{\mathbf{p}}(\mu, \mu, \mu)}{u} \int_{-u}^{0} \frac{du'}{2\pi} \int_{0}^{u} \frac{d\bar{u}}{2\pi} \int_{0}^{u} \frac{d\bar{u}'}{2\pi} \delta(1 + \frac{u' - \bar{u} - \bar{u}'}{u}),$$
(13.23)

where in the last equality we approximate F with its value at the three frequencies equal to μ since F is smooth. Changing the integration variables $u'/u = x$, $\bar{u}/u = \bar{x}$, and $\bar{u}'/u = \bar{x}'$

we get a factor u^3 from $du'd\bar{u}d\bar{u}'$ and an integral which is independent of u. Thus we conclude that for ω very close to μ

$$\Sigma_c^>(\mathbf{p}, \omega) = -\mathrm{i}\, C_\mathbf{p}\, \theta(\omega - \mu)\, (\omega - \mu)^2,$$

where $C_\mathbf{p}$ is a real constant. In a similar way we can derive that $\Sigma_c^<(\mathbf{p}, \omega) = \mathrm{i}\, C_\mathbf{p}\, \theta(\mu - \omega)\, (\omega - \mu)^2$. Therefore the rate operator

$$\Gamma(\mathbf{p}, \omega) = C_\mathbf{p}\, (\omega - \mu)^2 \qquad (13.24)$$

vanishes quadratically as $\omega \to \mu$ *for all* $\mathbf{p}$. This result is due to Luttinger [101] who also showed that self-energy diagrams with three or more interaction lines give contributions to the rates $\Sigma_c^{\lessgtr}$ that vanish even faster as $\omega \to \mu$. In particular these contributions vanish as $(\omega - \mu)^{2m}$ where m are integers larger than 1. Consequently (13.24) provides the *exact* leading term of the Taylor expansion of $\Gamma(\mathbf{p}, \omega)$ in powers of $(\omega - \mu)$. The constant of proportionality

$$C_\mathbf{p} = \lim_{\omega \to \mu} \frac{\Gamma(\mathbf{p}, \omega)}{(\omega - \mu)^2} > 0,$$

since in a fermionic system Γ is positive. From this result we can understand why the low-energy excitations of metals can be described in an effective single-particle picture. Although the Coulomb interaction is strong it is very ineffective in scattering particles close to the Fermi surface due to phase space restrictions. This explains many of the low-energy excitations of the electron gas. Needless to say, this is not true anymore at higher energies. Moreover the calculation of the details of the properties of the low energy spectrum (such as life-times) is still very much a many-body problem. In Section 15.5.4 we calculate $\Sigma_c^{\lessgtr}(\mathbf{p}, \omega)$ explicitly as well as the rate function $\Gamma(\mathbf{p}, \omega)$, the constant $C_\mathbf{p}$, the life-time $\tau_\mathbf{p}$, etc. within the GW approximation.

In the same paper Luttinger observed (in a seminal footnote) that the above phase-space argument needs to be modified in one dimension. We can understand his observation by using the noninteracting spectral function $A(\mathbf{p}, \omega) = 2\pi\delta(\omega - p^2/2)$. In this case the δ-function of momentum conservation appearing in (13.21) is sufficient to determine the energy, say, $\bar{\omega}' = \bar{p}'^2/2$ in terms of $\omega' = p'^2/2$ and $\bar{\omega} = \bar{p}^2/2$, and hence we can no longer treat $\bar{\omega}'$ as an independent variable. In other words the function $F_\mathbf{p}(\omega', \bar{\omega}, \bar{\omega}')$ is no longer smooth but rather it is proportional to a δ-function. As a consequence (13.23) contains one integral less and $\Gamma(\mathbf{p}, \omega) \propto |\omega - \mu|$. This fact has profound consequences on the nature of the interacting gas. For instance, the sharpness of the Fermi surface (discontinuity in the momentum distribution) requires that $\Gamma(\mathbf{p}, \omega)$ vanishes faster than $|\omega - \mu|$ as $\omega \to \mu$, see Appendix K. In fact, in one dimension there is no Fermi surface and the interacting electron gas is said to be a *Luttinger liquid*. The interested reader should consult Ref. [102] for a detailed discussion on the properties of Luttinger liquids.

13.3.1 Polarization effects

In this section we discuss an equivalent physical interpretation of the second-Born self-energy. The first diagram in (13.15) suggests that a particle in $\bar{1}$ hits a particle in $\bar{2}$ which, as a consequence, moves away from its original position leaving a hole behind it. A particle-hole pair is formed in $\bar{2}$ and propagates until 2 at which space–spin–time point it recombines and disappears. We can say that the particle in $\bar{1}$ polarizes the medium by pushing other particles away from it, i.e., by creating particle–hole pairs around it. The name "polarization diagram" commonly used for the bubble diagram in (13.15) is related to this effect. In fact, the bubble can be seen as the propagator of a particle–hole pair. The inclusion of polarization effects in Σ is, in general, relevant in systems with low-energy particle–hole excitations like open-shell molecules or metals. Below we show the importance of including the second-Born polarization diagram in the calculation of the spectral function of the system discussed in Section 6.3.4.

We rewrite the Hamiltonian (6.97) in a form suitable for MBPT:

$$\hat{H} = \underbrace{T\sum_{j=1}^{N-1}(\hat{d}_j^\dagger \hat{d}_{j+1} + \hat{d}_{j+1}^\dagger \hat{d}_j) - U\hat{N}_r}_{\hat{H}'_{\mathrm{met}}} + \underbrace{\sum_{\alpha=H,L}(\epsilon_\alpha - U\bar{N}_r)\hat{c}_\alpha^\dagger \hat{c}_\alpha}_{\hat{H}'_{\mathrm{mol}}} + U\sum_{\substack{\alpha=H,L\\ j\leq r}}\hat{c}_\alpha^\dagger \hat{d}_j^\dagger \hat{d}_j \hat{c}_\alpha.$$

(13.25)

The constant part $U\bar{N}_r$ has been neglected since the Green's function is not affected by a constant shift of all eigenvalues of $\hat{H}$. The goal is to calculate $\Sigma \to G \to A$ in different approximations and to assess the quality of these approximations by benchmarking the results against the exact spectral function calculated in Section 6.3.3.

The first important observation is that the Green's function (no matter in which approximation) is block diagonal, i.e., $G_{\alpha j} = G_{j\alpha} = 0$ and $G_{\alpha\alpha'} = 0$ if $\alpha \neq \alpha'$, since the occupation operator of the HOMO and LUMO states commutes with the Hamiltonian (the number of electrons in these states is a conserved quantity). From the Dyson equation we then infer that also the self-energy is block diagonal. The second observation pertains to the specific form of the interaction. Except for the Hartree–Fock diagrams, all other diagrams for the metallic self-energy $\Sigma_{jj'}$ have the form

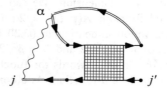

where $\alpha = H, L$ and the square grid is the reducible kernel K_r, see (12.39) and (12.40). If an electron is in α then we can only destroy it, while if it is not in α then we can only create it. In other words we cannot generate a particle–hole excitation in α because $\hat{n}_\alpha = \hat{c}_\alpha^\dagger \hat{c}_\alpha$ is conserved. This implies that all diagrams for $\Sigma_{jj'}$ beyond the Hartree–Fock approximation vanish. It is easy to realize that the Fock diagram also vanishes since there is no interaction line connecting j to j'. As the only surviving diagram is the Hartree diagram we can write (for electrons the charge $q = -1$) $\Sigma_{jj'}(z, z') = -\delta_{jj'}\delta(z, z')V_{\mathrm{H},j}(z)$ with $V_{\mathrm{H},j}(z) = 0$ for $j > r$ and

$$V_{\mathrm{H},j}(z) = iU\sum_\alpha G_{\alpha\alpha}(z, z^+) = -U,$$

for $j \leq r$. In the last equality we use the fact that the density $-iG_{\alpha\alpha}(z, z^+)$ on level α is unity for the HOMO and zero for the LUMO, independently of the approximation. We conclude that $G_{jj'}$ is the Hartree Green's function with Hartree Hamiltonian

$$\hat{h}'_{\text{met,H}} = T \sum_{j=1}^{N-1} \left(|j\rangle\langle j+1| + |j+1\rangle\langle j| \right) - \sum_{j \leq r} (U + V_{\text{H},j}) |j\rangle\langle j|.$$

$$= \hat{h}_{\text{met}}.$$

This Hamiltonian can easily be diagonalized to construct $G_{jj'}$ using the formulas of Section 6.2.

An alternative way to understand that the metallic Green's function coincides with that of the isolated noninteracting metal is through the Lehmann representation of Section 6.3.3. The Green's function is completely determined by the knowledge of the eigenstates and eigenvalues of the system with one particle more and with one particle less. Since the number of particles in the molecule is conserved the only eigenstates contributing to $G_{jj'}$ are those with one particle more or less in the metal. These eigenstates are independent of U since $N_{\text{mol}} = 1$ and hence the interaction Hamiltonian (6.96) is zero in this eigenspace.

Next we consider the molecular Green's function. We start with $G_{\alpha\alpha}$ in the Hartree-Fock approximation. Also in this case the Fock self-energy diagram vanishes since there is no interaction line connecting α to itself. The Hartree self-energy reads $\Sigma_{\alpha\alpha'}(z, z') = -\delta_{\alpha\alpha'}\delta(z, z')V_{\text{H},\alpha}(z)$ with

$$V_{\text{H},\alpha}(z) = iU \sum_{j \leq r} G_{jj}(z, z^+) = -U\bar{N}_r.$$

This Hartree potential cancels exactly the shift $U\bar{N}_r$ in (13.25) and the spectral function turns out to be independent of the interaction

$$A_{\alpha\alpha}(\omega) = 2\pi\delta(\omega - \epsilon_\alpha). \tag{13.26}$$

In Fig. 13.1 we compare the Hartree-Fock spectral function of the molecule $A_{\text{mol}} = A_{HH} + A_{LL}$ with the exact spectral function for a chain of $N = 15$ sites with $M = 5$ electrons, $\epsilon_L - \epsilon_H = 2$ and different values of U (all energies are measured in units of $|T|$). The exact HOMO–LUMO gap reduces as U increases due to the image-charge effect discussed in Section 6.3.3. The Hartree-Fock approximation is not able to capture this effect since the HOMO electron couples directly to the metallic density. The formation of an image charge is caused by the excitation of particle–hole pairs and these excitations are simply not accounted for in a mean field approximation [60–62].

To include polarization effects to some degree we evaluate the Green's function in the second-Born approximation. Let us start by calculating the self-energy using the Hartree-Fock Green's function (non-self-consistent approximation). As we shall see this is already enough to observe the tendency of the HOMO (LUMO) peak to move rightward (leftward). The exchange diagram in (13.15) vanishes since G is block diagonal and the interaction connects only sites in different blocks (this remains true for all approximations to G). Thus

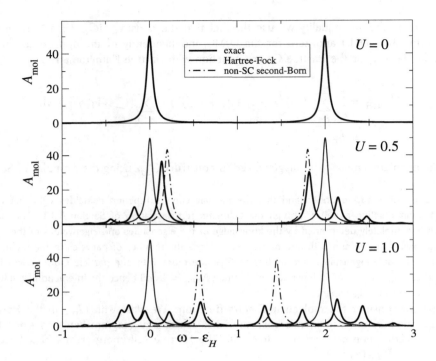

Figure 13.1 Spectral function of the molecule in the Hartree–Fock and in the non-self-consistent (non-SC) second-Born approximation compared with the exact spectral function as obtained from (6.101) and (6.102). The chain has $N = 15$ sites and $M = 5$ electrons ($M/N = 3$), the bare HOMO–LUMO gap is $\epsilon_L - \epsilon_H = 2$ and the infinitesimally small positive constant $\eta = 0.04$. All energies are in units of $|T|$.

we need only to calculate the bubble diagram. From (13.16) we have

$$\Sigma^{>}_{c,\alpha\alpha}(\omega) = -i^2 U^2 \sum_{jj'\leq r} \int \frac{d\omega_1}{2\pi} \frac{d\omega_2}{2\pi} \, G^{>}_{\alpha\alpha}(\omega - \omega_1 + \omega_2) G^{>}_{jj'}(\omega_1) G^{<}_{j'j}(\omega_2), \quad (13.27)$$

$$\Sigma^{<}_{c,\alpha\alpha}(\omega) = -i^2 U^2 \sum_{jj'\leq r} \int \frac{d\omega_1}{2\pi} \frac{d\omega_2}{2\pi} \, G^{<}_{\alpha\alpha}(\omega - \omega_1 + \omega_2) G^{<}_{jj'}(\omega_1) G^{>}_{j'j}(\omega_2). \quad (13.28)$$

The greater/lesser Hartree–Fock G of the molecule can be obtained using the fluctuation-dissipation theorem (6.94) with spectral function (13.26). The Fermi function $f(\omega - \mu)$ is zero for $\omega = \epsilon_L$ and one for $\omega = \epsilon_H$ and hence

$$G^{>}_{HH}(\omega) = 0, \qquad G^{<}_{HH}(\omega) = 2\pi i \delta(\omega - \epsilon_H),$$

$$G^{<}_{LL}(\omega) = 0, \qquad G^{>}_{LL}(\omega) = -2\pi i \delta(\omega - \epsilon_L),$$

from which it follows that $\Sigma^{>}_{c,HH} = \Sigma^{<}_{c,LL} = 0$. The greater/lesser G of the metal has the form (6.48) and (6.49). Using the eigenstates $|\lambda\rangle$ of $\hat{h}_{\mathrm{met}}$ with eigenvalues ϵ_λ and taking into account that we are at zero temperature

$$G^<_{jj'}(\omega) = 2\pi i \sum_\lambda^{\text{occ}} \langle j|\lambda\rangle\langle\lambda|j'\rangle \, \delta(\omega - \epsilon_\lambda), \quad G^>_{jj'}(\omega) = -2\pi i \sum_\lambda^{\text{unocc}} \langle j|\lambda\rangle\langle\lambda|j'\rangle \, \delta(\omega - \epsilon_\lambda),$$

where the sum in $G^<$ runs over the M occupied eigenstates (M = number of electrons in the metal) and the sum in $G^>$ runs over the $N - M$ unoccupied states. Inserting these Green's functions into the self-energies we find

$$\Sigma^<_{c,HH}(\omega) = 2\pi i U^2 \sum_{jj'\leq r} \sum_\lambda^{\text{occ}} \sum_\mu^{\text{unocc}} \langle j|\mu\rangle\langle\mu|j'\rangle\langle j'|\lambda\rangle\langle\lambda|j\rangle \, \delta(\omega + \epsilon_\mu - \epsilon_\lambda - \epsilon_H),$$

$$\Sigma^>_{c,LL}(\omega) = -2\pi i U^2 \sum_{jj'\leq r} \sum_\lambda^{\text{occ}} \sum_\mu^{\text{unocc}} \langle j|\lambda\rangle\langle\lambda|j'\rangle\langle j'|\mu\rangle\langle\mu|j\rangle \, \delta(\omega + \epsilon_\lambda - \epsilon_\mu - \epsilon_L).$$

Having the correlation self-energy we can construct the rates

$$\Gamma_{HH} = i[\Sigma^>_{c,HH} - \Sigma^<_{c,HH}] = -i\Sigma^<_{c,HH} \, ,$$

$$\Gamma_{LL} = i[\Sigma^>_{c,LL} - \Sigma^<_{c,LL}] = i\Sigma^>_{c,LL} \, ,$$

from which to extract $\Sigma^{R/A}_c$ using (9.57)

$$\Sigma^{R/A}_{c,HH}(\omega) = \int \frac{d\omega'}{2\pi} \frac{\Gamma_{HH}(\omega')}{\omega - \omega' \pm i\eta} = U^2 \sum_{jj'\leq r} \sum_\lambda^{\text{occ}} \sum_\mu^{\text{unocc}} \frac{\langle j|\mu\rangle\langle\mu|j'\rangle\langle j'|\lambda\rangle\langle\lambda|j\rangle}{\omega + \epsilon_\mu - \epsilon_\lambda - \epsilon_H \pm i\eta},$$

$$\Sigma^{R/A}_{c,LL}(\omega) = \int \frac{d\omega'}{2\pi} \frac{\Gamma_{LL}(\omega')}{\omega - \omega' \pm i\eta} = U^2 \sum_{jj'\leq r} \sum_\lambda^{\text{occ}} \sum_\mu^{\text{unocc}} \frac{\langle j|\lambda\rangle\langle\lambda|j'\rangle\langle j'|\mu\rangle\langle\mu|j\rangle}{\omega + \epsilon_\lambda - \epsilon_\mu - \epsilon_L \pm i\eta}.$$

We now have all ingredients to calculate the spectral function in the non-self-consistent second-Born approximation, see (13.3),

$$A_{\alpha\alpha}(\omega) = \frac{\Gamma_{\alpha\alpha}(\omega)}{|\omega - \epsilon_\alpha - \Sigma^R_{\alpha\alpha}(\omega)|^2}.$$

In Fig. 13.1 we show this spectral function with a dot-dashed curve. As anticipated, the effect of the polarization diagram is to close the gap between the HOMO and the LUMO levels.

For a self-consistent treatment of the second-Born approximation we should use the above Σ to calculate a new G, plug this G into (13.27) and (13.28) to get a new Σ and so on and so forth until convergence. At every iteration the number of δ-peaks in the spectral function increases and at self-consistency we end up with a smooth function of ω.[5] Thus the *self-consistent* spectral function has broadened peaks while the *exact* spectral function has δ-peaks. This is a feature of every approximate self-consistent Σ and is due to a lack of cancellations [103]. As discussed in Chapter 10 an approximate skeletonic self-energy $\Sigma[G]$

---

[5]The curves in Fig. 13.1 have been *artificially* broadened using a finite η.

contains an infinite subset of diagrams and to high order in the interaction these diagrams contain a large number of particle–hole excitations. The finite width of the peaks in an approximate self-consistent spectral function is caused by the infinite number of particle–hole excitations with which the added/removed particle is coupled, see Section 6.1.1. In finite systems, however, the number of particle–hole excitations is finite since it is not possible to excite more electrons than those present in the ground state. To cancel the nonphysical excitations of an approximate treatment one has to include diagrams not present in the original subset of diagrams. The exact result is recovered only when all diagrams are included. In accordance with our discussion in Section 13.1, the spurious broadening of a self-consistent approximation is also responsible for a spurious damping in the time domain [103].[6] These facts are conceptually important since they elucidate some aspects of self-consistency in Green's function theory. It should be stressed, however, that there is not such a self-consistency problem in bulk systems (where the number of particle–hole excitations is infinite) and that the spurious broadening/damping is appreciable only in very small and strongly interacting systems.

Exercise 13.3. Show that the exact self-energy $\Sigma_{\alpha\alpha}$ with $\alpha = H, L$ is the sum of diagrams containing one single metallic bubble connected in all possible ways to a $G_{\alpha\alpha}$-line by an arbitrary number of interaction lines.

13.4   Ground-state energy and correlation energy

Let us consider a system of identical and interacting particles in equilibrium at a given temperature and chemical potential. In this section we derive a formula for the ground-state energy which involves either the couple Σ, G or the couple P, W. The main advantage of these formulas over the Galitskii–Migdal formula derived in Section 6.4 lies in the possibility of separating the total energy into a noncorrelated and correlated part, thus highlighting the dependence on the many-body approximation. As we shall see, these formulas are similar to the dressed expansion (11.26) of the grand potential.

Let $\hat{\rho}$ be the equilibrium density matrix of $\hat{H} = \hat{H}_0 + \hat{H}_{\text{int}}$. Then the energy of the system is

$$E = \text{Tr}\left[\hat{\rho}\hat{H}_0\right] + \text{Tr}\left[\hat{\rho}\hat{H}_{\text{int}}\right] = E_{\text{one}} + E_{\text{int}}. \qquad (13.29)$$

The one-body part of the total energy, E_{one}, can easily be expressed in terms of $G^<$; the result is the first term on the r.h.s. of (8.17). Since the system is in equilibrium, the single particle Hamiltonian $\hat{h}_{\text{S}}(t) = \hat{h}$ is time-independent and $G^<$ depends only on the time difference. Then, Fourier transforming $G^<$ we find

$$\boxed{E_{\text{one}} = \pm i \int \frac{d\omega}{2\pi} \int d\mathbf{x} \, \langle \mathbf{x}|\hat{h}\,\hat{\mathcal{G}}^<(\omega)|\mathbf{x}\rangle} \qquad (13.30)$$

---

[6]In finite systems the sudden switch-on of a constant perturbation causes an *undamped* oscillatory behavior.

The interaction energy is given in terms of Σ and G in (9.19). It is convenient to separate the singular Hartree–Fock self-energy from the correlation self-energy. We write

$$\hat{\Sigma}(z_1, z_2) = \delta(z_1, z_2) q \hat{V}_{\mathrm{HF}} + \hat{\Sigma}_{\mathrm{c}}(z_1, z_2),$$
(13.31)

where the Hartree–Fock potential operator has matrix elements [see (7.39)]

$$\langle \mathbf{x}_1 | q \hat{V}_{\mathrm{HF}} | \mathbf{x}_2 \rangle = \delta(\mathbf{x}_1 - \mathbf{x}_2) \int d\mathbf{x}\, v(\mathbf{x}_1, \mathbf{x}) n(\mathbf{x}) \pm v(\mathbf{x}_1, \mathbf{x}_2) n(\mathbf{x}_1, \mathbf{x}_2),$$

with $n(\mathbf{x}_1, \mathbf{x}_2)$ the one-particle density matrix and $n(\mathbf{x}) = n(\mathbf{x}, \mathbf{x})$ the density. Inserting (13.31) into (9.19) we get

$$E_{\mathrm{int}} = E_{\mathrm{int,HF}} + E_{\mathrm{int,c}}.$$

In this equation the Hartree–Fock part of the interaction energy is given by

$$\boxed{E_{\mathrm{int,HF}} = \underbrace{\frac{1}{2} \int d\mathbf{x} d\mathbf{x}'\, v(\mathbf{x}, \mathbf{x}') n(\mathbf{x}) n(\mathbf{x}')}_{E_{\mathrm{int,H}}} \pm \underbrace{\frac{1}{2} \int d\mathbf{x} d\mathbf{x}'\, v(\mathbf{x}, \mathbf{x}') n(\mathbf{x}, \mathbf{x}') n(\mathbf{x}', \mathbf{x})}_{E_{\mathrm{int,x}}}}$$
(13.32)

which is the sum of a Hartree term $E_{\mathrm{int,H}}$ and a Fock (or exchange) term $E_{\mathrm{int,x}}$. Instead the correlation part reads

$$E_{\mathrm{int,c}} = \pm \frac{\mathrm{i}}{2} \int d\mathbf{x}_1 \langle \mathbf{x}_1 | \left[\hat{\Sigma}_{\mathrm{c}}^{\rceil} \star \hat{\mathcal{G}}^{\lceil} \right] (t_0, t_0) | \mathbf{x}_1 \rangle,$$
(13.33)

as follows directly from (9.19) with, e.g., $z_1 = t_{0-}$ ($E_{\mathrm{int,c}}$ is independent of time and hence we can choose any z_1).[7] Equation (13.33) is particularly interesting. If we set $t_1 = t_2 = t_0$ in (9.60) then the first two terms on the l.h.s. vanish while the third term is exactly what appears in (13.33). Therefore, we can use (9.60) to rewrite the correlation part of the interaction energy as

$$E_{\mathrm{int,c}} = \pm \frac{\mathrm{i}}{2} \int \frac{d\omega}{2\pi} \int d\mathbf{x}_1 \langle \mathbf{x}_1 | \hat{\Sigma}_{\mathrm{c}}^{\lessgtr}(\omega) \hat{\mathcal{G}}^{\mathrm{A}}(\omega) + \hat{\Sigma}_{\mathrm{c}}^{\mathrm{R}}(\omega) \hat{\mathcal{G}}^{\lessgtr}(\omega) | \mathbf{x}_1 \rangle.$$
(13.34)

In this formula we can choose either the greater or the lesser components of $\hat{\Sigma}_{\mathrm{c}}$ and $\hat{\mathcal{G}}$; the result is independent of the choice. To show it we use the relations

$$\hat{\mathcal{G}}^{\mathrm{R/A}}(\omega) = \mathrm{i} \int \frac{d\omega'}{2\pi} \frac{\hat{\mathcal{G}}^{>}(\omega') - \hat{\mathcal{G}}^{<}(\omega')}{\omega - \omega' \pm \mathrm{i}\eta}, \qquad \hat{\Sigma}_{\mathrm{c}}^{\mathrm{R/A}}(\omega) = \mathrm{i} \int \frac{d\omega'}{2\pi} \frac{\hat{\Sigma}_{\mathrm{c}}^{>}(\omega') - \hat{\Sigma}_{\mathrm{c}}^{<}(\omega')}{\omega - \omega' \pm \mathrm{i}\eta},$$

already derived in (6.93) and (9.57). Independently of the Keldysh component, the insertion of the above relations in (13.34) yields the following unique result

$$\boxed{E_{\mathrm{int,c}} = \mp \frac{1}{2} \int \frac{d\omega\, d\omega'}{2\pi\, 2\pi} \int d\mathbf{x}_1 \frac{\langle \mathbf{x}_1 | \hat{\Sigma}_{\mathrm{c}}^{<}(\omega) \hat{\mathcal{G}}^{>}(\omega') - \hat{\Sigma}_{\mathrm{c}}^{>}(\omega) \hat{\mathcal{G}}^{<}(\omega') | \mathbf{x}_1 \rangle}{\omega - \omega'}}$$
(13.35)

---

[7]For $z_1 = t_{0-}$ the integral over z_3 along the forward branch cancels the integral along the backward branch and only the vertical track remains.

In this formula we have removed the infinitesimal imaginary part in the denominator since for $\omega = \omega'$ the numerator vanishes.[8]

An alternative formula for the interaction energy involves the polarizability P and the screened interaction W. The starting point is the identity (11.52) which we write again below

$$\boxed{\text{tr}_\gamma\left[\Sigma_{\text{xc}}G\right] = \pm\text{tr}_\gamma\left[PW\right]} \tag{13.36}$$

Separating out the Fock (exchange) part

$$\Sigma_{\text{x}}(1;2) = iv(1;2)G(1;2^+) = \pm\delta(z_1,z_2)v(\mathbf{x}_1,\mathbf{x}_2)n(\mathbf{x}_1,\mathbf{x}_2),$$

from Σ_{xc} we can rewrite (13.36) as

$$\text{tr}_\gamma\left[\Sigma_{\text{c}}G\right] \mp 2\beta\, E_{\text{int,x}} = \pm\text{tr}_\gamma\left[PW\right], \tag{13.37}$$

where β is the inverse temperature[9] and $E_{\text{int,x}}$ is the exchange part of the Hartree–Fock interaction energy (13.32). Furthermore, since the system is in equilibrium

$$\text{tr}_\gamma\left[\Sigma_{\text{c}}G\right] = -i\beta\int d\mathbf{x}_1\langle\mathbf{x}_1|\left[\hat{\Sigma}_{\text{c}}^\rceil \star \hat{\mathcal{G}}^\lceil\right](t_0,t_0)|\mathbf{x}_1\rangle$$
$$= \mp 2\beta E_{\text{int,c}}. \tag{13.38}$$

Similarly we can evaluate the r.h.s. of (13.37). We separate out the singular part v from W by defining $W = v + \delta W$ so that

$$\text{tr}_\gamma\left[PW\right] = -i\beta\int d\mathbf{x}_1\langle\mathbf{x}_1|\hat{P}(z,z)\hat{v}|\mathbf{x}_1\rangle + \text{tr}_\gamma\left[P\delta W\right].$$

The first term on the r.h.s. requires the calculation of the polarizability at equal times. We recall that the many-body expansion of $P(1;2)$ starts with $\pm iG(1;2)G(2;1)$. For $z_1 = z_2$ this is the only ambiguous term of the expansion. The correct way of removing the ambiguity consists in using $\pm iG(1;2^+)G(2;1^+)$, in accordance with the discussion below (11.53). If we write

$$P(1;2) = \pm iG(1;2^+)G(2;1^+) + \delta P(1;2),$$

then

$$-i\beta\int d\mathbf{x}_1\langle\mathbf{x}_1|\hat{P}(z,z)\hat{v}|\mathbf{x}_1\rangle = -2\beta\, E_{\text{int,x}} - i\beta\int\frac{d\omega}{2\pi}\int d\mathbf{x}_1\langle\mathbf{x}_1|\delta\hat{P}^{\lessgtr}(\omega)\hat{v}|\mathbf{x}_1\rangle.$$

As expected, the zeroth order polarizability generates the exchange energy. The second term is independent of whether we choose the greater or lesser component since $v(\mathbf{x}_1,\mathbf{x}_2)$ is

---

[8]From the fluctuation–dissipation theorem for Σ and G the numerator is proportional to $[f(\omega)\bar{f}(\omega') - \bar{f}(\omega)f(\omega')]\hat{\Gamma}(\omega)\hat{A}(\omega')$ which vanishes for $\omega = \omega'$.

[9]We take into account that $\int_\gamma dz = -i\beta$.

symmetric under $\mathbf{x}_1 \leftrightarrow \mathbf{x}_2$ and from (10.26) we have $\delta P^>(1;2) = \delta P^<(2;1)$.[10] Taking into account (13.38) we can cast (13.37) in the following form:

$$E_{\text{int,c}} = \frac{\mathrm{i}}{2} \int \frac{d\omega}{2\pi} \int d\mathbf{x}_1 \langle \mathbf{x}_1 | \delta \hat{\mathcal{P}}^{\lessgtr}(\omega) \hat{v} | \mathbf{x}_1 \rangle - \frac{1}{2\beta} \mathrm{tr}_\gamma \left[P \delta W \right]. \tag{13.39}$$

Let us now manipulate the last term of this equation. We have

$$\mathrm{tr}_\gamma \left[P \delta W \right] = -\mathrm{i}\beta \int d\mathbf{x}_1 \langle \mathbf{x}_1 | \left[\hat{\mathcal{P}}^\rceil \star \delta \hat{\mathcal{W}}^\lceil \right] (t_0, t_0) | \mathbf{x}_1 \rangle$$

$$= -\mathrm{i}\beta \int \frac{d\omega}{2\pi} \int d\mathbf{x}_1 \langle \mathbf{x}_1 | \hat{\mathcal{P}}^\lessgtr(\omega) \delta \hat{\mathcal{W}}^{\mathrm{A}}(\omega) + \hat{\mathcal{P}}^{\mathrm{R}}(\omega) \delta \hat{\mathcal{W}}^\lessgtr(\omega) | \mathbf{x}_1 \rangle, \tag{13.40}$$

where in the last equality we use (13.13) with $t = t' = t_0$ (for these times the first two terms on the l.h.s. vanish). This equation now has the same structure as (13.34). By definition, the Fourier transform of the retarded/advanced polarizability is

$$\hat{\mathcal{P}}^{\mathrm{R/A}}(\omega) = \mathrm{i} \int \frac{d\omega'}{2\pi} \frac{\hat{\mathcal{P}}^>(\omega') - \hat{\mathcal{P}}^<(\omega')}{\omega - \omega' \pm \mathrm{i}\eta},$$

whereas the Fourier transform of the retarded/advanced $\delta \hat{\mathcal{W}}$ is

$$\delta \hat{\mathcal{W}}^{\mathrm{R/A}}(\omega) = \mathrm{i} \int \frac{d\omega'}{2\pi} \frac{\hat{\mathcal{W}}^>(\omega') - \hat{\mathcal{W}}^<(\omega')}{\omega - \omega' \pm \mathrm{i}\eta},$$

where we take into account that $\hat{v}^\lessgtr = 0$ and hence $\hat{\mathcal{W}}^\lessgtr = \delta \hat{\mathcal{W}}^\lessgtr$. Inserting these Fourier transforms into the trace we find

$$\mathrm{tr}_\gamma \left[P \delta W \right] = \beta \int \frac{d\omega}{2\pi} \frac{d\omega'}{2\pi} \int d\mathbf{x}_1 \langle \mathbf{x}_1 | \frac{\hat{\mathcal{P}}^<(\omega) \hat{\mathcal{W}}^>(\omega') - \hat{\mathcal{P}}^>(\omega) \hat{\mathcal{W}}^<(\omega')}{\omega - \omega'} | \mathbf{x}_1 \rangle.$$

This result can be inserted in (13.39) to obtain a formula for the interaction energy in terms of the polarizability and the screened interaction

$$\boxed{\begin{aligned} E_{\text{int,c}} = {}& \frac{\mathrm{i}}{2} \int \frac{d\omega}{2\pi} \int d\mathbf{x}_1 \langle \mathbf{x}_1 | \delta \hat{\mathcal{P}}^\lessgtr(\omega) \hat{v} | \mathbf{x}_1 \rangle \\ & - \frac{1}{2} \int \frac{d\omega}{2\pi} \frac{d\omega'}{2\pi} \int d\mathbf{x}_1 \langle \mathbf{x}_1 | \frac{\hat{\mathcal{P}}^<(\omega) \hat{\mathcal{W}}^>(\omega') - \hat{\mathcal{P}}^>(\omega) \hat{\mathcal{W}}^<(\omega')}{\omega - \omega'} | \mathbf{x}_1 \rangle \end{aligned}} \tag{13.41}$$

---

[10]More explicitly we have

$$\int d\mathbf{x}_1 \langle \mathbf{x}_1 | \hat{\mathcal{P}}^<(t,t) \hat{v} | \mathbf{x}_1 \rangle = \int d\mathbf{x}_1 d\mathbf{x}_2 \langle \mathbf{x}_1 | \hat{\mathcal{P}}^<(t,t) | \mathbf{x}_2 \rangle v(\mathbf{x}_2, \mathbf{x}_1)$$

$$= \int d\mathbf{x}_1 d\mathbf{x}_2 \langle \mathbf{x}_2 | \hat{\mathcal{P}}^>(t,t) | \mathbf{x}_1 \rangle v(\mathbf{x}_2, \mathbf{x}_1)$$

$$= \int d\mathbf{x}_1 d\mathbf{x}_2 \langle \mathbf{x}_1 | \hat{\mathcal{P}}^>(t,t) | \mathbf{x}_2 \rangle v(\mathbf{x}_2, \mathbf{x}_1) = \int d\mathbf{x}_1 \langle \mathbf{x}_1 | \hat{\mathcal{P}}^>(t,t) \hat{v} | \mathbf{x}_1 \rangle,$$

where in the third line we rename $\mathbf{x}_1 \leftrightarrow \mathbf{x}_2$ and use the symmetry of v.

The formulas (13.30), (13.32), and (13.35) [or equivalently (13.41)] allow us to calculate the ground-state energy of an interacting system for any given many-body approximation. While it is easy to separate the Hartree–Fock part from the correlation part in E_{int} the same cannot be said of E_{one}. It is in general quite useful to write the *total energy* E as the sum of the Hartree–Fock (noncorrelated) energy plus a correlation energy. The latter provides a measure of how much the system is correlated. A standard trick to perform this separation consists in using the Hellmann–Feynman theorem. The Hellmann–Feynman theorem, however, applies only to density matrices $\hat{\rho}$ which are a mixture of eigenstates with *fixed* coefficients. We therefore specialize the discussion to zero temperature and assume that the degeneracy of the ground-state multiplet of $\hat{H}_\lambda^M \equiv \hat{H}_0^M + \lambda \hat{H}_{int}$ does not change with λ. Then the zero-temperature density matrix reads

$$\hat{\rho}(\lambda) = \frac{1}{d} \sum_{g=1}^{d} |\Psi_g(\lambda)\rangle\langle\Psi_g(\lambda)|,$$

where d is the degeneracy and the sum runs over the components of the ground-state multiplet. As the weights in $\hat{\rho}(\lambda)$ are λ-independent we can use the Hellmann–Feynman theorem and find

$$\frac{d}{d\lambda} \text{Tr}\left[\hat{\rho}(\lambda)\hat{H}_\lambda\right] = \text{Tr}\left[\hat{\rho}(\lambda)\hat{H}_{int}\right]. \tag{13.42}$$

Integrating this equation between $\lambda = 0$ and $\lambda = 1$ we get

$$E = E_0 + \int_0^1 \frac{d\lambda}{\lambda} \text{Tr}\left[\hat{\rho}(\lambda)\hat{H}_{int}\right] = E_0 + \int_0^1 \frac{d\lambda}{\lambda} E_{int}(\lambda),$$

with E the interacting energy and E_0 the energy of the noninteracting system. The *noncorrelated* part of the total energy is defined from the above equation when the interaction energy is evaluated in the Hartree–Fock approximation, $E_{int}[\Sigma] \to E_{int}[\Sigma_{HF}] \equiv E_{int}^{HF}$:

$$E^{HF} = E_0 + \int_0^1 \frac{d\lambda}{\lambda} E_{int}^{HF}(\lambda).$$

The correlation energy is therefore the difference between the total energy and the Hartree–Fock energy

$$\boxed{E_{corr} \equiv E - E^{HF} = \int_0^1 \frac{d\lambda}{\lambda}\left(E_{int}(\lambda) - E_{int}^{HF}(\lambda)\right)} \tag{13.43}$$

It is important to appreciate the difference between $E_{int}^{HF} = E_{int}[\Sigma_{HF}]$ and $E_{int,HF} = E_{int,HF}[\Sigma]$. The quantity $E_{int,HF}[\Sigma]$ given in (13.32) is evaluated with a one-particle density matrix $n(\mathbf{x}_1, \mathbf{x}_2)$ [and density $n(\mathbf{x}) = n(\mathbf{x}, \mathbf{x})$] which comes from a Green's function $G = G_0 + G_0\Sigma G$, where Σ can be the second-Born, GW, or any other approximate self-energy. As such $E_{int,HF}[\Sigma]$ contains some correlation as well. Instead $E_{int}[\Sigma_{HF}]$ is the interaction energy evaluated with a Hartree–Fock Green's function. From (13.35) we see that $E_{int,c}[\Sigma_{HF}] = 0$ and hence $E_{int}[\Sigma_{HF}] = E_{int,HF}[\Sigma_{HF}]$. In the next section we calculate the correlation energy of an electron gas in the GW approximation.

Exercise 13.4. Verify that $E_{\text{one}} + E_{\text{int,HF}}$ evaluated with the Hartree–Fock Green's function yields exactly the result (7.50).

Exercise 13.5. Prove (13.42).

13.5 GW correlation energy of a Coulombic electron gas

In the electron gas the interparticle interaction $v(\mathbf{x}_1, \mathbf{x}_2) = v(\mathbf{r}_1 - \mathbf{r}_2)$ is independent of spin and hence the screened interaction $W(\mathbf{r}_1, z_1; \mathbf{r}_2, z_2)$ is also independent of spin. The sum over spin in (10.24) gives

$$W(\mathbf{r}_1, z_1; \mathbf{r}_2, z_2) = v(\mathbf{r}_1, \mathbf{r}_2) + \int d\bar{\mathbf{r}}_1 d\bar{\mathbf{r}}_2 d\bar{z} \, v(\mathbf{r}_1, \bar{\mathbf{r}}_1) P(\bar{\mathbf{r}}_1, z_1; \bar{\mathbf{r}}_2, \bar{z}) W(\bar{\mathbf{r}}_2, \bar{z}; \mathbf{r}_2, z_2),$$

(13.44)

where the spin-independent polarizability is defined as

$$P(\mathbf{r}_1, z_1; \mathbf{r}_2, z_2) \equiv \sum_{\sigma_1 \sigma_2} P(\mathbf{x}_1, z_1; \mathbf{x}_2, z_2).$$

For an electron gas all quantities in (13.44) depend only on the difference between the spatial coordinates. In momentum space (13.41) reads

$$
\begin{aligned}
\frac{E_{\text{int,c}}}{V} = {} & \frac{i}{2} \int \frac{d\mathbf{p} d\omega}{(2\pi)^4} \, \delta P^{\lessgtr}(\mathbf{p}, \omega) \tilde{v}_{\mathbf{p}} \\
& - \frac{1}{2} \int \frac{d\omega}{2\pi} \frac{d\omega'}{2\pi} \int \frac{d\mathbf{p}}{(2\pi)^3} \frac{P^<(\mathbf{p}, \omega) W^>(\mathbf{p}, \omega') - P^>(\mathbf{p}, \omega) W^<(\mathbf{p}, \omega')}{\omega - \omega'},
\end{aligned}
$$
(13.45)

with V the system volume and $\tilde{v}_{\mathbf{p}}$ the Fourier transform of the interaction. This equation can be further manipulated by expressing W in terms of v and P. In momentum space the retarded/advanced component of the screened interaction can easily be derived from (13.44) and reads

$$W^{\text{R/A}}(\mathbf{p}, \omega) = \frac{\tilde{v}_{\mathbf{p}}}{1 - \tilde{v}_{\mathbf{p}} P^{\text{R/A}}(\mathbf{p}, \omega)}.$$

Using the fluctuation–dissipation theorem (13.10) for P and (13.11) for W we have (omitting the explicit dependence on frequency and momentum)

$$W^> = \bar{f}\,[W^{\text{R}} - W^{\text{A}}] = |W^{\text{R}}|^2 \, P^>,$$

where

$$|W^{\text{R}}|^2 = |W^{\text{A}}|^2 = \frac{\tilde{v}^2}{(1 - \tilde{v} P^{\text{R}})(1 - \tilde{v} P^{\text{A}})}.$$

Similarly, for the lesser component of W we find

$$W^< = |W^{\text{R}}|^2 \, P^<.$$

Substituting these relations into (13.45) we obtain an equivalent formula for the correlation part of the interaction energy

$$\frac{E_{\text{int,c}}}{V} = \frac{i}{2} \int \frac{d\mathbf{p} d\omega}{(2\pi)^4} \delta P^{\lessgtr}(\mathbf{p}, \omega) \tilde{v}_{\mathbf{p}} - \frac{1}{2} \int \frac{d\omega}{2\pi} \frac{d\omega'}{2\pi} \int \frac{d\mathbf{p}}{(2\pi)^3} |W^{\text{R}}(\mathbf{p}, \omega')|^2$$

$$\times \frac{P^{<}(\mathbf{p}, \omega) P^{>}(\mathbf{p}, \omega') - P^{>}(\mathbf{p}, \omega) P^{<}(\mathbf{p}, \omega')}{\omega - \omega'}. \quad (13.46)$$

We stress that no approximations have been done so far: equation (13.46) is an exact rewriting of (13.45).

In the following we calculate the correlation energy (13.43) of an electron gas with Coulomb interaction $v(\mathbf{x}_1, \mathbf{x}_2) = 1/|\mathbf{r}_1 - \mathbf{r}_2|$, and in the presence of an external potential $V(\mathbf{r}) = n\tilde{v}_0$ generated by a uniform positive background charge with the same density of electrons. Let us start with the evaluation of the λ-integral of $E_{\text{int}}^{\text{HF}}(\lambda)$. This quantity is given in (13.32) with rescaled one-particle density matrix $n \to n_\lambda$ calculated in the Hartree–Fock approximation and, of course, with $v \to \lambda v$. Since the number of particles does not change when λ is varied the density $n_\lambda(\mathbf{x}) = n/2$ is independent of λ. For the one-particle density matrix we have [see (7.52)]

$$n_\lambda(\mathbf{x}_1, \mathbf{x}_2) = -iG_\lambda^{\text{M}}(\mathbf{x}_1, \tau; \mathbf{x}_2, \tau^+) = \delta_{\sigma_1 \sigma_2} \int \frac{d\mathbf{k}}{(2\pi)^3} e^{i\mathbf{k} \cdot (\mathbf{r}_1 - \mathbf{r}_2)} \theta(p_{\text{F}, \lambda} - k),$$

where $p_{\text{F}, \lambda}$ is the Fermi momentum of the system with rescaled interaction. The Fermi momentum cannot depend on λ for otherwise $n_\lambda(\mathbf{x}, \mathbf{x}) = n_\lambda(\mathbf{x})$ would be λ-dependent. Therefore the one-particle density matrix is independent of λ as well. We conclude that for an electron gas $E_{\text{int}}^{\text{HF}}(\lambda)$ is linear in λ and hence its λ-integral is trivial,

$$\int_0^1 \frac{d\lambda}{\lambda} E_{\text{int}}^{\text{HF}}(\lambda) = V \left[\frac{1}{2} \tilde{v}_0 n^2 - \int \frac{d\mathbf{p}}{(2\pi)^3} \frac{d\mathbf{k}}{(2\pi)^3} \tilde{v}_{\mathbf{p} - \mathbf{k}} \theta(p_{\text{F}} - k) \theta(p_{\text{F}} - p) \right]. \quad (13.47)$$

The reader can check that by adding E_0 to this energy one recovers the Hartree–Fock energy (7.54). We further encourage the reader to verify that we get the same result by using the one-particle density matrix in the Hartree approximation. Indeed, both the Hartree and Hartree–Fock eigenstates are plane-waves and since the density is constant in space $n_\lambda(\mathbf{x}_1, \mathbf{x}_2)$ is the same in both approximations (and independent of λ).

To calculate the correlation energy we need an approximation to the self-energy. We consider here the GW approximation according to which $\Sigma_{\text{xc}}(1; 2) = iG(1; 2) W(1; 2)$ with $W = v + vPW$ and $P = \chi_0 = -iGG$. It is easy to show that $E_{\text{int,c}}$ evaluated with (13.35) using a GW self-energy or evaluated with (13.41) using $P = \chi_0$ (equivalently $\delta P = 0$) yields the same result. The diagram which represent $\text{tr}_\gamma[\Sigma_{\text{xc}} G]$ in the GW approximation is

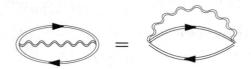

and it is topologically equivalent to the diagram in the r.h.s. which is represented by $\mathrm{tr}_\gamma[\chi_0 W]$.

To carry on the calculation with pencil and paper we take the Green's function in χ_0 to be the Hartree Green's function.[11] Then, according to the observation below (13.43), we have $E_{\mathrm{int,HF}}(\lambda) = E_{\mathrm{int}}^{\mathrm{HF}}(\lambda)$ since $n_\lambda(\mathbf{x}_1, \mathbf{x}_2)$ is the same in the Hartree and Hartree–Fock approximation. Thus the correlation energy (13.43) reduces to

$$E_{\mathrm{corr}} = \int_0^1 \frac{d\lambda}{\lambda} E_{\mathrm{int,c}}(\lambda). \tag{13.48}$$

The GW correlation energy corresponds to the resummation of the following (infinite) subset of diagrams (also called ring diagrams):

$$\tag{13.49}$$

as can easily be checked from (13.39) with $P = -iGG$ and $W = v + vPW$.

In Fourier space the greater/lesser component of the spin-independent and noninteracting polarizability $P^{\lessgtr}(\mathbf{r}_1, t_1; \mathbf{r}_2, t_2) = -i \sum_{\sigma_1 \sigma_2} G^{\lessgtr}(1;2) G^{\gtrless}(2;1)$ reads

$$P^{\lessgtr}(\mathbf{p}, \omega) = -2i \int \frac{d\bar{\mathbf{p}} d\bar{\omega}}{(2\pi)^4} G^{\lessgtr}(\mathbf{p} + \bar{\mathbf{p}}, \omega + \bar{\omega}) G^{\gtrless}(\bar{\mathbf{p}}, \bar{\omega}), \tag{13.50}$$

where we take into account that G is diagonal in spin space and hence the factor of 2 comes from spin. Inserting this polarizability into (13.46) the formula for $E_{\mathrm{int,c}}$ becomes (after some renaming of the integration variables)

$$
\begin{aligned}
\frac{E_{\mathrm{int,c}}}{V} = -2 \int & \frac{dp d\omega}{(2\pi)^4} \frac{dp' d\omega'}{(2\pi)^4} \frac{d\bar{p} d\bar{\omega}}{(2\pi)^4} \int \frac{d\bar{\omega}'}{2\pi} \frac{|W^{\mathrm{R}}(\mathbf{p} - \bar{\mathbf{p}}, \omega - \bar{\omega})|^2}{\omega + \omega' - \bar{\omega} - \bar{\omega}'} \\
& \times \left[G^{>}(\mathbf{p}, \omega) G^{>}(\mathbf{p}', \omega') G^{<}(\bar{\mathbf{p}}, \bar{\omega}) G^{<}(\mathbf{p} + \mathbf{p}' - \bar{\mathbf{p}}, \bar{\omega}') \right. \\
& \left. - G^{<}(\mathbf{p}, \omega) G^{<}(\mathbf{p}', \omega') G^{>}(\bar{\mathbf{p}}, \bar{\omega}) G^{>}(\mathbf{p} + \mathbf{p}' - \bar{\mathbf{p}}, \bar{\omega}') \right].
\end{aligned}
\tag{13.51}
$$

Next we use the explicit expression of the Hartree Green's function [see (6.48) and (6.49)]

$$G^{<}(\mathbf{p}, \omega) = 2\pi i\, \theta(p_{\mathrm{F}} - p) \delta\left(\omega - \frac{p^2}{2}\right)$$

$$G^{>}(\mathbf{p}, \omega) = -2\pi i\, \bar{\theta}(p_{\mathrm{F}} - p) \delta\left(\omega - \frac{p^2}{2}\right),$$

---

[11]In accordance with the discussion below (7.54), the Hartree Green's function of a Coulombic electron gas in a uniform positive background charge is equal to the Green's function of a noninteracting electron gas *without* the positive background charge.

with $\bar{\theta}(x) = 1 - \theta(x)$, and perform the integral over all frequencies. Renaming the momenta as $\mathbf{p} = \mathbf{q} + \mathbf{k}/2$, $\mathbf{p}' = \mathbf{q}' - \mathbf{k}/2$, and $\bar{\mathbf{p}} = \mathbf{q} - \mathbf{k}/2$ we find

$$\frac{E_{\text{int,c}}}{V} = -4 \int \frac{d\mathbf{q}}{(2\pi)^3} \frac{d\mathbf{q}'}{(2\pi)^3} \frac{d\mathbf{k}}{(2\pi)^3} \frac{|W^{\text{R}}(\mathbf{k}, \mathbf{q} \cdot \mathbf{k})|^2}{(\mathbf{q} - \mathbf{q}') \cdot \mathbf{k}}$$

$$\times \bar{\theta}(p_{\text{F}} - |\mathbf{q} + \frac{\mathbf{k}}{2}|) \bar{\theta}(p_{\text{F}} - |\mathbf{q}' - \frac{\mathbf{k}}{2}|) \theta(p_{\text{F}} - |\mathbf{q} - \frac{\mathbf{k}}{2}|) \theta(p_{\text{F}} - |\mathbf{q}' + \frac{\mathbf{k}}{2}|). \quad (13.52)$$

In deriving (13.52) we use the fact that the contributions coming from the four Green's functions in the second and third line of (13.51) are identical, so we only include the second line and multiply by 2. To prove that they are identical one has to use the symmetry property (10.27), which implies $W^{\text{R}}(\mathbf{k}, \omega) = [W^{\text{R}}(-\mathbf{k}, -\omega)]^*$, and the invariance of the system under rotations, which implies that $W^{\text{R}}(\mathbf{k}, \omega)$ depends only on the modulus $k = |\mathbf{k}|$.

Before continuing with the evaluation of (13.52) an important observation is in order. By replacing $W(\mathbf{k}, \omega)$ with $\tilde{v}_{\mathbf{k}} = 4\pi/k^2$ in (13.52) we get the $E_{\text{int,c}}$ of the second-order ring diagram [first term of the series (13.49)]. This quantity contains the integral $\int \frac{d\mathbf{k}}{(2\pi)^3} \tilde{v}_{\mathbf{k}}^2 \ldots = \int \frac{d\mathbf{k}}{(2\pi)^3} \frac{1}{k^4} \ldots$ and is therefore divergent! In fact, each diagram of the series is divergent. As anticipated at the end of Section 5.3 this is a typical feature of bulk systems with long-range interparticle interactions. For these systems the resummation of the same class of diagrams to infinite order is absolutely essential to get meaningful results. We could compare this mathematical behavior to that of the Taylor expansion of the function $f(x) = 1 - 1/(1 - x)$. This function is finite for $x \to \infty$ but each term of the Taylor expansion $f(x) = x + x^2 + \ldots$ diverges in the same limit. For bulk systems with long-range interactions MBPT can be seen as a Taylor series with vanishing convergence radius and hence we must resum the series (or pieces of it) to get a finite result. The GW approximation is the leading term in the G- and W-skeletonic formulation.

In Section 15.5.2 we calculate the screened interaction $W = v + v\chi_0 W$ appearing in (13.52) and show that at zero frequency and for small k

$$W^{\text{R}}(\mathbf{k}, 0) = \frac{4\pi}{k^2 + \lambda_{\text{TF}}^{-2}}, \quad (13.53)$$

with $\lambda_{\text{TF}} = \sqrt{\pi/(4p_{\text{F}})}$ the so called *Thomas–Fermi screening length*. Consequently W^{R} does not diverge for small momenta and the integral (13.52) is finite. The effect of the screening is to transform the original (long-range) Coulomb interaction into a (short-range) Yukawa-type of interaction since Fourier transforming (13.53) back to real space one finds

$$W^{\text{R}}(\mathbf{r}_1, \mathbf{r}_2; \omega = 0) = \frac{e^{-|\mathbf{r}_1 - \mathbf{r}_2|/\lambda_{\text{TF}}}}{|\mathbf{r}_1 - \mathbf{r}_2|}.$$

We can move forward with the analytic calculations if we consider the high density limit, i.e., $r_s \to 0$. Recalling that $r_s = (9\pi/4)^{1/3}/p_{\text{F}}$, in this limit $\lambda_{\text{TF}}^{-1} \ll p_{\text{F}}$ and therefore the dominant contribution to the integral (13.52) comes from the region $k \ll p_{\text{F}}$. For small k we can approximate $W^{\text{R}}(\mathbf{k}, \mathbf{q} \cdot \mathbf{k})$ with (13.53) since $W^{\text{R}}(\mathbf{k}, \omega)$ depends smoothly on ω for small ωs, see Section 15.5.2. Then the integral (13.52) simplifies to

$$\frac{E_{\text{int,c}}}{V} = -4 \int \frac{d\mathbf{q}}{(2\pi)^3} \frac{d\mathbf{q}'}{(2\pi)^3} \int_{k<p_F} \frac{d\mathbf{k}}{(2\pi)^3} \frac{(4\pi)^2}{(k^2 + \lambda_{\text{TF}}^{-2})^2 (\mathbf{q}-\mathbf{q}') \cdot \mathbf{k}}$$
$$\times \bar{\theta}(p_F - |\mathbf{q}+\frac{\mathbf{k}}{2}|)\bar{\theta}(p_F - |\mathbf{q}'-\frac{\mathbf{k}}{2}|)\theta(p_F - |\mathbf{q}-\frac{\mathbf{k}}{2}|)\theta(p_F - |\mathbf{q}'+\frac{\mathbf{k}}{2}|). \quad (13.54)$$

Next we observe that for $k \ll p_F$ the product $\bar{\theta}(p_F - |\mathbf{q}+\frac{\mathbf{k}}{2}|)\theta(p_F - |\mathbf{q}-\frac{\mathbf{k}}{2}|)$ is nonzero only for $\mathbf{q}$ close to the Fermi sphere. Denoting by c the direction cosine between $\mathbf{q}$ and $\mathbf{k}$ we then have $\mathbf{q} \cdot \mathbf{k} \sim p_F k c$ and $|\mathbf{q} \pm \frac{\mathbf{k}}{2}| \sim q \pm \frac{kc}{2}$. Further approximating

$$\int \frac{d\mathbf{q}}{(2\pi)^3} \sim \left(\frac{p_F}{2\pi}\right)^2 \int_0^\infty dq \int_{-1}^1 dc$$

we get

$$\frac{E_{\text{int,c}}}{V} = -4 \left(\frac{p_F}{2\pi}\right)^4 \frac{(4\pi)^2}{2\pi^2} \frac{1}{p_F} \int_0^\infty dq dq' \int_{-1}^1 dc dc' \int_0^{p_F} dk \frac{k}{(k^2 + \lambda_{\text{TF}}^{-2})^2 (c - c')}$$
$$\times \bar{\theta}(p_F - q - \frac{kc}{2})\bar{\theta}(p_F - q' + \frac{kc'}{2})\theta(p_F - q + \frac{kc}{2})\theta(p_F - q' - \frac{kc'}{2}). \quad (13.55)$$

The integral over q and q' is now straightforward since

$$\int dq \, \bar{\theta}(p_F - q - \frac{kc}{2})\theta(p_F - q + \frac{kc}{2}) = \int dx \, \bar{\theta}(x - \frac{kc}{2})\theta(x + \frac{kc}{2}) = \theta(c)kc.$$

Inserting this result into (13.55) we find

$$\frac{E_{\text{int,c}}}{V} = -2 \left(\frac{p_F}{\pi^2}\right)^2 p_F \int_0^{p_F} dk \frac{k^3}{(k^2 + \lambda_{\text{TF}}^{-2})^2} \underbrace{\int_0^1 dc \int_{-1}^0 dc' \frac{-cc'}{c - c'}}_{\frac{2}{3}(1-\ln 2)}. \quad (13.56)$$

The integral over k can be performed analytically since the integrand is a simple rational function. The result is

$$\int_0^{p_F} dk \frac{k^3}{(k^2 + \lambda_{\text{TF}}^{-2})^2} = \frac{1}{2}\left(-1 + \frac{\lambda_{\text{TF}}^{-2}}{\lambda_{\text{TF}}^{-2} + p_F^2} + \log(1 + \lambda_{\text{TF}}^2 p_F^2)\right).$$
$$\xrightarrow[r_s \to 0]{} \log(p_F \lambda_{\text{TF}})$$

The correlation energy (13.48) is given by the λ-integral of (13.56) with rescaled quantities. From (13.53) we see that $W^R \to \lambda W^R$ since p_F is independent of λ. Therefore $E_{\text{int,c}}(\lambda)$ is given by the r.h.s. of (13.56) multiplied by λ^2. Then the λ-integral produces the extra factor $\int_0^1 \frac{d\lambda}{\lambda}\lambda^2 = 1/2$ and the correlation energy reads

$$\boxed{E_{\text{corr}} = -n V \frac{2}{\pi^2}(1 - \ln 2) \ln(p_F \lambda_{\text{TF}})} \quad (13.57)$$

where we use the relation $n = p_F^3/(3\pi^2)$ between the density and p_F. This result was first derived by Gell-Mann and Brueckner in 1957 [104]. By analogy with the Hartree–Fock energy (7.62) we can express (13.57) in terms of the radius r_s. Adding the resulting expression to the Hartree–Fock energy (7.63) we obtain

$$\frac{E_{\text{tot}}}{n\text{V}} = \frac{1}{2}\left[\frac{2.21}{r_s^2} - \frac{0.916}{r_s} + 0.0622\ln r_s\right] \quad \text{(in atomic units)} \tag{13.58}$$

This formula contains the three leading contributions to E_{tot} in the high density limit $r_s \to 0$.

At the end of Section 5.3 we mentioned that the MBPT expansion is an asymptotic expansion in the interaction strength.[12] Equation (13.58) provides a nice example of this statement. Reinserting all the fundamental constants in (13.58) the prefactor

$$\frac{1}{2} \to \frac{e^2}{2a_{\text{B}}},$$

with the Bohr radius $a_{\text{B}} = \frac{\hbar^2}{m_e e^2}$ and e, m_e the electron charge and mass respectively. Taking into account that

$$r_s = \left(\frac{9\pi}{4}\right)^{\frac{1}{3}}\frac{1}{\hbar a_{\text{B}} p_F} \sim e^2, \tag{13.59}$$

we see that the first term $\propto e^0$, the second term $\propto e^2$ while the third term $\propto e^4\ln e^2$. After many years of study of the Coulombic electron gas the community feels reasonably confident that the correlation energy has the following asymptotic expansion

$$\frac{E_{\text{corr}}}{n\text{V}} = a_0\ln r_s + a_1 + a_2 r_s\ln r_s + a_3 r_s + a_4 r_s^2\ln r_s + a_5 r_s^2 + \ldots \tag{13.60}$$

We have just shown that the coefficient a_0 originates from the sum of ring diagrams evaluated with the Hartree Green's function. This approximation is called the G_0W_0 approximation to distinguish it from the GW approximation where G and W are self-consistently calculated from $G = G_0 + G_0(iGW)G$ and $W = v + v(-iGG)W$. The coefficients a_1 and a_2 can be extracted from the second-order exchange diagram and from a more precise evaluation of the sum of ring diagrams. The many-body origin of the coefficient a_3 is more complicated and we refer the interested reader to Ref. [105]. We write below the value of the first four coefficients in atomic units [106]

$$a_0 = +0.03109$$
$$a_1 = -0.04692$$
$$a_2 = +0.00923$$
$$a_3 = -0.010.$$

In Fig. 13.2 we show the correlation energy per particle as obtained by truncating the sum (13.60) to the first m terms, with $m = 1, 2, 3, 4$. For comparison we also report

---

[12]For a brief introduction to asymptotic expansions we refer the reader to Appendix F.

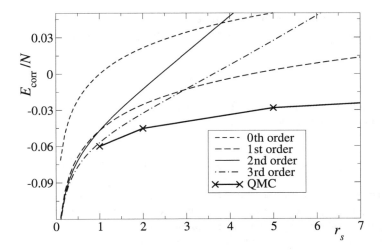

Figure 13.2 Correlation energy per particle (in atomic units) of the Coulombic electron gas.

the correlation energy per particle as obtained numerically [107] using a popular stochastic method, namely the Quantum Monte Carlo (QMC) method [108, 109]. A close inspection of the figure reveals the typical feature of an asymptotic expansion. We see that for $r_s = 1$ the 3rd order approximation is the most accurate. Going to larger values of r_s, however, the agreement deteriorates faster the higher the order of the approximation; for example the 1st order approximation is more accurate than the 2nd and 3rd order approximations for $r_s = 5$.

Finally we should note that the correlation energy calculated using the fully self-consistent GW approximation is as accurate as that from QMC calculations [110]. On the other hand, for spectral properties the $G_0 W_0$ approximation seems to perform better than the GW approximation.[13] This is typically attributed to a lack of cancellations. In Section 15.4 we show that conserving approximations in combination with gauge invariance lead to an equation between the vertex and the self-energy known as the *Ward identity*. The Ward identity suggests that self-consistent corrections are partially cancelled by vertex corrections. Since the GW vertex is zero this cancellation does not occur in the GW approximation [111–115]. The fact that self-consistent calculations of total energies are very accurate even without these cancellations is still not completely understood and the topic of ongoing research.

13.6 T-**matrix approximation**

Let us consider a system at zero temperature and with *zero* particles, and ask what physical processes G_2 should contain. Since there are no particles in the ground state we can only add two particles (we cannot destroy particles) and these two added particles can only

---

[13]In systems with a gap like semiconductors, insulators, or molecules the situation is much more unclear.

interact directly an arbitrary number of times. This is particularly clear in the Lippmann–Schwinger equation (M.6), see Appendix M, for the two-particle scattering state

$$|\psi\rangle = |\mathbf{k}\rangle + \frac{1}{E - \hat{\boldsymbol{p}}^2/m \pm i\eta} v(\hat{\mathbf{r}})|\psi\rangle, \tag{13.61}$$

of energy E and relative momentum $\mathbf{k}$. Iterating (13.61) we find the following expansion of $|\psi\rangle$ in powers of the interaction v

$$|\psi\rangle = |\mathbf{k}\rangle + \frac{1}{E - \hat{\boldsymbol{p}}^2/m \pm i\eta} v(\hat{\mathbf{r}})|\mathbf{k}\rangle + \frac{1}{E - \hat{\boldsymbol{p}}^2/m \pm i\eta} v(\hat{\mathbf{r}}) \frac{1}{E - \hat{\boldsymbol{p}}^2/m \pm i\eta} v(\hat{\mathbf{r}})|\mathbf{k}\rangle + \dots \tag{13.62}$$

The first term is a free scattering state. In the second term the particles interact once and then continue to propagate freely.[14] In the third term we have a first interaction, then a free propagation followed by a second interaction and then again a free propagation. In Section 13.3 we saw that the second-Born approximation for G_2 corresponds to truncating the expansion (13.62) to the term which is first order in v. We rewrite its diagrammatic representation as follows:

$$\tag{13.63}$$

where "exchange" represents the diagrams obtained from the explicitly drawn diagrams by exchanging c and d. The reader can easily verify that (13.63) is topologically equivalent to (12.7). The T-matrix approximation for G_2 goes beyond the second-Born approximation as it takes into account all multiple scatterings in the expansion (13.62). In the T-matrix approximation the two added particles interact directly an arbitrary number of times. The diagrammatic representation of this G_2 is then

+ exchange.

$$\tag{13.64}$$

---

[14]The operator $(E - \hat{\boldsymbol{p}}^2/m \pm i\eta)^{-1}$ is the Fourier transform of the evolution operator of two noninteracting particles.

The name T-matrix approximation is given after a very useful quantity known as the *transfer matrix*. In scattering theory the transfer matrix $\hat{T}$ is the operator which transforms the free scattering-state $|\mathbf{k}\rangle$ into the interacting scattering-state $|\psi\rangle$: $\hat{T}|\mathbf{k}\rangle = v(\hat{\mathbf{r}})|\psi\rangle$. From (13.62) we see that

$$\hat{T} = v(\hat{\mathbf{r}}) + v(\hat{\mathbf{r}})\frac{1}{E - \hat{\mathbf{p}}^2/m \pm i\eta}v(\hat{\mathbf{r}}) + \ldots = v(\hat{\mathbf{r}}) + \hat{T}\frac{1}{E - \hat{\mathbf{p}}^2/m \pm i\eta}v(\hat{\mathbf{r}}). \quad (13.65)$$

In MBPT the transfer matrix is defined in a similar way,

$$T(1,2;1',2') = \delta(1;1')\delta(2;2')v(1';2') + i\int d3d4\, T(1,2;3,4)G(3;1')G(4;2')v(1';2'). \quad (13.66)$$

In Section 5.2 we found that this quantity emerges in a natural way as the first nontrivial approximation to G_2, although in that case the Green's function was noninteracting. We can represent the transfer matrix with the following diagrammatic series

$$(13.67)$$

where an interaction v is associated with a wiggly line and a Green's function is associated with an oriented double line. The analogy between (13.67) and (13.65) is evident if we consider that the product $G(3;1')G(4;2')$ is the propagator of two independent particles.

From a knowledge of the transfer matrix we can determine G_2 in the T-matrix approximation since

$$v(1;2)G_2(1,2;1',2') = \int d3d4\, T(1,2;3,4)\left[G(3;1')G(4;2') \pm G(3;2')G(4;1')\right]. \quad (13.68)$$

We expect this approximation to be accurate in systems with a low density of particles (since it is exact for vanishing densities) and short-range interactions v (since it originates from the ideas of scattering theory where two particles at large enough distance are essentially free). If these two requirements are met then the T-matrix approximation provides a reliable description of the many-particle system independently of the interaction strength. In this respect the T-matrix approximation is very different from the second-Born or the GW approximations. The latter perform well in the high density limit $r_s \to 0$ or, equivalently, in the weak coupling limit $e^2 \to 0$, see (13.59).

The l.h.s. of (13.68) with $2' = 2^+$ also appears in the definition (9.1) of the self-energy. Comparing these equations it is straightforward to find[15]

$$\boxed{\Sigma(1;2) = \pm i\int d3d4\left[T(1,3;2,4) \pm T(1,3;4,2)\right]G(4;3^+)} \quad (13.69)$$

---

[15]Compare with (5.24) where a zeroth order version of the T-matrix was derived by truncating the Martin–Schwinger hierarchy. The equations become identical after relabeling $2 \leftrightarrow 3$.

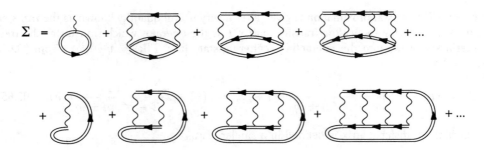

Figure 13.3 Self-energy in the T-matrix approximation.

Is this a conserving self-energy? The diagrammatic representation of Σ in shown in Fig. 13.3. The reader can easily recognize that the diagrams up to second order are topologically equivalent to the second-Born approximation, as it should be. Each diagram of the expansion in Fig. 13.3 is the functional derivative of a well-defined Φ-diagram and therefore the T-matrix self-energy is conserving. In particular the nth order diagram in the first row is the functional derivative of the Φ-diagram:

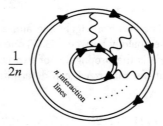

while the nth order diagram in the second row is the functional derivative of the Φ-diagram below:

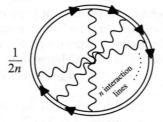

Like G_2 the transfer matrix is a difficult object to manipulate since it depends on four variables. Nevertheless the dependence on the time arguments is rather simple. From (13.67) we see that each term of the expansion starts and ends with an interaction v which is local in time. Therefore

$$T(1,2;1',2') = \delta(z_1,z_2)\delta(z_1',z_2')T(\mathbf{x}_1,\mathbf{x}_2,\mathbf{x}_1',\mathbf{x}_2';z_1,z_1').$$

To highlight the dependence on z_1 and z_1' we omit the position-spin variables of the reduced density matrix $T(\mathbf{x}_1, \mathbf{x}_2, \mathbf{x}_1', \mathbf{x}_2'; z_1, z_1') \rightarrow T(z_1, z_1')$. Then, from (13.67) we have

$$T(z_1, z_1') = \delta(z_1, z_1')v + iv\, G^2(z_1, z_1')\, v + i^2 v \int d\bar{z}\, G^2(z_1, \bar{z})\, v\, G^2(\bar{z}, z_1')\, v + \dots \quad (13.70)$$

This function belongs to the Keldysh space and contains a singular part. For systems in equilibrium the real-time Keldysh components (retarded, advanced, lesser, etc.) depend only on the time difference and can be Fourier transformed. If we define

$$\mathcal{T}(\omega) \equiv i\left[T^{>}(\omega) - T^{<}(\omega)\right] = i\left[T^{\mathrm{R}}(\omega) - T^{\mathrm{A}}(\omega)\right], \quad (13.71)$$

then, by the definition of retarded and advanced functions, we can write

$$T^{\mathrm{R/A}}(\omega) = v + \int \frac{d\omega'}{2\pi} \frac{\mathcal{T}(\omega')}{\omega - \omega' \pm i\eta},$$

or, reinserting the dependence on the position–spin variables,

$$T^{\mathrm{R/A}}(\mathbf{x}_1, \mathbf{x}_2, \mathbf{x}_1', \mathbf{x}_2'; \omega) = v(\mathbf{x}_1, \mathbf{x}_2)\delta(\mathbf{x}_1, \mathbf{x}_1')\delta(\mathbf{x}_2, \mathbf{x}_2') + \int \frac{d\omega'}{2\pi} \frac{\mathcal{T}(\mathbf{x}_1, \mathbf{x}_2, \mathbf{x}_1', \mathbf{x}_2'; \omega')}{\omega - \omega' \pm i\eta}.$$

The reduced transfer matrix also obeys its own fluctuation–dissipation theorem. The proof goes along the same lines as for the self-energy, polarizability, and screened interaction. Since $T(z_1, z_1')$ starts with $G^2(z_1, \dots)$ and ends with $G^2(\dots, z_1')$ the starting external vertices produce the factor $e^{2\mu\beta}$, whereas the ending external vertices produce the factor $e^{-2\mu\beta}$. Thus, it is easy to derive

$$T^{>}(\omega) = e^{\beta(\omega - 2\mu)}T^{<}(\omega). \quad (13.72)$$

Combining (13.72) with the definition (13.71) we can determine $T^{\lessgtr}$ from a sole knowledge of $\mathcal{T}$ since

$$\boxed{\begin{aligned} T^{<}(\omega) &= -if(\omega - 2\mu)\, \mathcal{T}(\omega) \\[2mm] T^{>}(\omega) &= -i\bar{f}(\omega - 2\mu)\, \mathcal{T}(\omega) \end{aligned}} \quad (13.73)$$

with $f(\omega) = 1/(e^{\beta\omega} - 1)$, the Bose function, and $\bar{f}(\omega) = 1 + f(\omega)$. This is the fluctuation-dissipation theorem for T.

The T-matrix approximation has been used in several contexts of condensed matter physics and nuclear physics. As the implementation of the fully self-consistent T-matrix self-energy is computationally demanding, different flavors of this approximation have been studied over the years. From (13.66) we see that the transfer matrix satisfies an equation with two Green's functions. If we here set $G = G_0$ we get the T_0-matrix approximation of Section 5.2. The transfer matrix T_0 can then be multiplied by G_0 to generate a self-energy according to (13.69). This $T_0 G_0$ approximation was employed by Thouless to describe the superconducting instability of a normal metal with attractive interactions [116]. The $T_0 G_0$ approximation, however, gives rise to unphysical results in two dimensions [117]. A different approximation consists in constructing the transfer matrix T from the self-consistent G, which solves the Dyson equation with self-energy $\Sigma = \pm iTG_0$ [118]. Also this scheme is not fully self-consistent and has some problems. The advantage of working with the self-consistent $\Sigma = \pm iTG$ is that several exact properties, like the conservation laws, are preserved [103, 119]. It should be said, however, that many spectral properties can already be described in partially self-consistent or non-self-consistent treatments (and sometimes even more accurately). For instance the non-self-consistent second-Born approximation correctly captures the gap shrinkage due to polarization effects, see Section 13.3.1. In the electron gas the G_0W_0 approximation produces better quasi-particle renormalization factors than the fully self-consistent GW approximation [115]. Below we discuss another example of non-self-consistent treatments. The example is concerned with the formation of a Cooper pair in a superconductor.

13.6.1 Formation of a Cooper pair

As a pedagogical example of the accuracy of the T-matrix approximation in systems where the repeated interaction between two electrons plays a crucial role we consider the formation of a bound state in the BCS model. Consider the BCS Hamiltonian (2.37) and suppose that there is only one electron, say of spin down, in the system. Let us order the single particle energies as $\epsilon_0 < \epsilon_1 < \ldots < \epsilon_N$ and assume no degeneracy. The ground state is $|\Psi_0\rangle = \hat{c}_{0\downarrow}^\dagger |0\rangle$ with ground-state energy ϵ_0. We are interested in calculating the spectral function for an electron of spin up. With only one spin-down electron in the ground state it is not possible to remove a spin-up electron and therefore the spectral function contains only addition energies. Furthermore, the interaction couples pairs of electrons with the same k label and opposite spin. This implies that if we add an electron of momentum $k \neq 0$ the resulting two-particle state $\hat{c}_{k\uparrow}^\dagger |\Psi_0\rangle = \hat{c}_{k\uparrow}^\dagger \hat{c}_{0\downarrow}^\dagger |0\rangle$ is an eigenstate with energy $\epsilon_k + \epsilon_0$. Consequently the spectral function exhibits only one peak at the addition energy ϵ_k. The situation is much more interesting if we add a spin-up electron with $k = 0$ since $\hat{c}_{0\uparrow}^\dagger \hat{c}_{0\downarrow}^\dagger |0\rangle$ is not an eigenstate. Below we calculate the exact spectral function $A(\omega) \equiv A_{0\uparrow 0\uparrow}(\omega)$ and then compare the results with the spectral function in the T-matrix and second-Born approximation. As we shall see the nonperturbative nature of the T-matrix approximation gives very accurate results while the second-Born approximation performs rather poorly in this context.

Let us start with the exact solution. The spectral function is $A(\omega) = iG_{0\uparrow 0\uparrow}^>(\omega)$ since the lesser Green's function vanishes. To obtain the greater Green's function we have to calculate the energies $E_{2,m}$ of the two-electron eigenstates $|\Psi_{2,m}\rangle$ as well as the overlaps

$$P_m = \langle \Psi_0 | \hat{c}_{0\uparrow} | \Psi_{2,m} \rangle = \langle 0 | \hat{c}_{0\downarrow} \hat{c}_{0\uparrow} | \Psi_{2,m} \rangle$$

between the state resulting from the addition of a particle of spin up in level 0 and $|\Psi_{2,m}\rangle$, see (6.88). The two-electron eigenstates are either of the form $\hat{c}_{k\uparrow}^\dagger \hat{c}_{p\downarrow}^\dagger |0\rangle$ with $k \neq p$ (singly occupied levels are blocked) or linear combinations of doubly occupied levels. The former give zero overlaps, $P_m = 0$, and it is therefore sufficient to calculate the latter. In agreement with the notation of Section 2.6 we denote these eigenstates by $|\Psi_i^{(1)}\rangle$ where the superscript

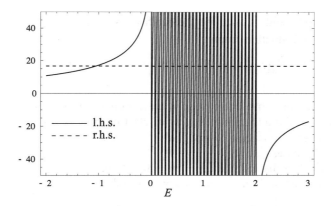

Figure 13.4 Graphical solution of the eigenvalue equation (13.74) for a single electron pair in the BCS model. The numerical parameters are $N = 30$ levels, equidistant energies $\epsilon_k = (\mathcal{E}/N)k$ with $k = 0, \ldots, N - 1$, and $v = 1.8\mathcal{E}/N$ (all energies are measured in units of $\mathcal{E}$).

"(1)" indicates that there is one pair. The generic form of $|\Psi_i^{(1)}\rangle$ is

$$|\Psi_i^{(1)}\rangle = \sum_k \alpha_k^{(i)} b_k^\dagger |0\rangle,$$

where $b_k^\dagger = \hat{c}_{k\uparrow}^\dagger \hat{c}_{k\downarrow}^\dagger$ is the creation operator for a pair of electrons in level k. We have seen that the eigenvalues E_i of $|\Psi_i^{(1)}\rangle$ are the solutions of the algebraic equation

$$\sum_k \frac{1}{2\epsilon_k - E_i} = \frac{1}{v}, \tag{13.74}$$

and that the coefficients

$$\alpha_k^{(i)} = \frac{C}{2\epsilon_k - E_i},$$

with C a normalization constant. The l.h.s. of (13.74) approaches $+\infty$ for $E_i \to 2\epsilon_k$ from the left and $-\infty$ for $E_i \to 2\epsilon_k$ from the right. Thus for $E_i = \epsilon_0 - \delta$, $\delta > 0$, the l.h.s. is always positive, diverges when $\delta \to 0$, and goes to zero when $\delta \to \infty$. This means that for positive v (attractive interaction) we have a bound state below the energy band $\{\epsilon_k\}$. In the context of the BCS model this bound state is known as *Cooper pair*. The graphical solution of (13.74) is shown in Fig. 13.4 for $N = 30$ levels, equidistant energies $\epsilon_k = (\mathcal{E}/N)k$ with $k = 0, \ldots, N - 1$ and $v = 1.8\mathcal{E}/N$. We clearly see an intersection at energy ~ -1, which corresponds to a Cooper pair. The full set of solutions E_i can easily be determined numerically and then used to construct the coefficients $\alpha_k^{(i)}$. We actually do not need these coefficients for all k since the overlap

$$P_i = \langle 0|\hat{c}_{0\downarrow}\hat{c}_{0\uparrow}|\Psi_i^{(1)}\rangle = \alpha_0^{(i)},$$

and it is therefore sufficient to calculate the coefficients with $k = 0$. Having the E_i and the $\alpha_0^{(i)}$ the exact spectral function reads

$$A(\omega) = 2\pi \sum_i |\alpha_0^{(i)}|^2 \,\delta(\omega - [E_i - \epsilon_0]). \qquad (13.75)$$

This spectral function has an isolated peak in correspondence with the Cooper-pair energy. We can then say that the addition of a spin-up electron in $k = 0$ excites the system in a combination of states of energy E_i among which there is one state (the Cooper pair) with an infinitely long life-time.

Let us now calculate the spectral function in the T-matrix approximation. The first task is to rewrite the interaction Hamiltonian in a form suitable for MBPT, i.e.,

$$\hat{H}_{\text{int}} = -v \sum_{kk'} \hat{c}_{k\uparrow}^\dagger \hat{c}_{k\downarrow}^\dagger \hat{c}_{k'\downarrow} \hat{c}_{k'\uparrow} = \frac{1}{2} \sum_{ijmn} v_{ijmn} \hat{c}_i^\dagger \hat{c}_j^\dagger \hat{c}_m \hat{c}_n,$$

where $i = (k_i, \sigma_i)$, $j = (k_j, \sigma_j)$ etc. are collective indices for the orbital and spin quantum numbers. The BCS interaction is nonvanishing only for $k_i = k_j$, $k_m = k_n$, and $\sigma_i = \bar{\sigma}_j = \bar{\sigma}_m = \sigma_n$, where $\bar{\sigma}$ is the spin opposite to σ. Therefore

$$v_{ijmn} = -v\,\delta_{k_i k_j}\delta_{k_m k_n}\delta_{\sigma_i \sigma_n}\delta_{\sigma_j \sigma_m}\delta_{\sigma_i \bar{\sigma}_j},$$

meaning that the only possible scattering processes are

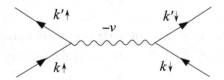

Since the interaction preserves the spin orientation, after every scattering the Green's function and the self-energy are diagonal in spin space. However, since the ground state is not symmetric under a spin flip, the Green's function for spin-up electrons is different from the Green's function for spin-down electrons. Another interesting feature of the interaction is that if we add or remove an electron of label k, let the system evolve, and then remove or add an electron of label k', we find zero unless $k = k'$. Thus, the Green's function is also diagonal in k-space: $G_{k\sigma\,k'\sigma'} = \delta_{\sigma\sigma'}\delta_{kk'} G_{k\sigma}$. For a ground state with only one spin-down electron in level 0 the greater and lesser *noninteracting* Green's function G_0 has diagonal matrix elements

spin up

$$G_{0,k\uparrow}^>(\omega) = -2\pi i\,\delta(\omega - \epsilon_k),$$
$$G_{0,k\uparrow}^<(\omega) = 0,$$

spin down

$$G_{0,k\downarrow}^>(\omega) = -2\pi i\,(1 - \delta_{k0})\delta(\omega - \epsilon_k),$$
$$G_{0,k\downarrow}^<(\omega) = 2\pi i\,\delta_{k0}\delta(\omega - \epsilon_0).$$

Like the Green's function, the self-energy is also diagonal in k- and spin-space, $\Sigma_{k\sigma\,k'\sigma'} = \delta_{\sigma\sigma'}\delta_{kk'}\Sigma_{k\sigma}$. In the T-matrix approximation Σ has the diagrammatic expansion of Fig. 13.3, and taking into account the explicit form of the BCS interaction we can easily assign the labels of each internal vertex. For instance the self-energy for a spin-up electron looks like from which it is evident that the self-energy vanishes unless $k = k'$ (the Green's function in the lower part of all these diagrams starts with k and ends with k'). We also see that all exchange diagrams (second row) are zero since after the scattering the particle cannot change its spin. The mathematical expression of the self-energy is therefore (remember that the interaction is $-v$):

$$\Sigma_{k\uparrow}(z_1, z_2) = i\left[v\delta(z_1, z_2) - iv^2 B(z_1, z_2) + i^2 v^3 \int_\gamma d\bar{z} B(z_1, \bar{z})B(\bar{z}, z_2) + \ldots\right]G_{k\downarrow}(z_2, z_1),$$

where B is the Cooper pair propagator

$$B(z_1, z_2) = \sum_p G_{p\uparrow}(z_1, z_2)G_{p\downarrow}(z_1, z_2).$$

The spectral function $A(\omega) = A_{0\uparrow 0\uparrow}$ is given in terms of the self-energy in agreement with (13.3):

$$A(\omega) = \frac{\Gamma_{0\uparrow}(\omega)}{|\omega - \epsilon_0 - \Sigma_{0\uparrow}^R(\omega)|^2}, \qquad \text{where} \qquad \Gamma_{0\uparrow}(\omega) = -2\text{Im}\left[\Sigma_{0\uparrow}^R(\omega)\right].$$

To calculate $\Sigma_{0\uparrow}$ we define the reduced T-matrix as the quantity in the square brackets,

$$T(z_1, z_2) = v\delta(z_1, z_2) - iv^2 B(z_1, z_2) + i^2 v^3 \int_\gamma d\bar{z} B(z_1, \bar{z})B(\bar{z}, z_2) + \ldots$$

$$= v\delta(z_1, z_2) - iv \int_\gamma d\bar{z}\, T(z_1, \bar{z})B(\bar{z}, z_2),$$

so that $\Sigma_{0\uparrow}(z_1, z_2) = iT(z_1, z_2)G_{0\downarrow}(z_2, z_1)$. Using the Langreth rules of Table 5.1 the retarded component of the self-energy in frequency space reads

$$\Sigma_{0\uparrow}^R(\omega) = i \int \frac{d\omega'}{2\pi}\left[T^R(\omega + \omega')G_{0\downarrow}^<(\omega') + T^<(\omega + \omega')G_{0\downarrow}^A(\omega')\right]. \qquad (13.76)$$

To extract the Keldysh components of the reduced T-matrix we use the simplified Langreth rules for equilibrium systems (see end of Section 13.2). Then in frequency space $T^< = -iv[T^R B^< + T^< B^A]$ from which it follows that $T^< = -ivT^R B^</(1 + ivB^A) = 0$. Indeed the lesser Cooper pair propagator is

$$B^<(t_1, t_2) = \sum_p G_{p\uparrow}^<(t_1, t_2)G_{p\downarrow}^<(t_1, t_2) = 0,$$

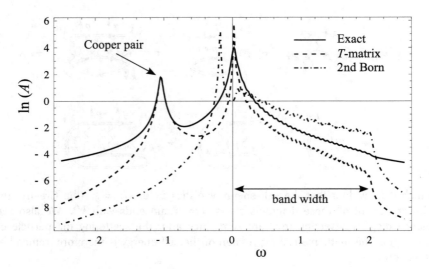

Figure 13.5 Logarithm of the spectral function $A(\omega)$ for the same parameters as Fig. 13.4, as obtained from the exact formula (13.75), the T-matrix approximation (13.77), and the second-Born approximation. The parameter $\eta = 0.3$ and for the exact formula we approximate $\delta(\omega) = \frac{1}{\pi} \frac{\eta}{\omega^2 + \eta^2}$ (all energies are measured in units of $\mathcal{E}$).

since there are no electrons of spin up in the ground state, i.e., $G^<_{p\uparrow} = 0$. It remains to calculate the first term in (13.76). This can be done analytically if we use the noninteracting Green's function worked out before (non self-consistent T-matrix approximation). We find

$$\Sigma^R_{0\uparrow}(\omega) = \mathrm{i} \int \frac{d\omega'}{2\pi} T^R(\omega + \omega') \times 2\pi\mathrm{i}\, \delta(\omega' - \epsilon_0) = -T^R(\omega + \epsilon_0).$$

The retarded component of the reduced T-matrix satisfies $T^R = v - \mathrm{i}v\, T^R B^R$ and hence $T^R = v/(1 + \mathrm{i}vB^R)$. We are then left with the evaluation of the retarded Cooper pair propagator. Using the noninteracting Green's function

$$B^>(\omega) = \sum_p \int \frac{d\omega'}{2\pi} G^>_{0,p\uparrow}(\omega - \omega') G^>_{0,p\downarrow}(\omega')$$

$$= \sum_p \int \frac{d\omega'}{2\pi} 2\pi\mathrm{i}\, \delta(\omega - \omega' - \epsilon_p) \times 2\pi\mathrm{i}\,(1 - \delta_{p0})\delta(\omega' - \epsilon_p)$$

$$= -2\pi \sum_{p \neq 0} \delta(\omega - 2\epsilon_p),$$

and hence

$$B^R(\omega) = \mathrm{i} \int \frac{d\omega'}{2\pi} \frac{B^>(\omega') - B^<(\omega')}{\omega - \omega' + \mathrm{i}\eta} = -\mathrm{i} \sum_{p \neq 0} \frac{1}{\omega - 2\epsilon_p + \mathrm{i}\eta}.$$

In conclusion, the spectral function in the non self-consistent T-matrix approximation reads

$$A(\omega) = 2\frac{\text{Im}[T^{\text{R}}(\omega + \epsilon_0)]}{|\omega - \epsilon_0 + T^{\text{R}}(\omega + \epsilon_0)|^2}, \qquad T^{\text{R}}(\omega) = \frac{v}{1 + v\sum_{p \neq 0}\dfrac{1}{\omega - 2\epsilon_p + i\eta}}. \qquad (13.77)$$

In Fig. 13.5 we show the exact spectral function as well as the T-matrix spectral function for the same parameters as Fig. 13.4. We see that the position of the Cooper pair binding energy in the T-matrix approximation is in excellent agreement with the exact result. The T-matrix approximation overestimates the height of the onset of the continuum and predicts a too sharp transition at the band edge $\omega = 2$. Overall, however, the agreement is rather satisfactory. For comparison we also display the spectral function in the second-Born approximation. The second-Born self-energy is given by the first two terms of the T-matrix expansion and it is therefore perturbative in the interaction v. The second-Born approximation severely underestimates the binding energy of the Cooper pair, and the continuum part of the spectrum decays far too slowly.

One final remark before concluding the section: it might look surprising that a second-order approximation like the second-Born approximation is able to describe the formation of a bound state, since a bound state is a highly nonperturbative (in v) solution of the eigenvalue problem. We should keep in mind, however, that what is perturbative here is the self-energy and *not* the Green's function. We have already stressed in Chapter 10 that the Green's function calculated from a self-energy with a finite number of diagrams corresponds to summing an infinite number of Green's function diagrams.

Exercise 13.6. Show that the number of symmetries N_{S} of the nth order Φ-diagram of the T-matrix approximation is $N_{\text{S}} = 2n$.

Exercise 13.7. Extract the greater/lesser component of the generic term of the expansion (13.70) taking into account that for systems in equilibrium the Langreth rules can be modified, i.e., the vertical track can be ignored provided that $t_0 \to -\infty$. Fourier transforming, show that each term of the expansion fulfills (13.72).

14

Linear response theory: preliminaries

14.1 Introduction

In the previous chapters we have developed a method to approximate the correlation between particles in both equilibrium and nonequilibrium systems. All approximations coincide with the exact solution when the interaction is turned off. In this chapter we investigate whether or not the general problem can be simplified by treating the external time-dependent fields perturbatively. To first order the corresponding theory is known as *linear response theory* (or simply *linear response*) and, of course, it is an approximate theory even for noninteracting systems. Through a careful selection of examples we first explore for what systems and external fields, and under which conditions we can trust the linear response results. Then, having clear in mind the limitations of the theory, in Chapter 15 we develop a general formalism based on nonequilibrium Green's functions.

The problem that we want to solve can be formulated as follows. Consider a system of interacting particles described by the time-dependent Hamiltonian

$$\hat{H}_{\text{tot}}(t) = \hat{H} + \lambda \hat{H}'(t),$$

where $\hat{H}$ is time independent, $\lambda \ll 1$ is a small dimensionless parameter, and $\hat{H}'(t)$ is a time dependent perturbation which we assume to vanish for times $t < t_0$. If the system is initially in the state $|\Psi\rangle$, how does $|\Psi\rangle$ evolve to first order in λ?

To answer this question we need an approximate expression for the evolution operator. The exact evolution operator obeys the differential equation

$$i\frac{d}{dt}\hat{U}_{\text{tot}}(t, t_0) = \left[\hat{H} + \lambda \hat{H}'(t)\right]\hat{U}_{\text{tot}}(t, t_0), \tag{14.1}$$

with boundary condition $\hat{U}_{\text{tot}}(t_0, t_0) = \hat{\mathbb{1}}$. We look for a solution of the form

$$\hat{U}_{\text{tot}}(t, t_0) = \hat{U}(t, t_0)\hat{F}(t), \tag{14.2}$$

with $\hat{U}(t, t_0) = \exp[-i\hat{H}(t - t_0)]$ the evolution operator of the unperturbed system. Substituting (14.2) into (14.1) we find a differential equation for $\hat{F}$

$$i\frac{d}{dt}\hat{F}(t) = \lambda \hat{H}'_H(t)\hat{F}(t), \tag{14.3}$$

385

where the subscript "H" denotes the operator in the Heisenberg picture, i.e.,

$$\hat{H}'_H(t) \equiv e^{i\hat{H}(t-t_0)}\hat{H}'(t)e^{-i\hat{H}(t-t_0)} = \hat{U}(t_0,t)\hat{H}'(t)\hat{U}(t,t_0). \tag{14.4}$$

The operator $\hat{H}'_H(t)$ is self-adjoint for all times t. Equation (14.3) must be solved with boundary condition $\hat{F}(t_0) = \hat{\mathbb{1}}$ and hence an integration between t_0 and t leads to

$$\hat{F}(t) = \hat{\mathbb{1}} - i\lambda \int_{t_0}^{t} dt'\, \hat{H}'_H(t')\hat{F}(t') = \hat{\mathbb{1}} - i\lambda \int_{t_0}^{t} dt'\, \hat{H}'_H(t') + \mathcal{O}(\lambda^2),$$

where in the last equality we replace $\hat{F}(t')$ with the whole r.h.s. (iterative solution). We conclude that to first order in λ the evolution operator reads

$$\hat{U}_{\text{tot}}(t,t_0) = \hat{U}(t,t_0)\left[\hat{\mathbb{1}} - i\lambda \int_{t_0}^{t} dt'\, \hat{H}'_H(t')\right]. \tag{14.5}$$

Let us comment on this result. We see at once that the operator (14.5) is not unitary. This implies that if we start from a normalized ket $|\Psi\rangle$ then the time evolved ket $|\Psi(t)\rangle = \hat{U}_{\text{tot}}(t,t_0)|\Psi\rangle$ is no longer normalized. The correction is, however, of second order in λ since

$$\langle\Psi(t)|\Psi(t)\rangle = 1 - i\lambda \int_{t_0}^{t} dt'\, \langle\Psi|\hat{H}'_H(t') - \hat{H}'_H(t')|\Psi\rangle + \mathcal{O}(\lambda^2) = 1 + \mathcal{O}(\lambda^2).$$

Nevertheless, the nonunitarity of the evolution operator represents a warning sign that we should not ignore, especially if we are interested in knowing the long-time behavior of the system. For instance in the simplest case $\hat{H}'(t) = \theta(t - t_0)E\hat{\mathbb{1}}$ is a uniform shift, the evolution operator (14.5) becomes $\hat{U}_{\text{tot}}(t,t_0) = e^{-i\hat{H}(t-t_0)}[1 - i\lambda E(t - t_0)]$ and for times $t \gg t_0 + 1/(\lambda E)$ the correction is far from being small. In the next section we investigate the consequences of this and related problems and we establish the domain of applicability of linear response theory.

14.2 Shortcomings of the linear response theory

To establish the domain of applicability of linear response theory we solve *exactly* some paradigmatic examples and then use the exact solution to benchmark the quality of the linear response results. As we shall see, many of the conclusions that we draw in this section have a completely general character.

We denote by $|\Psi_k\rangle$ the orthonormal eigenkets of the Hamiltonian $\hat{H}$ with eigenvalues E_k:

$$\hat{H} = \sum_k E_k |\Psi_k\rangle\langle\Psi_k|, \qquad \langle\Psi_k|\Psi_{k'}\rangle = \delta_{kk'}.$$

The most general form of the perturbing Hamiltonian in the basis $\{|\Psi_k\rangle\}$ is

$$\hat{H}'(t) = \sum_{kk'} T_{kk'}(t)|\Psi_k\rangle\langle\Psi_{k'}|, \tag{14.6}$$

with $T_{kk'} = T^*_{k'k}$. If we expand the time-evolved ket $|\Psi(t)\rangle$ as

$$|\Psi(t)\rangle = \sum_k c_k(t) e^{-iE_k t} |\Psi_k\rangle, \tag{14.7}$$

solving the time-dependent Schrödinger equation $i\frac{d}{dt}|\Psi(t)\rangle = [\hat{H} + \lambda\hat{H}'(t)]|\Psi(t)\rangle$ is equivalent to solving the following linear system of coupled differential equations:

$$i\frac{d}{dt}c_k(t) = \lambda \sum_{k'} T_{kk'}(t) e^{i\omega_{kk'}t} c_{k'}(t), \tag{14.8}$$

with $\omega_{kk'} \equiv E_k - E_{k'}$ the Bohr frequencies. We emphasize that (14.8) is an exact reformulation of the original Schrödinger equation. It is easy to show that the coefficients $c_k(t)$ which solve (14.8) preserve the normalization of the state, i.e.,[1]

$$\sum_k |c_k(t)|^2 = \sum_k |c_k(t_0)|^2 = \langle\Psi(t_0)|\Psi(t_0)\rangle. \tag{14.9}$$

From this exact result it follows that if the sum over k is restricted to a subset $\mathcal{S}$ of eigenkets then

$$\sum_{k\in\mathcal{S}} |c_k(t)|^2 \leq \langle\Psi(t_0)|\Psi(t_0)\rangle. \tag{14.10}$$

Equation (14.10) provides a useful benchmark for assessing the accuracy of linear response theory. Let us denote by $c_k^{(n)}(t)$ the nth order coefficient of the expansion of $c_k(t)$ in powers of λ. In most cases the initial state is an eigenstate of $\hat{H}$, say $|\Psi(t_0)\rangle = |\Psi_i\rangle$, and therefore $c_k^{(0)}(t) = 0$ for all $k \neq i$. This implies that if the subset $\mathcal{S}$ includes all states with the exception of $|\Psi_i\rangle$ then the lowest order contribution to the sum in (14.10) is obtained by replacing $c_k(t)$ with $c_k^{(1)}(t)$. This leads to the following upper bound for the linear response coefficients:

$$\sum_{k\neq i} |c_k^{(1)}(t)|^2 \leq \langle\Psi(t_0)|\Psi(t_0)\rangle. \tag{14.11}$$

The inequality (14.11) is obviously fulfilled at $t = t_0$ since $c_k^{(1)}(t_0) = 0$ for $k \neq i$. As we shall see, however, there exist situations for which (14.11) is violated at sufficiently long times. In these cases the probabilistic interpretation of the coefficients $|c_k^{(1)}(t)|^2$ breaks down and the results of linear response theory become unreliable.

14.2.1 Discrete-discrete coupling

We start our analysis by considering the simplest possible case, i.e., a perturbation that couples only two states, see Fig 14.1(a). Let $k = i$ and $k = f$ be the label of these states and

---

[1]To prove (14.9) we multiply both sides of (14.8) by $c_k^*(t)$, take the complex conjugate of the resulting equation and subtract it from the original one. Then, summing over all k we find that the time derivative of the l.h.s. of (14.9) is zero.

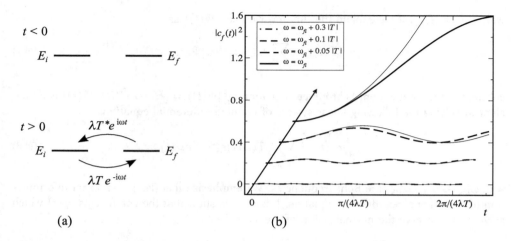

Figure 14.1 (a) Representation of the time-dependent perturbation. (b) The coefficient $|c_f(t)|^2$ for $\lambda = 0.01$ and different values of ω. The thin curves are the linear response results.

let us assume, for simplicity, that the time-dependent perturbation is monochromatic with frequency ω. The most general $\hat{H}'$ fulfilling these requirements is

$$\hat{H}'(t) = Te^{-i\omega t}|\Psi_f\rangle\langle\Psi_i| + T^* e^{i\omega t}|\Psi_i\rangle\langle\Psi_f|.$$

This Hamiltonian is a special case of (14.6) where all $T_{kk'}(t) = 0$ except for $T_{fi}(t) = T^*_{if}(t) = Te^{-i\omega t}$. If the system is in $|\Psi_i\rangle$ at time t_0 the differential equations (14.8) must be solved with boundary conditions $c_k(t_0) = 0$ if $k \neq i$ and $c_i(t_0) = 1$. Without loss of generality we take $t_0 = 0$. As the perturbation couples $|\Psi_i\rangle$ to $|\Psi_f\rangle$ only, the coefficients $c_k(t)$ with $k \neq i, f$ remain zero at all times. On the other hand the coefficients c_i and c_f are time dependent since they are solutions of

$$i\frac{d}{dt}c_i(t) = \lambda T^* e^{i(\omega-\omega_{fi})t}c_f(t), \qquad (14.12)$$

$$i\frac{d}{dt}c_f(t) = \lambda T e^{-i(\omega-\omega_{fi})t}c_i(t). \qquad (14.13)$$

This coupled system of equations can easily be solved. We define the frequency $\bar{\omega} = \omega - \omega_{fi}$. Then by multiplying (14.13) by $e^{i\bar{\omega}t}$, differentiating with respect to t and using (14.12) we find

$$\frac{d^2}{dt^2}c_f(t) + i\bar{\omega}\frac{d}{dt}c_f(t) + \lambda^2|T|^2 c_f(t) = 0,$$

the general solution of which is $c_f(t) = Ae^{i\alpha_+ t} + Be^{i\alpha_- t}$ with

$$\alpha_\pm = \frac{1}{2}\left(-\bar{\omega} \pm \sqrt{\bar{\omega}^2 + 4\lambda^2|T|^2}\right).$$

The boundary condition $c_f(0) = 0$ implies $A = -B$. To determine the constant A we calculate $c_i(t)$ from (14.13) and impose that $c_i(0) = 1$ (boundary condition). The final result is

$$c_f(t) = -\frac{2\mathrm{i}\lambda T}{\sqrt{\bar{\omega}^2 + 4\lambda^2|T|^2}} e^{-\mathrm{i}\bar{\omega}t/2} \sin\left(\frac{\sqrt{\bar{\omega}^2 + 4\lambda^2|T|^2}}{2}t\right).$$

The probability $|c_f(t)|^2$ of finding the system in $|\Psi_f\rangle$ at time t oscillates in time and the amplitude of the oscillations is correctly bounded between 0 and 1; it is easy to verify that $|c_i(t)|^2 + |c_f(t)|^2 = 1$ for all t. Note that the amplitude reaches the maximum value of 1 when the frequency of the perturbing field is equal to the Bohr frequency ω_{fi}, in which case $\bar{\omega} = 0$. This is the so-called *resonance phenomenon*: $|c_f(t)|^2$ increases *quadratically* up to a time $t_{\max} \sim \pi/(4\lambda|T|)$, at which time it reaches the value $\sim 1/2$. For frequencies ω much larger or smaller than ω_{fi} the amplitude of the oscillations is of the order of $(\lambda T/\bar{\omega})^2$, see Fig. 14.1(b), and hence the eigenket $|\Psi_f\rangle$ has a very small probability of being excited.

The linear response result is obtained by approximating $c_f(t)$ to first order in λ and reads

$$c_f(t) \sim c_f^{(1)}(t) = -\mathrm{i}\lambda T e^{-\mathrm{i}\bar{\omega}t/2} \frac{\sin\left(\frac{\bar{\omega}}{2}t\right)}{\left(\frac{\bar{\omega}}{2}\right)}. \tag{14.14}$$

The comparison between the exact and the linear response solution is displayed in Fig. 14.1(b). We can distinguish two regimes:

- $|\bar{\omega}| \gg \lambda|T|$: the linear response theory is accurate for short times but it eventually breaks down at times $t > |\bar{\omega}|/(\lambda|T|)^2$ since $c_f(t)$ and $c_f^{(1)}(t)$ start to oscillate out of phase. The amplitude of the oscillations of $|c_f^{(1)}(t)|^2$ is, however, in good agreement with the exact solution.

- $|\bar{\omega}| \ll \lambda|T|$: at the frequency of the resonance phenomenon ($\bar{\omega} = 0$) we have $c_f^{(1)}(t) = -\mathrm{i}\lambda T t$, which steadily increases with time. The square modulus $|c_f^{(1)}(t)|^2$ becomes larger than 1 for $t > 1/(\lambda T)$, see Fig. 14.1(b), and the inequality (14.11) breaks down. More generally, close to a resonance neither the frequency nor the amplitude of the oscillations is well reproduced in linear response.

Even though the above conclusions have been drawn from a rather simple model they are valid for all systems with a discrete spectrum. Integrating the exact equation (14.8) between 0 and t we find

$$c_k(t) = c_k(0) - \mathrm{i}\lambda \sum_{k'} \int_0^t dt' T_{kk'}(t') e^{\mathrm{i}\omega_{kk'}t'} c_{k'}(t').$$

If the system is initially in one of the eigenstates of $\hat{H}$, say the one with $k = i$, the boundary conditions are $c_k(0) = \delta_{ki}$. Then to first order in λ the coefficients $c_k(t)$ have the following form

$$c_k(t) \sim \delta_{ki} + c_k^{(1)}(t) = \delta_{ki} - \mathrm{i}\lambda \int_0^t dt' T_{ki}(t') e^{\mathrm{i}\omega_{ki}t'}. \tag{14.15}$$

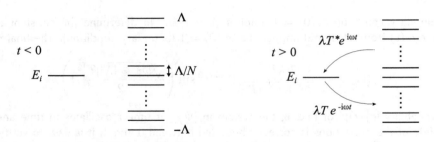

Figure 14.2 A single eigenstate with energy E_i is coupled at time $t_0 = 0$ to a macroscopic number $2N + 1$ of eigenstates with equally spaced energies $E_q = q(\Lambda/N) = q\Delta E$. In the limit $N \to \infty$ the spectrum becomes a continuum between $-\Lambda$ and Λ.

Assuming (as in the previous example) that for $k \neq i$ the perturbation is monochromatic[2] with frequency ω, $T_{ki}(t) = T_{ki}e^{-i\omega t}$, the coefficient $c_k(t)$ in (14.15) simplifies to

$$c_k(t) \sim c_k^{(1)}(t) = -i\lambda T_{ki}\, e^{-i(\omega-\omega_{ki})t/2}\, \frac{\sin\left(\frac{\omega-\omega_{ki}}{2}t\right)}{\left(\frac{\omega-\omega_{ki}}{2}\right)}, \qquad k \neq i, \qquad (14.16)$$

which agrees with (14.14) for $T_{ki} = \delta_{kf}T$. The coefficient $c_k^{(1)}(t)$ oscillates with frequency $(\omega - \omega_{ki})$; this frequency does not contain any information on the perturbation. Furthermore the probability $|c_k^{(1)}(t)|^2$ of finding the system in $|\Psi_k\rangle$ diverges when $\omega \to \omega_{ki}$ and $t \to \infty$, a typical signature of the breakdown of linear response theory according to the discussion below (14.11).

To summarize, the theory of linear response in systems with a discrete spectrum, like atoms and molecules, yields a good approximation for the coefficient $c_k(t)$ provided that the frequency ω of the perturbing Hamiltonian does not match the Bohr frequency ω_{ki} and provided that the time t is smaller than $|\omega - \omega_{ki}|/(\lambda T)^2$, where T is an energy of the order of magnitude of the $T_{kk'}$.

14.2.2 Discrete-continuum coupling

We now show that the performance of linear response theory improves when the perturbation $\hat{H}'$ couples the eigenket $|\Psi_i\rangle$ to a macroscopic number of eigenkets $|\Psi_q\rangle$ whose energies E_q form a continuum. A weak coupling between one or few discrete states and a continuum of states does not change the continuum of energies. Therefore we expect that if the system is initially in $|\Psi_i\rangle$ then the coefficients c_q oscillate in time with frequency ω_{iq} which is the same frequency predicted by linear response theory. To make this intuitive argument more rigorous we consider the system of Fig 14.2. An eigenket $|\Psi_i\rangle$ is coupled to $(2N+1)$ other eigenkets which we denote by $|\Psi_q\rangle$, $q = -N, \ldots, N$. The eigenkets $|\Psi_q\rangle$ are instead not coupled by the perturbation. We are interested in studying how the evolution of the system changes with increasing N and eventually in the limit $N \to \infty$. For simplicity we take the energies $E_q = (\Lambda/N)q$ equally spaced and the energy E_i in the range $(-\Lambda, \Lambda)$.

---

[2]Arbitrary perturbations can always be written as a linear combination of monochromatic ones.

The unperturbed Hamiltonian of the system reads

$$\hat{H} = E_i|\Psi_i\rangle\langle\Psi_i| + \sum_q E_q|\Psi_q\rangle\langle\Psi_q|. \tag{14.17}$$

In the large N limit the energy E_q becomes a continuous variable that we call E and the spectrum becomes a continuum of energies between $-\Lambda$ and Λ. The sum over q in (14.17) can then be converted into an integral. This is done by introducing kets that are orthonormalized to the Dirac δ-function rather than to the δ of Kronecker. Denoting the energy spacing by $\Delta E = \Lambda/N$, we define these kets as

$$|\Psi_E\rangle \equiv \frac{|\Psi_q\rangle}{\sqrt{\Delta E}} \quad \Rightarrow \quad \langle\Psi_E|\Psi_{E'}\rangle = \frac{\delta_{qq'}}{\Delta E} \xrightarrow{N\to\infty} \delta(E-E').$$

In terms of the kets $|\Psi_E\rangle$ the unperturbed Hamiltonian has the correct scaling properties since

$$\hat{H} = E_i|\Psi_i\rangle\langle\Psi_i| + \Delta E \sum_q E_q|\Psi_E\rangle\langle\Psi_E| \xrightarrow{N\to\infty} E_i|\Psi_i\rangle\langle\Psi_i| + \int_{-\Lambda}^{\Lambda} dE\, E\, |\Psi_E\rangle\langle\Psi_E|.$$

The kets $|\Psi_E\rangle$ are also useful to expand a generic ket $|\Psi\rangle$ when $N\to\infty$. Indeed

$$|\Psi\rangle = c_i|\Psi_i\rangle + \sum_q c_q|\Psi_q\rangle \xrightarrow{N\to\infty} c_i|\Psi_i\rangle + \int_{-\Lambda}^{\Lambda} dE\, c_E|\Psi_E\rangle, \tag{14.18}$$

where we define the coefficients

$$c_E \equiv \frac{c_q}{\sqrt{\Delta E}} = \langle\Psi_E|\Psi\rangle. \tag{14.19}$$

The square modulus of c_E is the *probability density* of finding the system in $|\Psi_E\rangle$. For $N\to\infty$ it is more convenient (and physically meaningful) to work with the c_E than with the c_q, see also discussion in Section 1.1. We stress that the probability density is not bounded between 0 and 1 but varies in the range $(0,\infty)$.

As in the example of Section 14.2.1, we specialize the discussion to monochromatic perturbations of frequency ω. For $\hat{H}'(t)$ to have the correct scaling properties we take

$$T_{qi}(t) = T_{iq}^*(t) = \sqrt{\Delta E}\, S_E\, e^{-i\omega t},$$

and $T_{ii} = T_{qq'} = 0$. The quantity S_E has the physical dimension of the square root of an energy since T_{qi} has the physical dimension of an energy. The scaling factor $\sqrt{\Delta E}$ in T_{qi} can easily be understood from the form of $\hat{H}'(t)$ written in terms of the $|\Psi_E\rangle$

$$\hat{H}'(t) = \Delta E \sum_q \left(S_E e^{-i\omega t}|\Psi_E\rangle\langle\Psi_i| + S_E^* e^{i\omega t}|\Psi_i\rangle\langle\Psi_E|\right),$$

which correctly becomes an integral over E in the limit $N\to\infty$.

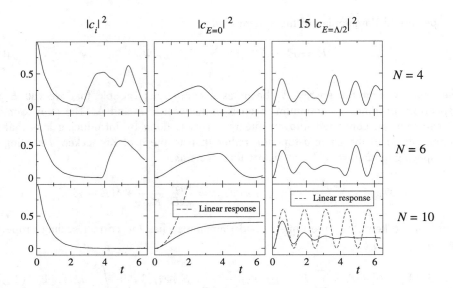

Figure 14.3 Time dependent modulus square of the coefficient c_i and of the coefficients $c_E = c_q/\sqrt{\Delta E}$ with energy $E = 0$ and $E = \Lambda/2$ (scaled up by a factor 15) for different numbers $(2N + 1)$ of $|\Psi_q\rangle$ states. Time is measured in units of $1/S^2$.

At this point a crucial observation about the normalization of the initial state $|\Psi(0)\rangle$ is in order. If the initial state is discrete, then $|\Psi(0)\rangle = |\Psi_i\rangle$ and the normalization is $\langle\Psi(0)|\Psi(0)\rangle = 1$. On the other hand, if the initial state is a continuum state, then $|\Psi(0)\rangle = |\Psi_E\rangle$ and the normalization is $\langle\Psi(0)|\Psi(0)\rangle = \lim_{N\to\infty} 1/\Delta E = \delta(0)$. We study here the evolution of the system when the initial state $|\Psi(0)\rangle = |\Psi_i\rangle$ and defer the problem of starting from a continuum state to the next section. With the parameters specified above, the linear system of equations (14.8) becomes

$$i\frac{d}{dt}c_i(t) = \lambda\sqrt{\Delta E}\sum_q S_E^* \, e^{i(\omega-\omega_{qi})t}c_q(t), \tag{14.20}$$

$$i\frac{d}{dt}c_q(t) = \lambda\sqrt{\Delta E}\, S_E \, e^{-i(\omega-\omega_{qi})t}c_i(t). \tag{14.21}$$

In Fig. 14.3 we display the numerical solution of (14.20) and (14.21) for different numbers $(2N + 1) = 9, 13, 21$ of $|\Psi_q\rangle$ states and frequency $\omega = 0$. We have taken $S_E = S$ independent of E and real, a large bandwidth $2\Lambda = 20S^2$ and the parameter $\lambda = 0.5$. For $E_i = 0$ the square modulus of $c_i(t)$ is shown in the left panel. It starts from 1 and decreases exponentially until a critical time $t_c(N)$, after which it behaves quite irregularly. Note that the exponential decay is the same for all N but the larger is N the longer is $t_c(N)$. Eventually when $N \to \infty$ the coefficient $|c_i(t)|^2 \sim e^{-t/\tau}$ for all t. The same kind of behavior is observed for the coefficients $c_E = c_q/\sqrt{\Delta E}$; they all approach some limiting function for $N \to \infty$, see middle and right panels of Fig. 14.3.

The limit $N \to \infty$ can easily be worked out analytically. Integrating (14.21) between 0 and t with boundary condition $c_q(0) = 0$, substituting the result in (14.20) and taking the limit $N \to \infty$ we find

$$i\frac{d}{dt}c_i(t) = -i(\lambda S)^2 \int_0^t dt' \int_{-\Lambda}^{\Lambda} dE\, e^{i(\omega - E + E_i)(t - t')}c_i(t').$$ (14.22)

If the quantity $|\omega + E_i| \ll \Lambda$ (as it is in our case since $\omega = E_i = 0$) we can extend the energy integral between $-\infty$ and ∞ since the contributions with large E are rapidly varying in time and the integral over t' averages them to zero (Riemann–Lebesgue theorem). When $\Lambda \to \infty$ the energy integral of the exponential function yields $2\pi\delta(t - t')$ and the integro-differential equation reduces to[3]

$$i\frac{d}{dt}c_i(t) = -i(\lambda S)^2 2\pi\frac{1}{2}c_i(t) \quad \Rightarrow \quad c_i(t) = e^{-\pi(\lambda S)^2 t}.$$ (14.23)

We find the anticipated exponential behavior with an inverse damping time $\tau^{-1} = \pi(\lambda S)^2$ quadratic in λ. As expected the exponential decay cannot be captured in linear response theory which predicts $c_i(t) = 1$ at all times. With the analytic solution (14.23) we can calculate the time-dependent coefficient $c_E(t) = c_q(t)/\sqrt{\Delta E}$ by direct integration of (14.21). The result is

$$c_E(t) = i\lambda S\frac{e^{-i(\omega - E + E_i)t - t/\tau} - 1}{i(\omega - E + E_i) + \tau^{-1}}.$$ (14.24)

This formula (strictly valid for $N \to \infty$) reproduces with remarkable accuracy the curves $|c_{E=0}(t)|^2$ and $|c_{E=\Lambda/2}(t)|^2$ in the middle and right panels of Fig. 14.3 up to the critical time $t_c(N)$ (not shown here).

The main difference between the discrete and the continuum coefficients is that $|c_E(t)|^2$ is finite in the long-time limit

$$\lim_{t \to \infty}|c_E(t)|^2 = \frac{(\lambda S)^2}{(\omega - E + E_i)^2 + \tau^{-2}} = \frac{1}{\pi}\frac{\tau^{-1}}{(\omega - E + E_i)^2 + \tau^{-2}}.$$ (14.25)

The r.h.s. of (14.25) can vary between 0 and ∞. We know that this is, in general, not a problem since $|c_E(t)|^2$ is a probability density and not a probability. The problem would exist if the normalization of the ket $|\Psi(t)\rangle$ was different from the normalization of the ket $|\Psi(0)\rangle$ which is unity in our case. It can easily be verified, however, that $|c_i(t)|^2 + \int dE|c_E(t)|^2 = 1$ at all times and hence that there is nothing wrong with our solution.[4]

For $E_i = \omega = 0$ the analytic result (14.24) predicts that $|c_{E=0}(t)|^2$ has the largest asymptotic value and no transient oscillations; this is nicely confirmed by the numerical simulations in Fig. 14.3 (middle panel). We can interpret the curve $|c_{E=0}(t)|^2$ as being the continuum version of the resonance phenomenon.

With the exact solution at our disposal we can now analyze the performance of linear response theory. To first order in λ the solution (14.24) reads

$$c_E(t) \sim c_E^{(1)}(t) = i\lambda S\frac{e^{-i(\omega - E + E_i)t} - 1}{i(\omega - E + E_i)} = -i\lambda S\, e^{-i(\omega - E + E_i)t/2}\frac{\sin\left(\frac{\omega - E + E_i}{2}t\right)}{\left(\frac{\omega - E + E_i}{2}\right)},$$ (14.26)

---

[3]To obtain (14.23) we use $\int_0^t dt'\delta(t - t')f(t') = f(t)/2$.

[4]In particular for $t \to \infty$ we have $c_i = 0$ and the integral of (14.25) between $-\infty$ and ∞ is 1.

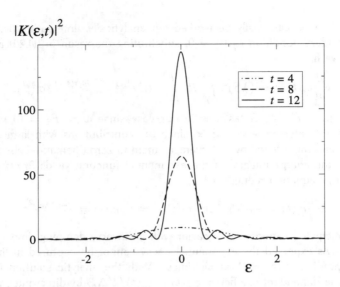

Figure 14.4 Square modulus $|K(\epsilon, t)|^2$ as a function of ϵ (arbitrary units) for different values of the time parameter t (arbitrary units).

which agrees with the general result (14.16). As in the discrete–discrete coupling, the coefficient $c_E^{(1)}$ for which $\omega = E - E_i$ (resonance) increases linearly in time (see middle-bottom panel of Fig. 14.3) and hence diverges when $t \to \infty$. Again linear response theory fails at the resonant frequency. This failure is also signaled by the breakdown of the upper bound (14.11) which again establishes itself as a very useful criterion for determining the range of validity of the linear response results. In the limit $N \to \infty$ the l.h.s. of (14.11) reads $\lim_{N \to \infty} \sum_q |c_q^{(1)}(t)|^2 = \int dE \, |c_E^{(1)}(t)|^2$. To evaluate this quantity we introduce the function

$$K(\epsilon, t) = e^{-i\epsilon t/2} \frac{\sin\left(\frac{\epsilon}{2}t\right)}{\left(\frac{\epsilon}{2}\right)}, \tag{14.27}$$

in terms of which the linear response probability density reads

$$|c_E^{(1)}(t)|^2 = (\lambda S)^2 |K(\omega - E + E_i, t)|^2. \tag{14.28}$$

The square modulus $|K(\epsilon, t)|^2$ is displayed in Fig. 14.4 as a function of ϵ for different values of t. It has a peak in $\epsilon = 0$ whose height increases as t^2 and whose width decreases as $2\pi/t$. For $t \to \infty$ we can approximate $|K(\epsilon, t)|^2$ as a "square barrier" of height t^2, width $2\pi/t$, centered in $\epsilon = 0$, i.e.,[5]

$$\lim_{t \to \infty} |K(\epsilon, t)|^2 \sim \lim_{t \to \infty} t^2 \, \theta\left(\frac{\pi}{t} - |\epsilon|\right) = \lim_{t \to \infty} 2\pi t \, \delta(\epsilon), \tag{14.29}$$

where in the last step we use the representation of the δ-function

$$\delta(\epsilon) = \lim_{\xi \to 0} \theta(\xi - |\epsilon|)/(2\xi).$$

---

[5]A rigorous derivation of (14.29) can be found in any textbook of quantum mechanics, see, e.g., Ref. [120].

At first sight (14.29) does not seem to be correct since from the definition (14.27)

$$\lim_{\epsilon \to 0} |K(\epsilon, t)|^2 = t^2, \tag{14.30}$$

which grows *quadratically* in t, as opposed to the linear dependence in (14.29). This apparent contradiction can easily be explained by the fact that $\delta(0)$ is proportional to the time volume, i.e., $\delta(0) \sim \lim_{t \to \infty} t$. Indeed taking the limit $\epsilon \to 0$ in (14.29) before the limit $t \to \infty$ we see that

$$\lim_{\epsilon \to 0} \delta(\epsilon) = \delta(0) = \lim_{t \to \infty} \lim_{\epsilon \to 0} \frac{t}{2\pi} \theta \left(\frac{\pi}{t} - |\epsilon| \right) = \lim_{t \to \infty} \frac{t}{2\pi}, \tag{14.31}$$

and hence (14.29) and (14.30) agree for large t. Using (14.29) in (14.28) we find

$$\lim_{t \to \infty} \int_{-\infty}^{\infty} dE \, |c_E^{(1)}(t)|^2 = \lim_{t \to \infty} (\lambda S)^2 2\pi \, t = \lim_{t \to \infty} \frac{2t}{\tau} \to \infty. \tag{14.32}$$

Thus the inequality (14.11) is not fulfilled when t becomes larger than $\tau/2$. We conclude that the mere fact of having a continuum of states does not eliminate the problems of linear response theory at long times. We could have anticipated this conclusion using the following intuitive argument. Since $\hat{H}'$ does not couple the continuum states the only possible transitions are $|\Psi_i\rangle \to |\Psi_E\rangle \to |\Psi_i\rangle \to |\Psi_{E'}\rangle \to \ldots$ If the probability density for the transition $|\Psi_i\rangle \to |\Psi_E\rangle$ is peaked around some energy E_r [as it is in our case with $E_r = E_i + \omega$, see (14.25)] the dynamics of the discrete–continuum system is very similar to the dynamics of the resonant two level system (with energies E_i and E_r) analyzed in the previous section. From this argument we expect that a coupling between the continuum states is necessary for linear response theory to be reliable at all times. This is the topic of the next section.

Before moving to the next section, however, we still have to check whether the linear response results are as bad also for the off-resonance coefficients. In the right panel of Fig. 14.3 we display $|c_E(t)|^2$ for $E = \Lambda/2 = 5S^2$ (we recall that $\Lambda = 10S^2$ in the figure). Before the critical time $t_c(N)$ the oscillations have a period T_p of about $1.25/S^2$, which is consistent with the frequency $|\omega - E + E_i| = E = 2\pi/T_p \sim 5S^2$ of the solution (14.24). For large N the probability density $|c_E(t)|^2$ approaches a limiting function whose asymptotic value is smaller than that for $E = 0$, again in agreement with (14.24). The linear response solution (14.26) is accurate up to times $t < \tau$, see right-bottom panel. The loss of accuracy is solely due to the fact that the amplitude of the oscillations in $|c_E^{(1)}|^2$ is not damped, as the frequency of the oscillations in c_E and $c_E^{(1)}$ is the *same*. This behavior is qualitatively different from the behavior of systems with a discrete spectrum where 1) *both* the exact and the linear response probabilities are not damped, and 2) the frequency of the oscillations in c_E and $c_E^{(1)}$ is *different* and hence $c_E^{(1)}$ eventually goes out of phase.

To summarize, in systems with a continuum spectrum coupled to discrete states, linear response theory fails in reproducing the coefficients $c_E(t)$ at the resonant energies, misses the damping and violates the inequality (14.11) when $t \to \infty$. As we shall see, the coupling between continuum states alleviates all these problems.

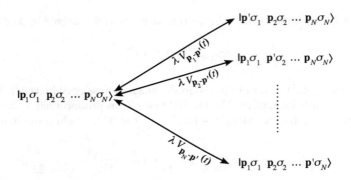

Figure 14.5 Coupling between continuum states.

14.2.3 Continuum–continuum coupling

For systems with a continuum spectrum there exist several physical situations in which the perturbation $\hat{H}'$ couples the states of the continuum. An example is the perturbation $\hat{H}'(t) = \int d\mathbf{x}\, V(\mathbf{r}, t)\hat{\psi}^\dagger(\mathbf{x})\hat{\psi}(\mathbf{x})$ added to the free-particle Hamiltonian $\hat{H} = -\frac{1}{2}\int d\mathbf{x}\,\hat{\psi}^\dagger(\mathbf{x})$ $\nabla^2\hat{\psi}(\mathbf{x})$ for particles with unit mass and charge. This Hamiltonian is diagonal in the momentum–spin basis

$$\hat{H} = \sum_\sigma \int \frac{d\mathbf{p}}{(2\pi)^3} \frac{p^2}{2}\, \hat{d}^\dagger_{\mathbf{p}\sigma}\hat{d}_{\mathbf{p}\sigma},$$

where $\hat{d}^\dagger_{\mathbf{p}\sigma}$ $(\hat{d}_{\mathbf{p}\sigma})$ creates (annihilates) a particle with momentum $\mathbf{p}$ and spin σ [see also (1.65)]. Expressing $\hat{H}'$ in terms of the $\hat{d}$-operators we find

$$\hat{H}'(t) = \sum_\sigma \int \frac{d\mathbf{p}}{(2\pi)^3} \frac{d\mathbf{p}'}{(2\pi)^3} V_{\mathbf{p}-\mathbf{p}'}(t)\hat{d}^\dagger_{\mathbf{p}\sigma}\hat{d}_{\mathbf{p}'\sigma},$$

with $V_{\mathbf{p}}(t) = \int d\mathbf{r}\, e^{-i\mathbf{p}\cdot\mathbf{r}} V(\mathbf{r}, t)$, the Fourier transform of the potential. Thus a generic continuum eigenket $|\mathbf{p}_1\sigma_1 \ldots \mathbf{p}_N\sigma_N\rangle = \hat{d}^\dagger_{\mathbf{p}_N\sigma_N}\ldots\hat{d}^\dagger_{\mathbf{p}_1\sigma_1}|0\rangle$ is coupled to infinitely many eigenkets of the continuum, as illustrated in Fig. 14.5. The coupling between two eigenkets is $\lambda V_{\mathbf{p}_i-\mathbf{p}'}(t)$ with $\mathbf{p}_{i=1,\ldots,N}$, one of the momenta of the initial state, and $\mathbf{p}'$ an arbitrary other momentum. The more localized is the potential in real space, the more spread is its Fourier transform; in the limiting case $V(\mathbf{x}, t) = V(t)\delta(\mathbf{r})$, the amplitude $V_{\mathbf{p}-\mathbf{p}'}(t) = V(t)$ for all $\mathbf{p}$ and $\mathbf{p}'$. In the following we investigate the time evolution of a system that can be considered as a simplified version of this example.

Let us consider a Hamiltonian $\hat{H}$ with only continuum eigenkets $|\Psi_q\rangle$ coupled by the perturbation $\hat{H}'(t)$. For simplicity we take the eigenenergies $E_q = q(\Lambda/N) = q\Delta E$, $q = -N, \ldots, N$, equally spaced as illustrated in the left panel of Fig. 14.6. We consider a perturbation of the form

$$\hat{H}'(t) = \Delta E\, f(t) \sum_{qq'} |\Psi_q\rangle\langle\Psi_{q'}| \quad \Rightarrow \quad T_{qq'}(t) = \Delta E\, f(t), \tag{14.33}$$

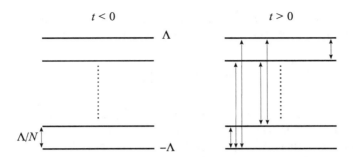

Figure 14.6 Representation of the energy levels of the equilibrium system (left panel) and of the time-dependent perturbation (right panel).

with $f(t)$ a real function of time, see right panel of Fig. 14.6. The prefactor ΔE guarantees that $\hat{H}'$ has the correct scaling properties when $N \to \infty$. Indeed, if we define the continuum kets $|\Psi_E\rangle = |\Psi_q\rangle/\sqrt{\Delta E}$ (as we did in the previous section) orthonormalized to the δ-function we have

$$\lim_{N \to \infty} \hat{H}'(t) = f(t) \int dE\,dE' |\Psi_E\rangle\langle\Psi_{E'}|.$$

We can think of $|\Psi_E\rangle$ as the eigenkets of a free particle in one dimension or as the eigenkets of N-particles, $N-1$ of which are blocked while one is free to move in a one-dimensional continuum.[6]

The linear system of equations (14.8) for the coefficients c_q with $T_{qq'}$ from (14.33) reads

$$i\frac{d}{dt}c_q(t) = \lambda\Delta E\,f(t)\sum_{q'} e^{i\omega_{qq'}t}c_{q'}(t) = \lambda\Delta E\,e^{iE_qt}f(t)C(t), \tag{14.34}$$

where we define the function

$$C(t) \equiv \sum_{q'} e^{-iE_{q'}t}c_{q'}(t).$$

To solve (14.34) we need a boundary condition for the c_q. We assume that the ket at time $t_0 = 0$ is one of the eigenkets of $\hat{H}$, and in accordance with the discussion of the previous section we normalize it as $|\Psi(0)\rangle = |\Psi_{q_0}\rangle/\sqrt{\Delta E}$, which implies

$$\langle\Psi(0)|\Psi(0)\rangle = \frac{1}{\Delta E} \xrightarrow[N \to \infty]{} \delta(0). \tag{14.35}$$

This choice of the initial state implies the boundary condition $c_q(0) = \delta_{qq_0}/\sqrt{\Delta E}$. Then the integral of (14.34) between 0 and t yields

$$c_q(t) - \frac{\delta_{qq_0}}{\sqrt{\Delta E}} = -i\lambda\Delta E \int_0^t dt'\, e^{iE_qt'} f(t')C(t'). \tag{14.36}$$

---

[6]These are, for instance, the eigenkets of a single ionized atom or molecule in which the free electron has radial momentum p and fixed angular momentum quantum numbers l, m.

From (14.36) it is possible to derive a simple integral equation for the function $C(t)$. Multi-plying both sides by $e^{-iE_q t}$ and summing over q we get

$$C(t) - \frac{e^{-iE_0 t}}{\sqrt{\Delta E}} = -i\lambda \int_0^t dt' \Delta E \sum_q e^{-iE_q(t-t')} f(t')C(t'), \qquad (14.37)$$

where $E_0 = E_{q_0}$. This equation can be solved numerically for any finite N, and the solution can be inserted in (14.36) to extract the coefficients $c_q(t)$. Since we have made no approximations, the normalization $\sum_q |c_q(t)|^2$ remains equal to $1/\Delta E$ for all times.

The analytic calculations can be pushed further when N is very large (or equivalently ΔE very small). For $N \gg 1$ the exponential $e^{-iE_q(t-t')}$ is a slowly varying function of q up to times $t \lesssim 2\pi/\Delta E$, and hence the sum over q in (14.37) is well approximated by an integral over energy,

$$C(t) - \frac{e^{-iE_0 t}}{\sqrt{\Delta E}} = -i\lambda \int_0^t dt' \int_{-\Lambda}^{\Lambda} dE \, e^{-iE(t-t')} f(t')C(t'), \qquad t \lesssim \frac{2\pi}{\Delta E}. \qquad (14.38)$$

Equation (14.38) is an iterative integral equation for $C(t)$. The first iteration consists in replacing $C(t')$ in the r.h.s. with $e^{-iE_0 t'}/\sqrt{\Delta E}$. If $|E_0| \ll \Lambda$ the contributions to the time integral with $|E| \gg |E_0|$ are negligible, since $e^{i(E-E_0)t'}$ is a rapidly varying function of t' and the integral over t' of $e^{i(E-E_0)t'} f(t') \sim 0$ (Riemann–Lebesgue theorem). The reader can easily check that the same reasoning applies to all the subsequent terms of the iteration. Therefore we can safely extend the domain of integration from $(-\Lambda, \Lambda)$ to $(-\infty, \infty)$. Then the energy integral yields $2\pi\delta(t-t')$ and the integral over t' becomes trivial. The final result for $C(t)$ is

$$C(t) = \frac{1}{\sqrt{\Delta E}} \frac{e^{-iE_0 t}}{1 + i\pi\lambda f(t)}, \qquad t \lesssim \frac{2\pi}{\Delta E}. \qquad (14.39)$$

Inserting (14.39) in (14.36) we also obtain the solution for the coefficients c_q,

$$c_q(t) = \frac{1}{\sqrt{\Delta E}} \left[\delta_{qq_0} - i\lambda\Delta E \int_0^t dt' e^{i(E_q - E_0)t'} \frac{f(t')}{1 + i\pi\lambda f(t')} \right], \qquad t \lesssim \frac{2\pi}{\Delta E}. \qquad (14.40)$$

Before continuing an observation is in order. Let $Q(t)$ denote the quantity in the square bracket. Then $|Q(t)|$ must be less than unity for all t for otherwise the coefficient c_q would have modulus larger than $1/\sqrt{\Delta E}$ and hence the normalization of the state would become larger than $1/\Delta E$. It is instructive to calculate $Q(t)$ in some simple case like, e.g., $f(t) = f_0\theta(t)$:

$$Q(t) = \delta_{qq_0} - i\lambda\Delta E \frac{f_0}{1 + i\pi\lambda f_0} e^{i\omega_{qq_0} t/2} \frac{\sin\left(\frac{\omega_{qq_0} t}{2}\right)}{\left(\frac{\omega_{qq_0}}{2}\right)}. \qquad (14.41)$$

Now note that for energies E_q close to E_0 the ratio $\sin(\omega_{qq_0}t/2)/(\omega_{qq_0}/2) \sim t$ and hence $|Q(t)| \to \infty$ for large t. Is there anything wrong? No. The coefficient $c_q(t)$ is given by the r.h.s. of (14.40) only up to times $t \sim 2\pi/\Delta E$, meaning that we cannot take the limit $t \to \infty$ before the limit $\Delta E \to 0$.

As already pointed out several times for continuous spectra, it makes more sense to work with the coefficient $c_E = c_q/\sqrt{\Delta E}$ whose square modulus represents the *probability density* of finding the state at energy E. Multiplying (14.40) by $1/\sqrt{\Delta E}$ and taking the limit $\Delta E \rightarrow 0$ we find the following nice result:

$$c_E(t) = \delta(E - E_0) - i\lambda \int_0^t dt' e^{i(E-E_0)t'} \frac{f(t')}{1 + i\pi\lambda f(t')}, \tag{14.42}$$

which is valid *for all times*. The exact solution (14.42) contains plenty of information of general character and we therefore discuss it in some detail below.

Normalization: Let us verify that the normalization $\int dE |c_E(t)|^2$ is equal to $\delta(0)$, see (14.35), for all times t and all values of λ. The modulus square of $c_E(t)$ is the sum of four terms one of which is $\delta(E - E_0)^2$. This term gives $\delta(0)$ upon integration. Thus the integral over E of the remaining three terms must be identically zero. The integral of the cross products contains $\delta(E - E_0)$ and can easily be evaluated. We find

$$\int dE |c_E(t)|^2 = \delta(0) - 2\pi \int_0^t dt' \frac{(\lambda f(t'))^2}{1 + (\pi\lambda f(t'))^2}$$
$$+ \int dE \int_0^t dt' dt'' e^{i(E-E_0)t'} \frac{\lambda f(t')}{1 + i\pi\lambda f(t')} e^{-i(E-E_0)t''} \frac{\lambda f(t'')}{1 - i\pi\lambda f(t'')}.$$

The integral over E in the last term on the r.h.s. yields $2\pi\delta(t' - t'')$ and upon integration over t'' we recognize that it exactly cancels the second term on the r.h.s.. The solution (14.42) preserves the normalization for all t!

Photon absorption/emission: Let us calculate $c_E(t)$ when $f(t)$ is a dichromatic perturbation of the form, e.g., $f(t) = \theta(t) f_0 \cos(\omega t)$.[7] We expand (14.42) in powers of λ and denote by $c_E^{(n)}$ the nth order term of the expansion. Taking into account that

$$\frac{-i\lambda f(t)}{1 + i\pi\lambda f(t)} = \sum_{n=0}^{\infty} \pi^n [-i\lambda f(t)]^{n+1} = \sum_{n=0}^{\infty} \pi^n \left(-\frac{i\lambda f_0}{2}\right)^{n+1} \left(e^{i\omega t} + e^{-i\omega t}\right)^{n+1}$$
$$= \sum_{n=0}^{\infty} \pi^n \left(-\frac{i\lambda f_0}{2}\right)^{n+1} \sum_{k=0}^{n+1} \binom{n+1}{k} e^{-i(n+1-2k)\omega t},$$

the quantity $c_E^{(n)}$ can be expressed in terms of the function $K(\epsilon, t)$ defined in (14.27):

$$c_E^{(n)}(t) = \pi^{n-1} \left(-\frac{i\lambda f_0}{2}\right)^n \sum_{k=0}^{n} \binom{n}{k} K((n-2k)\omega - E + E_0, t); \tag{14.43}$$

for $n = 0$ we have to add $\delta(E - E_0)$ to the r.h.s.. Let us comment on this result. Since $K(\epsilon, t)$ has a peak in $\epsilon = 0$ whose width decreases in time, (14.43) tells us that only states with energy E such that $E - E_0$ is an integer multiple of ω have a nonvanishing probability

---

[7]In our example we cannot consider purely monochromatic perturbations since we chose $T_{qq'}(t)$ independent of q and q' which implies that $T_{qq'}(t)$ must be real. To form a real function with monochromatic perturbations the minimum number of frequencies is two, i.e., ω and $-\omega$.

of being excited when $t \to \infty$. The well known quantum mechanical interpretation of this result is that the transition to a state with energy $E = E_0 + m\omega$ occurs via absorption or emission of $|m|$ quanta (photon) of energy ω. We also see that to describe a process involving the absorption/emission of $|m|$ quanta it is necessary to include at least the $|m|$th order term in the perturbative expansion. Therefore linear response theory gives us access to processes with the absorption/emission of just one photon. This is a completely general result.

Linear response solution: To first order in λ the solution (14.42) reads

$$c_E(t) \sim c_E^{(0)}(t) + c_E^{(1)}(t) = \delta(E - E_0) - i\lambda \int_0^t dt' e^{i(E - E_0)t'} f(t'). \qquad (14.44)$$

If $\lambda \ll 1$ then the approximate result (14.44) is in excellent agreement with the exact result (14.42) for all times and energies. We have thus found that for systems with a continuous spectrum and for perturbations that couple the continuum states the theory of linear response works well _even at long times_. In the example of the quantum discharge of a capacitor considered in Section 7.2.3 we were in exactly this situation: the time-dependent bias in (7.29) couples the eigenstates of the equilibrium Hamiltonian in a similar way as illustrated in Fig. 14.5.

We can learn something more about the performance of linear response theory by studying the coefficient c_E for the dichromatic perturbation considered above. From (14.43) we have

$$c_E^{(1)}(t) = -\frac{i\lambda f_0}{2} \{K(\omega - E + E_0, t) + K(-\omega - E + E_0, t)\},$$

and hence for $E \neq E_0$

$$|c_E(t)|^2 \sim |c_E^{(1)}(t)|^2 = \left(\frac{\lambda f_0}{2}\right)^2 \{|K(\omega - E + E_0, t)|^2 + |K(-\omega - E + E_0, t)|^2$$
$$+ 2\mathrm{Re}\,[K(\omega - E + E_0, t)K^*(-\omega - E + E_0, t)]\}. \quad (14.45)$$

In Fig. 14.7 we show the quantity in the curly bracket as a function of E for different values of t. As expected it has two main peaks located at $E = E_0 \pm \omega$ whose height increases as t^2 and whose width narrows as $2\pi/t$. In the limit $t \to \infty$ the cross product between the two K functions [second row of (14.45)] approaches zero (in a distribution sense) and using (14.29) we obtain the following asymptotic behavior:

$$\lim_{t \to \infty} |c_E^{(1)}(t)|^2 = \lim_{t \to \infty} \left(\frac{\lambda f_0}{2}\right)^2 2\pi t\,[\delta(\omega - E + E_0) + \delta(-\omega - E + E_0)]. \qquad (14.46)$$

This is our first encounter with the so called _Fermi golden rule:_ the probability density is nonvanishing provided that $E = E_0 \pm \omega$ (energy conservation) and increases linearly in time. In the previous section (discrete–continuum coupling) the same asymptotic behavior $|c_E^{(1)}(t)|^2 \sim t \times \delta$-function led to the violation of the upper bound (14.11). Why is it not so in the present case? We can answer by evaluating the l.h.s. of (14.11). Using (14.46) we find

$$\lim_{t \to \infty} \int_{E \neq E_0} dE\,|c_E^{(1)}(t)|^2 = \lim_{t \to \infty} 2\left(\frac{\lambda f_0}{2}\right)^2 2\pi t = 2\left(\frac{2\pi\lambda f_0}{2}\right)^2 \times \delta(0), \qquad (14.47)$$

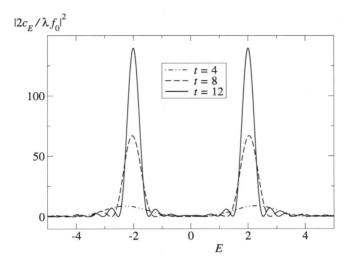

Figure 14.7 The function in the curly bracket of (14.45) versus E (in a.u.) for $E_0 = 0$, $\omega = 2$ a.u., and for different values of t (in a.u.).

where in the last equality we take (14.31) into account. If $\lambda \ll 1$ the r.h.s. of this equation is smaller than $\delta(0) = \langle \Psi(0) | \Psi(0) \rangle$ and (14.11) is fulfilled. The difference between this case and the one of the previous section should now be evident. In (14.47) the term proportional to t must be compared to $\delta(0)$ while in (14.32) it must be compared to 1.

14.3 Fermi golden rule

We are now ready to study the general case of a Hamiltonian $\hat{H}$ with both discrete and continuum eigenkets and a perturbation $\hat{H}'(t)$ which couples discrete states to discrete and continuum states as well as continuum states to continuum states. We denote by $|\Psi_j\rangle$ the discrete eigenkets of $\hat{H}$ with energies E_j. The continuum eigenkets $|\Psi_\alpha\rangle$ of $\hat{H}$ are labeled with the collective index $\alpha = (E, Q)$ where E is the continuum energy and Q is a set of discrete or continuum quantum numbers which specify the physical state uniquely. In order to highlight the typical structure of the linear response formulas we find it convenient to introduce a notation that treats discrete and continuum states on an equal footing. We define the label a which runs over both discrete states, $a = j$, and continuum states, $a = \alpha$, and the shorthand notation $\int da \equiv \sum_j + \int d\alpha$. The eigenkets of the Hamiltonian $\hat{H}$ are therefore represented by the set $\{|\Psi_a\rangle\}$ with inner product

$$\langle \Psi_a | \Psi_{a'} \rangle = \delta(a - a'), \tag{14.48}$$

where $\delta(a-a') = \delta_{jj'}$ if $a = j$ and $a' = j'$, and $\delta(a-a') = \delta(\alpha-\alpha') = \delta(E-E')\delta(Q-Q')$ if $a = \alpha$ and $a' = \alpha'$. In the examples of Section 14.2 there was no Q quantum number since the energy E was enough to specify the continuum eigenkets.

One advantage of the new notation is that $\hat{H}$ and $\hat{H}'(t)$ take the compact form

$$\hat{H} = \int da\, E_a |\Psi_a\rangle\langle\Psi_a|, \qquad \hat{H}'(t) = \int da\, da'\, T_{aa'}(t)|\Psi_a\rangle\langle\Psi_{a'}|.$$

Similarly, the expansion of the ket $|\Psi(t)\rangle$ reads

$$|\Psi(t)\rangle = \int da\, c_a(t) e^{-iE_a t}|\Psi_a\rangle.$$

Using the inner product (14.48), the time-dependent Schrödinger equation $i\frac{d}{dt}|\Psi(t)\rangle = (\hat{H} + \lambda\hat{H}'(t))|\Psi(t)\rangle$ can be shown to be equivalent to

$$i\frac{d}{dt}c_a(t) = \lambda \int da'\, T_{aa'}(t) e^{i\omega_{aa'}t} c_{a'}(t), \qquad (14.49)$$

with the obvious notation $\omega_{aa'} = E_a - E_{a'}$. If $|\Psi(0)\rangle = |\Psi_{a_0}\rangle$ is an eigenket at time $t_0 = 0$, then $c_a(0) = \delta(a - a_0)$ and to first order in λ the solution of (14.49) reads

$$c_a(t) \sim c_a^{(0)}(t) + c_a^{(1)}(t) = \delta(a - a_0) - i\lambda \int_0^t dt'\, T_{aa_0}(t') e^{i\omega_{aa_0}t'}. \qquad (14.50)$$

The analytic calculations can be carried out a little further when the perturbation is monochromatic. This is not a serious restriction since, as we have already observed, any perturbation can be written as a linear combination (Fourier transform) of monochromatic perturbations. For simplicity we set the diagonal element $T_{a_0 a_0}(t) = 0$; its inclusion is straightforward and is left as an exercise for the reader. As for the off-diagonal elements, we take the monochromatic dependence $T_{aa_0}(t) = T_{aa_0}e^{-i\omega t}$. Inserting this form into (14.50) we obtain

$$c_a^{(1)}(t) = -i\lambda T_{aa_0} K(\omega - \omega_{aa_0}, t),$$

from which we can extract the long-time limit of the probability (if $a = j$) or of the probability density (if $a = \alpha$) for the transition $a_0 \to a$ in the usual manner,

$$\lim_{t\to\infty} |c_a^{(1)}(t)|^2 = |\lambda T_{aa_0}|^2 2\pi\, t\, \delta(\omega - \omega_{aa_0}). \qquad (14.51)$$

Below we study the solution (14.51) for $|\Psi(t_0)\rangle = |\Psi_{a_0}\rangle$ which is either a discrete state or a continuum state.

First case: $|\Psi(t_0)\rangle$ *is a discrete state.* Let $|\Psi(0)\rangle = |\Psi_i\rangle$ be one of the discrete eigenkets of $\hat{H}$. Then the normalization is $\langle\Psi(0)|\Psi(0)\rangle = 1$. From (14.50) and (14.51) the probability for the transition to another discrete state $j \neq i$ is

$$\lim_{t\to\infty} |c_j(t)|^2 \sim \lim_{t\to\infty} |c_j^{(1)}(t)|^2 = |\lambda T_{ji}|^2 2\pi\, t\, \delta(\omega - \omega_{ji}).$$

This result tells us that for $\omega = \omega_{ji}$ the probability $|c_j^{(1)}(t)|^2$ grows as $t\delta(0) \sim t^2$ (unless $T_{ji} = 0$) and eventually becomes larger than 1. Consequently the probability overcomes the upper bound (14.11) and the theory of linear response breaks down. Regarding the probability density for the transition to a continuum state α we have

$$\lim_{t\to\infty} |c_\alpha(t)|^2 \sim \lim_{t\to\infty} |c_\alpha^{(1)}(t)|^2 = |\lambda T_{\alpha i}|^2 2\pi\, t\, \delta(\omega - \omega_{\alpha i}). \tag{14.52}$$

Even though the probability density can vary between 0 and ∞, the continuum states too are responsible for the breakdown of the upper bound (14.11). Indeed (14.52) implies

$$\lim_{t\to\infty} \int d\alpha\, |c_\alpha^{(1)}(t)|^2 = \lim_{t\to\infty} 2\pi\, t\, \lambda^2 \int dQ\, |T_{\alpha i}|^2_{E=E_i+\omega} \to \infty,$$

where in the last step we split the integral over $\alpha = (E, Q)$ into an integral over energy E and quantum number Q. It is noteworthy, however, that the divergence of the continuum states is milder than that of the discrete states. The former goes as $\sim t$ independent of the value of ω while the latter goes as $t\delta(0) \sim t^2$ when the frequency ω matches a Bohr frequency.

Second case: $|\Psi(t_0)\rangle$ *is a continuum state.* Let us study the linear response solution when $|\Psi(0)\rangle = |\Psi_{\alpha_0}\rangle$ is one of the continuum eigenkets of $\hat{H}$. In this case the normalization is $\langle\Psi(0)|\Psi(0)\rangle = \delta_\alpha(0)$. The notation $\delta_\alpha(0) = \delta(\alpha - \alpha)$ is used to distinguish the δ-function in α-space from the δ-function in energy space $\delta(0) = \delta(E - E)$. We may write that $\delta_\alpha(0) = \delta(0)\delta_Q(0)$ with $\delta_Q(0) = \delta(Q - Q)$, the δ-function in Q space. The quantity $\delta_Q(0)$ is unity only provided that Q is a discrete quantum number; if Q is an angle (or a set of angles) or a momentum (or a set of momenta) or any other kind of continuous quantum number then the quantity $\delta_Q(0)$ is infinite.

The probability for the transition to a discrete state j can be extracted from (14.51) by setting $a = j$ and $a_0 = \alpha_0$:

$$\lim_{t\to\infty} |c_j(t)|^2 \sim \lim_{t\to\infty} |c_j^{(1)}(t)|^2 = |\lambda T_{j\alpha_0}|^2 2\pi\, t\, \delta(\omega - \omega_{j\alpha_0}). \tag{14.53}$$

For $\omega = \omega_{j\alpha_0}$ we obtain the (by now) familiar divergence $t\delta(0)$. This divergence, however, does not necessarily imply that the inequality (14.11) is violated when $t \to \infty$. Indeed for large t we can write $t\delta(0) \sim \delta(0)\delta(0)$ which must be compared with the initial normalization $\delta_\alpha(0) = \delta(0)\delta_Q(0)$. Only if Q is a discrete quantum number or if $\delta(0)$ "diverges faster" than $\delta_Q(0)$ when Q is a continuum quantum number, then the probability $|c_j^{(1)}(t \to \infty)|^2$ exceeds the upper bound in (14.11). In all other cases (14.11) is fulfilled and other criteria must be found to assess the quality of the linear response approximation.

Next we consider the probability density for the transition to another continuum state $\alpha \neq \alpha_0$. From (14.50) and (14.51) we find

$$\lim_{t\to\infty} |c_\alpha(t)|^2 \sim \lim_{t\to\infty} |c_\alpha^{(1)}(t)|^2 = |\lambda T_{\alpha\alpha_0}|^2 2\pi\, t\, \delta(\omega - \omega_{\alpha\alpha_0}). \tag{14.54}$$

We can check the reliability of this result by replacing the sum over k in (14.11) with the integral over $\alpha \neq \alpha_0$. We have

$$\lim_{t\to\infty} \int_{\alpha\neq\alpha_0} d\alpha\, |c_\alpha^{(1)}(t)|^2 = \left[(2\pi\lambda)^2 \int dQ\, |T_{\alpha\alpha_0}|^2_{E=E_0+\omega} \right] \times \delta(0), \tag{14.55}$$

where we replace t with $2\pi\delta(0)$ and define E_0 as the eigenenergy of $|\Psi_{\alpha_0}\rangle$. Excluding pathological perturbations, the integral over Q is either finite (when Q is a discrete quantum number[8]) or at most divergent like $\delta_Q(0)$ (when the coupling is extremely localized in Q-space as, e.g., $T_{\alpha\alpha_0} = T_{EE_0}\delta(Q - Q_0)$). Thus, for sufficiently small λ the r.h.s. of (14.55) is always smaller than $\delta_\alpha(0) = \delta(0)\delta_Q(0)$ and the inequality (14.11) is fulfilled. This result is a generalization of what we found in Section 14.2.3, see (14.47).

General remarks: From the previous analysis we can conclude that if the frequency of the perturbation is far off the Bohr frequencies $\omega_{j\alpha_0} = E_j - E_0$, then the linear response coefficients fulfill the upper bound (14.11). We stress, however, that (14.11) is not the only condition to fulfill for the long-time results to be accurate and reliable. For instance the damping discussed in Section 14.2.2 cannot be captured in linear response. Things considerably improve when there are only continuum states. In most of these cases the linear response formulas are accurate also when $t \to \infty$ and, admittedly, very elegant. Considering again (14.54) we see that the probability density $|c_\alpha(t)|^2$ grows linearly in time provided that the energy $E = E_0 + \omega$. Therefore it makes sense to define a probability density per unit time and cast (14.54) in the form

$$\boxed{\lim_{t\to\infty} \frac{d}{dt}|c_\alpha(t)|^2 \sim 2\pi|\lambda T_{\alpha\alpha_0}|^2\delta(E - E_0 - \omega), \qquad \alpha \neq \alpha_0}\tag{14.56}$$

This equation is known as the *Fermi golden rule* and it is of great practical use.[9] For example the power P fed into the system is equal to the integral of the probability density per unit time to excite a state of energy E times the energy $E - E_0$ of the transition. In the long time limit we can write

$$\lim_{t\to\infty} P(t) = \lim_{t\to\infty} \int d\alpha \,(E - E_0)\frac{d}{dt}|c_\alpha(t)|^2 \sim 2\pi\omega \int dQ\, |\lambda T_{\alpha\alpha_0}|^2_{E=E_0+\omega}.$$

Thus to lowest order in λ the system can absorb energy only via transitions to states with energy $E = E_0 + \omega$. This is in agreement with the discussion in Section 14.2.3, where we observed that the linear response theory can only describe the absorption or emission of a single photon.

14.4 Kubo formula

The coefficients of the expansion of $|\Psi(t)\rangle$ constitute the basic ingredients to calculate the time-dependent quantum average of any operator $\hat{O}(t)$. We have

$$\langle\Psi(t)|\hat{O}(t)|\Psi(t)\rangle = \int da\,da'\, c_a^*(t)c_{a'}(t)e^{i(E_a - E_{a'})t}\langle\Psi_a|\hat{O}(t)|\Psi_{a'}\rangle$$

$$= \int da\,da'\, c_a^*(t)c_{a'}(t)\langle\Psi_a|\hat{O}_H(t)|\Psi_{a'}\rangle,\tag{14.57}$$

---

[8]In this case the integral over Q is actually a sum.

[9]Despite its name the Fermi golden rule was derived the first time by Dirac in 1927 [121]. In 1949 Fermi gave a course on Nuclear Physics at the University of Chicago (his lectures have been collected in a book titled *Nuclear Physics*, see Ref. [122]) in which he rederived (14.56) and coined for it the name "Golden rule No. 2." This name was so appealing that people started to refer to (14.56) initially as "Golden rule No. 2, by E. Fermi" and eventually as the "Fermi golden rule," see also Ref. [123].

with $\hat{O}_H(t) = e^{i\hat{H}t}\hat{O}(t)e^{-i\hat{H}t}$ the operator in the Heisenberg picture (we remind the reader that we set $t_0 = 0$). To first order in λ the coefficients $c_a(t)$ are given in (14.50). Since the coupling $T_{aa_0}(t) = \langle\Psi_a|\hat{H}'(t)|\Psi_{a_0}\rangle$, we can write

$$c_a(t) \sim \langle\Psi_a|\hat{\mathbb{1}} - i\lambda \int_0^t dt'\, \hat{H}'_H(t')|\Psi(0)\rangle, \qquad |\Psi(0)\rangle = |\Psi_{a_0}\rangle. \tag{14.58}$$

This result can also be seen as a direct consequence of (14.5). Indeed, $c_a(t)e^{-iE_a t} = \langle\Psi_a|\Psi(t)\rangle = \langle\Psi_a|\hat{U}_{\text{tot}}(t,0)|\Psi(0)\rangle$, and using the first order approximation (14.5) for the evolution operator we recover (14.58). Inserting (14.58) into (14.57) and performing the integral over a and a' ($\int da\,|\Psi_a\rangle\langle\Psi_a| = \hat{\mathbb{1}}$) we obtain

$$\delta O(t) = -i\lambda \int_0^t dt'\, \langle\Psi(0)|\left[\hat{O}_H(t), \hat{H}'_H(t')\right]_-|\Psi(0)\rangle, \tag{14.59}$$

where

$$\delta O(t) \equiv \left[\langle\Psi(t)|\hat{O}(t)|\Psi(t)\rangle - \langle\Psi(0)|\hat{O}(t)|\Psi(0)\rangle\right]_{\text{first order in } \lambda}$$

represents the change of the time-dependent quantum average induced by the external perturbation to first order in λ. Equation (14.59) is known as the *Kubo formula*. The Kubo formula lends itself to be generalized to time-dependent ensemble averages. We simply have to calculate $\delta O(t)$ for $|\Psi(0)\rangle = |\Psi_a\rangle$ and then average over all a with probabilities ρ_a that the system is in $|\Psi_a\rangle$:

$$\delta O(t) = -i\lambda \int da\, \rho_a \int_0^t dt'\, \langle\Psi_a|\left[\hat{O}_H(t), \hat{H}'_H(t')\right]_-|\Psi_a\rangle$$
$$= -i\lambda \int_0^t dt'\, \frac{\text{Tr}\left[e^{-\beta\hat{H}^{\text{M}}}\left[\hat{O}_H(t), \hat{H}'_H(t')\right]_-\right]}{\text{Tr}\left[e^{-\beta\hat{H}^{\text{M}}}\right]} = -i\lambda \int_0^t dt'\, \langle\left[\hat{O}_H(t), \hat{H}'_H(t')\right]_-\rangle, \tag{14.60}$$

where we use the short-hand notation $\langle\ldots\rangle$ to denote ensemble averages. Equation (14.60) is the starting point for our subsequent derivations. In Chapter 15 we lay down the basis for a formulation of linear response theory in terms of Green's functions.

Before concluding we should make an important remark about the possibility of expanding in powers of λ the change of a time-dependent average. To first order we see that (14.60) involves an average of operators in the Heisenberg picture with the *unperturbed* Hamiltonian $\hat{H}$. Therefore at zero temperature one might be tempted to use the zero-temperature formalism (see Section 5.4) to evaluate this average. Unfortunately, however, the zero-temperature formalism gives us access only to *time-ordered* averages. In the special case of (14.60) there exists a simple trick (see for instance Ref. [124]) to extract the average of the commutator from the average of $T\{\hat{O}_H(t)\hat{H}'_H(t')\}$. Going beyond linear response this is not possible anymore. For instance to second order in λ the change of the time-dependent average involves a double commutator $\left[\left[\hat{O}_H(t), \hat{H}'_H(t')\right]_-, \hat{H}'_H(t'')\right]_-$ and there is no trick to calculate this quantity from the average of $T\{\hat{O}_H(t)\hat{H}'_H(t')\hat{H}'_H(t'')\}$. Second order changes are relevant every time the linear order change vanishes. This is the case of photoemission spectra where the current is quadratic in the radiative coupling $\hat{H}' = \hat{H}_{l-e}$. In Section 6.3.4 we developed a simple theory of photoemission based on the assumption that the photoelectron was suddenly kicked out of the sample. A more microscopic theory consists in calculating the second-order change of the average of the occupation operator $\hat{O} = \hat{d}_\epsilon^\dagger \hat{d}_\epsilon$ for electrons of energy ϵ [125, 126]. In this context MBPT has to be formulated on the contour [127] despite the fact that all averages are equilibrium averages.

15

Linear response theory: many-body formulation

15.1 Current and density response function

In the previous chapter we learned how to approximate the evolution operator to first order in the time-dependent perturbation. We separated the total Hamiltonian into a time-independent part $\hat{H}$ plus a "small" time-dependent perturbation $\lambda\hat{H}'(t)$, and derived the result (14.5). The careful reader might have realized that the idea of Section 14.1 can be exploited in much more general situations. In fact, the derivation remains valid also when $\lambda\hat{H}'(t)$ is added to a *time-dependent* Hamiltonian. Let us make this point clear. Consider a system described by the time-dependent Hamiltonian $\hat{H}(t)$ and subject to a further, but "small," time-dependent perturbation $\lambda\hat{H}'(t)$. The full evolution operator obeys the differential equation

$$\mathrm{i}\frac{d}{dt}\hat{U}_{\mathrm{tot}}(t,t_0) = \left[\hat{H}(t) + \lambda\hat{H}'(t)\right]\hat{U}_{\mathrm{tot}}(t,t_0),$$

with boundary condition $\hat{U}_{\mathrm{tot}}(t_0,t_0) = \hat{\mathbb{1}}$. As in Section 14.1 we look for solutions of the form $\hat{U}_{\mathrm{tot}}(t,t_0) = \hat{U}(t,t_0)\hat{F}(t)$ where $\hat{U}(t,t_0)$ is the evolution operator for the system with Hamiltonian $\hat{H}(t)$. Following the same steps as in Section 14.1 we soon arrive at the result (14.5), the only difference being that the operators are in the Heisenberg picture with time-dependent Hamiltonian $\hat{H}(t)$. Since (14.5) is the only ingredient in the derivation of the Kubo formula, (14.60) also does not change. It is instructive to rederive the Kubo formula in this more general context. We will follow an alternative (and shorter) path from which the generality and the simplicity of (14.60) can be better appreciated. The time-dependent ensemble average of the operator $\hat{O}(t)$ is

$$O(t) = \langle\hat{U}_{\mathrm{tot}}(t_0,t)\hat{O}(t)\hat{U}_{\mathrm{tot}}(t,t_0)\rangle$$
$$\sim \langle\left[\hat{\mathbb{1}} + \mathrm{i}\lambda\int_{t_0}^t dt'\,\hat{H}'_H(t')\right]\hat{U}(t_0,t)\hat{O}(t)\hat{U}(t,t_0)\left[\hat{\mathbb{1}} - \mathrm{i}\lambda\int_{t_0}^t dt'\,\hat{H}'_H(t')\right]\rangle$$
$$= \langle\hat{O}_H(t)\rangle - \mathrm{i}\lambda\int_{t_0}^t dt'\,\langle\left[\hat{O}_H(t),\hat{H}'_H(t')\right]_-\rangle + \mathcal{O}(\lambda^2).$$

In this equation the symbol $\langle\ldots\rangle$ denotes the ensemble average as in Section 14.4. Letting $\delta O(t) = O(t) - \langle\hat{O}_H(t)\rangle$ be the change of the time-dependent average of $\hat{O}(t)$ to first order in λ we find again the Kubo formula

$$\delta O(t) = -\mathrm{i}\lambda \int_{t_0}^t dt' \left\langle \left[\hat{O}_H(t), \hat{H}'_H(t')\right]_-\right\rangle \tag{15.1}$$

In this section we use the linear response theory to study a system of interacting particles with mass m and charge q moving under the influence of an external vector potential $\mathbf{A}(\mathbf{x}, t)$ and scalar potential $V(\mathbf{x}, t)$.[1] The time-dependent Hamiltonian is therefore $\hat{H}(t) = \hat{H}_0(t) + \hat{H}_{\mathrm{int}}$ with

$$\hat{H}_0(t) = \frac{1}{2m}\int d\mathbf{x}\,\hat{\psi}^\dagger(\mathbf{x})\left(-\mathrm{i}\boldsymbol{\nabla} - \frac{q}{c}\mathbf{A}(\mathbf{x}, t)\right)^2 \hat{\psi}(\mathbf{x}) + q\int d\mathbf{x}\,V(\mathbf{x}, t)\hat{n}(\mathbf{x}), \tag{15.2}$$

and $\hat{n}(\mathbf{x}) = \hat{\psi}^\dagger(\mathbf{x})\hat{\psi}(\mathbf{x})$ the density operator.[2] To be concrete we take the system in thermodynamic equilibrium at times $t < t_0$. The equilibrium Hamiltonian $\hat{H}^{\mathrm{M}} = \hat{H}_0 + \hat{H}_{\mathrm{int}} - \mu\hat{N}$ has the one-body part $\hat{H}_0$ given by (15.2) but with static vector and scalar potentials $\mathbf{A}(\mathbf{x})$ and $V(\mathbf{x})$. The question we ask is: how does the time evolution change under a change of the external potentials $\mathbf{A} \to \mathbf{A} + \delta\mathbf{A}$ and $V \to V + \delta V$ for times $t > t_0$? To first order in $\delta\mathbf{A}$ and δV the change of $\hat{H}_0$ can most easily be worked out from (3.29) and reads[3]

$$\hat{H}_0(t) \to \hat{H}_0(t) - \frac{q}{c}\int d\mathbf{x}\,\hat{\mathbf{j}}(\mathbf{x})\cdot\delta\mathbf{A}(\mathbf{x}, t) + \int d\mathbf{x}\,\hat{n}(\mathbf{x})\left(q\delta V(\mathbf{x}, t) + \frac{q^2}{mc^2}\mathbf{A}(\mathbf{x}, t)\cdot\delta\mathbf{A}(\mathbf{x}, t)\right).$$

The perturbation on the r.h.s. of this equation is the explicit form of $\lambda\hat{H}'(t)$. Let us introduce a few definitions and rewrite the perturbation in a more transparent way. We first note that the terms linear in $\delta\mathbf{A}$ can be grouped to form the (gauge-invariant) current density operator $\hat{\mathbf{J}}$ defined in (3.31). It is therefore natural to define the four-dimensional vector of operators $(\hat{n}, \hat{\mathbf{J}}) = (\hat{n}, \hat{J}_x, \hat{J}_y, \hat{J}_z)$ with components $\hat{J}_\mu$, $\mu = 0, 1, 2, 3$, so that $\hat{J}_0 = \hat{n}$ and $\hat{J}_1 = \hat{J}_x$, etc. Similarly, we define the four-dimensional vector $(\delta V, -\delta\mathbf{A}/c)$ with components δA^μ, $\mu = 0, 1, 2, 3$, so that $\delta A^0 = V$ and $\delta A^1 = -\delta A_x/c$, etc. Then the perturbation takes the following compact form:

$$\lambda\hat{H}'(t) = q\int d\mathbf{x}\,\hat{J}_\mu(\mathbf{x}, t)\,\delta A^\mu(\mathbf{x}, t),$$

with the Einstein convention of summing over repeated upper and lower indices.

With the explicit form of $\lambda\hat{H}'(t)$ we can try to calculate some physical quantity like, e.g., the change in the time-dependent density n and current density $\mathbf{J}$. From the Kubo formula (15.1) these changes are given by

$$\delta J_\mu(\mathbf{x}, t) = -\mathrm{i}q\int_{t_0}^t dt' \int d\mathbf{x}'\left\langle\left[\hat{J}_{\mu,H}(\mathbf{x}, t), \hat{J}_{\nu,H}(\mathbf{x}', t')\right]_-\right\rangle \delta A^\nu(\mathbf{x}', t'). \tag{15.3}$$

---

[1]Even though the vector and the scalar potentials do not depend on spin we use the variable $\mathbf{x}$ for notational convenience.

[2]For simplicity we do not include the Pauli coupling between the spin of the particles and the magnetic field.

[3]The last term of (3.29) has been ignored since it is a total divergence.

To make the link with quantities already introduced in the previous chapters we replace the current operators in the above commutator with the deviation current operators

$$\Delta \hat{J}_{\mu,H}(\mathbf{x},t) \equiv \hat{J}_{\mu,H}(\mathbf{x},t) - \langle \hat{J}_{\mu,H}(\mathbf{x},t) \rangle,$$

and rewrite (15.3) as

$$\delta J_\mu(\mathbf{x},t) = -iq \int_{t_0}^t dt' \int d\mathbf{x}' \, \langle \left[\Delta \hat{J}_{\mu,H}(\mathbf{x},t), \Delta \hat{J}_{\nu,H}(\mathbf{x}',t') \right]_- \rangle \, \delta A^\nu(\mathbf{x}',t'). \tag{15.4}$$

It is clear that (15.4) is equivalent to (15.3) since a number commutes with all operators. The structure of (15.4) prompts us to define the following correlator on the contour[4]

$$\boxed{\chi_{\mu\nu}(\mathbf{x},z;\mathbf{x}',z') \equiv -i \langle \mathcal{T} \left\{ \Delta \hat{J}_{\mu,H}(\mathbf{x},z) \Delta \hat{J}_{\nu,H}(\mathbf{x}',z') \right\} \rangle} \tag{15.5}$$

which has the symmetry property

$$\boxed{\chi_{\mu\nu}(1;2) = \chi_{\nu\mu}(2;1)} \tag{15.6}$$

The Keldysh function $\chi_{\mu\nu}$ can be converted into standard time-dependent averages by choosing the contour arguments on different branches. For example the greater/lesser components of $\chi_{\mu\nu}$ are

$$\chi_{\mu\nu}^>(\mathbf{x},t;\mathbf{x}',t') = \chi_{\mu\nu}(\mathbf{x},t_+;\mathbf{x}',t'_-) = -i \langle \Delta \hat{J}_{\mu,H}(\mathbf{x},t) \Delta \hat{J}_{\nu,H}(\mathbf{x}',t') \rangle, \tag{15.7}$$

$$\chi_{\mu\nu}^<(\mathbf{x},t;\mathbf{x}',t') = \chi_{\mu\nu}(\mathbf{x},t_-;\mathbf{x}',t'_+) = -i \langle \Delta \hat{J}_{\nu,H}(\mathbf{x}',t') \Delta \hat{J}_{\mu,H}(\mathbf{x},t) \rangle$$
$$= - \left[\chi_{\mu\nu}^>(\mathbf{x},t;\mathbf{x}',t') \right]^*. \tag{15.8}$$

Since $\chi_{\mu\nu}$ has no singular part, i.e., $\chi_{\mu\nu}^\delta = 0$, the retarded component reads

$$\chi_{\mu\nu}^R(\mathbf{x},t;\mathbf{x}',t') = \theta(t-t') \left[\chi_{\mu\nu}^>(\mathbf{x},t;\mathbf{x}',t') - \chi_{\mu\nu}^<(\mathbf{x},t;\mathbf{x}',t') \right]$$
$$= -i\theta(t-t') \langle \left[\Delta \hat{J}_{\mu,H}(\mathbf{x},t), \Delta \hat{J}_{\nu,H}(\mathbf{x}',t') \right]_- \rangle, \tag{15.9}$$

which is exactly the kernel of (15.4). Thus we can rewrite (15.4) as

$$\boxed{\delta J_\mu(\mathbf{x},t) = q \int_{-\infty}^\infty dt' \int d\mathbf{x}' \, \chi_{\mu\nu}^R(\mathbf{x},t;\mathbf{x}',t') \, \delta A^\nu(\mathbf{x}',t')} \tag{15.10}$$

where we take into account that δA^ν vanishes for times smaller than t_0. The function $\chi_{\mu\nu}^R(\mathbf{x},t;\mathbf{x}',t')$ is a real function. This can be seen directly from the second line of (15.9) or, equivalently, from the first line of (15.9) using the relation (15.8) between $\chi^<$ and the complex conjugate of $\chi^>$. The correlator $\chi_{\mu\nu}$ has different names depending on the values of μ and ν. For $\mu = \nu = 0$ we have the component $\chi_{00} = \chi$ which is the *density*

---

[4]In (15.5) the operators are in the Heisenberg picture on the contour, see (4.38).

response function already encountered in (11.44) and subsequently in (12.18), see below. The components χ_{0j} and χ_{j0}, with $j = 1, 2, 3$, are called the *current-density response functions*, and the components χ_{jk} with $j, k = 1, 2, 3$ are called the *current response functions*. All these components can be expressed in terms of the two-particle XC function (12.16) that we rewrite here for convenience

$$
\begin{aligned}
L(1, 2; 1', 2') &= \pm [G_2(1, 2; 1', 2') - G(1; 1')G(2; 2')] \\
&= \mp \langle \mathcal{T} \{ \hat{\psi}_H(1)\hat{\psi}_H(2)\hat{\psi}_H^\dagger(2')\hat{\psi}_H^\dagger(1') \} \rangle \\
&\quad \pm \langle \mathcal{T} \{ \hat{\psi}_H(1)\hat{\psi}_H^\dagger(1') \} \rangle \langle \mathcal{T} \{ \hat{\psi}_H(2)\hat{\psi}_H^\dagger(2') \} \rangle,
\end{aligned} \tag{15.11}
$$

where, as usual, the upper sign is for bosons and the lower sign for fermions. For instance the density response function in (15.5) is explicitly given by

$$
\chi_{00}(1; 2) = -\mathrm{i} \langle \mathcal{T} \{ \hat{n}_H(1)\hat{n}_H(2) \} \rangle + \mathrm{i}n(1)n(2) = \pm \mathrm{i}L(1, 2; 1^+, 2^+), \tag{15.12}
$$

which agrees with the result (12.18) derived diagrammatically. Similarly the current-density response function and the current response function are related to L by

$$
\begin{aligned}
\chi_{0j}(1; 2) &= \pm \mathrm{i} \left[\left(\frac{\partial_{2,j} - \partial_{2',j}}{2mi} - \frac{q}{mc} A_j(2) \right) L(1, 2; 1^+, 2') \right]_{2'=2^+} \\
&= \pm \mathrm{i} \left[\left(\frac{D_{2,j} - D_{2',j}^*}{2mi} \right) L(1, 2; 1^+, 2') \right]_{2'=2^+}, \\
\chi_{j0}(1; 2) &= \pm \mathrm{i} \left[\left(\frac{D_{1,j} - D_{1',j}^*}{2mi} \right) L(1, 2; 1', 2^+) \right]_{1'=1^+}, \\
\chi_{jk}(1; 2) &= \pm \mathrm{i} \left[\left(\frac{D_{1,j} - D_{1',j}^*}{2mi} \right) \left(\frac{D_{2,k} - D_{2',k}^*}{2mi} \right) L(1, 2; 1', 2') \right]_{\substack{1'=1^+ \\ 2'=2^+}}, \tag{15.13}
\end{aligned}
$$

with $\mathbf{D}_1 = \boldsymbol{\nabla}_1 - \mathrm{i}\frac{q}{c}\mathbf{A}(1)$ the (gauge-invariant) derivative already introduced in (8.2). Due to the symmetry property $L(1, 2; 1', 2') = L(2, 1; 2', 1')$ of the exact (as well as of any conserving approximation to the) two-particle XC function the symmetry property (15.6) is satisfied.

The result (15.10) as well as the connection between $\chi_{\mu\nu}$ and L are certainly nice results, but what are the advantages of using these formulas? Ultimately they only give us access to the linear response change δJ_μ. Why don't we instead calculate the *full* J_μ using the Green's function G, an object much easier than $\chi_{\mu\nu}$ or L? The point here is that we are interested in changes with respect to *equilibrium* quantities. If we specialize the above formulas to systems in equilibrium ($\hat{H}_0(t)$ independent of time) then it may be more advantageous to calculate equilibrium correlators, like $\chi_{\mu\nu}$ and L, rather than the nonequilibrium Green's function G. Indeed, in the presence of time-dependent external fields G depends on two time coordinates whereas the equilibrium response functions $\chi_{\mu\nu}$ depend only on the time difference.

In Chapter 12 we learned that to calculate L and hence $\chi_{\mu\nu}$ we must solve the Bethe–Salpeter equation. In practice this is done by approximating the kernel K in some way.

We then encounter another question: how are the δJ_μ calculated from an approximate L via (15.10) and the δJ_μ calculated directly from an approximate G related? This question is of fundamental importance since we would like to export everything that we derived on conserving approximations into the linear response world, so that all basic conservation laws are automatically built in. In Section 15.3 we show that the δJ_μ coming from a conserving G with self-energy Σ is the same as the δJ_μ coming from (15.10) where L is the solution of the Bethe–Salpeter equation with kernel $K = \pm\delta\Sigma/\delta G$. The proof is carried out in the general case of a time-dependent perturbation added to a pre-existing time-dependent electromagnetic field, since no extra complications arise from this. Another motivation for keeping the formulas so general is related to an important identity between L and G that we prove in Section 15.4. Let us, however, first see what physics is contained in the equilibrium response functions.

15.2 Lehmann representation

When the system is not perturbed by external fields, the time evolution operator is simply the exponential $\hat{U}(t, t_0) = \exp[-\mathrm{i}\hat{H}(t - t_0)]$. Let us see what this simplification leads to. Consider for example the greater component of the response function given in (15.7)

$$\chi_{\mu\nu}^{>}(\mathbf{x}, t; \mathbf{x}', t') = -\mathrm{i}\frac{\mathrm{Tr}\left[e^{-\beta(\hat{H}-\mu\hat{N})}e^{\mathrm{i}\hat{H}(t-t_0)}\Delta\hat{J}_\mu(\mathbf{x})e^{-\mathrm{i}\hat{H}(t-t')}\Delta\hat{J}_\nu(\mathbf{x}')e^{-\mathrm{i}\hat{H}(t'-t_0)}\right]}{\mathrm{Tr}\left[e^{-\beta(\hat{H}-\mu\hat{N})}\right]}$$

$$= -\mathrm{i}\int da\,\rho_a\,\langle\Psi_a|\Delta\hat{J}_\mu(\mathbf{x})e^{-\mathrm{i}(\hat{H}-E_a)(t-t')}\Delta\hat{J}_\nu(\mathbf{x}')|\Psi_a\rangle, \tag{15.14}$$

where the integral is over the (continuum and/or discrete) quantum number a of the eigenkets $|\Psi_a\rangle$ of $\hat{H}$ with energy E_a and number of particles N_a. The weights ρ_a are therefore

$$\rho_a = \frac{e^{-\beta(E_a-\mu N_a)}}{\mathrm{Tr}\left[e^{-\beta(\hat{H}-\mu\hat{N})}\right]}. \tag{15.15}$$

As expected $\chi^>$ depends only on the time difference $t - t'$. A similar result can be worked out for the lesser component and hence we can define the Fourier transform according to

$$\chi_{\mu\nu}^{\gtrless}(\mathbf{x}, t; \mathbf{x}', t') = \int\frac{d\omega}{2\pi}e^{-\mathrm{i}\omega(t-t')}\chi_{\mu\nu}^{\gtrless}(\mathbf{x}, \mathbf{x}'; \omega).$$

Using the same trick that led to (6.79) it is easy to find the following exact relation:

$$\chi_{\mu\nu}^{>}(\mathbf{x}, \mathbf{x}'; \omega) = e^{\beta\omega}\chi_{\mu\nu}^{<}(\mathbf{x}, \mathbf{x}'; \omega)$$

between the greater and lesser components.[5] As for the equilibrium Green's function, self-energy, polarizability, etc. we then have a fluctuation–dissipation theorem for the equilibrium $\chi_{\mu\nu}$. Omitting the position–spin variables we can write

$$\chi_{\mu\nu}^{>}(\omega) = \bar{f}(\omega)\left[\chi_{\mu\nu}^{\mathrm{R}}(\omega) - \chi_{\mu\nu}^{\mathrm{A}}(\omega)\right],$$

---

[5]This identity for $\chi_{00} = \chi$ was proved in Section 13.2 using the diagrammatic expansion.

$$\chi_{\mu\nu}^<(\omega) = f(\omega) \left[\chi_{\mu\nu}^R(\omega) - \chi_{\mu\nu}^A(\omega)\right],$$

with $f(\omega) = 1/(e^{\beta\omega} - 1)$ the Bose function and $\bar{f}(\omega) = 1 + f(\omega)$.

The Fourier transform of $\chi_{\mu\nu}^{\lessgtr}$ can be used to calculate the Fourier transform of $\chi_{\mu\nu}^R$ since from (15.9)

$$\chi_{\mu\nu}^R(\mathbf{x}, t; \mathbf{x}', t') = \int \frac{d\omega}{2\pi} e^{-i\omega(t-t')} \chi_{\mu\nu}^R(\mathbf{x}, \mathbf{x}'; \omega)$$

$$= \theta(t - t') \int \frac{d\omega'}{2\pi} e^{-i\omega'(t-t')} \left[\chi_{\mu\nu}^>(\mathbf{x}, \mathbf{x}'; \omega') - \chi_{\mu\nu}^<(\mathbf{x}, \mathbf{x}'; \omega')\right]. \quad (15.16)$$

Taking into account the representation (6.57) of the Heaviside function we get

$$\chi_{\mu\nu}^R(\mathbf{x}, \mathbf{x}'; \omega) = i \int \frac{d\omega'}{2\pi} \frac{\chi_{\mu\nu}^>(\mathbf{x}, \mathbf{x}'; \omega') - \chi_{\mu\nu}^<(\mathbf{x}, \mathbf{x}'; \omega')}{\omega - \omega' + i\eta} \quad (15.17)$$

with η an infinitesimal positive constant. Similarly one can show that

$$\chi_{\mu\nu}^A(\mathbf{x}, \mathbf{x}'; \omega) = i \int \frac{d\omega'}{2\pi} \frac{\chi_{\mu\nu}^>(\mathbf{x}, \mathbf{x}'; \omega') - \chi_{\mu\nu}^<(\mathbf{x}, \mathbf{x}'; \omega')}{\omega - \omega' - i\eta}. \quad (15.18)$$

The Fourier transforms (15.17) or (15.18) contain plenty of physical information and they play a central role in this chapter.

A nice warm-up exercise which illustrates some of the physics contained in $\chi_{\mu\nu}^R$ is the calculation of the total energy dissipated by a system which is invariant under translations (like the electron gas) and which is spin unpolarized (meaning that the ensemble average of the spin density is zero everywhere). In such systems $\chi_{\mu\nu}^R(\mathbf{x}, \mathbf{x}'; \omega)$ depends only on the difference $\mathbf{r} - \mathbf{r}'$. We define its Fourier transform in momentum space according to

$$\sum_{\sigma\sigma'} \chi_{\mu\nu}^R(\mathbf{x}, \mathbf{x}'; \omega) = \int \frac{d\mathbf{p}}{(2\pi)^3} e^{i\mathbf{p}\cdot(\mathbf{r}-\mathbf{r}')} \chi_{\mu\nu}^R(\mathbf{p}, \omega). \quad (15.19)$$

The dissipated energy E_{diss} is the integral over time of (8.16)

$$E_{\text{diss}} = q \int_{-\infty}^{\infty} dt \int d\mathbf{x}\, \mathbf{J}(\mathbf{x}, t) \cdot \mathbf{E}_{\text{ext}}(\mathbf{r}, t) = q \int \frac{d\omega d\mathbf{p}}{(2\pi)^4} \mathbf{J}(\mathbf{p}, \omega) \cdot \mathbf{E}_{\text{ext}}(-\mathbf{p}, -\omega), \quad (15.20)$$

where q is the charge of the particles and $\mathbf{J}(\mathbf{p}, \omega)$ is the Fourier transform of the current $\mathbf{J}(\mathbf{r}, t) = \sum_\sigma \mathbf{J}(\mathbf{x}, t)$ summed over all spin projections.[6] For simplicity we consider a longitudinal external electric field and choose the gauge in which the vector potential $\mathbf{A} = 0$. In linear response we then have $\mathbf{E}_{\text{ext}}(\mathbf{r}, t) = -\boldsymbol{\nabla}\delta V(\mathbf{x}, t)$ and the current $\mathbf{J}$ in (15.20) is the first order change $\delta\mathbf{J}$ calculated in the previous section. Let us now manipulate these quantities a little. In Fourier space $\mathbf{E}_{\text{ext}}(-\mathbf{p}, -\omega) = \mathbf{E}_{\text{ext}}^*(\mathbf{p}, \omega) = i\mathbf{p}\delta V^*(\mathbf{p}, \omega)$.[7] Substitution of this result into (15.20) generates the scalar product $\mathbf{p} \cdot \delta\mathbf{J}(\mathbf{p}, \omega)$ which can be expressed in terms of the density change using the continuity equation in Fourier space, i.e., $-i\omega\delta n(\mathbf{p}, \omega) + i\mathbf{p} \cdot \delta\mathbf{J}(\mathbf{p}, \omega) = 0$, with $\delta n(\mathbf{p}, \omega)$ the Fourier transform of the density $n(\mathbf{r}, t) = \sum_\sigma n(\mathbf{x}, t)$. The density change is in turn given in (15.10) and reads $\delta n(\mathbf{p}, \omega) =$

---

[6]The Fourier transform of a function $f(\mathbf{r}, t)$ is defined in the usual manner $f(\mathbf{r}, t) = \int \frac{d\omega d\mathbf{p}}{(2\pi)^4} e^{i\mathbf{p}\cdot\mathbf{r} - i\omega t} f(\mathbf{p}, \omega)$.

[7]For real functions $f(\mathbf{r}, t)$ the Fourier transform $f(\mathbf{p}, \omega) = f^*(-\mathbf{p}, -\omega)$.

$q\chi_{00}^{R}(\mathbf{p},\omega)\delta V(\mathbf{p},\omega)$ since in our gauge $\delta\mathbf{A} = 0$. Collecting these results we find the following expression for the dissipated energy of a gas:

$$E_{\text{diss}} = q^2 \int \frac{d\omega d\mathbf{p}}{(2\pi)^4} \frac{i\omega}{p^2} \chi^{R}(\mathbf{p},\omega) |\mathbf{E}_{\text{ext}}(\mathbf{p},\omega)|^2, \tag{15.21}$$

with $\chi^{R} = \chi_{00}^{R}$, in agreement with our notation. Clearly in (15.21) only the imaginary part of χ^{R} contributes since E_{diss} is a real quantity.[8] It is then convenient to define the so called *energy-loss function* $\mathcal{L}$ as

$$\mathcal{L}(\mathbf{p},\omega) \equiv -\frac{4\pi}{p^2}\text{Im}[\chi^{R}(\mathbf{p},\omega)] \quad \Rightarrow \quad E_{\text{diss}} = \frac{q^2}{4\pi} \int \frac{d\omega d\mathbf{p}}{(2\pi)^4} \omega \mathcal{L}(\mathbf{p},\omega) |\mathbf{E}_{\text{ext}}(\mathbf{p},\omega)|^2.$$

Thus $\mathcal{L}$ tells us for which frequencies and momenta the system can dissipate energy. Since our system is initially in thermodynamic equilibrium the dissipated energy must be positive for positive frequencies (absorption) and negative for negative frequencies (emission). This implies that

$$\text{Im}[\chi^{R}(\mathbf{p},\omega)] \quad \begin{matrix} < 0 & \text{for } \omega > 0 \\ > 0 & \text{for } \omega < 0 \end{matrix}. \tag{15.22}$$

Any approximation to χ that violates the sign property (15.22) also violates the energy conservation law. Equation (15.21) has been obtained without any assumption on the form of the interparticle interaction. In the special case of Coulombic interactions $\tilde{v}_{\mathbf{p}} = 4\pi/p^2$, and the Fourier transform of the retarded inverse dielectric function, see (L.18), is

$$\varepsilon^{-1,R}(\mathbf{p},\omega) = 1 + \tilde{v}_{\mathbf{p}}\chi^{R}(\mathbf{p},\omega) = 1 + \frac{4\pi}{p^2}\chi^{R}(\mathbf{p},\omega). \tag{15.23}$$

Hence the energy loss function becomes

$$\mathcal{L}(\mathbf{p},\omega) = -\text{Im}\left[\varepsilon^{-1,R}(\mathbf{p},\omega)\right].$$

The formulas for the response functions at finite temperature are a bit more cumbersome than those at zero temperature. Thus we discuss here only the zero-temperature case and leave the generalization to finite temperature as an exercise for the curious reader. At zero temperature only the ground state $|\Psi_0\rangle$ contributes to the integral over a in (15.14) and, again for simplicity, we assume that $|\Psi_0\rangle$ is nondegenerate and normalized to 1. Then (15.14) simplifies to

$$\chi_{\mu\nu}^{>}(\mathbf{x},t;\mathbf{x}',t') = -i\langle\Psi_0|\Delta\hat{J}_\mu(\mathbf{x})e^{-i(\hat{H}-E_0)(t-t')}\Delta\hat{J}_\nu(\mathbf{x}')|\Psi_0\rangle$$
$$= -i\int db\, e^{-i(E_b-E_0)(t-t')} f_{\mu,b}(\mathbf{x})f_{\nu,b}^*(\mathbf{x}'), \tag{15.24}$$

where in the second equality we have inserted the completeness relation $\int db|\Psi_b\rangle\langle\Psi_b| = \hat{\mathbb{1}}$ and defined the so called *excitation amplitudes*

$$\boxed{f_{\mu,b}(\mathbf{x}) \equiv \langle\Psi_0|\Delta\hat{J}_\mu(\mathbf{x})|\Psi_b\rangle}$$

---

[8]As we shall see the real part of $\chi^{R}(\mathbf{p},\omega)$ is even in ω and therefore does not give any contribution to the r.h.s. of (15.21).

We see that the excitation amplitudes $f_{\mu,b}$ vanish if the number of particles in the excited state $|\Psi_b\rangle$ differs from the number of particles in the ground state $|\Psi_0\rangle$. From (15.24) and the like for $\chi^<_{\mu\nu}$ we can readily extract the Fourier transforms

$$\chi^>_{\mu\nu}(\mathbf{x},\mathbf{x}';\omega) = -i\int db\, 2\pi\delta(\omega - \Omega_b)f_{\mu,b}(\mathbf{x})f^*_{\nu,b}(\mathbf{x}') = -[\chi^>_{\nu\mu}(\mathbf{x}',\mathbf{x};\omega)]^*,$$

$$\chi^<_{\mu\nu}(\mathbf{x},\mathbf{x}';\omega) = -i\int db\, 2\pi\delta(\omega + \Omega_b)f^*_{\mu,b}(\mathbf{x})f_{\nu,b}(\mathbf{x}') = -[\chi^<_{\nu\mu}(\mathbf{x}',\mathbf{x};\omega)]^*,$$

with $\Omega_b = E_b - E_0 > 0$ the excitation energies. Substituting these expressions into (15.17) we find the *Lehmann representation* of the retarded response function

$$\chi^R_{\mu\nu}(\mathbf{x},\mathbf{x}';\omega) = \int db\left[\frac{f_{\mu,b}(\mathbf{x})f^*_{\nu,b}(\mathbf{x}')}{\omega - \Omega_b + i\eta} - \frac{f^*_{\mu,b}(\mathbf{x})f_{\nu,b}(\mathbf{x}')}{\omega + \Omega_b + i\eta}\right] = [\chi^A_{\nu\mu}(\mathbf{x}',\mathbf{x};\omega)]^*. \qquad (15.25)$$

The Lehmann representation allows us to understand what physical information is contained in χ^R. In the remainder of this section we explore and discuss this result.

15.2.1 Analytic structure

For real frequencies ω the retarded response function (15.25) has the property

$$\boxed{\chi^R_{\mu\nu}(\mathbf{x},\mathbf{x}';-\omega) = \chi^R_{\mu\nu}(\mathbf{x},\mathbf{x}';\omega)^*}$$

which implies that the real part is an even function of ω whereas the imaginary part is an odd function of ω.[9] For complex ω we see from (15.25) that χ^R is analytic in the upper half plane. This analyticity property together with the fact that $\chi^R_{\mu\nu} \to 0$ for large ω implies that $\chi^R(\mathbf{x},t;\mathbf{x}',t')$ vanishes for t' larger than t due to the Cauchy residue theorem. Hence the densities $\delta J_\mu(\mathbf{x},t)$ are sensitive to changes in the potentials $\delta A^\nu(\mathbf{x}',t')$ only if these changes occur at times $t' < t$. We then say that χ^R has the *causality property* or that χ^R is causal. In the lower half plane, instead, χ^R has either simple poles in $\pm\Omega_b - i\eta$ when b is a discrete quantum number or the branch cuts along the real axis when b is a continuum quantum number. Consequently the plot of the imaginary part of χ^R as a function of the real frequency ω exhibits δ-like peaks in correspondence with the discrete excitation energies, but is a smooth curve in the continuum of excitations and is zero everywhere else.

For positive frequencies the Lehmann representation (15.25) yields

$$\text{Im}\left[\chi^R_{\mu\nu}(\mathbf{x},\mathbf{x}';\omega)\right] = -\pi\int db\, f_{\mu,b}(\mathbf{x})f^*_{\nu,b}(\mathbf{x}')\delta(\omega - \Omega_b), \qquad \omega > 0.$$

If we think of $\text{Im}\left[\chi^R_{\mu\nu}\right]$ as a matrix with indices $\mathbf{x},\mathbf{x}'$ the above result implies that the matrices $\text{Im}\left[\chi^R_{\mu\mu}\right]$ are negative definite for $\omega > 0$, i.e.,

$$\alpha_\mu(\omega) \equiv \int d\mathbf{x}d\mathbf{x}'\alpha^*(\mathbf{x})\,\text{Im}\left[\chi^R_{\mu\mu}(\mathbf{x},\mathbf{x}';\omega)\right]\alpha(\mathbf{x}') < 0,$$

---

[9]This property is also a direct consequence of the fact that $\chi^R_{\mu\nu}(\mathbf{x},t;\mathbf{x}',t')$ is a real function.

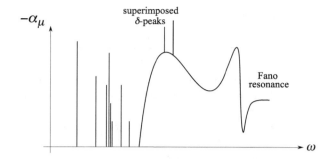

Figure 15.1 General form of a diagonal matrix element of $\mathrm{Im}\left[\chi^{\mathrm{R}}_{\mu\mu}(\omega)\right]$.

for any complex function $\alpha(\mathbf{x})$. This property, together with the fact that the imaginary part is an odd function of ω, represents the generalization of the sign property (15.22) to arbitrary systems. The general form of the function $\alpha_\mu(\omega)$ looks as in Fig. 15.1, where the sharp vertical lines represent δ-like peaks. In principle the δ-like peaks can also be superimposed on the continuum; this is actually a common feature of noninteracting systems. In interacting systems the discrete and continuum noninteracting excitations are coupled by the interaction (unless prohibited by symmetry) and the superimposed δ-like peaks transform into resonances with a characteristic asymmetric lineshape. This effect was pointed out by Fano [14] and is illustrated in Fig. 15.1. In general the continuum part can have different shapes depending on the nature of the elementary excitations: Fano resonances, excitons, plasmons, etc. These excitations are all charge neutral (or particle conserving) since the energies E_b that contribute to (15.25) correspond to eigenstates with the same number of particles as in the ground state. This should be contrasted with the excitations of the Green's function which are not charge conserving since they involve eigenstates with one particle more or less, see Section 6.3.2.

From the Kubo formula (15.10) we can establish a very nice link between the physics contained in $\chi^{\mathrm{R}}_{\mu\nu}$ and the results of a time propagation. This link is particularly relevant when there are discrete excitations in the spectrum. We therefore analyze the contribution to $\chi^{\mathrm{R}}_{\mu\nu}$ coming from the discrete excitations $b = j$ in detail. To be concrete let us consider the density response function $\chi_{00} = \chi$. Substitution of (15.25) into (15.16) yields

$$\chi^{\mathrm{R}}(\mathbf{x}, t; \mathbf{x}', t') = -\mathrm{i}\theta(t - t') \sum_j \left[e^{-\mathrm{i}\Omega_j(t-t')} f_j(\mathbf{x}) f_j^*(\mathbf{x}') - e^{\mathrm{i}\Omega_j(t-t')} f_j^*(\mathbf{x}) f_j(\mathbf{x}') \right]$$
$$+ \chi^{\mathrm{R}}_{\mathrm{cont}}(\mathbf{x}, t; \mathbf{x}', t'), \tag{15.26}$$

where we drop the subscript 0 in the excitation amplitudes and where $\Omega_j = E_j - E_0$ are the discrete excitation energies. The last term on the r.h.s. of (15.26) is the part of the response function that comes from the integral over the continuum of excitations. If the vector potential $\mathbf{A}(\mathbf{x})$ (of the equilibrium system) vanishes then the excitation amplitudes can be chosen real. We assume here that this is the case, i.e., $f_j = f_j^*$, even though the general conclusion remains valid regardless of this simplification. We now perturb the system with, e.g., a scalar potential,

$$\delta V(\mathbf{x}, t) = \theta(t - t_0) \int \frac{d\omega}{2\pi} e^{-i\omega t} \delta V(\mathbf{x}, \omega),$$

and calculate the time-dependent density induced by this perturbation. Without loss of generality we take $t_0 = 0$. From the Kubo formula (15.10) the first order change in the density can be written as

$$\delta n(\mathbf{x}, t) = \int \frac{d\omega}{2\pi} \delta n_\omega(\mathbf{x}, t) + \delta n_{\text{cont}}(\mathbf{x}, t),$$

with δn_{cont} the contribution due to $\chi^{\text{R}}_{\text{cont}}$ and

$$\delta n_\omega(\mathbf{x}, t) = -i \sum_j f_j(\mathbf{x}) \delta V_j(\omega) \left[e^{-i\frac{(\omega+\Omega_j)}{2}t} \frac{\sin\left(\frac{\omega-\Omega_j}{2}t\right)}{\frac{\omega-\Omega_j}{2}} - e^{-i\frac{(\omega-\Omega_j)}{2}t} \frac{\sin\left(\frac{\omega+\Omega_j}{2}t\right)}{\frac{\omega+\Omega_j}{2}} \right],$$

where the quantities $\delta V_j(\omega) \equiv \int d\mathbf{x}' \, f_j(\mathbf{x}') \delta V(\mathbf{x}', \omega)$. As expected the linear response density $\delta n_\omega(\mathbf{x}, t)$ oscillates with frequencies $\omega \pm \Omega_j$. In accordance with the results of Chapter 14 these frequencies do not depend on the perturbation and hence the exact and the linear response densities eventually go out of phase. A view of this result from a different perspective is that the frequencies of the *exact* density change $\delta n(\mathbf{x}, t) = n(\mathbf{x}, t) - n(\mathbf{x})$ provide an approximation (of first order in the perturbing field) to the particle-conserving excitation energies of the *unperturbed* system.[10] Reading out these frequencies from the exact time-dependent solution is not always an easy task since the oscillatory behavior may get damped very fast as the time passes and only a finite number of periods may be clearly visible. Nevertheless, we have established a link which is conceptual in nature: the linear response theory can be used to interpret the exact time-dependent results in terms of particle-conserving excitations and, *vice versa*, from the exact time-dependent results we can extract information on the particle-conserving excitation spectrum. We come back to this link in Chapter 16.

15.2.2 The f-sum rule

From the Lehmann representation (15.25) we can derive an important sum rule for the density response function. The *Thomas–Reiche–Kuhn sum rule* [128–130] or simply the f-sum rule relates the first momentum of the retarded density response function χ^{R} of a system in equilibrium to the corresponding equilibrium density. To prove the f-sum rule we consider the large ω limit of (15.25) with $\mu = \nu = 0$. Dropping, as before, the subscript 0 from the excitation functions we have

$$\chi^{\text{R}}(\mathbf{x}, \mathbf{x}'; \omega) = \frac{1}{\omega} \int db \, [f_b(\mathbf{x}) f_b^*(\mathbf{x}') - f_b^*(\mathbf{x}) f_b(\mathbf{x}')]$$

$$+ \frac{1}{\omega^2} \int db \, \Omega_b \, [f_b(\mathbf{x}) f_b^*(\mathbf{x}') + f_b^*(\mathbf{x}) f_b(\mathbf{x}')] + \mathcal{O}(\frac{1}{\omega^3}). \qquad (15.27)$$

---

[10] For instance we have seen in the example of Section 14.2.1 that the *exact* frequency of the oscillations is $\omega_{\text{exact}} = \frac{1}{2}\sqrt{(\omega - \omega_{fi})^2 + 4\lambda^2 |T|^2}$ and therefore the quantity $\omega - 2\omega_{\text{exact}}$ provides an approximation (in this case of second order in λ) to the true excitation energy ω_{fi}.

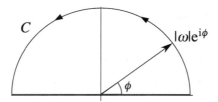

Figure 15.2 The closed contour C going from $-\infty$ to $+\infty$ along a line just above the real axis and from $+\infty$ to $-\infty$ along a semicircle of infinite radius in the upper half plane.

The definition $f_b(\mathbf{x}) = \langle \Psi_0 | \hat{n}(\mathbf{x}) | \Psi_b \rangle$ of the excitation functions and the completeness relation $\hat{1} = \int db \, | \Psi_b \rangle \langle \Psi_b |$ allow us to recognize in the first term on the r.h.s. the ground-state average of the commutator $[\hat{n}(\mathbf{x}), \hat{n}(\mathbf{x}')]_-$, which is zero. Similarly, recalling that $\Omega_b = E_b - E_0$, the second term on the r.h.s. of (15.27) can be written as the ground-state average of a double commutator

$$\chi^{\mathrm{R}}(\mathbf{x}, \mathbf{x}'; \omega) = \frac{1}{\omega^2} \langle \Psi_0 | \big[[\hat{n}(\mathbf{x}), \hat{H}]_-, \hat{n}(\mathbf{x}') \big]_- | \Psi_0 \rangle + \mathcal{O}(\frac{1}{\omega^3})$$
$$= \frac{1}{\omega^2} \left[-\frac{1}{m} \boldsymbol{\nabla} \cdot \big(n(\mathbf{x}) \boldsymbol{\nabla} \delta(\mathbf{x} - \mathbf{x}') \big) \right] + \mathcal{O}(\frac{1}{\omega^3}). \tag{15.28}$$

In the second equality we use the fact that $[\hat{n}(\mathbf{x}), \hat{H}]_- = -i\boldsymbol{\nabla} \cdot \hat{\mathbf{J}}(\mathbf{x})$ and that the commutator between the current density and the density is (see Exercise 3.5)

$$\big[\hat{\mathbf{J}}(\mathbf{x}), \hat{n}(\mathbf{x}') \big]_- = -\frac{i}{m} \hat{n}(\mathbf{x}) \boldsymbol{\nabla} \delta(\mathbf{x} - \mathbf{x}'). \tag{15.29}$$

Since χ^{R} is analytic in the upper half of the complex ω-plane the integral of $\omega \chi^{\mathrm{R}}$ along the contour C of Fig. 15.2 is zero. Then we can write

$$0 = \oint_C d\omega \, \omega \chi^{\mathrm{R}}(\mathbf{x}, \mathbf{x}'; \omega) = \int_{-\infty}^{\infty} d\omega \, \omega \chi^{\mathrm{R}}(\mathbf{x}, \mathbf{x}'; \omega) + i \lim_{|\omega| \to \infty} \int_0^{\pi} d\phi \, |\omega|^2 e^{2i\phi} \chi^{\mathrm{R}}(\mathbf{x}, \mathbf{x}'; |\omega| e^{i\phi}).$$

Using (15.28) to evaluate the integral over ϕ we obtain a relation between the first momentum of χ^{R} and the equilibrium density

$$\boxed{ \int_{-\infty}^{\infty} d\omega \, \omega \, \mathrm{Im}[\chi^{\mathrm{R}}(\mathbf{x}, \mathbf{x}'; \omega)] = \frac{\pi}{m} \boldsymbol{\nabla} \cdot \big(n(\mathbf{x}) \boldsymbol{\nabla} \delta(\mathbf{x} - \mathbf{x}') \big) } \tag{15.30}$$

where we take into account that only the imaginary part contributes to the integral since $\mathrm{Re}[\chi^{\mathrm{R}}]$ is even in ω. Equation (15.30) is known as the frequency- or f-sum rule for χ^{R}. It is easy to show that the f-sum rule is valid also at finite temperature.

The f-sum rule is alternatively written in terms of the density operator in momentum space

$$\hat{n}_{\mathbf{p}\sigma} \equiv \int d\mathbf{r} \, e^{-i\mathbf{p} \cdot \mathbf{r}} \, \hat{n}(\mathbf{x}) \qquad \Rightarrow \qquad \hat{n}(\mathbf{x}) = \int \frac{d\mathbf{p}}{(2\pi)^3} e^{i\mathbf{p} \cdot \mathbf{r}} \hat{n}_{\mathbf{p}\sigma}.$$

The average of $\hat{n}_{\mathbf{p}\sigma}$ is simply the Fourier transform of the density $n(\mathbf{x})$. The correlator between two of these operators is defined as

$$\Xi_{\sigma\sigma'}(\mathbf{p}, z, z') \equiv -\mathrm{i}\,\langle\mathcal{T}\left\{\hat{n}_{\mathbf{p}\sigma, H}(z)\hat{n}_{-\mathbf{p}\sigma', H}(z')\right\}\rangle.$$

The retarded component is directly related to the retarded component of the density response function via[ll]

$$\Xi_{\sigma\sigma'}^{\mathrm{R}}(\mathbf{p}, \omega) = \int d\mathbf{r}d\mathbf{r}'\, e^{-\mathrm{i}\mathbf{p}\cdot(\mathbf{r}-\mathbf{r}')}\chi^{\mathrm{R}}(\mathbf{x}, \mathbf{x}'; \omega). \tag{15.31}$$

Multiplying (15.30) by $e^{-\mathrm{i}\mathbf{p}\cdot(\mathbf{r}-\mathbf{r}')}$ and integrating over all $\mathbf{r}$ and $\mathbf{r}'$ we then find a sum rule for $\Xi_{\sigma\sigma'}^{\mathrm{R}}$

$$\int_{-\infty}^{\infty} d\omega\,\omega\,\Xi_{\sigma\sigma'}^{\mathrm{R}}(\mathbf{p}, \omega) = -\mathrm{i}\delta_{\sigma\sigma'}\frac{\pi p^2}{m}\int d\mathbf{r}\,n(\mathbf{x}) = -\mathrm{i}\delta_{\sigma\sigma'}\frac{\pi p^2}{m}N_\sigma,$$

with N_σ the total number of particles of spin σ.

15.2.3 Noninteracting fermions

We now wish to discuss the response function of a system of noninteracting fermions at zero temperature. There are three reasons for choosing this special case. First, we can acquire some more familiarity with the physics contained in the response function. Second, the noninteracting χ is needed to calculate the interacting χ from the Bethe–Salpeter equation. And third, it is possible to derive an important analytic formula for χ which is used later in our examples.

Let $\hat{d}_n$, $\hat{d}_n^\dagger$ be the annihilation and creation operators which diagonalize $\hat{H} = \hat{H}_0 = \sum_n \epsilon_n \hat{d}_n^\dagger \hat{d}_n$. For a given chemical potential μ the ground state $|\Psi_0\rangle$ has all levels with energy smaller than μ occupied whereas those with energy larger than μ are empty. The charge-neutral excited states $|\Psi_b\rangle$ (with the same number of fermions as in $|\Psi_0\rangle$) are obtained by "moving" one, two, three, etc. fermions from the occupied ground-state levels to some empty levels or, in other words, by creating *electron–hole pairs*. If we introduce the convention of labeling the occupied levels with indices n and the unoccupied levels with barred indices $\bar{n}$ then a generic charge-neutral excited state has the form

$$|\Psi_b\rangle = |\Psi_{n_1\ldots n_N \bar{n}_1\ldots\bar{n}_N}\rangle = \hat{d}_{\bar{n}_1}^\dagger \ldots \hat{d}_{\bar{n}_N}^\dagger \hat{d}_{n_1} \ldots \hat{d}_{n_N}|\Psi_0\rangle. \tag{15.32}$$

This state describes a system in which N fermions have been excited from their original levels $n_1, \ldots, n_N$ to the empty levels $\bar{n}_1, \ldots, \bar{n}_N$ and its energy is

$$E_b = E_0 + \sum_{j=1}^{N}(\epsilon_{\bar{n}_j} - \epsilon_{n_j}),$$

with E_0 the ground-state energy. The excitation amplitude $f_{\mu,b}$ is the matrix element of the one-body operator $\Delta\hat{J}_\mu$ between $|\Psi_0\rangle$ and $|\Psi_b\rangle$. It is then clear that the excitation

---

[ll]For a system invariant under translations the sum over σ, σ' of the r.h.s. of (15.31) yields $\mathrm{V}\chi^{\mathrm{R}}(\mathbf{p}, \omega)$, where $\mathrm{V} = \int d\mathbf{r}$ is the volume of the system and $\chi^{\mathrm{R}}(\mathbf{p}, \omega)$ is the response function defined in (15.19).

amplitudes vanish if b contains more than one electron–hole pair. Consequently the zero-temperature response function (15.25) for a system of noninteracting fermions simplifies to

$$\chi_{\mu\nu}^{R}(\mathbf{x}, \mathbf{x}'; \omega) = \sum_{n\bar{n}} \left[\frac{f_{\mu,n\bar{n}}(\mathbf{x}) f_{\nu,n\bar{n}}^{*}(\mathbf{x}')}{\omega - (\epsilon_{\bar{n}} - \epsilon_{n}) + i\eta} - \frac{f_{\mu,n\bar{n}}^{*}(\mathbf{x}) f_{\nu,n\bar{n}}(\mathbf{x}')}{\omega + (\epsilon_{\bar{n}} - \epsilon_{n}) + i\eta} \right], \qquad (15.33)$$

where only excitations with one electron–hole pair contribute. The excitation amplitudes for the density response function

$$f_{n\bar{n}}(\mathbf{x}) \equiv f_{0,n\bar{n}}(\mathbf{x}) = \langle \Psi_0 | \Delta\hat{n}(\mathbf{x}) | \Psi_{n\bar{n}} \rangle = \langle \Psi_0 | \hat{n}(\mathbf{x}) | \Psi_{n\bar{n}} \rangle = \langle \Psi_0 | \hat{\psi}^{\dagger}(\mathbf{x}) \hat{\psi}(\mathbf{x}) \hat{d}_{\bar{n}}^{\dagger} \hat{d}_{n} | \Psi_0 \rangle$$
$$= \langle \mathbf{x} | \bar{n} \rangle \langle n | \mathbf{x} \rangle = \varphi_{\bar{n}}(\mathbf{x}) \varphi_{n}^{*}(\mathbf{x}) \qquad (15.34)$$

are the product of one occupied and one unoccupied single-particle eigenfunction of $\hat{H}_0$. Similarly the excitation amplitudes for the current-density and the current response functions contain products of the form $\varphi_{\bar{n}} \boldsymbol{\nabla} \varphi_{n}^{*}$, $(\boldsymbol{\nabla} \varphi_{\bar{n}}) \varphi_{n}^{*}$, etc. We also note that for real eigenfunctions φ (the φ can always be chosen real in the absence of a magnetic field) the density response function takes the simple form

$$\chi^{R}(\mathbf{x}, \mathbf{x}'; \omega) = \sum_{n\bar{n}} (\epsilon_{\bar{n}} - \epsilon_{n}) \frac{\varphi_{\bar{n}}(\mathbf{x}) \varphi_{n}(\mathbf{x}) \varphi_{\bar{n}}(\mathbf{x}') \varphi_{n}(\mathbf{x}')}{(\omega + i\eta)^2 - (\epsilon_{\bar{n}} - \epsilon_{n})^2}.$$

It is also instructive to derive (15.33) from the many-body formulation. We consider here the density response function only; the interested reader can easily generalize the steps below to the current-density and the current response functions. From (15.12) we know that $\chi(1; 2) = -iL(1, 2; 1^{+}, 2^{+})$ and therefore

$$\chi(\mathbf{x}, z; \mathbf{x}', z') = -i\, L(\mathbf{x}, z, \mathbf{x}', z'; \mathbf{x}, z, \mathbf{x}', z') \equiv -i\, L(\mathbf{x}, \mathbf{x}', \mathbf{x}, \mathbf{x}'; z, z'),$$

where in the last identity we use the definition (12.20). In the noninteracting case this L was calculated in (12.28). Hence

$$\chi^{R}(\mathbf{x}, \mathbf{x}'; \omega) = \sum_{nm} f_{n} \bar{f}_{m} \left[\frac{f_{nm}(\mathbf{x}) f_{nm}^{*}(\mathbf{x}')}{\omega - (\epsilon_{m} - \epsilon_{n}) + i\eta} - \frac{f_{nm}^{*}(\mathbf{x}) f_{nm}(\mathbf{x}')}{\omega + (\epsilon_{m} - \epsilon_{n}) + i\eta} \right] \qquad (15.35)$$

in which we have extended the definition of the excitation amplitudes $f_{nm}(\mathbf{x}) = \varphi_{m}(\mathbf{x}) \varphi_{n}^{*}(\mathbf{x})$ to arbitrary couples (not necessarily unoccupied–occupied) of eigenfunctions. The response function in (15.35) generalizes the result (15.33) to finite temperature and reduces to it in the zero temperature limit.[12] Of course the finite temperature extension of χ^{R} can also be worked out starting from its definition. To do it we must use the weights ρ_{a} of (15.15), generalize the discussion leading to (15.25), and then specialize the new formulas to the noninteracting case. The reader should appreciate that the many-body formulation provides a much faster way to get to (15.35).

To summarize, only excited states with one electron–hole pair contribute to the noninteracting χ. We can try to understand how and why things change in the interacting case,

---

[12] At zero temperature f_n is unity if n is occupied and is zero otherwise.

at least for small interactions. If the interaction is weak then the interacting ground state $|\Psi_0\rangle$ differs from the noninteracting ground state by a small correction. This correction can be written as a linear combination of the states in (15.32). Similarly the interacting excited eigenstates $|\Psi_b\rangle$ have a dominant term of the form (15.32) plus a small correction; we say that the interacting $|\Psi_b\rangle$ is a *double excited state* if the dominant term of the form (15.32) has $N = 2$, a *triple excited state* if the dominant term has $N = 3$, and so on. In this complicated situation the excitation amplitude $f_{\mu,b}$ with b a double excitation, a triple excitation, etc., is generally nonzero. For instance if $|\Psi_0\rangle$ has a component proportional to $|\Psi_{n\bar{n}}\rangle$ then this component contributes to $f_{0,b}(\mathbf{x})$ with the overlap $\langle\Psi_{n\bar{n}}|\Delta\hat{n}(\mathbf{x})|\Psi_b\rangle$, and this overlap is large when the dominant term of $|\Psi_b\rangle$ has the form $|\Psi_{n_1 n_2 \bar{n}_1 \bar{n}_2}\rangle$, i.e., when b is a double excitation. The interacting response function changes by developing new poles in the discrete part of the spectrum and/or by modifying the continuum line-shape, see Section 16.8.

15.3 Bethe–Salpeter equation from the variation of a conserving G

We consider again here the system of Section 15.1 in which interacting particles are subject to a time-dependent vector potential $\mathbf{A} + \delta\mathbf{A}$ and scalar potential $V + \delta V$. The interaction between the particles is accounted for by some Φ-derivable self-energy Σ. The equation of motion for the Green's function G_{tot} in the total fields is, to first order in $\delta\mathbf{A}$ and δV, given by

$$\left[\mathrm{i}\frac{d}{dz_1} + \frac{1}{2m}D_1^2 - qV(1) - q\left(\delta V(1) - \frac{1}{c}\frac{\mathbf{D}_1\cdot\delta\mathbf{A}(1) + \delta\mathbf{A}(1)\cdot\mathbf{D}_1}{2mi}\right)\right] G_{\text{tot}}(1;3)$$

$$- \int d2\,\Sigma[G_{\text{tot}}](1;2)G_{\text{tot}}(2;3) = \delta(1;3), \quad (15.36)$$

where the functional dependence of Σ on the Green's function is made explicit. In this equation the perturbing potentials with the time variable on the contour are defined similarly to (8.3) and (8.4) [and in agreement with the more general definition (4.6)]

$$\delta\mathbf{A}(\mathbf{x}, z = t_\pm) = \delta\mathbf{A}(\mathbf{x}, t), \qquad \delta V(\mathbf{x}, z = t_\pm) = \delta V(\mathbf{x}, t), \quad (15.37)$$

and $\delta\mathbf{A} = \delta V = 0$ for z on the vertical track. The Green's function G of the system with $\delta\mathbf{A} = \delta V = 0$ obeys the equation of motion

$$\left[\mathrm{i}\frac{d}{dz_1} + \frac{1}{2m}D_1^2 - qV(1)\right] G(1;3) - \int d3\,\Sigma[G](1;2)G(2;3) = \delta(1;3), \quad (15.38)$$

with the same functional form of the self-energy Σ (otherwise the two systems would be treated at a different level of approximation). For later convenience we introduce the shorthand notation

$$\overrightarrow{G}^{-1}(1;2) \equiv \left[\left(\mathrm{i}\frac{\overrightarrow{d}}{dz_1} + \frac{\overrightarrow{D}_1^2}{2m} - qV(1)\right)\delta(1;2) - \Sigma(1;2)\right],$$

$$\overleftarrow{G}^{-1}(1;2) \equiv \left[\delta(1;2)\left(-\mathrm{i}\frac{\overleftarrow{d}}{dz_2} + \frac{(\overleftarrow{D}_2^2)^*}{2m} - qV(2)\right) - \Sigma(1;2)\right],$$

where, as usual, the left/right arrows specify that the derivatives act on the quantity to their left/right. In terms of these differential operators the equations of motion read

$$\int d2\, \overrightarrow{G}^{-1}(1;2)G(2;3) = \int d2\, G(1;2)\overleftarrow{G}^{-1}(2;3) = \delta(1;3). \tag{15.39}$$

We now derive an equation for the difference $\delta G = G_{\text{tot}} - G$. Adding and subtracting $\Sigma[G](1;2)G_{\text{tot}}(2;3)$ to and from the integrand of (15.38) and then subtracting (15.36) we find

$$\int d2\, \overrightarrow{G}^{-1}(1;2)\delta G(2;3) = q\delta h(1)G_{\text{tot}}(1;3) + \int d2\, \delta\Sigma(1;2)G_{\text{tot}}(2;3), \tag{15.40}$$

where we define the perturbation

$$\delta h(1) \equiv \left(\delta V(1) - \frac{1}{c} \frac{\mathbf{D}_1 \cdot \delta \mathbf{A}(1) + \delta \mathbf{A}(1) \cdot \mathbf{D}_1}{2mi} \right), \tag{15.41}$$

as well as the self-energy variation

$$\delta\Sigma(1;2) \equiv \Sigma[G_{\text{tot}}](1;2) - \Sigma[G](1;2).$$

Taking into account that G fulfills (15.39) we can rewrite (15.40) in the integral form:

$$\delta G(1;3) = q\int d2\, G(1;2)\delta h(2)G_{\text{tot}}(2;3) + \int d2 d4\, G(1;2)\delta\Sigma(2;4)G_{\text{tot}}(4;3), \tag{15.42}$$

which can easily be verified by applying $\overrightarrow{G}^{-1}$ to both sides and by checking that both sides satisfy the KMS boundary conditions. The integral equation (15.42) with $\Sigma = 0$ has the same structure as the integral equation derived in Section 7.2.3 for the quantum discharge of a capacitor, see (7.31); (15.42) can be regarded as the proper extension of (7.31) to approximations beyond the Hartree approximation. We observe that if we had included the (discarded) term proportional to δA^2 in δh then (15.42) would have been *exact* for all $\delta \mathbf{A}$ and δV. Since, however, we are interested in calculating δG to first order in the perturbing fields it is enough to consider the δh of (15.41). The next step is to evaluate the self-energy variation $\delta\Sigma$ to first order. Consider a diagram for $\Sigma[G]$ containing n Green's functions. The variation of this diagram when $G \to G + \delta G$ is the sum of n diagrams obtained by replacing a G with a δG for all the Gs of the diagram. An example is the variation of the second-order bubble diagram below:

In accordance with the observation in Section 9.3 we see that the variation of Σ is the result of the operation $\delta\Sigma/\delta G$ (which cuts one Green's function line from Σ in all possible ways) multiplied by the variation of the Green's function. Thus, to first order we have

$$\delta\Sigma(2;4) = \Sigma[G+\delta G](2;4) - \Sigma[G](2;4) = \int d5d6\,\frac{\delta\Sigma(2;4)}{\delta G(5;6)}\delta G(5;6)$$

$$= \pm\int d5d6\,K(2,6;4,5)\delta G(5;6),\tag{15.43}$$

with K the kernel of the Bethe–Salpeter equation defined in Section 12.2. Equation (15.43) has the following diagrammatic representation (remember the rule for K: the first two variables label opposite vertices with outgoing lines and the third variable labels the vertex to the right of an imaginary oriented line connecting diagonally the first to the second variable):

Since δh and $\delta\Sigma$ are of first order in the perturbing potentials we can replace G_{tot} with G in (15.42), thus obtaining

$$\delta G(1;3) = q\int d2\,G(1;2)\delta h(2)G(2;3)$$

$$\pm\int d2d4d5d6\,G(1;2)G(4;3)K(2,6;4,5)\delta G(5;6).\tag{15.44}$$

This is a recursive equation for δG. Its solution provides the first-order change in G for a given Φ-derivable approximation to the self-energy. To make a link with the discussion in Section 15.1 we manipulate the first term on the r.h.s.. Using the explicit form of δh we can write

$$\int d2\,G(1;2)\delta h(2)G(2;3) = \int d2\,G(1;2)\left(\delta V(2) - \frac{1}{c}\frac{-\overleftarrow{\mathbf{D}}_2^*\cdot\delta\mathbf{A}(2) + \delta\mathbf{A}(2)\cdot\mathbf{D}_2}{2mi}\right)G(2;3)$$

$$= \int d2d4\left[\delta(2;4)\underbrace{\left(\delta V(2) - \frac{\delta\mathbf{A}(2)}{c}\cdot\frac{\mathbf{D}_2 - \mathbf{D}_4^*}{2mi}\right)}_{\delta h(4;2)}G(1;4)G(2;3)\right],\tag{15.45}$$

where in the first equality we perform an integration by parts (the arrow over $\mathbf{D}_2^*$ indicates that the derivative acts on the left) and use the identity (9.20). Substituting this result into (15.44) we find

$$\delta G(1;3) = q\int d2d4\,\delta h(4;2)\,G(1;4)G(2;3)$$

$$\pm\int d2d4d5d6\,G(1;2)G(4;3)K(2,6;4,5)\delta G(5;6),\tag{15.46}$$

where $\delta h(4; 2)$ is the differential operator implicitly defined in (15.45). To visualize the structure of this equation we give to $q\,\delta h(4; 2)$ the diagrammatic representation of a dashed line joining the points 2 and 4. In this way (15.46) is represented like this:

where every oriented line is a G-line and where the $\pm$ sign in the first term stems from the Feynman rules for G_2, see Section 12.1. Iterating this equation we see that at every iteration the diagram with δG contains one more structure GGK. Therefore, the solution is

with

The grey blob solves the Bethe–Salpeter equation (12.17) for the two-particle XC function with kernel $K = \pm\delta\Sigma/\delta G$ and hence it must be identified with L. Converting the diagrammatic solution into a formula we obtain the following important result for the variation of a conserving Green's function

$$\delta G(1; 3) = q \int d2d4\ \delta h(4; 2)\, L(1, 2; 3, 4)$$
$$= q \int d2 \left[\left(\delta V(2) - \frac{\delta \mathbf{A}(2)}{c}\frac{\mathbf{D}_2 - \mathbf{D}_4^*}{2mi} \right) L(1, 2; 3, 4) \right]_{4=2}. \qquad (15.47)$$

From this equation we can calculate the density and current variations

$$\delta n(1) = \pm i\delta G(1; 1^+), \qquad \delta \mathbf{J}(1) = \pm i \left(\frac{\mathbf{D}_1 - \mathbf{D}_2^*}{2mi}\delta G(1; 2) \right)_{2=1^+},$$

and it is a simple exercise to show that these variations can be written in the following compact form[13]

$$\delta J_\mu(1) = q \int d2 \, \chi_{\mu\nu}(1;2) \delta A^\nu(2), \tag{15.48}$$

with $\chi_{\mu\nu}$ the conserving response function defined, for a given L, as in (15.12) and (15.13). It is important to stress that in Section 15.1 the relations (15.12) and (15.13) have been *derived* starting from the definitions of L and $\chi_{\mu\nu}$ in terms of the field operators. Here, instead, the response function $\chi_{\mu\nu}$ has been *defined* as in (15.12) and (15.13).[14] Equation (15.48) is an equation between quantities in Keldysh space and contains an integral along the contour. To convert (15.48) into an equation on the real time axis we take, e.g., $z_1 = t_{1-}$ on the forward branch, use the result of Exercise 5.5, and find

$$\delta J_\mu(\mathbf{x}_1, t_1) = q \int_{-\infty}^{\infty} dt_2 \int d\mathbf{x}_2 \, \chi_{\mu\nu}^{\mathrm{R}}(\mathbf{x}_1, t_1; \mathbf{x}_2, t_2) \delta A^\nu(\mathbf{x}_2, t_2),$$

which is identical to (15.10) for the exact quantities.

To summarize, we have proved that the current and the density variations produced by a conserving G with self-energy Σ are, to first order, obtained from (15.10) with a response function that satisfies the Bethe–Salpeter equation with kernel $K = \pm \delta\Sigma/\delta G$. In the next section we use (15.47) to prove an important identity between the vertex function and the self-energy.

15.4   Ward identity and the f-sum rule

There exist very special variations $\delta\mathbf{A}$ and δV of the external potentials for which the variation of a conserving Green's function is trivial, and these are the variations induced by a gauge transformation. For an infinitesimal gauge transformation $\Lambda(\mathbf{x}, z)$ we have [see, e.g., (3.24)]

$$\delta\mathbf{A}(1) = \boldsymbol{\nabla}_1 \Lambda(1), \qquad \delta V(1) = -\frac{1}{c}\frac{d}{dz_1}\Lambda(1). \tag{15.49}$$

Under this gauge transformation $h \rightarrow h + \delta h$ and consequently $G \rightarrow G + \delta G$. The transformed G is the Green's function that satisfies the Kadanoff–Baym equations with single-particle Hamiltonian $h + \delta h$. As already observed below (9.40), this Green's function is simply $G[\Lambda](1;2) = e^{i\frac{q}{c}\Lambda(1)} G(1;2) e^{-i\frac{q}{c}\Lambda(2)}$ for any Φ-derivable self-energy. Therefore to first order in Λ

$$\delta G(1;2) = \left[e^{i\frac{q}{c}\Lambda(1)} G(1;2) e^{-i\frac{q}{c}\Lambda(2)} - G(1;2) \right] = i\frac{q}{c}[\Lambda(1) - \Lambda(2)] G(1;2). \tag{15.50}$$

Inserting (15.49) and (15.50) into (15.47) we find

$$i[\Lambda(1) - \Lambda(2)]G(1;2) = -\int d3 \left[\left(\frac{d\Lambda(3)}{dz_3} + (\boldsymbol{\nabla}_3 \Lambda(3)) \cdot \frac{\mathbf{D}_3 - \mathbf{D}_4^*}{2mi} \right) L(1,3;2,4^+) \right]_{4=3}.$$

---

[13] If L is defined as the solution of the Bethe–Salpeter equation then $L(1,2;3,2) = L(1,2;3^+,2^+)$; this is due to the fact that in the Bethe–Salpeter equation the Hartree term in the definition (15.11) is already subtracted and hence there is no ambiguity in setting $2^+ = 2$ and $3^+ = 3$.

[14] For an approximate response function there is no definition in terms of the field operators. If L is exact then the response function calculated as in (15.12) and (15.13) coincides with the response function (15.5).

We stress once more that the replacement $L(1, 4; 2, 3) \rightarrow L(1, 4; 2, 3^+)$ on the r.h.s. is irrelevant if L comes from the solution of the Bethe–Salpeter equation, while it matters if L is calculated directly from (15.11) due to the presence of the Hartree term. If we now integrate the r.h.s. by parts and take into account that $\Lambda(\mathbf{x}, t_{0-}) = \Lambda(\mathbf{x}, t_0 - \mathrm{i}\beta) = 0$ (the perturbing field is switched on at times $t > t_0$) we get

$$\frac{d}{dz_3} L(1, 3; 2, 3^+) + \boldsymbol{\nabla}_3 \cdot \left(\frac{\mathbf{D}_3 - \mathbf{D}_4^*}{2m\mathrm{i}} L(1, 3; 2, 4^+) \right)_{4=3} = \mathrm{i} \left[\delta(1; 3) - \delta(2; 3) \right] G(1; 2), \tag{15.51}$$

which is valid for the exact L and G as well as for any conserving approximation to L and G. This relation is known as the *Ward identity* and guarantees the gauge invariance of the theory. The Ward identity is usually written in terms of the scalar vertex function Λ, defined in Section 12.5, and the vector vertex function $\boldsymbol{\Lambda}$:

$$\int d4 d5 \, G(1; 4) G(5; 2) \Lambda(4, 5; 3) \equiv L(1, 3; 2, 3^+),$$

$$\int d4 d5 \, G(1; 4) G(5; 2) \boldsymbol{\Lambda}(4, 5; 3) \equiv \left(\frac{\mathbf{D}_3 - \mathbf{D}_4^*}{2m\mathrm{i}} L(1, 3; 2, 4^+) \right)_{4=3}.$$

With these definitions (15.51) becomes

$$\int d4 d5 \, G(1; 4) G(5; 2) \left[\frac{d}{dz_3} \Lambda(4, 5; 3) + \boldsymbol{\nabla}_3 \cdot \boldsymbol{\Lambda}(4, 5; 3) \right] = \mathrm{i} \left[\delta(1; 3) - \delta(2; 3) \right] G(1; 2). \tag{15.52}$$

We can eliminate the Green's functions on the l.h.s. using the equations of motion. Acting on (15.52) with $\overrightarrow{G}^{-1}$ from the left and then with $\overleftarrow{G}^{-1}$ from the right we arrive at the standard form of the Ward identity

$$\boxed{ \frac{d}{dz_3} \Lambda(1, 2; 3) + \boldsymbol{\nabla}_3 \cdot \boldsymbol{\Lambda}(1, 2; 3) = \mathrm{i} \left[\overrightarrow{G}^{-1}(1; 3) \delta(3; 2) - \delta(1; 3) \overleftarrow{G}^{-1}(3; 2) \right] } \tag{15.53}$$

This equation relates the vertex to the self-energy and is valid for any conserving approximation. It was derived in 1950 by Ward [131] (who published it in a letter of less than a column) in the context of quantum electrodynamics and it was used by the author to demonstrate an exact cancellation between divergent quantities. Later in 1957 the Ward identity was generalized by Takahashi [132] to higher order correlators and, especially in textbooks on quantum field theory [29, 30], these generalized identities are called the *Ward–Takahashi identities*. In the theory of Fermi liquids the Ward identity provides a relation between the vertex and the quasi-particle renormalization factor Z [124, 133]. We emphasize that our derivation of the Ward identity does not require that the system is in the ground state; (15.53) is an identity between out-of-equilibrium correlators in the Keldysh space and reduces to the standard Ward identity for systems in equilibrium.

The satisfaction of the Ward identity implies that the response functions fulfill several exact relations. Among them there is the f-sum rule discussed in Section 15.2. In the remainder of this section we prove the f-sum rule for conserving approximations. We start by taking the limit $2 \rightarrow 1$ in (15.51); using the relations (15.12) and (15.13) we can write this limit as

$$\frac{d}{dz_3}\chi_{00}(1;3) + \partial_{3,k}\chi_{0k}(1;3) = 0, \tag{15.54}$$

where the sum over $k = 1, 2, 3$ is understood (Einstein convention). Similarly, we can apply the operator $(\mathbf{D}_1 - \mathbf{D}_2^*)/(2mi)$ to (15.51) and then take the limit $2 \rightarrow 1^+$. Using again the relations (15.12) and (15.13) we arrive at

$$\frac{d}{dz_3}\chi_{j0}(1;3) + \partial_{3,k}\chi_{jk}(1;3) = \frac{n(1)}{m}\partial_{1,j}\delta(1;3). \tag{15.55}$$

The response function $\chi_{\mu\nu}(1;3)$ belongs to the Keldysh space and has a vanishing singular part. Therefore its structure is

$$\chi_{\mu\nu}(1;3) = \theta(z_1, z_3)\chi_{\mu\nu}^>(1;3) + \theta(z_3, z_1)\chi_{\mu\nu}^<(1;3). \tag{15.56}$$

Inserting (15.56) into (15.54) we find

$$-\delta(z_1, z_3)\left[\chi_{00}^>(1;3) - \chi_{00}^<(1;3)\right] + \theta(z_1, z_3)\left[\frac{d}{dz_3}\chi_{00}^>(1;3) + \partial_{3,k}\chi_{0k}^>(1;3)\right]$$
$$+ \theta(z_3, z_1)\left[\frac{d}{dz_3}\chi_{00}^<(1;3) + \partial_{3,k}\chi_{0k}^<(1;3)\right] = 0,$$

from which it follows that

$$\chi_{00}^>(\mathbf{x}_1, t_1; \mathbf{x}_3, t_1) - \chi_{00}^<(\mathbf{x}_1, t_1; \mathbf{x}_3, t_1) = 0, \tag{15.57}$$

and

$$\frac{d}{dt_3}\chi_{00}^\lessgtr(\mathbf{x}_1, t_1; \mathbf{x}_3, t_3) + \partial_{3,k}\chi_{0k}^\lessgtr(\mathbf{x}_1, t_1; \mathbf{x}_3, t_3) = 0. \tag{15.58}$$

In a similar way insertion of (15.56) into (15.55) leads to

$$\chi_{j0}^>(\mathbf{x}_1, t_1; \mathbf{x}_3, t_1) - \chi_{j0}^<(\mathbf{x}_1, t_1; \mathbf{x}_3, t_1) = -\frac{n(\mathbf{x}_1, t_1)}{m}\partial_{1,j}\delta(\mathbf{x}_1 - \mathbf{x}_3), \tag{15.59}$$

and

$$\frac{d}{dt_3}\chi_{j0}^\lessgtr(\mathbf{x}_1, t_1; \mathbf{x}_3, t_3) + \partial_{3,k}\chi_{jk}^\lessgtr(\mathbf{x}_1, t_1; \mathbf{x}_3, t_3) = 0. \tag{15.60}$$

The equations (15.58) and (15.60) are gauge conditions on the response functions. They guarantee that the switching of a pure gauge does not change the density and the current in the system. As for (15.57) and (15.59), we observe that in the exact case they are a direct consequence of the commutators $[\hat{n}(\mathbf{x}_1), \hat{n}(\mathbf{x}_3)]_- = 0$ and of the commutator (15.29). We now combine (15.58) and (15.59) using the symmetry property (15.6) which implies $\chi_{0k}^\lessgtr(\mathbf{x}_1, t_1; \mathbf{x}_3, t_3) = \chi_{k0}^\gtrless(\mathbf{x}_3, t_3; \mathbf{x}_1, t_1)$. We define the difference $\Delta_{\mu\nu} = \chi_{\mu\nu}^> - \chi_{\mu\nu}^<$ and write

$$\frac{d}{dt_3}\Delta_{00}(\mathbf{x}_1, t_1; \mathbf{x}_3, t_3)\big|_{t_3=t_1} = -\partial_{3,k}\Delta_{0k}(\mathbf{x}_1, t_1; \mathbf{x}_3, t_1) = \partial_{3,k}\Delta_{k0}(\mathbf{x}_3, t_1; \mathbf{x}_1, t_1)$$

$$= -\frac{1}{m}\boldsymbol{\nabla}_3 \cdot [n(\mathbf{x}_3, t_1)\boldsymbol{\nabla}_3\delta(\mathbf{x}_3 - \mathbf{x}_1)]. \tag{15.61}$$

Let us specialize this formula to equilibrium situations, i.e., $\mathbf{A}(\mathbf{x}, t) = \mathbf{A}(\mathbf{x})$ and $V(\mathbf{x}, t) = V(\mathbf{x})$. Then the density $n(\mathbf{x}_3, t_1) = n(\mathbf{x}_3)$ is the equilibrium density and $\Delta_{\mu\nu}$ depends only on the time difference. Denoting by $\Delta_{\mu\nu}(\mathbf{x}_1, \mathbf{x}_3, \omega)$ its Fourier transform, (15.61) becomes

$$\mathrm{i} \int \frac{d\omega'}{2\pi} \, \omega' \Delta_{00}(\mathbf{x}_1, \mathbf{x}_3, \omega') = -\frac{1}{m} \boldsymbol{\nabla}_3 \cdot [n(\mathbf{x}_3) \boldsymbol{\nabla}_3 \delta(\mathbf{x}_3 - \mathbf{x}_1)]. \tag{15.62}$$

Similarly, in equilibrium (15.57) can be rewritten as

$$\int \frac{d\omega'}{2\pi} \Delta_{00}(\mathbf{x}_1, \mathbf{x}_3, \omega') = 0. \tag{15.63}$$

The relation between Δ_{00} and χ^{R} is provided by (15.17), that for large ω reads

$$\chi^{\mathrm{R}}(\mathbf{x}_1, \mathbf{x}_3, \omega) = \frac{\mathrm{i}}{\omega} \int \frac{d\omega'}{2\pi} \Delta_{00}(\mathbf{x}_1, \mathbf{x}_3, \omega') + \frac{\mathrm{i}}{\omega^2} \int \frac{d\omega'}{2\pi} (\omega' - \mathrm{i}\eta) \Delta_{00}(\mathbf{x}_1, \mathbf{x}_3, \omega') + \mathcal{O}(\frac{1}{\omega^3}).$$

Using (15.62) and (15.63) we see that the large ω behavior of any *conserving* density response function is the same as the behavior (15.28) of the *exact* density response function. Furthermore, *assuming* that the conserving response function is analytic in the upper half of the complex ω plane, we could apply the same trick of Section 15.2.2 and integrate along the contour C of Fig. 15.2 to find the f-sum rule again. The reason why we say "assuming" is that conserving approximations do not necessarily preserve the correct analytic structure of the response functions [134].

15.5 Time-dependent screening in an electron gas

In this section we present a nice application of linear response theory in the Coulombic electron gas. Suppose that a very high-energy photon is absorbed by the gas and that, as a consequence, an electron is instantaneously expelled by the system. How does the density of the gas rearrange in order to screen the suddenly created hole? And how long does it take? To answer these questions we need some preliminary results on the equilibrium density response function. We first derive an analytic result for the noninteracting χ, that we denote by χ_0. Then, we approximate the interacting χ as the solution of the Bethe–Salpeter equation with kernel $K \sim K_{\mathrm{H}} = -\delta\Sigma_{\mathrm{H}}/\delta G$ at the Hartree level (the minus sign is because we are dealing with fermions). This approximation for the response function is known as the *Random Phase Approximation* (RPA) and, like the Hartree approximation for G, it becomes exact in the high density limit $r_s \to 0$.[15] Lastly, we use the Kubo formula to calculate the time-dependent density in the neighborhood of the suddenly created hole. The formula for the RPA density response function is also used to discuss the spectral properties of the electron gas in the $G_0 W_0$ approximation.

---

[15]Remember that r_s is related to the density n of the gas by $\frac{1}{n} = \frac{4\pi}{3}(a_{\mathrm{B}} r_s)^3$, see Section 7.3.2. In the Hartree approximation the energy of the electron gas is given by the first term in (13.58) which is the dominant term for $r_s \to 0$.

15.5.1 Noninteracting density response function

The general formula for the noninteracting density response function χ_0 is given in (15.33). For an electron gas the single-particle energy eigenkets $|\mathbf{p}\tau\rangle$ are the momentum–spin kets and therefore the excitation amplitudes (15.34) read

$$f_{\mathbf{p}\tau\bar{\mathbf{p}}\bar{\tau}}(\mathbf{r}\sigma) = \langle \mathbf{r}\sigma|\bar{\mathbf{p}}\bar{\tau}\rangle\langle \mathbf{p}\tau|\mathbf{r}\sigma\rangle = \delta_{\sigma\bar{\tau}}\delta_{\sigma\tau}e^{i(\bar{\mathbf{p}}-\mathbf{p})\cdot\mathbf{r}}.$$

Substitution of these excitation amplitudes into (15.35) and summation over the spin indices τ and $\bar{\tau}$ leads to

$$\chi_0^{R}(\mathbf{x},\mathbf{x}';\omega) = \delta_{\sigma\sigma'}\int \frac{d\mathbf{p}\,d\bar{\mathbf{p}}}{(2\pi)^6}\,f_{\mathbf{p}}\bar{f}_{\bar{\mathbf{p}}}\left[\frac{e^{i(\bar{\mathbf{p}}-\mathbf{p})\cdot(\mathbf{r}-\mathbf{r}')}}{\omega - (\epsilon_{\bar{\mathbf{p}}} - \epsilon_{\mathbf{p}}) + i\eta} - \frac{e^{-i(\bar{\mathbf{p}}-\mathbf{p})\cdot(\mathbf{r}-\mathbf{r}')}}{\omega + (\epsilon_{\bar{\mathbf{p}}} - \epsilon_{\mathbf{p}}) + i\eta}\right],$$

with the standard notation $\mathbf{x} = \mathbf{r}\sigma$ and $\mathbf{x}' = \mathbf{r}'\sigma'$, and the zero-temperature Fermi function f and $\bar{f} = 1 - f$ for the occupied and unoccupied states respectively. We rename the integration variables as $\mathbf{p} = \mathbf{k}$, $\bar{\mathbf{p}} = \mathbf{k} + \mathbf{q}$ in the first integral, and $\bar{\mathbf{p}} = \mathbf{k}$, $\mathbf{p} = \mathbf{k} + \mathbf{q}$ in the second integral; in this way χ_0^{R} becomes

$$\chi_0^{R}(\mathbf{x},\mathbf{x}';\omega) = \delta_{\sigma\sigma'}\int \frac{d\mathbf{q}\,d\mathbf{k}}{(2\pi)^6}\,e^{i\mathbf{q}\cdot(\mathbf{r}-\mathbf{r}')}\left[\frac{f_{\mathbf{k}}\bar{f}_{\mathbf{k}+\mathbf{q}} - f_{\mathbf{k}+\mathbf{q}}\bar{f}_{\mathbf{k}}}{\omega - (\epsilon_{\mathbf{k}+\mathbf{q}} - \epsilon_{\mathbf{k}}) + i\eta}\right]. \tag{15.64}$$

Due to the translational invariance of the system it is convenient to work with the Fourier transform $\chi_0(\mathbf{k},\omega)$, which we define similarly to (15.19), i.e.,

$$\sum_{\sigma\sigma'}\chi_0^{R}(\mathbf{x},\mathbf{x}';\omega) = \int \frac{d\mathbf{q}}{(2\pi)^3}e^{i\mathbf{q}\cdot(\mathbf{r}-\mathbf{r}')}\chi_0^{R}(\mathbf{q},\omega). \tag{15.65}$$

Comparing (15.65) with (15.64) and taking into account that the combination of Fermi functions $f_1\bar{f}_2 - f_2\bar{f}_1 = f_1 - f_2$, we obtain

$$\chi_0^{R}(\mathbf{q},\omega) = 2\int \frac{d\mathbf{k}}{(2\pi)^3}\,f_{\mathbf{k}}\left[\frac{1}{\omega - (\epsilon_{\mathbf{k}+\mathbf{q}} - \epsilon_{\mathbf{k}}) + i\eta} - \frac{1}{\omega - (\epsilon_{\mathbf{k}} - \epsilon_{\mathbf{k}-\mathbf{q}}) + i\eta}\right]. \tag{15.66}$$

The evaluation of this integral is a bit tedious but doable. For free electrons the energy dispersion is $\epsilon_{\mathbf{p}} = p^2/2$. Introducing the dimensionless variables

$$x = \frac{q}{p_{\mathrm{F}}}, \qquad y = \frac{k}{p_{\mathrm{F}}}, \qquad \nu = \frac{\omega}{\epsilon_{p_{\mathrm{F}}}} = \frac{2\omega}{p_{\mathrm{F}}^2}, \tag{15.67}$$

with $\epsilon_{p_{\mathrm{F}}}$ the noninteracting energy with Fermi momentum p_{F}, we can rewrite (15.66) as

$$\chi_0^{R}(\mathbf{q},\omega) = \frac{4p_{\mathrm{F}}}{(2\pi)^2}\int_0^1 dy\,y^2\int_{-1}^1 dc\left[\frac{1}{\nu - x^2 - 2xyc + i\eta} - \frac{1}{\nu + x^2 - 2xyc + i\eta}\right], \tag{15.68}$$

where c is the cosine of the angle between the vectors $\mathbf{k}$ and $\mathbf{q}$. Due to the invariance of the system under rotations, χ_0^{R} correctly depends only on the modulus q of the vector $\mathbf{q}$. We now calculate the real and the imaginary part of χ_0^{R}. The resulting final form of $\mathrm{Re}\left[\chi_0^{R}\right]$

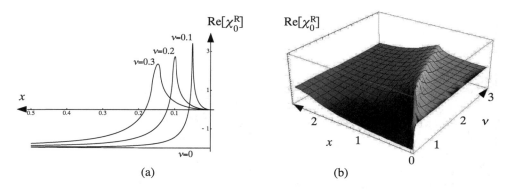

Figure 15.3 Real part of χ_0^R in units of $2p_F/(2\pi)^2$. (a) For frequencies $\nu = 0, 0.1, 0.2, 0.3$ as a function of x. (b) 3D plot as a function of ν and x.

and $\text{Im}\left[\chi_0^R\right]$ was worked out for the first time by Lindhard in a classic paper from 1954 [135]. Today it is common to refer to χ_0^R as the *Lindhard function*.

The real part: To calculate the real part we simply set to zero the infinitesimal $i\eta$ in (15.68). Performing the integral over c we are left with the evaluation of the logarithmic integral I defined in (7.58). It is then straightforward to arrive at

$$\text{Re}\left[\chi_0^R(\mathbf{q}, \omega)\right] = \frac{2p_F}{(2\pi)^2} \frac{I(2x, \nu - x^2) - I(2x, \nu + x^2)}{x}.$$

This function is discontinuous at the origin since the limit $x \to 0$ does not commute with the limit $\nu \to 0$. Using the explicit form of I, at zero frequency we find

$$\text{Re}\left[\chi_0^R(\mathbf{q}, 0)\right] = -\frac{4p_F}{(2\pi)^2} F(x/2), \tag{15.69}$$

where F is the same function (7.60) that appears in the Hartree–Fock eigenvalues of the electron gas. At any finite frequency ν the small-x behavior is instead parabolic:

$$\text{Re}\left[\chi_0^R(\mathbf{q} \to 0, \omega)\right] \to \frac{8p_F}{(2\pi)^2} \frac{2}{3} \frac{x^2}{\nu^2}, \tag{15.70}$$

and the real part approaches zero when $x \to 0$. This is illustrated in Fig. 15.3(a) where the discontinuous behavior for $\nu = 0$ is clearly visible. For $\nu \to 0$ the function $\text{Re}\left[\chi_0^R\right]$ versus x exhibits a peak in $x_{\max} \sim \nu/2$ whose width decreases as $\sim \nu$ and whose height increases as $\sim \ln \nu$. The 3D plot of $\text{Re}\left[\chi_0^R\right]$ as a function of x and ν is displayed in Fig. 15.3(b); we see a maximum along the curve $\nu = x^2 + 2x$ and otherwise a rather smooth function away from the origin.

The imaginary part: The calculation of the imaginary part of χ_0^R is simpler. Using the identity $1/(x + i\eta) = P(1/x) - i\pi\delta(x)$ in (15.68) we find

$$\text{Im}\left[\chi_0^R(\mathbf{q}, \omega)\right] = -\frac{4\pi p_F}{(2\pi)^2} \int_0^1 dy\, y^2 \int_{-1}^1 dc \left[\delta(\nu - x^2 - 2xyc) - \delta(\nu + x^2 - 2xyc)\right].$$

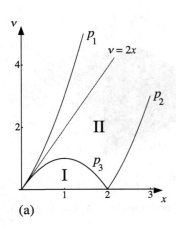
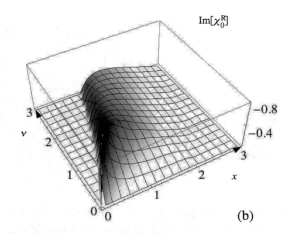

Figure 15.4 (a) Domain of the (ν, x) plane where the imaginary part of χ_0^R is nonvanishing. (b) 3D plot of $\mathrm{Im}\left[\chi_0^R\right]$ in units of $2\pi p_F/(2\pi)^2$ as a function of ν and x.

The integral over c between -1 and 1 of the δ-function $\delta(\alpha - \beta c)$ yields $1/|\beta|$ if $|\alpha/\beta| < 1$ and zero otherwise. Then, the integral over y can easily be performed also, since it is of the form $\int_0^1 dy\, y\, \theta(y - \gamma) = \frac{1}{2}(1 - \gamma^2)\,\theta(1 - \gamma)$ with γ a positive constant. The final result is

$$\mathrm{Im}\left[\chi_0^R(\mathbf{q}, \omega)\right] = -\frac{2\pi p_F}{(2\pi)^2}\frac{1}{x}\left[\frac{1 - P_-^2(\nu, x)}{2}\theta(1 - P_-(\nu, x)) \right.$$
$$\left. - \frac{1 - P_+^2(\nu, x)}{2}\theta(1 - P_+(\nu, x))\right], \qquad (15.71)$$

where we define

$$P_\pm(\nu, x) = \frac{|\nu \pm x^2|}{2x},$$

and where we take into account that both x and y are, by definition, positive quantities. In the quarter of the (ν, x) plane with $\nu > 0$ and $x > 0$ the first θ-function is nonvanishing in the area delimited by the parabolas $\nu = x^2 + 2x$, which we call p_1, and $\nu = x^2 - 2x$, which we call p_2, whereas the second θ-function is nonvanishing in the area below the parabola p_3 with equation $\nu = -x^2 + 2x$, see Fig. 15.4(a).[16] We define region I as the area where both the θ-functions are different from zero and region II as the area where the second θ-function

---

[16] It is intuitively clear that the imaginary part of χ_0^R is nonvanishing in this region. For instance at zero frequency the only particle–hole excitations which contribute to $\mathrm{Im}[\chi_0^R]$ are those in which both the particle and the hole are on the Fermi surface since in this case their energy difference is zero. Then the maximum distance between the momentum $\mathbf{k}$ of the particle and the momentum $\mathbf{k} + \mathbf{q}$ of the hole is $q = 2p_F$, which is obtained for $\mathbf{q} = -2\mathbf{k}$ (hole-momentum opposite to electron-momentum). Therefore at zero frequency $\mathrm{Im}[\chi_0^R]$ is nonzero for all $q \in (0, 2p_F)$. With similar geometric considerations one can generalize the above argument to finite frequencies.

does instead vanish. Then, from (15.71) we have

$$\text{Im}\left[\chi_0^{\text{R}}(\mathbf{q},\omega)\right] = -\frac{2\pi p_{\text{F}}}{(2\pi)^2}\begin{cases}\frac{\nu}{2x} & \text{region I} \\ \frac{1}{2x}\left[1-\left(\frac{\nu-x^2}{2x}\right)^2\right] & \text{region II.}\end{cases} \tag{15.72}$$

The limits $\nu \to 0$ and $x \to 0$ must be taken with care. If we approach the origin along the parabolas p_1 and p_3 that delimit region II we find

$$\text{Im}\left[\chi_0^{\text{R}}(\mathbf{q},\omega)\right] \xrightarrow[\substack{\nu,x\to0 \\ \text{along } p_1}]{} 0, \qquad \text{Im}\left[\chi_0^{\text{R}}(\mathbf{q},\omega)\right] \xrightarrow[\substack{\nu,x\to0 \\ \text{along } p_3}]{} -\frac{2\pi p_{\text{F}}}{(2\pi)^2}. \tag{15.73}$$

On the other hand, if we approach the origin from region I along the line $\nu = mx$ with $m < 2$ we find

$$\text{Im}\left[\chi_0^{\text{R}}(\mathbf{q},\omega)\right] \xrightarrow[\substack{\nu=mx,x\to0 \\ \text{(from region I)}}]{} -\frac{2\pi p_{\text{F}}}{(2\pi)^2}\frac{m}{2}, \tag{15.74}$$

which for $m \to 2$ reduces to the value in the second limit of (15.73). For any finite frequency ν the imaginary part $\text{Im}\left[\chi_0^{\text{R}}\right]$ is a continuous function of x with cusps (discontinuity of the first derivative) at the points where regions I and II start or end. The 3D plot of $\text{Im}\left[\chi_0^{\text{R}}\right]$ is shown in Fig. 15.4(b) where it is evident that the function is everywhere nonpositive, in agreement with our discussion on the energy-loss function in Section 15.2. The noninteracting electron gas can dissipate energy only via the creation or annihilation of electron–hole pairs. As we see in the next sections, the physics becomes much more interesting in the interacting case.

15.5.2 RPA density response function

In the electron gas the noninteracting response function χ_0 constitutes the basic ingredient to calculate the linear response density δn in the Hartree approximation. Let us consider the Bethe–Salpeter equation for L corresponding to the variation of the Hartree Green's function G_{H} [see (12.17)],

$$L_{\text{H}}(1,2;3,4) = G_{\text{H}}(1;4)G_{\text{H}}(2;3)$$
$$- \int d5d6d7d8\, G_{\text{H}}(1;5)G_{\text{H}}(7;3)K_{\text{H}}(5,6;7,8)L_{\text{H}}(8,2;6,4). \tag{15.75}$$

In this equation $K_{\text{H}} \equiv -\delta\Sigma_{\text{H}}/\delta G$ is the kernel that gives the linear response change of the Hartree self-energy $\Sigma_{\text{H}}(1;2) = -\mathrm{i}\delta(1;2)\int d3\, v(1;3)G_{\text{H}}(3;3^+)$ in accordance with (15.43). Equation (15.75) defines the two-particle XC function L_{H} in the Hartree approximation. The explicit form of the Hartree kernel is

$$K_{\text{H}}(5,6;7,8) = -\frac{\delta\Sigma_{\text{H}}(5;7)}{\delta G(8;6)} = \mathrm{i}\delta(5;7)\delta(6;8)v(5;8).$$

Inserting this result into (15.75) and taking the limit $3 \to 1$ and $4 \to 2$ we find

$$\chi(1;2) = -\mathrm{i}G_{\text{H}}(1;2)G_{\text{H}}(2;1) - \mathrm{i}\int d5d6\, G_{\text{H}}(5;1)G_{\text{H}}(1;5)v(5;6)\chi(6;4), \tag{15.76}$$

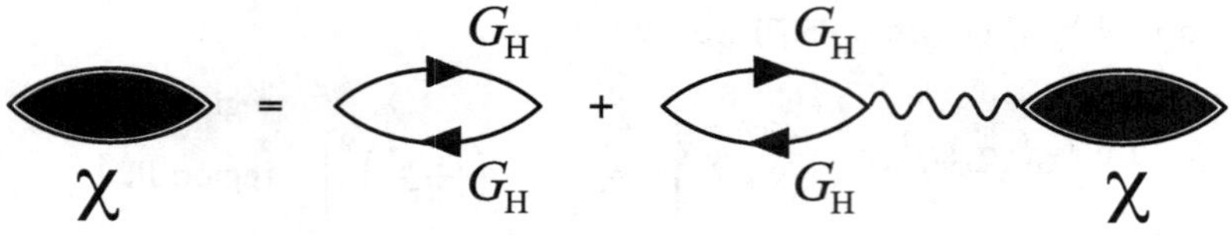

Figure 15.5 Representation of the RPA response function.

where $\chi(1;2) = -iL_H(1,2;1,2)$ is the density response function (we omit the subscript "H" in χ in order to simplify the notation), see (15.12). Thus in the Hartree approximation χ is given by a Dyson-like equation whose diagrammatic representation is shown in Fig. 15.5. For historical reasons this approximation is called the *Random Phase Approximation* (RPA). In 1953 Bohm and Pines [136] found a very ingenious way to map the Hamiltonian of a gas of electrons interacting via a (long-range) Coulomb interaction into the Hamiltonian of a gas of electrons *plus* collective excitations interacting via a screened short-range Coulomb interaction. The RPA is equivalent to neglecting the interaction between the electrons and the collective excitations, and it becomes exact in the limit of very high densities ($r_s \to 0$). In Appendix N we present a simplified version of the original treatment by Bohm and Pines. Going beyond the Hartree approximation the kernel K is no longer proportional to a product of δ-functions and if we take the limit $3 \to 1$ and $4 \to 2$ in the Bethe–Salpeter equation we find that L (under the integral sign) does not reduce to χ.

Let us now go back to (15.76) and consider the structure $-iG_H G_H$ that appears in it. For an electron gas subject to a uniform potential V_0 (positive background charge) the momentum–spin kets $|\mathbf{p}\sigma\rangle$ are eigenkets of the noninteracting Hamiltonian $\hat{h} = \hat{p}^2/2 + V_0\hat{1}$ as well as of the Hartree Hamiltonian $\hat{h}_H = \hat{p}^2/2 - V_0\hat{1} - \hat{V}_H$ with $\hat{V}_H = -n\tilde{v}_0\hat{1}$, see (7.14). This means that for $V_0 = n\tilde{v}_0$ the Hartree Green's function G_H is the same as the Green's function G_0 of a system of noninteracting electrons with energy dispersion $\epsilon_{\mathbf{p}} = p^2/2$. In Section 15.2.3 we showed that $-iG_0 G_0$ is the noninteracting density response function χ_0, and in the previous section we calculated this χ_0 just for electrons with energy dispersion $\epsilon_{\mathbf{p}} = p^2/2$. Therefore, in an electron gas the RPA density response function is the solution of

$$\chi(1;2) = \chi_0(1;2) + \int d3d4\, \chi_0(1;3)v(3;4)\chi(4;2), \qquad (15.77)$$

with χ_0 given in (15.68). It is worth noting that the Coulomb interaction v is spin independent and hence χ is not diagonal in spin space even though χ_0 is.[17] Equation (15.77) is an integral equation for two-point correlators in Keldysh space. As Keldysh functions the Coulomb interaction $v(3;4) = \delta(z_3, z_4)/|\mathbf{r}_3 - \mathbf{r}_4|$ has only a singular part while χ and χ_0 have the structure (15.56) and do not contain any singular part. Using the Langreth rules of Table 5.1 to extract the retarded component and Fourier transforming to frequency space we obtain the equation:

$$\chi^R(\mathbf{x}_1, \mathbf{x}_2; \omega) = \chi_0^R(\mathbf{x}_1, \mathbf{x}_2; \omega) + \int d\mathbf{x}_3 d\mathbf{x}_4\, \chi_0^R(\mathbf{x}_1, \mathbf{x}_3; \omega)v(\mathbf{x}_3, \mathbf{x}_4)\chi^R(\mathbf{x}_4, \mathbf{x}_2; \omega).$$

---

[17]This is simply due to the fact that in the interacting gas a change of, say, the spin-up density will affect both the spin-up and the spin-down densities.

Since both χ_0^R and v depend only on the difference of the spatial arguments, χ^R also depends only on the difference of the spatial arguments. If we define the Fourier transform of χ^R by analogy with (15.19),

$$\sum_{\sigma\sigma'} \chi^R(\mathbf{x}, \mathbf{x}'; \omega) = \int \frac{d\mathbf{p}}{(2\pi)^3} e^{i\mathbf{p}\cdot(\mathbf{r}-\mathbf{r}')} \chi^R(\mathbf{p}, \omega),$$

then the RPA equation for χ^R becomes a simple algebraic equation whose solution is

$$\boxed{\chi^R(\mathbf{q}, \omega) = \frac{\chi_0^R(\mathbf{q}, \omega)}{1 - \tilde{v}_\mathbf{q} \chi_0^R(\mathbf{q}, \omega)}, \qquad \tilde{v}_\mathbf{q} = \frac{4\pi}{q^2}} \tag{15.78}$$

In the remainder of the section we discuss this result.

The static Thomas–Fermi screening: The first remark is about the effective Coulomb interaction $W = v + v\chi v$, see (11.45), between two electrons in the static limit or, equivalently, in the zero frequency limit. Approximating χ as in (15.78) the retarded component of W in Fourier space reads

$$W^R(\mathbf{q}, \omega) = \frac{\tilde{v}_\mathbf{q}}{1 - \tilde{v}_\mathbf{q} \chi_0^R(\mathbf{q}, \omega)}.$$

For $\omega = 0$ the imaginary part of χ_0^R vanishes, see (15.74), while the real part approaches $-4p_F/(2\pi)^2$ for $\mathbf{q} \to 0$, see (15.69). The small q behavior of the static effective interaction is therefore

$$W^R(\mathbf{q} \to 0, 0) \to \frac{4\pi}{q^2 + \frac{4p_F}{\pi}}. \tag{15.79}$$

This is exactly the screened interaction of the Thomas–Fermi theory [93,137,138]. If we Fourier transform $W^R(\mathbf{q}, 0)$ back to real space, (15.79) implies that at large distances the effective interaction has the Yukawa form

$$W(\mathbf{x}_1, \mathbf{x}_2; 0) \simeq \frac{e^{-|\mathbf{r}_1-\mathbf{r}_2|/\lambda_{\mathrm{TF}}}}{|\mathbf{r}_1 - \mathbf{r}_2|},$$

with $\lambda_{\mathrm{TF}} = \sqrt{\pi/4p_F}$ the *Thomas–Fermi screening length*.[18] The physical meaning of the static effective interaction is discussed in the next section.

Plasmons: The second remark is more relevant to our subsequent discussion and concerns the aforementioned collective excitations found by Bohm and Pines. For an electron gas subject to an external scalar potential δV the Kubo formula in Fourier space reads $\delta n(\mathbf{q}, \omega) = -\chi^R(\mathbf{q}, \omega)\delta V(\mathbf{q}, \omega)$ with $\delta n(\mathbf{q}, \omega)$ the Fourier transform of $\delta n(\mathbf{r}, t) = \sum_\sigma \delta n(\mathbf{x}, t)$ (for an electron the charge $q = -1$). Therefore

$$\frac{\delta n(\mathbf{q}, \omega)}{\chi^R(\mathbf{q}, \omega)} = -\delta V(\mathbf{q}, \omega). \tag{15.80}$$

---

[18] In the Fourier transform back to real space, $W^R(\mathbf{q}, 0)$ can be approximated with its small q limit (15.79) if the distance $|\mathbf{r}_1 - \mathbf{r}_2| \to \infty$.

Suppose that there exist points in the (q, ω) plane for which $1/\chi^R(\mathbf{q}, \omega) = 0$. Then (15.80) is compatible with a scenario in which the electron density oscillates in space and time without any driving field, i.e., $\delta n \neq 0$ even though $\delta V = 0$. These persistent (undamped) density oscillations are the collective excitations of Bohm and Pines who gave them the name of *plasmons*. From our previous analysis of the noninteracting response function it is evident that there are no points in the (q, ω) plane for which $1/\chi_0^R(\mathbf{q}, \omega) = 0$. As we see below, the existence of plasmons is intimately connected to the long-range nature of the Coulomb interaction.

For an intuitive understanding of the self-sustained density oscillations in an electron gas we can use the following classical picture. Let $\mathbf{u}(\mathbf{r}, t)$ be the displacement of an electron from its equilibrium position in $\mathbf{r}$ at a certain time t. Consider the special case in which the displacement is of the form

$$\mathbf{u}(\mathbf{r}, t) = \mathbf{u}_0 \cos(\mathbf{q} \cdot \mathbf{r} - \omega t), \tag{15.81}$$

with $\mathbf{u}_0$ parallel to $\mathbf{q}$ (longitudinal displacement). The polarization associated with this displacement is $\mathbf{P}(\mathbf{r}, t) = -n\mathbf{u}(\mathbf{r}, t)$ with n the equilibrium density, and therefore the electric field $\mathbf{E} = -4\pi\mathbf{P} = 4\pi n\mathbf{u}$ is longitudinal as well. The classical equation of motion for the electron with equilibrium position in $\mathbf{r}$ reads

$$\ddot{\mathbf{u}}(\mathbf{r}, t) = -\mathbf{E}(\mathbf{r}, t) = -4\pi n\mathbf{u}(\mathbf{r}, t).$$

For (15.81) to be a solution of this differential equation the frequency must be

$$\omega = \omega_p = \sqrt{4\pi n} = \sqrt{\frac{3}{r_s^3}}. \tag{15.82}$$

This frequency is called the *plasma frequency*.[19]

Let us investigate the possibility that the inverse of the RPA response function is zero. From (15.78) we see that $1/\chi^R(\mathbf{q}, \omega) = 0$ implies

$$(a) \qquad \mathrm{Im}[\chi_0^R(\mathbf{q}, \omega)] = 0, \tag{15.83}$$

$$(b) \qquad 1 - \tilde{v}_\mathbf{q} \mathrm{Re}[\chi_0^R(\mathbf{q}, \omega)] = 0. \tag{15.84}$$

We look for solutions with $x \ll 1$, i.e., to the left of the regions I and II in Fig. 15.4(a), so that $\mathrm{Im}[\chi_0^R] = 0$. We rewrite χ_0^R in (15.68) in a slightly different form. Changing the variable $c \to -c$ in the first integral we find the equivalent expression

$$\chi_0^R(\mathbf{q}, \omega) = \frac{8p_F}{(2\pi)^2} \int_0^1 dy\, y^2 \int_{-1}^1 dc\, \frac{x^2 - 2xyc}{(\nu + i\eta)^2 - (x^2 - 2xyc)^2}.$$

To calculate the real part when $x \ll 1$ we set $\eta = 0$ and expand the integrand in powers of $(x^2 - 2xyc)$,

$$\mathrm{Re}\left[\chi_0^R(\mathbf{q}, \omega)\right] = \frac{8p_F}{(2\pi)^2} \int_0^1 dy\, y^2 \int_{-1}^1 dc\, \frac{x^2 - 2xyc}{\nu^2}\left[1 + \frac{(x^2 - 2xyc)^2}{\nu^2} + \dots\right].$$

---

[19]Expressing the plasma frequency in terms of the fundamental constants we find $\omega_p = \sqrt{\frac{4\pi e^2 n}{m_e}}$. Typical values of the plasma frequency are in the range 10–20 eV.

Taking into account that terms with odd powers of c do not contribute, we obtain

$$
\begin{aligned}
\text{Re}\left[\chi_0^{\text{R}}(\mathbf{q},\omega)\right] &= \frac{8p_{\text{F}}}{(2\pi)^2}\frac{2}{3}\left(\frac{x}{\nu}\right)^2\left[1+\frac{12}{5}\left(\frac{x}{\nu}\right)^2+\mathcal{O}(x^4)\right] \\
&= \frac{p_{\text{F}}^3}{3\pi^2}\left(\frac{q}{\omega}\right)^2\left[1+\frac{3}{5}\left(\frac{qp_{\text{F}}}{\omega}\right)^2+\mathcal{O}(x^4)\right],
\end{aligned}
$$

where in the second equality we reintroduce the physical momentum and frequency. Inserting this result into (15.84) and taking into account the relation $n=p_{\text{F}}^3/(3\pi^2)$ between the density and the Fermi momentum, as well as the definition (15.82) of the plasma frequency, we get

$$
1-\tilde{v}_{\mathbf{q}}\text{Re}[\chi_0^{\text{R}}(\mathbf{q},\omega)] = 1-\frac{\omega_{\text{p}}^2}{\omega^2}\left[1+\frac{3}{5}\left(\frac{qp_{\text{F}}}{\omega}\right)^2+\mathcal{O}(x^4)\right]=0. \tag{15.85}
$$

From this equation it follows that for $q=0$ the electron gas supports undamped density oscillations with frequency ω_{p} in agreement with the classical picture. For small q, however, the frequency changes according to

$$
\omega=\omega_{\text{p}}(q)\simeq\omega_{\text{p}}\sqrt{1+\frac{3}{5}\left(\frac{qp_{\text{F}}}{\omega_{\text{p}}}\right)^2}, \tag{15.86}
$$

which follows directly from (15.85) by replacing ω with ω_{p} in the square brackets.

An important consequence of the existence of plasmons is that the imaginary part of $\chi^{\text{R}}(\mathbf{q},\omega)$ consists of a continuum of electron–hole excitations (as in the noninteracting electron gas) and a plasmon peak. To show it let us introduce the real functions B_1,B_2 and A_1,A_2 that depend only on the modulus of $\mathbf{q}$ and on the frequency ω according to

$$
\begin{aligned}
B_1(q,\omega)+\text{i}B_2(q,\omega) &= \tilde{v}_{\mathbf{q}}\chi^{\text{R}}(\mathbf{q},\omega), \\
A_1(q,\omega)+\text{i}A_2(q,\omega) &= \tilde{v}_{\mathbf{q}}\chi_0^{\text{R}}(\mathbf{q},\omega).
\end{aligned}
$$

If we multiply both sides of (15.78) by $\tilde{v}_{\mathbf{q}}$ we find that B_2 is given by

$$
B_2=\frac{A_2}{(1-A_1)^2+A_2^2},
$$

where the dependence on (q,ω) has been dropped. From this equation we see that $A_2\neq 0$ implies $B_2\neq 0$ and therefore that the imaginary part of the RPA response function is certainly nonvanishing in the regions I and II of Fig. 15.4(a). Outside these regions A_2 is an infinitesimally small function proportional to η that vanishes only in the limit $\eta\to 0$. We then see that B_2 is not necessarily zero outside regions I and II since

$$
\lim_{A_2\to\pm 0}B_2=\lim_{A_2\to\pm 0}\frac{A_2}{(1-A_1)^2+A_2^2}=\pm\pi\delta(1-A_1), \tag{15.87}
$$

where $A_2\to\pm 0$ signifies that A_2 approaches zero from negative/positive values. The r.h.s. of this equation is nonvanishing when the argument of the δ-function is zero. Taking into account that $1-A_1(q,\omega)=1-\tilde{v}_{\mathbf{q}}\text{Re}\left[\chi_0^{\text{R}}(\mathbf{q},\omega)\right]$ is exactly the function that establishes the

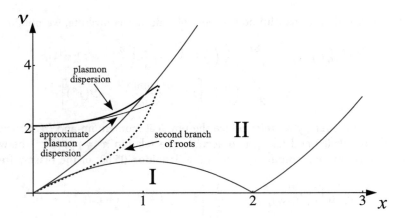

Figure 15.6 Domain of the (ν, x) plane where the imaginary part of the RPA response function is nonvanishing. Besides the regions I and II we also have a δ-peak along the plasmon branch (thick line). The dashed line corresponds to a second solution of the equation $1 - A_1 = 0$. The thin line is the function in (15.86). In this plot $r_s = 5$.

existence of plasmons, see (15.84), we conclude that the imaginary part of the RPA response function also has a δ-like peak along the plasmon curve in the (q, ω) plane.

From the discussion in Section 15.2 about energy dissipation we can say that the plasmon peak must be related to a strong absorption of light at the plasma frequency, a result that has been experimentally confirmed in all simple metals. This absorption can also be deduced from the Maxwell equations. In an electron gas the Fourier transform of the external electric field obeys the equation

$$q^2 \mathbf{E}_{\text{ext}}(\mathbf{q}, \omega) - \frac{\omega^2}{c^2} \varepsilon^{\text{R}}(\mathbf{q}, \omega) \mathbf{E}_{\text{ext}}(\mathbf{q}, \omega) = 0,$$

with ε^{R} the dielectric function. Thus an electromagnetic wave with wave-vector $\mathbf{q}$ and energy ω can penetrate the medium provided that the dispersion relation $q^2 = \frac{\omega^2}{c^2} \varepsilon^{\text{R}}(\mathbf{q}, \omega)$ is satisfied. In (15.23) we saw that $\varepsilon^{-1,\text{R}}(\mathbf{q}, \omega) = 1 + \tilde{v}_{\mathbf{q}} \chi^{\text{R}}(\mathbf{q}, \omega)$ and therefore $\varepsilon^{\text{R}}(\mathbf{q}, \omega) = 1 - \tilde{v}_{\mathbf{q}} P^{\text{R}}(\mathbf{q}, \omega)$ with P the polarizability. In the RPA $P = \chi_0$ and using (15.85) we deduce that for small q the dispersion relation becomes $q^2 = \frac{1}{c^2}(\omega^2 - \omega_{\text{p}}^2)$. For energies larger than ω_{p} there always exist real wave-vectors q for which light can penetrate and be absorbed. On the contrary for $\omega < \omega_{\text{p}}$ the wave-vector is complex and the amplitude of the electric field decays exponentially in the medium. For such energies there cannot be any absorption by plasmons and hence the light is fully reflected. Visible light has energy below the typical plasma frequency and it is indeed common experience that it is reflected by most metals.

In Fig. 15.6 we show the domain of the (ν, x) plane where the imaginary part of χ^{R} is nonvanishing. The figure shows the regions I and II, which are in common with the noninteracting response function, as well as the numerical solution of the equation $1 - A_1(q, \omega) = 0$ for $r_s = 5$ (thick line). The plasmon branch is well approximated by (15.86) (thin line) for small x. Besides the plasmon curve the function $1 - A_1$ also vanishes along a second curve (dashed line) which is, however, entirely contained in the region where $\text{Im} \left[\chi^{\text{R}} \right] \neq 0$. Inside region II the two solutions of $1 - A_1 = 0$ approach each other until they touch, and thereafter $1 - A_1$ is always different from zero. Just after the critical value x_c at which the plasmon branch crosses region II the plasmon peak gets broadened by the particle–hole excitations and the lifetime of the corresponding plasmon excitation becomes finite. This phenomenon is known as the *Landau damping*. As a matter of fact the plasmon lifetime is finite at any finite temperature as well as at zero temperature if we go beyond RPA.

Exercise 15.1. Show that the RPA response function satisfies the f-sum rule.

15.5.3 Sudden creation of a localized hole

In this section we study the time-dependent density δn induced by the sudden creation of a charge Q at the origin. As already mentioned, this study is relevant to the description of the transient screening of a core-hole in simple metals. It also constitutes a very pedagogical application of the linear response theory since the calculations can be carried out analytically. We closely follow the derivation of Canright in Ref. [139]. The potential δV generated by a charge Q suddenly created at time $t = 0$ in $\mathbf{r} = 0$ is

$$\delta V(\mathbf{x}, t) = \theta(t)\frac{Q}{r} = \int \frac{d\mathbf{q}}{(2\pi)^3} \int \frac{d\omega}{2\pi} e^{i\mathbf{q}\cdot\mathbf{r} - i\omega t} \, \delta V(\mathbf{q}, \omega),$$

with

$$\delta V(\mathbf{q}, \omega) = \frac{4\pi Q}{q^2} \frac{i}{\omega + i\eta} = \tilde{v}_\mathbf{q} \, Q \frac{i}{\omega + i\eta}.$$

From the linear response equation (15.80) we then have

$$\delta n(\mathbf{r}, t) = \sum_\sigma \delta n(\mathbf{x}, t) = -\int \frac{d\mathbf{q}}{(2\pi)^3} \int \frac{d\omega}{2\pi} e^{i\mathbf{q}\cdot\mathbf{r} - i\omega t} \, B(q, \omega) \, Q\frac{i}{\omega + i\eta},$$

where the function $B = B_1 + iB_2 = \tilde{v}\chi^{\mathrm{R}}$ was defined in the previous section. Since B depends only on the modulus q we can easily perform the angular integration and find

$$\delta n(\mathbf{r}, t) = -\frac{4\pi Q}{(2\pi)^4} \frac{1}{r} \int_0^\infty dq \, q \sin(qr) \int_{-\infty}^\infty d\omega B(q, \omega)\frac{ie^{-i\omega t}}{\omega + i\eta}. \tag{15.88}$$

The function B is analytic in the upper-half of the complex ω plane and goes to zero as $1/\omega^2$ when $\omega \to \infty$, see Section 15.4. Therefore B has the required properties for the Kramers–Kronig relations (these relations are derived in Appendix O). Accordingly we write B in terms of the imaginary part B_2,

$$B(q, \omega) = -\frac{1}{\pi} \int_{-\infty}^\infty d\omega' \frac{B_2(q, \omega')}{\omega - \omega' + i\eta}.$$

Inserting this relation into (15.88) and performing the integral over ω we obtain

$$\delta n(\mathbf{r}, t) = -\frac{16\pi Q}{(2\pi)^4} \frac{1}{r} \int_0^\infty dq \, q \sin(qr) \int_0^\infty d\omega B_2(q, \omega)\frac{1 - \cos(\omega t)}{\omega}, \tag{15.89}$$

where we rename the integration variable ω' with ω and where we exploit the fact that B_2 is odd in ω, see Section 15.2.1. From Fig. 15.6 we see that this integral naturally splits into two terms of physically different nature. At fixed q the function B_2 is a smooth function between the values $\omega_{\min}(q)$ and $\omega_{\max}(q)$ that delimit the electron–hole continuum (regions I+II) with

$$\omega_{\min}(q) = \epsilon_{p_F} \begin{cases} 0 & \text{if } x < 2 \\ x^2 - 2x & \text{if } x > 2 \end{cases}, \qquad \omega_{\max}(q) = \epsilon_{p_F}(x^2 + 2x),$$

and $x = q/p_F$ as in (15.67). For $\omega \notin (\omega_{\min}(q), \omega_{\max}(q))$ the function B_2 is zero everywhere except along the plasmon branch $\omega_p(q)$. Denoting by q_c the momentum at which the plasmon branch crosses region II, we have that for $q < q_c$ and for $\omega > \omega_{\max}(q)$ the function B_2 is given in (15.87), i.e.,

$$B_2(q,\omega) = -\pi\delta(1 - A_1(q,\omega)) = -\pi \frac{\delta(\omega - \omega_p(q))}{\left| \frac{\partial A_1}{\partial \omega}(q, \omega_p(q)) \right|},$$

where we take into account that for positive frequencies A_2 is negative, see (15.22). From this analysis it is natural to write the density δn in (15.89) as the sum $\delta n_{eh} + \delta n_p$, where δn_{eh} is generated by the excitation of electron–hole pairs whereas δn_p is generated by the excitation of plasmons. Specifically, we have

$$\delta n_{eh}(\mathbf{r}, t) = -\frac{16\pi Q}{(2\pi)^4} \frac{1}{r} \int_0^\infty dq\, q \sin(qr) \int_{\omega_{\min}(q)}^{\omega_{\max}(q)} d\omega\, B_2(q, \omega) \frac{1 - \cos(\omega t)}{\omega},$$

$$\delta n_p(\mathbf{r}, t) = \frac{16\pi^2 Q}{(2\pi)^4} \frac{1}{r} \int_0^{q_c} dq\, q \sin(qr) \frac{1 - \cos(\omega_p(q)t)}{\omega_p(q) \left| \frac{\partial A_1}{\partial \omega}(q, \omega_p(q)) \right|}.$$

In Fig. 15.7 we show the 3D plot of δn_{eh} [panel (a), left], δn_p [panel (b), left] as well as the total density δn [panel (c), left] as a function of time t and distance r from the origin. These plots have been generated by calculating the above integrals numerically for $Q = 1$ and $r_s = 3$. In the right part of the figure the densities are multiplied by $4\pi(rp_F)^2$ so as to highlight the large r behavior. As pointed out by Canright [139] the partitioning into electron–hole pairs and plasmons is "somewhat illuminating but also unphysical." The density δn_{eh}, in contrast with the density δn_p, clearly loses phase coherence at long times [panel (a) and (b), right]. We also notice that at small r the screening is essentially due to δn_{eh} which builds up in a time $t \sim 1/\omega_p$ [panel (a), left]; the plasmon contribution is an order of magnitude smaller [panel (b), left] and adds a small, damped ringing to the total density. The unphysical aspect of the partitioning is evident in the plots to the right of panels (a) and (b). The hole is suddenly created in $r = 0$ and therefore, as for a stone tossed into a pond, the density in r should change only after the shock-wave has had the time to propagate till r. We instead see that at small t the densities δn_{eh} and δn_p decay very slowly for large r. This unphysical behavior is absent in the physical sum $\delta n = \delta n_{eh} + \delta n_p$ [panel (c), right] where an almost perfect cancellation occurs. We clearly see a density front that propagates at a speed of about the Fermi velocity $v_F = p_F$, which is $\simeq 0.65$ in our case. Thus the physical density $\delta n(r, t)$ is essentially zero up to a time $t \sim r/v_F$; after this time δn changes and exhibits damped oscillations around its steady-state value.

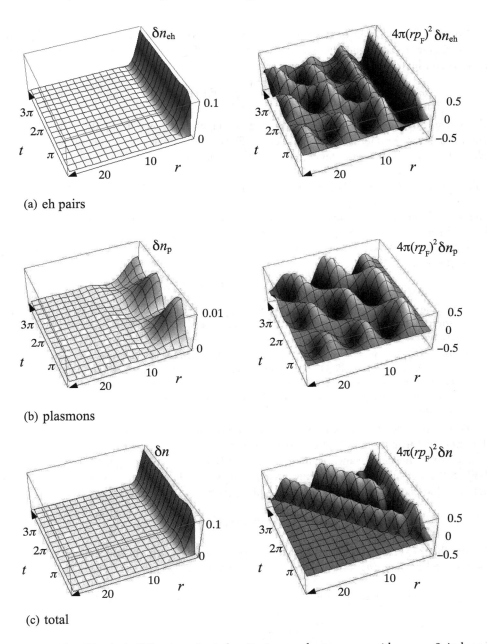

(a) eh pairs

(b) plasmons

(c) total

Figure 15.7 The 3D plot of the transient density in an electron gas with $r_s = 3$ induced by the sudden creation of a point-like positive charge $Q = 1$ in the origin at $t = 0$. The contribution due to the excitation of electron–hole pairs (a), and plasmons (b), is, for clarity, multiplied by $4\pi(rp_F)^2$ in the plots to the right. Panel (c) is simply the sum of the two contributions. Units: r is in units of $1/p_F$, t is in units of $1/\omega_p$, and all densities are in units of p_F^3.

The steady-state limit of the RPA density δn can easily be worked out from (15.89). Due to the Riemann–Lebesgue theorem the term containing $\cos(\omega t)$ integrates to zero when $t \to \infty$ and hence

$$\delta n_s(\mathbf{r}) \equiv \lim_{t \to \infty} \delta n(\mathbf{r}, t) = -\frac{Q}{2\pi^3} \frac{1}{r} \int_0^\infty dq\, q \sin(qr) \int_{-\infty}^\infty d\omega \frac{B_2(q, \omega)}{\omega}.$$

The integral over ω can be interpreted as a principal part since $B_2(q, 0) = 0$, see (15.72). Then from the Kramers–Kronig relation (O.5) we can replace the integral over ω with $B(q, 0) = \tilde{v}_\mathbf{q} \chi^\mathrm{R}(\mathbf{q}, 0)$, and hence recover the well known result

$$\delta n_s(\mathbf{r}) = -\frac{Q}{2\pi^2} \frac{1}{r} \int_0^\infty dq\, q \sin(qr) \tilde{v}_\mathbf{q} \chi^\mathrm{R}(\mathbf{q}, 0) = -Q \int \frac{d\mathbf{q}}{(2\pi)^3} e^{i\mathbf{q}\cdot\mathbf{r}} \tilde{v}_\mathbf{q} \chi^\mathrm{R}(\mathbf{q}, 0). \quad (15.90)$$

This formula yields the *static* screening in an electron gas. We refer the reader to the classic book of Fetter and Walecka [45] for a thorough analysis of (15.90) with χ^R in the RPA.

Among the most interesting consequences of (15.90) with χ^R in the RPA we should mention:

- The induced charge

$$\delta Q(R) \equiv -4\pi \int_0^R dr\, r^2 \delta n_s(\mathbf{r})$$

  is close to minus the positive charge Q for $R \gtrsim \lambda_\mathrm{TF}$, meaning that the electron gas screens an external charge within a distance of a few λ_TF. At very large distances the screening is perfect, i.e., $\delta Q(R \to \infty) = Q$.

- For $r \gg 1/p_\mathrm{F}$ the density $\delta n_s(\mathbf{r})$ goes to zero as $\cos(2p_\mathrm{F}r)/r^3$, a result which was derived by Langer and Vosko in 1960 [140]. However, it was Friedel who first pointed out that these damped spatial oscillations are a general consequence of the discontinuity of the Fermi function at zero temperature [141]. For this reason, in the scientific literature they are known as the *Friedel oscillations*.

Equation (15.90) allows us to give a physical interpretation to the effective interaction of the Thomas–Fermi theory. Suppose that the charge $Q = q = -1$ is the same as the electron charge. Then the total change in the charge density is $q\,\delta n_\mathrm{tot}(\mathbf{r}) = q[\delta(\mathbf{r}) + \delta n_s(\mathbf{r})]$, where $\delta(\mathbf{r})$ is the density of the suddenly created charge at the origin. The interaction energy between the charge distribution $q\,\delta n_\mathrm{tot}$ and a generic electron in position $\mathbf{r}$ is

$$e_\mathrm{int}(\mathbf{r}) = \int d\mathbf{r}' v(\mathbf{r}, \mathbf{r}') \delta n_\mathrm{tot}(\mathbf{r}').$$

We see that $e_\mathrm{int}(\mathbf{r})$ correctly reduces to $v(\mathbf{r}, 0) = 1/r$ in an empty space since in this case the induced density $\delta n_s = 0$. Instead, in the electron gas we have

$$e_\mathrm{int}(\mathbf{r}) = \int d\mathbf{r}' v(\mathbf{r}, \mathbf{r}') \left[\delta(\mathbf{r}') + \int \frac{d\mathbf{q}}{(2\pi)^3} e^{i\mathbf{q}\cdot\mathbf{r}'} \tilde{v}_\mathbf{q} \chi^\mathrm{R}(\mathbf{q}, 0)\right]$$

$$= \int \frac{d\mathbf{q}}{(2\pi)^3} e^{i\mathbf{q}\cdot\mathbf{r}} \left[\tilde{v}_\mathbf{q} + \tilde{v}_\mathbf{q}^2 \chi^\mathrm{R}(\mathbf{q}, 0)\right]$$

$$= \int \frac{d\mathbf{q}}{(2\pi)^3} e^{i\mathbf{q}\cdot\mathbf{r}} W^\mathrm{R}(\mathbf{q}, 0) \xrightarrow[r \to \infty]{} \frac{e^{-r/\lambda_\mathrm{TF}}}{r}.$$

Thus the effective Yukawa interaction is the interaction between a "test" electron and a statically screened electron.

15.5.4 Spectral properties in the $G_0 W_0$ approximation

In the previous sections we have seen that a sudden creation of a positive charge in the electron gas yields a density response that can be physically interpreted as the sum of two contributions: it consists of a piece in which we excite particle–hole pairs and a piece in which we excite plasmons. We may therefore expect that the sudden addition or removal of an electron as described by the greater and lesser Green's functions induces similar excitations which appear as characteristic structures in the spectral function. This has indeed been found in photoemission and inverse photoemission experiments on metals such as sodium or aluminum [142] which strongly resemble the electron gas. To calculate the spectral function we must first determine the self-energy. Since $W = v + vPW = v + v\chi v$ already contains the physics of both single-particle excitations and plasmons it is natural to consider the expansion of the self-energy $\Sigma_{ss,\mathrm{xc}}[G, W]$ in terms of the dressed Green's function and screened interaction as described in Section 10.7. In the following we restrict ourselves to the lowest order term of the expansion which is the GW approximation $\Sigma_{\mathrm{xc}}(1; 2) = iG(1; 2)W(2; 1)$, where the screened interaction is calculated from the polarizability $P(1; 2) = -iG(1; 2)G(2; 1)$.

In Chapter 7 we learned that the Hartree self-energy Σ_{H} cancels with the external potential of the positive background charge and can therefore be disregarded. Since the electron gas is translationally invariant and in equilibrium we can Fourier transform to momentum-energy space and write

$$\Sigma^{\mathrm{R}}(\mathbf{p}, \omega) = \Sigma_{\mathrm{x}}(\mathbf{p}) + \underbrace{i \int \frac{d\omega'}{2\pi} \frac{\Sigma^{>}(\mathbf{p}, \omega') - \Sigma^{<}(\mathbf{p}, \omega')}{\omega - \omega' + i\eta}}_{\Sigma_{\mathrm{c}}^{\mathrm{R}}(\mathbf{p}, \omega)}, \tag{15.91}$$

where $\Sigma_{\mathrm{x}}(\mathbf{p}) = -\frac{2p_{\mathrm{F}}}{\pi} F(\frac{p}{p_{\mathrm{F}}})$ is the time-local exchange (or Fock) self-energy, see (7.61), whereas $\Sigma_{\mathrm{c}}^{\mathrm{R}}$ is the correlation self-energy, see (9.57). Remember that in equilibrium the Green's function and the self-energy are diagonal in spin space, since v, and hence W, are independent of spin. In accordance with our notation the quantities $G(\mathbf{p}, \omega)$ and $\Sigma(\mathbf{p}, \omega)$ refer to the Fourier transform of the diagonal (and σ-independent) matrix elements $G(\mathbf{r}_1\sigma, t_1; \mathbf{r}_2\sigma, t_2)$ and $\Sigma(\mathbf{r}_1\sigma, t_1; \mathbf{r}_2\sigma, t_2)$. Since the electron gas is also rotationally invariant all Fourier transformed quantities depend only on the modulus of the momentum. To lighten the notation we therefore denote $G(\mathbf{p}, \omega)$ by $G(p, \omega)$, $\Sigma(\mathbf{p}, \omega)$ by $\Sigma(p, \omega)$, $W(\mathbf{p}, \omega)$ by $W(p, \omega)$, etc. In Fourier space the GW self-energy becomes a convolution between G and W:

$$\Sigma^{\lessgtr}(p, \omega) = i \int \frac{d\mathbf{k} d\omega'}{(2\pi)^4} G^{\lessgtr}(\mathbf{k}, \omega') W^{\gtrless}(|\mathbf{k} - \mathbf{p}|, \omega' - \omega)$$

$$= i \int \frac{d\omega'}{(2\pi)^4} 2\pi \int_{-1}^{1} dc \int_0^{\infty} dk\, k^2\, G^{\lessgtr}(k, \omega') W^{\gtrless}(\sqrt{k^2 + p^2 - 2kpc}, \omega' - \omega),$$

where we use the fact that $\Sigma^{\lessgtr}(1; 2) = iG^{\lessgtr}(1; 2)W^{\gtrless}(2; 1)$. Instead of integrating over the cosine c we perform the substitution $q = \sqrt{k^2 + p^2 - 2kpc}$ such that $q\,dq = -kp\,dc$, and rewrite the lesser/greater self-energy as

$$\Sigma^{\lessgtr}(p,\omega) = \frac{i}{(2\pi)^3 p} \int d\omega' \int_0^\infty dk\, k\, G^{\lessgtr}(k,\omega') \int_{|k-p|}^{k+p} dq\, q\, W^{\gtrless}(q,\omega'-\omega). \qquad (15.92)$$

We have therefore reduced the calculation of the self-energy to a three-dimensional integral. Let us give a physical interpretation to $\Sigma^{\lessgtr}$. We have seen in Section 13.3 that $\Sigma^{>}(p,\omega)$ is the decay rate for an added particle with momentum p and energy ω. According to (15.92) this particle is scattered into momentum–energy state (k,ω'), thereby creating a particle–hole pair of momentum–energy $(|\mathbf{p} - \mathbf{k}|, \omega - \omega')$. If the energy ω is just above the chemical potential μ then, since all states below the chemical potential are occupied, there is very little phase-space left to excite a particle–hole pair. We therefore expect that the scattering rate goes to zero as a function of $\omega - \mu$. According to our interpretation one factor $\omega - \mu$ comes from the phase-space requirement on $G^{>}$ (available states after the scattering), and another one from the phase-space requirement on $W^{<}$ (density of particle–hole pairs). Together this leads to the $(\omega - \mu)^2$ behavior that we already deduced in Section 13.3 for the second-Born approximation, and that we derive more rigorously below for the approximation (15.92).[20] Similar considerations apply to $\Sigma^{<}$.

Equation (15.92) together with the equation for the screened interaction W and the Dyson equation

$$G^{\mathrm{R}}(k,\omega) = \frac{1}{\omega - \epsilon_k - \Sigma^{\mathrm{R}}(k,\omega)}, \qquad \epsilon_k = k^2/2, \qquad (15.93)$$

form a self-consistent set of equations which can only be solved with some considerable numerical effort [110]. However, our goal is to get some insight into the properties of the spectral function. Therefore, rather than solving the equations self-consistently we insert into them a physically motivated Green's function. The simplest Green's function we can consider is, of course, the noninteracting Green's function $G_0^{\mathrm{R}}(k,\omega) = (\omega - \epsilon_k + i\eta)^{-1}$. However, this has some objections. First of all, the choice of the GW-diagram came from a skeletonic expansion in dressed Green's functions. We, therefore, at least want to include some kind of self-energy renormalization. The simplest of such a Green's function, which still allows for analytical manipulations, is given by

$$G^{\mathrm{R}}(k,\omega) = \frac{1}{\omega - \epsilon_k - \Delta + i\eta}, \qquad (15.94)$$

with Δ a real number [143]. In this poor man's choice we approximate the self-energy with a real number (as in the Hartree–Fock approximation, but this time k-independent). The next question is then how to choose Δ. We know from Section 13.3 that the exact as well as any approximate Green's function has a rate $\Gamma(k,\omega) \sim (\omega - \mu)^2$, and therefore that $G^{\mathrm{R}}(k,\mu)$ has a pole in $k = p_{\mathrm{F}}$ provided that p_{F} is chosen as the solution of

$$0 = \mu - \epsilon_{p_{\mathrm{F}}} - \Sigma^{\mathrm{R}}(p_{\mathrm{F}},\mu). \qquad (15.95)$$

We then require that our approximate self-consistent $G^{\mathrm{R}}(k,\mu)$ in (15.94) has a pole at the same spot. This implies

$$0 = \mu - \epsilon_{p_{\mathrm{F}}} - \Delta$$

---

[20] The fact that the second-Born and the GW approximations predict a rate proportional to $(\omega - \mu)^2$ is a direct consequence of the fact that both approximations contain the single bubble self-energy diagram.

and yields $\Delta = \mu - \epsilon_{p_F}$. As we shall see this shift considerably simplifies the analytical manipulations. In practice the Fermi momentum $p_F = (3\pi^2 n)^{1/3}$ is determined by the density of the gas and therefore rather than fixing μ and calculating p_F we fix p_F and calculate μ. In this way our G^R has a pole at the desired p_F.[21] Thus the strategy is: first evaluate Σ^R using the μ-dependent Green's function (15.94) and subsequently fix the chemical potential using (15.95). This is the so called $G_0 W_0$ approximation for the self-energy.

The Green's function in (15.94) has the form of a noninteracting G and therefore using (6.80) and (6.81) we get

$$G^<(k,\omega) = 2\pi i f(\omega - \mu)\delta(\omega - \epsilon_k - \Delta),$$
$$G^>(k,\omega) = -2\pi i \bar{f}(\omega - \mu)\delta(\omega - \epsilon_k - \Delta).$$

At zero temperature $f(\omega - \mu) = \theta(\mu - \omega)$ and due to the δ-function we can use $\theta(\mu - \epsilon_k - \Delta) = \theta(\epsilon_{p_F} - \epsilon_k) = \theta(p_F - k)$. Similarly in $G^>$ we can replace $\bar{f}(\omega - \mu)$ with $\theta(k - p_F)$. Inserting $G^<$ into (15.92) we find

$$\Sigma^<(p,\omega) = \frac{-1}{4\pi^2 p} \int_0^{p_F} dk\, k \int_{|k-p|}^{k+p} dq\, q\, W^>(q, \epsilon_k + \Delta - \omega).$$

If we define the new variable $\omega' = \epsilon_k + \Delta - \omega$ so that $d\omega' = k\,dk$, then the integral becomes

$$\Sigma^<(p,\omega) = \frac{-1}{4\pi^2 p} \int_{\mu-\omega-\epsilon_{p_F}}^{\mu-\omega} d\omega' \int_{q_-}^{q_+} dq\, q\, W^>(q, \omega'),$$

where $q_- = |\sqrt{2(\omega' - \Delta + \omega)} - p|$ and $q_+ = \sqrt{2(\omega' - \Delta + \omega)} + p$. In the same way it is easy to show that

$$\Sigma^>(p,\omega) = \frac{1}{4\pi^2 p} \int_{\mu-\omega}^{\infty} d\omega' \int_{q_-}^{q_+} dq\, q\, W^<(q, \omega').$$

The calculation of the self-energy is now reduced to a two-dimensional integral. It only remains to calculate W from the polarizability $P = -iGG$. We have $W = v + vPW = v + v\chi v$ where $\chi = P + PvP + \dots$ If we calculate P using the Green's function (15.94) then $P = \chi_0$ is the Lindhard function. Indeed our approximate G only differs from the noninteracting G_0 by a constant shift of the one-particle energies, and hence the particle-hole excitation energies $\epsilon_{k+q} - \epsilon_k$ in χ_0 are unchanged, see (15.66). Since $P = \chi_0$ then χ is exactly the RPA expression of Section 15.5.2. To extract the lesser/greater screened interaction we can use the fluctuation–dissipation theorem (13.11). Due to the property (15.25) we have $W^A(q,\omega) = [W^R(q,\omega)]^*$ and hence

$$W^<(q,\omega) = 2if(\omega)\mathrm{Im}[W^R(q,\omega)], \quad W^>(q,\omega) = 2i\bar{f}(\omega)\mathrm{Im}[W^R(q,\omega)],$$

with f the Bose function and $\bar{f} = 1 + f$. At zero temperature $f(\omega) = -\theta(-\omega)$ and $\bar{f}(\omega) = 1 + f(\omega) = 1 - \theta(-\omega) = \theta(\omega)$. Furthermore $W^R(q,\omega) = \tilde{v}_q + \tilde{v}_q^2 \chi^R(q,\omega)$ and

---

[21]In general the comparison between Green's functions in different approximations makes sense only if these Green's functions have a pole at the same Fermi momentum. This means that different approximations to Σ yield different values of the chemical potential according to (15.95).

therefore the lesser/greater self-energies assume the form

$$i\Sigma^<(p,\omega) = \frac{1}{2\pi^2 p} \int_{\mu-\omega-\epsilon_{p_F}}^{\mu-\omega} d\omega'\, \theta(\omega') \int_{q_-}^{q_+} dq\, q\, \tilde{v}_q^2\, \text{Im}[\chi^R(q,\omega')], \qquad (15.96)$$

$$i\Sigma^>(p,\omega) = \frac{1}{2\pi^2 p} \int_{\mu-\omega}^{\infty} d\omega'\, \theta(-\omega') \int_{q_-}^{q_+} dq\, q\, \tilde{v}_q^2\, \text{Im}[\chi^R(q,\omega')]. \qquad (15.97)$$

Since $\text{Im}[\chi^R(q,\omega')]$ is negative (positive) for positive (negative) frequencies ω', see (15.22), from these equations we infer that $i\Sigma^>$ is only nonvanishing for $\omega > \mu$ in which case it is positive and $i\Sigma^<$ is only nonvanishing for $\omega < \mu$, in which case it is negative. Hence the rate function $\Gamma = i(\Sigma^> - \Sigma^<)$ is positive as it should be. We also observe that both $\Sigma^<$ and $\Sigma^>$ depend only on $\omega - \mu$ and therefore (15.95) provides an explicit expression for μ:

$$\mu = \epsilon_{p_F} + \Sigma^R(p_F, \mu) = \epsilon_{p_F} + \Sigma_x(p_F) + \Sigma_c^R(p_F, \mu), \qquad (15.98)$$

since the correlation self-energy $\Sigma_c^R(p_F, \mu)$ does not depend on μ. Any other shift Δ would have led to a more complicated equation for μ.

This is how far we can get with pencil and paper. What we have to do next is to calculate numerically $\Sigma^{\lessgtr}$, insert them into (15.91) to obtain Σ_c^R and then extract the spectral function from (13.5). First, however, it is instructive to analyze the behavior of the self-energy for energies close to μ and for energies close to the plasma frequency ω_p. This analysis facilitates the interpretation of the numerical results and at the same time provides us with useful observations for the actual implementation.

Self-energy for energies close to the chemical potential

Let us analyze the integrals (15.96) and (15.97) for ω close to μ. We take, for example, the derivative of $i\Sigma^<$. For ω close to μ we have that $|\omega - \mu| < \epsilon_{p_F}$ and the lower limit of the integral in (15.96) can be set to zero. Differentiation with respect to ω then gives three terms, one term stemming from the differentiation of the upper limit of the ω' integral and two more terms stemming from the differentiation of q_- and q_+. These last two terms vanish when $\omega \to \mu^-$ since the interval of the ω' integration goes to zero in this limit. Therefore

$$i\frac{\partial \Sigma^<(p,\omega)}{\partial \omega}\bigg|_{\omega=\mu^-} = -\frac{1}{2\pi^2 p}\int_{|p_F-p|}^{p_F+p} dq\, q\, \tilde{v}_q^2\, \text{Im}[\chi^R(q,0)] = 0,$$

since $\text{Im}[\chi^R(q,0)] = 0$. In a similar way one can show that the derivative of $\Sigma^>(p,\omega)$ with respect to ω vanishes for $\omega \to \mu^+$. Thus for $\omega \sim \mu$ we have

$$i\Sigma^<(p,\omega) = -C_p^< \theta(\mu - \omega)\,(\omega - \mu)^2,$$

$$i\Sigma^>(p,\omega) = C_p^> \theta(\omega - \mu)\,(\omega - \mu)^2,$$

where $C_p^{\lessgtr} > 0$, see discussion below (15.97). The constants $C_p^{\lessgtr}$ can easily be determined by taking the second derivative of $\Sigma^{\lessgtr}$ with respect to ω in $\omega = \mu^{\mp}$. It is a simple exercise for the reader to show that $C_p^> = C_p^< = C_p$ with

$$C_p = -\frac{1}{4\pi^2 p}\int_{|p_F-p|}^{p_F+p} dq\, q\, \frac{\tilde{v}_q^2}{|1 - \tilde{v}_q\chi_0^R(q,0)|^2}\, \frac{\partial\, \text{Im}[\chi_0^R(q,\omega)]}{\partial\omega}\bigg|_{\omega=0}, \qquad (15.99)$$

where we have used the RPA expression (15.78) for χ^R in terms of the Lindhard function χ_0^R. The fact that $C_p^> = C_p^<$ agrees with the discussion in Section 13.3 and allows us to write the rate operator as in (13.24). Let us comment in more detail on the explicit formula for C_p in the $G_0 W_0$ approximation. Since the derivative of $\text{Im}[\chi_0^R(q,\omega)]$ with respect to ω is negative in $\omega = 0$ for $q \leq 2p_F$ and zero otherwise, see Fig. 15.4(b), we find that the constant $C_p > 0$ for $|p_F - p| \leq 2p_F$ (i.e. for $p \leq 3p_F$) and zero otherwise. This is physically related to the fact that an injected particle with momentum $3p_F$ and energy μ can scatter into a particle with momentum p_F and energy μ and an electron–hole pair with momentum $2p_F$ and energy zero at the Fermi surface. Any particle with higher momentum than $3p_F$ (and energy μ) can only excite electron–hole pairs with nonzero energy.

From the knowledge of C_p we can calculate the lifetime τ_p of a quasi-particle (or quasi-hole) with momentum p. We recall that $\tau_p^{-1} = Z_p \Gamma(p, E_p)$ where Z_p is the quasi-particle renormalization factor and E_p is the quasi-particle energy defined as the solution of $\omega - \epsilon_p - \Sigma_x(p) - \Lambda(p,\omega) = 0$, see Section 13.1 (in Section 13.1 Σ_x was absorbed in ϵ_p). For $p \sim p_F$ the quasi-particle energy $\epsilon_p \sim \mu$ and hence $\Gamma(p, E_p) \sim C_{p_F}(E_p - \mu)^2$. Expanding in powers of $p - p_F$ we then find

$$\frac{1}{\tau_p} = Z C_{p_F} \left(\frac{p_F}{m^*}\right)^2 (p - p_F)^2,$$

where $Z = Z_{p_F}$ and the *effective mass* m^* of the quasi-particle is defined by

$$\frac{1}{m^*} \equiv \frac{1}{p_F} \frac{dE_p}{dp}\bigg|_{p=p_F}.$$

Let us calculate a more explicit form of C_p. From (15.69) and (15.72) we have that $\chi_0^R(q,0) = -(p_F/\pi^2) F(q/2p_F)$ and $\partial \text{Im}[\chi_0^R(q,\omega)]/\partial\omega|_{\omega=0} = -\theta(2p_F - q)/(2\pi q)$ (we use the expression in region I). Inserting these results into (15.99) we get

$$C_p = \frac{1}{8\pi^3 p} \int_{|p_F - p|}^{p_F + p} dq\, \theta(2p_F - q) W^R(q,0)^2, \quad \text{with} \quad W^R(q,0) = \frac{4\pi}{q^2 + \frac{4p_F}{\pi} F(\frac{q}{2p_F})}.$$

We now evaluate this formula for $p = p_F$:

$$C_{p_F} = \frac{1}{8\pi^3 p_F} \int_0^{2p_F} dq \left(\frac{4\pi}{q^2 + \frac{4p_F}{\pi} F(\frac{q}{2p_F})}\right)^2 = \frac{\pi}{4p_F^2} \xi(r_s),$$

where we define

$$\xi(r_s) = \int_0^1 dx\, \frac{1}{\left[F(x) + \left(\frac{9\pi}{4}\right)^{\frac{1}{3}} \frac{\pi}{r_s} x^2\right]^2}$$

and use the fact that $p_F = (9\pi/4)^{\frac{1}{3}}/r_s$. Thus the quasi-particle lifetime becomes [67]

$$\frac{1}{\tau_p} = \frac{Z\pi}{4m^{*2}} \xi(r_s) (p - p_F)^2.$$

As an exercise we can analyze the high density $r_s \to 0$ behavior of $\xi(r_s)$. Since $F(0) = 1$ and $F(1) = 0$ the integrand of $\xi(r_s)$ is equal to 1 in $x = 0$ and is proportional to r_s^2 in $x = 1$, which is small in the limit $r_s \to 0$. The largest contribution to the integral therefore comes from the integration region around $x = 0$ and we can then simply approximate $F(x) \approx F(0) = 1$. Denoting by $\alpha = (9\pi/4)^{\frac{1}{6}} \sqrt{\pi/r_s}$ the square root of the constant which multiplies x^2 we have

$$\xi(r_s) = \frac{1}{\alpha} \int_0^\alpha dy \frac{1}{(1+y^2)^2} = \frac{1}{2\alpha} \arctan \alpha + \frac{1}{2(1+\alpha^2)} \sim \frac{\pi}{4\alpha} \quad (\alpha \to \infty).$$

So we see that $\xi(r_s) \sim \sqrt{r_s}$ for $r_s \to 0$ and hence the lifetime increases by increasing the density of the electron gas. Strictly speaking also m^* and Z are functions of r_s but they are weakly dependent on r_s and approach 1 as $r_s \to 0$ [67]. This limiting behavior of m^* and Z is somehow intuitive since for $r_s \to 0$ the dominant self-energy diagram is the Hartree diagram, which is, however, cancelled by the external potential. Therefore the high-density interacting gas behaves like a noninteracting gas.

The constant C_p pertains to the quasi-particle lifetime. What about the quasi-particle energy E_p? To answer this question we have to study the behavior of $\Lambda = \text{Re}[\Sigma_c^R]$, since E_p is the solution of $\omega - \epsilon_p - \Sigma_x(p) - \Lambda(p, \omega) = 0$, see Section 13.1. This is not easily done on paper. As we discuss in more detail below, we find from numerical calculations for the electron gas at metallic densities that $E_p - \mu \approx \epsilon_p - \epsilon_{p_F}$. This implies that

$$\Sigma_x(p) + \Lambda(p, E_p) = E_p - \epsilon_p \approx \mu - \epsilon_{p_F}$$

and hence the real part of the self-energy does not change much when we move along the quasi-particle branch in the (p, ω)-plane. With hindsight our poor man's choice for G is more accurate than the Hartree–Fock G, since in the Hartree–Fock approximation $\epsilon_p = p^2/2$ is renormalized by the exchange self-energy $\Sigma_x(p)$ and hence the Hartree–Fock $E_p = \epsilon_p + \Sigma_x(p)$ deviates considerably from a parabola, see middle panel of Fig. 7.6 (note that in Section 7.3.2 we used the symbol ϵ_p instead of E_p).

Self-energy for energies close to the plasma frequency

So far we have only been looking at the behavior of the self-energy close to the Fermi surface. However, an added or removed particle at sufficiently high energy may also excite plasmons. For instance, for an added particle we would expect this to happen at energies $\omega \geq \mu + \omega_p$ where ω_p is the plasmon frequency. Let us study how these plasmon excitations appear from the equations. We observe that the function under the q-integral in (15.96) and (15.97) is $q \tilde{v}_q B_2(q, \omega')$ where B_2 is the imaginary part of the function $B = \tilde{v} \chi^R$ defined in the previous section. As pointed out in (15.87), the function B_2 has δ-like singularities outside regions I and II (where $A_2(q, \omega') \sim -\eta \, \text{sgn}(\omega')$) when $1 - A_1(q, \omega') = 0$, i.e., when $\omega' = \pm\omega_p(q)$. Thus for $\omega' \sim \pm\omega_p(q)$ and for $q < q_c$ (q_c momentum at which the plasmon branch crosses the particle–hole continuum), we can write

$$B_2(q, \omega') = -\pi \frac{\delta(\omega' - \omega_p(q))}{\left| \frac{\partial A_1}{\partial \omega'}(q, \omega_p(q)) \right|} + \pi \frac{\delta(\omega' + \omega_p(q))}{\left| \frac{\partial A_1}{\partial \omega'}(q, \omega_p(q)) \right|},$$

where we take into account that A_2 is negative (positive) for positive (negative) frequencies. We thus see that the lesser and greater self-energy can be split into a particle–hole part and a plasmon part $\Sigma^{\lessgtr} = \Sigma_{ph}^{\lessgtr} + \Sigma_p^{\lessgtr}$ where the particle–hole part is given by (15.96) and (15.97)

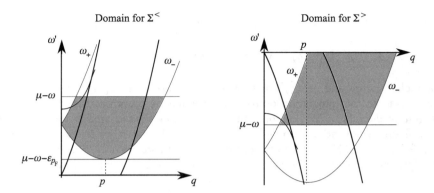

Figure 15.8 (Left panel) Integration domain for $\Sigma^<(p, \omega')$ in the case that $\mu - \omega > \omega_\mathrm{p}$ and $\mu - \omega - \epsilon_{p_\mathrm{F}} > 0$ and $p < 2p_\mathrm{F}$. (Right panel) Integration domain for $\Sigma^>(p, \omega')$ in the case that $\mu - \omega < -\omega_\mathrm{p}$ and $p < 2p_\mathrm{F}$. The figure further shows with thick lines the regions I and II of the particle–hole continuum and the plasmon branch, compare with Fig. 15.6.

integrated over regions I and II, and the plasmon part is given by

$$i\Sigma_\mathrm{p}^<(p, \omega) = -\frac{1}{2\pi p} \int_{\mu - \omega - \epsilon_{p_\mathrm{F}}}^{\mu - \omega} d\omega' \, \theta(\omega') \int_{q_-}^{q_+} dq \, \frac{q \, \tilde{v}_q \, \theta(q_\mathrm{c} - q)}{\left| \frac{\partial A_1}{\partial \omega'}(q, \omega_\mathrm{p}(q)) \right|} \, \delta(\omega' - \omega_\mathrm{p}(q)),$$

$$i\Sigma_\mathrm{p}^>(p, \omega) = \frac{1}{2\pi p} \int_{\mu - \omega}^{\infty} d\omega' \, \theta(-\omega') \int_{q_-}^{q_+} dq \, \frac{q \, \tilde{v}_q \, \theta(q_\mathrm{c} - q)}{\left| \frac{\partial A_1}{\partial \omega'}(q, \omega_\mathrm{p}(q)) \right|} \, \delta(\omega' + \omega_\mathrm{p}(q)).$$

When can we expect a contribution from the plasmon excitations? Let us first consider $\Sigma_\mathrm{p}^<$. Since the upper limit of the ω'-integration is given by $\mu - \omega$ we can only get a contribution from the integral when $\mu - \omega \geq \omega_\mathrm{p}(q) \geq \omega_\mathrm{p}$, with $\omega_\mathrm{p} = \omega_\mathrm{p}(0)$. The allowed q values are still determined by the functions q_+ and q_-, but it is at least clear that there can be no contribution unless $\omega \leq \mu - \omega_\mathrm{p}$. This means physically that a removed particle can only excite plasmons when its energy is at least ω_p below the chemical potential. A similar story applies to $\Sigma_\mathrm{p}^>$: from the lower limit of the integral we see that there can be no contribution unless $\mu - \omega \leq -\omega_\mathrm{p}(q) \leq -\omega_\mathrm{p}$. Therefore an added particle can only excite plasmons when its energy is at least ω_p above the chemical potential.

We can deduce the energies for which $\Sigma_\mathrm{p}^{\lessgtr}$ are maximal by analyzing in more detail the integration domain. For $\Sigma_\mathrm{p}^<$ the ω'-integral goes from $\max(0, \mu - \omega - \epsilon_{p_\mathrm{F}})$ to $\mu - \omega$ (which must be positive for otherwise $\Sigma_\mathrm{p}^< = 0$). The left panel of Fig. 15.8 shows an example of the ω'-domain (the region between the two horizontal lines) in the case that $\mu - \omega > \omega_\mathrm{p}$ and $\mu - \omega - \epsilon_{p_\mathrm{F}} > 0$. For any fixed ω' we have to integrate over q between q_- and q_+, i.e.,

$$q_- = \left| \sqrt{2(\omega' - \mu + \omega + \epsilon_{p_\mathrm{F}})} - p \right| \leq q \leq \sqrt{2(\omega' - \mu + \omega + \epsilon_{p_\mathrm{F}})} + p = q_+$$

where we use the fact that $\Delta = \mu - \epsilon_{p_F}$. This amounts to integrating over all qs such that[22]

$$\omega_-(q) \le \omega' \le \omega_+(q), \quad \text{where} \quad \omega_\pm(q) = \frac{1}{2}(p \pm q)^2 + \mu - \omega - \epsilon_{p_F}.$$

The parabolas $\omega_+(q)$ and $\omega_-(q)$ are shown in Fig. 15.8 and, together with the previous constraint, lead to the integration domain represented by the grey-shaded area. Similarly the integration domain of $\Sigma_p^>$ is represented by the grey-shaded area in the right panel of the same figure. Here the horizontal lines are at $\omega' = 0$ and $\omega' = \mu - \omega$ (which must be negative, for otherwise $\Sigma_p^> = 0$). In the example of the figure we have chosen $\mu - \omega < -\omega_p$.

If the plasmon branch lies in the integration domain then the integral over ω' leads to

$$i\Sigma_p^<(p, \omega) = -\frac{1}{2\pi p} \int_{q_0^<}^{q_1^<} dq \, \frac{q \, \tilde{v}_q \, \theta(q_c - q)}{\left| \frac{\partial A_1}{\partial \omega'}(q, \omega_p(q)) \right|}, \tag{15.100}$$

$$i\Sigma_p^>(p, \omega) = \frac{1}{2\pi p} \int_{q_0^>}^{q_1^>} dq \, \frac{q \, \tilde{v}_q \, \theta(q_c - q)}{\left| \frac{\partial A_1}{\partial \omega'}(q, \omega_p(q)) \right|}, \tag{15.101}$$

where $q_0^<$ and $q_1^<$ ($q_0^>$ and $q_1^>$) are the momenta at which the plasmon branch $\omega_p(q)$ $(-\omega_p(q))$ enters and leaves the grey-shaded area. Consider now the integrand for small q. We have

$$A_1(q, \omega') = \tilde{v}_q \, \text{Re}[\chi_0^R(q, \omega')] \xrightarrow[q \to 0]{} \tilde{v}_q \, n \, \frac{q^2}{\omega'^2} = \tilde{v}_q \, \frac{\omega_p^2}{4\pi} \frac{q^2}{\omega'^2},$$

where we use (15.70) and the formula $\omega_p = \sqrt{4\pi n}$ for the plasma frequency. Thus we see that for small momenta

$$\frac{\partial A_1}{\partial \omega'}(q, \omega_p(q)) = -\tilde{v}_q \, \frac{q^2}{2\pi \omega_p}$$

and the integrand of $\Sigma_p^{\lessgtr}$ becomes $2\pi\omega_p/q$. This leads to logarithmic behavior of $\Sigma_p^{\lessgtr}$ when the lower integration limit $q_0^{\lessgtr}$ gets very small. Consequently the rate function $\Gamma(p, \omega) = i[\Sigma^>(p, \omega) - \Sigma^<(p, \omega)]$ also has logarithmic behavior and its Hilbert transform $\Lambda(p, \omega)$ has a discontinuous jump at the same point.[23]

Let us see for which values of p and ω the logarithmic singularity occurs. For $q = 0$ the parabolas $\omega_+(0) = \omega_-(0) = \epsilon_p + \mu - \omega - \epsilon_{p_F} \equiv \omega_0$ have the same value. For $\Sigma_p^<$ the point $(0, \omega_0)$ is in the integration domain when $\omega_0 \le \mu - \omega$, i.e., when $p \le p_F$, whereas for $\Sigma_p^>$ this point is in the integration domain when $\omega_0 \ge \mu - \omega$, i.e., when $p \ge p_F$. In the examples of Fig. 15.8 the point $(0, \omega_0)$ is inside the integration domain for $\Sigma_p^<$ and outside for $\Sigma_p^>$. It is now clear that the integrals in (15.100) and (15.101) become logarithmically divergent when the plasmon branches enter the integration domain exactly at point $(0, \omega_0)$, i.e., when $\omega_0 = \pm\omega_p$, since in this case $q_0^{\lessgtr} = 0$. This happens exactly at the frequency

$$\omega = \epsilon_p - \epsilon_{p_F} + \mu - \omega_p, \quad (\omega = \epsilon_p - \epsilon_{p_F} + \mu + \omega_p),$$

---

[22]We can easily derive this equivalent form of the integration domain. Let $x = \omega' - \mu + \omega + \epsilon_{p_F} > 0$. Then $|\sqrt{2x} - p| \le q$ implies $-q \le \sqrt{2x} - p \le q$ or equivalently $p - q \le \sqrt{2x} \le p + q$. Denoting $M_1 = \max(0, p - q)$ we can square this inequality and get $\frac{1}{2}M_1^2 \le x \le \frac{1}{2}(p + q)^2$. Next consider $q \le \sqrt{2x} + p$ which implies $q - p \le \sqrt{2x}$. Denoting $M_2 = \max(0, q - p)$ we can square this inequality and get $\frac{1}{2}M_2^2 \le x$. Since the maximum value between M_1^2 and M_2^2 is $(p - q)^2$ we find $\frac{1}{2}(p - q)^2 \le x \le \frac{1}{2}(p + q)^2$.

[23]If $\Lambda(p, \omega)$ has a discontinuity D in $\omega = \omega_D$ then $\Gamma(p, \omega) = -2P \int \frac{d\omega'}{\pi} \frac{\Lambda(p, \omega')}{\omega - \omega'} \sim -\frac{2D}{\pi} \ln \omega_D$.

in $\Sigma_{\mathrm{p}}^<$ ($\Sigma_{\mathrm{p}}^>$) for $p \leq p_{\mathrm{F}}$ ($p \geq p_{\mathrm{F}}$). More generally, if ω_0 is close to $\pm\omega_{\mathrm{p}}$ the lower limit $q_0^{\lessgtr}$ is given by the solution of $\omega_-(q) = \pm\omega_{\mathrm{p}}(q) \sim \pm\omega_{\mathrm{p}}$ or $\omega_+(q) = \pm\omega_{\mathrm{p}}(q) \sim \pm\omega_{\mathrm{p}}$ (depending on whether ω_0 is larger or smaller than ω_{p}) and it is an easy exercise to check that $q_0^{\lessgtr} \sim \frac{1}{p}|\omega_0 \mp \omega_{\mathrm{p}}|$. Then the integrals (15.100) and (15.101) can be split into a singular contribution coming from the small momentum region and a remaining finite contribution. The singular part behaves as

$$i\Sigma_{\mathrm{p}}^{\lessgtr}(p,\omega) = \pm\frac{\omega_{\mathrm{p}}}{p} \ln \frac{|\omega_0 \mp \omega_{\mathrm{p}}|}{p} = \pm\frac{\omega_{\mathrm{p}}}{p} \ln \frac{|\epsilon_p + \mu - \omega - \epsilon_{p_{\mathrm{F}}} \mp \omega_{\mathrm{p}}|}{p} \qquad (15.102)$$

for $p \leq p_{\mathrm{F}}$ ($p \geq p_{\mathrm{F}}$) for $\Sigma_{\mathrm{p}}^<$ ($\Sigma_{\mathrm{p}}^>$). In the limit $p \to 0$ this analysis must be done more carefully since then also the integration domain itself goes to zero as the two parabolas approach each other. The reader can work out this case for him/herself. From (15.102) we see that the rate function $\Gamma(p,\omega)$ has a logarithmic singularity of strength $-\omega_{\mathrm{p}}/p$. Therefore $\Lambda(p,\omega)$ has a discontinuous jump of $\pi\omega_{\mathrm{p}}/(2p)$ at the same point. Strictly speaking this is true only if p is smaller (larger) than p_{F} for $\Sigma_{\mathrm{p}}^<$ ($\Sigma_{\mathrm{p}}^>$). At exactly $p = p_{\mathrm{F}}$ the plasmon branch enters the integration domain at the critical point $(0, \pm\omega_{\mathrm{p}})$ when $\mu - \omega = \pm\omega_{\mathrm{p}}$ and is outside the integration domains of $\Sigma_{\mathrm{p}}^{\lessgtr}$ for $|\mu - \omega| \leq \omega_{\mathrm{p}}$. Therefore the logarithmic singularity in (15.102) becomes one-sided only and as a consequence the Hilbert transform of the rate function induces a logarithmic singularity at $\mu - \omega = \pm\omega_{\mathrm{p}}$ in $\Lambda(p,\omega)$.

Let us now interpret our results. We have seen that the quasi-particle energies satisfy the property $E_p - \mu \sim \epsilon_p - \epsilon_{p_{\mathrm{F}}}$. Therefore $\Sigma_{\mathrm{p}}^<$ exhibits a logarithmic peak at $\omega = E_p - \omega_{\mathrm{p}}$ for $p \leq p_{\mathrm{F}}$, whereas $\Sigma_{\mathrm{p}}^>$ exhibits a logarithmic peak at $\omega = E_p + \omega_{\mathrm{p}}$ for $p \geq p_{\mathrm{F}}$. As we show below this leads to sharp features in the real part of Σ^{R} and peak structures at $\omega \sim E_p \pm \omega_{\mathrm{p}}$ in the spectral function in addition to the quasi-particle peak at $\omega = E_p$. This means that if we add a particle with momentum $p \geq p_{\mathrm{F}}$ then the created state evolves in time as a superposition of many-body eigenstates with energy components mainly centered around E_p and to a lesser extent around $E_p + \omega_{\mathrm{p}}$. The interpretation in terms of an inverse photoemission experiment is that the system after injection of the particle is most likely to be found having the energy E_p or $E_p + \omega_{\mathrm{p}}$ relative to its initial energy. A similar interpretation can be given for the removal of a particle of momentum $p \leq p_{\mathrm{F}}$. After the removal the system is most likely to be found at the quasi-hole energy E_p or at the energy $E_p - \omega_{\mathrm{p}}$ where we have created a quasi-hole state and removed a plasmon.

Numerical results

A practical calculation starts by specifying the density of the electron gas by choosing a value of r_s and hence the Fermi momentum p_{F}. In our example we take an electron gas of density characterized by the Wigner–Seitz radius $r_s = 4$, which corresponds to a density that is close to the valence density of the sodium crystal. Subsequently the chemical potential can be determined from (15.98) and evaluation of the function $\Sigma^{\mathrm{R}}(p_{\mathrm{F}}, \mu)$. We find that $\Sigma_{\mathrm{x}}(p_{\mathrm{F}}) = -1.33\,\epsilon_{p_{\mathrm{F}}}$ and $\Sigma_{\mathrm{c}}^{\mathrm{R}}(p_{\mathrm{F}}, \mu) = -0.44\,\epsilon_{p_{\mathrm{F}}}$, yielding a chemical potential of $\mu = -0.77\epsilon_{p_{\mathrm{F}}}$, which is quite different from the value $\mu = \epsilon_{p_{\mathrm{F}}}$ of the noninteracting electron gas. The plasma frequency at $r_s = 4$ is $\omega_{\mathrm{p}} = 1.88\,\epsilon_{p_{\mathrm{F}}}$ and the plasmon branch enters the particle–hole continuum at the critical momentum $q_c = 0.945\,p_{\mathrm{F}}$. Then all parameters of the problem are determined and we can calculate the lesser/greater self-energy, the rate function and the spectral function.

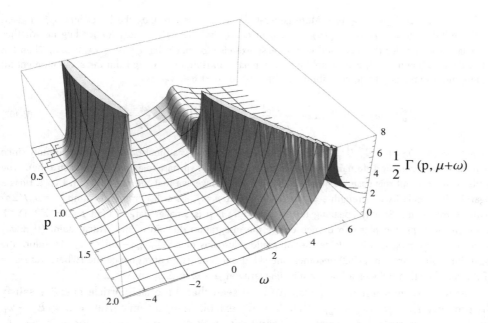

Figure 15.9 The imaginary part of the retarded self-energy $-\mathrm{Im}[\Sigma^{\mathrm{R}}(p, \omega + \mu)] = \Gamma(p, \omega + \mu)/2$ for an electron gas at $r_s = 4$ within the $G_0 W_0$ approximation as a function of the momentum and energy. The momentum p is measured in units of p_{F} and the energy ω and the self-energy in units of $\epsilon_{p_{\mathrm{F}}} = p_{\mathrm{F}}^2/2$.

Let us start with a discussion of the self-energy. In Fig. 15.9 we show $-\mathrm{Im}[\Sigma^{\mathrm{R}}(p, \omega + \mu)] = \Gamma(p, \omega + \mu)/2$ relative to the chemical potential μ in the (p, ω)-plane. The first distinctive feature of this graph is the valley around the chemical potential where $\Gamma \sim (\omega - \mu)^2$ for $\omega \to \mu$ for all values of the momentum p. As discussed in this chapter and in Section 13.3 this behavior is completely determined by phase-space restrictions. Away from the chemical potential at energies $|\omega - \mu| \geq \omega_{\mathrm{p}}$ there is a possibility for the system with a particle of momentum p added or removed to decay into plasmons as well. As a consequence the rate function is large for hole states with momentum $p < p_{\mathrm{F}}$ and for particle states with momentum $p > p_{\mathrm{F}}$. This boundary at $p = p_{\mathrm{F}}$ is not strict. Unlike the noninteracting electron gas, for an interacting gas it is possible to add a particle with momentum $p \leq p_{\mathrm{F}}$ or to remove a particle with momentum $p \geq p_{\mathrm{F}}$. As we see below, the momentum distribution of the interacting gas is not a strict Heaviside step function anymore.

Let us now turn our attention to the real part of the self-energy and the spectral function. In Fig. 15.10 we show the real and imaginary parts of the self-energy and the spectral function as a function of the energy relative to μ for the momentum values $p/p_{\mathrm{F}} = 0.5, 1, 1.5$. For $p = 0.5\,p_{\mathrm{F}}$ (upper left panel) we see that $\mathrm{Im}[\Sigma^{\mathrm{R}}]$ (solid line) has a logarithmic singularity due to plasmons below the chemical potential. At the same point $\mathrm{Re}[\Sigma^{\mathrm{R}}]$ (dashed line) jumps discontinuously in accordance with the previous analysis. The

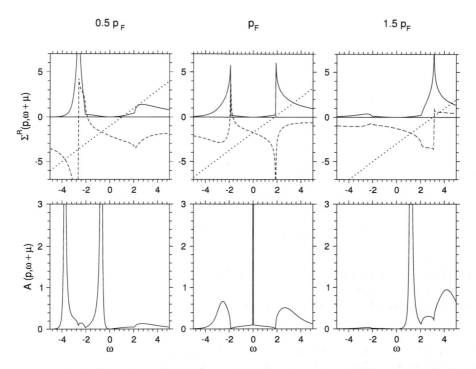

Figure 15.10 The self-energy and the spectral function in the $G_0 W_0$ approximation for $p/p_{\mathrm{F}} = 0.5$, 1.0, 1.5 at density $r_s = 4$ with ω and self-energies in units of $\epsilon_{p_{\mathrm{F}}} = p_{\mathrm{F}}^2/2$. In the top row the solid lines represent $-\mathrm{Im}[\Sigma^{\mathrm{R}}(p,\omega + \mu)] = \Gamma(p,\omega + \mu)/2$ and the dashed lines represent $\mathrm{Re}[\Sigma^{\mathrm{R}}(p,\omega + \mu)] = \Sigma_{\mathrm{x}}(p) + \Lambda(p,\omega + \mu)$. The dotted line represents the curve $\omega + \mu - \epsilon_p$. The crossings of this line with $\mathrm{Re}[\Sigma^{\mathrm{R}}(p,\omega + \mu)]$ determine the position of the peaks in the spectral function $A(p,\omega + \mu)$ displayed in the bottom row. The δ-function in the spectral function $A(p_{\mathrm{F}}, \mu + \omega)$ at $\omega = 0$ is indicated by a vertical line and has strength $Z = 0.64$.

structure of the spectral function is determined by the zeros of $\omega - \epsilon_p = \mathrm{Re}[\Sigma^{\mathrm{R}}(p,\omega)]$. These zeros occur at the crossings of the dotted line with the dashed line. At two of the crossings $\mathrm{Im}[\Sigma^{\mathrm{R}}]$ is small and as consequence the spectral function has two pronounced peaks (lower left panel). The peak on the right close to μ represents the quasi-hole peak whereas the peak to the left tells us that there is considerable probability of exciting a plasmon. For $p = 1.5\, p_{\mathrm{F}}$ (upper right panel) the plasmon contribution occurs at positive energies. There are again three crossings between the dashed and dotted lines but this time only one crossing (the quasi-particle one) occurs in an energy region where $-\mathrm{Im}[\Sigma^{\mathrm{R}}] \ll 1$. Thus the spectral function has only one main quasi-particle peak (lower right panel). There are still some plasmonic features visible at larger energies but they are much less prominent than for the case of $p = 0.5\, p_{\mathrm{F}}$. This can be understood from the fact that plasmons are mainly excited at low momenta since for $p > q_c$ the plasmon branch enters the particle–hole

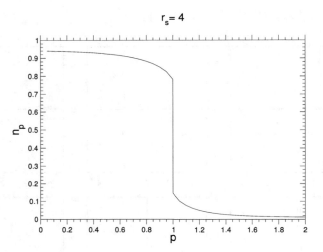

Figure 15.11 Momentum distribution n_p of an electron gas with $r_s = 4$ within the G_0W_0 approximation. The momentum p is measured in units of p_{F}. The momentum distribution jumps with a value of $Z = 0.64$ at the Fermi surface.

continuum. For an added particle at high momentum to excite a plasmon at low momentum it should at the same time transfer a large momentum to a particle–hole excitation. However, if the particle thereby loses most of its energy to a plasmon then the particle–hole pair should at the same time have large momentum and low energy, which is unlikely. Finally, in the upper middle panel we display the self-energy for $p = p_{\mathrm{F}}$. In this case $\mathrm{Im}[\Sigma^{\mathrm{R}}]$ displays a one-sided logarithmic singularity at $\omega - \mu = \pm\omega_{\mathrm{p}}$ (the numerics does not show this infinity exactly due to the finite energy grid) and as a consequence $\mathrm{Re}[\Sigma^{\mathrm{R}}]$ displays a logarithmic singularity at these points. These singularities would be smeared out to finite peaks if we had used a more advanced approximation for the screened interaction than the RPA in which we also allow for plasmon broadening. We further see that $\mathrm{Re}[\Sigma^{\mathrm{R}}]$ crosses the dotted line in only one point. At this point the derivative of $\mathrm{Re}[\Sigma^{\mathrm{R}}]$ is negative, in accordance with the general result (13.9), and $\mathrm{Im}[\Sigma^{\mathrm{R}}] = 0$. The quasi-particle peak becomes infinitely sharp and develops into a δ-function (vertical line in the lower middle panel). For $r_s = 4$ the strength of the δ-function is $Z = 0.64$.

From the behavior of the spectral function we can also deduce the structure of the momentum distribution $n_p = \int_{-\infty}^{\mu} \frac{d\omega}{2\pi} A(p, \omega)$. For $p \leq p_{\mathrm{F}}$ we integrate over the quasi-hole peak below the chemical potential. When p approaches p_{F} from below, the quasi-hole peak develops into a δ-function, while for p immediately above p_{F} the quasi-particle peak appears at an energy above μ and does not contribute to the integral. We thus expect that the momentum distribution suddenly jumps at $p = p_{\mathrm{F}}$. This is indeed shown rigorously in Appendix K where we prove that the discontinuity has exactly the size of the quasi-particle renormalization factor Z. This jump is directly related to the fact that the imaginary part of self-energy vanishes as $(\omega - \mu)^2$ close to the chemical potential. The formula for n_p in terms of the spectral function is awkward to use in practice since we need to integrate over a very

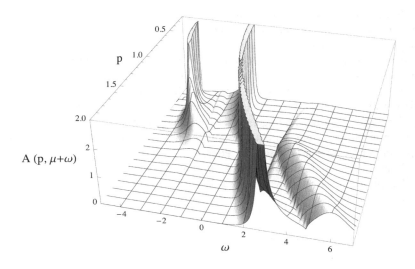

Figure 15.12 The spectral function $A(p, \mu+\omega)$ as a function of the momentum and energy for an electron gas at $r_s = 4$ within the $G_0 W_0$ approximation. The momentum p is measured in units of p_F and the energy ω and the spectral function in units of $\epsilon_{p_F} = p_F^2/2$.

spiky function, which is not easily done numerically. We therefore use for the numerics the equivalent formula (K.2) where the self-energy for complex frequencies is calculated from the rate function according to

$$\Sigma^R(p, \mu + i\omega) = \Sigma_x(p) + \int \frac{d\omega'}{2\pi} \frac{\Gamma(p, \omega')}{i\omega + \mu - \omega' + i\eta}.$$

The result is given in Fig. 15.11. We clearly see the sudden jump of magnitude Z at $p = p_F$. Therefore the interacting electron gas preserves the sharp Fermi surface of the noninteracting system. The quasi-particle renormalization factor Z is smaller for lower densities, whereas for high densities it approaches unity [67]. The discontinuity of n_p and hence the sharpness of the Fermi surface are experimentally observable by means of Compton scattering on electron-gas-like metals such as sodium. For more details on this issue we refer the reader to Ref. [75]. The sharpness of the Fermi surface applies to a large class of fermionic systems, not only metallic crystals but also other quantum systems such as liquid ^{3}He. In fact, theoretically it is used to define an important class of physical systems known as *Fermi liquids*, the low-energy behavior (excitations close to the Fermi surface) of which was first successfully described by Landau in a phenomenological way [144].

Let us finally study the overall behavior of the spectral function in the (p, ω)-plane as displayed in Fig. 15.12. The quasi-particle peak appears in the middle of the figure and broadens when we move away from the Fermi momentum p_F. Its position at $E_p - \mu$ is approximately given by $p^2/2 - p_F^2/2$, or in units of the figure $(E_p - \mu)/\epsilon_{p_F} \approx (p/p_F)^2 - 1$. An important improvement of the dispersion curve E_p compared to the Hartree–Fock dispersion curve of Section 7.3.2 is that it does not exhibit the logarithmic divergence at $p = p_F$. The disappearance of the logarithmic divergence can be shown from a further analysis of the $G_0 W_0$-equations. The interested reader can find the details and much more information on the spectral properties of the electron gas in papers by Lundqvist, and Lundqvist and Samathiyakanit [145, 146]. Physically, the logarithmic divergence is removed by the fact that the screened interaction W does not have the divergent behavior of the bare Coulomb interaction v at low momenta. Apart from the quasi-particle structure we clearly see in Fig. 15.12 plasmonic features at energies below the chemical potential for $p \leq p_F$ (the addition of a hole can excite plasmons or, equivalently, the hole can decay into plasmon excitations). Similar plasmonic features are present for energies above the chemical potential for momenta $p \geq p_F$. These features are, however, less pronounced. As explained before, this is due to the fact that plasmons are mainly excited at low momenta.

16

Applications of MBPT to nonequilibrium problems

In this chapter we solve the Kadanoff–Baym equations in finite interacting systems, possibly connected to macroscopic reservoirs. We have already encountered examples of these kinds of system. For instance, the Hamiltonian for the quantum discharge of a capacitor introduced in Section 7.2.3 describes a molecule (finite system) connected to two capacitor plates (macroscopic reservoirs). The difference here is that we include an interaction in the finite system and, for simplicity, we discard the interaction in the reservoirs. In fact, no complication arises if the interaction in the reservoirs is treated at the Hartree level, see below.

The general Hamiltonian that we have in mind is

$$\hat{H} = \hat{H}_0 + \hat{H}_{\text{int}},$$

where the noninteracting part reads

$$\hat{H}_0 = \underbrace{\sum_{\substack{k\alpha \\ \sigma}} \epsilon_{k\alpha} \hat{n}_{k\alpha\sigma}}_{\text{reservoirs}} + \underbrace{\sum_{\substack{mn \\ \sigma}} T_{mn} \hat{d}^\dagger_{m\sigma} \hat{d}_{n\sigma}}_{\text{molecule}} + \underbrace{\sum_{\substack{m,k\alpha \\ \sigma}} (T_{m\,k\alpha} \hat{d}^\dagger_{m\sigma} \hat{d}_{k\alpha\sigma} + T_{k\alpha\,m} \hat{d}^\dagger_{k\alpha\sigma} \hat{d}_{m\sigma})}_{\text{coupling}}. \quad (16.1)$$

As in Section 7.2.3, the index $k\alpha$ refers to the kth eigenfunction of the αth reservoir whereas the indices m, n refer to basis functions of the finite system, which from now on we call the molecule, see Fig. 16.1. We also add the spin degree of freedom σ in order to model realistic interactions. As usual $\hat{n}_{s\sigma} = \hat{d}^\dagger_{s\sigma} \hat{d}_{s\sigma}$ is the occupation number operator for the basis function $s = k\alpha$ or $s = m$. The interacting part of the Hamiltonian is taken of the general form

$$\hat{H}_{\text{int}} = \frac{1}{2} \sum_{\substack{ijmn \\ \sigma\sigma'}} v_{ijmn} \hat{d}^\dagger_{i\sigma} \hat{d}^\dagger_{j\sigma'} \hat{d}_{m\sigma'} \hat{d}_{n\sigma}. \quad (16.2)$$

The sum is restricted to indices in the molecule so that there is no interaction between a particle in the molecule and a particle in the reservoirs or between two particles in the

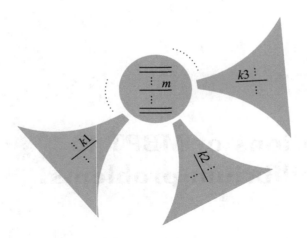

Figure 16.1 Representation of the molecule in contact with many reservoirs.

reservoirs.[1] All applications and examples discussed in this chapter are for a system of electrons. To the best of our knowledge there are still no implementations of the Kadanoff–Baym equations for systems of bosons.

We consider the system initially in thermodynamic equilibrium at inverse temperature β and chemical potential μ. The initial density matrix is then $\hat{\rho} = \exp[-\beta(\hat{H} - \mu\hat{N})]/Z$, which corresponds to having $\hat{H}^{\mathrm{M}} = \hat{H} - \mu\hat{N}$ on the vertical track of the contour. At a certain time t_0 we drive the system out of equilibrium by changing the parameters of the Hamiltonian. We could change the interaction or the couplings between the molecule and the reservoirs or the energies $\epsilon_{k\alpha}$. The formalism is so general that we can deal with all sorts of combinations. The Hamiltonian at times $t > t_0$ is therefore given by $\hat{H}$ in which the parameters are time-dependent, i.e., $\epsilon_{k\alpha} \to \epsilon_{k\alpha}(t)$, $T_{mk\alpha} \to T_{mk\alpha}(t)$, $v_{ijmn} \to v_{ijmn}(t)$, etc. The goal is to calculate the Green's function from the Kadanoff–Baym equations and then to extract quantities of physical interest like the density, current, screened interaction, polarizability, nonequilibrium spectral functions, etc. This is in general a very difficult task since the system of Fig. 16.1 has infinitely many degrees of freedom and does not have any special symmetry.[2] In the next section we show that for our particular choice of the interaction an important simplification occurs: the equations to be solved involve only a finite number of matrix elements of the Green's function.

---

[1]The form (16.2) of the interaction is identical to (1.88) and it should not be confused with the interaction in (1.83) where the spin indices are not spelled out. Thus the i, j, m, n indices in this chapter are orbital indices as opposed to Section 1.6 where they were orbital-spin indices.

[2]Continuous symmetries of the Hamiltonian can simplify the calculations considerably. In the electron gas the invariance under translations allowed us to derive several analytic results.

16.1 Kadanoff–Baym equations for open systems

The first observation is that the Green's function and the self-energy are diagonal in spin space since the interaction preserves the spin orientation at every vertex and the noninteracting part of the Hamiltonian is spin diagonal. The inclusion of terms in the Hamiltonian which flip the spin, like the spin-orbit interaction or the coupling with a noncollinear magnetic field, is straightforward and does not introduce conceptual complications. We denote by G_{sr} and Σ_{sr} the matrix elements of Green's function and self-energy for a given spin projection, where the indices s, r refer either to the reservoirs or to the molecule. Similarly, we denote by h_{sr} the matrix elements of the first quantized Hamiltonian $\hat{h}$. The matrices G, Σ, and h are the representation of the operators (in first quantization) $\hat{\mathcal{G}}$, $\hat{\Sigma}$, and $\hat{h}$ in the one-particle basis $\{k\alpha, m\}$. The equations of motion for the Green's function matrix G read

$$\left[i\frac{d}{dz} - h(z) \right] G(z, z') = \delta(z, z') + \int_\gamma d\bar{z}\, \Sigma(z, \bar{z}) G(\bar{z}, z'),$$

$$G(z, z') \left[-i\frac{\overleftarrow{d}}{dz'} - h(z') \right] = \delta(z, z') + \int_\gamma d\bar{z}\, G(z, \bar{z})\Sigma(\bar{z}, z'),$$

and must be solved with the KMS boundary conditions $G(t_{0-}, z') = -G(t_0 - i\beta, z')$ and the like for the second argument. At this point we find it convenient to introduce the "blocks" forming the matrices h, G, and Σ. A block of h, G, and Σ is the projection of these matrices onto the subspace of the reservoirs or of the molecule. Thus, for instance, $h_{\alpha\alpha'}$ is the block of h with matrix elements $[h_{\alpha\alpha'}]_{kk'} = \delta_{\alpha\alpha'}\delta_{kk'}\epsilon_{k\alpha}$, $h_{\alpha M}$ is the block of h with matrix elements $[h_{\alpha M}]_{km} = T_{k\alpha\,m}$ and h_{MM} is the block of h with matrix elements $[h_{MM}]_{mn} = T_{mn}$. If we number the reservoirs as $\alpha = 1, 2, 3 \ldots$, the block form of the matrix h is

$$h = \begin{pmatrix} h_{11} & 0 & 0 & \ldots & h_{1M} \\ 0 & h_{22} & 0 & \ldots & h_{2M} \\ 0 & 0 & h_{33} & \ldots & h_{3M} \\ \vdots & \vdots & \vdots & \vdots & \vdots \\ h_{M1} & h_{M2} & h_{M3} & \ldots & h_{MM} \end{pmatrix},$$

where each entry (including the 0s) is a block matrix. For the time being we do not specify the number of reservoirs. Similarly to h we can write the matrices G and Σ in a block form. The matrix G has nonvanishing blocks everywhere. Physically there is no reason to expect that $G_{\alpha\alpha'} \propto \delta_{\alpha\alpha'}$, since an electron is free to go from reservoir α to reservoir α'. On the contrary the self-energy Σ *has only one nonvanishing block*, that is Σ_{MM}. This is a direct consequence of the fact that the diagrammatic expansion of the self-energy starts and ends with an interaction line which in our case is confined to the molecular region. This also implies that $\Sigma_{MM}[G_{MM}]$ is a functional of G_{MM} only. The fact that Σ_{MM} depends only on G_{MM} is very important since it allows us to close the equations of motion for the molecular Green's function. Let us project the equation of motion onto regions MM and

αM:

$$\left[i\frac{d}{dz} - h_{MM}(z) \right] G_{MM}(z, z') = \delta(z, z') + \sum_\alpha h_{M\alpha}(z) G_{\alpha M}(z, z')$$

$$+ \int_\gamma d\bar{z}\, \Sigma_{MM}(z, \bar{z}) G_{MM}(\bar{z}, z'), \qquad (16.3)$$

$$\left[i\frac{d}{dz} - h_{\alpha\alpha}(z) \right] G_{\alpha M}(z, z') = h_{\alpha M}(z) G_{MM}(z, z').$$

The latter equation can be solved for $G_{\alpha M}$ and yields

$$G_{\alpha M}(z, z') = \int_\gamma d\bar{z}\, g_{\alpha\alpha}(z, \bar{z}) h_{\alpha M}(\bar{z}) G_{MM}(\bar{z}, z'), \qquad (16.4)$$

where we define the Green's function $g_{\alpha\alpha}$ of the isolated αth reservoir as the solution of

$$\left[i\frac{d}{dz} - h_{\alpha\alpha}(z) \right] g_{\alpha\alpha}(z, z') = \delta(z, z')$$

with KMS boundary conditions. Again we stress the importance of solving the equation for $g_{\alpha\alpha}$ with KMS boundary conditions so that $G_{\alpha M}$ too fulfills the KMS relations. Any other boundary conditions for $g_{\alpha\alpha}$ would lead to unphysical time-dependent results. It is also important to realize that $g_{\alpha\alpha}$ is the Green's function of a noninteracting system and is therefore very easy to calculate. In particular, since the block Hamiltonian $h_{\alpha\alpha}(z)$ is diagonal with matrix elements $\epsilon_{k\alpha}(z)$, we simply have $[g_{\alpha\alpha'}]_{kk'}(z, z') = \delta_{\alpha\alpha'}\delta_{kk'}g_{k\alpha}(z, z')$ with

$$g_{k\alpha}(z, z') = -i\left[\theta(z, z')\bar{f}(\epsilon_{k\alpha}^M) - \theta(z', z)f(\epsilon_{k\alpha}^M) \right] e^{-i\int_{z'}^z d\bar{z}\, \epsilon_{k\alpha}(\bar{z})},$$

where in accordance with our notation $\epsilon_{k\alpha}^M = \epsilon_{k\alpha}(t_0 - i\tau) = \epsilon_{k\alpha} - \mu$. The reader can verify the correctness of this result using (6.39). Inserting the solution for $G_{\alpha M}$ in the second term on the r.h.s. of (16.3) we get

$$\sum_\alpha h_{M\alpha}(z) G_{\alpha M}(z, z') = \int d\bar{z}\, \Sigma_{em}(z, \bar{z}) G_{MM}(\bar{z}, z'),$$

where we define the *embedding self-energy* in the usual way,

$$\Sigma_{em}(z, z') = \sum_\alpha \Sigma_\alpha(z, z'), \qquad \Sigma_\alpha(z, z') = h_{M\alpha}(z) g_{\alpha\alpha}(z, z') h_{\alpha M}(z'). \qquad (16.5)$$

Like $g_{\alpha\alpha}$, the embedding self-energy is also independent of the electronic interaction, and hence of G_{MM}, and it is completely specified by the reservoir Hamiltonian $h_{\alpha\alpha}$ and by the contact Hamiltonian $h_{\alpha M}$. In conclusion, the equation of motion for the molecular Green's function becomes

$$\left[i\frac{d}{dz} - h_{MM}(z) \right] G_{MM}(z, z') = \delta(z, z') + \int_\gamma d\bar{z}\, [\Sigma_{MM}(z, \bar{z}) + \Sigma_{em}(z, \bar{z})] G_{MM}(\bar{z}, z'). \qquad (16.6)$$

The adjoint equation of motion can be derived similarly.

Equation (16.6) is an *exact* closed equation for G_{MM} provided that the exact many-body self-energy Σ_{MM} as a functional of G_{MM} is inserted. This is a very nice result since G_{MM} is the Green's function of a finite system and therefore (16.6) can be implemented numerically using a finite basis. In practical implementations (16.6) is converted into a set of coupled real-time equations, i.e., the Kadanoff–Baym equations, which are then solved by means of time-propagation techniques [147]. We come back to this point in the following sections. From the knowledge of G_{MM} we have access to the time-dependent ensemble average of all one-body operators of the molecular region. We also have access to the total current flowing between the molecule and the reservoir α. This current is a simple generalization of (7.30) and reads

$$I_\alpha(t) = q \frac{d}{dt} N_\alpha(t) = 2q \times 2\,\mathrm{Re}\left\{ \mathrm{Tr}_M \left[h_{M\alpha}(t) G^<_{\alpha M}(t,t) \right] \right\},$$

where Tr_M denotes the trace over the single-particle states of the molecular basis, q is the charge of the fermions and the extra factor of 2 comes from spin. Extracting the lesser component of $G_{\alpha M}$ from (16.4) we can rewrite I_α in terms of the Green's function of the molecular region only

$$\boxed{I_\alpha(t) = 4q\,\mathrm{Re}\left\{ \mathrm{Tr}_M \left[\Sigma^<_\alpha \cdot G^A_{MM} + \Sigma^R_\alpha \cdot G^<_{MM} + \Sigma^\rceil_\alpha \star G^\lceil_{MM} \right](t,t) \right\}} \qquad (16.7)$$

The last term in (16.7) explicitly accounts for the effects of initial correlations and initial-state dependence. If one assumes that both dependencies are washed out in the limit $t \to \infty$, then for large times we can discard the imaginary time convolution. The resulting formula is known as the *Meir–Wingreen formula* [148, 149]. Equation (16.7) provides a generalization of the Meir–Wingreen formula to the transient time-domain. Furthermore, if in the same limit ($t \to \infty$) we reach a steady state then the Green's function, and hence the self-energy, become a function of the times difference only. In this case we can Fourier transform with respect to the relative time and write the steady-state value of the current as

$$\boxed{I_\alpha^{(S)} \equiv \lim_{t\to\infty} I_\alpha(t) = 2iq \int \frac{d\omega}{2\pi} \mathrm{Tr}_M \left[\Sigma^<_\alpha(\omega) A_{MM}(\omega) - \Gamma_\alpha(\omega) G^<_{MM}(\omega) \right]} \qquad (16.8)$$

where $\Gamma_\alpha(\omega) = i\left[\Sigma^R_\alpha(\omega) - \Sigma^A_\alpha(\omega) \right]$ is the nonequilibrium embedding rate operator and $A_{MM}(\omega) = i\left[G^R_{MM}(\omega) - G^A_{MM}(\omega) \right]$ is the nonequilibrium spectral function.[3] Physically we expect that a necessary condition for the development of a steady-state is that the Hamiltonian $\hat{H}(t)$ is independent of t in the long-time limit. In general, however, we do not expect that this condition is also sufficient. We have already encountered examples where the presence of bound states in noninteracting systems prevents the formation of a steady state, see Section 6.3.1 and Refs. [55, 150]. In interacting systems the issue is more difficult to address due to the absence of exact solutions. What we can say is that in all

---

[3]To derive (16.8) one uses the fact that $\mathrm{Re}\{\mathrm{Tr}[B]\} = \frac{1}{2}\mathrm{Tr}[B + B^\dagger]$ for any matrix B and then the properties of Section 9.4 for the transpose conjugate of the Green's functions and self-energies as well as the cyclic property of the trace.

numerical simulations performed so far the current reaches a steady value in the long time limit provided that the self-energy contains some correlation. Instead, if we approximate Σ_{MM} at the Hartree or Hartree–Fock level then the development of a steady state is not guaranteed [62,151]

In the following sections we discuss the solution of (16.6) using different approximations to the self-energy Σ_{MM}. First, however, we get acquainted with (16.6) by solving it analytically in the absence of interactions, $\Sigma_{MM} = 0$. The noninteracting case is particularly instructive. On one hand it allows us to introduce the procedure for solving the more complicated case of interacting systems. On the other hand it provides us with reference results to interpret the effects of the interactions.

Exercise 16.1. Prove the formula (16.7) for the current $I_\alpha(t)$ by calculating the commutator between the total number of particle operators $\hat{N}_\alpha = \sum_{k\sigma} \hat{d}^\dagger_{k\alpha\sigma} \hat{d}_{k\alpha\sigma}$ for reservoir α and the Hamiltonian $\hat{H}(t)$. Note that $[\hat{N}_\alpha, \hat{H}_{\text{int}}]_- = 0$.

Exercise 16.2. Prove the formula (16.8) for the steady-state current.

16.2 Time-dependent quantum transport: an exact solution

We consider the same situation as in Section 7.2.3 in which an external electric field is suddenly switched on to bring the reservoirs to different potentials, thereby inducing a current through the molecule [54,152]. Since the particles are electrons we set the charge $q = -1$. In this section the interaction is completely neglected so that $\hat{H}^M = \hat{H}_0 - \mu\hat{N}$ and $\hat{H}(t)$ is obtained from $\hat{H}_0$ by replacing $\epsilon_{k\alpha} \to \epsilon_{k\alpha}(t) = \epsilon_{k\alpha} + V_\alpha(t)$, where $V_\alpha(t)$ is the external bias. We mention that this kind of external perturbation models the classical potential inside a metal. When a metal is exposed to an external field V the effect of the Hartree potential V_H is to screen V in such a way that the classical (external plus Hartree) potential is uniform in the interior of the metal. Consequently the potential drop is entirely confined to the molecular region, as shown in the schematic representation below:

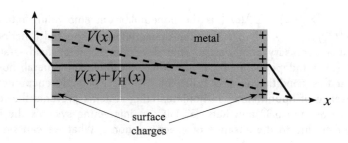

The uniform bias $V_\alpha(t)$ in reservoir α can then be interpreted as the sum of the external and Hartree potential. This effectively means that the reservoirs are treated at a Hartree mean field level. Of course, for the classical potential to be uniform *at all times* we must vary the external potential on a time scale much longer than the characteristic time for the formation of surface charges. A reasonable estimate of this characteristic time is the inverse of the plasma frequency, which in a bulk metal is of the order of ~ 10 eV $\sim 10^{-18}$ s, see also the discussion in Section 15.5.3.

To calculate the time-dependent ensemble average of one-body operators, e.g., the occupation of the molecular levels or the current flowing through the molecule, we must determine $G^<_{MM}(t,t)$. To lighten the notation we drop the subscript "MM" since all quantities in (16.6) are matrices with indices in the molecular region and therefore there is no risk of misunderstandings. Taking into account that the many-body self-energy is zero we have

$$\left[i\frac{d}{dz} - h(z) \right] G(z,z') = \delta(z,z') + \int_\gamma d\bar{z}\, \Sigma_{em}(z,\bar{z}) G(\bar{z},z')$$

and the adjoint equation

$$G(z,z') \left[-i\overleftarrow{\frac{d}{dz'}} - h(z') \right] = \delta(z,z') + \int_\gamma d\bar{z}\, G(z,\bar{z})\Sigma_{em}(\bar{z},z'),$$

with $h = h_{MM}$ the finite matrix Hamiltonian of the isolated molecule. We can generate an equation for $G^<(t,t)$ by setting $z = t_-$, $z' = t'_+$, subtracting the equation of motion from its adjoint and then setting $t = t'$. The result is

$$i\frac{d}{dt}G^<(t,t) - \left[h(t), G^<(t,t) \right]_- = \left[\Sigma^<_{em} \cdot G^A - G^R \cdot \Sigma^<_{em} + \Sigma^R_{em} \cdot G^< - G^< \cdot \Sigma^A_{em} \right](t,t)$$

$$+ \left[\Sigma^\rceil_{em} \star G^\lceil - G^\rceil \star \Sigma^\lceil_{em} \right](t,t). \tag{16.9}$$

It is easy to see that the terms with the plus sign on the r.h.s. are the transpose conjugate of the terms with the minus sign. For instance from (6.24) and (9.45)

$$\int_{t_0}^\infty d\bar{t}\, \Sigma^<_{em}(t,\bar{t}) G^A(\bar{t},t) = -\int_{t_0}^\infty d\bar{t}\, [\Sigma^<_{em}(\bar{t},t)]^\dagger [G^R(t,\bar{t})]^\dagger = -\left[G^R \cdot \Sigma^<_{em} \right]^\dagger(t,t),$$

and similarly from (9.42) and (9.46)

$$-i\int_0^\beta d\tau\, \Sigma^\rceil_{em}(t,\tau) G^\lceil(\tau,t) = -i\int_0^\beta d\tau\, [\Sigma^\lceil_{em}(\beta-\tau,t)]^\dagger [G^\rceil(t,\beta-\tau)]^\dagger = -\left[G^\rceil \star \Sigma^\lceil_{em} \right]^\dagger(t,t).$$

Therefore we can rewrite (16.9) as

$$i\frac{d}{dt}G^<(t,t) - \left[h(t), G^<(t,t) \right]_- = -\left[G^R \cdot \Sigma^<_{em} + G^< \cdot \Sigma^A_{em} + G^\rceil \star \Sigma^\lceil_{em} \right](t,t) + \text{H.c.}, \tag{16.10}$$

where H.c. stands for Hermitian conjugate. In order to solve this differential equation we must calculate the Keldysh components of the embedding self-energy as well as the Green's function G^R and $G^\rceil$. G^R can easily be derived from the sole knowledge of Σ^R_{em}. To calculate $G^\rceil$ we must first calculate the Matsubara Green's function G^M, see (9.35). We then proceed as follows. First we determine the embedding self-energy, then we calculate G^{M} from (9.34) and use it to calculate $G^{\rceil}$, which we then use to solve (16.10). This way of proceeding is completely general. One always starts from the Matsubara Green's function since it is the only Kadanoff–Baym equation which is not coupled to the others. The difference in the interacting case is that the Matsubara self-energy is a functional of G^{M} and therefore cannot be determined a priori like the embedding self-energy. In the interacting case the Matsubara Kadanoff–Baym equation must be solved self-consistently.

For simplicity we take $V_{\alpha}(t) = V_{\alpha}$ independent of time i.e. we consider a sudden switch on of the bias. Then the Hamiltonian is time independent on the horizontal branches of the contour, and according to the results of Section 6.2 the retarded and advanced Green's functions $g^{\mathrm{R/A}}$ depend only on the time difference. The advanced embedding self-energy for reservoir α is then

$$\Sigma^{\mathrm{A}}_{\alpha,mn}(t,t') = \int \frac{d\omega}{2\pi} e^{-i\omega(t-t')} \underbrace{\sum_{k} T_{m\,k\alpha}\, g^{\mathrm{A}}_{k\alpha}(\omega)\, T_{k\alpha\,n}}_{\Sigma^{\mathrm{A}}_{\alpha,mn}(\omega)}.$$

To simplify the analytic calculations we take the eigenvalues of the molecular Hamiltonian h well inside the continuum spectrum of reservoir α and make the WBLA according to which

$$\Sigma^{\mathrm{A}}_{\alpha,mn}(\omega) = \sum_{k} T_{m\,k\alpha} \frac{1}{\omega - \epsilon_{k\alpha} - V_{\alpha} - i\eta} T_{k\alpha\,n} = \frac{i}{2}\Gamma_{\alpha,mn}$$

is a purely imaginary constant. The quantity

$$\Gamma_{\alpha,mn} = 2\pi \sum_{k} T_{m\,k\alpha}\, \delta(\omega - \epsilon_{k\alpha} - V_{\alpha})\, T_{k\alpha\,n} = \Gamma^{*}_{\alpha,nm}$$

can be seen as the (m,n) matrix element of a self-adjoint matrix Γ_{α}. Consequently

$$\Sigma^{\mathrm{A}}_{\alpha}(t,t') = \frac{i}{2}\Gamma_{\alpha}\delta(t - t').$$

In the WBLA $\Sigma^{\mathrm{A}}_{\alpha}(\omega)$, and hence $\Sigma^{\mathrm{R}}_{\alpha}(\omega) = [\Sigma^{\mathrm{A}}_{\alpha}(\omega)]^{\dagger}$, is the same as in equilibrium since the dependence on V_{α} drops out. Then we can use (9.58) and write

$$\Sigma^{\mathrm{M}}_{\alpha}(\omega_{q}) = \frac{i}{2}\left\{ \begin{array}{ll} -\Gamma_{\alpha} & \text{if } \mathrm{Im}[\omega_{q}] > 0 \\ +\Gamma_{\alpha} & \text{if } \mathrm{Im}[\omega_{q}] < 0 \end{array} \right..$$

The right and left self-energy can now be derived very easily. Without loss of generality we choose the time at which the bias is switched on to be $t_{0} = 0$. Using the relation (6.53),

$$g^{\rceil}_{k\alpha}(t,\tau) = i g^{\mathrm{R}}_{k\alpha}(t,0) g^{\mathrm{M}}_{k\alpha}(0,\tau) = e^{-i(\epsilon_{k\alpha}+V_{\alpha})t} g^{\mathrm{M}}_{k\alpha}(0,\tau).$$

Expanding $g^{\mathrm{M}}_{k\alpha}(0,\tau)$ in Matsubara frequencies, see (6.17), we find

$$\Sigma^{\rceil}_{\alpha,mn}(t,\tau) = \frac{1}{-i\beta} \sum_{q} e^{\omega_{q}\tau} \sum_{k} T_{m\,k\alpha} \frac{e^{-i(\epsilon_{k\alpha}+V_{\alpha})t}}{\omega_{q} - \epsilon_{k\alpha} + \mu} T_{k\alpha\,n}$$

$$= \frac{1}{-i\beta} \sum_{q} e^{\omega_{q}\tau} \int \frac{d\omega}{2\pi} \underbrace{2\pi \sum_{k} T_{m\,k\alpha}\delta(\omega - \epsilon_{k\alpha}) T_{k\alpha\,n}}_{\Gamma_{\alpha,mn}} \frac{e^{-i(\omega+V_{\alpha})t}}{\omega_{q} - \omega + \mu}.$$

This trick to convert the sum over k into an integral over ω is used over and over in the following derivations. We observe that in the original definition of Γ_α the argument of the δ-function was $\omega - \epsilon_{k\alpha} - V_\alpha$. However, since Γ_α is independent of ω we can shift ω as we like without changing the value of the sum. In conclusion

$$\Sigma_\alpha^{\rceil}(t, \tau) = \Gamma_\alpha \times \frac{1}{-i\beta} \sum_q e^{\omega_q \tau} \int \frac{d\omega}{2\pi} \frac{e^{-i(\omega + V_\alpha)t}}{\omega_q - \omega + \mu}.$$

In a completely analogous way the reader can check that the expression for $\Sigma_\alpha^{\lceil}$ is

$$\Sigma_\alpha^{\lceil}(\tau, t) = \Gamma_\alpha \times \frac{1}{-i\beta} \sum_q e^{-\omega_q \tau} \int \frac{d\omega}{2\pi} \frac{e^{i(\omega + V_\alpha)t}}{\omega_q - \omega + \mu}.$$

It is also straightforward to verify that $\Sigma_\alpha^{\rceil}(t, \tau) = [\Sigma_\alpha^{\lceil}(\beta - \tau, t)]^\dagger$, in agreement with (9.46). Finally, the lesser self-energy is obtained from

$$\Sigma_{\alpha,mn}^<(t, t') = \sum_k T_{m\,k\alpha} \underbrace{i f(\epsilon_{k\alpha} - \mu) e^{-i(\epsilon_{k\alpha} + V_\alpha)(t - t')}}_{g_{k\alpha}^<(t, t')} T_{k\alpha\,n}$$

$$= i\,\Gamma_{\alpha,mn} \int \frac{d\omega}{2\pi} f(\omega - \mu) e^{-i(\omega + V_\alpha)(t - t')}.$$

We are now in the position to solve the Kadanoff–Baym equations. The Matsubara Green's function is simply

$$G^M(\omega_q) = \frac{1}{\omega_q - h - \Sigma_{em}^M(\omega_q) + \mu} = \begin{cases} \dfrac{1}{\omega_q - h + i\Gamma/2 + \mu} & \text{if } \mathrm{Im}[\omega_q] > 0 \\[2mm] \dfrac{1}{\omega_q - h - i\Gamma/2 + \mu} & \text{if } \mathrm{Im}[\omega_q] < 0 \end{cases}, \qquad (16.11)$$

with $\Gamma = \sum_\alpha \Gamma_\alpha$. The combination $h \pm i\Gamma/2$ appears very often below and it is therefore convenient to introduce an effective non-Hermitian Hamiltonian

$$h_{\text{eff}} = h - \frac{i}{2}\Gamma \quad \Rightarrow \quad h_{\text{eff}}^\dagger = h + \frac{i}{2}\Gamma$$

to shorten the formulas. To calculate $G^{\rceil}$ we use its equation of motion

$$\left[i\frac{d}{dt} - h\right] G^{\rceil}(t, \tau) = \int_0^\infty d\bar{t}\, \Sigma_{em}^R(t, \bar{t}) G^{\rceil}(\bar{t}, \tau) - i \int_0^\beta d\bar{\tau}\, \Sigma_{em}^{\rceil}(t, \bar{\tau}) G^M(\bar{\tau}, \tau).$$

Taking into account that $\Sigma_{em}^R(t, \bar{t}) = [\Sigma_{em}^A(\bar{t}, t)]^\dagger = -\frac{i}{2}\Gamma\delta(\bar{t} - t)$, the first term on the r.h.s. is simply $-\frac{i}{2}\Gamma G^{\rceil}(t, \tau)$. Moving this term to the l.h.s. we form the combination $h - \frac{i}{2}\Gamma = h_{\text{eff}}$, and hence

$$G^{\rceil}(t, \tau) = e^{-i h_{\text{eff}} t} \left[G^M(0, \tau) - \int_0^t dt'\, e^{i h_{\text{eff}} t'} \int_0^\beta d\bar{\tau}\, \Sigma_{em}^{\rceil}(t', \bar{\tau}) G^M(\bar{\tau}, \tau) \right], \qquad (16.12)$$

where we take into account that $G^M(0, \tau) = G^\rceil(0, \tau)$. In order to solve (16.10) we must calculate one remaining quantity, which is the retarded Green's function. Again we observe that the Hamiltonian is time independent on the horizontal branches of the contour and therefore $G^R(t, t')$ depends only on the time difference $t - t'$,

$$G^R(t, t') = \int \frac{d\omega}{2\pi} e^{-i\omega(t-t')} G^R(\omega).$$

Using the retarded Dyson equation we find

$$G^R(\omega) = \frac{1}{\omega - h + i\eta} \left[1 + \Sigma_{em}^R(\omega) G^R(\omega) \right] \quad \Rightarrow \quad G^R(\omega) = \frac{1}{\omega - h_{eff}} \tag{16.13}$$

and hence

$$G^R(t, t') = -i\theta(t - t') e^{-ih_{eff}(t-t')}.$$

We can now evaluate the three terms in the square brackets of (16.10). Let us start with the first term,

$$\left[G^R \cdot \Sigma_{em}^< \right](t, t) = -ie^{-ih_{eff}t} \int_0^t dt' e^{ih_{eff}t'} i \sum_\alpha \Gamma_\alpha \int \frac{d\omega}{2\pi} f(\omega - \mu) e^{-i(\omega+V_\alpha)t'} e^{i(\omega+V_\alpha)t}$$

$$= i \sum_\alpha \int \frac{d\omega}{2\pi} f(\omega - \mu) \left(1 - e^{i(\omega+V_\alpha-h_{eff})t} \right) G^R(\omega + V_\alpha) \Gamma_\alpha. \tag{16.14}$$

The second term involves Σ^A which is proportional to a δ-function and therefore

$$\left[G^< \cdot \Sigma^A \right](t, t) = \frac{i}{2} G^<(t, t) \Gamma. \tag{16.15}$$

The third term is more complicated even though the final result is rather simple. Using (16.12) we find

$$\left[G^\rceil \star \Sigma_{em}^\lceil \right](t, t) = e^{-ih_{eff}t} \left\{ \left[G^M \star \Sigma_{em}^\lceil \right](0, t) - i \int_0^t dt' e^{ih_{eff}t'} \left[\Sigma_{em}^\rceil \star G^M \star \Sigma_{em}^\lceil \right](t', t) \right\}.$$

We now show that the second term on the r.h.s. vanishes. Taking into account the explicit form of the left and right embedding self-energies,

$$\left[\Sigma_{em}^\rceil \star G^M \star \Sigma_{em}^\lceil \right](t', t) = \int \frac{d\omega \, d\omega'}{2\pi \, 2\pi} \sum_{\alpha\alpha'} \Gamma_\alpha \frac{1}{-i\beta} \sum_q \frac{e^{-i(\omega+V_\alpha)t'}}{\omega_q - \omega + \mu} G^M(\omega_q) \frac{e^{i(\omega'+V_{\alpha'})t}}{\omega_q - \omega' + \mu} \Gamma_{\alpha'},$$

where we use $\int_0^\beta d\tau \, e^{(\omega_q - \omega_{q'})\tau} = \beta \delta_{qq'}$. Since both t and t' are positive the integral over ω can be performed by closing the contour in the lower half of the complex ω-plane, whereas the integral over ω' can be performed by closing the contour in the upper half of the complex ω'-plane. We then see that the integral over ω is nonzero only for $\text{Im}[\omega_q] < 0$, whereas the integral over ω' is nonzero only for $\text{Im}[\omega_q] > 0$. This means that for every ω_q

the double integral vanishes. We are left with the calculation of the convolution between G^M and $\Sigma^\lceil$. We have

$$\left[G^M \star \Sigma_{em}^\lceil \right](0,t) = \int \frac{d\omega}{2\pi} \frac{1}{-i\beta} \sum_q \frac{G^M(\omega_q)e^{\eta\omega_q}}{\omega_q - \omega + \mu} \sum_\alpha \Gamma_\alpha e^{i(\omega + V_\alpha)t},$$

where factor $e^{\eta\omega_q}$ stems from the fact that in the limit $t \to 0$ the quantity $\left[G^M \star \Sigma_{em}^\lceil \right](0,t)$ must be equal to $\left[G^M \star \Sigma_{em}^M \right](0,0^+)$. To evaluate the sum over the Matsubara frequencies we use the trick (6.19) and then deform the contour from Γ_a to Γ_b since $G^M(\zeta)$ is analytic in the upper and lower half of the complex ζ-plane, see (16.11). We then have

$$\left[G^M \star \Sigma_{em}^\lceil \right](0,t) = \int \frac{d\omega}{2\pi} \frac{d\omega'}{2\pi} f(\omega') \left[\frac{G^M(\omega' - i\delta)}{\omega' - \omega + \mu - i\delta} - \frac{G^M(\omega' + i\delta)}{\omega' - \omega + \mu + i\delta} \right] \sum_\alpha \Gamma_\alpha e^{i(\omega + V_\alpha)t},$$

where, as usual, the limit $\delta \to 0$ is understood. As $t > 0$ we can close the integral over ω in the upper half of the complex ω-plane. Then only the second term in the square brackets contributes and we find

$$\left[G^M \star \Sigma_{em}^\lceil \right](0,t) = i \int \frac{d\omega'}{2\pi} f(\omega') G^M(\omega' + i\delta) \sum_\alpha \Gamma_\alpha e^{i(\omega' + \mu + V_\alpha)t}$$

$$= i \int \frac{d\omega}{2\pi} f(\omega - \mu) G^R(\omega) \sum_\alpha \Gamma_\alpha e^{i(\omega + V_\alpha)t},$$

where in the last equality we change the integration variable $\omega' = \omega - \mu$ and observe that $G^M(\omega - \mu + i\delta) = G^R(\omega)$. Therefore the last term on the r.h.s. of (16.10) reads

$$\left[G^\lceil \star \Sigma_{em}^\lceil \right](t,t) = i \int \frac{d\omega}{2\pi} f(\omega - \mu) \sum_\alpha e^{i(\omega + V_\alpha - h_{eff})t} G^R(\omega) \Gamma_\alpha. \qquad (16.16)$$

We now have all ingredients to solve the equation of motion for $G^<(t,t)$. Using the results (16.14), (16.15), and (16.16),

$$i\frac{d}{dt} G^<(t,t) - h_{eff} G^<(t,t) + G^<(t,t) h_{eff}^\dagger = -i \int \frac{d\omega}{2\pi} f(\omega - \mu) \sum_\alpha$$

$$\times \left\{ G^R(\omega + V_\alpha) + e^{i(\omega + V_\alpha - h_{eff})t} \left[G^R(\omega) - G^R(\omega + V_\alpha) \right] \right\} \Gamma_\alpha + \text{H.c.}$$

It is natural to make the transformation

$$G^<(t,t) = e^{-ih_{eff}t} \tilde{G}^<(t,t) e^{ih_{eff}^\dagger t}, \qquad (16.17)$$

so that the differential equation for $\tilde{G}^<(t,t)$ reads

$$i\frac{d}{dt} \tilde{G}^<(t,t) = -i \int \frac{d\omega}{2\pi} f(\omega - \mu) \sum_\alpha e^{ih_{eff}t} \left[G^R(\omega + V_\alpha) \Gamma_\alpha - \Gamma_\alpha G^A(\omega + V_\alpha) \right] e^{-ih_{eff}^\dagger t}$$

$$- i \int \frac{d\omega}{2\pi} f(\omega - \mu) \sum_\alpha V_\alpha \left[G^R(\omega) G^R(\omega + V_\alpha) \Gamma_\alpha e^{i(\omega + V_\alpha - h_{eff}^\dagger)t} - \text{H.c.} \right],$$

where in the second line we use the Dyson-like identity

$$G^{\mathrm{R}}(\omega + V_\alpha) = G^{\mathrm{R}}(\omega) - V_\alpha G^{\mathrm{R}}(\omega) G^{\mathrm{R}}(\omega + V_\alpha) \tag{16.18}$$

and the symbol "H.c." now refers to the transpose conjugate of the quantity in the square bracket, i.e., H.c. $= e^{-\mathrm{i}(\omega + V_\alpha - h_{\mathrm{eff}})t} \Gamma_\alpha G^A(\omega + V_\alpha) G^A(\omega)$. The quantity $\tilde{G}^<(t, t)$ can be calculated by direct integration of the r.h.s.. The integral over time between 0 and t of the second line is easy since the dependence on t is all contained in the exponential. To integrate the first line we use the identity

$$\int_0^t dt'\, e^{\mathrm{i}At'} \left[\frac{1}{x-A} B - B \frac{1}{x-A^\dagger} \right] e^{-\mathrm{i}A^\dagger t'} = -\mathrm{i} e^{\mathrm{i}At'} \frac{1}{x-A} B \frac{1}{x-A^\dagger} e^{-\mathrm{i}A^\dagger t'} \Big|_0^t,$$

which is valid for arbitrary matrices A and B and can be verified by direct differentiation of the r.h.s. after the replacements $e^{\mathrm{i}At'} \to e^{\mathrm{i}(A-x)t'}$ and $e^{-\mathrm{i}A^\dagger t'} \to e^{-\mathrm{i}(A^\dagger - x)t'}$ (clearly this replacement does not change the r.h.s.). Finally the integration between 0 and t of the l.h.s. yields the difference $\tilde{G}^<(t, t) - \tilde{G}^<(0, 0)$. The matrix $\tilde{G}^<(0, 0) = G^<(0, 0) = G^{\mathrm{M}}(0, 0^+)$ and

$$G^{\mathrm{M}}(0, 0^+) = \frac{1}{-\mathrm{i}\beta} \sum_q e^{\eta \omega_q} G^{\mathrm{M}}(\omega_q) = \int \frac{d\omega}{2\pi} f(\omega)[G^{\mathrm{M}}(\omega - \mathrm{i}\delta) - G^{\mathrm{M}}(\omega + \mathrm{i}\delta)]$$

$$= \mathrm{i} \int \frac{d\omega}{2\pi} f(\omega - \mu) G^{\mathrm{R}}(\omega) \Gamma G^A(\omega). \tag{16.19}$$

This result generalizes the formula (6.20) to molecules with an arbitrary number of levels. Equation (16.19) could also have been derived from the fluctuation–dissipation theorem, since $G^{\mathrm{M}}(0, 0^+) = G^<(0, 0) = \int \frac{d\omega}{2\pi} G^<(\omega)$ with $G^<(\omega) = \mathrm{i} f(\omega - \mu) A(\omega)$ and the spectral function

$$A(\omega) = \mathrm{i} \left[G^{\mathrm{R}}(\omega) - G^A(\omega) \right] = G^{\mathrm{R}}(\omega) \Gamma G^A(\omega).$$

Interestingly, the spectral function can be written as the sum of partial spectral functions corresponding to different reservoirs, i.e.,

$$A(\omega) = \sum_\alpha A_\alpha(\omega) \quad \text{where} \quad A_\alpha(\omega) = G^{\mathrm{R}}(\omega) \Gamma_\alpha G^A(\omega).$$

Collecting the terms coming from the integration and taking into account (16.17) and (16.18), after some algebra we arrive at

$$\boxed{\begin{aligned} -\mathrm{i}G^<(t, t) = \int \frac{d\omega}{2\pi} f(\omega - \mu) \sum_\alpha \Big\{ & A_\alpha(\omega + V_\alpha) \\ & + V_\alpha \left[e^{\mathrm{i}(\omega + V_\alpha - h_{\mathrm{eff}})t} G^{\mathrm{R}}(\omega) A_\alpha(\omega + V_\alpha) + \text{H.c.} \right] \\ & + V_\alpha^2\, e^{-\mathrm{i}h_{\mathrm{eff}}t} G^{\mathrm{R}}(\omega) A_\alpha(\omega + V_\alpha) G^A(\omega)\, e^{\mathrm{i}h_{\mathrm{eff}}^\dagger t} \Big\} \end{aligned}} \tag{16.20}$$

Considering the original complication of the problem, equation (16.20) is an extremely compact result and, as we see below, contains a lot of physics. This formula can easily be generalized to situations where the molecular Hamiltonian has terms which flip the spin [153], e.g., $\sum_{mn}(T^{sf}_{mn}d^{\dagger}_{m\uparrow}d_{n\downarrow} + T^{sf}_{nm}d^{\dagger}_{n\uparrow}d_{m\downarrow})$, ferromagnetic leads where $\epsilon_{k\alpha\uparrow} \neq \epsilon_{k\alpha\downarrow}$, and arbitrary time-independent perturbations in the molecular region, $T_{mn} \rightarrow T_{mn}(t) = T'_{mn}$. The only crucial ingredient to obtain a close analytic formula for $G^{<}(t,t)$ is the WBLA.

Let us come back to (16.20). We can easily check that $G^{<}(0,0) = G^{M}(0,0^{+})$ and that for zero bias, $V_{\alpha} = 0$, one has $G^{<}(t,t) = G^{<}(0,0)$ at all times, as it should be. Due to the non-hermiticity of $h_{\text{eff}} = h - i\Gamma/2$ the terms in the second and third lines vanish exponentially fast in the limit $t \rightarrow \infty$. Consequently, $G^{<}(t,t)$ approaches the steady-state value

$$\lim_{t\to\infty} G^{<}(t,t) = i \int \frac{d\omega}{2\pi} f(\omega - \mu) \sum_{\alpha} A_{\alpha}(\omega + V_{\alpha}).$$

This result should be compared with the equilibrium result (16.19) in which the partial spectral functions are all calculated at the same frequency ω. Instead, out of equilibrium, the partial spectral functions are calculated at frequencies shifted by the applied bias, $\omega \rightarrow \omega + V_{\alpha}$.

The transient behavior of $G^{<}(t,t)$ is described by the last two lines of (16.20). These terms have different physical origin. At sufficiently low temperatures the Fermi function has a sharp step at $\omega = \mu$ and the second line gives rise to transient oscillations with frequencies $\omega_{j} \sim |\mu + V_{\alpha} - h_{j}|$ with damping times τ_{j}, where $h_{j} - i\tau_{j}^{-1}/2$ are the eigenvalues of the effective Hamiltonian $h - i\Gamma/2$. These oscillations originate from virtual transitions between the resonant levels of the molecule and the Fermi level of the biased reservoirs. The third line of (16.20) describes intramolecular trasitions. If the effective Hamiltonian h_{eff} does not commute with Γ_{α} (and hence with A_{α}) this term produces oscillations of frequency $\omega_{ij} = h_{i} - h_{j}$ which are exponentially damped over a time scale $\tau_{ij} = 1/(\tau_{i}^{-1} + \tau_{j}^{-1})$. It is worth stressing that if h_{eff} and Γ_{α} commute no intramolecular frequencies are observed in the transient behavior of $G^{<}(t,t)$ since $e^{-ih_{\text{eff}}t+ih^{\dagger}_{\text{eff}}t} = e^{-\Gamma t}$. It is also interesting to observe that intramolecular transitions are damped on a faster time scale than the transitions between the Fermi level of the reservoirs and the molecular levels.

16.2.1 Landauer–Büttiker formula

Having all the self-energies and the Green's functions we can also calculate the time-dependent current $I_{\alpha}(t)$ in (16.7). The mathematical steps are very similar to the ones leading to (16.20) and here we write the final result directly (take into account that the electron charge is $q = -1$):

$$
\begin{aligned}
I_\alpha(t) = 2 \int \frac{d\omega}{2\pi} f(\omega - \mu) \sum_\beta \mathrm{Tr}_M \Big\{ &\Gamma_\alpha G^{\mathrm{R}}(\omega + V_\alpha) \Gamma_\beta G^{\mathrm{A}}(\omega + V_\alpha) \\
&- \Gamma_\alpha G^{\mathrm{R}}(\omega + V_\beta) \Gamma_\beta G^{\mathrm{A}}(\omega + V_\beta) \\
&- V_\beta \left[\Gamma_\alpha e^{\mathrm{i}(\omega + V_\beta - h_{\mathrm{eff}})t} G^{\mathrm{R}}(\omega) \left(-\mathrm{i}\delta_{\alpha\beta} G^{\mathrm{R}}(\omega + V_\beta) + A_\beta(\omega + V_\beta) \right) + \mathrm{H.c.} \right] \\
&- V_\beta^2\, \Gamma_\alpha e^{-\mathrm{i}h_{\mathrm{eff}}t}\, G^{\mathrm{R}}(\omega) A_\beta(\omega + V_\beta) G^{\mathrm{A}}(\omega)\, e^{\mathrm{i}h_{\mathrm{eff}}^\dagger t} \Big\}
\end{aligned}
$$

$$(16.21)$$

where the sum over β is a sum over all reservoirs.

The analysis of this formula is very similar to the analysis for the time-dependent behavior of $G^<(t,t)$. The last two lines vanish exponentially fast in the long time limit and the steady-state value of the current is given by the first two terms in the curly bracket. Let us play around with (16.21). Consider the simple case of a 2×2 Hamiltonian h for the molecule with matrix elements

$$
h = \begin{pmatrix} \epsilon_0 + \Delta & 0 \\ 0 & \epsilon_0 - \Delta \end{pmatrix}
$$

and two reservoirs that we call left, $\alpha = L$, and right, $\alpha = R$. We further assume that $\Gamma_L = \Gamma_R$. We study two different cases: in the first case the matrix Γ_L is proportional to the 2×2 identity matrix, while in the second case it is proportional to a 2×2 matrix with all identical entries,

$$
\text{first case}: \quad \Gamma_L = \Gamma_0 \begin{pmatrix} 1 & 0 \\ 0 & 1 \end{pmatrix} \quad \Big| \quad \text{second case}: \quad \Gamma_L = \Gamma_0 \begin{pmatrix} 1 & 1 \\ 1 & 1 \end{pmatrix}. \quad (16.22)
$$

The system is initially in equilibrium at zero temperature and chemical potential $\mu = 0$. At time $t = 0$ we suddenly switch on a bias $V_L = 2\epsilon_0$ in the left reservoir while we keep the right reservoir at zero bias, $V_R = 0$. If we take $\epsilon_0 > 0$ and $\Delta \ll \epsilon_0$, and if the molecule is weakly coupled to the reservoirs, $\Gamma_0 \ll \epsilon_0$, then the initial number of electrons on the molecule is approximately zero. The effect of the bias is to raise the levels of the left reservoir above the molecular levels so that the molecular levels end up in the bias window, see schematic representation below:

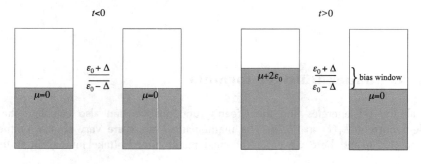

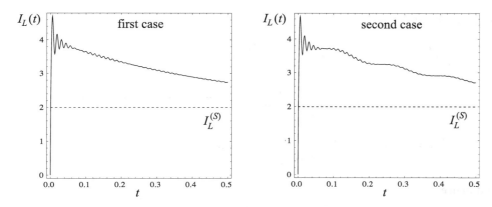

Figure 16.2 Left current $I_L(t)$ for the two-level molecular system described in the main text with parameters $\epsilon_0 = 500$, $\Delta = 20$, $V_L = 2\epsilon_0$ and $V_R = 0$. The left (right) panel corresponds to the first (second) case of (16.22). Energies and current are in units of Γ_0 and time is in units of Γ_0^{-1}.

In this situation a net current starts flowing. In the first case of (16.22) the transient current only exhibits oscillations of frequency $\omega = |V_L - \epsilon_0 \pm \Delta| = |\epsilon_0 \pm \Delta|$ and $\omega = |V_R - \epsilon_0 \pm \Delta| = |\epsilon_0 \mp \Delta|$ corresponding to transitions between the molecular levels and the Fermi energy of the biased reservoirs. No intramolecular transitions are possible since Γ_α commutes with h. This is nicely illustrated in the left panel of Fig. 16.2 where we plot the function $I_L(t)$ of (16.21). The only two visible frequencies $\omega_\pm = |\epsilon_0 \pm \Delta|$ produce coherent quantum beats in the current. This phenomenon has been observed in the context of spin transport where the energy spacing Δ should be seen as the Zeeman splitting induced by an external magnetic field [153–156]. In the second case of (16.22) the matrices Γ_α do not commute with h and according to our previous analysis we should observe an additional frequency $\omega = 2\Delta$ corresponding to the intramolecular transition. In the right panel of Fig. 16.2 we show the left current $I_L(t)$ for the second case. We clearly see an additional oscillation of frequency $\omega = 2\Delta$ superimposed on the oscillations of the first case.

We have already observed that in the long-time limit the last two lines of (16.21) vanish. Let us check that the steady-state value of the current agrees with the formula (16.8) when we specialize it to noninteracting situations and to WBLA reservoirs. The formula (16.8) was derived under the assumptions that initial correlations and initial-state dependencies are washed out in the limit $t \to \infty$ and that, in the same limit, the invariance under time translations is recovered, i.e., the Green's function and the self-energy depend only on the time difference. Then the lesser Green's function of the molecular region can be calculated from (9.71) where, for the noninteracting case considered here, the self-energy is only the embedding self-energy. In Fourier space the general form of the retarded and lesser embedding self-energy is

$$\Sigma_\alpha^{R}(\omega) = h_{M\alpha}\frac{1}{\omega - h_{\alpha\alpha} - V_\alpha + i\eta}h_{\alpha M}$$

$$= \underbrace{P\, h_{M\alpha}\frac{1}{\omega - h_{\alpha\alpha} - V_\alpha}h_{\alpha M}}_{\Lambda_\alpha(\omega)} \underbrace{-i\pi\, h_{M\alpha}\delta(\omega - h_{\alpha\alpha} - V_\alpha)h_{\alpha M}}_{\Gamma_\alpha(\omega)/2},$$

$$\Sigma_\alpha^{<}(\omega) = 2\pi i\, h_{M\alpha}f(h_{\alpha\alpha} - \mu)\delta(\omega - h_{\alpha\alpha} - V_\alpha)h_{\alpha M}$$

$$= i\, f(\omega - V_\alpha - \mu)\Gamma_\alpha(\omega),$$

and hence

$$G^{<}(\omega) = G^{R}(\omega)\Sigma_{em}^{<}(\omega)G^{A}(\omega) = i\sum_\alpha f(\omega - V_\alpha - \mu)G^{R}(\omega)\Gamma_\alpha(\omega)G^{A}(\omega).$$

In contrast to the WBLA, the rate operator is frequency dependent and Σ_α^{R} also has a real part.[4] The retarded Green's function is easily obtained from the Dyson equation, $G^{R}(\omega) = 1/(\omega - h - \Sigma_{em}^{R}(\omega))$, and from it we can calculate the nonequilibrium spectral function

$$A(\omega) = i[G^{R}(\omega) - G^{A}(\omega)] = \sum_\alpha G^{R}(\omega)\Gamma_\alpha(\omega)G^{A}(\omega).$$

Inserting these results into (16.8) we obtain the *Landauer–Büttiker* formula [157, 158] for the steady-state current

$$\boxed{I_\alpha^{(S)} = 2\int\frac{d\omega}{2\pi}\sum_\beta [f(\omega - V_\alpha - \mu) - f(\omega - V_\beta - \mu)]\,\mathrm{Tr}_M\left[\Gamma_\alpha(\omega)G^{R}(\omega)\Gamma_\beta(\omega)G^{A}(\omega)\right]}$$

(16.23)

In the WBLA the rate operators Γ_α are frequency independent, the retarded Green's functions simplify as in (16.13) and the Landauer–Büttiker formula reduces to the first two terms of (16.21), as it should. The quantities

$$\mathcal{T}_{\alpha\beta} = \mathrm{Tr}_M\left[\Gamma_\alpha(\omega)G^{R}(\omega)\Gamma_\beta(\omega)G^{A}(\omega)\right]$$

are related to the probability for an electron of energy ω to be transmitted from reservoir α to reservoir β. We then see that the current $I_\alpha^{(S)}$ is the sum of the probabilities for electrons in α (and hence with energy below $\mu + V_\alpha$) to go from α to β minus the sum of the probabilities for electrons in β (and hence with energy below $\mu + V_\beta$) to go from β to α. As expected the steady-state current vanishes if the biases $V_\alpha = V$ are all the same. It is important to realize that the derivation of the Landauer–Büttiker formula does not apply in the interacting case since $G^{<} = G^{R}(\Sigma^{<} + \Sigma_{em}^{<})G^{A}$ and there is no simple relation between the *nonequilibrium* many-body self-energy $\Sigma^{<}$ and the *nonequilibrium* many-body rate operator $\Gamma = i[\Sigma^{>} - \Sigma^{<}]$.

---

[4]Remember that Λ_α and Γ_α are related by a Hilbert transformation.

For the case of only two reservoirs $\alpha = L, R$ the Landauer–Büttiker formula simplifies to

$$I_L^{(S)} = 2 \int \frac{d\omega}{2\pi} \left[f(\omega - V_L - \mu) - f(\omega - V_R - \mu) \right] \mathrm{Tr}_M \left[\Gamma_L(\omega) G^{\mathrm{R}}(\omega) \Gamma_R(\omega) G^{\mathrm{A}}(\omega) \right]$$

and $I_R^{(S)} = -I_L^{(S)}$. This formula generalizes the result (7.37) to finite bias and to molecules with an arbitrary number of levels, and it is not restricted to the WBLA. We can easily check that for one single level and small biases, and within the WBLA we reobtain (7.37). In this case

$$\mathrm{Tr}_M \left[\Gamma_L(\omega) G^{\mathrm{R}}(\omega) \Gamma_R(\omega) G^{\mathrm{A}}(\omega) \right] = \frac{\Gamma_L \Gamma_R}{(\omega - \epsilon_0)^2 + (\Gamma/2)^2}.$$

Furthermore for $V_R = 0$ and small biases, $V_L = qV_0 = -V_0$ and we can Taylor expand $f(\omega + V_0 - \mu) \sim f(\omega - \mu) + f'(\omega - \mu)V_0$. Then

$$I_R^{(S)} = -2 \int \frac{d\omega}{2\pi} f'(\omega - \mu) \frac{\Gamma_L \Gamma_R}{(\omega - \epsilon_0)^2 + (\Gamma/2)^2} V_0.$$

Integrating by parts we obtain (7.37) up to a factor of 2 due to spin which was neglected in Section 7.2.3.

Exercise 16.3. Show that for any bias V_α the $G^<(0,0)$ in (16.20) reduces to the equilibrium Green's function $G^{\mathrm{M}}(0, 0^+)$ in (16.19).

Exercise 16.4. Prove (16.21) and show that (a) $I_\alpha(t = 0) = 0$, and (b) $I_\alpha^{(S)} = 0$ when the biases $V_\alpha = V$ are the same in all reservoirs.

16.3 Implementation of the Kadanoff–Baym equations

For open systems the Kadanoff–Baym equations of Section 9.4 must be modified by adding to the many-body self-energy the embedding self-energy, see again (16.6). Dropping the subscript "MM" as we did in the previous section, we denote by

$$\Sigma_{\mathrm{tot}} = \Sigma + \Sigma_{\mathrm{em}}$$

the sum of the two self-energies. In this section we discuss how to implement the Kadanoff–Baym equations for the Green's function of the molecular region. The presentation follows very closely the one of Ref. [147]. Before entering into the details of the procedure we observe that the Kadanoff–Baym equations are first-order differential equations in time. The numerical technique to solve these kinds of equation is the so called *time-stepping technique*. Consider the differential equation

$$\frac{d}{dt} f(t) = g(t),$$

where $g(t)$ is a known function. From the value of f at the initial time t_0 we can calculate the value of f at one time step later, i.e., at $t_0 + \Delta_t$, by approximating the derivative with a finite difference

$$\frac{f(t_0 + \Delta_t) - f(t_0)}{\Delta_t} = g(t_0).$$

Of course $f(t_0 + \Delta_t)$ calculated in this way provides a good approximation if the time step Δ_t is much smaller than the typical time scale over which $g(t)$ varies. From $f(t_0 + \Delta_t)$ we can then calculate $f(t_0 + 2\Delta_t)$ by approximating again the derivative with a finite difference. In general, from the value of f at time $t_0 + n\Delta_t$ we can calculate the value of f at time $t_0 + (n+1)\Delta_t$ from

$$\frac{f(t_0 + (n+1)\Delta_t) - f(t_0 + n\Delta_t)}{\Delta_t} = g(t_0 + n\Delta_t). \tag{16.24}$$

This is the basic idea of the time-stepping technique. There are of course dozens (or maybe hundreds) of refinements and variants to make the iterative procedure more stable and accurate. For example one can replace $g(t_0 + n\Delta_t)$ in the r.h.s. of (16.24) with the average $\frac{1}{2}[g(t_0 + n\Delta_t) + g(t_0 + (n+1)\Delta_t)]$ or with the value of g at half the time step, i.e., $g(t_0 + (n+\frac{1}{2})\Delta_t))$. The advantage of these variants is clear: if we use the same scheme to go backward in time then we recover exactly $f(t_0)$ from $f(t_0 + n\Delta_t)$. The variant of choice depends on the physical problem at hand. In Appendix P we describe the time-stepping algorithm employed to produce the results of the following sections. Here we explain how to solve the Kadanoff–Baym equations using a generic time-stepping technique, and then give details on how to calculate the self-energy in the second-Born and GW approximation. We choose these two self-energies because most of the results discussed later have been obtained within these approximations. However, at this point the reader should be able to work out the Keldysh components of any other approximate self-energy.

16.3.1   Time-stepping technique

Let us assume that we have calculated the Matsubara Green's function, and hence the Matsubara self-energy, using some suitable self-consistent method. We now show that the lesser and greater Green's functions can be calculated by solving the following three coupled Kadanoff–Baym equations

$$\left[i\frac{d}{dt} - h(t)\right] G^{\rceil}(t, \tau) = \left[\Sigma^{\mathrm{R}}_{\mathrm{tot}} \cdot G^{\rceil} + \Sigma^{\rceil}_{\mathrm{tot}} \star G^{\mathrm{M}}\right](t, \tau),$$

$$\left[i\frac{d}{dt} - h(t)\right] G^{>}(t, t') = \left[\Sigma^{\mathrm{R}}_{\mathrm{tot}} \cdot G^{>} + \Sigma^{>}_{\mathrm{tot}} \cdot G^{\mathrm{A}} + \Sigma^{\rceil}_{\mathrm{tot}} \star G^{\lceil}\right](t, t'),$$

$$G^{<}(t, t')\left[-i\frac{\overleftarrow{d}}{dt'} - h(t')\right] = \left[G^{\mathrm{R}} \cdot \Sigma^{<}_{\mathrm{tot}} + G^{<} \cdot \Sigma^{\mathrm{A}}_{\mathrm{tot}} + G^{\rceil} \star \Sigma^{\lceil}_{\mathrm{tot}}\right](t, t'),$$

together with the equation for the time-diagonal Green's function

$$i\frac{d}{dt} G^{<}(t, t) - \left[h(t), G^{<}(t, t)\right]_{-} = -\left[G^{\mathrm{R}} \cdot \Sigma^{<}_{\mathrm{tot}} + G^{<} \cdot \Sigma^{\mathrm{A}}_{\mathrm{tot}} + G^{\rceil} \star \Sigma^{\lceil}_{\mathrm{tot}}\right](t, t) + \mathrm{H.c.},$$

which can be derived similarly to (16.10). In other words, we do not need the equation for $G^\lceil$ nor the equation for $G^>(t, t')$ with the derivative with respect to t' nor the equation for $G^<(t, t')$ with the derivative with respect to t.

Without loss of generality we choose $t_0 = 0$. If we use the simple minded time-stepping method (16.24), then the collision integral (i.e., the r.h.s.) of the first equation for $t = 0$ is $[\Sigma^\lceil_{\text{tot}} \star G^{\text{M}}](0, \tau) = [\Sigma^{\text{M}}_{\text{tot}} \star G^{\text{M}}](0, \tau)$, which is known; we can make one time step to calculate $G^\lceil(\Delta_t, \tau)$ since $G^\lceil(0, \tau) = G^{\text{M}}(0, \tau)$, which is also known. Similarly, the collision integral of the second and third equations is known for $t = t' = 0$ and we can make one time step to calculate $G^>(\Delta_t, 0)$ and $G^<(0, \Delta_t)$, since we know $G^>(0, 0) = G^{\text{M}}(0^+, 0)$ and $G^<(0, 0) = G^{\text{M}}(0, 0^+)$. Using the symmetry properties (9.41) and (9.42) we then also know $G^\lceil(\tau, \Delta_t)$ as well as $G^>(0, \Delta_t)$ and $G^<(\Delta_t, 0)$. In order to calculate $G^<(\Delta_t, \Delta_t)$ we use the fourth equation. The value of $G^>(\Delta_t, \Delta_t)$ can be computed from

$$G^>(t, t) = \mathrm{i}\hat{\mathbb{1}} + G^<(t, t),$$

which is a direct consequence of the (anti)commutation rules of the field operators. In more refined time-stepping techniques one needs the collision integrals not only at $t = t' = 0$ but also at the first time step, see again the discussion below (16.24). In these cases one typically implements a *predictor corrector* scheme. The idea is essentially to make the first time step as described above, use the result to improve the collision integrals (for instance by taking an average) and again make the first time step. In principle we can do more than one predictor corrector, i.e., we can use the result of the first predictor corrector to improve further the collision integrals and make again the first time step. In most cases, however, one predictor corrector is enough.

It should now be clear how to proceed. Having the Green's functions with real times up to Δ_t we can compute the collision integrals up to the same time. Then, we can use the first equation to calculate $G^\lceil(2\Delta_t, \tau)$ and the second and third equations to calculate $G^>(2\Delta_t, n\Delta_t)$ and $G^<(n\Delta_t, 2\Delta_t)$ with $n = 0, 1$. The quantities $G^\lceil(\tau, 2\Delta_t)$, $G^>(n\Delta_t, 2\Delta_t)$, and $G^<(2\Delta_t, n\Delta_t)$ follow directly from the symmetry properties. Finally we use the fourth equation to calculate $G^<(2\Delta_t, 2\Delta_t)$. At the end of the second time step we have the Green's function with real times smaller or equal to $2\Delta_t$. In Fig. 16.3 we illustrate how the procedure works from time step m to time step $m + 1$. The first equation is used to evaluate $G^\lceil((m+1)\Delta_t, \tau)$, the second and third equations to evaluate $G^>((m+1)\Delta_t, n\Delta_t)$ and $G^<(n\Delta_t, (m+1)\Delta_t)$ for all $n \leq m$. The symmetry properties are used to extract the Green's functions with interchanged time arguments and finally the fourth equation is used to calculate $G^<((m+1)\Delta_t, (m+1)\Delta_t)$.

In the next section we work out explicitly the self-energy in terms of the Green's function. The derivation is very instructive since it confirms with two examples the validity of the observation made in Section 5.3 [see also discussion below (9.33)] according to which, in order to perform the $(m+1)$th time step, we do not need the self-energy with time-arguments larger than $m\Delta_t$.

16.3.2 Second-Born and GW self-energies

The main purpose of this section is to study how the Keldysh components of the self-energy are expressed in terms of the Keldysh components of the Green's function and to highlight

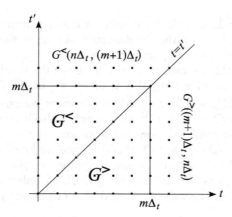

Figure 16.3 Time stepping technique in the (t, t') plane. $G^>(t, t')$ is calculated for $t > t'$ and $G^<(t, t')$ is calculated for $t \leq t'$.

the dependence on the time variables. No matter what the approximation to the self-energy is, the four differential equations of the previous section tell us that in order to evaluate the time derivative with respect to T of $G^\rceil(T, \tau)$, $G^>(T, \bar{t})$, and $G^<(\bar{t}, T)$ with $\bar{t} \leq T$, and of $G^<(T, T)$ we only need to know $\Sigma^\rceil(\bar{t}, \tau)$ for $\bar{t} \leq T$, $\Sigma^>(\bar{t}, \bar{t}')$ for $\bar{t}' \leq \bar{t} \leq T$, and $\Sigma^<(\bar{t}, \bar{t}')$ for $\bar{t} \leq \bar{t}' \leq T$. We now show that to calculate these self-energies in the second-Born and GW approximation it is enough to know the Green's function with real times up to T, see also discussion below (9.33).

The second-Born self-energy is the sum of the Hartree–Fock self-energy, Σ_{HF}, and the second order correlation self-energy Σ_c. The Hartree–Fock self-energy is local in time and can be calculated from the sole knowledge of $G^<(\mathbf{x}, t; \mathbf{x}', t)$. The second-order correlation self-energy is given in (13.16), which we rewrite below for fermions of spin $1/2$

$$\Sigma_c(1; 2) = -\mathrm{i}^2 \int d3d4 \, v(1; 3)v(2; 4) \left[G(1; 2)G(3; 4)G(4; 3) - G(1; 4)G(4; 3)G(3; 2) \right].$$

For convenience of notation we suppress the position–spin variables. Taking into account that v is local in time the right component of Σ_c is

$$\Sigma_c^\rceil(t, \tau) = -\mathrm{i}^2 \int d\mathbf{x}_3 d\mathbf{x}_4 \, v \, v \left[G^\rceil(t, \tau)G^\rceil(t, \tau)G^\lceil(\tau, t) - G^\rceil(t, \tau)G^\lceil(\tau, t)G^\rceil(t, \tau) \right].$$

The left component $\Sigma_c^\lceil$ can be worked out similarly and, like $\Sigma_c^\rceil$, contains only $G^\rceil(t, \tau)$ and $G^\lceil(\tau, t)$. The lesser and greater components read

$$\Sigma_c^\lessgtr(t, t') = -\mathrm{i}^2 \int d\mathbf{x}_3 d\mathbf{x}_4 \, v \, v \left[G^\lessgtr(t, t')G^\lessgtr(t, t')G^\gtrless(t', t) - G^\lessgtr(t, t')G^\gtrless(t', t)G^\lessgtr(t, t') \right].$$

Thus we see that in order to calculate the self-energy with real times up to T we only need the Green's function with real times up to T.

In the GW approximation the XC part of the self-energy is the product of the Green's function G and the screened interaction W, see (12.48). The screened interaction satisfies the Dyson equation

$$W(1;2) = v(1;2) + \int d3d4 \, v(1;3)P(3;4)W(4;2),$$

where the polarizability is approximated as $P(1;2) = -iG(1;2)G(2;1)$. In order to split the self-energy into a Hartree–Fock part and a correlation part we write $W = v + \delta W$ so that the GW self-energy reads $\Sigma = \Sigma_{\mathrm{HF}} + \Sigma_{\mathrm{c}}$ with

$$\Sigma_{\mathrm{c}}(1;2) = -iG(1;2)\delta W(2;1).$$

Understanding again the dependence on the position–spin variables we have

$$\Sigma_{\mathrm{c}}^{\rceil}(t,\tau) = -iG^{\rceil}(t,\tau)\delta W^{\lceil}(\tau,t) \quad ; \quad \Sigma_{\mathrm{c}}^{\lceil}(\tau,t) = -iG^{\lceil}(\tau,t)\delta W^{\rceil}(t,\tau)$$

and

$$\Sigma_{\mathrm{c}}^{\lessgtr}(t,t') = -iG^{\lessgtr}(t,t')\delta W^{\gtrless}(t',t).$$

As $W(1;2) = W(2;1)$ is symmetric, see property (10.27), we only need to calculate $\delta W^{\rceil}(t,\tau)$ and $\delta W^{<}(t,t')$ for $t \leq t'$ since[5]

$$\delta W^{\lceil}(\tau,t) = \delta W^{\rceil}(t,\tau) \quad \text{and} \quad \delta W^{>}(t,t') = \delta W^{<}(t',t).$$

From the Dyson equation for the screened interaction we find

$$\delta W^{\rceil}(t,\tau) = vP^{\rceil}(t,\tau)v + vX^{\rceil}(t,\tau)$$

and

$$\delta W^{<}(t,t') = vP^{<}(t,t')v + vX^{<}(t,t'),$$

where

$$P^{\rceil}(t,\tau) = -iG^{\rceil}(t,\tau)G^{\lceil}(\tau,t) \quad \text{and} \quad P^{<}(t,t') = -iG^{<}(t,t')G^{>}(t',t),$$

and

$$X^{\rceil}(t,\tau) = \int_0^t d\bar{t}\, P^{\mathrm{R}}(t,\bar{t})\delta W^{\rceil}(\bar{t},\tau) - i\int_0^\beta d\bar{\tau}\, P^{\rceil}(t,\bar{\tau})\delta W^{\mathrm{M}}(\bar{\tau},\tau),$$

$$X^{<}(t,t') = \int_0^t d\bar{t}\, P^{\mathrm{R}}(t,\bar{t})\delta W^{<}(\bar{t},\tau) + \int_0^{t'} d\bar{t}\, P^{<}(t,\bar{t})\delta W^{\mathrm{A}}(\bar{t},\tau)$$

$$- i\int_0^\beta d\bar{\tau}\, P^{\rceil}(t,\bar{\tau})\delta W^{\lceil}(\bar{\tau},t').$$

---

[5]We recall that we only need $\Sigma^{>}(t,t')$ for $t \geq t'$ and $\Sigma^{<}(t,t')$ for $t \leq t'$.

Let us assume that we propagated up to a maximum time T and see if we can calculate $\delta W^\rceil$ and $\delta W^<$ with real times up to T. The first term in $\delta W^\rceil$ and $\delta W^<$ is known since the polarizability is given in terms of Green's functions with real times smaller than T. The same is true for the second term since the time integrals in X have an upper limit which never exceeds T. For instance, for the first time step we need

$$X^\rceil(0,\tau) = -\mathrm{i}\int_0^\beta d\bar\tau\, P^{\mathrm{M}}(0,\bar\tau)\delta W^{\mathrm{M}}(\bar\tau,\tau)$$

and $X^<(0;0) = X^\rceil(0,0^+)$. These quantities are known from the solution of the equilibrium problem. After the first time step we can calculate P and δW with real times up to Δ_t and hence X with real times up to Δ_t. We can then perform the second time step and so on and so forth. In more refined time-stepping techniques we also need to implement a predictor corrector. This means that to go from time step m to time step $m+1$ we also need δW with real times equal to $(m+1)\Delta_t$. In this case we can use an iterative scheme similar to the one previously described. As a first guess for, e.g., $\delta W^<(n\Delta_t,(m+1)\Delta_t)$ we take $\delta W^<(n\Delta_t,m\Delta_t)$. Similarly we approximate $\delta W^\rceil((m+1)\Delta_t,\tau) \sim \delta W^\rceil(m\Delta_t,\tau)$ and $\delta W^<((m+1)\Delta_t,(m+1)\Delta_t) \sim \delta W^<(m\Delta_t,m\Delta_t)$. We then calculate X with real times equal to $(m+1)\Delta_t$ and use it to obtain a new value of δW with real times equal to $(m+1)\Delta_t$. This new value can then be used to improve X and in turn δW and the process is repeated until convergence is reached.

In the next sections we present the results of the numerical solution of the Kadanoff–Baym equations for open systems and for finite systems.

16.4 Initial-state and history dependence

The dependence on the initial state manifests itself in two ways in the Kadanoff–Baym equations. First, the initial values of the time-dependent Green's functions are determined by the equilibrium Green's function at $t = t_0$, and this value is obtained by solving the Matsubara Kadanoff–Baym equation (9.32) or the Matsubara Dyson equation (9.34). Second, the Kadanoff–Baym equations contain terms that describe the memory of the initial state during the time propagation. These terms depend on the self-energies $\Sigma_{\mathrm{tot}}^{\rceil,\lceil}(z,z')$ with mixed real- and imaginary-time arguments. In this section we investigate these two initial-state dependencies separately by either setting the MBPT self-energy $\Sigma(z,z')$ to zero for z and/or z' on the vertical track of the contour (the initial state is noninteracting) or by setting $\Sigma_{\mathrm{em}}^{\rceil,\lceil}$ and/or $\Sigma^{\rceil,\lceil}$ to zero (the initial state is interacting but the memory effects due to embedding and/or electron correlations are neglected).

We consider the system illustrated below in which a molecular region described with only two basis functions is connected to a left and a right noninteracting one-dimensional reservoir (semi-infinite chains):

The bonds between the chain sites denote the nonvanishing off-diagonal matrix elements of the one-particle Hamiltonian. So we have a link T in the reservoirs, a link T_α between the $\alpha = L, R$ reservoir and the molecular basis functions a, b and a link T_M in the molecule. The diagonal matrix elements in lead α are all equal to ϵ_α whereas they are equal to ϵ_m, $m = a, b$, in the molecule. The site basis $|j\alpha\rangle$ for the reservoirs is not the basis of the eigenfunctions of the reservoir Hamiltonian. For a reservoir α with N sites the reservoir Hamiltonian in the site basis reads

$$\hat{H}_\alpha = \sum_{\substack{j=1 \\ \sigma}}^{N-1} T \left(\hat{c}^\dagger_{j\alpha\sigma} \hat{c}_{j+1\alpha\sigma} + \hat{c}^\dagger_{j+1\alpha\sigma} \hat{c}_{j\alpha\sigma} \right).$$

The operators $\hat{d}_{k\alpha\sigma}$ that diagonalize $\hat{H}_\alpha$ can easily be worked out, see Exercise 2.3, and are given by $\hat{d}_{k\alpha\sigma} = \sum_{j=1}^{N} \langle k\alpha|j\alpha\rangle \hat{c}_{j\alpha\sigma}$ with $\langle j\alpha|k\alpha\rangle = \sqrt{\frac{2}{N+1}} \sin \frac{\pi k j}{N+1}$. In the $k\alpha$-basis $\hat{H}_\alpha = \sum_{k\sigma} \epsilon_{k\alpha} \hat{d}^\dagger_{k\alpha\sigma} \hat{d}_{k\alpha\sigma}$ has the same form as in (16.1) with the eigenvalues $\epsilon_{k\alpha} = \epsilon_\alpha + 2T \cos \frac{\pi k j}{N+1}$. In a similar way we can deduce the coupling Hamiltonian in the $k\alpha$-basis. The coupling Hamiltonian between the molecule and the left reservoir in the site basis is

$$\sum_\sigma T_L \left(\hat{c}^\dagger_{1L\sigma} \hat{d}_{a\sigma} + \hat{d}^\dagger_{a\sigma} \hat{c}_{1L\sigma} \right) = \sum_{k\sigma} T_{kL} \left(\hat{d}^\dagger_{kL\sigma} \hat{d}_{a\sigma} + \hat{d}^\dagger_{a\sigma} \hat{d}_{kL\sigma} \right),$$

where on the r.h.s. we define $T_{kL} = T_L \langle kL|1L\rangle$. Finally we choose the Coulomb integrals $v_{ijmn} = \delta_{in}\delta_{jm} v_{mn}$ so that the interaction Hamiltonian takes the diagonal form

$$\hat{H}_{\text{int}} = \frac{1}{2} \sum_{\substack{mn \\ \sigma\sigma'}} v_{mn} \, \hat{d}^\dagger_{m\sigma} \hat{d}^\dagger_{n\sigma'} \hat{d}_{n\sigma'} \hat{d}_{m\sigma},$$

where the sum over m, n is restricted to the molecular basis functions a and b.

The one-particle Hamiltonian of the molecule in the basis $\{a, b\}$ reads

$$h = \begin{pmatrix} \epsilon_a & T_M \\ T_M & \epsilon_b \end{pmatrix}$$

and the embedding self-energy Σ_α in the same basis has only one nonvanishing entry,

$$\Sigma_L = \begin{pmatrix} \sigma_L & 0 \\ 0 & 0 \end{pmatrix}, \qquad \Sigma_R = \begin{pmatrix} 0 & 0 \\ 0 & \sigma_R \end{pmatrix}.$$

To calculate σ_α we use the following trick. From the definition (16.5) we have $\Sigma_\alpha = h_{M\alpha} g_{\alpha\alpha} h_{\alpha M}$ and therefore

$$\sigma_\alpha = \sum_k T_{k\alpha} \, g_{k\alpha} \, T_{k\alpha} = T_\alpha^2 \sum_k \langle 1\alpha|k\alpha\rangle g_{k\alpha} \langle k\alpha|1\alpha\rangle.$$

Denoting by $\hat{g}_{\alpha\alpha} = \sum_k g_{k\alpha} |k\alpha\rangle\langle k\alpha|$ the noninteracting Green's function operator (in first quantization) of the isolated αth reservoir we see that

$$\sigma_\alpha = T_\alpha^2 \, \langle 1\alpha|\hat{g}_{\alpha\alpha}|1\alpha\rangle.$$

The retarded component of $\hat{g}_{\alpha\alpha}$ in frequency space reads $\hat{g}_{\alpha\alpha}^{R}(\omega) = 1/(\omega - \hat{h}_{\alpha\alpha} + i\eta)$ where $\hat{h}_{\alpha\alpha}$ is the single-particle Hamiltonian (in first quantization) of the αth reservoir. We write $\hat{h}_{\alpha\alpha} = \hat{h}_{\alpha\alpha}^{0} + \hat{T}$ where $\hat{T} = T(|1\alpha\rangle\langle 2\alpha| + |2\alpha\rangle\langle 1\alpha|)$ is responsible for the link between sites 1 and 2. Thus $\hat{h}_{\alpha\alpha}^{0}$ is the one-particle Hamiltonian of the isolated site 1 and of the chain starting from site 2. In this way $|1\alpha\rangle$ is an eigenket of $\hat{h}_{\alpha\alpha}^{0}$ with eigenvalue ϵ_{α}. From the Dyson equation

$$\hat{g}_{\alpha\alpha}^{R}(\omega) = \frac{1}{\omega - \hat{h}_{\alpha\alpha}^{0} + i\eta} + \frac{1}{\omega - \hat{h}_{\alpha\alpha}^{0} + i\eta} \hat{T}\,\hat{g}_{\alpha\alpha}^{R}(\omega)$$

we have

$$\langle 1\alpha|\hat{g}_{\alpha\alpha}^{R}(\omega)|1\alpha\rangle = \frac{1}{\omega - \epsilon_{\alpha} + i\eta}\left[1 + T\,\langle 2\alpha|\hat{g}_{\alpha\alpha}^{R}(\omega)|1\alpha\rangle\right],$$

$$\langle 2\alpha|\hat{g}_{\alpha\alpha}^{R}(\omega)|1\alpha\rangle = \langle 2\alpha|\frac{1}{\omega - \hat{h}_{\alpha\alpha}^{0} + i\eta}|2\alpha\rangle\,T\,\langle 1\alpha|\hat{g}_{\alpha\alpha}^{R}(\omega)|1\alpha\rangle.$$

Solving this system for $\langle 1\alpha|\hat{g}_{\alpha\alpha}^{R}(\omega)|1\alpha\rangle$ we get

$$\langle 1\alpha|\hat{g}_{\alpha\alpha}^{R}(\omega)|1\alpha\rangle = \frac{1}{\omega - \epsilon_{\alpha} - T^2\langle 2\alpha|\frac{1}{\omega - \hat{h}_{\alpha\alpha}^{0} + i\eta}|2\alpha\rangle + i\eta}. \tag{16.25}$$

Now the crucial observation is that for an infinitely long reservoir $\langle 2\alpha|\frac{1}{\omega - \hat{h}_{\alpha\alpha}^{0} + i\eta}|2\alpha\rangle = \langle 1\alpha|\frac{1}{\omega - \hat{h}_{\alpha\alpha} + i\eta}|1\alpha\rangle = \langle 1\alpha|\hat{g}_{\alpha\alpha}^{R}(\omega)|1\alpha\rangle$ since both matrix elements are the terminal site Green's function of a semi-infinite chain with the same Hamiltonian. Then (16.25) becomes a simple quadratic equation, the two solutions of which yield two solutions for the embedding self-energy:

$$\sigma_{\alpha}^{R}(\omega) = \frac{T_{\alpha}^2}{2T^2}\left[(\omega - \epsilon_{\alpha} + i\eta) \pm \sqrt{(\omega - \epsilon_{\alpha} + i\eta)^2 - 4T^2}\right].$$

Which solution should we pick? To decide we observe that for $|\omega| \to \infty$ we have $g_{1\alpha,1\alpha}^{R}(\omega) \sim 1/\omega$ and hence $\sigma_{\alpha}^{R}(\omega) \sim T_{\alpha}^2/\omega$. Expanding the square root in powers of $4T^2/(\omega - \epsilon_{\alpha})^2$ it is easy to see that for $\omega \to +\infty$ we must take the solution with the minus sign whereas for $\omega \to -\infty$ we must take the solution with the plus sign. In this way we fix the sign for all frequencies such that $|\omega - \epsilon_{\alpha}| > |2T|$. For $|\omega - \epsilon_{\alpha}| < |2T|$ the argument of the square root becomes negative and the embedding self-energy acquires an imaginary part. By construction the sign of the imaginary part must be negative.[6] To summarize

$$\sigma_{\alpha}^{R}(\omega) = \frac{T_{\alpha}^2}{2T^2}\begin{cases} \omega - \epsilon_{\alpha} - \sqrt{(\omega - \epsilon_{\alpha})^2 - 4T^2} & \omega - \epsilon_{\alpha} > |2T| \\ \omega - \epsilon_{\alpha} - i\sqrt{4T^2 - (\omega - \epsilon_{\alpha})^2} & |\omega - \epsilon_{\alpha}| < |2T| \\ \omega - \epsilon_{\alpha} + \sqrt{(\omega - \epsilon_{\alpha})^2 - 4T^2} & \omega - \epsilon_{\alpha} < -|2T| \end{cases},$$

where we take the limit $\eta \to 0$ since $\sigma_{\alpha}^{R}(\omega)$ is nonsingular everywhere. The real and imaginary parts of $\sigma_{\alpha}^{R}(\omega)$ are shown in Fig. 16.4. All other Keldysh components of σ_{α} can be

---

[6]From the Cauchy relation we have $\text{Im}[\sigma_{\alpha}^{R}(\omega)] = -\pi \sum_{k} T_{k\alpha}^2 \delta(\omega - \epsilon_{k\alpha})$.

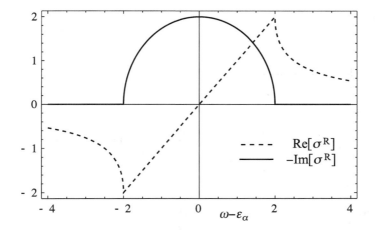

Figure 16.4 Real and imaginary parts of the embedding self-energy for a reservoir which is a semi-infinite chain. Frequencies are in units of T and σ_α^R is in units of $T_\alpha^2/(2T)$.

derived as we did in Section 16.2. With Σ_α at our disposal we can solve the Kadanoff–Baym equations for the open molecular region and in particular we can address the initial state dependence.

We start by considering a system prepared in a noninteracting state so that the many-body self-energy is zero when one or both time arguments are on the vertical track of the contour. At time $t = t_0 = 0$ we switch on the interaction and propagate until $t = t_1$, at which time we also switch on a sudden symmetric bias, i.e., $\epsilon_\alpha \to \epsilon_\alpha + V_\alpha(t)$, with $V_L(t) = -V_R(t) = V$. Since electron correlations are taken into account in the time propagation but not in the initial state, there is a charge redistribution for times $t > 0$. The result is compared to an initially correlated Kadanoff–Baym propagation. For the molecule we use the parameters $\epsilon_a = \epsilon_b = 0$, $T_M < 0$ and $v_{aa} = v_{bb} = 2$ and $v_{ab} = v_{ba} = 1$ (all energies are in units of $|T_M|$). The chemical potential μ is set at the middle of the Hartree–Fock gap and is determined by an equilibrium Hartree–Fock calculation of the uncontacted but interacting molecule which yields the value $\mu = 2$. For the reservoirs we use $\epsilon_\alpha = \mu$ and $T = -1.5$. The coupling between the molecule and the reservoirs is set to $T_L = T_R = 0.5$ (weak coupling). Finally we consider the low temperature regime and take $\beta = 90$. The Hamiltonian of the noninteracting and isolated molecule has eigenvalues $\epsilon_\pm = \pm 1$ which are both below the chemical potential. As a result of the sudden electron interaction for $t > 0$ a charge of about 2 electrons is pushed into the reservoirs. The corresponding current $I_L(t) = -dN_L/dt$ (the charge of the particles is $q = -1$) is shown in Fig. 16.5 for the second-Born approximation. $I_L(t)$ is saturated before the bias voltage $V = 1$ is switched on at time $t_1 = 20$ (times are in units of $1/|T_M|$). For later times $t > t_1$ the transient currents with inclusion and with neglect of initial correlations are indistinguishable. Therefore the initially uncorrelated system has *thermalized* to a correlated state when the bias is switched on. It is important to emphasize that this behavior is not general. In our case the interaction is confined in the molecular region and therefore represents a

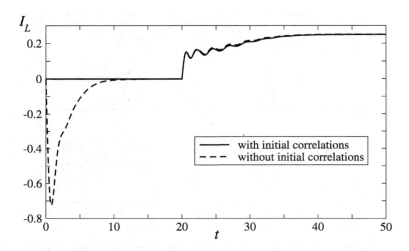

Figure 16.5 Time-dependent current $I_L(t)$ in the second-Born approximation with and without initial correlations, i.e., starting from an initial interacting and noninteracting state.

local perturbation in the language of Section 6.3.1. If the interaction were present also in the leads (global perturbation) then the steady-state current would depend on whether the system is initially interacting or not [159]. This effect is known as *thermalization breakdown* [160–163] and has recently attracted considerable attention due to the possibility of studying it experimentally [164].

To study the memory of the initial state we compare the full solution of the Kadanoff–Baym equations to those in which we neglect the terms $\Sigma_{\text{emb}}^{\rceil,\lceil}$ and/or $\Sigma^{\rceil,\lceil}$. However, at the initial time $t = t_0 = 0$ we still employ the fully correlated and embedded equilibrium Green's function. The results are displayed in Fig. 16.6 for the Hartree–Fock and second-Born approximations. We see that neglect of the memory terms has an important effect on the transient current. In the Hartree–Fock case these terms only contain the embedding self-energy (there is no correlation self-energy) and therefore here we can investigate only the memory of the initial contacting to the reservoirs. Neglect of this term leads to the curve labeled HF1 in the left panel of Fig. 16.6. For the second-Born case there is also a dependence on the many-body self-energy. We therefore have two curves, one in which we neglect only $\Sigma^{\rceil,\lceil}$ (labeled 2B1), and one in which we neglect both $\Sigma^{\rceil,\lceil}$ and $\Sigma_{\text{em}}^{\rceil,\lceil}$ (labeled 2B2). From the right panel of Fig. 16.6 we observe that the neglect of the embedding self-energy has a larger impact than the neglect of the correlation self-energy in the transient regime. We further see that the same steady-state current develops as with the memory terms included, indicating that these terms eventually die out in the long-time limit. This is in agreement with our expectation on the existence of *relaxation* in macroscopic and interacting systems, see Section 6.1.3.

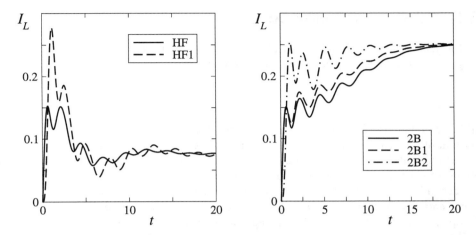

Figure 16.6 Time-dependent current $I_L(t)$ in the Hartree–Fock (left panel) and second-Born (right panel) approximation with and without the memory terms $\Sigma^{\rceil,\lceil}$ and $\Sigma_{\text{em}}^{\rceil,\lceil}$, see the main text.

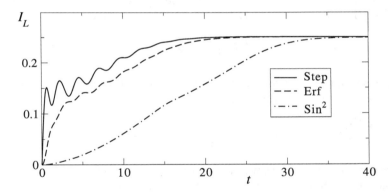

Figure 16.7 Time-dependent current $I_L(t)$ in the second-Born approximation for different switch-ons of the external bias.

We conclude this section by investigating the dependence of the transient currents on various forms of the time-dependent bias. In Fig. 16.7 we show the second-Born transient currents generated by different time-dependent biases $V_L(t) = -V_R(t) = V(t)$. We take $V(t) = \theta(t)V$, $V(t) = V\,\mathrm{Erf}(\omega_1 t)$, and $V(t) = V\sin^2(\omega_2 t)$ for $t \leq \pi/(2\omega_2)$ and $V(t) = V$ for $t > \pi/(2\omega_2)$, with $V = 1$, $\omega_1 = 0.5$, and $\omega_2 = 0.1$. We observe that rapid switch-ons produce large oscillations as compared to slower switch-ons. The steady-state currents, however, are the same for all three cases. The same is true for the equal time $G^<(t,t)$ of the molecular region. Thus, the steady-state values depend only on the steady-state Hamiltonian and not on its history.[7]

---

[7]For noninteracting electrons the proof of the history independence can be found in Ref. [54].

16.5   Charge conservation

The Hartree–Fock, second-Born and GW self-energies are all examples of Φ-derivable self-energies. From Chapter 9 we know that if the self-energy is Φ-derivable, i.e.,

$$\Sigma_{mn}(z,z') = \frac{\delta \Phi[G]}{\delta G_{nm}(z',z^+)},$$

and if the equations of motion for the Green's function are solved fully self-consistently for this form of the self-energy, then basic conservation laws are satisfied. For an open system, like our molecule, charge conservation does not imply that the time derivative of the total number of electrons $N_M(t)$ in the molecule is constant in time. It rather implies that the time derivative of $N_M(t)$, also known as the *displacement current*, is equal to the sum of the currents that flow into the leads. It is instructive to specialize the discussion of Chapter 9 to open systems and clarify again the importance of the Φ-derivability in this context.

The total number of electrons in the molecule is given by

$$N_M(t) = -2\mathrm{i}\,\mathrm{Tr}_M\left[G^<(t,t)\right],$$

where the factor of 2 comes from spin. Subtracting the equation of motion (16.6) from its adjoint and setting $z = t_-$, $z' = t_+$ we obtain the generalization of (16.10) to interacting systems. The only difference is that $\Sigma_{\mathrm{em}} \to \Sigma_{\mathrm{tot}} = \Sigma_{\mathrm{em}} + \Sigma$ and hence

$$\frac{dN_M(t)}{dt} = -4\mathrm{Re}\left\{\mathrm{Tr}_M\left[\int_\gamma d\bar{z}\,\Sigma_{\mathrm{tot}}(t_-,\bar{z})G(\bar{z},t_+)\right]\right\}.$$

Comparing this result with the formula (16.7) for the current (the charge $q = -1$ in our case) we can rewrite the displacement current as

$$\frac{dN_M(t)}{dt} = \sum_\alpha I_\alpha(t) - 4\mathrm{Re}\left\{\mathrm{Tr}_M\left[\int_\gamma d\bar{z}\,\Sigma(t_-,\bar{z})G(\bar{z},t_+)\right]\right\}. \tag{16.26}$$

Charge conservation, or equivalently the satisfaction of the continuity equation, implies that the integral in (16.26) vanishes. We know that this is a direct consequence of the invariance of the functional Φ under gauge transformations. Indeed, changing the external potential in the molecule by an arbitrary gauge function $\Lambda_m(z)$ [with boundary conditions $\Lambda_m(t_0) = \Lambda_m(t_0 - \mathrm{i}\beta)$] changes the Green's function according to [see (9.25)]

$$G_{mn}(z,z') \to e^{\mathrm{i}\Lambda_m(z)}G_{mn}(z,z')e^{-\mathrm{i}\Lambda_n(z')} \equiv G_{mn}[\Lambda](z,z'). \tag{16.27}$$

Since the Φ functional is a linear combination of vacuum diagrams and since the interaction is diagonal, i.e., $v_{ijmn} = \delta_{in}\delta_{jm}v_{mn}$, Φ does not change under a gauge transformation and we find

$$0 = \frac{\delta \Phi}{\delta \Lambda_q(z)} = \sum_{mn}\int d\bar{z}d\bar{z}'\,\frac{\delta \Phi}{\delta G_{mn}[\Lambda](\bar{z}',\bar{z}^+)}\frac{\delta G_{mn}[\Lambda](\bar{z}',\bar{z}^+)}{\delta \Lambda_q(z)}$$

$$= \sum_{mn}\int d\bar{z}d\bar{z}'\,\Sigma_{nm}[\Lambda](\bar{z},\bar{z}')\frac{\delta G_{mn}[\Lambda](\bar{z}',\bar{z}^+)}{\delta \Lambda_q(z)}. \tag{16.28}$$

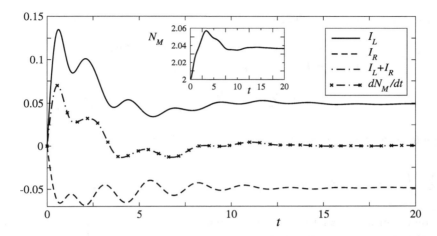

Figure 16.8 Time-dependent left and right current and their sum. The inset shows the plot of $N_M(t)$, the time derivative of which is in excellent agreement with $I_L(t) + I_R(t)$.

Here we explicitly use the Φ-derivability condition of the self-energy. If we now insert the derivative of the Green's function with respect to Λ_q, sum over q, and evaluate the resulting expression in $z = t_{\pm}$ and $\Lambda = 0$ we obtain the integral in (16.26). Therefore the last term in (16.26) vanishes and the time derivative of the number of particles $N_M(t)$ in the molecule is equal to the sum of the currents that flow into the leads.

Conserving approximations guarantee that important physical requirements, like the conservation of charge, are automatically built into the MBPT framework. This is nicely illustrated in Fig. 16.8: here we consider the same system as in the previous section and drive the system out of equilibrium by an asymmetric sudden bias $V_L = 0.9$ and $V_R = -0.4$. The plot shows that the left/right currents and the time derivative of the number of electrons $N_M(t)$ satisfy the continuity equation [165]. We should mention that in the long-time limit the number of particles in the molecule is constant provided that the system attains a steady state. In this case $\sum_\alpha I_\alpha = 0$, i.e., the current that flows in the molecule equals the current that flows out of the molecule. The importance of using conserving approximations in steady-state transport calculations (see end of Section 9.7) has been carefully addressed in Ref. [166].

Exercise 16.5. For a nondiagonal interaction v_{ijmn} the Φ functional is invariant under a gauge transformation provided that G changes according to (16.27) *and* the interaction changes according to

$$v_{ijmn}(z, z') \to v_{ijmn}(z, z')\, e^{i(\Lambda_i(z) + \Lambda_j(z') - \Lambda_m(z') - \Lambda_n(z))} \equiv v_{ijmn}[\Lambda](z, z').$$

In this case (16.28) must be modified as

$$0 = \frac{\delta \Phi}{\delta \Lambda_q(z)} = \sum_{mn} \int d\bar{z} d\bar{z}' \frac{\delta \Phi}{\delta G_{mn}[\Lambda](\bar{z}', \bar{z}^+)} \frac{\delta G_{mn}[\Lambda](\bar{z}', \bar{z}^+)}{\delta \Lambda_q(z)}$$

$$+ \sum_{ijmn} \int d\bar{z} d\bar{z}' \frac{\delta \Phi}{\delta v_{ijmn}[\Lambda](\bar{z}', \bar{z})} \frac{\delta v_{ijmn}[\Lambda](\bar{z}', \bar{z})}{\delta \Lambda_q(z)}.$$

Taking into account that $v_{ijmn}(z, z') = v_{ijmn}\delta(z, z')$ is local in time, show that

$$\mathrm{Re}\left[\sum_n \int_\gamma d\bar{z} \, \Sigma_{qn}(t_-, \bar{z}) G_{nq}(\bar{z}, t_+) \right] + \sum_{ijmn} \frac{\delta \Phi}{\delta v_{ijmn}} v_{ijmn} \left[\delta_{iq} + \delta_{jq} - \delta_{mq} - \delta_{nq} \right] = 0.$$

The second term vanishes for diagonal interactions $v_{ijmn} = \delta_{in}\delta_{jm}v_{mn}$, as expected. In the general case of a four-index interaction the second term does not vanish and the "continuity equation" for the average of the level occupation contains a term proportional to v. However, the equation for the total number of particles does not change since the sum over q of the second term vanishes.

16.6 Time-dependent GW approximation in open systems

In this section we analyze a more complex system consisting of two-dimensional reservoirs and a larger molecular region. We present results obtained within the Hartree–Fock, second-Born as well as the GW approximation. The molecular region is modeled by a chain of four localized basis functions 1, 2, 3, 4 coupled through 1 to a left reservoir and through 4 to a right reservoir as illustrated below:

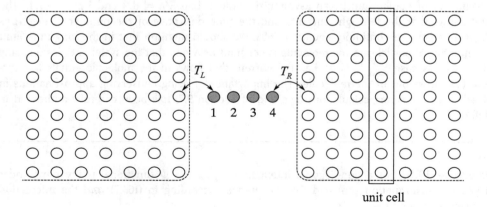

unit cell

The reservoirs are two-dimensional crystals with nine transverse channels. We can think of them as one-dimensional crystals with a unit cell consisting of nine localized basis functions.

The matrix elements of the Hamiltonian $h_{\alpha\alpha}$ for reservoir $\alpha = L, R$ are nonvanishing only for nearest neighbor basis functions and along the diagonal. We choose the off-diagonal elements of $h_{\alpha\alpha}$ to be all equal to T whereas the diagonal elements ϵ_α are set equal to the chemical potential. The reservoirs are therefore half-filled. The end-points of the molecular chain are coupled only to the terminal basis function of the central row of the reservoirs with a link $T_L = T_R$. Finally the 4×4 Hamiltonian in the basis 1, 2, 3, 4 is modeled as

$$h = \begin{pmatrix} 0 & T_M & 0 & 0 \\ T_M & 0 & T_M & 0 \\ 0 & T_M & 0 & T_M \\ 0 & 0 & T_M & 0 \end{pmatrix},$$

with $T_M < 0$. As for the interaction Hamiltonian in the molecular region, we use

$$\hat{H}_{\text{int}} = \frac{1}{2} \sum_{\substack{mn \\ \sigma\sigma'}} v_{mn}\, \hat{d}^\dagger_{m\sigma} \hat{d}^\dagger_{n\sigma'} \hat{d}_{n\sigma'} \hat{d}_{m\sigma},$$

where the sum over m, n goes from 1 to 4 and the Coulomb integrals have the form

$$v_{mn} = \begin{cases} v & m = n \\ \dfrac{v}{2|m-n|} & m \neq n \end{cases}.$$

The choice $v_{ijmn} = \delta_{in}\delta_{jm}v_{mn}$ for the Coulomb integrals is appropriate for molecular-like systems weakly coupled to leads, as it is commonly used in the study of isolated molecules based on the Pariser–Parr–Pople model, see Section 2.2. This type of interaction appears naturally from a calculation of the Coulomb integrals v_{ijmn} in localized basis states. The use of the full set of Coulomb integrals v_{ijmn} is not conceptually more complicated but the computational effort grows (see Ref. [98] for a first step in this direction).

Measuring all energies in units of $|T_M|$ we choose $T = -2$, $T_L = T_R = -0.5$ and $v = 1.5$. For these parameters the equilibrium Hartree–Fock levels of the isolated molecule lie at $\epsilon_1 = 0.39$, $\epsilon_2 = 1.32$, $\epsilon_3 = 3.19$, $\epsilon_4 = 4.46$. In the simulations below, the chemical potential is fixed between the HOMO ϵ_2 and the LUMO ϵ_3 levels at $\mu = 2.26$ and the inverse temperature is set to $\beta = 90$. This effectively corresponds to the zero temperature limit since the results do not change for higher values of β. We start from the interacting system initially in thermal equilibrium, and then we suddenly apply a constant bias at an initial time $t_0 = 0$, i.e., $\epsilon_\alpha \to \epsilon_\alpha + V_\alpha$ with $V_L = -V_R = V$ independent of time.

16.6.1 Keldysh Green's functions in the double-time plane

The calculation of any physical quantity requires the calculation of all Keldysh components of G. Due to their importance we wish to present the behavior of the lesser Green's function $G^<$ as well as of the right Green's function $G^\rceil$ in the double-time plane for the Hartree–Fock approximation. The Green's functions corresponding to the second-Born and GW approximations are qualitatively similar but show more strongly damped oscillations.

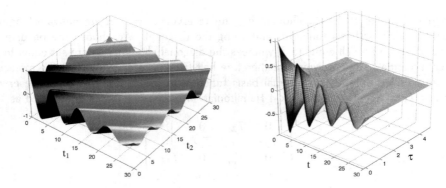

Figure 16.9 The imaginary part of the lesser Green's function $G^<_{HH}(t_1, t_2)$ (left), and Green's function $G^\rceil_{HH}(t, \tau)$ (right), of the molecular region projected onto the HOMO level. Bias voltage $V = 1.2$, Hartree–Fock approximation.

In the left panel of Fig. 16.9 we display the imaginary part of $G^<_{HH}(t, t')$ in the basis of the initial Hartree–Fock molecular orbitals, for an applied bias $V = 1.2$. This matrix element corresponds to the HOMO level of the molecular chain. The value of the Green's function on the time diagonal, i.e., $n_H(t) = -iG^<_{HH}(t, t)$ gives the level occupation per spin. We see that $n_H(t)$ decays from a value of 1.0 at the initial time to a value of 0.5 at time $t = 30$ (times are measured in units of $1/|T_M|$). An analysis of the LUMO level occupation $n_L(t)$ shows that almost all the charge is transferred to this level. When we move away from the time-diagonal we consider the time-propagation of holes in the HOMO level. We observe here a damped oscillation whose frequency corresponds to the removal energy of an electron from the HOMO level. As we see in Section 16.6.2, this oscillation leads to a distinct peak in the spectral function. The damping of the oscillation instead indicates the occurrence of relaxation.

The imaginary part of $G^\rceil_{HH}(t, \tau)$ within the Hartree–Fock approximation is displayed in the right panel of Fig. 16.9 for real times between $t = 0$ and $t = 30$ and imaginary times from $\tau = 0$ to $\tau = 5$. The right Green's function accounts for initial correlations as well as initial embedding effects (within the Hartree–Fock approximation only the latter). At $t = 0$ we have the ground-state Matsubara Green's function, and as the real time t increases all elements of $G^\rceil(t, \tau)$ approach zero independently of the value of τ. Thus, initial correlation effects and the initial-state dependence die out in the long-time limit in agreement with the results of Section 16.4. A very similar behavior is found within the second-Born and GW approximation but with a stronger damping of the oscillations. The Green's function $G^\rceil(t, \tau)$ is antiperiodic in τ with period β. However, since in our case $\beta = 90$ the antiperiodicity cannot be observed from Fig. 16.9.

16.6.2   Time-dependent current and spectral function

The time-dependent current at the left interface between the chain and the two-dimensional reservoir is shown in the top panels of Fig. 16.10 for the Hartree–Fock (HF), second-Born

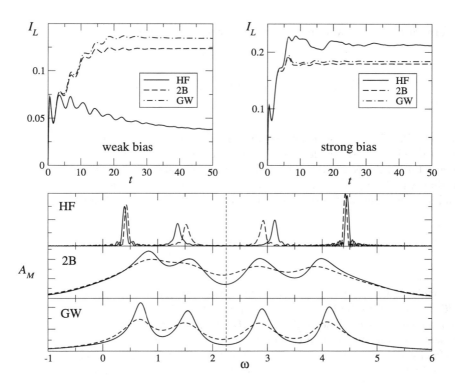

Figure 16.10 Top panels: Time-dependent current $I_L(t)$ in the Hartree–Fock (HF), second-Born (2B) and GW approximations with the applied bias $V = 0.8$ (left) and $V = 1.2$ (right). Bottom panel: Spectral functions $A_M(\omega)$ for the HF (top), 2B (middle) and GW (bottom) approximations with the applied bias $V = 0.8$ (solid line) and $V = 1.2$ (dashed line). The spectral function is in arbitrary units. A vertical dashed line that intersects the ω-axis at the chemical potential $\mu = 2.26$ is also shown.

(2B) and GW approximations and for two different values of the applied bias $V = 0.8$ (weak bias) and 1.2 (strong bias). The first remarkable feature is that the 2B and GW results are in excellent agreement *at all times* both in the weak and strong bias regime while the HF current already deviates from the correlated results after a few time units. This result indicates that a chain of four atoms is already long enough for screening effects to play a crucial role. The 2B and GW approximations have in common the first bubble diagram (besides the Hartree–Fock diagrams). We thus conclude that the first-order exchange diagram (Fock) with an interaction screened by a single polarization bubble (with fully dressed Green's functions) already captures the essential physics of the problem. The 2B approximation also includes the second-order exchange diagram. Even though this diagram contains only two interaction lines (like the first bubble diagram), we find numerically that it is less relevant.

The steady-state value of the current in the various approximations can be nicely related to the spectral function. We define the time-dependent spectral function as the Fourier transform of $G^> - G^<$ with respect to the relative time coordinate, i.e.,

$$A(T, \omega) = \mathrm{i} \int \frac{dt}{2\pi} \, e^{\mathrm{i}\omega t} [G^> - G^<] (T + \frac{t}{2}, T - \frac{t}{2}). \qquad (16.29)$$

For values of T after the transients have died out the spectral function becomes independent of T. For these times we denote the spectral function by $A(\omega)$. This function is the generalization of the equilibrium spectral function to nonequilibrium steady-state situations. By analogy with its equilibrium counterpart, $A(\omega)$ displays peaks that correspond to removal and addition energies of the system at the steady-state. The spectral function

$$A_M(\omega) = \mathrm{Tr}_M [A(\omega)]$$

of the molecular system is displayed in the bottom panels of Fig. 16.10. At weak bias the HOMO–LUMO gap in the HF approximation is fairly similar to the equilibrium gap, whereas the 2B and GW gaps collapse, and the HOMO and LUMO levels move in the bias window. As a consequence the steady-state HF current is notably smaller than the 2B and GW currents. This effect was first observed in Ref. [167] by performing a steady-state calculation according to the scheme outlined at the end of Section 9.7. The explicit solution of the Kadanoff–Baym equations confirms that the assumptions implicit in the steady-state scheme, i.e., the occurence of relaxation and the washing out of initial correlations, are satisfied.

The physics changes considerably in the strong bias regime. The HF HOMO and LUMO levels move into the bias window and lift the steady-state current above the corresponding 2B and GW values. This can be explained by observing that the peaks of the HF spectral function are very sharp compared to the rather broadened structures in the 2B and GW approximations, see again Fig. 16.10. In the correlated case the HOMO and LUMO levels are only partially available to electrons with energy below $\mu + V$ and we thus observe a suppression of the current with respect to the HF case. From a mathematical point of view, the steady-state current is roughly proportional to the integral of $A_M(\omega)$ over the bias window, see (16.23), and this integral is larger in the HF approximation.

The time-evolution of the spectral function $A_M(T, \omega)$ as a function of T is illustrated in Fig. 16.11 for the case of the HF and the 2B approximation. For these results, the ground state system was propagated without bias up to a time $T = 40$, after which a bias was suddenly switched on. The HF peaks remain rather sharp during the entire evolution and the HOMO–LUMO levels near each other linearly in time until a critical distance is reached. On the contrary, the broadening of the 2B peaks remains small during the initial transient regime (up to $T = 70$) but then increases dramatically. This behavior indicates that there is a critical charging time after which an enhanced renormalization of the quasi-particle states takes place, causing a substantial reshaping of the spectral function with respect to its equilibrium value.

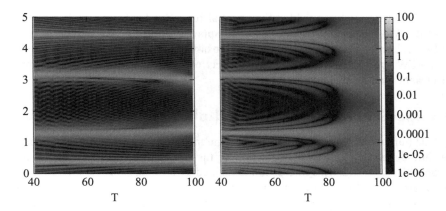

Figure 16.11 Contour plot of the real-time evolution of the molecular spectral function $A_M(T,\omega)$ for the HF (left panel) and the 2B (right panel) approximation for an applied bias of $V = 1.2$. On the horizontal axis is plotted the time T, and on the vertical axis the frequency ω.

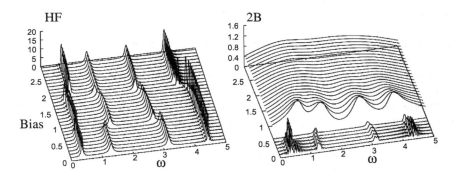

Figure 16.12 Steady-state spectral function $A_M(\omega)$ for the HF (left panel), and 2B (right panel) approximation, as a function of the bias voltage. For the 2B approximation the spectral functions corresponding to $V < 0.6$ were divided by a factor of 30.

In Fig. 16.12 we also show the steady-state spectral function $A_M(\omega)$ in the HF and 2B approximation for different applied bias. In the HF case the width of the peaks is fairly constant with increasing V. This is an expected result since the only source of broadening comes from the embedding self-energy. What is instead more interesting is that there exist values of the bias at which the positions of the HF peaks suddenly get shifted. Even more striking, however, is the behavior of $A_M(\omega)$ in the 2B approximation. When the bias voltage $V \sim 0.7$ we observe a substantial broadening of the spectral peaks. To see this effect clearly, in Fig. 16.12 we have divided the spectral functions for biases up to $V = 0.6$ by a factor of 30. We further notice that in the 2B approximation the gap closing as a function of the bias voltage is faster than in the HF approximation.[8] Very similar results are obtained within the GW approximation. In conclusion, electronic correlations have a major impact on both transient and steady-state currents.

16.6.3   Screened interaction and physical interpretation

In Fig. 16.13 (top panel) we show the trace over the molecular region of the lesser component of the time-dependent screened interaction of the GW approximation in the double-time plane for a bias $V = 1.2$. This interaction is defined as $W = v + v\,P\,W$ where $P = -i\,GG$ is the polarization bubble (with self-consistent GW Green's functions) of the connected and correlated system. Since W has the same symmetry properties as the response function χ, we have for the diagonal elements

$$\mathrm{Re}[W_{ii}^{\lessgtr}(t,t')] = -\mathrm{Re}[W_{ii}^{\lessgtr}(t',t)]; \qquad \mathrm{Im}[W_{ii}^{\lessgtr}(t,t')] = \mathrm{Im}[W_{ii}^{\lessgtr}(t',t)],$$

i.e., the real part is antisymmetric while the imaginary part is symmetric under the interchange $t \leftrightarrow t'$. Since $W_{ii}^{>}(t,t') = -[W_{ii}^{<}(t,t')]^*$ we do not plot the greater component of W. The good agreement between the 2B and GW approximations suggests that the dominant contribution to the screening comes from the first bubble diagram, that is $W^{<} \sim vP^{<}v$. In the bottom panel of Fig. 16.13 we show the trace over the molecular region of the retarded component of $\delta W = W - v$, which is a real function like the retarded component of χ. The retarded interaction between the electrons is mainly attractive. In particular $\sum_i \delta W_{ii}^{\mathrm{R}}(t, t + \Delta_t) \sim -3$ for $\Delta_t \sim 0.1$. How can we interpret this result?

For a physical interpretation of $\delta W^{\mathrm{R}}(t, t')$ let us consider again the electron gas with interparticle interaction $v(\mathbf{r}_1, \mathbf{r}_2) = q^2/|\mathbf{r}_1 - \mathbf{r}_2|$ where $q = -1$ is the charge of the electrons. The system is in equilibrium when at time $t = t_0$ we suddenly put and remove a particle of charge q in $\mathbf{r} = \mathbf{r}_0$. This particle generates an external potential

$$V(\mathbf{r}, t) = \frac{q}{|\mathbf{r} - \mathbf{r}_0|}\delta(t - t_0) = \frac{1}{q}v(\mathbf{r}, \mathbf{r}_0)\delta(t - t_0). \tag{16.30}$$

This perturbation is different from the one considered in Section 15.5.3 where we put a particle in the system but we did not remove it. The linear-response density change induced by (16.30) is

$$\delta n(\mathbf{r}, t) = q \int_{-\infty}^{\infty} dt' \int d\mathbf{r}' \chi^{\mathrm{R}}(\mathbf{r}, t; \mathbf{r}', t') V(\mathbf{r}', t') = \int d\mathbf{r}' \chi^{\mathrm{R}}(\mathbf{r}, t; \mathbf{r}', t_0) v(\mathbf{r}', \mathbf{r}_0),$$

where $\chi^{\mathrm{R}}(\mathbf{r}, t; \mathbf{r}', t') = \sum_{\sigma\sigma'} \chi^{\mathrm{R}}(\mathbf{x}, t; \mathbf{x}', t')$ is the density response function summed over spin and $\delta n(\mathbf{r}, t) = \sum_{\sigma} \delta n(\mathbf{x}, t)$ is the total density change. This density change generates

---

[8] This gap closing has nothing to do with the gap closing discussed in Section 13.3.1. There the HOMO–LUMO gap was studied as a function of the interaction strength, whereas here it is studied as a function of bias. Furthermore, here the only interaction is between electrons in the molecule and hence there is no interaction between an electron in the molecule and an electron in the reservoirs.

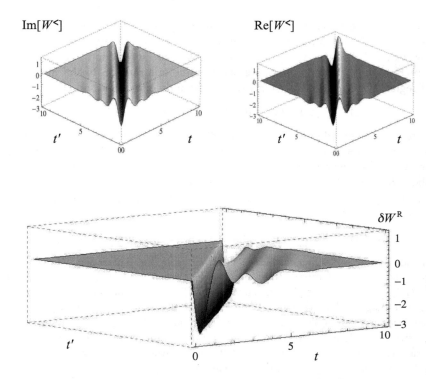

Figure 16.13 Imaginary part and real part of the trace of $W^<(t,t')$ (top panel) and $W^R(t,t')$ in the GW approximation.

a classical (or Hartree) potential given by

$$V_H(\mathbf{r},t) = \frac{1}{q} \int d\mathbf{r}'\, v(\mathbf{r},\mathbf{r}')\delta n(\mathbf{r}',t)$$

$$= \frac{1}{q} \int d\mathbf{r}' d\mathbf{r}''\, v(\mathbf{r},\mathbf{r}')\chi^R(\mathbf{r}',t;\mathbf{r}'',t_0)v(\mathbf{r}'',\mathbf{r}_0)$$

$$= \frac{1}{q} \delta W^R(\mathbf{r},t;\mathbf{r}_0,t_0),$$

where in the last equality we use the fact that $\delta W = vPW = v\chi v$. Therefore $\delta W^R(\mathbf{r},t; \mathbf{r}_0,t_0) = qV_H(\mathbf{r},t)$ can be seen as the classical interaction energy between a point-like charge q in $\mathbf{r}$ and a charge distribution $q\delta n(\mathbf{r}',t)$ induced by the sudden switch-on/switch-off of a charge q in $\mathbf{r}_0$ at time t_0. This equivalence allows us to draw a few conclusions. At time $t = t_0^+$ the system has no time to respond to the external perturbation and hence $\delta n(\mathbf{r}',t_0^+) = 0$, which implies

$$\delta W^R(\mathbf{r},t_0^+;\mathbf{r}_0,t_0) = 0. \tag{16.31}$$

In Section 15.5.3 we have seen that the response time of an electron gas is roughly proportional to the inverse of the plasma frequency ω_p. We therefore expect that at times

$t \sim 1/\omega_p$ the system has had time to respond by pushing the electrons away from $\mathbf{r}_0$. The density change $\delta n(\mathbf{r}', t)$ is negative for $\mathbf{r}' \sim \mathbf{r}_0$ and positive further away in such a way that $\int d\mathbf{r}' \delta n(\mathbf{r}', t) = 0$. If we calculate the interaction energy $qV_H(\mathbf{r}, t) = q^2 \int d\mathbf{r}' \delta n(\mathbf{r}', t)/|\mathbf{r} - \mathbf{r}'|$ for $\mathbf{r} \sim \mathbf{r}_0$ we see that the contribution to the integral coming from the region $\mathbf{r}' \sim \mathbf{r}_0$ (where $\delta n(\mathbf{r}', t) < 0$) is the most important and hence

$$\delta W^R(\mathbf{r}, t; \mathbf{r}_0, t_0) < 0 \qquad \text{for } \mathbf{r} \sim \mathbf{r}_0 \text{ and } t \sim 1/\omega_p. \tag{16.32}$$

Thus the retarded interaction is attractive at sufficiently short distances for times $t \sim 1/\omega_p$. Finally we consider the limit of large times. In this case we expect that in the neighborhood of $\mathbf{r}_0$ the density has relaxed back to its equilibrium value, i.e., $\delta n(\mathbf{r}', t) \sim 0$ for $|\mathbf{r}' - \mathbf{r}_0| < R(t)$, where $R(t)$ is the extension of the relaxed region. $R(t)$ increases with time and the interaction energy $qV_H(\mathbf{r}, t) = q^2 \int d\mathbf{r}' \delta n(\mathbf{r}', t)/|\mathbf{r} - \mathbf{r}'|$ approaches zero for any $|\mathbf{r} - \mathbf{r}_0|$ much smaller than $R(t)$:

$$\delta W^R(\mathbf{r}, t; \mathbf{r}_0, t_0) \sim 0 \qquad \text{for } |\mathbf{r} - \mathbf{r}_0| \ll R(t) \text{ and } t \gg 1/\omega_p. \tag{16.33}$$

Even though our molecular system is different from the electron gas we see that the above conclusions have a quite general character. From Fig. 16.13 (bottom panel) we have $\text{Tr}[\delta W^R(t, t')] = 0$ along the time-diagonal $t = t'$, in agreement with (16.31). Moving away from the time-diagonal $\text{Tr}[\delta W^R(t, t')]$ goes negative very fast, in agreement with (16.32). It should be said that our molecular region is too small to develop well-defined plasmonic-like collective excitations. Nevertheless the response time is very short as compared to the typical relaxation time scale, see Fig. 16.9. The relaxation time is mainly related to the hybridization and goes as $1/\Gamma$ whereas the response time is mainly related to the interaction between the particles. Finally we see that for large $|t - t'|$ the on-site W_{ii}^R approaches zero in agreement with (16.33).

16.7 Inbedding technique: how to explore the reservoirs

So far we have discussed the molecular Green's function and physical quantities of the molecule. Is it possible from a knowledge of the molecular Green's function to calculate physical quantities of the reservoirs? The answer is affirmative and we now show how to do it. The Green's function $G_{\alpha\alpha}$ projected onto reservoir α satisfies the equation of motion

$$\left[i\frac{d}{dz} - h_{\alpha\alpha}(z) \right] G_{\alpha\alpha}(z, z') = \delta(z, z') + h_{\alpha M}(z) G_{M\alpha}(z, z'),$$

where we take into account that there is no interaction in the reservoirs. Inserting into this equation the adjoint of (16.4) we obtain the following integro-differential equation:

$$\left[i\frac{d}{dz} - h_{\alpha\alpha}(z) \right] G_{\alpha\alpha}(z, z') = \delta(z, z') + \int_\gamma d\bar{z} \, \Sigma_{\text{in},\alpha}(z, \bar{z}) g_{\alpha\alpha}(\bar{z}, z'),$$

where we define the *inbedding self-energy* as

$$\Sigma_{\text{in},\alpha}(z, z') = h_{\alpha M}(z) G_{MM}(z, z') h_{M\alpha}(z').$$

The inbedding self-energy is completely known once we have solved the Kadanoff–Baym equations for the open molecular system, i.e., once we know G_{MM}. The equation for $G_{\alpha\alpha}$ can be integrated using the Green's function $g_{\alpha\alpha}$ and the result is

$$G_{\alpha\alpha}(z, z') = g_{\alpha\alpha}(z, z') + \int d\bar{z} d\bar{z}' \, g_{\alpha\alpha}(z, \bar{z}) \Sigma_{\text{in},\alpha}(\bar{z}, \bar{z}') g_{\alpha\alpha}(\bar{z}', z'). \qquad (16.34)$$

From this equation we can obtain physical quantities like density, current, energy, etc. in the reservoirs.[9] Below we use the inbedding technique to calculate the time-dependent occupation $n_{i\alpha}(t) = -i[G^<_{\alpha\alpha}(t, t)]_{ii}$ of the basis function located on site i of the crystal. This study is of special importance since it challenges an assumption implicitly made when we approximated the reservoirs as noninteracting and when we modeled the bias as a uniform constant shift, namely the assumption that the reservoirs remain in thermodynamic equilibrium during the entire evolution.

In Fig. 16.14 we show the evolution of the density in the two-dimensional nine rows wide crystal after the sudden switch-on of a bias voltage. We display snapshots of the crystal densities up to ten layers deep inside the crystal for times $t = 0.0$, 1.7, 3.6, and 10.0. In order to improve the visibility we have interpolated the density between the sites. The molecular chain is connected to the central terminal site and acts as an impurity, similarly to an external electric charge in the electron gas. We observe density oscillations with a cross-shaped pattern; they are the crystal analog of the continuum Friedel oscillations in the electron gas, see Section 15.5.3. We can understand the cross-shaped pattern by considering the linear-response density change caused by an external impurity in a truly two-dimensional crystal. According to the Bloch theorem of Section 2.3.1 the one-particle eigenstates are Bloch waves $e^{i\mathbf{k}\cdot\mathbf{n}}$ and their eigenenergy is $\epsilon_{\mathbf{k}} = \mu + 2T(\cos k_x + \cos k_y)$ where k_x and k_y vary between $-\pi$ and π. Let $\delta n(\mathbf{n}, \omega)$ be the linear-response density change (per spin) in unit cell $\mathbf{n}$ caused by the sudden switch-on of a potential $V_{\text{imp}}(\mathbf{n}, \omega)$ at time t_0. We have

$$\delta n(\mathbf{n}, \omega) = -\sum_{\mathbf{n}'} \chi^{\text{R}}(\mathbf{n}, \mathbf{n}'; \omega) V_{\text{imp}}(\mathbf{n}', \omega),$$

where the minus sign comes from the electron charge $q = -1$. Fourier transforming the response function as

$$\chi^{\text{R}}(\mathbf{n}, \mathbf{n}'; \omega) = \int_{-\pi}^{\pi} \frac{d\mathbf{q}}{(2\pi)^2} \, e^{i\mathbf{q}\cdot(\mathbf{n}-\mathbf{n}')} \chi^{\text{R}}(\mathbf{q}, \omega)$$

we find

$$\delta n(\mathbf{n}, \omega) = -\int_{-\pi}^{\pi} \frac{d\mathbf{q}}{(2\pi)^2} \, e^{i\mathbf{q}\cdot\mathbf{n}} \, \chi^{\text{R}}(\mathbf{q}, \omega) \tilde{V}_{\text{imp}}(\mathbf{q}, \omega),$$

---

[9]Since the integral in (16.34) is a contour integral, in order to extract, say, $G^<_{\alpha\alpha}$ we must convert using the Langreth rules.

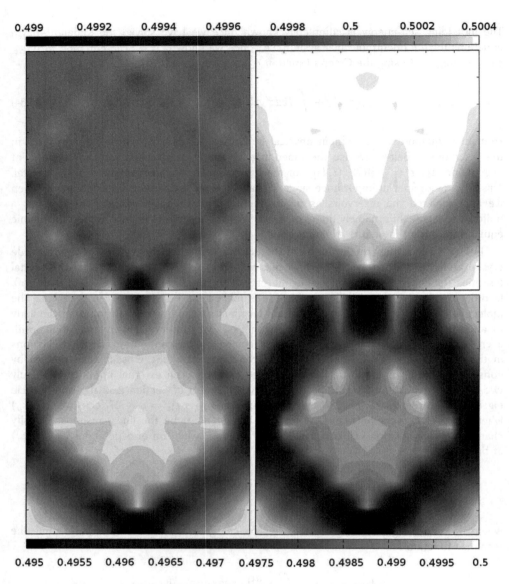

Figure 16.14 Snapshots of the HF density in the left reservoir after a bias $V = 1.2$ is suddenly switched-on. On the horizontal axes is shown the transverse dimension of the crystal (nine rows wide, with the site connected to the chain in the center) and ten layers deep. Upper left panel: initial density. Upper right panel: density at time $t = 1.7$. Lower left panel: density at time $t = 3.6$. Lower right panel: density at time $t = 10$. The upper grey-scale-bar refers to the initial density in the upper left panel. The lower grey-scale-bar refers to the other panels.

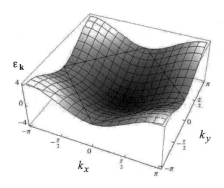
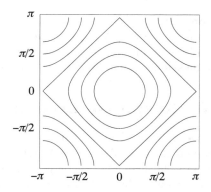

Figure 16.15 Left panel: Energy dispersion $\epsilon_\mathbf{k} = 2T(\cos k_x + \cos k_y)$ with k_x and k_y between $-\pi$ and π in units of $|T|$ for negative T. Right panel: Contour plot of the energy dispersion.

where $\tilde{V}_{\mathrm{imp}}(\mathbf{q}, \omega)$ is the Fourier transform of the impurity potential. The noninteracting response function for our two-dimensional crystal reads [see (15.66)]

$$\chi(\mathbf{q}, \omega) = \int_{-\pi}^{\pi} \frac{d\mathbf{k}}{(2\pi)^2} f(\epsilon_\mathbf{k} - \mu) \left[\frac{1}{\omega - (\epsilon_{\mathbf{k}+\mathbf{q}} - \epsilon_\mathbf{k}) + i\eta} - \frac{1}{\omega - (\epsilon_\mathbf{k} - \epsilon_{\mathbf{k}-\mathbf{q}}) + i\eta} \right].$$

We are interested in the value of χ at zero frequency since, in the long-time limit, the density change $\delta n(\mathbf{n}, t)$ approaches a constant value. In Fig. 16.15 we show the energy dispersion $\epsilon_\mathbf{k} - \mu$ (left panel) as well as its contour plot (right panel). At zero temperature the Fermi surface is a square with vertices in $(0, \pm\pi)$ and $(\pm\pi, 0)$. Then, the dominant contribution to the above integral comes from the values of $\mathbf{k}$ close to these vertices where the single-particle density of states diverges.[10] For zero frequency and $\mathbf{k}$ close to the vertices of the Fermi surface the response function is large when $\mathbf{q}$ is approximately equal to one of the *nesting vectors* $\mathbf{Q} = (\pi, \pi)$, $(\pi, -\pi)$, $(-\pi, \pi)$ and $(-\pi, -\pi)$. The name nesting vectors comes from the fact that by adding to an occupied state $\mathbf{k}$ a nesting vector we get an unoccupied state $\mathbf{k} + \mathbf{Q}$ and vice versa. Now we see that for $\mathbf{k}$ close to a vertex $\mathbf{k} + \mathbf{Q}$ is also close to a vertex and therefore $\epsilon_\mathbf{k} \sim \epsilon_{\mathbf{k}+\mathbf{Q}}$ and $\chi^{\mathrm{R}}(\mathbf{Q}, 0)$ is large. Each of these nesting vectors contributes to the density change $\delta n(\mathbf{n}, \omega = 0)$ with a function $\sim e^{i\mathbf{Q}\cdot\mathbf{n}}$. Since at zero frequency the product $\chi^{\mathrm{R}} \tilde{V}_{\mathrm{imp}}$ is real then $\delta n(\mathbf{n}, \omega = 0) \propto \cos \mathbf{Q}\cdot\mathbf{n} = \cos[\pi(n_x \pm n_y)]$. Therefore a single impurity in a two-dimensional half-filled crystal induces a cross-shaped density pattern. Due to the fact that in our case the crystal is semi-infinite, we only observe two arms of this cross.

The results of Fig. 16.14 also allow us to test the assumption of thermal equilibrium in the leads. The equilibrium density (per spin), see top-left panel, is essentially the same as its equilibrium bulk value at half-filling, i.e., 0.5. After the switching of the bias a density corrugation with the shape of a cross starts to propagate deep into the reservoir. The largest deviation from the bulk density value occurs at the corners of the cross and is about 2% at the junction while it reduces to about 1% after ten layers. It can be numerically verified that this deviation is about 3 times larger for reservoirs with only three transverse channels.

---

[10]It is easy to check that $\partial\epsilon_\mathbf{k}/\partial k_x = \partial\epsilon_\mathbf{k}/\partial k_y = 0$ for $\mathbf{k}$ at one of the vertices of the Fermi surface.

We conclude that the density deviation is inversely proportional to the cross-section of the reservoir. These results suggest that for a true mean field description of two-dimensional reservoirs with nine transverse channels it is enough to include few atomic layers for an accurate self-consistent time-dependent calculation of the Hartree potential.

16.8   Response functions from time-propagation

We have seen in Chapter 15 that the spectral properties of neutral (particle conserving) excitations can be read out from the two-particle XC function L or, equivalently, the response function χ. In Section 15.3 we derived the important result that the change δG of the Green's function induced by a change in the external potential can be obtained from the two-particle XC function which satisfies the Bethe–Salpeter equation with kernel $K = \pm \delta\Sigma/\delta G$, where $\Sigma = \Sigma[G]$ is the self-energy that determines G through $G = G_0 + G_0 \Sigma G$. The direct implementation of the Bethe–Salpeter equation has proven to be computationally challenging and, in practice, often requires a number of additional approximations, such as neglect of self-consistency, kernel diagrams and/or frequency dependence. Instead, obtaining the response function by time propagation of the Green's function does not require any of the aforementioned approximations [97–99]. Moreover, the resulting response function automatically satisfies the f-sum rule.

Suppose that we are interested in calculating the change $\delta n_s(t) = \delta n_{s\uparrow}(t) + \delta n_{s\downarrow}(t) = -2i\,\delta G_{ss}^<(t,t)$ in the occupation of orbital s due to a change $T_{ss} \to T_{ss} + \delta\epsilon_s(t)$ of a diagonal element of the single-particle Hamiltonian h. Then from (15.10) we have

$$\delta n_s(t) = \int_{-\infty}^{\infty} dt' \, \chi_{ss'}^{\mathrm{R}}(t,t')\delta\epsilon_{s'}(t'),$$

with density response function

$$\chi_{ss'}(z,z') = -i\sum_{\sigma\sigma'}\langle \mathcal{T}\{\Delta\hat{n}_{s\sigma,H}(z)\,\Delta\hat{n}_{s'\sigma',H}(z')\}\rangle$$

and $\Delta\hat{n}_{s\sigma,H}(t) = \hat{n}_{s\sigma,H}(t) - \langle\hat{n}_{s\sigma,H}(t)\rangle$. Choosing to perturb the system with a δ-like time-dependent energy $\delta\epsilon_{s'}(t) = \delta\epsilon_{s'}\delta(t)$, we get

$$\frac{\delta n_s(t)}{\delta\epsilon_{s'}} = \chi_{ss'}^{\mathrm{R}}(t,0). \tag{16.35}$$

This equation tells us that if the amplitude $\delta\epsilon_{s'}$ of the δ-function is small then we can solve the Kadanoff–Baym equations with self-energy $\Sigma[G]$, calculate the occupations $\delta n_s(t) = n_s(t) - n_s(0)$, divide by $\delta\epsilon_{s'}$ and extract the retarded response function $\chi_{ss'}^{\mathrm{R}}(t,0)$. Although nonlinear effects are always present, they are easily reduced by ensuring that the magnitude of the perturbation lies well in the linear response region. In practice, this is verified by doubling the amplitude $\delta\epsilon_{s'}$ and checking that also $\delta n_s(t)$ doubles to a sufficient accuracy. The resulting response function χ solves the Bethe–Salpeter equation with kernel $K = \pm\delta\Sigma/\delta G$ and therefore already a relatively simple self-energy accounts for quite a sophisticated approximation for χ.

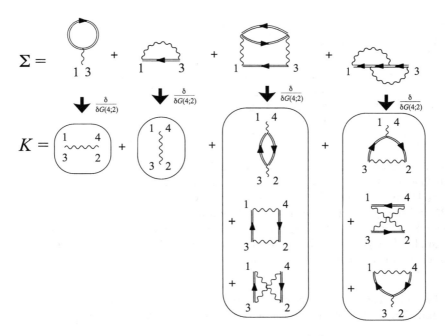

Figure 16.16 In the frame below every self-energy diagram $\Sigma^{(D)}$ we report the kernel diagrams generated from $\pm\delta\Sigma^{(D)}/\delta G$.

The HF self-energy yields the *frequency independent* kernel K_{HF}, see Section 12.4, used in several state-of-the-art implementations of the Bethe–Salpeter equation. The diagrammatic expansion of the HF response function is illustrated in Fig. 12.3(b) where the solid lines are HF Green's functions. In Fig. 16.16 we illustrate the diagrammatic content of the 2B self-energy and kernel. In this figure the double lines are 2B Green's functions and below each self-energy diagram $\Sigma^{(D)}$ we display the kernel diagrams generated from $\pm\delta\Sigma^{(D)}/\delta G$. It is instructive to expand the 2B response function in terms of the bare interaction and Hartree–Fock Green's function. To highlight the number of particle–hole excitations in every diagram we imagine that the direction of time is from left to right so that all v-lines are drawn vertical (the interaction is instantaneous). Some representative diagrams of the infinite series are shown in Fig. 16.17. The series of bubbles and ladders in the first row corresponds to the HF approximation of Fig. 12.3. Here we clearly see that at every instant of time there is at most one particle–hole excitation. Thus the HF kernel renormalizes the single-particle excitations of the underlying noninteracting system but cannot describe excitations of multiple-particle character like double-excitations, triple-excitations, etc., see last paragraph of Section 15.2.3. These interaction-induced excitations require either the use of a Green's function beyond the Hartree–Fock approximation or the use of kernel diagrams beyond the Hartree–Fock approximation, i.e., *frequency dependent* kernels. This is clearly illustrated in the diagrams of the second row of Fig. 16.17. Every diagram contains at least two particle–hole excitations and their origin is due to either 2B self-energy insertions [see diagrams (a), (b), and (c)] or 2B kernel diagrams [see (d) and (e)], or both [see (f)].

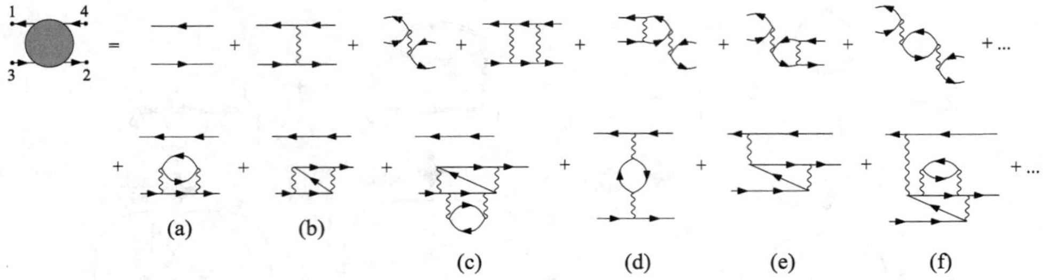

(a)   (b)

(c)   (d)   (e)   (f)

Figure 16.17 Expansion of the two-particle XC function in terms of the bare interaction v (wiggly line) and Hartree–Fock Green's function G_{HF} (oriented solid line). In the first row we have the HF diagrams of Fig. 12.3. In the second row we have examples of HF diagrams with one 2B self-energy insertion [see (a) and (b)], and with two 2B self-energy insertions [see (c)]. Then we have examples of a diagrams with 2B kernel-diagrams [see (d) and (e)] and finally an example of diagram with both 2B self-energy insertion and 2B kernel-diagrams [see (f)]. The direction of time is from left to right.

Below we present results on simple model systems obtained by solving the Kadanoff–Baym equations with a HF and 2B self-energy. We consider a Hubbard ring with Hamiltonian

$$\hat{H} = -T \sum_{s=1}^{N} \sum_{\sigma} \left(\hat{d}_{s\sigma}^{\dagger} \hat{d}_{s+1\sigma} + \hat{d}_{s+1\sigma}^{\dagger} \hat{d}_{s\sigma} \right) + U \sum_{s=1}^{N} \hat{n}_{s\uparrow} \hat{n}_{s\downarrow}, \tag{16.36}$$

where $\hat{n}_{s\sigma} = \hat{d}_{s\sigma}^{\dagger} \hat{d}_{s\sigma}$ is the occupation operator for site s and spin σ. In this equation the operator $\hat{d}_{N+1\sigma} \equiv \hat{d}_{1\sigma}$, so that the nearest-neighbor sites of 1 are 2 and N. In the Hubbard model the interaction is spin-dependent, since if we want to rewrite the last term of (16.36) in the canonical form

$$\frac{1}{2} \sum_{s\sigma, s'\sigma'} v_{s\sigma\,s'\sigma'} \hat{d}_{s\sigma}^{\dagger} \hat{d}_{s'\sigma'}^{\dagger} \hat{d}_{s'\sigma'} \hat{d}_{s\sigma}$$

the interaction must be

$$v_{s\sigma\,s'\sigma'} = U \delta_{ss'} \delta_{\sigma\bar{\sigma}'},$$

where $\bar{\sigma}'$ is the spin opposite to σ'. Consequently the exchange (Fock) self-energy diagram vanishes whereas the Hartree self-energy is simply

$$[\Sigma_H(z, z')]_{s\sigma\,s'\sigma'} = \delta_{ss'} \delta_{\sigma\sigma'} \delta(z, z') U \langle n_{s\bar{\sigma}}(z) \rangle.$$

The average occupation $\langle n_{s\sigma}(z) \rangle$ is independent of z if we are in equilibrium, independent of s due to the invariance of the Hubbard ring under discrete rotations, and independent of σ if the number of spin-up and spin-down particles is the same. In this case the HF eigenstates are identical to the noninteracting eigenstates (like in the electron gas) and their energy is $\epsilon_k = -2T \cos(2\pi k/N) + Un/2$, with $n/2 = n_\sigma$ the average occupation per spin. In Fig. 16.18 we show a Hubbard ring with six sites as well as the HF single-particle energies and degeneracies when the ring is filled with two electrons of opposite spin, i.e., $n = 1/3$,

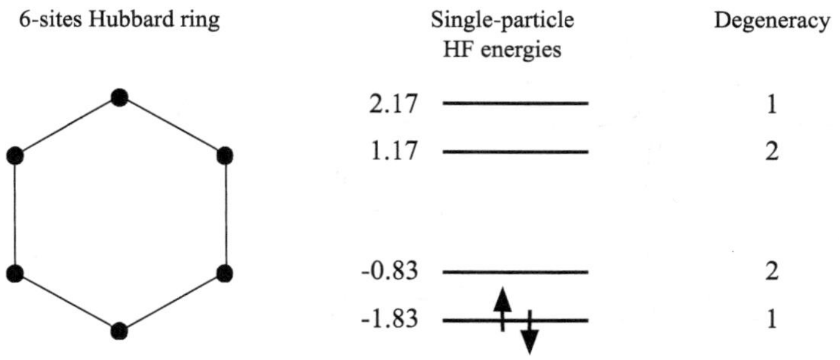

Single-particle
HF energies

Degeneracy

2.17 ──────── 1

1.17 ──────── 2

-0.83 ──────── 2

-1.83 ──────── 1

Figure 16.18 (Left) Hubbard ring with six sites. (Right) HF energy levels for $U = 1$. All energies are in units of $T > 0$.

and $U = T$. The single particle excitations of the response function $\chi_0 = -iG_{\mathrm{HF}}G_{\mathrm{HF}}$ correspond to the energy difference between an unoccupied level and the only occupied level.

For the system of Fig. 16.18 we have obtained the eigenfunctions $|\Psi_k\rangle$ and eigenenergies E_k via exact diagonalization (ED) methods (Lanczos method [168,169]), and hence the excitation energies $\Omega_k = E_k - E_0$ for the transitions $|\Psi_0\rangle \leftrightarrow |\Psi_k\rangle$ between the ground state $|\Psi_0\rangle$ and the kth excited state $|\Psi_k\rangle$. Additionally we used the $|\Psi_k\rangle$ and E_k to calculate the time evolution of the system after a δ-like perturbation $\delta\epsilon_1\delta(t)$ of the onsite energy of site 1. According to (16.35) the ratio $\delta n_1(t)/\delta\epsilon_1$ is the response function $\chi_{11}^{\mathrm{R}}(t,0)$. As we can only propagate for a finite time ΔT we must approximate the Fourier transform with[II]

$$\chi_{11}^{\mathrm{R}}(\omega) \sim \int_0^{\Delta T} dt\, e^{i\omega t}\, \chi_{11}^{\mathrm{R}}(t,0). \tag{16.37}$$

To have an idea of how this function looks like let us write $\chi_{11}^{\mathrm{R}}(t,0)$ in the form (15.26), i.e.,

$$\chi_{11}^{\mathrm{R}}(t,0) = -i\theta(t)\sum_{k\neq 0}\left[e^{-i\Omega_k t}|f_k(1)|^2 - e^{i\Omega_k t}|f_k(1)|^2\right], \tag{16.38}$$

where

$$f_k(1) = \sum_\sigma \langle\Psi_0|\hat{n}_{1\sigma}|\Psi_k\rangle$$

are the excitation amplitudes on site 1. Inserting (16.38) into (16.37) we get

$$\chi_{11}^{\mathrm{R}}(\omega) \sim -i\sum_k |f_k(1)|^2\left[e^{i(\omega-\Omega_k)\Delta T/2}\frac{\sin\frac{\omega-\Omega_k}{2}\Delta T}{\frac{\omega-\Omega_k}{2}} - e^{i(\omega+\Omega_k)\Delta T/2}\frac{\sin\frac{\omega+\Omega_k}{2}\Delta T}{\frac{\omega+\Omega_k}{2}}\right].$$

The imaginary part of this function exhibits peaks of width $\sim 1/\Delta T$ instead of sharp δ-peaks. These broadened peaks, however, are located exactly in $\pm\Omega_k$. We also observe

----
[II]This integral should be evaluated numerically since we know $\chi_{11}^{\mathrm{R}}(t,0)$ only at the discrete times $t = k\Delta_t$, with Δ_t the time step.

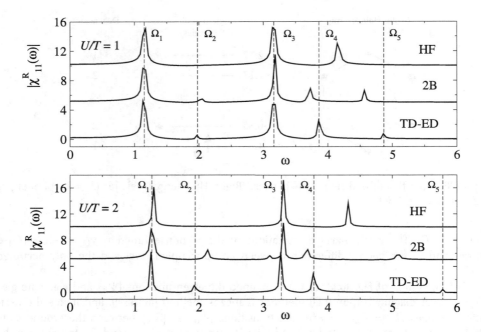

Figure 16.19 Modulus of the response function $\chi_{11}^{R}(\omega)$ for $U/T = 1, 2$ as obtained from the exact time-propagation (TD-ED) and the solution of the Kadanoff–Baym equations in the 2B and HF approximation. The dashed vertical lines indicate the position of the ED excitation energies.

that for any finite ΔT the imaginary part has an oscillatory background "noise" that goes to zero as $1/\Delta T$. Thus peaks smaller than $1/\Delta T$ are hardly visible. If we are only interested in the position and relative strength of the peaks a better way to process the time-dependent information consists in performing a discrete Fourier transform, i.e.,

$$\chi_{11}^{R}(\omega_m) \sim \Delta_t \sum_{k=1}^{N_t} e^{i\omega_m k\Delta_t} \chi_{11}^{R}(k\Delta_t, 0),$$

where Δ_t is the time-step, N_t is the total number of time steps and the frequency is sampled according to $\omega_m = 2\pi m/(N_t\Delta_t)$. The discrete Fourier transform corresponds to extending periodically the time-dependent results between 0 and $\Delta T = N_t\Delta_t$ so that, e.g., $\chi_{11}^{R}(N_t\Delta_t + p\Delta_t, 0) = \chi_{11}^{R}(p\Delta_t, 0)$. Furthermore, it is aesthetically nicer to plot $|\chi_{11}^{R}(\omega_m)|$ rather than the imaginary part of $\chi_{11}^{R}(\omega_m)$, since the latter is not always positive for finite ΔT. The list-plot of $|\chi_{11}^{R}(\omega_m)|$ is always positive and has maxima in $\pm\Omega_k$ whose height is proportional to the square of the excitation amplitudes. Thus also the modulus of the discrete Fourier transform contains the information of interest. In Fig. 16.19 we plot the time-dependent ED (TD-ED) results for $|\chi_{11}^{R}(\omega)|$. The position of the ED excitation energies (dashed vertical lines) matches exactly the position of the TD-ED peaks, in accordance with

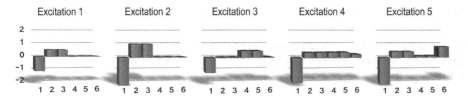

Figure 16.20 Histograms of the HF level occupation difference $\Delta n_l^{\mathrm{HF}}(\Omega)$ versus $l = 1, \ldots, 6$ for the excitation energies $\Omega = \Omega_1, .., \Omega_5$.

the previous discussion. In the same figure the TD-ED response function is compared with the HF and 2B response function as obtained from the solution of the Kadanoff–Baym equations for two different values $U = T$ and $U = 2T$ of the interaction.

As expected, not all excitations are reproduced within the HF approximation. The HF response function shows only three peaks corresponding to the (renormalized) single-excitations in which an electron is promoted from the lowest HF level to one of the three excited HF levels of Fig. 16.18. Which of these HF excitations can be considered as an approximation to the exact excitations? Following Ref. [99] we calculated the average of the occupation operator $\hat{n}_l^{\mathrm{HF}} \equiv \sum_\sigma \hat{c}_{l\sigma}^\dagger \hat{c}_{l\sigma}$ over the excited states $|\Psi_k\rangle$. Here $\hat{c}_{l\sigma}$ annihilates an electron of spin σ on the lth HF level. The difference

$$\Delta n_l^{\mathrm{HF}}(k) \equiv \langle \Psi_k | \hat{n}_l^{\mathrm{HF}} | \Psi_k \rangle - \langle \Psi_0 | \hat{n}_l^{\mathrm{HF}} | \Psi_0 \rangle$$

tells us how the occupation of the lth HF level changes in the transition $|\Psi_0\rangle \to |\Psi_k\rangle$. If we number the HF levels from the lowest to the highest in energy (thus, e.g., $l = 1$ is the lowest, $l = 2, 3$ the first excited, etc.) then single-excitations are characterized by $\Delta n_1^{\mathrm{HF}}(k) \sim -1$ whereas double-excitations by $\Delta n_1^{\mathrm{HF}}(k) \sim -2$. In Fig. 16.20 we show the histogram of

$$\Delta n_l^{\mathrm{HF}}(\Omega) = \frac{1}{d_\Omega} \sum_{k:\Omega_k=\Omega} \Delta n_l^{\mathrm{HF}}(k), \qquad d_\Omega = \text{degeneracy of excitation } \Omega,$$

for the first five excitations $\Omega_1, \ldots, \Omega_5$ and for interaction $U = T$. We see that only Ω_1 and Ω_3 are single-excitations and therefore only the first and the second HF peaks correspond to physical excitations. The third HF peak between Ω_4 and Ω_5 is instead unphysical and indeed quite far from the TD-ED peaks.

The qualitative agreement between TD-ED and MBPT results improves considerably in the 2B approximation. The 2B response function captures both single- and double-excitations! Unfortunately the position of the double-excitation peaks Ω_2, Ω_4, and Ω_5 is not as good as that of the single-excitation peaks and tends to worsen the higher we go in energy. Is this error due to an inaccurate estimation of the ground-state energy E_0, or of the excited-state energy E_k, or both? This question can be answered by calculating the ground-state energy either using the Galitskii–Migdal formula or the Luttinger–Ward functional,[12] and

---

[12]For the self-consistent 2B Green's function the two approaches yield the same result, see end of Section 11.9.

confronting the result with the exact diagonalization result E_0. What one finds is that the discrepancy between the 2B ground-state energy and E_0 is about 0.2% for $U = T$ and 1.2% for $U = 2T$. Thus the 2B ground-state energy is extremely accurate and we must attribute the mismatch between the position of the 2B and TD-ED peaks to an erroneous description of the (doubly) excited states.

Appendices

A

From the *N* roots of 1 to the Dirac δ-function

The roots of the equation

$$z = \sqrt[N]{1} = 1^{1/N}$$

are the complex numbers $z_k = \exp[\frac{2\pi i}{N} k]$ with integers $k = 1, \ldots, N$, as can be readily verified by taking the Nth power of z_k. In the complex plane these roots lie at the vertices of a regular polygon inscribed in a unit circle. In Fig. A.1(a) we show the location of the roots for $N = 3$.

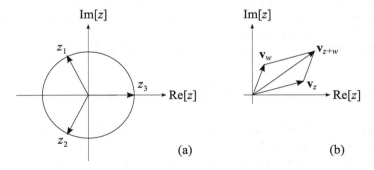

(a)   (b)

Figure A.1 (a) Location of the cubic roots of 1 in the complex plane. (b) Representation of the sum of two complex numbers.

Which complex number do we get when we sum all roots? The addition of complex numbers is equivalent to the addition of vectors. If $\mathbf{v}_z$ is the vector going from the origin to z then $\mathbf{v}_z + \mathbf{v}_w = \mathbf{v}_{z+w}$, as shown in Fig. A.1(b). It is then clear that

$$\sum_{k=1}^{N} z_k = \sum_{k=1}^{N} e^{\frac{2\pi i}{N} k} = 0.$$

We now ask ourselves how this result changes if we replace $z_k \to z_k^n$ for some integer n. The operation of taking the nth power of a complex number z on the unit circle corresponds to rotating the vector $\mathbf{v}_z$ by an angle $n\theta_z$, with θ_z the angle between $\mathbf{v}_z$ and the real axis. In the example of Fig. A.1(a) we see that by taking, e.g., the square of each root we get $z_1 \to z_1^2 = z_2$, $z_2 \to z_2^2 = z_1$ and $z_3 \to z_3^2 = z_3$. Of course this is a general property: the set $\{z_k\}$ and the set $\{z_k^n\}$ contain the same elements since the angle $n\theta_k = n(2\pi k/N)$ is equal to the angle θ_{nk}. There is however one exception! If $n = N$ then, by definition, $z_k^N = 1$ and hence all roots collapse in 1. This is clearly true also if n is a multiple of N, i.e., $n = mN$. Therefore, we conclude that

$$\frac{1}{N} \sum_{k=1}^{N} z_k^n = \frac{1}{N} \sum_{k=1}^{N} e^{\frac{2\pi i}{N} kn} = \ldots + \delta_{n,-N} + \delta_{n,0} + \delta_{n,N} + \delta_{n,2N} + \ldots, \qquad \text{(A.1)}$$

where $\delta_{n,m}$ is the Kronecker delta which is 1 for $n = m$ and zero otherwise. This important identity can also be derived algebraically. The sum is a geometric sum since $z_k^n = (z_1^n)^k$ and hence

$$\frac{1}{N} \sum_{k=1}^{N} z_k^n = \frac{1}{N} \sum_{k=1}^{N} (z_1^n)^k = \frac{(z_1^n)}{N} \frac{1 - (z_1^n)^N}{1 - (z_1^n)}. \qquad \text{(A.2)}$$

The numerator in the r.h.s. is always zero since $(z_1^n)^N = (z_1^N)^n = 1$, while the denominator is zero only provided that n is a multiple of N. In this case $z_k^n = 1$ for all k and we recover (A.1).

Next we observe that the result (A.1) does not change if we sum over k between M and $M + N$, i.e., over N arbitrary consecutive integers, rather than between 1 and N. This is obvious from the geometrical interpretation of (A.1) since it amounts to a sum over all roots starting from M instead of 1. Let us prove it also with some algebra. We have

$$\frac{1}{N} \sum_{k=M}^{M+N-1} z_k^n = \frac{1}{N} \sum_{k=1}^{N} z_{k+M-1}^n = \frac{z_{M-1}^n}{N} \sum_{k=1}^{N} z_k^n,$$

where we use the fact that $z_{k_1+k_2} = z_{k_1} z_{k_2}$. The statement follows from the fact that the sum vanishes unless n is multiple of N, in which case the sum equals N and $z_{M-1}^n = 1$.

Let us consider the case that N is, say, even and choose $M = -N/2$. Then we can write

$$\boxed{\frac{1}{N} \sum_{k=-N/2}^{N/2-1} e^{\frac{2\pi i}{N} kn} = \ldots + \delta_{n,-N} + \delta_{n,0} + \delta_{n,N} + \delta_{n,2N} + \ldots} \qquad \text{(A.3)}$$

Now we define the variable $y_k = 2\pi k/N$. The distance between two consecutive y_k is $\Delta_y = y_{k+1} - y_k = 2\pi/N$. Therefore taking the limit $N \to \infty$ we get

$$\lim_{N \to \infty} \frac{1}{N} \sum_{k=-N/2}^{N/2-1} e^{\frac{2\pi i}{N} kn} = \lim_{\Delta_y \to 0} \sum_{k=-N/2}^{N/2-1} \frac{\Delta_y}{2\pi} e^{i y_k n} = \int_{-\pi}^{\pi} \frac{dy}{2\pi} e^{i y n}, \qquad \text{(A.4)}$$

where in the last equality we have transformed the sum into an integral in accordance with the standard definition of integrals as the limit of a Riemann sum. Setting $n = m - m'$ and taking into account that for $N \to \infty$ only the Kronecker $\delta_{n,0}$ remains in the r.h.s. of (A.3) we obtain the following important identity:

$$\int_{-\pi}^{\pi} \frac{dy}{2\pi} \, e^{iy(m-m')} = \delta_{m,m'} \tag{A.5}$$

In (A.3) we could, alternatively, think of the variable $n/N = y_n$ as a continuum variable when $N \to \infty$. The infinitesimal increment is then $\Delta_y = y_{n+1} - y_n = 1/N$ and taking the limit $N \to \infty$ we find

$$\sum_{k=-\infty}^{\infty} e^{2\pi i k y} = \lim_{N \to \infty} \frac{1}{\Delta_y}(\ldots + \delta_{n,-N} + \delta_{n,0} + \delta_{n,N} + \delta_{n,2N} + \ldots). \tag{A.6}$$

For $N \to \infty$ and hence $\Delta_y \to 0$ the quantity $\delta_{n,mN}/\Delta_y$ is zero unless $n = y_n N = mN$, i.e., $y_n = m$, in which case it diverges like $1/\Delta_y$. Since for any function $f(y)$ we have

$$f(m) = \sum_{n=-\infty}^{\infty} \delta_{n,mN} f(y_n) = \sum_{n=-\infty}^{\infty} \Delta_y \frac{\delta_{n,mN}}{\Delta_y} f(y_n)$$

$$\xrightarrow[\Delta_y \to 0]{} \int_{-\infty}^{\infty} dy \left(\lim_{\Delta_y \to 0} \frac{\delta_{n,mN}}{\Delta_y} \right) f(y),$$

we can identify the Dirac δ-function

$$\lim_{\Delta_y \to 0} \frac{\delta_{n,mN}}{\Delta_y} = \delta(y - m). \tag{A.7}$$

Therefore, taking the limit $\Delta_y \to 0$ in (A.6) we obtain a second important identity

$$\sum_{k=-\infty}^{\infty} e^{2\pi i k y} = \ldots + \delta(y+1) + \delta(y) + \delta(y-1) + \delta(y-2) + \ldots$$

Lastly, we again consider (A.5) and divide both sides by an infinitesimal Δ_p

$$\frac{1}{\Delta_p} \int_{-\pi}^{\pi} \frac{dy}{2\pi} \, e^{iy(m-m')} = \frac{\delta_{m,m'}}{\Delta_p}.$$

In the limit $\Delta_p \to 0$ we can define the continuous variables $p = m\Delta_p$ and $p' = m'\Delta_p$, and using (A.7) we get

$$\lim_{\Delta_p \to 0} \frac{1}{\Delta_p} \int_{-\pi}^{\pi} \frac{dy}{2\pi} \, e^{iy(m-m')} = \delta(p - p').$$

The product in the exponential can be rewritten as $y(m - m') = y(p - p')/\Delta_p$. Thus, if we change variable $x = y/\Delta_p$ the above equation becomes

$$\int_{-\infty}^{\infty} \frac{dx}{2\pi} \, e^{ix(p-p')} = \delta(p - p')$$

which is one of the possible representations of the Dirac δ-function. In conclusion the Dirac δ-function is intimately related to the sum of the N roots of 1 when $N \to \infty$.

B

Graphical approach to permanents and determinants

The quantum states of a set of identical particles are either symmetric or antisymmetric in the single-particle labels. An orthonormal basis for the space spanned by these quantum states is formed by the complete set of (anti)symmetric products of single-particle orthonormal states, and their manipulation leads naturally to the consideration of permanents and determinants. The standard algebraic derivation of identities for permanents and determinants usually involves several steps with much relabeling of permutations, which makes the derivations often long and not very insightful. In this appendix we instead give a simple and intuitive graphical derivation of several of these identities. The basic ingredient of the graphical approach is the *permutation graph* that keeps track of how the permutation moves around the elements on which it acts. A permutation P is defined as a one-to-one mapping from the set of integers $(1, \ldots, n)$ to itself,

$$P(1, \ldots, n) = (P(1), \ldots, P(n)) = (1', \ldots, n'),$$

where $j' = P(j)$ denotes the image of j under the permutation P. The permutation graph is then defined by drawing a figure with the numbers $(1, \ldots, n)$ as dots ordered from top to bottom along a vertical line on the left and with the images $(1', \ldots, n')$ as dots also ordered from top to bottom along a vertical line on the right, and by connecting the dots j with the images j'. For example, if $n = 5$, we can consider the permutation

$$P(1, 2, 3, 4, 5) = (2, 1, 5, 3, 4) = (1', 2', 3', 4', 5'). \tag{B.1}$$

The permutation graph is then given by

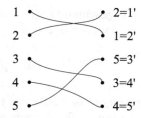

The permutation graph thus simply provides a graphical representation of how the numbers $(1, \ldots, n)$ are moved around by the permutation. An important feature of the permutation graph is that, for a given permutation, the number of crossings of the lines is always even or odd, no matter how we deform the lines in the diagram, provided that we do not deform the lines outside the left and right boundaries. For example we have

It should be noted that for this to be true we also need to exclude drawings in which the lines touch each other in one point and drawings in which multiple lines cross in exactly the same point. The crossing rule is intuitively clear since any deformation of the lines that creates new crossing points always creates two of them. Hence the parity of the number of crossings is preserved. We can therefore divide the set of permutations into two classes, those with an even number of crossings and those with an odd number of crossings. These are naturally referred to as even and odd permutations. Another way to characterize this property is by defining the *sign* of a permutation P to be

$$\text{sign}\, P \equiv (-)^{n_c}, \tag{B.2}$$

where n_c is the number of crossings in a permutation graph and $(-)^{n_c}$ is a short-hand notation for $(-1)^{n_c}$. So even permutations have sign 1 and odd permutations have sign -1. It is also intuitively clear that any permutation can be built up from successive interchanges of two numbers. This is what one, for instance, would do when given the task of reordering numbered balls by each time swapping two of them. Each such a swap is called a transposition. If a transposition interchanges labels i and j then we write it as $(i\,j)$. For instance, the permutation (B.1) can be constructed from the identity permutation by the subsequent transpositions $(4\,5)$, $(3\,5)$, and $(1\,2)$. This is described by the following four permutation graphs (to be read from left to right)

We can thus write

$$P = (1\,2)\,(3\,5)\,(4\,5), \tag{B.3}$$

in which the subsequent transpositions in this expression are carried out from right to left. We therefore need three transpositions to reorder numbered balls from $(1, 2, 3, 4, 5)$ to $(2, 1, 5, 3, 4)$. From these graphs we further see the interesting fact that every transposition changes the number of crossings by an odd number (which is equal to 1 in the example above). This fact is easily deduced from the following graph that displays the very right hand side of a permutation graph

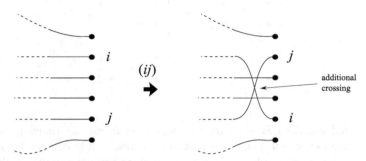

By interchanging i and j on the right hand side we introduce one additional crossing plus an even number of crossings due to the upward and downward running lines induced by the swap. Hence a transposition always generates an odd number of crossings. It therefore follows that even permutations are built up from an even number of transpositions and odd permutations are built up from an odd number of transpositions. For example, the permutation P of our example (B.3) is odd and indeed built up from three transpositions. As experience with reordering objects tells us, the way to achieve a given reordering is not unique. However, as we have just proved, the parity of the number of transpositions is unique, and therefore the permutation in our example can only be decomposed using an odd number of transpositions. For instance, it is not difficult to check that permutation (B.3) can also be written as

$$P = (2\,3)\,(2\,5)\,(4\,5)\,(1\,3)\,(2\,3),$$

which indeed again has an odd number of transpositions, as it should. From our considerations we thus conclude that if $|P|$ is the number of transpositions in a given decomposition of a permutation P, then the sign of this permutation is given by $(-)^{|P|}$. By comparing with (B.2) we find the useful relation

$$(-)^{n_c} = (-)^{|P|}.$$

Alternatively this relation is usually written as

$$(-)^{n_c} = (-)^P,$$

where $(-)^P$ simply signifies the sign of the permutation. Before we continue we further give another useful way of looking at the permutation graphs. In the graphs above we thought of the labels j as numbered balls and imagined the balls being moved around. This image is, for instance, very useful when thinking about operator orderings as we did in Chapter 4. However, for deriving identities for permanents and determinants it is useful to think about the permutation graphs in a slightly different way. In the permutation P the ball numbered $P(i)$ acquires the new position i. For instance, the example (B.1) tells us that after the permutation P the balls labeled $(2, 1, 5, 3, 4)$ can be found at positions $(1, 2, 3, 4, 5)$. We can express this information in a new permutation graph where on the left hand side we put the positions i and on the right hand side we put the ball numbers $P(i)$ directly opposite to i and then connect them with horizontal lines. In our example this gives

Position		Ball number
1	———————	2
2	———————	1
3	———————	5
4	———————	3
5	———————	4

For instance we have the line $(3, 5)$ which tells us that ball 5 is found at position 3 after the permutation. Now we can reorder the balls on the right hand side of the graph to their original positions by subsequent transpositions:

Since each transposition leads to an odd number of crossings the sign of the permutation is again given by $(-1)^{n_c}$ with n_c the number of crossings, but now for a graph in which both sides are labeled 1 to n from top to bottom and where we connect i to $P(i)$. This is an alternative way to calculate the sign of a permutation from a permutation graph, and it proves to be very useful in deriving some important identities for permanents and determinants.

The permanent or determinant of an $n \times n$ matrix A_{ij} is defined as

$$|A|_{\pm} \equiv \sum_P (\pm)^P \prod_{i=1}^n A_{i\,P(i)},$$

where the "+" sign refers to the permanent and the "−" sign to the determinant (in this formula $(+)^P = (+1)^{|P|} = 1$ for all P). For instance, for a 2×2-matrix we only have the even permutation $P(1,2) = (1,2)$ and the odd permutation $P(1,2) = (2,1)$ and therefore we have that the permanent/determinant is

$$|A|_\pm = A_{11} A_{22} \pm A_{12} A_{21}. \tag{B.4}$$

The permanent or determinant can be visualized using permutation graphs. First draw the permutation graph which connects with lines points i to points $P(i)$ and then associate with every line the matrix element $A_{i\,P(i)}$. For instance, the permanent/determinant in (B.4) can be written as

and for a 6×6 permanent/determinant we have a term of the form

$$= (\pm)^3 A_{11} A_{25} A_{32} A_{43} A_{54} A_{66}.$$

For permanents and determinants we can now prove the following Laplace expansion theorems:

$$|A|_\pm = \sum_{j=1}^{n} (\pm)^{i+j} A_{ij}\, \tilde{D}_{ij}$$

$$|A|_\pm = \sum_{i=1}^{n} (\pm)^{i+j} A_{ij}\, \tilde{D}_{ij}$$

in which $\tilde{D}_{ij}$ is the minor of $|A|_\pm$, i.e., the permanent/determinant of matrix A in which row i and column j are removed. The first expression gives the expansion of $|A|_\pm$ along row i whereas the second one gives its expansion along column j. These identities can be proven almost immediately from drawing a single diagram. We consider a particular permutation that moves ball numbered j in $(1, \ldots, n)$ just before ball numbered i. If $j > i$ this permutation is explicitly given by

$$P(1, \ldots, n) = (1, \ldots, i-1, j, i, \ldots, j-1, j+1, \ldots, n),$$

with a very similar expression when $j < i$. Since there are $|j-i|$ crossings in the permutation graph the sign of this permutation graph is $(\pm)^{i-j} = (\pm)^{i+j}$. For example, for $n = 6$ we can consider the permutation with $j = 5$ and $i = 2$,

$$P(1,2,3,4,5,6) = (1,5,2,3,4,6),$$

which is graphically given by the permutation graph in the previous figure. Let us now in the figure fix the line $(i,j) = (2,5)$ and subsequently connect the remaining lines in all possible ways. In this way we are exactly constructing the minor $\tilde{D}_{25}$, i.e.

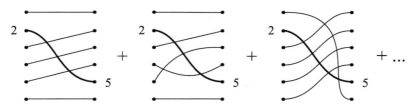

$$= (\pm)^{2+5} A_{25} \tilde{D}_{25} = (\pm)^{2+5} A_{25} \begin{vmatrix} A_{11} & A_{12} & A_{13} & A_{14} & A_{16} \\ A_{31} & A_{32} & A_{33} & A_{34} & A_{36} \\ A_{41} & A_{42} & A_{43} & A_{44} & A_{46} \\ A_{51} & A_{52} & A_{53} & A_{54} & A_{56} \\ A_{61} & A_{62} & A_{63} & A_{64} & A_{66} \end{vmatrix}_{\pm}.$$

Indeed the sign of the generic permutation graph in the above expansion is given by $(\pm)^{2+5} \times (\pm)^{n_c^{(25)}}$ where $n_c^{(25)}$ is the number of crossings *without* the line $(2,5)$.[1] It is clear that we obtain the full permanent/determinant by doing the same after singling out the lines $(2,j)$ for $j = 1,2,3,4,6$ and adding the results, since in that case we have summed over all possible connections exactly once. This yields

$$|A|_{\pm} = \sum_{j=1}^{6} (\pm)^{2+j} A_{2j} \tilde{D}_{2j}.$$

This is the Laplace formula for expanding the permanent/determinant along row 2. Alternatively we could have singled out the lines $(i,5)$ for $i = 1, \ldots, 6$ and obtained

$$|A|_{\pm} = \sum_{i=1}^{6} (\pm)^{i+5} A_{i5} \tilde{D}_{i5}$$

which gives the expansion along row 5. It is clear that there is nothing special about this example and that the general proof obviously goes the same way.

This graphical proof also gives an idea on how to generalize the Laplace formula. Consider a permutation graph in which we single out three lines (i_1, j_1), (i_2, j_2), and (i_3, j_3),

---

[1] The reader can easily check that the line $(2,5)$ is always crossed an odd number of times (3 times in the second graph and 5 times in the third graph).

where $i_i < i_2 < i_3$ and $j_1 < j_2 < j_3$ and where the remaining lines are drawn in a noncrossing way, i.e.

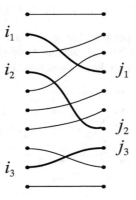

The sign of this permutation graph is clearly given by

$$(\pm)^{j_1-i_1}(\pm)^{j_2-i_2}(\pm)^{j_3-i_3} = (\pm)^{i_1+i_2+i_3+j_1+j_2+j_3}.$$

If we now fix the lines (i_1, j_1), (i_2, j_2), and (i_3, j_3), i.e., we fix the position of ball numbered j_1 to be i_1, etc., and make a permutation Q of the remaining balls (so that their position is different from the position they have in the above permutation graph) we get a new permutation graph. As in the case of a single line, the total number of crossings for the lines (i_1, j_1), (i_2, j_2), and (i_3, j_3) always has the same parity (even or odd) for all Q and therefore the sign of this new permutation graph is

$$(\pm)^{i_1+i_2+i_3+j_1+j_2+j_3} \times (\pm)^{n_c^{(i_1 j_1),(i_2 j_2),(i_3 j_3)}},$$

where $n_c^{(i_1 j_1),(i_2 j_2),(i_3 j_3)}$ is the number of crossings with the lines (i_1, j_1), (i_2, j_2), and (i_3, j_3) removed from the graph. This observation allows us to derive an important identity. Let $\Pi_{i_1 i_2 i_3, j_1 j_2 j_3}(Q)$ be this new permutation graph. Then we can write

$$\sum_Q \Pi_{i_1 i_2 i_3, j_1 j_2 j_3}(Q) = (\pm)^{i_1+i_2+i_3+j_1+j_2+j_3} A_{i_1 j_1} A_{i_2 j_2} A_{i_3 j_3} \tilde{D}_{i_1 i_2 i_3, j_1 j_2 j_3}. \qquad \text{(B.5)}$$

In this equation $\tilde{D}_{i_1 i_2 i_3, j_1 j_2 j_3}$ is a generalized minor, i.e., the permanent/determinant of the matrix A in which we have removed rows i_1, i_2, i_3 and columns j_1, j_2, j_3, i.e.,

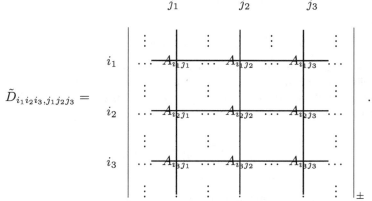

Subsequently in (B.5) we sum over all the permutations P of the integers j_1, j_2, and j_3. The sign of the permutation graph $\Pi_{i_1 i_2 i_3, P(j_1) P(j_2) P(j_3)}(Q)$ differs from the sign of $\Pi_{i_1 i_2 i_3, j_1 j_2 j_3}(Q)$ by a factor $(\pm)^{n_c}$ with n_c the number of crossings of the lines $(i_1, P(j_1))$, $(i_2, P(j_2))$ and $(i_3, P(j_3))$. Therefore we have

$$\sum_P \sum_Q \Pi_{i_1 i_2 i_3, P(j_1) P(j_2) P(j_3)}(Q) = (\pm)^{i_1 + i_2 + i_3 + j_1 + j_2 + j_3} D_{i_1 i_2 i_3, j_1 j_2 j_3} \tilde{D}_{i_1 i_2 i_3, j_1 j_2 j_3},$$

where $D_{i_1 i_2 i_3, j_1 j_2 j_3}$ is the permanent/determinant consisting of the rows i_1, i_2, i_3 and columns j_1, j_2, j_3 of matrix A, i.e.

$$D_{i_1 i_2 i_3, j_1 j_2 j_3} = \begin{vmatrix} A_{i_1 j_1} & A_{i_1 j_2} & A_{i_1 j_3} \\ A_{i_2 j_1} & A_{i_2 j_2} & A_{i_2 j_3} \\ A_{i_3 j_1} & A_{i_3 j_2} & A_{i_3 j_3} \end{vmatrix}_{\pm}.$$

If we finally sum over all triples (j_1, j_2, j_3) with $j_1 < j_2 < j_3$ then we clearly obtain the full permanent/determinant of matrix A since we have summed exactly once over all possible connections of lines in the permutation graph. We therefore find that

$$|A|_\pm = \sum_{j_1 < j_2 < j_3}^n (\pm)^{i_1 + i_2 + i_3 + j_1 + j_2 + j_3} D_{i_1 i_2 i_3, j_1 j_2 j_3} \tilde{D}_{i_1 i_2 i_3, j_1 j_2 j_3}.$$

It is clear that the choice of the initial three lines in this example was arbitrary. We might as well have chosen $m < n$ lines instead. Then nothing essential would change in the derivation. In that case we obtain the formula

$$\boxed{|A|_\pm = \sum_J^n (\pm)^{|I+J|} D_{IJ} \tilde{D}_{IJ}} \tag{B.6}$$

where we sum over the ordered m-tuple

$$J = (j_1, \dots, j_m) \quad \text{with} \quad j_1 < \dots < j_m$$

for a fixed ordered m-tuple

$$I = (i_1, \ldots, i_m) \quad \text{with} \quad i_1 < \ldots < i_m,$$

and defined

$$|I + J| = i_1 + \ldots + i_m + j_1 + \ldots + j_m.$$

Further D_{IJ} is the permanent/determinant containing rows I and columns J whereas $\tilde{D}_{IJ}$ is the generalized minor, not containing the rows I and columns J. The result (B.6) is known as the *generalized Laplace formula* for permanents/determinants. Alternatively, in the derivation we could have summed over the rows rather than the columns and written

$$|A|_{\pm} = \sum_{I}^{n} (\pm)^{|I+J|} D_{IJ} \tilde{D}_{IJ} \tag{B.7}$$

The number of terms in the summation in the generalized Laplace formula is equal to the number of ordered m-tuples that can be chosen from n indices and is therefore equal to $\binom{n}{m}$. The use of the permutation graphs leads to a very compact derivation of the identities (B.6) and (B.7). As an example of (B.6) we can expand the permanent/determinant of a 4×4 matrix into products of 2×2 permanents/determinants. Taking, e.g., $(i_1, i_2) = (1, 2)$ we have

$$
\begin{aligned}
|A|_{\pm} &= \sum_{j_1 < j_2} (\pm)^{1+2+j_1+j_2} D_{12,j_1 j_2} \tilde{D}_{12,j_1 j_2} \\
&= \begin{vmatrix} A_{11} & A_{12} \\ A_{21} & A_{22} \end{vmatrix}_{\pm} \begin{vmatrix} A_{33} & A_{34} \\ A_{43} & A_{44} \end{vmatrix}_{\pm} \pm \begin{vmatrix} A_{11} & A_{13} \\ A_{21} & A_{23} \end{vmatrix}_{\pm} \begin{vmatrix} A_{32} & A_{34} \\ A_{42} & A_{44} \end{vmatrix}_{\pm} \\
&+ \begin{vmatrix} A_{11} & A_{14} \\ A_{21} & A_{24} \end{vmatrix}_{\pm} \begin{vmatrix} A_{32} & A_{33} \\ A_{42} & A_{43} \end{vmatrix}_{\pm} + \begin{vmatrix} A_{12} & A_{13} \\ A_{22} & A_{23} \end{vmatrix}_{\pm} \begin{vmatrix} A_{31} & A_{34} \\ A_{41} & A_{44} \end{vmatrix}_{\pm} \\
&\pm \begin{vmatrix} A_{12} & A_{14} \\ A_{22} & A_{24} \end{vmatrix}_{\pm} \begin{vmatrix} A_{31} & A_{33} \\ A_{41} & A_{43} \end{vmatrix}_{\pm} + \begin{vmatrix} A_{13} & A_{14} \\ A_{23} & A_{24} \end{vmatrix}_{\pm} \begin{vmatrix} A_{31} & A_{32} \\ A_{41} & A_{42} \end{vmatrix}_{\pm}.
\end{aligned}
$$

Finally we derive a useful formula for the permanent/determinant of the sum of two matrices. Let us start with an example. If A and B are two 2×2 matrices then we can readily calculate that

$$
\begin{vmatrix} A_{11} + B_{11} & A_{12} + B_{12} \\ A_{21} + B_{21} & A_{22} + B_{22} \end{vmatrix}_{\pm} =
$$

$$
\begin{vmatrix} A_{11} & A_{12} \\ A_{21} & A_{22} \end{vmatrix}_{\pm} + A_{11}B_{22} \pm A_{12}B_{21} \pm A_{21}B_{12} + A_{22}B_{11} + \begin{vmatrix} B_{11} & B_{12} \\ B_{21} & B_{22} \end{vmatrix}_{\pm}.
$$

This can be written in a compact way as

$$|A + B|_{\pm} = |A|_{\pm} + \sum_{i_1, j_1}^{2} (\pm)^{i_1 + j_1} A_{i_1 j_1} B_{\check{i}_1, \check{j}_1} + |B|_{\pm},$$

where $\check{i}$ is the element complementary to i in the set $(1, 2)$, i.e., if $i = 1$ then $\check{i} = 2$ and vice versa. We now derive the following generalization of this equation:

$$|A + B|_{\pm} = |A|_{\pm} + \sum_{l=1}^{n-1} \sum_{IJ} (\pm)^{|I+J|} |A|_{l,\pm}(I, J) |B|_{n-l,\pm}(\check{I}, \check{J}) + |B|_{\pm} \qquad \text{(B.8)}$$

where A and B are $n \times n$ matrices. In (B.8) I is an ordered l-tuple $(i_1, \ldots, i_l)$ with $i_1 < \ldots < i_l$ and similarly for J. Further $\check{I}$ is the set of ordered complementary indices to I in the set $(1, \ldots, n)$ and similarly for $\check{J}$. For example, if $n = 5$ and $I = (1, 3, 4)$ then $\check{I} = (2, 5)$. The quantity $|A|_{l,\pm}(I, J)$ denotes the $l \times l$ permanent/determinant of A with rows I and columns J. If $l = n$ the only possible n-tuple is $I = J = (1, \ldots, n)$ and therefore $|A|_{n,\pm}(I, J) = |A|_{\pm}$. Analogously $|B|_{n-l,\pm}(\check{I}, \check{J})$ denotes the $(n - l) \times (n - l)$ permanent/determinant of B with rows $\check{I}$ and columns $\check{J}$. To prove (B.8) we start with the example of the 2×2 matrix $C = A + B$. Let the thick line (i, j) represent the matrix element $C_{ij} = A_{ij} + B_{ij}$:

where the thin line represents matrix element A_{ij} and the dashed line matrix element B_{ij}. The graphic expression of the 2×2 permanent/determinant $|C|_{\pm}$ is then given by

In the general case that C is an $n \times n$ matrix we can consider a given graph containing l solid lines and $n - l$ dashed lines. Let the thin lines run from the ordered set I to the ordered set J and the remaining dashed lines from $\check{I}$ to $\check{J}$. Since the sets are ordered there is no crossing between two or more thin lines and between two or more dashed lines. The graph representing this situation looks like this:

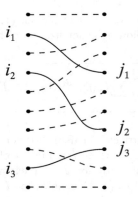

The sign of the graph is $(\pm)^{|I+J|}$. If we sum over all the dashed connections obtained by considering all permutations of the elements in the $(n-l)$-tuple $\check{J}$ and multiply by the prefactor $(\pm)^{n_c}$ for each crossing of the dashed lines we build the permanent/determinant $|B|_{n-l,\pm}(\check{I},\check{J})$. The number of additional crossings of the dashed lines with the thin lines that are caused by these permutations is always even and we therefore do not need to add additional signs. Subsequently summing over all permutations in the l-tuples J and multiplying by the prefactor $(\pm)^{n_c}$ for each crossing of the thin lines we build the permanent/determinant $|A|_{l,\pm}(I,J)$. As before the number of additional crossings of the solid lines with the dashed ones caused by these permutations is always even so we do not need to count these crossings to get the correct sign. When we finally sum over all ordered l-tuples I and J then we sum over all graphs that have l solid lines and $n-l$ dashed lines. It is then clear that all possible graphs are obtained by summing over l from 1 to $n-1$ and adding $|A|_\pm$ and $|B|_\pm$. This proves (B.8).

To make the formula (B.8) explicit we give an example for the case of 3×3 matrices A and B:

$$|A+B|_\pm = |A|_\pm + \sum_{i_1,j_1}^{3} (\pm)^{i_1+j_1} |A|_{1,\pm}(i_1,j_1)|B|_{2,\pm}(\check{i}_1,\check{j}_1)$$

$$+ \sum_{\substack{i_1<i_2 \\ j_1<j_2}} (\pm)^{i_1+i_2+j_1+j_2} |A|_{2,\pm}(i_{12},j_{12})|B|_{1,\pm}(\check{i}_{12},\check{j}_{12}) + |B|_\pm,$$

where we use the short notation $i_{12}=(i_1,i_2)$ and $j_{12}=(j_1,j_2)$. As a final remark we note that for the generalized Laplace formulas and the formula for the permanent/determinant of the sum of two matrices the symmetric (permanent) and anti-symmetric (determinant) cases can be treated on equal footing. This is not anymore the case for the product of matrices. Whereas it is true that for a product AB of two matrices one has $|AB|_- = |A|_-|B|_-$ (the determinant of a product is the product of the determinants) we have that $|AB|_+ \neq |A|_+|B|_+$ in general.

C

Density matrices and probability interpretation

In this appendix we discuss the probability interpretation of the density matrices Γ_n introduced in Section 1.7. Let us, for simplicity, consider a system of spinless fermions in one dimension. Then the states $|x_1 \ldots x_N\rangle$ with $x_1 > \ldots > x_N$ form a basis in the N-particle Hilbert space $\mathcal{H}_N$. Any state $|\Psi\rangle \in \mathcal{H}_N$ can be expanded as

$$|\Psi\rangle = \int_{x_1 > \ldots > x_N} dx_1 \ldots dx_N |x_1 \ldots x_N\rangle \underbrace{\langle x_1 \ldots x_N | \Psi\rangle}_{\Psi(x_1, \ldots, x_N)},$$

and the normalization condition reads

$$1 = \langle \Psi | \Psi \rangle = \int_{x_1 > \ldots > x_N} dx_1 \ldots dx_N \langle \Psi | x_1 \ldots x_N \rangle \langle x_1 \ldots x_N | \Psi \rangle$$

$$= \frac{1}{N!} \int dx_1 \ldots dx_N |\Psi(x_1, \ldots, x_N)|^2. \tag{C.1}$$

Let us now ask what is the probability density $p(z)$ for finding a particle at position z. We start with the first nontrivial case of $N = 2$ particles. The configuration space is given by the grey area in the figure below:

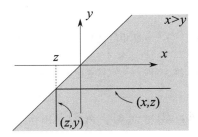

We then have

$$p(z) = \int_z^\infty dx \, |\Psi(x, z)|^2 + \int_{-\infty}^z dy \, |\Psi(z, y)|^2 = \int_{-\infty}^\infty dy |\Psi(z, y)|^2,$$

where in the last equality we use the antisymmetry of the wavefunction. We observe, however, that

$$\int_{-\infty}^{\infty} dz\, p(z) = \int_{-\infty}^{\infty} dz\, dy\, |\Psi(z,y)|^2 = 2,$$

as follows directly from (C.1). Mathematically the fact that $p(z)$ does not integrate to unity is not surprising since $p(z) = \Gamma_1(z; z) = n(z)$ is the density of particles, see (1.93), and hence correctly integrates to 2. Intuitively, however, it seems strange that a probability does not integrate to unity and therefore an explanation is needed.

The point is that the event of finding a particle in z and the event of finding a particle in z' are not independent events. To calculate $p(z)$ we have to integrate $|\Psi(z,y)|^2$ over all y and this includes the point $y = z'$. Thus among the events with a particle in z there is an event in which there is a particle in z'. This fact can most easily be illustrated by a simple discrete example. Consider two bosons that can either occupy state $|1\rangle$ or state $|2\rangle$. Then $\mathcal{H}_2 = \{|11\rangle, |12\rangle, |22\rangle\}$. A general state describing these two particles is a linear combination of the basis states in $\mathcal{H}_2$,

$$|\Psi\rangle = \Psi(1,1)|11\rangle + \Psi(1,2)|12\rangle + \Psi(2,2)|22\rangle,$$

and if $|\Psi\rangle$ is normalized we have

$$|\Psi(1,1)|^2 + |\Psi(1,2)|^2 + |\Psi(2,2)|^2 = 1.$$

Now the probability of finding a particle in state 1 is

$$p(1) = |\Psi(1,1)|^2 + |\Psi(1,2)|^2.$$

Similarly the probability of finding a particle in state 2 is

$$p(2) = |\Psi(1,2)|^2 + |\Psi(2,2)|^2.$$

However, the probability of finding a particle in state 1 or in state 2 is not the sum of the two probabilities since

$$p(1) + p(2) = |\Psi(1,1)|^2 + 2|\Psi(1,2)|^2 + |\Psi(2,2)|^2 = 1 + |\Psi(1,2)|^2.$$

In this way we double count the state $|12\rangle$. The proper probability formula for overlapping event sets A and B is

$$p(A \cup B) = p(A) + p(B) - p(A \cap B).$$

The fact that we do find a particle in state 1 does not exclude the fact that another particle can be found in state 2. This joint probability is $|\Psi(1,2)|^2$ and needs to be subtracted from $p(1) + p(2)$. In this way the probability of finding a particle either in $|1\rangle$ or $|2\rangle$ is unity, as it should be.

Let us now calculate the probability density $p(z)$ for N fermions in one dimension. We have

$$p(z) = \int_{z>x_2>\ldots>x_N} dx_2 \ldots dx_N |\Psi(z, x_2, \ldots, x_N)|^2$$

$$+ \int_{x_1>z>x_3>\ldots>x_N} dx_1 dx_3 \ldots dx_N |\Psi(x_1, z, x_3, \ldots, x_N)|^2$$

$$+ \ldots + \int_{x_1>\ldots>x_{N-1}>z} dx_1 \ldots dx_{N-1} |\Psi(x_1, \ldots, x_{N-1}, z)|^2$$

$$= \int_{x_2>\ldots>x_N} dx_2 \ldots dx_N |\Psi(z, x_2, \ldots, x_N)|^2$$

$$= \frac{1}{(N-1)!} \int dx_2 \ldots dx_N |\Psi(z, x_2, \ldots, x_N)|^2,$$

where we use the antisymmetry of the wavefunction. As in the previous case, $p(z) = \Gamma_1(z; z) = n(z)$ is the density of particles in z and does not integrate to unity but to the total number of particles N. The reason is the same as before. Clearly the above derivation can readily be generalized to particles with spin, dimensions higher than 1, and to the bosonic case. Thus the physical interpretation of the density $n(\mathbf{x})$ is the probability of finding a particle in $\mathbf{x}$.

Coming back to our one-dimensional system of fermions, we now consider the joint probability $p(z, z')$ of finding a particle at z and another at z'. Since $p(z, z') = p(z', z)$ we can assume that $z > z'$. Then

$$p(z, z') = \int_{z>z'>x_3>\ldots>x_N} dx_3 \ldots dx_N |\Psi(z, z', x_3, \ldots, x_N)|^2$$

$$+ \int_{z>x_2>z'>\ldots>x_N} dx_2 dx_4 \ldots dx_N |\Psi(z, x_2, z', \ldots, x_N)|^2$$

$$+ \ldots + \int_{x_1>z>z'>\ldots>x_N} dx_1 dx_4 \ldots dx_N |\Psi(x_1, z, z', \ldots, x_N)|^2 + \ldots$$

$$= \int_{x_3>\ldots>x_N} dx_3 \ldots dx_N |\Psi(z, z', x_3, \ldots, x_N)|^2$$

$$= \frac{1}{(N-2)!} \int dx_3 \ldots dx_N |\Psi(z, z', x_3, \ldots, x_N)|^2,$$

where again the antisymmetry of the wavefunction has been used. We thus see that

$$p(z, z') = \Gamma_2(z, z'; z, z')$$

is the two-particle density matrix defined in (1.96). The above derivation is readily seen to be valid also for bosons and also in the case of arbitrary spin and spatial dimensions. Thus, more generally, we have $p(\mathbf{x}, \mathbf{x}') = \Gamma_2(\mathbf{x}, \mathbf{x}'; \mathbf{x}, \mathbf{x}')$. The probability interpretation of Γ_2 gives a nice interpretation of the interaction energy (1.97):

$$E_{\text{int}} = \langle \Psi | \hat{H}_{\text{int}} | \Psi \rangle = \frac{1}{2} \int d\mathbf{x} d\mathbf{x}' v(\mathbf{x}, \mathbf{x}') p(\mathbf{x}, \mathbf{x}').$$

The interaction energy is simply the product of the probability of finding a particle at $\mathbf{x}$ and another at $\mathbf{x}'$ times their interaction; the factor $1/2$ makes sure that we count each pair only once, see discussion before (1.80).

If points $\mathbf{r}$ and $\mathbf{r}'$ are far from each other then it is reasonable to expect that the probability $p(\mathbf{x}, \mathbf{x}')$ becomes the independent product $n(\mathbf{x})n(\mathbf{x}')$ for any physically relevant state $|\Psi\rangle$. We therefore define the *pair correlation function*

$$g(\mathbf{x}, \mathbf{x}') = \frac{p(\mathbf{x}, \mathbf{x}')}{n(\mathbf{x})n(\mathbf{x}')}. \tag{C.2}$$

It is easy to show that $g \to 1$ for $|\mathbf{r} - \mathbf{r}'| \to \infty$ when the state $|\Psi\rangle$ is the ground state of some homogeneous system so that $n(\mathbf{x}) = n(\mathbf{r}\sigma) = n_\sigma$ is independent of position and $p(\mathbf{x}, \mathbf{x}') = p_{\sigma\sigma'}(\mathbf{r} - \mathbf{r}')$ depends only on the coordinate difference. If the system is in a cubic box of volume $\mathbf{V} = L^3$ then

$$p_{\sigma\sigma'}(\mathbf{r} - \mathbf{r}') = \frac{1}{\mathbf{V}} \sum_{\mathbf{k}} e^{i\mathbf{k}\cdot(\mathbf{r}-\mathbf{r}')} \tilde{p}_{\sigma\sigma'}(\mathbf{k}).$$

For large $|\mathbf{r} - \mathbf{r}'|$ we can restrict the sum over $\mathbf{k}$ to those wavevectors with $|\mathbf{k}| \lesssim 2\pi/|\mathbf{r} - \mathbf{r}'|$ since the sum over the wavevectors with $|\mathbf{k}| \gtrsim 2\pi/|\mathbf{r} - \mathbf{r}'|$ gives approximately zero. Thus in the limit $|\mathbf{r} - \mathbf{r}'| \to \infty$ only the term with $\mathbf{k} = 0$ survives,

$$\lim_{|\mathbf{r}-\mathbf{r}'|\to\infty} p_{\sigma\sigma'}(\mathbf{r} - \mathbf{r}') = \frac{1}{\mathbf{V}} \tilde{p}_{\sigma\sigma'}(0). \tag{C.3}$$

To calculate $\tilde{p}_{\sigma\sigma'}(0)$ we observe that by definition

$$p_{\sigma\sigma'}(\mathbf{r} - \mathbf{r}') = \langle\Psi|\hat{\psi}^\dagger(\mathbf{x})\hat{\psi}^\dagger(\mathbf{x}')\hat{\psi}(\mathbf{x}')\hat{\psi}(\mathbf{x})|\Psi\rangle = \langle\Psi|\hat{n}(\mathbf{x})\hat{n}(\mathbf{x}')|\Psi\rangle \pm \delta(\mathbf{x} - \mathbf{x}')n(\mathbf{x}). \tag{C.4}$$

Therefore

$$\tilde{p}_{\sigma\sigma'}(0) = \int d\mathbf{r}\, p_{\sigma\sigma'}(\mathbf{r}) = N_\sigma n_{\sigma'} \pm \delta_{\sigma\sigma'} n_\sigma,$$

where we use the fact that $|\Psi\rangle$ is an eigenstate of the total number of particles of spin σ operator $\hat{N}_\sigma = \int d\mathbf{r}\, n(\mathbf{x})$ with eigenvalue N_σ. Inserting this result into (C.3), taking the limit $\mathbf{V} \to \infty$ and using $N_\sigma/\mathbf{V} = n_\sigma$ we get

$$\lim_{|\mathbf{r}-\mathbf{r}'|\to\infty} p_{\sigma\sigma'}(\mathbf{r} - \mathbf{r}') = n_\sigma n_{\sigma'},$$

which implies that $g \to 1$ in the same limit.

A formula that is commonly found in the literature is

$$E_{\text{int}} = \frac{1}{2} \int d\mathbf{x}d\mathbf{x}'v(\mathbf{x}, \mathbf{x}')n(\mathbf{x})n(\mathbf{x}') + \frac{1}{2} \int d\mathbf{x}d\mathbf{x}'v(\mathbf{x}, \mathbf{x}')n(\mathbf{x})n(\mathbf{x}') \left[g(\mathbf{x}, \mathbf{x}') - 1\right].$$

The first term represents the classical (Hartree) interaction between two densities $n(\mathbf{x})$ and $n(\mathbf{x}')$ whereas the second term denotes the exchange–correlation part of the interaction.

In fact g is important also because it can be measured using X-ray scattering [75]. Theoretically the quantity $p(\mathbf{x}, \mathbf{x}')$ is closely related to the density response function since in equilibrium systems we can rewrite (C.4) as

$$p(\mathbf{x}, \mathbf{x}') = \langle\Psi|\hat{n}_H(\mathbf{x}, t)\hat{n}_H(\mathbf{x}', t)|\Psi\rangle \pm \delta(\mathbf{x} - \mathbf{x}')n(\mathbf{x}).$$

In the first term we recognize the greater part of the density response function $\chi(\mathbf{x}, z; \mathbf{x}', z')$, i.e.,

$$\langle \Psi | \hat{n}_H(\mathbf{x}, t) \hat{n}_H(\mathbf{x}', t) | \Psi \rangle = n(\mathbf{x}) n(\mathbf{x}') + i \chi^>(\mathbf{x}, t; \mathbf{x}', t)$$
$$= n(\mathbf{x}) n(\mathbf{x}') + i \int \frac{d\omega}{2\pi} \chi^>(\mathbf{x}, \mathbf{x}'; \omega),$$

where we use (15.12). This means that $p(\mathbf{x}, \mathbf{x}')$ can be calculated from diagrammatic perturbation theory as explained in Chapter 15.

As an example we consider the electron gas. In momentum space the relation between p and χ reads

$$\tilde{p}_{\sigma\sigma'}(\mathbf{k}) = n_\sigma n_{\sigma'} (2\pi)^3 \delta(\mathbf{k}) - \delta_{\sigma\sigma'} n_\sigma + i \int \frac{d\omega}{2\pi} \chi^>_{\sigma\sigma'}(\mathbf{k}, \omega). \tag{C.5}$$

We define $p = \sum_{\sigma\sigma'} p_{\sigma\sigma'}$ and similarly $\chi = \sum_{\sigma\sigma'} \chi_{\sigma\sigma'}$ and $n = \sum_\sigma n_\sigma$. Then, summing (C.5) over σ and σ' and using the fluctuation–dissipation theorem for $\chi^>$ (see Section 15.2) we get

$$\tilde{p}(\mathbf{k}) = n^2 (2\pi)^3 \delta(\mathbf{k}) - n + i \int \frac{d\omega}{2\pi} \bar{f}(\omega) \left[\chi^R(\mathbf{k}, \omega) - \chi^A(\mathbf{k}, \omega) \right].$$

In this equation the difference inside the square brackets is $2i \operatorname{Im}[\chi^R(\mathbf{k}, \omega)]$ since the property (15.25) implies that $\chi^A(\mathbf{k}, \omega) = [\chi^R(\mathbf{k}, \omega)]^*$. In the zero-temperature limit $\beta \to \infty$, the Bose function $\bar{f}(\omega) = 1/(1 - e^{-\beta\omega})$ vanishes for negative ωs and is unity for positive ωs. Therefore

$$\tilde{p}(\mathbf{k}) = n^2 (2\pi)^3 \delta(\mathbf{k}) - n - \frac{1}{\pi} \int_0^\infty d\omega \, \operatorname{Im}\left[\chi^R(\mathbf{k}, \omega) \right].$$

Let us evaluate the frequency integral when χ is the Lindhard density response function worked out in Section 15.5.1. According to Fig. 15.4(a) and to equation (15.72) we have for $k = |\mathbf{k}| < 2p_F$

$$\int_0^\infty d\omega \, \operatorname{Im}\left[\chi^R(\mathbf{k}, \omega) \right] = \epsilon_{p_F} \int_0^{-x^2+2x} d\nu \left(-\frac{p_F}{2\pi} \right) \frac{\nu}{2x}$$
$$+ \epsilon_{p_F} \int_{-x^2+2x}^{x^2+2x} d\nu \left(-\frac{p_F}{2\pi} \right) \frac{1}{2x} \left[1 - \left(\frac{\nu - x^2}{2x} \right)^2 \right]$$
$$= -\frac{p_F^3}{4\pi} \left(x - \frac{x^3}{12} \right),$$

whereas for $k > 2p_F$

$$\int_0^\infty d\omega \, \operatorname{Im}\left[\chi^R(\mathbf{k}, \omega) \right] = \epsilon_{p_F} \int_{x^2-2x}^{x^2+2x} d\nu \left(-\frac{p_F}{2\pi} \right) \frac{1}{2x} \left[1 - \left(\frac{\nu - x^2}{2x} \right)^2 \right]$$
$$= -\frac{p_F^3}{4\pi} \frac{4}{3}.$$

In these equations $x = k/p_F$, $\nu = \omega/\epsilon_{p_F}$, and $\epsilon_{p_F} = p_F^2/2$, see (15.67). Taking into account that $p_F^3 = 3\pi^2 n$, see (7.57), we conclude that

$$\tilde{p}(\mathbf{k}) = \begin{cases} n^2 (2\pi)^3 \delta(\mathbf{k}) + n(-1 + \frac{3x}{4} - \frac{x^3}{16}) & |\mathbf{k}| < 2p_F \\ 0 & |\mathbf{k}| > 2p_F \end{cases}.$$

Fourier transforming back to real space we find

$$
\begin{aligned}
p(\mathbf{r}) &= \int \frac{d\mathbf{k}}{(2\pi)^3} e^{i\mathbf{k}\cdot\mathbf{r}} \, \tilde{p}(\mathbf{k}) \\
&= n^2 + \frac{n}{2\pi^2 r} \int_0^{2p_F} dk \left(-1 + \frac{3x}{4} - \frac{x^3}{16} \right) k \sin kr \qquad \text{(setting } \alpha = p_F r) \\
&= n^2 + \frac{3n^2}{2\alpha} \int_0^2 dx \left(-x + \frac{3x^2}{4} - \frac{x^4}{16} \right) \sin x\alpha \\
&= n^2 - \frac{9n^2}{2} \left(\frac{\sin\alpha - \alpha\cos\alpha}{\alpha^3} \right)^2.
\end{aligned}
\tag{C.6}
$$

In the electron gas the pair-correlation function (C.2) reads

$$g(\mathbf{x}, \mathbf{x}') = g_{\sigma\sigma'}(\mathbf{r} - \mathbf{r}') = \frac{p_{\sigma\sigma'}(\mathbf{r} - \mathbf{r}')}{n_\sigma n_{\sigma'}}.$$

If we define $g = \frac{1}{4}\sum_{\sigma\sigma'} g_{\sigma\sigma'}$ and use the fact that $n_\sigma = n/2$, then (C.6) implies

$$g(\mathbf{r}) = \frac{p(\mathbf{r})}{n^2} = 1 - \frac{9}{2} \left(\frac{\sin p_F r - p_F r \cos p_F r}{(p_F r)^3} \right)^2.$$

As expected $g \to 1$ for $r \to \infty$. For small r we have $\sin\alpha - \alpha\cos\alpha = \alpha^3/3 + \ldots$ and hence $g(0) = 1/2$. This result can easily be interpreted. Since we use the noninteracting density response function we have only incorporated antisymmetry. Therefore like-spin electrons are correlated through the Pauli exclusion principle, but unlike-spin electrons are not. If we go beyond the simple noninteracting case then we find that $g(0) < 1/2$. In fact, if we use the RPA approximation for χ we find that $g(0)$ can even become negative at small enough densities. Going beyond RPA repairs this situations. For a thorough discussion on these topics see Refs. [67,75,76].

D

Thermodynamics and statistical mechanics

We start by considering a system with a given number N of particles whose volume V and temperature T can change. The first law of thermodynamics establishes a relation between the change of energy dE of the system, the work $\delta\mathcal{L}$ done by the system and the heat δQ absorbed by the system:

$$dE = \delta Q - \delta\mathcal{L}. \tag{D.1}$$

In (D.1) the infinitesimals δQ and $\delta\mathcal{L}$ are not exact differentials since the total absorbed heat or the total work done depend on the path of the transformation which brings the system from one state to another and not only on the initial and final states. The second law of thermodynamics establishes that

$$\delta Q \leq TdS,$$

where S is the entropy and the equality holds only for reversible processes. In what follows we consider only reversible processes and write $\delta Q = TdS$. As the only work that the system can do is to increase its volume we have $\delta\mathcal{L} = PdV$, where P is the internal pressure. Therefore (D.1) takes the form

$$dE = TdS - PdV.$$

From this relation we see that the internal energy $E = E(S, V, N)$ is a function of the entropy, the volume, and the number of particles.

If we now also allow the number of particles to change, then dE acquires the extra term $(\partial E/\partial N)_{S,T}\, dN$. The quantity

$$\mu \equiv \left(\frac{\partial E}{\partial N}\right)_{S,T}$$

is known as the chemical potential and represents the energy cost for adding a particle to the system. In conclusion, the change in energy of a system that can exchange heat and particles, and that can expand its volume reads

$$dE = TdS - PdV + \mu dN. \tag{D.2}$$

The internal energy E is not the most convenient quantity to work with due to its explicit dependence on S, a difficult quantity to control and measure. The only situation for which we do not need to bother about S is that of a system at zero temperature since then $TdS = 0$. In fact, the energy is mostly used in this circumstance. Experimentally, it is much easier to control and measure the temperature, the pressure and the volume. We can then introduce more convenient quantities by a Legendre transformation. The *free energy* or *Helmholtz energy* is defined as

$$F = E - TS \qquad \Rightarrow \qquad dF = -SdT - PdV + \mu dN \tag{D.3}$$

and depends on T, V, N. From it we can also define the *Gibbs energy*

$$G = F + PV \qquad \Rightarrow \qquad dG = -SdT + VdP + \mu dN, \tag{D.4}$$

which instead depends on T, P, N. We see that in G the only extensive variable is the number of particles and therefore

$$G(T, P, N) = N\mu(T, P), \tag{D.5}$$

according to which the chemical potential is the Gibbs energy per particle. For later purposes we also define another important thermodynamic quantity known as the *grand potential*,

$$\Omega = F - \mu N \qquad \Rightarrow \qquad d\Omega = -SdT - PdV - Nd\mu, \tag{D.6}$$

which depends explicitly on the chemical potential μ. From a knowledge of these thermodynamic energies we can extract all thermodynamic quantities by differentiation. For instance from (D.2) we have

$$\left(\frac{\partial E}{\partial S} \right)_{V,N} = T, \qquad \left(\frac{\partial E}{\partial V} \right)_{S,N} = -P, \qquad \left(\frac{\partial E}{\partial N} \right)_{S,V} = \mu,$$

which can be conveniently shortened as

$$(\partial_S, \partial_V, \partial_N) E = (T, -P, \mu).$$

In this compact notation it is evident that E depends on the independent variables (S, V, N). Similarly we can write

$$(\partial_T, \partial_V, \partial_N) F = (-S, -P, \mu),$$

$$(\partial_T, \partial_P, \partial_N) G = (-S, V, \mu),$$

$$(\partial_T, \partial_V, \partial_\mu) \Omega = (-S, -P, -N).$$

Let us briefly summarize what we have seen so far. By combining the first and second law of thermodynamics we were able to express the *differential* of the various energies in terms of the basic thermodynamic variables (S, T, P, V, N, μ). Using (D.5) we can *integrate* dG and obtain $G = \mu N$. This result is particularly important since we can now give an explicit form to all other energies. If we insert (D.5) into (D.4) we find

$$F = -PV + \mu N$$

and similarly we can construct the energy and the grand potential

$$E = TS - PV + \mu N, \qquad \Omega = -PV. \tag{D.7}$$

The grand potential is simply the product of pressure and volume, and recalling that its independent variables are T, V, μ, we have

$$d\Omega = -\left(\frac{\partial P}{\partial T}\right)_{V,\mu} V dT - \left(\frac{\partial P}{\partial V}\right)_{T,\mu} V dV - \left(\frac{\partial P}{\partial \mu}\right)_{T,V} V d\mu - PdV.$$

Comparing this result with (D.6) we find the important relations

$$S = \left(\frac{\partial P}{\partial T}\right)_{V,\mu} V, \qquad N = \left(\frac{\partial P}{\partial \mu}\right)_{T,V} V, \tag{D.8}$$

along with the obvious one $(\partial P/\partial V)_{T,\mu} = 0$ (the pressure of a system with a given temperature and density is the same for all volumes). The second of (D.8) was used in Section 7.2.2 to calculate the pressure of an electron gas in the Hartree approximation.

All relations found so far are useless without a microscopic way to calculate the energies. Statistical mechanics, in its classical or quantum version, provides the bridge between the microscopic laws governing the motion of the particles and the macroscopic quantities of thermodynamics. The connection between statistical mechanics and thermodynamics is due to Boltzmann. Let E_n be one of the possible energies of the system with N_n particles and let us consider $\mathcal{M}$ identical copies of the same system. This hypersystem is usually referred to as an *ensemble*. Suppose that in the ensemble there are M_1 systems with energy E_1 and number of particles N_1, M_2 systems with energy E_2 and number of particles N_2, etc. with, of course, $M_1 + M_2 + \ldots = \mathcal{M}$. The total energy of the ensemble is therefore $\mathcal{E} = \sum_n M_n E_n$ while the total number of particles is $\mathcal{N} = \sum_n M_n N_n$. Let us calculate the degeneracy of level $\mathcal{E}$. This is a simple combinatorial problem. Consider for instance an ensemble of 12 copies like the one illustrated in the table below:

E_1	E_2	E_3	E_2
E_3	E_3	E_1	E_2
E_2	E_1	E_2	E_3

We have $M_1 = 3$, $M_2 = 5$ and $M_3 = 4$ and the total energy of the ensemble is $\mathcal{E} = 3E_1 + 5E_2 + 4E_3$. The degeneracy $d_{\mathcal{E}}$ of level $\mathcal{E}$ is

$$d_{\mathcal{E}} = \frac{12!}{3!\,5!\,4!} = 27720.$$

In general the number of ways to have M_1 systems with energy E_1, M_2 systems with energy E_2, etc. over an ensemble of $M_1 + M_2 + \ldots = \mathcal{M}$ copies is given by

$$d_{\mathcal{E}} = \frac{\mathcal{M}!}{M_1!\,M_2!\ldots}.$$

If the ensemble is formed by a very large number of copies, $\mathcal{M} \gg 1$, then also the $M_k \gg 1$ and it is more convenient to work with the logarithm of the degeneracy. For a large number P we can use the *Stirling formula*

$$\ln P! \sim P \ln P - P$$

so that

$$\ln d_{\mathcal{E}} \sim \mathcal{M} \ln \mathcal{M} - \mathcal{M} - \sum_n (M_n \ln M_n - M_n)$$

$$= -\mathcal{M} \sum_n w_n \ln w_n,$$

where $w_n \equiv M_n/\mathcal{M}$ can be interpreted as the probability of finding one of the systems of the ensemble in a state of energy E_n. It is now reasonable to expect that an ensemble in equilibrium is an ensemble in the most probable configuration, i.e., an ensemble which maximizes the degeneracy $d_{\mathcal{E}}$. Let us then study what we get if we maximize the quantity

$$S[\{w_n\}] \equiv -K_B \sum_n w_n \ln w_n, \tag{D.9}$$

under the constraints that the probabilities w_n sum up to 1, $\sum_n w_n = 1$, that the average energy is E, $\sum_n w_n E_n = E$, and that the average number of particles is N, $\sum_n w_n N_n = N$. The constant K_B in (D.9) is completely irrelevant for our purpose and its presence is justified below. With the help of three Lagrange multipliers we must find the unconstrained maximum of the function

$$\frac{1}{K_B} \tilde{S}[\{w_n\}, \lambda_1, \lambda_2, \lambda_3] = -\sum_n w_n \ln w_n - \lambda_1 \left(\sum_n w_n - 1 \right)$$

$$- \lambda_2 \left(\sum_n w_n E_n - E \right) - \lambda_3 \left(\sum_n w_n N_n - N \right)$$

with respect to all w_ns and the three λs. Setting to zero the derivative of $\tilde{S}$ with respect to w_n we find

$$\frac{\partial \tilde{S}}{\partial w_n} = -\ln w_n - 1 - \lambda_1 - \lambda_2 E_n - \lambda_3 N_n = 0,$$

from which it follows that

$$w_n = e^{-(1+\lambda_1) - \lambda_2 E_n - \lambda_3 N_n}.$$

This solution is a maximum since $\frac{\partial^2 \tilde{S}}{\partial w_n \partial w_m} = -\frac{\delta_{mn}}{w_n} < 0$. To find the Lagrange multipliers we use the constraints. The first constraint $\sum_n w_n = 1$ yields

$$e^{1+\lambda_1} = \sum_n e^{-\lambda_2 E_n - \lambda_3 N_n} \equiv Z, \quad \Rightarrow \quad w_n = \frac{e^{-\lambda_2 E_n - \lambda_3 N_n}}{Z}. \tag{D.10}$$

Choosing λ_2 and λ_3 so as to satisfy the constraints

$$\sum_n \underbrace{\frac{e^{-\lambda_2 E_n - \lambda_3 N_n}}{Z}}_{w_n} E_n = E, \qquad \sum_n \underbrace{\frac{e^{-\lambda_2 E_n - \lambda_3 N_n}}{Z}}_{w_n} N_n = N,$$

then $\tilde{S}$ evaluated at the w_ns of (D.10) is equal to

$$S = K_B \left(\lambda_2 \sum_n w_n E_n + \lambda_3 \sum_n w_n N_n + \ln Z \right) = K_B (\lambda_2 E + \lambda_3 N + \ln Z). \quad \text{(D.11)}$$

The groundbreaking idea of Boltzmann was to identify this quantity with the entropy. The constant $K_B = 8.3 \times 10^{-5}$ eV/K, duly called the *Boltzmann constant*, was chosen to fit the first thermodynamic relation in (D.7). In fact, from the comparison between (D.11) and (D.7) we see that the Boltzmann idea is very sound and sensible. The physical meaning of λ_2, λ_3, and Z can be deduced by equating these two formulas; what we find is

$$\lambda_2 = \beta = \frac{1}{K_B T}, \qquad \lambda_3 = -\beta\mu, \qquad \ln Z = \beta P V = -\beta\Omega. \quad \text{(D.12)}$$

In quantum mechanics the possible values of the energies E_k are the eigenvalues of the Hamiltonian operator $\hat{H}$. Then the quantity Z, also called the partition function, can be written as

$$Z = \sum_k e^{-\beta(E_k - \mu N_k)} = \text{Tr}\left[e^{-\beta(\hat{H} - \mu\hat{N})} \right].$$

It is possible to show that all the thermodynamics derivatives agree with the Boltzmann idea. For instance, from (D.6) we have

$$-\left(\frac{\partial}{\partial\beta} \beta\Omega \right)_{V,\mu} = -\Omega - \beta \left(\frac{\partial\Omega}{\partial T} \right)_{V,\mu} \frac{\partial T}{\partial\beta} = PV - TS = -E + \mu N,$$

and the same result follows from (D.12) since

$$-\left(\frac{\partial}{\partial\beta} \beta\Omega \right)_{V,\mu} = \left(\frac{\partial\ln Z}{\partial\beta} \right)_{V,\mu} = -\frac{\text{Tr}\left[e^{-\beta(\hat{H} - \mu\hat{N})}(\hat{H} - \mu\hat{N}) \right]}{\text{Tr}\left[e^{-\beta(\hat{H} - \mu\hat{N})} \right]} = -E + \mu N.$$

Similarly, from (D.6) we have

$$\left(\frac{\partial\Omega}{\partial\mu} \right)_{V,T} = -N,$$

and the same result follows from (D.12) since

$$\left(\frac{\partial\Omega}{\partial\mu} \right)_{V,T} = -\frac{1}{\beta} \left(\frac{\partial\ln Z}{\partial\mu} \right)_{V,T} = -\frac{\text{Tr}\left[e^{-\beta(\hat{H} - \mu\hat{N})}\hat{N} \right]}{\text{Tr}\left[e^{-\beta(\hat{H} - \mu\hat{N})} \right]} = -N.$$

The statistical approach to thermodynamics also allows us to understand the positivity of the temperature from a microscopic point of view. By increasing the average energy E the degeneracy $d_{\mathcal{E}}$ increases, and hence also the entropy increases, $\partial S/\partial E > 0$. From (D.2) we see that $\partial S/\partial E = 1/T$ and hence the temperature must be positive.

In conclusion, the procedure to extract thermodynamic quantities from a system with Hamiltonian $\hat{H}$ whose eigenstates $|\Psi_k\rangle$ have energies E_k and number of particles N_k is:

- Calculate the partition function

$$Z = \sum_k e^{-\beta(E_k - \mu N_k)},$$

which depends on μ, T, V (the dependence on V comes from the Hamiltonian).

- Calculate the grand potential

$$\Omega = -\frac{1}{\beta} \ln Z.$$

- To calculate the energy we can use

$$E = -\frac{\partial}{\partial \beta} \ln Z.$$

In this way, however, the energy depends on μ, T, V. To eliminate μ in favour of N we must invert

$$N = \frac{1}{\beta} \frac{\partial}{\partial \mu} \ln Z.$$

- To find the equation of state of the system we can use the third of (D.12) and eliminate μ in favour of N as described above.

- Other thermodynamics quantities of interest to the reader can be obtained with similar techniques.

E

Green's functions and lattice symmetry

In Section 2.3.1 we introduced several important physical systems with periodic symmetry such as carbon nanotubes and the graphene lattice. There the discussion was simplified by the fact that we dealt with noninteracting electrons. In this appendix we work out the consequence of lattice symmetry for interacting systems and ask ourselves in which way the many-body states can be labeled by the crystal momentum vectors. The fact that this should be possible is suggested by (inverse) photoemission experiments on crystals which clearly measure the band structure. The deeper physical meaning of band structure of interacting systems should therefore be apparent from a calculation of the spectral function, see Section 6.3.4. To elucidate these aspects we must study the symmetry properties of the interacting Green's function under lattice translations. For concreteness we consider the electronic system of a three-dimensional crystal with lattice vectors $\mathbf{v}_1$, $\mathbf{v}_2$, and $\mathbf{v}_3$. This means that every unit cell is repeated periodically along $\mathbf{v}_1$, $\mathbf{v}_2$, and $\mathbf{v}_3$.

The starting point of our discussion is the Hamiltonian (the electron charge is $q = -1$)

$$\hat{H} = \int d\mathbf{x}\, \hat{\psi}^\dagger(\mathbf{x}) \left(-\frac{\nabla^2}{2} - V(\mathbf{r}) \right) \hat{\psi}(\mathbf{x}) + \frac{1}{2} \int d\mathbf{x} d\mathbf{x}'\, v(\mathbf{x}, \mathbf{x}') \hat{\psi}^\dagger(\mathbf{x}) \hat{\psi}^\dagger(\mathbf{x}') \hat{\psi}(\mathbf{x}') \hat{\psi}(\mathbf{x}), \quad \text{(E.1)}$$

where we take the interaction to be $v(\mathbf{x}, \mathbf{x}') = v(\mathbf{r} - \mathbf{r}')$ and the potential V generated by the atomic nuclei in the crystal to be spin-independent and lattice-periodic,

$$V(\mathbf{r} + \mathbf{v}_j) = V(\mathbf{r}) \qquad j = 1, 2, 3.$$

This symmetry is, strictly speaking, only valid in a truly infinite system. The mathematical treatment of a Hamiltonian of an infinite number of particles is ill-defined as the Schrödinger equation becomes a differential equation for a wavefunction with infinitely many coordinates. A real piece of solid is, of course, finite and the symmetry is an approximation that is very good for bulk electrons. Since for the theoretical treatment the full periodic symmetry is a clear advantage we impose periodic boundary conditions on a large box (BvK boundary conditions), as we did in Section 2.3.1. For simplicity we choose the box with edges given by the vectors $\mathbf{V}_j = N_j \mathbf{v}_j$, each radiating from, say, the origin of our reference frame. In the language of Section 2.3.1 this choice corresponds to taking $\mathbf{N}_1 = (N_1, 0, 0)$, $\mathbf{N}_2 = (0, N_2, 0)$, and $\mathbf{N}_3 = (0, 0, N_3)$. The requirement of periodicity for a many-body state $|\Psi\rangle$ with N particles implies that $|\Psi\rangle$ satisfies

$$\langle \mathbf{x}_1 \ldots (\mathbf{r}_i + \mathbf{V}_j)\sigma_i \ldots \mathbf{x}_N | \Psi \rangle = \langle \mathbf{x}_1 \ldots \mathbf{x}_i \ldots \mathbf{x}_N | \Psi \rangle \tag{E.2}$$

for all $j = 1, 2, 3$. In this way we have as many conditions as spatial differential operators in the Schrödinger equation (since the kinetic energy operator is a second order differential operator we also need to put periodic boundary conditions on the derivatives of the many-body wave functions). Although these conditions seem natural they are not trivial in the presence of many-body interactions. This is because the many-body interaction is invariant under the simultaneous translation of all particles, i.e., $v(\mathbf{r} - \mathbf{r}') = v((\mathbf{r} + \mathbf{V}_j) - (\mathbf{r}' + \mathbf{V}_j))$ but not under the translation of a single particle. The Hamiltonian does not therefore have a symmetry compatible with the boundary conditions (E.2). To solve this problem we replace the two-body interaction by

$$v(\mathbf{r} - \mathbf{r}') = \frac{1}{V} \sum_{\mathbf{k}} \tilde{v}_{\mathbf{k}} \, e^{i\mathbf{k} \cdot (\mathbf{r} - \mathbf{r}')}, \tag{E.3}$$

where V is the volume of the box, $\tilde{v}_{\mathbf{k}}$ is the Fourier transform of v, and the sum runs over all vectors such that $\mathbf{k} \cdot \mathbf{V}_j = 2\pi m_j$ for $j = 1, 2, 3$, and m_j integers. The interaction (E.3) satisfies $v(\mathbf{r} + \mathbf{V}_j - \mathbf{r}') = v(\mathbf{r} - \mathbf{r}')$ and becomes equal to the original interaction when we take the limit $V \to \infty$. With this replacement and the BvK boundary conditions (E.2) the eigenvalue equation for the Hamiltonian $\hat{H}$ becomes well-defined (of course the spatial integrations in (E.1) must be restricted to the box).

Let us now explore the lattice symmetry. We consider the total momentum operator of the system defined in (3.32):

$$\hat{\mathbf{P}} \equiv \frac{1}{2i} \int d\mathbf{x} \left[\hat{\psi}^\dagger(\mathbf{x}) \left(\boldsymbol{\nabla} \hat{\psi}(\mathbf{x}) \right) - \left(\boldsymbol{\nabla} \hat{\psi}^\dagger(\mathbf{x}) \right) \hat{\psi}(\mathbf{x}) \right].$$

This operator has the property that

$$-i\boldsymbol{\nabla}\hat{\psi}(\mathbf{x}) = \left[\hat{\psi}(\mathbf{x}), \hat{\mathbf{P}} \right]_- \; ; \quad -i\boldsymbol{\nabla}\hat{\psi}^\dagger(\mathbf{x}) = \left[\hat{\psi}^\dagger(\mathbf{x}), \hat{\mathbf{P}} \right]_- .$$

From these equations we deduce that

$$\hat{\psi}(\mathbf{r} + \mathbf{r}'\sigma) = e^{-i\hat{\mathbf{P}} \cdot \mathbf{r}} \, \hat{\psi}(\mathbf{r}'\sigma) \, e^{i\hat{\mathbf{P}} \cdot \mathbf{r}}, \tag{E.4}$$

with a similar equation with $\hat{\psi}^\dagger$ replacing $\hat{\psi}$. Equation (E.4) is readily checked by taking the gradient with respect to $\mathbf{r}$ on both sides and verifying the condition that both sides are equal for $\mathbf{r} = \mathbf{r}'$. It is now readily seen that the unitary operator $e^{-i\hat{\mathbf{P}} \cdot \mathbf{r}}$ is the operator that shifts all particle coordinates over the vector $\mathbf{r}$. This follows from the adjoint of (E.4) if we evaluate

$$
\begin{aligned}
|\mathbf{r}_1 + \mathbf{r}\,\sigma_1, \ldots, \mathbf{r}_N + \mathbf{r}\,\sigma_N\rangle &= \hat{\psi}^\dagger(\mathbf{r}_N + \mathbf{r}\,\sigma_N) \ldots \hat{\psi}^\dagger(\mathbf{r}_1 + \mathbf{r}\,\sigma_1)|0\rangle \\
&= e^{-i\hat{\mathbf{P}} \cdot \mathbf{r}} \hat{\psi}^\dagger(\mathbf{r}_N \sigma_N) \ldots \hat{\psi}^\dagger(\mathbf{r}_1 \sigma_1) e^{i\hat{\mathbf{P}} \cdot \mathbf{r}}|0\rangle \\
&= e^{-i\hat{\mathbf{P}} \cdot \mathbf{r}}|\mathbf{x}_1, \ldots, \mathbf{x}_N\rangle,
\end{aligned}
$$

where we use the fact that the action of $\hat{\mathbf{P}}$ on the empty ket $|0\rangle$ is zero. It therefore follows in particular that the many-body states that satisfy the BvK boundary conditions (E.2) also satisfy

$$e^{i\hat{\mathbf{P}}\cdot\mathbf{V}_j}|\Psi\rangle = |\Psi\rangle. \tag{E.5}$$

Let us now consider the Hamiltonian (E.1) with the spatial integrals restricted to the box. We first demonstrate that

$$\hat{H} = e^{-i\hat{\mathbf{P}}\cdot\mathbf{v}_j}\hat{H}e^{i\hat{\mathbf{P}}\cdot\mathbf{v}_j}, \qquad j = 1, 2, 3. \tag{E.6}$$

The one-body operator $\hat{V}$ of the potential energy is

$$\hat{V} = \int_V d\mathbf{x}\, V(\mathbf{r})\hat{\psi}^\dagger(\mathbf{x})\hat{\psi}(\mathbf{x}) = \int_V d\mathbf{x}\, V(\mathbf{r})\hat{\psi}^\dagger(\mathbf{r} + \mathbf{v}_j\,\sigma)\hat{\psi}(\mathbf{r} + \mathbf{v}_j\,\sigma)$$

$$= \int_V d\mathbf{x}\, e^{-i\hat{\mathbf{P}}\cdot\mathbf{v}_j}\hat{\psi}^\dagger(\mathbf{x})\hat{\psi}(\mathbf{x})e^{i\hat{\mathbf{P}}\cdot\mathbf{v}_j}V(\mathbf{r}) = e^{-i\hat{\mathbf{P}}\cdot\mathbf{v}_j}\hat{V}e^{i\hat{\mathbf{P}}\cdot\mathbf{v}_j}.$$

Similarly we can check the same transform law for the kinetic energy operator and the interaction operator with the interaction (E.3). We have therefore proven (E.6) or equivalently

$$\left[\hat{H}, e^{i\hat{\mathbf{P}}\cdot\mathbf{v}_j}\right]_- = 0, \qquad j = 1, 2, 3.$$

The set of unitary operators $e^{i\hat{\mathbf{P}}\cdot\mathbf{v}}$ with $\mathbf{v} = n_1\mathbf{v}_1 + n_2\mathbf{v}_2 + n_3\mathbf{v}_3$ all commute with the Hamiltonian and we can thus find a common set of eigenstates. These eigenstates satisfy

$$e^{i\hat{\mathbf{P}}\cdot\mathbf{v}_j}|\Psi\rangle = e^{i\alpha_j}|\Psi\rangle,$$

where the eigenvalues are a pure phase factor. The conditions (E.5) on the eigenstates imply that (recall that $\mathbf{V}_j = N_j\mathbf{v}_j$)

$$\left(e^{i\hat{\mathbf{P}}\cdot\mathbf{v}_j}\right)^{N_j}|\Psi\rangle = |\Psi\rangle. \tag{E.7}$$

Then we see that the phases α_j are not arbitrary but have to fulfill $e^{i\alpha_j N_j} = 1$ and hence $\alpha_j = 2\pi m_j/N_j$ with m_j an integer. Note that the dimensionless vectors $\tilde{\mathbf{k}} = (\alpha_1, \alpha_2, \alpha_3)$ are exactly the vectors of Section 2.3.1, see (2.12). Equation (E.7) tells us that we can label the eigenstates of $\hat{H}$ with a crystal momentum $\mathbf{k}$ and a remaining quantum number l since for an arbitrary translation $\mathbf{v} = n_1\mathbf{v}_1 + n_2\mathbf{v}_2 + n_3\mathbf{v}_3$ we have

$$e^{i\hat{\mathbf{P}}\cdot\mathbf{v}}|\Psi_{\mathbf{k}\,l}\rangle = e^{in_1\alpha_1 + n_2\alpha_2 + n_3\alpha_3}|\Psi_{\mathbf{k}\,l}\rangle = e^{i\tilde{\mathbf{k}}\cdot\mathbf{n}}|\Psi_{\mathbf{k}\,l}\rangle = e^{i\mathbf{k}\cdot\mathbf{v}}|\Psi_{\mathbf{k}\,l}\rangle, \tag{E.8}$$

where $\mathbf{k}$ is such that $\mathbf{k}\cdot\mathbf{v}_j = \alpha_j$.[1] Similarly to the $\tilde{\mathbf{k}}$ vectors, the $\mathbf{k}$ vectors which differ by a vector $\mathbf{K}$ or multiples thereof with the property that $\mathbf{K}\cdot\mathbf{v}_j = 2\pi$ must be identified. The

---

[1] The vectors $\mathbf{k}$ have the physical dimension of the inverse of a length and should not be confused with the dimensionless vectors $\tilde{\mathbf{k}}$ of Section 2.3.1. Let us derive the relation between the two. The vectors $\tilde{\mathbf{k}}$ can be written as $\tilde{\mathbf{k}} = \sum_j \alpha_j \mathbf{e}_j$. On the other hand the solution of $\mathbf{k}\cdot\mathbf{v}_j = \alpha_j$ is $\mathbf{k} = \sum_j \alpha_j \mathbf{b}_j$ where

$$\mathbf{b}_1 = \frac{\mathbf{v}_2 \times \mathbf{v}_3}{\mathbf{v}_1\cdot(\mathbf{v}_2 \times \mathbf{v}_3)}, \qquad \mathbf{b}_2 = \frac{\mathbf{v}_3 \times \mathbf{v}_1}{\mathbf{v}_1\cdot(\mathbf{v}_2 \times \mathbf{v}_3)}, \qquad \mathbf{b}_3 = \frac{\mathbf{v}_1 \times \mathbf{v}_2}{\mathbf{v}_1\cdot(\mathbf{v}_2 \times \mathbf{v}_3)}.$$

In the special case that $\mathbf{v}_1 = (a_1, 0, 0)$, $\mathbf{v}_2 = (0, a_2, 0)$, and $\mathbf{v}_3 = (0, 0, a_3)$ are orthogonal we simply have $\mathbf{b}_j = \mathbf{e}_j/a_j$.

set of inequivalent $\mathbf{k}$ vectors is called the *first Brillouin zone*. The vectors $\mathbf{K}$ are known as reciprocal lattice vectors. With this convention (E.8) represents the many-body generalization of the Bloch theorem for single-particle states. It tells us that a simultaneous translation of all particles over a lattice vector $\mathbf{v}$ changes the many-body states by a phase factor.

We are now ready to study the symmetry properties of the Green's function. Let us, for example, consider the expression (6.88) for $G^>$ in position basis. The labels m in our case are $\mathbf{k}\,l$ and we have

$$G^>(\mathbf{x}, \mathbf{x}'; \omega) = -2\pi i \sum_{\mathbf{k}\,l} P_{\mathbf{k}\,l}(\mathbf{x}) P^*_{\mathbf{k}\,l}(\mathbf{x}') \delta(\omega - [E_{N+1,\mathbf{k}\,l} - E_{N,0}])$$

with quasi-particle wavefunctions

$$P_{\mathbf{k}\,l}(\mathbf{x}) = \langle \Psi_{N,0} | \hat{\psi}(\mathbf{x}) | \Psi_{N+1,\mathbf{k}\,l} \rangle.$$

Let us now see how the quasi-particle wave functions change under a lattice translation:

$$\begin{aligned} P_{\mathbf{k}\,l}(\mathbf{x} + \mathbf{v}) &= \langle \Psi_{N,0} | \hat{\psi}(\mathbf{x} + \mathbf{v}) | \Psi_{N+1,\mathbf{k}\,l} \rangle \\ &= \langle \Psi_{N,0} | e^{-i\hat{\mathbf{P}}\cdot\mathbf{v}} \hat{\psi}(\mathbf{x}) e^{i\hat{\mathbf{P}}\cdot\mathbf{v}} | \Psi_{N+1,\mathbf{k}\,l} \rangle = e^{i\mathbf{k}\cdot\mathbf{v}} P_{\mathbf{k}\,l}(\mathbf{x}), \end{aligned} \qquad (\text{E.9})$$

where we assume that $e^{i\hat{\mathbf{P}}\cdot\mathbf{v}} | \Psi_{N,0} \rangle = | \Psi_{N,0} \rangle$, i.e., we assume that the ground state with N particles has $\mathbf{k} = 0$. In the limit of vanishing interactions the quasi-particle wavefunctions become equal to the single-particle eigenstates and we recover the well-known symmetry property of the single-particle Bloch orbitals under lattice translations. From (E.9) we can derive

$$G^>(\mathbf{x} + \mathbf{v}, \mathbf{x}' + \mathbf{v}'; \omega) = \sum_{\mathbf{k}} e^{i\mathbf{k}\cdot(\mathbf{v} - \mathbf{v}')} G^>(\mathbf{x}, \mathbf{x}'; \mathbf{k}, \omega),$$

where we define

$$G^>(\mathbf{x}, \mathbf{x}'; \mathbf{k}, \omega) = -2\pi i \sum_{l} P_{\mathbf{k}\,l}(\mathbf{x}) P^*_{\mathbf{k}\,l}(\mathbf{x}') \delta(\omega - [E_{N+1,\mathbf{k}\,l} - E_{N,0}]).$$

Thus if we know the Green's function $G^>(\mathbf{x}, \mathbf{x}'; \mathbf{k}, \omega)$ for $\mathbf{x}$ and $\mathbf{x}'$ in a given unit cell, then the Green's function $G^>(\mathbf{x}, \mathbf{x}'; \omega)$ in all units cells is readily calculated. The problem is thus reduced to calculating $G^>(\mathbf{x}, \mathbf{x}'; \mathbf{k}, \omega)$ in a single unit cell. So far we have worked in the position basis, but in general we can use any basis we like inside this unit cell. If we expand the field operators in a unit cell in a localized basis as

$$\hat{\psi}(\mathbf{x}) = \sum_{s\tau} \varphi_{s\tau}(\mathbf{x}) \hat{d}_{s\tau},$$

where s is a label for basis functions in the unit cell and τ is a spin index, then in the new basis we have

$$G^>_{s\tau\,s'\tau'}(\mathbf{k}, \omega) = -2\pi i \sum_{l} P_{\mathbf{k}\,l}(s\tau) P^*_{\mathbf{k}\,l}(s'\tau') \delta(\omega - [E_{N+1,\mathbf{k}\,l} - E_{N,0}]).$$

Here

$$P_{\mathbf{k}l}(s\tau) = \langle \Psi_{N,0}|\hat{d}_{s\tau}|\Psi_{N+1,\mathbf{k}l}\rangle = \int_V d\mathbf{x}\, P_{\mathbf{k}l}(\mathbf{x})\varphi_{s\tau}^*(\mathbf{x}).$$

A commonly used orthonormal basis in the unit cell is the set of functions

$$\varphi_{\mathbf{K}\tau}(\mathbf{x}) = \frac{\delta_{\sigma\tau}}{\sqrt{v}}\, e^{i\mathbf{K}\cdot\mathbf{r}}, \tag{E.10}$$

where v is the volume of the unit cell and where $\mathbf{K}$ is a reciprocal lattice vector. An advantage of using this basis is that the set of plane waves $e^{i(\mathbf{k}+\mathbf{K})\cdot\mathbf{r}}$ with $\mathbf{k}$ in the first Brillouin zone and $\mathbf{K}$ a reciprocal lattice vector is the same set used to expand the interaction in (E.3). Therefore the Coulomb integrals are especially simple in this basis.

Let us now describe the typical outcome of a calculation of $G^>$. We assume for simplicity that $G^>$ is diagonal in spin space with matrix elements $G^>_{ss'}(\mathbf{k},\omega)$. The quantity $iG^>_{ss'}(\mathbf{k},\omega) = A_{ss'}(\mathbf{k},\omega)$ is the (s,s') matrix element of the spectral function for $\omega > \mu$. The spectral function is a positive semi-definite self-adjoint matrix for any fixed $\mathbf{k}$ and ω, see Section 6.3.2. If we plot the eigenvalues $a_\nu(\mathbf{k},\omega)$ of this matrix as a function of ω we find that they are peaked around some value $\omega = \epsilon_{\mathbf{k}\nu}$ and zero otherwise. This result can easily be interpreted in terms of an inverse photoemission experiment. If in the experiment we measure both the energy and the momentum of the photons then we find that the photon intensity as a function of energy has peaks in $\epsilon_{\mathbf{k}\nu}$. Analogous considerations apply to $G^<$. In this way it is possible to measure the band structure of a crystal. Similarly to what we saw for the electron gas the peaks are not infinitely sharp. The real band structure corresponds to long-living states produced by addition and removal of electrons with appropriate energy and momentum.

F

Asymptotic expansions

Very often we are interested in calculating functions $f(x)$ depending on a variable x when x is close to some value x_0 or when x approaches x_0 if $x_0 = \infty$. If no exact solution exists (or if the exact solution is exceedingly complicated) it is useful to devise "perturbative" methods to approximate f. These perturbative methods lead to an expansion of $f(x)$ in terms of functions $\varphi_n(x)$ that for $x \to x_0$ approach zero faster and faster with increasing n. MBPT is one such method. The set of functions $\varphi_n(x)$, $n = 0, 1, 2, \ldots$, are called an asymptotic set if

$$\lim_{x \to x_0} \frac{\varphi_{n+1}(x)}{\varphi_n(x)} = 0,$$

and the expansion

$$f(x) \sim \sum_{n=0}^{\infty} a_n \varphi_n(x) \tag{F.1}$$

is called an *asymptotic expansion* if

$$\lim_{x \to x_0} \frac{f(x) - \sum_{n=0}^{N} a_n \varphi_n(x)}{\varphi_N(x)} = 0 \quad \text{for all } N. \tag{F.2}$$

Obviously a Taylor series is an asymptotic expansion while a Fourier series is not. In (F.1) we use the symbol "$\sim$" because the series need not be convergent! In many interesting physical situations the first terms of an asymptotic expansion decrease rapidly for small $|x - x_0|$ while higher order terms increase wildly with increasing n (at fixed $|x - x_0|$). In these cases we can get a good approximation to f by summing just the first few terms of the expansion. This is why asymptotic expansions, even when divergent, are very useful in practice.

To summarize, an asymptotic expansion need not be convergent and a convergent series need not be an asymptotic expansion (a Fourier series is not asymptotic). Distinguishing between these two concepts, i.e., convergence versus asymptoticity, is crucial to appreciate the power of asymptotic expansions. In mathematical terms we can say that convergence pertains to the behavior of the partial sum

$$S_N(x) = \sum_{n=0}^{N} a_n \varphi_n(x)$$

534

for $N \to \infty$, while asymptoticity pertains to the behavior of S_N for $x \to x_0$.

From the above definitions it follows that the coefficient a_{N+1} is given by

$$a_{N+1} = \lim_{x \to x_0} \frac{f(x) - \sum_{n=0}^{N} a_n \varphi_n(x)}{\varphi_{N+1}(x)}.$$

Thus if a function has an asymptotic expansion then this expansion is unique given the φ_ns. The converse is not true since different functions can have the same asymptotic expansion. For example, for any constant c

$$\frac{1}{1-x} + c e^{-1/x^2} = 1 + x + x^2 + \dots$$

The r.h.s. is an asymptotic expansion around $x_0 = 0$ and is independent of c since $\lim_{x \to 0} e^{-1/x^2}/x^n = 0$ for all n. We conclude this appendix with a classical example of asymptotic expansions.

Let us consider the error function

$$\mathrm{erf}(x) = \frac{2}{\sqrt{\pi}} \int_0^x dt\, e^{-t^2} = 1 - \frac{2}{\sqrt{\pi}} \int_x^\infty dt\, e^{-t^2} = 1 - \frac{1}{\sqrt{\pi}} \int_{x^2}^\infty ds\, \frac{e^{-s}}{\sqrt{s}}.$$

For non-negative integers n we define the function

$$F_n(x) = \int_{x^2}^\infty ds\, s^{-n-1/2} e^{-s} = \frac{e^{-x^2}}{x^{2n+1}} - \left(n + \frac{1}{2}\right) \int_{x^2}^\infty ds\, s^{-n-1-1/2} e^{-s}$$

$$= \frac{e^{-x^2}}{x^{2n+1}} - \left(n + \frac{1}{2}\right) F_{n+1}(x).$$

The asymptotic expansion of $\mathrm{erf}(x)$ around $x_0 = \infty$ follows from the repeated use of the above recursive relation

$$\mathrm{erf}(x) = 1 - \frac{1}{\sqrt{\pi}} F_0(x)$$

$$= 1 - \frac{1}{\sqrt{\pi}} \left[\frac{e^{-x^2}}{x} - \frac{1}{2} F_1(x) \right]$$

$$= 1 - \frac{1}{\sqrt{\pi}} \left[\frac{e^{-x^2}}{x} - \frac{1}{2} \frac{e^{-x^2}}{x^3} + \frac{1}{2} \frac{3}{2} F_2(x) \right]$$

$$= 1 - \frac{e^{-x^2}}{\sqrt{\pi}} \sum_{n=0}^{\infty} (-)^n \frac{(2n-1)!!}{2^n x^{2n+1}}, \tag{F.3}$$

which is clearly a divergent series for all x. To show that this is an asymptotic expansion we have to prove that the functions $\varphi_n \sim e^{-x^2}/x^{2n+1}$ form an asymptotic set (which is obvious) and that (F.2) is fulfilled. Let us define

$$R_{N+1}(x) = \mathrm{erf}(x) - 1 + \frac{e^{-x^2}}{\sqrt{\pi}} \sum_{n=0}^{N} (-)^n \frac{(2n-1)!!}{2^n x^{2n+1}}$$

$$= -\frac{(-)^{N+1}}{\sqrt{\pi}} \frac{(2N+1)!!}{2^{N+1}} F_{N+1}(x).$$

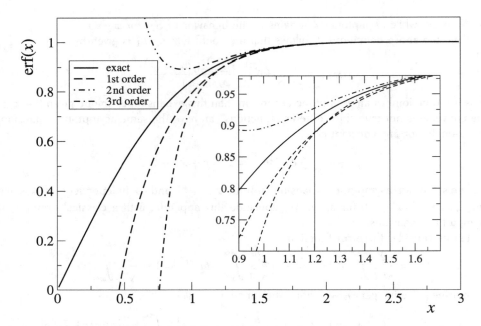

Figure F.1 Error function $\mathrm{erf}(x)$ and the partial sums $S_N(x)$ of the asymptotic expansion (F.3) for $N = 1, 2, 3$. The inset is a magnification of the same functions in a restricted domain of the x variable.

We have

$$|F_{N+1}(x)| = \left| \int_{x^2}^{\infty} ds \, s^{-N-3/2} e^{-s} \right| \leq \frac{1}{x^{2N+3}} \int_{x^2}^{\infty} ds \, e^{-s} = \frac{e^{-x^2}}{x^{2N+3}},$$

which goes to zero faster than $\varphi_n(x) \sim e^{-x^2}/x^{2N+1}$ for $x \to \infty$. Therefore (F.3) is an asymptotic expansion.

In Fig. F.1 we display the plot of $\mathrm{erf}(x)$ as well as the partial sums of the asymptotic expansion with $N = 0, 1, 2$, which corresponds to the 1st, 2nd and 3rd order approximation. We see that for $x < 1$ the best approximation is the 1st order one. The inset shows a magnification of the same curves in a narrower window of the x variable. For $x = 1.2$ we get closer to the exact curve by adding to the term with $n = 0$ the term with $n = 1$ (2nd order); the further addition of the term with $n = 3$ worsens the approximation. Increasing x even further, e.g., for $x = 1.5$, the 3rd order approximation is superior to the 1st and 2nd order approximations. Thus, given x there exists an optimal N for which the partial sum performs best whereas the inclusion of higher order terms brings the partial sum further away from the exact value. In this example the optimal value of N increases with increasing x.

G

Wick's theorem for general initial states

The discussion in this appendix deals with an alternative way of calculating the Green's functions of systems initially described by a density matrix $\hat{\rho}$. We present this appendix for completeness and to show that the formalism can in principle deal with very general situations. The presentation closely follows the one in Ref. [170].

The explicit form (5.2) and (5.3) of the Martin–Schwinger hierarchy assumes the fact that $\hat{H}^{\mathrm{M}}$ and $\hat{H}(t)$ contain at most two-body operators. The more general inclusion of m-body operators in $\hat{H}^{\mathrm{M}}$ would generate a coupling (for contour arguments along the vertical track) between G_n and G_{n+m-1}, thus rendering the system of equations even more complicated to solve. This is, however, what one should do for a general density matrix $\hat{\rho}$ (or in the case of pure states for a general initial state). For instance, if $\hat{\rho} = |\Psi\rangle\langle\Psi|$ with $|\Psi\rangle$ a general initial state then the operator $\hat{H}^{\mathrm{M}}$ for which $\hat{\rho} = e^{-\beta\hat{H}^{\mathrm{M}}}/\mathrm{Tr}[e^{-\beta\hat{H}^{\mathrm{M}}}]$ contains, in general, n-body operators of all orders n. It is clear that an alternative way of dealing with this problem would be most valuable. The approach that we follow in this appendix is not to introduce any additional contour, but to use the contour of Fig. 4.3 instead, which starts at t_{0-}, runs back and forth to infinity and returns in t_{0+}. On this contour the Martin–Schwinger hierarchy equations are still valid. However, the KMS boundary conditions cannot be used to solve these equations since the point $z_{\mathrm{f}} = t_0 - \mathrm{i}\beta$ does not belong to the contour. Consequently we must regard the problem as an initial value problem. Let us illustrate how it works.

We rewrite the n-particle Green's functions (5.1) with contour arguments on the horizontal branches as

$$
\begin{aligned}
G_n(1,\ldots,n;1',\ldots,n') &= \mathrm{Tr}\left[\hat{\rho}\,\hat{G}_n(1,\ldots,n;1',\ldots,n')\right] \\
&= \frac{1}{\mathrm{i}^n}\mathrm{Tr}\left[\hat{\rho}\,\mathcal{T}\left\{e^{-\mathrm{i}\int_\gamma d\bar{z}\hat{H}(\bar{z})}\hat{\psi}(1)\ldots\hat{\psi}(n)\hat{\psi}^\dagger(n')\ldots\hat{\psi}^\dagger(1')\right\}\right],
\end{aligned} \quad \text{(G.1)}
$$

where γ is the contour of Fig. 4.3 and $\hat{\rho}$ is a general density matrix. These Green's functions can be expanded in powers of the interaction v by simply taking into account that inside the $\mathcal{T}$ sign the Hamiltonians $\hat{H}_0(z)$ and $\hat{H}_{\mathrm{int}}(z)$ commute.[1] The result is (omitting the

---

[1]The same trick was used in (5.30) and (5.33).

arguments of G_n)

$$
G_n = \frac{1}{i^n} \sum_{k=0}^{\infty} \frac{(-i)^k}{k!} \int d\bar{z}_1 \ldots d\bar{z}_k
$$

$$
\times \text{Tr}\left[\hat{\rho}\, \mathcal{T}\left\{e^{-i\int_\gamma d\bar{z}\hat{H}_0(\bar{z})} \hat{H}_{\text{int}}(\bar{z}_1) \ldots \hat{H}_{\text{int}}(\bar{z}_k)\hat{\psi}(1) \ldots \hat{\psi}(n)\hat{\psi}^\dagger(n') \ldots \hat{\psi}^\dagger(1')\right\}\right]
$$

$$
= \frac{1}{i^n} \sum_{k=0}^{\infty} \left(-\frac{i}{2}\right)^k \frac{1}{k!} \int d\bar{1}d\bar{1}' \ldots d\bar{k}d\bar{k}'v(\bar{1};\bar{1}') \ldots v(\bar{k};\bar{k}')
$$

$$
\times i^{n+2k} g_{n+2k}(1,\ldots,n,\bar{1},\bar{1}',\ldots,\bar{k},\bar{k}';1',\ldots,n',\bar{1}^+,\bar{1}'^+,\ldots,\bar{k}^+,\bar{k}'^+), \qquad (G.2)
$$

where we define the noninteracting Green's functions g_n as in (G.1) but with $\hat{H}(\bar{z}) \to \hat{H}_0(\bar{z})$.[2] By definition, the g_n satisfy the noninteracting Martin–Schwinger hierarchy on the contour γ of Fig. 4.3. To solve this hierarchy we need an initial condition. For this purpose we observe that to specify $\hat{\rho}$ is the same as to specify the n-particle density matrices

$$
\Gamma_n(\mathbf{x}_1,\ldots,\mathbf{x}_n;\mathbf{x}_1',\ldots,\mathbf{x}_n') \equiv \text{Tr}\left[\hat{\rho}\,\hat{\psi}^\dagger(\mathbf{x}_1') \ldots \hat{\psi}^\dagger(\mathbf{x}_n')\hat{\psi}(\mathbf{x}_n) \ldots \hat{\psi}(\mathbf{x}_1)\right],
$$

which are the obvious generalization of the Γ_n of Section 1.7 to the case of ensemble averages. The link between the n-particle density matrices and the n-particle Green's functions follows directly from (G.1) and is

$$
\Gamma_n(\mathbf{x}_1,\ldots,\mathbf{x}_n;\mathbf{x}_1',\ldots,\mathbf{x}_n') = (\pm i)^n \lim_{\substack{z_i,z_i' \to t_{0-} \\ z_1' > \ldots > z_n' > z_n > \ldots > z_1}} G_n(1,\ldots,n;1',\ldots,n'), \qquad (G.3)
$$

where the prefactor $(\pm i)^n$ refers to bosons/fermions respectively and where the inequality symbol ">" between contour variables signifies "later than." In the case of fermions we use the fact that the sign of the permutation to put the operators in the same order as in Γ_n is $(-1)^n$, as can be readily checked. Now the crucial observation is that in the limit $z_i, z_i' \to t_{0-}$ the Green's functions G_n and g_n approach the same value since in both cases the exponential in (G.1) reduces to $\hat{U}(t_{0+},\infty)\hat{U}(\infty,t_{0-}) = \hat{1}$. Therefore, the initial condition to solve the noninteracting Martin–Schwinger hierarchy is

$$
\Gamma_n(\mathbf{x}_1,\ldots,\mathbf{x}_n;\mathbf{x}_1',\ldots,\mathbf{x}_n') = (\pm i)^n \lim_{\substack{z_i,z_i' \to t_{0-} \\ z_1' > \ldots > z_n' > z_n > \ldots > z_1}} g_n(1,\ldots,n;1',\ldots,n'). \qquad (G.4)
$$

Let us recapitulate what we have said so far. For a general initial density matrix $\hat{\rho}$ the initial conditions for the n-particle Green's functions are uniquely specified by the Γ_n, and the Green's functions at later times can be obtained by solving the Martin–Schwinger hierarchy on the contour γ of Fig. 4.3. If g_n solve the noninteracting hierarchy then the G_n expanded as in (G.2) solve the interacting hierarchy on the same contour.

It therefore remains to find the solution of the noninteracting hierarchy with initial conditions (G.4). We start by first considering the equations (5.17) and (5.18) for the noninteracting

---

[2]These g_n are different from the $G_{0,n}$ of Section 5.3 since they are averaged with a general density matrix instead of the noninteracting density matrix $\hat{\rho}_0 = e^{-\beta \hat{H}_0^M}/Z_0$.

one-particle Green's function $g \equiv g_1$. These equations must be solved with the initial condition

$$\Gamma_1(\mathbf{x}_1; \mathbf{x}_1') = \pm i \lim_{\substack{z_1, z_1' \to t_{0-} \\ z_1' > z_1}} g(1; 1'). \tag{G.5}$$

Due to the contour ordering the Green's function g has the structure

$$g(1; 1') = \theta(z_1, z_1') g^>(1; 1') + \theta(z_1', z_1) g^<(1; 1'), \tag{G.6}$$

where

$$g^>(1; 1') = -i \operatorname{Tr} \left[\hat{\rho} \, \hat{\psi}_H(1) \hat{\psi}_H^\dagger(1') \right],$$

$$g^<(1; 1') = \mp i \operatorname{Tr} \left[\hat{\rho} \, \hat{\psi}_H^\dagger(1') \hat{\psi}_H(1) \right].$$

The functions $g^{\lessgtr}$ have arguments on the contour γ but their value is independent of whether the arguments are on the forward or backward branch since there is no contour ordering operator in the definition of these functions. Thus,

$$g^{\lessgtr}(\mathbf{x}, t_+; \mathbf{x}', z') = g^{\lessgtr}(\mathbf{x}, t_-; \mathbf{x}', z') \qquad \text{for all } z',$$

and similarly

$$g^{\lessgtr}(\mathbf{x}, z; \mathbf{x}', t_+') = g^{\lessgtr}(\mathbf{x}, z; \mathbf{x}', t_-') \qquad \text{for all } z.$$

We see that (G.5) is an initial condition on the $g^<$ function

$$\Gamma_1(\mathbf{x}_1; \mathbf{x}_1') = \pm i \, g^<(\mathbf{x}_1, t_{0-}; \mathbf{x}_1', t_{0-}).$$

The initial condition on the $g^>$ function follows from the equal-time (anti)commutation relations of the field operators

$$-i\delta(\mathbf{x}_1 - \mathbf{x}_1') = -i \operatorname{Tr} \left[\hat{\rho} \left[\hat{\psi}_H(\mathbf{x}_1, t_{0-}), \hat{\psi}_H^\dagger(\mathbf{x}_1', t_{0-}) \right]_{\mp} \right]$$

$$= g^>(\mathbf{x}_1, t_{0-}; \mathbf{x}_1', t_{0-}) - g^<(\mathbf{x}_1, t_{0-}; \mathbf{x}_1', t_{0-}). \tag{G.7}$$

We can now solve (5.17) using as initial condition the value of $g^>$ in $z_1 = z_1' = t_{0-}$ and (5.18) using as initial condition the value of $g^<$ in $z_1 = z_1' = t_{0-}$. One can readily check that this property is completely general, i.e., due to the equal-time (anti)commutation relations the limit in (G.4) determines also the limit $z_i, z_i' \to t_{0-}$ taken for a different contour ordering.

Once we have the Green's function g, how can we calculate all the g_n? If we consider the permanent/determinant (5.27) in terms of this g then it is still true that this expression satisfies the noninteracting Martin–Schwinger hierarchy since in the derivation we only use the equations of motion. However, the permanent/determinant does not in general satisfy the required boundary conditions. From (G.5) we instead find that

$$(\pm i)^n \lim_{\substack{z_i, z_i' \to t_{0-} \\ z_1' > \dots > z_1}} \begin{vmatrix} g(1; 1') & \dots & g(1; n') \\ \vdots & & \vdots \\ g(n; 1') & \dots & g(n; n') \end{vmatrix}_{\pm} = \begin{vmatrix} \Gamma_1(\mathbf{x}_1; \mathbf{x}_1') & \dots & \Gamma_1(\mathbf{x}_1; \mathbf{x}_n') \\ \vdots & & \vdots \\ \Gamma_1(\mathbf{x}_n; \mathbf{x}_1') & \dots & \Gamma_1(\mathbf{x}_n; \mathbf{x}_n') \end{vmatrix}_{\pm},$$

which is, in general, different from Γ_n. The permanent/determinant is a particular solution to the noninteracting Martin–Schwinger hierarchy but with the wrong boundary conditions. In order to repair this problem the permanent/determinant must be supplied by the additional homogeneous solution. For pedagogical reasons we first consider the example of the two-particle Green's function. We write

$$g_2(1,2;1',2') = \begin{vmatrix} g(1;1') & g(1;2') \\ g(2;1') & g(2;2') \end{vmatrix}_{\pm} + \tilde{g}_2(1,2;1',2'), \tag{G.8}$$

where $\tilde{g}_2$ satisfies the four homogeneous equations

$$\left[\mathrm{i}\frac{d}{dz_1} - h(1)\right]\tilde{g}_2(1,2;1',2') = 0, \qquad \tilde{g}_2(1,2;1',2')\left[-\mathrm{i}\frac{\overleftarrow{d}}{dz_1'} - h(1')\right] = 0, \tag{G.9}$$

$$\left[\mathrm{i}\frac{d}{dz_2} - h(2)\right]\tilde{g}_2(1,2;1',2') = 0, \qquad \tilde{g}_2(1,2;1',2')\left[-\mathrm{i}\frac{\overleftarrow{d}}{dz_2'} - h(2')\right] = 0, \tag{G.10}$$

In these equations the r.h.s. are zero and the δ-functions of the Martin–Schwinger hierarchy have disappeared. We remind the reader that the δ-functions arose from the differentiations of the contour Heaviside functions. We thus conclude that $\tilde{g}_2$ is a continuous function of its contour arguments, i.e., there are no discontinuities when two contour arguments cross each other. Since

$$\tilde{g}_2(1,2;1',2') = g_2(1,2;1',2') - \begin{vmatrix} g(1;1') & g(1;2') \\ g(2;1') & g(2;2') \end{vmatrix}_{\pm},$$

this implies that a jump in g_2 is exactly compensated by a jump in the permanent/determinant. Accordingly, the limit

$$(\pm\mathrm{i})^2 \lim_{z_i, z_i' \to t_{0-}} \tilde{g}_2(1,2;1',2') = C_2(\mathbf{x}_1, \mathbf{x}_2; \mathbf{x}_1', \mathbf{x}_2') \tag{G.11}$$

is independent of the contour ordering of the z_i and z_i'. We call the quantity C_2 the two-particle initial-correlation function. Taking into account (G.4), C_2 can be written as

$$C_2(\mathbf{x}_1, \mathbf{x}_2; \mathbf{x}_1', \mathbf{x}_2') = \Gamma_2(\mathbf{x}_1, \mathbf{x}_2; \mathbf{x}_1', \mathbf{x}_2') - \begin{vmatrix} \Gamma_1(\mathbf{x}_1; \mathbf{x}_1') & \Gamma_1(\mathbf{x}_1; \mathbf{x}_2') \\ \Gamma_1(\mathbf{x}_2; \mathbf{x}_1') & \Gamma_1(\mathbf{x}_2; \mathbf{x}_2') \end{vmatrix}_{\pm}, \tag{G.12}$$

where Γ_1 and Γ_2 are the one- and two-particle density-matrices. We can find an explicit solution of the homogeneous equations (G.9) and (G.10) with initial condition (G.11) in terms of the one-particle Green's function g of (G.6). We define the function[3]

$$A(1;1') = \mathrm{Tr}\left[\hat{\rho}\left[\hat{\psi}_H(1), \hat{\psi}_H^\dagger(1')\right]_{\mp}\right] = \mathrm{i}[g^>(1;1') - g^<(1;1')],$$

---

[3]For systems in equilibrium the function $A(1;1')$ depends only on the time difference $t_1 - t_1'$, and its Fourier transform is the so called *spectral function*, see Chapter 6.

where, as usual, the upper/lower sign refers to bosons/fermions. This function satisfies the equations of motion

$$\left[i\frac{d}{dz_1} - h(1) \right] A(1;1') = 0, \qquad A(1;1') \left[-i\frac{\overleftarrow{d}}{dz_1'} - h(1') \right] = 0,$$

as follows directly from the equations of motion of the field operators in noninteracting systems, see (4.62) and (4.63). Furthermore (G.7) tells us that the function A satisfies the initial condition

$$A(\mathbf{x}_1, t_{0-}; \mathbf{x}_1', t_{0-}) = \delta(\mathbf{x}_1 - \mathbf{x}_1').$$

Consequently the function

$$\tilde{g}_2(1,2;1',2') = (\mp i)^2 \int d\mathbf{y}_1 d\mathbf{y}_2 d\mathbf{y}_1' d\mathbf{y}_2' \, A(1;\mathbf{y}_1,t_{0-})A(2;\mathbf{y}_2,t_{0-})$$
$$\times \, C_2(\mathbf{y}_1,\mathbf{y}_2;\mathbf{y}_1',\mathbf{y}_2')A(\mathbf{y}_1',t_{0-};1')A(\mathbf{y}_2',t_{0-};2')$$

is a solution to the homogeneous equations (G.9) and (G.10) and satisfies the initial condition (G.11). This solution can be rewritten in a more elegant form. We define the following linear combination of contour δ-functions

$$\delta(z, t_0) = \delta(t_0, z) \equiv \delta(z, t_{0-}) - \delta(z, t_{0+}).$$

Since the δ-functions appear at the edges of the integration interval we have to agree on a convention how to integrate them. Here we define this integral to be equal to one. With these definitions we can then derive the relation

$$\int_\gamma dz_1' \, g(1;1')\delta(z_1', t_0) = \int_\gamma dz_1' \, g(1;1')\big(\delta(z_1', t_{0-}) - \delta(z_1', t_{0+})\big)$$
$$= g^>(1;\mathbf{x}_1', t_{0-}) - g^<(1;\mathbf{x}_1', t_{0+})$$
$$= g^>(1;\mathbf{x}_1', t_{0-}) - g^<(1;\mathbf{x}_1', t_{0-})$$
$$= -i \, A(1;\mathbf{x}_1', t_{0-}),$$

where we use the fact that the functions $g^{\lessgtr}$ have the same value for arguments on the forward and backward branches of the contour. Similarly we have the relation

$$\int_\gamma dz_1 \, \delta(t_0, z_1)g(1;1') = i \, A(\mathbf{x}_1, t_{0-};1').$$

If we now define the two-particle initial-correlation function on the contour

$$C_2(1,2;1',2') \equiv \delta(z_1, t_0)\delta(z_2, t_0)C_2(\mathbf{x}_1, \mathbf{x}_2; \mathbf{x}_1', \mathbf{x}_2')\delta(t_0, z_1')\delta(t_0, z_2'), \qquad \text{(G.13)}$$

then the homogeneous solution can be rewritten as

$$\tilde{g}_2(1,2;1',2') = (\mp i)^2 \int d\bar{1}d\bar{2}d\bar{1}'d\bar{2}' \, g(1;\bar{1})g(2;\bar{2})C_2(\bar{1},\bar{2};\bar{1}',\bar{2}')g(\bar{1}';1')g(\bar{2}';2'). \qquad \text{(G.14)}$$

This result, together with (G.8), provides an explicit expression in terms of g for the solution of the equations of motion of g_2 with a given initial condition determined by a one- and two-particle density matrix. A diagrammatic representation of g_2 is given in Fig. G.1(a).

It is clear that the procedure to determine g_2 can be continued to higher order Green's functions. We illustrate the next step for the three-particle Green's function g_3 before we give the general result. The equations of motion for g_3 are

$$\left[i\frac{d}{dz_1} - h(1)\right]g_3(1,2,3;1',2',3') = \delta(1;1')g_2(2,3;2',3') \pm \delta(1;2')g_2(2,3;1',3')$$
$$+ \delta(1;3')g_2(2,3;1',2'),$$

$$\left[i\frac{d}{dz_2} - h(2)\right]g_3(1,2,3;1',2',3') = \pm\delta(2;1')g_2(1,3;2',3') + \delta(2;2')g_2(1,3;1',3')$$
$$\pm \delta(2;3')g_2(1,3;1',2'),$$

$$\left[i\frac{d}{dz_3} - h(3)\right]g_3(1,2,3;1',2',3') = \delta(3;1')g_2(1,2;2',3') \pm \delta(3;2')g_2(1,2;1',3')$$
$$+ \delta(3;3')g_2(1,2;1',2'),$$

plus the three equations with the derivative with respect to the primed arguments. If we insert (G.8) into the r.h.s. of these equations we can easily check that the general solution for the three-particle Green's function is

$$g_3(1,2,3;1',2',3') = \begin{vmatrix} g(1;1') & g(1;2') & g(1;3') \\ g(2;1') & g(2;2') & g(2;3') \\ g(3;1') & g(3;2') & g(3;3') \end{vmatrix}_{\pm}$$
$$+ g(1;1')\tilde{g}_2(2,3;2',3') \pm g(1;2')\tilde{g}_2(2,3;1',3') + g(1;3')\tilde{g}_2(2,3;1',2')$$
$$\pm g(2;1')\tilde{g}_2(1,3;2',3') + g(2;2')\tilde{g}_2(1,3;1',3') \pm g(2;3')\tilde{g}_2(1,3;1',2')$$
$$+ g(3;1')\tilde{g}_2(1,2;2',3') \pm g(3;2')\tilde{g}_2(1,2;1',3') + g(3;3')\tilde{g}_2(1,2;1',2')$$
$$+ \tilde{g}_3(1,2,3;1',2',3'), \tag{G.15}$$

where the function $\tilde{g}_3$ satisfies the homogeneous equations

$$\left[i\frac{d}{dz_1} - h(1)\right]\tilde{g}_3(1,2,3;1',2',3') = 0, \qquad \left[i\frac{d}{dz_2} - h(2)\right]\tilde{g}_3(1,2,3;1',2',3') = 0,$$

$$\left[i\frac{d}{dz_3} - h(3)\right]\tilde{g}_3(1,2,3;1',2',3') = 0,$$

as well as the three equations with the derivative with respect to the primed arguments. As for the two-particle case $\tilde{g}_3$ is a continuous function of the contour arguments since there are no δ-functions in the r.h.s. of its equations of motion. The precise form of $\tilde{g}_3$ must again be determined from the initial condition (G.4). More explicitly we have

$$(\pm i)^3 \lim_{z_i, z_i' \to t_{0-}} \tilde{g}_3(1,2,3;1',2',3') = C_3(\mathbf{x}_1, \mathbf{x}_2, \mathbf{x}_3; \mathbf{x}_1', \mathbf{x}_2', \mathbf{x}_3'),$$

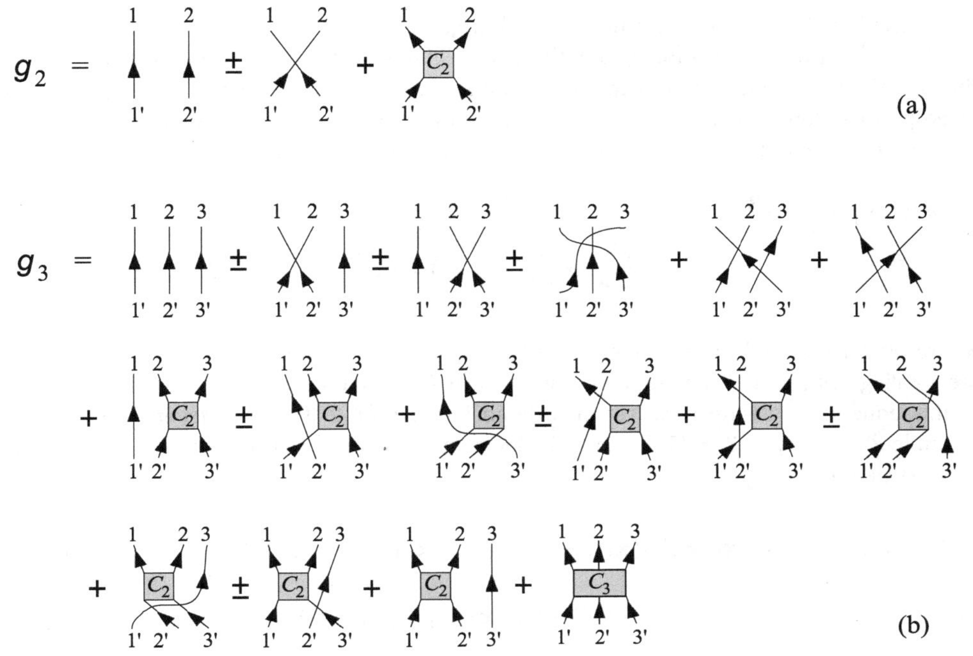

$$g_2 =$$ (a)

$$g_3 =$$ (b)

Figure G.1 Representation of the two- and three-particle Green's functions which solve the noninteracting Martin–Schwinger hierarchy with a general density matrix. With every oriented line going from j to i is associated the one-particle Green's function $g(i;j)$ and with every C_n-box is associated $(\mp i)^n C_n$.

where the three-particle initial-correlation function C_3 can be deduced from (G.15) to be

$$
\begin{aligned}
C_3(\mathbf{x}_1, \mathbf{x}_2, \mathbf{x}_3; \mathbf{x}_1', \mathbf{x}_2', \mathbf{x}_3') = {} & \Gamma_3(\mathbf{x}_1, \mathbf{x}_2, \mathbf{x}_3; \mathbf{x}_1', \mathbf{x}_2', \mathbf{x}_3') \\
& - \Gamma_1(\mathbf{x}_1; \mathbf{x}_1')C_2(\mathbf{x}_2, \mathbf{x}_3; \mathbf{x}_2', \mathbf{x}_3') \mp \Gamma_1(\mathbf{x}_1; \mathbf{x}_2')C_2(\mathbf{x}_2, \mathbf{x}_3; \mathbf{x}_1', \mathbf{x}_3') \\
& - \Gamma_1(\mathbf{x}_1; \mathbf{x}_3')C_2(\mathbf{x}_2, \mathbf{x}_3; \mathbf{x}_1', \mathbf{x}_2') \mp \Gamma_1(\mathbf{x}_2; \mathbf{x}_1')C_2(\mathbf{x}_1, \mathbf{x}_3; \mathbf{x}_2', \mathbf{x}_3') \\
& - \Gamma_1(\mathbf{x}_2; \mathbf{x}_2')C_2(\mathbf{x}_1, \mathbf{x}_3; \mathbf{x}_1', \mathbf{x}_3') \mp \Gamma_1(\mathbf{x}_2; \mathbf{x}_3')C_2(\mathbf{x}_1, \mathbf{x}_3; \mathbf{x}_1', \mathbf{x}_2') \\
& - \Gamma_1(\mathbf{x}_3; \mathbf{x}_1')C_2(\mathbf{x}_1, \mathbf{x}_2; \mathbf{x}_2', \mathbf{x}_3') \mp \Gamma_1(\mathbf{x}_3; \mathbf{x}_2')C_2(\mathbf{x}_1, \mathbf{x}_2; \mathbf{x}_1', \mathbf{x}_3') \\
& - \Gamma_1(\mathbf{x}_3; \mathbf{x}_3')C_2(\mathbf{x}_1, \mathbf{x}_2; \mathbf{x}_1', \mathbf{x}_2') \\
& - \left| \begin{array}{ccc}
\Gamma_1(\mathbf{x}_1; \mathbf{x}_1') & \Gamma_1(\mathbf{x}_1; \mathbf{x}_2') & \Gamma_1(\mathbf{x}_1; \mathbf{x}_3') \\
\Gamma_1(\mathbf{x}_2; \mathbf{x}_1') & \Gamma_1(\mathbf{x}_2; \mathbf{x}_2') & \Gamma_1(\mathbf{x}_2; \mathbf{x}_3') \\
\Gamma_1(\mathbf{x}_3; \mathbf{x}_1') & \Gamma_1(\mathbf{x}_3; \mathbf{x}_2') & \Gamma_1(\mathbf{x}_3; \mathbf{x}_3')
\end{array} \right|_{\pm} .
\end{aligned}
$$

Similarly to (G.14) the explicit solution for $\tilde{G}_3$ is given by

$$
\tilde{g}_3(1, 2, 3; 1', 2', 3') = (\mp i)^3 \int d\bar{1}\, d\bar{2}\, d\bar{3}\, d\bar{1}'\, d\bar{2}'\, d\bar{3}'\ g(1; \bar{1}) g(2; \bar{2}) g(3; \bar{3})
$$
$$
\times C_3(\bar{1}, \bar{2}, \bar{3}; \bar{1}', \bar{2}', \bar{3}') g(\bar{1}'; 1') g(\bar{2}'; 2') g(\bar{3}'; 3'),
$$

where the initial-correlation function C_3 with arguments on the contour is defined similarly to (G.13). This explicit form for $\tilde{g}_3$ together with (G.15) provides the solution in terms of g to the equations of motion of g_3 with a given initial condition determined by a one-, two-, and three-particle density matrix. A diagrammatic representation of g_3 is given in Fig. G.1(b).

A little inspection shows that (G.15) can be rewritten in a more compact form as

$$g_3(1,2,3;1',2',3') = |g|_{3,\pm}(1,2,3;1',2',3')$$
$$+ \sum_{j,k=1}^{3} (\pm)^{j+k} g(j;k') \tilde{g}_2(\breve{j};\breve{k}') + \tilde{g}_3(1,2,3;1',2',3'),$$

where we have added the subindex 3 on the permanent/determinant to remind us that we are dealing with a 3×3 matrix of one-particle Green's functions g as matrix elements. In this equation $\breve{j}$ denotes the set of ordered integers $(1,2,3)$ with integer j missing, for example $\breve{2} = (1,3)$ or $\breve{3} = (1,2)$. In a similar manner the initial-correlation function C_3 can be rewritten as

$$C_3(\mathbf{x}_1,\mathbf{x}_2,\mathbf{x}_3;\mathbf{x}_1',\mathbf{x}_2',\mathbf{x}_3') = \Gamma_3(\mathbf{x}_1,\mathbf{x}_2,\mathbf{x}_3;\mathbf{x}_1',\mathbf{x}_2',\mathbf{x}_3') - \sum_{j,k=1}^{3} (\pm)^{j+k} \Gamma_1(\mathbf{x}_j;\mathbf{x}_k') C_2(\breve{\mathbf{x}}_j;\breve{\mathbf{x}}_k')$$
$$- |\Gamma_1|_{3,\pm}(\mathbf{x}_1,\mathbf{x}_2,\mathbf{x}_3;\mathbf{x}_1',\mathbf{x}_2',\mathbf{x}_3'). \tag{G.16}$$

We now see a clear structure appearing. In the general case that we need to calculate the n-particle Green's function g_n given the initial density matrices Γ_k with $k = 1, \ldots, n$, we first need to determine $n-1$ initial-correlation functions $C_2, \ldots, C_n$ and subsequently construct $n-1$ solutions $\tilde{g}_2, \ldots, \tilde{g}_n$ of the homogeneous equations of motion. Below we first give the explicit prescription of how to construct g_n and subsequently the proof.

We define the m-particle initial-correlation function C_m from the recursive relations

$$C_m(X_M;X_M') = \Gamma_m(X_M;X_M') - \sum_{l=1}^{m-2} \sum_{K,J} (\pm)^{|K+J|} |\Gamma_1|_{l,\pm}(X_K;X_J') C_{m-l}(\breve{X}_K;\breve{X}_J')$$
$$- |\Gamma_1|_{m,\pm}(X_M;X_M'), \tag{G.17}$$

where $C_2 \equiv \Gamma_2 - |\Gamma_1|_{2,\pm}$. We need to explain the notation used in this equation. We introduce the collective coordinates $X_M = (\mathbf{x}_1, \ldots, \mathbf{x}_m)$ and $X_M' = (\mathbf{x}_1', \ldots, \mathbf{x}_m')$. For $l = 1, \ldots, m-2$ we further introduce the collective ordered indices $K = (k_1, \ldots, k_l)$ with $k_1 < \ldots < k_l$ and the collective coordinate $X_K = (\mathbf{x}_{k_1}, \ldots, \mathbf{x}_{k_l})$, and similarly the collective ordered index $J = (j_1, \ldots, j_l)$ with $j_1 < \ldots < j_l$ and the collective coordinate $X_J' = (\mathbf{x}_{j_1}', \ldots, \mathbf{x}_{j_l}')$. The sign of the various terms is determined by the quantity $|K+J|$ which is defined as

$$|K+J| = k_1 + \ldots + k_l + j_1 + \ldots + j_l.$$

The collective coordinate $\breve{X}_K \equiv X_{\breve{K}}$ denotes the set of coordinates with the complementary set $\breve{K}$ of ordered indices in $M = (1, \ldots, m)$. Finally, the quantity $|\Gamma_1|_{l,\pm}$ is the $l \times l$ permanent/determinant of one-particle density matrices with row indices K and column indices J. For example, let $m = 5$ and $l = 2$ with $K = (1,4)$ and $J = (2,4)$ and hence $\check{K} = (2,3,5)$ and $\check{J} = (1,3,5)$. We then have $X_K = (\mathbf{x}_1, \mathbf{x}_4)$, $X'_J = (\mathbf{x}'_2, \mathbf{x}'_4)$ and the complementary collective coordinates $\check{X}_K = (\mathbf{x}_2, \mathbf{x}_3, \mathbf{x}_5)$, $\check{X}'_J = (\mathbf{x}'_1, \mathbf{x}'_3, \mathbf{x}'_5)$. The corresponding term for C_5 under the summation sign in (G.17) is given by

$$(\pm)^{1+4+2+4} |\Gamma_1|_{2,\pm} (\mathbf{x}_1, \mathbf{x}_4; \mathbf{x}'_2, \mathbf{x}'_4) C_3 (\mathbf{x}_2, \mathbf{x}_3, \mathbf{x}_5; \mathbf{x}'_1, \mathbf{x}'_3, \mathbf{x}'_5)$$

$$= \pm \left| \begin{matrix} \Gamma_1(\mathbf{x}_1; \mathbf{x}'_2) & \Gamma_1(\mathbf{x}_1; \mathbf{x}'_4) \\ \Gamma_1(\mathbf{x}_4; \mathbf{x}'_2) & \Gamma_1(\mathbf{x}_4; \mathbf{x}'_4) \end{matrix} \right|_{\pm} C_3 (\mathbf{x}_2, \mathbf{x}_3, \mathbf{x}_5; \mathbf{x}'_1, \mathbf{x}'_3, \mathbf{x}'_5).$$

It can be readily checked that (G.17) gives back the expressions (G.12) and (G.16) for C_2 and C_3 that we derived before. The construction of the homogeneous solutions $\tilde{g}_m$ is readily guessed from $\tilde{g}_2$ and $\tilde{g}_3$. We define for all $m \geq 2$

$$C_m(M; M') = \left(\prod_{k=1}^{m} \delta(z_k, t_0) \right) C_m(X_M; X'_M) \left(\prod_{j=1}^{m} \delta(t_0, z'_j) \right)$$

and

$$\tilde{g}_m(M; M') = (\mp \mathrm{i})^m \int \left(\prod_{k=1}^{m} g(k; \bar{k}) \right) C_m(\bar{M}; \bar{M}') \left(\prod_{j=1}^{m} g(\bar{j}'; j') \right),$$

where the integral is over $\bar{1}, \ldots, \bar{m}, \bar{1}', \ldots, \bar{m}'$. These functions satisfy the homogeneous equations and at the initial times $z_i = z'_i = t_{0-}$ for all i are equal to the initial-correlation functions $C_m(X_M; X'_M)$. We are now ready to state the most important result of this appendix. The solution g_n of the noninteracting Martin–Schwinger hierarchy on the contour of Fig. 4.3 with the prescribed initial conditions (G.4) is

$$\boxed{\begin{aligned} g_n(N; N') &= |g|_{n, \pm}(N; N') \\ &\quad + \sum_{l=1}^{n-2} \sum_{K,J} (\pm)^{|K+J|} |g|_{l, \pm}(K; J') \tilde{g}_{n-l}(\check{K}; \check{J}') + \tilde{g}_n(N; N') \end{aligned}} \tag{G.18}$$

where $N = (1, \ldots, n)$ and the summation over K, J is a summation over all ordered subsets of N with l elements. We refer to this equation as the *generalized Wick's theorem*.

Before we prove the generalized Wick's theorem we note that (G.18) has the structure of the permanent/determinant of the sum of two matrices A and B since we have the Laplace formula (see Appendix B)

$$|A + B|_{\pm} = |A|_{\pm} + \sum_{l=1}^{n-1} \sum_{K,J} (\pm)^{|K+J|} |A|_{l, \pm}(K; J) |B|_{n-l, \pm}(\check{K}; \check{J}) + |B|_{\pm}, \tag{G.19}$$

where $|A|_{l, \pm}(K; J)$ is the permanent/determinant of the $l \times l$ matrix obtained with the rows K and the columns J of the matrix A. In the special case $l = n$ we have $K = J = N$ and hence $|A|_{n, \pm}(N; N) = |A|_{\pm}$. The same notation has been used for the matrix B. With the identification $A_{kj} = g(k; j')$ and $|B|_{n-l, \pm}(\check{K}; \check{J}) = \tilde{g}_{n-l}(\check{K}; \check{J}')$ for $l = 1 \ldots n-2$ and the definition $\tilde{g}_1 \equiv 0$ the equations (G.18) and (G.19) become identical. We can thus symbolically write the generalized Wick's theorem as

$$g_n = |g + \tilde{g}|_{n,\pm}$$

the precise meaning of which is given by (G.18).

It remains to prove (G.18). In the first step we check that the r.h.s. satisfies the Martin–Schwinger hierarchy and subsequently we show that it fulfills the required boundary conditions. Since the first term $|g|_{n,\pm}$ satisfies the hierarchy and the last term $\tilde{g}_n$ is a homogeneous solution it is enough to show that each function

$$g_n^{(l)}(N; N') = \sum_{K,J} (\pm)^{|K+J|} |g|_{l,\pm}(K; J') \tilde{g}_{n-l}(\check{K}; \check{J}') \tag{G.20}$$

appearing under the summation sign for the index l in (G.18) satisfies the Martin–Schwinger hierarchy. This function can be given a simple diagrammatic representation similar to Fig. G.1.[4] We represent $\tilde{g}_{n-l}(\check{K}; \check{J}')$ as a block with ingoing lines $\check{J}'$ and outgoing lines $\check{K}$, and $g(k; j')$ as an oriented line from j' to k. For instance, for $n = 5$ and $l = 2$ the term of the sum in (G.20) with $K = (1, 4)$ and $J = (2, 4)$ has the following diagrammatic representation:

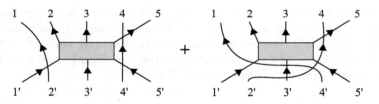

The sign of the first diagram is $(\pm)^{|K+J|} = \pm$ whereas the sign of the second diagram is $\pm(\pm)^{|K+J|} = +$. In Appendix B we showed that the sign of these diagrams can simply be determined from $(\pm)^{n_c}$ where n_c is the number of crossing lines (in the above example $n_c = 3$ in the first diagram and $n_c = 6$ in the second diagram). By summing over all possible diagrams of this kind we recover the function $g_n^{(l)}$.

If we now act with $[i\frac{d}{dz_k} - h(k)]$ on a particular diagram of $g_n^{(l)}$ the result is zero if k is an ingoing line to a $\tilde{g}_{n-l}$-block. This follows immediately from the fact that $\tilde{g}_{n-l}$ satisfies the homogeneous equations of motion. On the contrary, if the diagram contains a free line from j' to k then the result is $(\pm)^{k+j}\delta(k; j') \times$[a diagram for $g_{n-1}^{(l)}(\check{k}; \check{j}')$]. The sign $(\pm)^{k+j}$ follows immediately from the fact that the removal of a free line connecting j' to k reduces the number of crossing lines by a number with the parity of $k + j$. For instance, for the left diagram of the above example the removal of $g(1; 2')$ reduces the number of crossing lines by 1 which has the same parity as $1 + 2 = 3$. Similarly, the removal of $g(4; 4')$ reduces the number of crossing lines by 2 which has the same parity as $4 + 4 = 8$. The reader can verify that the same is true for the right diagram. This topological property is most easily explained in terms of "permutation graphs" introduced in Section 4.5. We refer the reader

---

[4]In contrast to Fig. G.1 the sign of the diagram is incorporated in the diagram itself.

to Appendix B for more details. In conclusion we can write

$$\left[i\frac{d}{dz_k} - h(k) \right] \text{[sum over all diagrams in } g_n^{(l)} \text{ containing } g(k;j')]$$

$$= (\pm)^{k+j}\delta(k;j')g_{n-1}^{(l)}(\check{k};\check{j}').$$

The satisfaction of the Martin–Schwinger hierarchy by $g_n^{(l)}$ follows from this relation when summing over all j. This proves our first statement.

Next we prove that g_n of (G.18) has the correct boundary conditions (G.4). Taking the limit $z_i, z_i' \to t_{0-}$ with $z_1' > ... > z_n' > z_n > ... > z_1$ we have $|g|_{l,\pm} \to (\mp i)^l |\Gamma_1|_{l,\pm}$ and $\tilde{g}_{n-l} \to (\mp i)^{n-l}C_{n-l}$ and therefore

$$\lim_{\substack{z_i,z_i' \to t_{0-} \\ z_1' > ... > z_n' > z_n > ... > z_1}} (\pm i)^n g_n(N;N') = |\Gamma_1|_{n,\pm}(X_N;X_N')$$

$$+ \sum_{l=1}^{n-2}\sum_{K,J}(\pm)^{|K+J|}|\Gamma_1|_{l,\pm}(X_K;X_J')C_{n-l}(\check{X}_K;\check{X}_J') + C_n(X_N;X_N').$$

Writing the last term C_n as in (G.17) we see by inspection that we recover the correct boundary conditions (G.4). This concludes the proof of the generalized Wick's theorem.

Note that if the system is prepared in a *noninteracting* configuration we have $\hat{\rho} = e^{-\beta\hat{H}_0^M}/Z_0$ and the Green's functions g_n coincide with the noninteracting Green's functions $G_{0,n}$ of Section 5.3. For these Green's functions the standard Wick's theorem (5.27) applies and therefore the n-particle density matrix $\Gamma_n = |\Gamma_1|_{n,\pm}$ is the permanent/determinant of one-particle density matrices. In this case the generalized Wick's theorem correctly reduces to the standard Wick's theorem since all the homogeneous solutions $\tilde{g}_m$ vanish. At zero temperature and in the case of fermions the ensemble average becomes a quantum average over the noninteracting ground state. This state is a product state of the form $|n_1 \ldots n_N\rangle$. Note that for *any* arbitrary choice of the quantum numbers $n_1, \ldots, n_N$ we can construct a noninteracting fermionic Hamiltonian for which $|n_1 \ldots n_N\rangle$ is the ground state. An example of such a Hamiltonian is $\hat{H}_0^M = -|E|\sum_{i=1}^N \hat{d}_{n_i}^\dagger \hat{d}_{n_i}$. This is not true in the case of bosons since at zero temperature the density matrix is either a mixture of states, see Exercise 6.3, or a pure state in which all bosons have the same quantum number (Bose condensate): $\hat{\rho} = |n_0n_0\ldots\rangle\langle n_0n_0\ldots|$. According to Wick's theorem we can then say that the noninteracting n-particle Green's function averaged over a product state $|n_0n_0\ldots\rangle$ in the case of condensed bosons or over a product state $|n_1\ldots n_N\rangle$ in the case of fermions is the permanent/determinant of one-particle Green's functions.

For systems prepared in an *interacting* configuration the generalized Wick's theorem offers an alternative route to expand the Green's functions. In this book we use the MBPT formulation on the contour of Fig. 4.5(a), and therefore include the information on the initial preparation in the Hamiltonian $\hat{H}^M$. Alternatively, however, we could use the generalized Wick's theorem and include the information on the initial preparation in the n-particle density matrices Γ_n. This MBPT formulation is based on the contour $\gamma_- \oplus \gamma_+$ which does not contain the vertical track. The two approaches are clearly complementary and the convenience of using one approach or the other depends on the information at hand. The relation between them is further discussed below. In order to continue reading, however, the reader should already be familiar with the concepts of self-energy and Dyson equation introduced in Chapters 10 and 11.

We intend to put forward the basic equations that relate the MBPT formulation on the Konstantinov–Perel' contour $\gamma_- \oplus \gamma_+ \oplus \gamma^M$ (in which the information on the initial preparation is encoded in $\hat{H}^M$) to the MBPT formulation on the Schwinger–Keldysh contour $\gamma_- \oplus \gamma_+$ (in which the information on the initial preparation is encoded in the n-particle density matrices Γ_n). We have seen that in the former approach the noninteracting Green's functions $G_{0,n}$ satisfy the noninteracting Martin–Schwinger hierarchy on the contour $\gamma_- \oplus \gamma_+ \oplus \gamma^M$ with KMS boundary conditions. These $G_{0,n}$ are given by (G.1) with $\rho \to \rho_0$ and $\hat{H}(z) \to \hat{H}_0$. On the other hand the noninteracting Green's functions g_n of the latter approach satisfy the noninteracting Martin–Schwinger hierarchy on the contour $\gamma_- \oplus \gamma_+$ with boundary conditions (G.4). These Green's functions are given by (G.1) with $\hat{H}(z) \to \hat{H}_0$.

If we expand the one-particle Green's function G in (G.1) in powers of the interaction $\hat{H}_{\text{int}}$ we get the formula (G.2) with $n = 1$, i.e.,

$$G(a,b) = \sum_{k=0}^{\infty} \frac{1}{k!} \left(\frac{i}{2} \right)^k \int v(1;1') \dots v(k;k') g_{2k+1}(a,1,1' \dots k, k'; b, 1^+, 1'^+ \dots k^+, k'^+),$$

where the integrals are over all variables with the exception of a and b. In contrast to (5.31), in this equation the contour integrals run on the contour $\gamma = \gamma_- \oplus \gamma_+$. In order to avoid confusion in the following we always specify the contour of integration in the integration symbol "$\int$". Using the generalized Wick's theorem (G.18) for the g_n we obtain a diagrammatic expansion for G. For example, to first order in the interaction we only need the g_3 of Fig. G.1. If we then take into account that the disconnected pieces vanish[5] we find that to first order in the interaction the Green's function is given by the connected diagrams displayed in Fig. G.2. In general the expansion of G in powers of $\hat{H}_{\text{int}}$ leads to a diagrammatic series which starts and ends with a g-line. The kernel of this expansion is therefore a *reducible self-energy* which we denote by σ_r to distinguish it from the reducible self-energy Σ_r of the MBPT formulation on the contour $\gamma_- \oplus \gamma_+ \oplus \gamma^M$, see (11.6). In the case of general initial states the σ_r-diagrams contain at most one initial-correlation function C_m and either begin and end with an interaction line (like the standard self-energy diagrams; we call their sum $\sigma_r^{(s)}$) or begin/end with an interaction line and end/begin with a C_m (we call their sum L/R reducible self-energy $\sigma_r^{L,R}$), see also Ref. [41]. This means that the general structure of the diagrammatic expansion is

$$G(1;2) = g(1;2) + \int_\gamma d\bar{1} d\bar{2} \; g(1;\bar{1}) \sigma_r(\bar{1};\bar{2}) g(\bar{2};2), \tag{G.21}$$

where

$$\sigma_r = \sigma_r^{(s)} + \sigma_r^L + \sigma_r^R \tag{G.22}$$

and

$$\sigma_r^L(1;2) = \sigma_r^L(1;\mathbf{x}_2)\delta(t_0,z_2),$$

---

[5]On the contour $\gamma_- \oplus \gamma_+$ all vacuum diagrams vanish since the integral along the forward branch cancels the integral along the backward branch and there is no vertical track.

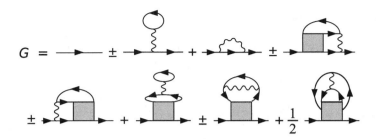

Figure G.2 First-order expansion of G for a two-body interaction. The last three diagrams vanish since both lines enter a C_m-block and hence the internal time-integration can be reduced to a point. This is a completely general feature.

$$\sigma_r^R(1;2) = \delta(z_1, t_0)\sigma_r^R(\mathbf{x}_1; 2).$$

The self-energy $\sigma_r^L(1; \mathbf{x}_2)$ has the same value for $z_1 = t_{1\pm}$ and similarly $\sigma_r^R(\mathbf{x}_1; 2)$ has the same value for $z_2 = t_{2\pm}$. Comparing (G.21) with (11.6) we see that the reducible self-energy is modified by diagrams with initial-correlation functions and by the addition of a L/R reducible self-energy which is nonzero only if its right/left time-argument is one of the end-points of γ. The diagrammatic expansion of these new self-energies is very similar to that of Chapter 10, as exemplified in Fig. G.2. The only extra ingredient is the appearance of the C_m-blocks which describe the initial m-body correlations.

Let us now consider the equation of motion for the Green's function on the contour $\gamma \oplus \gamma^M$,

$$\left[i\frac{d}{dz_1} - \hat{h}(z_1) \right] \hat{\mathcal{G}}(z_1, z_2) = \delta(z_1, z_2) + \int_{\gamma \oplus \gamma^M} d\bar{z}\, \hat{\Sigma}(z_1, \bar{z})\hat{\mathcal{G}}(\bar{z}, z_2),$$

with Σ the irreducible self-energy. We split the contour integral into an integral along γ and an integral along the vertical track γ^M,

$$\int_{\gamma \oplus \gamma^M} d\bar{z}\, \hat{\Sigma}(z_1, \bar{z})\hat{\mathcal{G}}(\bar{z}, z_2) = \int_\gamma d\bar{z}\, \hat{\Sigma}(z_1, \bar{z})\hat{\mathcal{G}}(\bar{z}, z_2) + \left[\hat{\Sigma}^\rceil \star \hat{\mathcal{G}}^\lceil \right](z_1, z_2), \qquad \text{(G.23)}$$

where we are implicitly assuming that z_1 and z_2 lie on the horizontal branches of γ. In this equation the left/right Keldysh components with argument on γ are defined in the most natural way, i.e.,

$$\hat{\Sigma}^\rceil(z = t_\pm, \tau) \equiv \hat{\Sigma}(z = t_\pm, t_0 - i\tau) = \hat{\Sigma}^\rceil(t, \tau)$$

and the like for $\hat{\Sigma}^\lceil$. Inserting in (G.23) the analogue of (9.68) for the left Green's function, i.e.,

$$\hat{\mathcal{G}}^\lceil(\tau, t) = -i\hat{\mathcal{G}}^M(\tau, 0)\hat{\mathcal{G}}^A(t_0, t) + \left[\hat{\mathcal{G}}^M \star \hat{\Sigma}^\lceil \cdot \hat{\mathcal{G}}^A \right](\tau, t),$$

and taking into account that

$$\int_\gamma dz'\, \delta(t_0, z')\hat{\mathcal{G}}(z', t_\pm) = \hat{\mathcal{G}}^A(t_0, t)$$

as well as [see also Exercise 5.5]

$$\int_\gamma dz' \,\hat{\Sigma}^\lceil(\tau, z')\hat{\mathcal{G}}(z', t_\pm) = \int_{t_0}^\infty dt' \,\hat{\Sigma}^\lceil(\tau, t')\hat{\mathcal{G}}^A(t', t),$$

we find

$$\int_{\gamma \oplus \gamma^M} d\bar{z} \,\hat{\Sigma}(z_1, \bar{z})\hat{\mathcal{G}}(\bar{z}, z_2) = \int_\gamma d\bar{z} \left[\hat{\Sigma} + \hat{\Sigma}^\lceil \star \hat{\mathcal{G}}^M \star \hat{\Sigma}^\lceil + \hat{\Sigma}^L\right](z_1, \bar{z})\hat{\mathcal{G}}(\bar{z}, z_2). \quad \text{(G.24)}$$

In (G.24) we have defined the irreducible L self-energy

$$\hat{\Sigma}^L(z_1, z_2) = -\mathrm{i}\,[\hat{\Sigma}^\lceil \star \hat{\mathcal{G}}^M](z_1, 0)\delta(t_0, z_2),$$

which has the same mathematical structure as σ_r^L. Similarly for the adjoint equation of motion we can easily derive

$$\int_{\gamma \oplus \gamma^M} d\bar{z}\, \hat{\mathcal{G}}(z_1, \bar{z})\hat{\Sigma}(\bar{z}, z_2) = \int_\gamma d\bar{z}\, \hat{\mathcal{G}}(z_1, \bar{z})\left[\hat{\Sigma} + \hat{\Sigma}^\rceil \star \hat{\mathcal{G}}^M \star \hat{\Sigma}^\lceil + \hat{\Sigma}^R\right](\bar{z}, z_2), \quad \text{(G.25)}$$

with

$$\hat{\Sigma}^R(z_1, z_2) = \mathrm{i}\delta(z_1, t_0)[\hat{\mathcal{G}}^M \star \hat{\Sigma}^\lceil](0, z_2),$$

which has the same mathematical structure as σ_r^R. Using (G.24) and (G.25) the integral over $\gamma \oplus \gamma^M$ appearing in the equations of motion for G is cast in terms of an integral over γ only. Thus, *given* the self-energy and the Green's function with arguments on the vertical track we can regard the equations of motion on $\gamma \oplus \gamma^M$ as equations of motion on γ for the unknown Green's function with arguments on γ. To integrate these equations we cannot use the noninteracting Green's function G_0 satisfying the KMS relations since the point $t_0 - \mathrm{i}\beta$ does not belong to γ. We must instead use the noninteracting Green's function g since it satisfies

$$\lim_{\substack{z_1, z_2 \to t_{0-} \\ z_2 > z_1}} g(1; 2) = \lim_{\substack{z_1, z_2 \to t_{0-} \\ z_2 > z_1}} G(1; 2) = \Gamma(\mathbf{x}_1; \mathbf{x}_2). \quad \text{(G.26)}$$

Thus if we define the total self-energy as

$$\hat{\Sigma}_{\text{tot}} = \hat{\Sigma} + \hat{\Sigma}^\rceil \star \hat{\mathcal{G}}^M \star \hat{\Sigma}^\lceil + \hat{\Sigma}^L + \hat{\Sigma}^R \quad \text{(G.27)}$$

we can write the following Dyson equation on γ:

$$G(1; 2) = g(1; 2) + \int_\gamma d\bar{1}d\bar{2}\, g(1; \bar{1})\Sigma_{\text{tot}}(\bar{1}; \bar{2})G(\bar{2}; 2). \quad \text{(G.28)}$$

Since the integral vanishes for $z_1, z_2 \to t_{0-}$ this G satisfies the equations of motion and the correct boundary conditions. Comparing (G.28) with (G.21) we now deduce an important relation between the reducible self-energy σ_r written in terms of initial-correlation functions and the self-energy Σ_{tot} written in terms of integrals along the vertical track,

$$\sigma_r = \Sigma_{\text{tot}} + \Sigma_{\text{tot}}g\Sigma_{\text{tot}} + \Sigma_{\text{tot}}g\Sigma_{\text{tot}}g\Sigma_{\text{tot}} + \dots \quad \text{(G.29)}$$

This relation constitutes a bridge between two different and complementary approaches [each built to optimize the nature (density matrix or Hamiltonian) of the initial information] and give a deep insight into the physics of initial correlations. Splitting σ_r as in (G.22) we can easily identify $\sigma_r^{(s)}(1;2)$, $\sigma_r^L(1;2)$, and $\sigma_r^R(1;2)$ as the terms in (G.29) which do not contain a δ-function, the terms which contain a $\delta(z_1, t_0)$ and the terms which contain a $\delta(t_0, z_2)$.

H

BBGKY hierarchy

In Section 5.1 we observed that from a knowledge of the equal-time n-particle Green's function we can calculate the time-dependent ensemble average of any n-body operator. Therefore it would be valuable if we could generate a close system of equations for the equal-time G_ns. Let us introduce the time-dependent generalization of the n-particle density matrix Γ_n defined in Section 1.7:

$$\Gamma_n(\mathbf{x}_1,...,\mathbf{x}_n;\mathbf{x}_1',...,\mathbf{x}_n'|t) = (\pm i)^n \lim_{\substack{z_i,z_i'\to t_- \\ z_1'>...>z_n'>z_n>...>z_1}} G_n(1,...,n;1',...,n'). \tag{H.1}$$

This definition reduces to (G.3) for $t = t_0$ and otherwise gives[1]

$$\Gamma_n(\mathbf{x}_1,...,\mathbf{x}_n;\mathbf{x}_1',...,\mathbf{x}_n'|t) = \mathrm{Tr}\left[\hat{\rho}\,\hat{\psi}_H^\dagger(\mathbf{x}_1',t)...\hat{\psi}_H^\dagger(\mathbf{x}_n',t)\hat{\psi}_H(\mathbf{x}_n,t)...\hat{\psi}_H(\mathbf{x}_1,t)\right].$$

Thus from Γ_n we can calculate the time-dependent ensemble average of any n-body operator. We can easily generate a hierarchy of equations for the Γ_ns starting from the Martin–Schwinger hierarchy. For notational convenience we introduce the collective coordinates $X_N = (\mathbf{x}_1,...,\mathbf{x}_n)$ and $X_N' = (\mathbf{x}_1',...,\mathbf{x}_n')$ in agreement with the notation of Appendix G:

$$\Gamma_n(\mathbf{x}_1,...,\mathbf{x}_n;\mathbf{x}_1',...,\mathbf{x}_n'|t) = \Gamma_n(X_N;X_N'|t).$$

Now we subtract (5.3) from (5.2), multiply by $(\pm i)^n$, sum over k between 1 and n and take the limit $z_i, z_i' \to t_-$ as in (H.1). The sum of all the time-derivatives of G_n yields the time-derivative of Γ_n,

$$(\pm i)^n \lim_{z_i,z_i'\to t_-} \sum_k \left[i\frac{d}{dz_k} + i\frac{d}{dz_k'}\right] G_n(1,...,n;1',...,n') = i\frac{d}{dt}\Gamma_n(X_N;X_N'|t).$$

The terms with the δ-functions cancel out since

$$\sum_{kj=1}^n (\pm)^{k+j}\left[\delta(k;j')G_{n-1}(1,...\overset{\sqcap}{k}...,n;1',...\overset{\sqcap}{j'}...,n')\right.$$

$$\left. - \delta(j;k')\,G_{n-1}(1,...\overset{\sqcap}{j}...,n;1',...\overset{\sqcap}{k'}...,n')\right] = 0,$$

---

[1]The reader can check that the same result would follow by choosing t_+ instead of t_- in (H.1).

as follows directly by renaming the indices $k \leftrightarrow j$ in, say, the second term. Interestingly this cancellation occurs independently of the choice of the contour arguments z_i and z_i'. The terms with the single-particle Hamiltonian h are also easy to handle since

$$(\pm i)^n \lim_{z_i, z_i' \to t_-} h(k) G_n(1, ..., n; 1', ..., n') = h(\mathbf{x}_k, t) \Gamma_n(X_N; X_N'|t),$$

and similarly for $h(k') G_n(1, ..., n; 1', ..., n')$, and hence

$$(\pm i)^n \lim_{z_i, z_i' \to t_-} \left[\sum_{k=1}^{n} h(k) - \sum_{k=1}^{n} h(k') \right] G_n(1, ..., n; 1', ..., n')$$

$$= \left[\sum_{k=1}^{n} h(\mathbf{x}_k, t) - \sum_{k=1}^{n} h(\mathbf{x}_k', t) \right] \Gamma_n(X_N; X_N'|t).$$

We are therefore left with the integrals $\int v \, G_{n+1}$. Consider for instance (5.2). By definition we have

$$G_{n+1}(1, ..., n, \bar{1}; 1', ..., n', \bar{1}^+)$$

$$= \frac{(\pm)}{i^{n+1}} \text{Tr} \left[\hat{\rho} \mathcal{T} \left\{ \hat{\psi}_H(1) ... \hat{n}_H(\bar{1}) \hat{\psi}_H(k) ... \hat{\psi}_H(n) \hat{\psi}_H^\dagger(n') ... \hat{\psi}_H^\dagger(1') \right\} \right].$$

As the interaction is local in time we have to evaluate this G_{n+1} when the time $\bar{z}_1 = z_k$. For these times the operator $\hat{n}_H(\bar{1}) \hat{\psi}_H(k) = \hat{n}_H(\bar{\mathbf{x}}_1, z_k) \hat{\psi}_H(\mathbf{x}_k, z_k)$ should be treated as a composite operator when we take the limit $z_i, z_i' \to t_-$. Consequently we can write

$$\lim_{z_i, z_i' \to t_-} G_{n+1}(1, ..., n, \bar{\mathbf{x}}_1, z_k; 1', ..., n', \bar{\mathbf{x}}_1, z_k^+)$$

$$= \frac{(\pm)^{n+1}}{i^{n+1}} \text{Tr} \left[\hat{\rho} \hat{\psi}_H^\dagger(\mathbf{x}_1', t) ... \hat{\psi}_H^\dagger(\mathbf{x}_n', t) \hat{\psi}_H(\mathbf{x}_n, t) ... \hat{n}_H(\bar{\mathbf{x}}_1, t) \hat{\psi}_H(\mathbf{x}_k, t) ... \hat{\psi}_H(\mathbf{x}_1, t) \right].$$

Next we want to move the density operator $\hat{n}_H(\bar{\mathbf{x}}_1, t)$ between $\hat{\psi}_H^\dagger(\mathbf{x}_n', t)$ and $\hat{\psi}_H(\mathbf{x}_n, t)$ so as to form the $(n+1)$-particle density matrix. This can easily be done since operators in the Heisenberg picture at equal time satisfy the same (anti)commutation relations as the original operators. From (1.48) we have

$$\hat{\psi}(\mathbf{x}_j) \hat{n}(\bar{\mathbf{x}}_1) = \hat{n}(\bar{\mathbf{x}}_1) \hat{\psi}(\mathbf{x}_j) + \delta(\mathbf{x}_j - \bar{\mathbf{x}}_1) \hat{\psi}(\mathbf{x}_j)$$

and therefore

$$(\pm i)^{n+1} \lim_{z_i, z_i' \to t_-} G_{n+1}(1, ..., n, \bar{\mathbf{x}}_1, z_k; 1', ..., n', \bar{\mathbf{x}}_1, z_k^+)$$

$$= \Gamma_{n+1}(X_N, \bar{\mathbf{x}}_1; X_N', \bar{\mathbf{x}}_1|t) + \sum_{j > k} \delta(\mathbf{x}_j - \bar{\mathbf{x}}_1) \Gamma_n(X_N; X_N'|t).$$

If we multiply this equation by $v(\mathbf{x}_k, \bar{\mathbf{x}}_1)$ and integrate over $\bar{\mathbf{x}}_1$ we get exactly the integral that we need since in the Martin–Schwinger hierarchy $\int v \, G_{n+1}$ is multiplied by $\pm i$ and we said before that we are multiplying the nth equation of the hierarchy by $(\pm i)^n$. We have

$$(\pm i)^{n+1} \lim_{z_i,z_i' \to t_-} \int d\bar{1}\, v(k;\bar{1})\, G_{n+1}(1,...,n,\bar{1};1',...,n',\bar{1}^+)$$

$$= \int d\bar{\mathbf{x}}_1\, v(\mathbf{x}_k,\bar{\mathbf{x}}_1)\, \Gamma_{n+1}(X_N,\bar{\mathbf{x}}_1;X_N',\bar{\mathbf{x}}_1|t) + \sum_{j>k} v(\mathbf{x}_k,\mathbf{x}_j)\, \Gamma_n(X_N;X_N'|t).$$

In a similar way we can prove that

$$(\pm i)^{n+1} \lim_{z_i,z_i' \to t_-} \int d\bar{1}\, v(k';\bar{1})\, G_{n+1}(1,...,n,\bar{1}^-;1',...,n',\bar{1})$$

$$= \int d\bar{\mathbf{x}}_1\, v(\mathbf{x}_k',\bar{\mathbf{x}}_1)\, \Gamma_{n+1}(X_N,\bar{\mathbf{x}}_1;X_N',\bar{\mathbf{x}}_1|t) + \sum_{j>k} v(\mathbf{x}_k',\mathbf{x}_j')\, \Gamma_n(X_N;X_N'|t).$$

Thus the interaction produces terms proportional to Γ_{n+1} and terms proportional Γ_n. Collecting all pieces we find the following equation:

$$\left[i\frac{d}{dt} - \sum_{k=1}^n \left(h(\mathbf{x}_k,t) + \sum_{j>k} v(\mathbf{x}_k,\mathbf{x}_j) \right) + \sum_{k=1}^n \left(h(\mathbf{x}_k',t) + \sum_{j>k} v(\mathbf{x}_k',\mathbf{x}_j') \right) \right] \Gamma_n(X_N;X_N'|t)$$

$$= \sum_{k=1}^n \int d\bar{\mathbf{x}}_1 \left(v(\mathbf{x}_k,\bar{\mathbf{x}}_1) - v(\mathbf{x}_k',\bar{\mathbf{x}}_1) \right) \Gamma_{n+1}(X_N,\bar{\mathbf{x}}_1;X_N',\bar{\mathbf{x}}_1|t)$$

This hierarchy of equations for the density matrices is known as the Born–Bogoliubov–Green-Kirkwood-Yvon (BBGKY) hierarchy [171-174]. Let us work out the lowest order equations. For $n = 1$ we have

$$\left[i\frac{d}{dt} - h(\mathbf{x}_1,t) + h(\mathbf{x}_1',t) \right] \Gamma_1(\mathbf{x}_1;\mathbf{x}_1'|t)$$

$$= \int d\bar{\mathbf{x}}_1 \left(v(\mathbf{x}_1,\bar{\mathbf{x}}_1) - v(\mathbf{x}_1',\bar{\mathbf{x}}_1) \right) \Gamma_2(\mathbf{x}_1,\bar{\mathbf{x}}_1;\mathbf{x}_1',\bar{\mathbf{x}}_1|t),$$

while for $n = 2$ we have

$$\left[i\frac{d}{dt} - h(\mathbf{x}_1,t) - h(\mathbf{x}_2,t) - v(\mathbf{x}_1,\mathbf{x}_2) + h(\mathbf{x}_1',t) + h(\mathbf{x}_2',t) + v(\mathbf{x}_1',\mathbf{x}_2') \right] \Gamma_2(\mathbf{x}_1,\mathbf{x}_2;\mathbf{x}_1',\mathbf{x}_2'|t)$$

$$= \int d\bar{\mathbf{x}}_1 \left(v(\mathbf{x}_1,\bar{\mathbf{x}}_1) + v(\mathbf{x}_2,\bar{\mathbf{x}}_1) - v(\mathbf{x}_1',\bar{\mathbf{x}}_1) - v(\mathbf{x}_2',\bar{\mathbf{x}}_1) \right) \Gamma_3(\mathbf{x}_1,\mathbf{x}_2,\bar{\mathbf{x}}_1;\mathbf{x}_1',\mathbf{x}_2',\bar{\mathbf{x}}_1|t).$$

For the BBGKY hierarchy to be useful one should devise suitable truncation schemes to express Γ_m in terms of Γ_ns with $n < m$. At present all standard truncation schemes give rise to several problems in the time domain and the practical use of the BBGKY is still limited to ground-state situations, see Ref. [175] and references therein.

I

From δ-like peaks to continuous spectral functions

Photocurrent experiments on bulk systems like metals and semiconductors reveal that the photocurrent $I_{ph}(\epsilon)$ and hence the spectral function $A(\epsilon)$, see Section 6.3.4, are continuous functions of ϵ. How can a sum of δ-like peaks transform into a continuous function? Understanding the transition from δ-like peaks to continuous functions not only satisfies a mathematical curiosity but is also important to make connection with experiments. In this appendix we explain the "mystery" with an example.

Consider a one-dimensional crystal like the one of Fig. 2.5 with zero onsite energy and nearest neighbor hoppings T. The atoms are labeled from 1 to N and 1 is nearest neighbor of N (BvK boundary conditions). The single-particle eigenkets $|k\rangle$ are Bloch waves whose amplitude on site j is

$$\langle j|k\rangle = \frac{e^{ikj}}{\sqrt{N}}$$

and whose eigenenergy is $\epsilon_k = 2T\cos k$, where $k = 2\pi m/N$ and $m = 1,\ldots,N$. We assume that the electrons are noninteracting and calculate the spectral function $A_{11}(\omega)$ projected on atom 1.[1] By definition

$$
\begin{aligned}
A_{11}(\omega) &= i\left[G_{11}^{R}(\omega) - G_{11}^{A}(\omega)\right] \\
&= -2\mathrm{Im}\left[G_{11}^{R}(\omega)\right] \\
&= -2\mathrm{Im}\,\langle 1|\frac{1}{\omega - \hat{h} + i\eta}|1\rangle \\
&= \frac{2}{N}\sum_{m=1}^{N}\frac{\eta}{(\omega - 2T\cos\frac{2\pi m}{N})^2 + \eta^2},
\end{aligned}
$$

where in the last equality we have inserted the completeness relation $\sum_k |k\rangle\langle k| = \hat{1}$. Since the eigenvalues with m and $m' = N - m$ are degenerate we can rewrite the above sum as

$$A_{11}(\omega) = \frac{4}{N}\sum_{m=1}^{N/2}\frac{\eta}{(\omega - 2T\cos\frac{2\pi m}{N})^2 + \eta^2}. \tag{I.1}$$

---

[1] Since $|1\rangle$ is not an eigenket we expect that $A_{11}(\omega)$ has peaks at different energies. Furthermore, due to the discrete translational invariance of the Hamiltonian, $A_{jj}(\omega)$ is independent of the atomic site j.

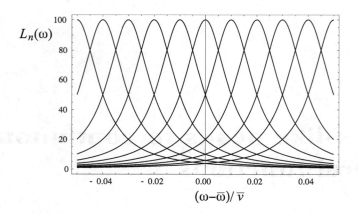

Figure I.1 Lorentzian functions $L_n(\omega) = \frac{\eta}{(\omega-\bar\omega-2\pi\bar v n/N)^2+\eta^2}$ for $n = -50, -40, \ldots, 40, 50$ in units of $1/(2\pi\bar v)$. The other parameters are $\eta/(2\pi\bar v) = 0.01$ and $N = 1000$. We see that $L_{\pm 50}(\bar\omega)$ is already very small and can be discarded.

Restricting the sum between 1 and $N/2$ guarantees that different ms correspond to different peaks. For every finite N the spectral function becomes a sum of δ-functions when the limit $\eta \to 0$ is taken. However, if we first take the limit $N \to \infty$ and then $\eta \to 0$ the situation is very different. For any arbitrary small but finite η the heights of the peaks go to zero like the prefactor $1/N$ while the position of the peaks moves closer and closer. When the energy spacing between two consecutive eigenvalues becomes much smaller than η many Lorentzians contribute to the sum for a given ω. Let us study the value of the sum for a frequency $\bar\omega = 2T \cos \frac{2\pi\bar m}{N}$ corresponding to the position of one of the peaks. For $m = \bar m + n$ with $n \ll N$ we can write

$$\epsilon_k = 2T \cos \frac{2\pi m}{N} = \bar\omega + 2\pi\bar v \frac{n}{N} + \mathcal{O}(n^2/N^2), \tag{I.2}$$

where $\bar v = -2T \sin \frac{2\pi\bar m}{N}$. We see that for $n = \pm N\eta/(2\pi\bar v) \ll N$ the r.h.s. of (I.2) is equal to $\bar\omega \pm \eta$. We can use this approximate expression for ϵ_k in $A_{11}(\bar\omega)$, since if the eigenvalues are much larger than $\bar\omega + \eta$ or much smaller than $\bar\omega - \eta$ the corresponding contribution is negligible. In Fig. I.1 we show the plot of several Lorentzians of width η centered in $\bar\omega + 2\pi\bar v n/N$ for $\eta/(2\pi\bar v) = 0.01$ and $N = 1000$. The value that they assume in $\bar\omega$ is rather small already for $2\pi\bar v n/N = \pm 5\eta$, i.e., $|n| = 5N\eta/(2\pi\bar v) = 50 \ll N$. Inserting the approximate expression of the eigenvalues in A_{11} and extending the sum over n between $-\infty$ and ∞ we find

$$A_{11}(\bar\omega) = \lim_{N\to\infty} \frac{4}{N} \sum_{n=-\infty}^{\infty} \frac{\eta}{(2\pi\bar v n/N)^2 + \eta^2} = 4 \int_{-\infty}^{\infty} dx \frac{\eta}{(2\pi\bar v x)^2 + \eta^2} = \frac{2}{|\bar v|}.$$

The final result is *finite and independent of* η. The limiting function can easily be expressed in terms of $\bar\omega$. Defining $\bar k = 2\pi\bar m/N$ we have $\bar v = -2T \sin \bar k$ and $\bar\omega = 2T \cos \bar k$. Solving

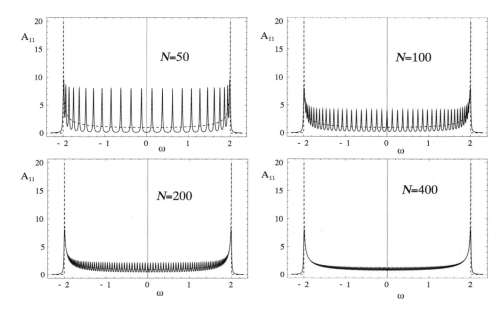

Figure I.2 Comparison between the limiting spectral function (dashed) and the spectral function (I.1) for $N = 50, 100, 200, 400$ (solid). The parameter $\eta = 0.01$ and all energies are in units of $|T|$.

the latter equation for $\bar{k}$ we get

$$A_{11}(\bar{\omega}) = \frac{2}{\left|2T \sin\left(\arccos \frac{\bar{\omega}-\epsilon}{2T}\right)\right|}.$$

In Fig. I.2 we show how fast the discrete spectral function (I.1) converges to the limiting spectral function as N increases. We can see that N must be larger than $\sim 5|T|/\eta$ for the peaks to merge and form a smooth function.

It is instructive to look at the transition $N \to \infty$ also in the time domain. For every finite N and at zero temperature the Green's function, say, $G_{11}^<(t,t')$ is an oscillatory function of the form

$$G_{11}^<(t,t') = \pm i \int \frac{d\omega}{2\pi} e^{-i\omega(t-t')} f(\omega - \mu) A_{11}(\omega) = \pm i \frac{1}{N} \sum_{\epsilon_k < \mu} e^{-i\epsilon_k(t-t')}.$$

In Fig. I.3 we plot the absolute value squared of $G_{11}^<(t,0)$, which corresponds to the probability of finding the system unchanged when removing a particle from site 1 at time 0 and then putting the particle back at time t. The probability initially decreases but for times $t \sim N/(2T)$ (roughly the inverse of the average energy spacing) there is a revival. The revival time becomes longer when N becomes larger. In the continuum limit $N \to \infty$ the revival time becomes infinite and the probability decays to zero.[2] If the decay is exponential,

---

[2]This follows from the fact that we can replace the sum by an integral and use the Riemann–Lebesgue theorem.

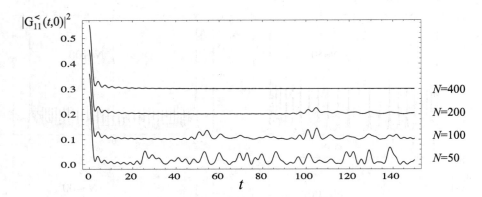

Figure I.3 Absolute value squared of $G_{11}^<(t,0)$ at zero temperature and $\mu = 0$. The curves with $N = 100, 200, 400$ are shifted upward by 0.1, 0.2, 0.3 respectively. Times are in units of $1/|T|$.

$e^{-\Gamma t}$, then the spectral function in frequency space is a Lorentzian of width Γ. This means that the limiting spectral function can have poles (or branch cuts) in the complex plane even though its discrete version has poles just above and below the real axis and is analytic everywhere else, see again (I.1). The appearance of poles in the complex plane is related to the infinite revival time which, in turn, is due to the negative interference of many oscillatory functions with different but very close energies.

J

Virial theorem for conserving approximations

We consider a system of identical particles with charge q and mass m, mutually interacting with a Coulomb interaction $v(\mathbf{r}_1, \mathbf{r}_2) = q^2/|\mathbf{r}_1 - \mathbf{r}_2|$ and subject to the static potential generated by nuclei of charge $\{-Z_i q\}$ in position $\{\mathbf{R}_i\}$. The single-particle Hamiltonian operator $\hat{h}$ reads

$$\hat{h} = \frac{\hat{p}^2}{2m} - q^2 \sum_i \frac{Z_i}{|\hat{\boldsymbol{r}} - \mathbf{R}_i|} \equiv \frac{\hat{p}^2}{2m} + qV(\hat{\boldsymbol{r}}; \{\mathbf{R}_i\}).$$

The total energy of the system

$$E_{\text{tot}} = E + E_{\text{nn}}$$

is the sum of the energy defined in (13.29),

$$E = E_{\text{one}} + E_{\text{int}},$$

and the classical nuclear energy

$$E_{\text{nn}} = \frac{1}{2} q^2 \sum_{ij} \frac{Z_j Z_i}{|\mathbf{R}_j - \mathbf{R}_i|}.$$

For later purposes it is convenient to split the one-body part of the energy into a kinetic and potential part:

$$E_{\text{one}} = E_{\text{kin}} + E_{\text{en}}(\{\mathbf{R}_i\})$$

with

$$E_{\text{kin}} = \pm i \int d\mathbf{x}_1 \left[-\frac{\nabla_1^2}{2m} G(1; 2) \right]_{2=1^+}$$

and

$$E_{\text{en}}(\{\mathbf{R}_i\}) = \pm i \int d\mathbf{x}_1 qV(\mathbf{r}_1; \{\mathbf{R}_i\}) G(1; 1^+) = \int d\mathbf{x}_1 qV(\mathbf{r}_1; \{\mathbf{R}_i\}) n(1), \qquad (J.1)$$

where $n(1) = \pm iG(1; 1^+)$ is the density. In (J.1) the dependence of E_{en} on the nuclear coordinates is explicitly reported since this quantity depends on $\{\mathbf{R}_i\}$ implicitly through G and explicitly through V.

In this appendix we show that the virial theorem

$$\boxed{2E_{\text{kin}} + E_{\text{en}}(\{\mathbf{R}_i\}) + E_{\text{int}} + E_{\text{nn}} = 0}$$

is satisfied provided that (1) G is self-consistently calculated from a conserving self-energy, and (2) the nuclei are in their equilibrium configuration, i.e.,

$$\frac{dE_{\text{tot}}}{d\mathbf{R}_i} = 0 \quad \text{for all } i. \tag{J.2}$$

This result was first derived in Ref. [176].

We begin by exploiting the important stationary property of the Klein functional

$$\frac{\delta \Omega_{\text{K}}[G, v]}{\delta G} = 0,$$

when G equals the self-consistent Green's function corresponding to a conserving self-energy [condition (1)]. We then define a Green's function G^λ which depends on the real parameter λ according to

$$G^\lambda(1; 2) = \lambda^3 G(\lambda \mathbf{r}_1\, \sigma_1, z_1; \lambda \mathbf{r}_2\, \sigma_2, z_2).$$

The Green's function $G^{\lambda=1} = G$ is the self-consistent Green's function and therefore

$$\left. \frac{d\Omega_{\text{K}}[G^\lambda, v]}{d\lambda} \right|_{\lambda=1} = 0. \tag{J.3}$$

Let us calculate this derivative from the expression (11.34) of the Klein functional. The term $\text{tr}_\gamma \left[\ln(-G^\lambda)\right] = \text{tr}_\gamma \left[\ln(-G)\right]$ is independent of λ since

$$\text{tr}_\gamma \left[(G^\lambda)^m \right] = \int d1 \ldots dm\, G^\lambda(1; 2) \ldots G^\lambda(m, 1^+) = \text{tr}_\gamma \left[G^m \right].$$

As a consequence (J.3) can be written as

$$\left. \frac{d\Phi[G^\lambda, v]}{d\lambda} \right|_{\lambda=1} - \frac{d}{d\lambda} \text{tr}_\gamma \left[G^\lambda \overleftarrow{G}_0^{-1} \right]_{\lambda=1} = 0. \tag{J.4}$$

To evaluate the derivative of Φ consider the expansion (11.18). The nth order term consists of n interaction lines, $2n$ Green's function lines, and integration over $2n$ spatial coordinates. Taking into account that the interaction v is Coulombic we have

$$\int d\mathbf{r}_1 \ldots d\mathbf{r}_{2n} \underbrace{v \ldots v}_{n \text{ times}} \underbrace{G^\lambda \ldots G^\lambda}_{2n \text{ times}} = \lambda^n \int d\mathbf{r}_1 \ldots d\mathbf{r}_{2n} \underbrace{v \ldots v}_{n \text{ times}} \underbrace{G \ldots G}_{2n \text{ times}}$$

and therefore

$$\Phi[G^\lambda, v] = \sum_{n=1}^{\infty} \frac{\lambda^n}{2n} \text{tr}_\gamma \left[\Sigma_s^{(n)} G \right].$$

The λ-derivative of this quantity is proportional to the interaction energy

$$\frac{d\Phi[G^\lambda, v]}{d\lambda}\bigg|_{\lambda=1} = \frac{1}{2}\text{tr}_\gamma\,[\Sigma G] = \frac{-i\beta}{2}\int dx_1 d2\,\Sigma(1;2)G(2;1^+) = \mp\beta E_{\text{int}},$$

where in the last equality we have used (9.19). Next we consider the second term of (J.4). We have

$$\text{tr}_\gamma\left[G^\lambda\overleftarrow{G_0^{-1}}\right] = \int d1\,G^\lambda(1;2)\left[-i\frac{\overleftarrow{d}}{dz_2} + \frac{\overleftarrow{\nabla_2^2}}{2m} - qV(r_2;\{R_i\})\right]_{2=1^+}$$

$$= \int d1\left[-i\frac{d}{dz_2}G^\lambda(1;2)\right]_{2=1^+} + \underbrace{\int d1\left[\frac{\nabla_2^2}{2m} - qV(r_2;\{R_i\})\right]G^\lambda(1;2)\bigg|_{2=1^+}}_{\pm\beta[\lambda^2 E_{\text{kin}}+\lambda E_{\text{en}}(\{\lambda R_i\})]}$$

and hence

$$\frac{d}{d\lambda}\text{tr}_\gamma\left[G^\lambda\overleftarrow{G_0^{-1}}\right]_{\lambda=1} = \pm\beta\left[2E_{\text{kin}} + E_{\text{en}}(\{R_i\}) + \int dx_1 q\frac{dV(r_1;\{\lambda R_i\})}{d\lambda}\bigg|_{\lambda=1} n(x_1)\right].$$

Substituting these results into (J.4) we find

$$2E_{\text{kin}} + E_{\text{en}}(\{R_i\}) + E_{\text{int}} + \int dx_1 q\frac{dV(r_1;\{\lambda R_i\})}{d\lambda}\bigg|_{\lambda=1} n(x_1) = 0, \qquad (J.5)$$

which is true for any conserving approximation.

The next step consists in evaluating the λ-derivative in (J.5). For this purpose we consider the variation of the energy E induced by a variation δR_j of the nuclear coordinates. In the zero temperature limit $\Omega = E - \mu N$ [see (D.7)] and the variations $\delta\Omega$ and δE are equal since the total number of particles is independent of the nuclear position, provided that there is a gap between the ground state and the first excited state. Then we can write

$$\delta E = \text{tr}_\gamma\left[\frac{\delta\Omega_K}{\delta G}\delta G\right]_{\{R_i\}} + \frac{d\Omega_K}{dR_i}\bigg|_G \cdot \delta R_i,$$

where we take into account that Ω_K depends on $\{R_i\}$ implicitly through G and explicitly through G_0. The first term on the r.h.s. vanishes for a conserving Green's function (stationary property). In the Klein functional the positions of the nuclei only enter as parameters in the term G_0^{-1}, while the Green's function is an independent variable. This means that

$$\frac{d\Omega_K}{dR_j}\bigg|_G = \int dx_1 q\frac{dV(r_1;\{R_i\})}{dR_j} n(x_1).$$

From these considerations we conclude that the variation of E induced by the variation $\{R_i\} \to \{\lambda R_i\}$ is

$$\frac{dE}{d\lambda}\bigg|_{\lambda=1} = \int dx_1 q\frac{dV(r_1;\{\lambda R_i\})}{d\lambda}\bigg|_{\lambda=1} n(x_1).$$

This is an important result since it establishes a relation between the derivative appearing in (J.5) and the derivative of E. This latter derivative can easily be worked out using condition (2). The nuclear positions must fulfill (J.2) and hence

$$\frac{dE}{d\lambda}\bigg|_{\lambda=1} = -\frac{d}{d\lambda}\frac{1}{2}q^2 \sum_{ij} \frac{Z_j Z_i}{\lambda|\mathbf{R}_j - \mathbf{R}_i|}\bigg|_{\lambda=1} = E_{nn}.$$

The virial theorem simply follows from (J.5) and the combination of these last two equations.

K

Momentum distribution and sharpness of the Fermi surface

The *momentum distribution* $n_{\mathbf{p}}$ of the electron gas at zero temperature is given by the equation

$$n_{\mathbf{p}} = \int_{-\infty}^{0} \frac{d\omega}{2\pi} A(\mathbf{p}, \mu + \omega), \tag{K.1}$$

where $A(\mathbf{p}, \mu + \omega)$ is the spectral function centered at the chemical potential $\mu = \epsilon_{\mathrm{F}}$. Interestingly, in terms of this quantity the total energy (6.106) for an electron gas takes the form

$$E_{\mathrm{S}} = \frac{1}{2} \int \frac{d\omega}{2\pi} \int \frac{d\mathbf{p}}{(2\pi)^3} \langle \mathbf{p} | (\omega + \hat{h}) f(\omega - \mu) \hat{\mathcal{A}}(\omega) | \mathbf{p} \rangle$$

$$= \frac{\mathrm{V}}{2} \int_{-\infty}^{\mu} \frac{d\omega}{2\pi} \int \frac{d\mathbf{p}}{(2\pi)^3} (\omega + \frac{p^2}{2}) A(\mathbf{p}, \omega) = \frac{\mathrm{V}}{2} \left[\int \frac{d\mathbf{p}}{(2\pi)^3} \frac{p^2}{2} n_{\mathbf{p}} + \int_{-\infty}^{\mu} \frac{d\omega}{2\pi} \omega D(\omega) \right],$$

where $D(\omega)$ is the density of states. It is easy to show that in the noninteracting case the two integrals in the square brackets are identical.

Equation (K.1) is difficult to use in practical numerical calculations since it has a very sharp quasi-particle peak close to the Fermi surface, the volume of which is hard to converge by integration. We therefore derive a different formula that also allows us to prove that the momentum distribution is sharp. First we note that since $G^{\mathrm{A}}(\mathbf{p}, \omega) = [G^{\mathrm{R}}(\mathbf{p}, \omega)]^*$ we have $A(\mathbf{p}, \omega) = -2 \, \mathrm{Im}[G^{\mathrm{R}}(\mathbf{p}, \omega)]$. The retarded Green's function is analytic in the upper-half of the complex ω-plane and therefore

$$0 = \int_{-\infty}^{0} d\omega \, G^{\mathrm{R}}(\mathbf{p}, \mu + \omega) + \mathrm{i} \int_{0}^{\infty} d\omega \, G^{\mathrm{R}}(\mathbf{p}, \mu + \mathrm{i}\omega) + \lim_{R \to \infty} \mathrm{i} \int_{\frac{\pi}{2}}^{\pi} d\theta \, Re^{\mathrm{i}\theta} G^{\mathrm{R}}(\mathbf{p}, \mu + Re^{\mathrm{i}\theta}),$$

where we first integrate along the real axis from $-\infty$ to 0, then along the positive imaginary axis to $\mathrm{i}\infty$ and finally along a quarter circle from $\mathrm{i}\infty$ to $-\infty$. Since for $|\omega| \to \infty$ we have $G^{\mathrm{R}}(\mathbf{p}, \mu + \omega) \sim 1/\omega$, see (6.78), the integral along the quarter circle becomes

$$\lim_{R \to \infty} \mathrm{i} \int_{\frac{\pi}{2}}^{\pi} d\theta \, Re^{\mathrm{i}\theta} G^{\mathrm{R}}(\mathbf{p}, \mu + Re^{\mathrm{i}\theta}) = \mathrm{i} \frac{\pi}{2}$$

and hence

$$\int_{-\infty}^{0} d\omega\, G^{\mathrm{R}}(\mathbf{p}, \mu + \omega) = -\mathrm{i}\frac{\pi}{2} - \mathrm{i}\int_{0}^{\infty} d\omega\, G^{\mathrm{R}}(\mathbf{p}, \mu + \mathrm{i}\omega).$$

Inserting this result into (K.1) then yields

$$n_{\mathbf{p}} = \frac{1}{2} + \frac{1}{\pi}\mathrm{Re}\int_{0}^{\infty} d\omega\, G^{\mathrm{R}}(\mathbf{p}, \mu + \mathrm{i}\omega). \tag{K.2}$$

This formula [110] is the basis for our subsequent derivations. We first write for the retarded Green's function centered at μ

$$G^{\mathrm{R}}(\mathbf{p}, \mu + \omega) = \frac{1}{\omega + \mu - \frac{p^2}{2} - \Sigma^{\mathrm{R}}(\mathbf{p}, \mu + \omega)} = \frac{1}{\omega + a_p - \sigma_p^{\mathrm{R}}(\omega)},$$

where we define

$$a_p = \mu - \frac{p^2}{2} - \Sigma_{\mathrm{x}}(p),$$

with the exchange self-energy $\Sigma_{\mathrm{x}}(p) = -\frac{2p_{\mathrm{F}}}{\pi}F(p/p_{\mathrm{F}})$, see (7.61), and $\sigma_p^{\mathrm{R}}(\omega) = \Sigma_{\mathrm{c}}^{\mathrm{R}}(\mathbf{p}, \mu + \omega)$, the correlation self-energy centered at μ.[1] Of course these quantities depend only on the modulus $p = |\mathbf{p}|$ due to rotational invariance. Let us now work out the integral in (K.2)

$$\mathrm{Re}\int_{0}^{\infty} d\omega\, G^{\mathrm{R}}(\mathbf{p}, \mu + \mathrm{i}\omega) = \mathrm{Re}\int_{0}^{\infty} \frac{d\omega}{\mathrm{i}\omega + a_p - \sigma_p^{\mathrm{R}}(\mathrm{i}\omega)}.$$

Since the denominator vanishes when $\omega \to 0$ and $p = p_{\mathrm{F}}$ according to the Luttinger–Ward theorem we add and subtract a term in which σ^{R} is expanded to first order in ω, i.e.,

$$\frac{1}{\mathrm{i}\omega + a_p - \sigma_p^{\mathrm{R}}(\mathrm{i}\omega)} = \underbrace{\left[\frac{1}{\mathrm{i}\omega + a_p - \sigma_p^{\mathrm{R}}(\mathrm{i}\omega)} - \frac{1}{\mathrm{i}\omega + a_p - \sigma_p^{\mathrm{R}}(0) - \mathrm{i}\omega\frac{\partial\sigma_p^{\mathrm{R}}(0)}{\partial\omega}}\right]}_{F_p(\omega)}$$
$$+ \frac{1}{\mathrm{i}\omega + a_p - \sigma_p^{\mathrm{R}}(0) - \mathrm{i}\omega\frac{\partial\sigma_p^{\mathrm{R}}(0)}{\partial\omega}}.$$

The last term can be integrated analytically. By definition

$$\sigma_p^{\mathrm{R}}(\omega) = \Lambda(\mathbf{p}, \mu + \omega) - \frac{\mathrm{i}}{2}\Gamma(\mathbf{p}, \mu + \omega). \tag{K.3}$$

The imaginary part of $\sigma_p^{\mathrm{R}}(\omega)$ vanishes quadratically as $\omega \to 0$, see (13.24), and therefore $\sigma_p^{\mathrm{R}}(0) = \Lambda(\mathbf{p}, \mu)$ and $\partial\sigma_p^{\mathrm{R}}(0)/\partial\omega = \partial\Lambda(\mathbf{p}, \mu)/\partial\omega$ are real numbers. In particular, from the Hilbert transform relation (13.4) we also deduce that

$$\frac{\partial\sigma_p^{\mathrm{R}}(0)}{\partial\omega} = -\int \frac{d\omega'}{2\pi}\frac{\Gamma(\mathbf{p}, \omega')}{(\omega' - \mu)^2} < 0, \tag{K.4}$$

---

[1]The Hartree self-energy cancels with the potential of the uniform positive background charge.

since the rate function $\Gamma(\mathbf{p}, \omega')$ is non-negative, see Exercise 13.1, and vanishes quadratically as $\omega' \to \mu$ (hence the integral is well defined even without the principal part). In conclusion we have

$$
\mathrm{Re} \int_0^\infty \frac{d\omega}{i\omega + a_p - \sigma_p^{\mathrm{R}}(0) - i\omega \frac{\partial \sigma_p^{\mathrm{R}}(0)}{\partial \omega}} = \int_0^\infty d\omega \, \frac{a_p - \sigma_p^{\mathrm{R}}(0)}{(1 - \frac{\partial \sigma_p^{\mathrm{R}}(0)}{\partial \omega})^2 \omega^2 + (a_p - \sigma_p^{\mathrm{R}}(0))^2}
$$

$$
= \frac{1}{1 - \frac{\partial \sigma_p^{\mathrm{R}}(0)}{\partial \omega}} \arctan \left(\frac{1 - \frac{\partial \sigma_p^{\mathrm{R}}(0)}{\partial \omega}}{a_p - \sigma_p^{\mathrm{R}}(0)} \, \omega \right) \Bigg|_0^\infty
$$

$$
= \frac{1}{1 - \frac{\partial \sigma_p^{\mathrm{R}}(0)}{\partial \omega}} \frac{\pi}{2} \, \mathrm{sgn}(a_p - \sigma_p^{\mathrm{R}}(0)),
$$

where in the last equality we take into account that $1 - \partial \sigma_p^{\mathrm{R}}(0)/\partial \omega$ is positive and larger than 1, see (K.4). The prefactor

$$
\tilde{Z}_p = \frac{1}{1 - \frac{\partial \sigma_p^{\mathrm{R}}(0)}{\partial \omega}} < 1
$$

coincides with the quasi-particle renormalization factor $Z_\mathbf{p}$ of (13.8) for $p = p_\mathrm{F}$ since in this case the quasi-particle energy $E_\mathbf{p} = \mu$. Inserting these results into (K.2) we get

$$
n_\mathbf{p} = \frac{1}{2} \left[1 + \tilde{Z}_p \, \mathrm{sgn}(\mu - \frac{p^2}{2} - \Sigma^{\mathrm{R}}(\mathbf{p}, \mu)) \right] + \frac{1}{\pi} \mathrm{Re} \int_0^\infty d\omega \, F_p(\omega). \tag{K.5}
$$

According to the Luttinger–Ward theorem (11.41) the first term in this expression jumps with a value $\tilde{Z}_{p_\mathrm{F}} = Z$ at the Fermi surface. The function $F_p(\omega)$ is the sum of two terms which are equally singular when $\omega \to 0$ and $p = p_\mathrm{F}$ in such a way that their difference is finite in this limit. Therefore $F_p(\omega)$ is suitable for numerical treatments. Thus the momentum distribution (K.5) is the sum of a discontinuous function and a smooth function. This allows us to define a Fermi surface as the set of $\mathbf{p}$ points where $n_\mathbf{p}$ is discontinuous (a sphere in the electron gas). In the noninteracting case $F_p(\omega) = 0$ and $Z = 1$ and $n_\mathbf{p}$ becomes the standard Heaviside function of a Fermi gas. The effect of the interaction is to reduce the size of the discontinuity and to promote electrons below the Fermi surface to states above the Fermi surface in such a way that $n_\mathbf{p} < 1$ for $p < p_\mathrm{F}$ and $n_\mathbf{p} > 0$ for $p > p_\mathrm{F}$, see Section 15.5.4.

We conclude by observing that in the one-dimensional electron gas $\partial \sigma_p^{\mathrm{R}}(0)/\partial \omega$ is not real since the rate function $\Gamma(\mathbf{p}, \omega)$ vanishes as $|\omega - \mu|$, see discussion in Section 13.3. In this system there is no Fermi surface, i.e., the momentum distribution is not discontinuous as p crosses p_F. The interacting electron gas in one dimension is called a *Luttinger liquid.*

L

Hedin equations from a generating functional

The idea behind the source field method is to construct a generating functional from which to obtain the quantities of interest by functional differentiation with respect to some "source" field S. Classical examples of generating functionals are the exponential e^{2xt-t^2} whose nth derivative with respect to the "source" variable t calculated in $t = 0$ gives the nth Hermite polynomial, or the function $1/\sqrt{1 - 2xt + t^2}$ whose nth derivative with respect to the "source" variable t calculated in $t = 0$ gives the nth Legendre polynomial multiplied by $n!$. Clearly there exist infinitely many formulations more or less complicated depending on which are the quantities of interest. In this appendix we follow the presentation of Strinati in Ref. [124], see also Ref. [177]. We define a generating functional which is an extension of the definition (5.1) of the Green's function[1]

$$G_S(1;2) = \frac{1}{i} \frac{\text{Tr}\left[\mathcal{T}\left\{e^{-i\int_\gamma d\bar{z}\hat{H}_S(\bar{z})}\hat{\psi}(1)\hat{\psi}^\dagger(2)\right\}\right]}{\text{Tr}\left[\mathcal{T}\left\{e^{-i\int_\gamma d\bar{z}\hat{H}_S(\bar{z})}\right\}\right]}, \qquad \text{(L.1)}$$

where $\hat{H}_S$ is given by the sum of the Hamiltonian $\hat{H}$ of the system and the coupling between a source field S and the density $\hat{n}(\mathbf{x}) = \hat{\psi}^\dagger(\mathbf{x})\hat{\psi}(\mathbf{x})$,

$$\hat{H}_S(z) = \hat{H}(z) + \int d\mathbf{x}\, S(\mathbf{x}, z)\hat{n}(\mathbf{x}).$$

The special feature of the source field S is that it can take different values on the two horizontal branches of the contour and, furthermore, it can vary along the imaginary track. This freedom allows for unconstrained variations of S in G_S, a necessary property if we want to take functional derivatives. In the special case $S(\mathbf{x}, t_+) = S(\mathbf{x}, t_-)$ and $S(\mathbf{x}, t_0 - i\tau) = S^M(\mathbf{x})$ the addition of the source field is equivalent to the addition of a scalar potential. In particular the generating functional G_S reduces to the standard Green's function G for $S = 0$. We also observe that G_S satisfies the KMS relations for any S.

---

[1] In (L.1) the field operators are *not* in the Heisenberg picture for otherwise we would have added the subscript "H". Their dependence on the contour time is fictitious and has the purpose only of specifying where the operators lie on the contour, see Section 4.2.

Let us derive the equations of motion for G_S.[2] We write the numerator in (L.1) as

$$\mathcal{T}\left\{e^{-i\int_\gamma d\bar{z}\hat{H}_S(\bar{z})}\hat{\psi}(1)\hat{\psi}^\dagger(2)\right\}$$

$$= \theta(z_1, z_2)\mathcal{T}\left\{e^{-i\int_{z_1}^{t_0-i\beta} d\bar{z}\hat{H}_S(\bar{z})}\right\}\hat{\psi}(1)\mathcal{T}\left\{e^{-i\int_{t_0-}^{z_1} d\bar{z}\hat{H}_S(\bar{z})}\hat{\psi}^\dagger(2)\right\}$$

$$\pm \theta(z_2, z_1)\mathcal{T}\left\{e^{-i\int_{z_1}^{t_0-i\beta} d\bar{z}\hat{H}_S(\bar{z})}\hat{\psi}^\dagger(2)\right\}\hat{\psi}(1)\mathcal{T}\left\{e^{-i\int_{t_0-}^{z_1} d\bar{z}\hat{H}_S(\bar{z})}\right\}.$$

Taking the derivative with respect to z_1 we find

$$i\frac{d}{dz_1}G_S(1;2) = \delta(1;2) + \frac{1}{i}\frac{\text{Tr}\left[\mathcal{T}\left\{e^{-i\int_\gamma d\bar{z}\hat{H}_S(\bar{z})}\left[\hat{\psi}(1), \hat{H}_S(z_1)\right]\hat{\psi}^\dagger(2)\right\}\right]}{\text{Tr}\left[\mathcal{T}\left\{e^{-i\int_\gamma d\bar{z}\hat{H}_S(\bar{z})}\right\}\right]}. \tag{L.2}$$

The commutator between the field operators and the Hamiltonian has been calculated several times [see for instance Section 3.3] and the reader should have no problems in rewriting (L.2) as

$$i\frac{d}{dz_1}G_S(1;2) - \int d3[h(1;3) + \delta(1;3)S(3)]G_S(3;1)$$

$$= \delta(1;2) \pm i\int d3\, v(1;3)G_{2,S}(1,3;2,3^+),$$

where we define the generalized two-particle Green's function

$$G_{2,S}(1,2;3,4) = \frac{1}{i^2}\frac{\text{Tr}\left[\mathcal{T}\left\{e^{-i\int_\gamma d\bar{z}\hat{H}_S(\bar{z})}\hat{\psi}(1)\hat{\psi}(2)\hat{\psi}^\dagger(4)\hat{\psi}^\dagger(3)\right\}\right]}{\text{Tr}\left[\mathcal{T}\left\{e^{-i\int_\gamma d\bar{z}\hat{H}_S(\bar{z})}\right\}\right]},$$

which reduces to G_2 for $S = 0$. We can also introduce a self-energy Σ_S as we did in (9.1),

$$\int d3\, \Sigma_S(1;3)G_S(3;2) = \pm i\int d3\, v(1;3)G_{2,S}(1,3;2,3^+), \tag{L.3}$$

and rewrite the equation of motion as

$$\int d3\, \overrightarrow{G}_S^{-1}(1;3)G_S(3;2) = \delta(1;2), \tag{L.4}$$

where the operator $\overrightarrow{G}_S^{-1}$ (as usual the arrow signifies that it acts on quantities to its right) is defined according to

$$\overrightarrow{G}_S^{-1}(1;3) = \delta(1;3)i\frac{\overrightarrow{d}}{dz_3} - \delta(1;3)S(3) - h(1;3) - \Sigma_S(1;3).$$

---

[2] In the special case $G_S = G$ the derivation below provides an alternative proof of the equations of motion for the Green's function.

Proceeding along the same lines it is possible to derive the adjoint equation of motion

$$\int d3\, G_S(1;3)\overleftarrow{G}_S^{-1}(3;2) = \delta(1;2),\qquad\text{(L.5)}$$

with

$$\overleftarrow{G}_S^{-1}(3;2) = -i\frac{\overleftarrow{d}}{dz_3}\delta(3;2) - S(3)\delta(3;2) - h(3;2) - \Sigma_S(3;2).\qquad\text{(L.6)}$$

Thus, no extra complication arises in deriving the equations of motion of the new Green's function provided that we properly extend the definition of the two-particle Green's function and self-energy.

We make extensive use of the operators $\overrightarrow{G}_S^{-1}$ and $\overleftarrow{G}_S^{-1}$ in what follows. These operators have the property that for any two functions $A(1;2)$ and $B(1;2)$ which fulfill the KMS boundary conditions

$$\int d3d4\, A(1;3)\overrightarrow{G}_S^{-1}(3;4)B(4;2) = \int d3d4\, A(1;3)\overleftarrow{G}_S^{-1}(3;4)B(4;2),$$

since an integration by parts with respect to the contour time does not produce any boundary term as a consequence of the KMS boundary conditions. Using (L.5) we can also solve (L.3) for Σ_S

$$\Sigma_S(1;4) = \pm i\int d2d3\, v(1;3)G_{2,S}(1,3;2,3^+)\overleftarrow{G}_S^{-1}(2;4),\qquad\text{(L.7)}$$

a result which we need later.

The starting point of our new derivation of the Hedin equations is a relation between $G_{2,S}$ and $\delta G_S/\delta S$. The generating functional G_S depends on S through the dependence on $\hat{H}_S$, which appears in the exponent of both the numerator and the denominator. Therefore

$$\frac{\delta G_S(1;2)}{\delta S(3)} = \frac{1}{i}\frac{\mathrm{Tr}\left[\mathcal{T}\left\{e^{-i\int_\gamma dz\hat{H}_S(\bar{z})}(-i)\hat{n}(3)\hat{\psi}(1)\hat{\psi}^\dagger(2)\right\}\right]}{\mathrm{Tr}\left[\mathcal{T}\left\{e^{-i\int_\gamma dz\hat{H}_S(\bar{z})}\right\}\right]}$$

$$-\frac{1}{i}\frac{\mathrm{Tr}\left[\mathcal{T}\left\{e^{-i\int_\gamma dz\hat{H}_S(\bar{z})}\hat{\psi}(1)\hat{\psi}^\dagger(2)\right\}\right]}{\left(\mathrm{Tr}\left[\mathcal{T}\left\{e^{-i\int_\gamma dz\hat{H}_S(\bar{z})}\right\}\right]\right)^2}\times\mathrm{Tr}\left[\mathcal{T}\left\{e^{-i\int_\gamma dz\hat{H}_S(\bar{z})}(-i)\hat{n}(3)\right\}\right]$$

$$= \pm\left[G_{2,S}(1;3;2,3^+) - G_S(1;2)G_S(3;3^+)\right].\qquad\text{(L.8)}$$

Inserting this result into (L.7) we find

$$\Sigma_S(1;4) = \underbrace{\pm i\delta(1;4)\int d3\, v(1;3)G_S(3;3^+)}_{\Sigma_{\mathrm{H},S}(1;4)}+\underbrace{i\int d2d3\, v(1;3)\frac{\delta G_S(1;2)}{\delta S(3)}\overleftarrow{G}_S^{-1}(2;4)}_{\Sigma_{\mathrm{xc},S}(1;4)}.\quad\text{(L.9)}$$

We recognize in the first term on the r.h.s. the Hartree self-energy $\Sigma_{\mathrm{H},S}(1;4) = \delta(1;4)qV_{\mathrm{H},S}(1)$ with Hartree potential

$$V_{\mathrm{H},S}(1) = \pm\frac{i}{q}\int d3\, v(1;3)G_S(3;3^+),$$

in agreement with (7.4). The sum of the source field S and the Hartree energy $qV_{\mathrm{H},S}$ defines the so-called classical (or total) energy

$$C(1) = S(1) + qV_{\mathrm{H},S}(1). \tag{L.10}$$

We recall that in extended systems, like an electron gas, the external potential and the Hartree potential are separately infinite while their sum is finite and meaningful, see Section 7.3.2. It is therefore more natural to study variations with respect to C rather than with respect to S. Regarding G_S as a functional of C instead of S we can apply the chain rule to the XC self-energy of (L.9) and obtain

$$\Sigma_{\mathrm{xc},S}(1;4) = \mathrm{i}\int d2d3d5\, v(1;3)\frac{\delta G_S(1;2)}{\delta C(5)}\frac{\delta C(5)}{\delta S(3)}\overleftarrow{G}_S^{-1}(2;4). \tag{L.11}$$

The derivative of the total field with respect to the source field in $S = 0$ is the *inverse (longitudinal) dielectric function* ε^{-1} defined as

$$\varepsilon^{-1}(1;2) = \left.\frac{\delta C(1)}{\delta S(2)}\right|_{S=0},$$

and it will be calculated shortly. The derivative $\delta G_S/\delta C$ is most easily worked out from the derivative $\delta\overleftarrow{G}_S^{-1}/\delta C$. Using the equations of motions (L.4) and (L.5) we have[3] (omitting the arguments)

$$\int\frac{\delta G_S}{\delta C}\overleftarrow{G}_S^{-1} + \int G_S\frac{\delta\overleftarrow{G}_S^{-1}}{\delta C} = 0 \quad\Rightarrow\quad \frac{\delta G_S}{\delta C} = -\int G_S\frac{\delta\overleftarrow{G}_S^{-1}}{\delta C}G_S, \tag{L.12}$$

and hence (L.11) can also be written as

$$\Sigma_{\mathrm{xc},S}(1;4) = -\mathrm{i}\int d2d3d5\, v(1;3)G_S(1;2)\frac{\delta\overleftarrow{G}_S^{-1}(2;4)}{\delta C(5)}\frac{\delta C(5)}{\delta S(3)}. \tag{L.13}$$

This result leads to the 2nd Hedin equation. In fact, all Hedin equations naturally follow from (L.13) when forcing on this equation the structure $\mathrm{i}GW\Gamma$.

Let us start by proving that the vertex function

$$\Gamma(1,2;3) \equiv -\left.\frac{\delta\overleftarrow{G}_S^{-1}(1;2)}{\delta C(3)}\right|_{S=0}, \tag{L.14}$$

appearing in (L.13) obeys the 5th Hedin equation. From (L.6) and the definition (L.10) of the classical energy we can rewrite $\overleftarrow{G}_S^{-1}$ as

$$\overleftarrow{G}_S^{-1}(1;2) = \left[-\mathrm{i}\frac{\overleftarrow{d}}{dz_1} - C(1)\right]\delta(1;2) - h(1;2) - \Sigma_{\mathrm{xc},S}(1;2).$$

---

[3]In principle we could add to $\delta G_S/\delta C$ any function F such that $\int F\,\overleftarrow{G}_S^{-1} = 0$. However this possibility is excluded in our case due to the satisfaction of the KMS relations, see also discussion in Section 9.1.

We thus see that

$$\Gamma(1,2;3) = \delta(1;2)\delta(1;3) + \left.\frac{\delta\Sigma_{\mathrm{xc},S}(1;2)}{\delta C(3)}\right|_{S=0}.$$

The XC self-energy depends on C only through its dependence on the generating functional G_S; using the chain rule we then find

$$\Gamma(1,2;3) = \delta(1;2)\delta(2;3) + \int d4d5\,\left.\frac{\delta\Sigma_{\mathrm{xc},S}(1;2)}{\delta G_S(4;5)}\frac{\delta G_S(4;5)}{\delta C(3)}\right|_{S=0}$$

$$= \delta(1;2)\delta(2;3) + \int d4d5d6d7\,\frac{\delta\Sigma_{\mathrm{xc}}(1;2)}{\delta G(4;5)}G(4,6)G(7;5)\Gamma(6,7;3),$$

where in the last step we use (L.12). This is exactly the Bethe–Salpeter equation for Γ, see (12.45), provided that we manually shift the contour time $z_2 \to z_2^+$ in $\delta(1;2)$.[4]

Next we observe that (L.8) implies

$$\left.\frac{\delta G_S(1;1^+)}{\delta S(3)}\right|_{S=0} = \pm\left[G_2(1;3;1^+,3^+) - G(1;1^+)G(3;3^+)\right]$$

$$= L(1,3;1,3)$$

$$= \mp i\chi(1;3). \tag{L.15}$$

If we define the polarizability P as

$$P(1;2) = \pm i\left.\frac{\delta G_S(1;1^+)}{\delta C(2)}\right|_{S=0}, \tag{L.16}$$

then (L.15) provides the following relation between χ and P:

$$\chi(1;2) = \int d3\,P(1;3)\left.\frac{\delta C(3)}{\delta S(2)}\right|_{S=0} = \int d3\,P(1;3)\varepsilon^{-1}(3;2). \tag{L.17}$$

Let us calculate the inverse dielectric function. Using the definition of the classical energy (L.10) we have

$$\varepsilon^{-1}(3;2) = \delta(3;2) \pm i\int d4\,v(3;4)\left.\frac{\delta G_S(4;4^+)}{\delta S(2)}\right|_{S=0}$$

$$= \delta(3;2) + \int d4\,v(3;4)\chi(4;2). \tag{L.18}$$

Inserting this result into (L.17) we get the familiar expansion, see (11.44), of the density response function in terms of the polarizability: $\chi = P + Pv\chi$. This suggests that we are on the right track in defining P according to (L.16). To make this argument more stringent

---

[4]This shift is crucial in order to get the correct Fock (or exchange) self-energy.

we evaluate the functional derivative (L.16). Using (L.12) and the definition (L.14) of the vertex function we find

$$P(1;2) = \mp i \int d3 d4 \, G(1;3) \left. \frac{\delta \overleftarrow{G}_s^{-1}(3;4)}{\delta C(2)} \right|_{S=0} G(4;1^+)$$

$$= \pm i \int d3 d4 \, G(1;3) G(4;1) \Gamma(3,4;2),$$

where in the last step we (safely) replaced 1^+ with 1 in the second Green's function. We have just obtained the 3rd Hedin equation.

The only remaining quantity to evaluate is $v \times \delta C/\delta S$, which appears in the XC self-energy (L.13). From (L.18) we see that

$$W(1;2) \equiv \int d3 \, v(1;3) \left. \frac{\delta C(2)}{\delta S(3)} \right|_{S=0} = v(1;2) + \int d3 d4 \, v(1;3) \chi(3;4) v(2;4).$$

Expanding χ in powers of the polarizability we get $W = v + vPv + vPvPv + \dots$ which is solved by the 4th Hedin equation

$$W(1;2) = v(1;2) + \int d3 d4 \, v(1;3) P(3;4) W(4;2).$$

Finally, taking into account the definition of Γ and W we can write the self-energy (L.13) in $S = 0$ as

$$\Sigma_{\mathrm{xc}}(1;2) \equiv \Sigma_{\mathrm{xc},S=0}(1;2) = i \int d3 d4 \, W(1;3) G(1;4) \Gamma(4,2;3),$$

which coincides with the 2nd Hedin equation. This concludes the alternative derivation of the Hedin equations using a generating functional.

M

Lippmann–Schwinger equation and cross-section

Let us consider two nonidentical particles with, for simplicity, the same mass m and described by the Hamiltonian (in first quantization)

$$\hat{\mathcal{H}} = \frac{\hat{\boldsymbol{p}}_1^2}{2m} + \frac{\hat{\boldsymbol{p}}_2^2}{2m} + v(\hat{\boldsymbol{r}}_1 - \hat{\boldsymbol{r}}_2), \tag{M.1}$$

where we use the standard notation $\hat{\boldsymbol{p}}_1 = \hat{\boldsymbol{p}} \otimes \hat{\mathbb{1}}$ for the momentum of the first particle, $\hat{\boldsymbol{p}}_2 = \hat{\mathbb{1}} \otimes \hat{\boldsymbol{p}}$ for the momentum of the second particle, and similarly for the position operators. For the time being we will ignore the spin of the particles. In the absence of the interaction v the eigenkets of $\hat{\mathcal{H}}$ are the momentum eigenkets $|\mathbf{k}_1\rangle|\mathbf{k}_2\rangle$. We are interested in calculating how the unperturbed eigenkets change due to the presence of the interaction. Since the interaction preserves the total momentum it is convenient to work in the reference frame of the center of mass. Let us then introduce a slightly different basis. The wavefunction

$$\Psi_{\mathbf{k}_1,\mathbf{k}_2}(\mathbf{r}_1, \mathbf{r}_2) = \langle \mathbf{r}_2 | \langle \mathbf{r}_1 | |\mathbf{k}_1\rangle|\mathbf{k}_2\rangle = e^{i\mathbf{k}_1 \cdot \mathbf{r}_1 + i\mathbf{k}_2 \cdot \mathbf{r}_2} \tag{M.2}$$

can also be written as

$$\Psi_{\mathbf{k}_1,\mathbf{k}_2}(\mathbf{r}_1, \mathbf{r}_2) = e^{i\mathbf{K} \cdot \mathbf{R} + i\mathbf{k} \cdot \mathbf{r}}, \tag{M.3}$$

with $\mathbf{K} = \mathbf{k}_1 + \mathbf{k}_2$ the total momentum, $\mathbf{k} = \frac{1}{2}(\mathbf{k}_1 - \mathbf{k}_2)$ the relative momentum, $\mathbf{R} = \frac{1}{2}(\mathbf{r}_1 + \mathbf{r}_2)$ the center of mass coordinate and $\mathbf{r} = \mathbf{r}_1 - \mathbf{r}_2$ the relative coordinate. We then see that we can define an equivalent basis $|\mathbf{K}\rangle|\mathbf{k}\rangle$ in which the first ket refers to the center of mass degree of freedom and the second ket to the relative coordinate degree of freedom. The ket $|\mathbf{K}\rangle|\mathbf{k}\rangle$ is an eigenket of the total momentum operator $\hat{\boldsymbol{\mathcal{P}}} = \hat{\boldsymbol{p}}_1 + \hat{\boldsymbol{p}}_2$ with eigenvalue $\mathbf{K}$ and of the relative momentum operator $\hat{\boldsymbol{p}} = \frac{1}{2}(\hat{\boldsymbol{p}}_1 - \hat{\boldsymbol{p}}_2)$ with eigenvalue $\mathbf{k}$. Similarly the ket $|\mathbf{R}\rangle|\mathbf{r}\rangle$ is an eigenket of the center-of-mass position operator $\hat{\boldsymbol{\mathcal{R}}} = \frac{1}{2}(\hat{\boldsymbol{r}}_1 + \hat{\boldsymbol{r}}_2)$ with eigenvalue $\mathbf{R}$ and of the relative coordinate operator $\hat{\boldsymbol{r}} = \hat{\boldsymbol{r}}_1 - \hat{\boldsymbol{r}}_2$ with eigenvalue $\mathbf{r}$. The inner products of these kets are

$$\langle \mathbf{r} | \langle \mathbf{R} | |\mathbf{K}\rangle|\mathbf{k}\rangle = e^{i\mathbf{K} \cdot \mathbf{R} + i\mathbf{k} \cdot \mathbf{r}},$$
$$\langle \mathbf{r} | \langle \mathbf{K}' | |\mathbf{K}\rangle|\mathbf{k}\rangle = (2\pi)^3 \delta(\mathbf{K} - \mathbf{K}') e^{i\mathbf{k} \cdot \mathbf{r}},$$

etc. In terms of the operators $\hat{\mathcal{P}}$, $\hat{p}$, $\hat{\mathcal{R}}$, and $\hat{r}$ the Hamiltonian in (M.1) takes the form

$$\hat{\mathcal{H}} = \frac{\hat{\mathcal{P}}^2}{4m} + \frac{\hat{p}^2}{m} + v(\hat{r}), \tag{M.4}$$

which is independent of the center-of-mass position operator $\hat{\mathcal{R}}$. We use this observation to write a generic eigenket of $\hat{\mathcal{H}}$ as $|\mathbf{K}\rangle|\psi\rangle$ where $|\psi\rangle$ fulfills

$$\hat{h}|\psi\rangle = \left[\frac{\hat{p}^2}{m} + v(\hat{r}) \right] |\psi\rangle = E|\psi\rangle. \tag{M.5}$$

The eigenenergy of $|\mathbf{K}\rangle|\psi\rangle$ is then $E + K^2/(4m)$. Let us take E in the continuum part of the spectrum of the operator $\hat{h}$. Assuming that $v(\mathbf{r})$ vanishes when $r \to \infty$ we have $E \geq 0$. Then the eigenfunction $\langle \mathbf{r}|\psi\rangle$ is a plane wave in this region of space, i.e., $\lim_{r \to \infty} \langle \mathbf{r}|\psi\rangle = \langle \mathbf{r}|\mathbf{k}\rangle = e^{i\mathbf{k}\cdot\mathbf{r}}$, with $E = k^2/m$. We calculate $|\psi\rangle$ using the *Lippmann–Schwinger equation*

$$|\psi\rangle = |\mathbf{k}\rangle + \frac{1}{E - \hat{p}^2/m \pm i\eta} v(\hat{r})|\psi\rangle. \tag{M.6}$$

The equivalence between (M.5) and (M.6) can easily be verified by multiplying both sides with $(E - \hat{p}^2/m \pm i\eta)$ and by letting $\eta \to 0$. The ket $|\mathbf{k}\rangle$ on the r.h.s. of (M.6) guarantees that for $v = 0$ we recover the noninteracting solution. In order to choose the sign in front of $i\eta$ we must understand what physical state $|\psi\rangle$ is. For this purpose we suppose that we prepare the system at time $t = 0$ in the state $|\mathbf{k}\rangle$ and then evolve the system in time according to the Hamiltonian $\hat{h}$. The state of the system at time t is therefore

$$|\psi(t)\rangle = e^{-i(\hat{h}-E)t}|\mathbf{k}\rangle, \tag{M.7}$$

where, for convenience, we have subtracted from $\hat{h}$ the eigenvalue E so that the noninteracting ket $|\psi(t)\rangle = |\mathbf{k}\rangle$ is independent of t (in the noninteracting case $v = 0$ and hence $|\mathbf{k}\rangle$ is an eigenket of $\hat{h}$ with eigenvalue $E = k^2/m$). Now for any $t > 0$ we can write

$$e^{-i(\hat{h}-E)t} = -i \int \frac{d\omega}{2\pi} e^{-i(\omega-E)t} \frac{1}{\omega - \hat{h} + i\eta}$$

$$= -i \int \frac{d\omega}{2\pi} e^{-i(\omega-E)t} \left[\frac{1}{\omega - \hat{p}^2/m + i\eta} + \frac{1}{\omega - \hat{p}^2/m + i\eta} v(\hat{r}) \frac{1}{\omega - \hat{h} + i\eta} \right],$$

where in the last equality we have expanded $1/(\omega - \hat{h} + i\eta)$ in a Dyson-like equation. Inserting this result into (M.7) we find

$$|\psi(t)\rangle = |\mathbf{k}\rangle - i \int \frac{d\omega}{2\pi} e^{-i(\omega-E)t} \frac{1}{\omega - \hat{p}^2/m + i\eta} v(\hat{r}) \frac{1}{\omega - \hat{h} + i\eta} |\mathbf{k}\rangle. \tag{M.8}$$

In the limit $t \to \infty$ the dominant contribution to the ω-integral comes from the region $\omega \sim E$ (Riemann–Lebesgue theorem). If we replace $\omega \to E$ in the first denominator (M.8) becomes

$$\lim_{t \to \infty} |\psi(t)\rangle = |\mathbf{k}\rangle + \lim_{t \to \infty} \frac{1}{E - \hat{p}^2/m + i\eta} v(\hat{r})|\psi(t)\rangle.$$

Comparing this with the Lippmann–Schwinger equation we see that $|\psi\rangle = \lim_{t\to\infty}|\psi(t)\rangle$ when the sign of $i\eta$ in (M.6) is plus. Similarly it is easy to show that $|\psi\rangle = \lim_{t\to-\infty}|\psi(t)\rangle$ when the sign of $i\eta$ in (M.6) is minus. As in an experiment we can only study the forward evolution we take the sign of $i\eta$ in (M.6) to be plus in the following discussion.

We multiply (M.6) from the left with $\langle\mathbf{r}|$ and insert the completeness relation $\hat{\mathbb{1}} = \int d\mathbf{r}'|\mathbf{r}'\rangle\langle\mathbf{r}'|$ to the left of v:

$$\psi(\mathbf{r}) = \langle\mathbf{r}|\psi\rangle = e^{i\mathbf{k}\cdot\mathbf{r}} + \int d\mathbf{r}'\langle\mathbf{r}|\frac{1}{E - \hat{p}^2/m + i\eta}|\mathbf{r}'\rangle v(\mathbf{r}')\psi(\mathbf{r}'). \qquad \text{(M.9)}$$

The kernel of this integral equation can be evaluated as follows:

$$\begin{aligned}
\langle\mathbf{r}|\frac{1}{E - \hat{p}^2/m + i\eta}|\mathbf{r}'\rangle &= \int \frac{d\mathbf{q}}{(2\pi)^3} \frac{e^{i\mathbf{q}\cdot(\mathbf{r}-\mathbf{r}')}}{E - q^2/m + i\eta} \\
&= \int_0^\infty \frac{dq}{(2\pi)^2} q^2 \int_{-1}^1 dc \frac{e^{iq|\mathbf{r}-\mathbf{r}'|c}}{E - q^2/m + i\eta} \\
&= -\frac{m}{8\pi^2}\frac{1}{i|\mathbf{r}-\mathbf{r}'|}\int_{-\infty}^\infty dq\, q \frac{e^{iq|\mathbf{r}-\mathbf{r}'|} - e^{-iq|\mathbf{r}-\mathbf{r}'|}}{q^2 - k^2 - i\eta} \\
&= -\frac{m}{4\pi}\frac{e^{ik|\mathbf{r}-\mathbf{r}'|}}{|\mathbf{r}-\mathbf{r}'|},
\end{aligned}$$

where in the last equality we take into account that the denominator has simple poles in $q = \pm(k+i\eta')$. If the interaction v is short-ranged and if we are only interested in calculating the wavefunction far away from the interacting region, we can approximate the kernel in (M.9) with its expansion for $r \gg r'$. We have $|\mathbf{r}-\mathbf{r}'| = (r^2 + r'^2 - 2\mathbf{r}\cdot\mathbf{r}')^{1/2} \sim r - \mathbf{r}\cdot\mathbf{r}'/r$. Let $\mathbf{k}' = k\mathbf{r}/r$ be the propagation vector pointing in the same direction as the vector $\mathbf{r}$ (where we want to compute ψ) and having the same modulus of the relative momentum. Then, we can approximate (M.9) as

$$\psi(\mathbf{r}) = e^{i\mathbf{k}\cdot\mathbf{r}} - \frac{m}{4\pi}\frac{e^{ikr}}{r}\int d\mathbf{r}' e^{-i\mathbf{k}'\cdot\mathbf{r}'} v(\mathbf{r}')\psi(\mathbf{r}') \equiv e^{i\mathbf{k}\cdot\mathbf{r}} + \frac{e^{ikr}}{r}f(\mathbf{k}',\mathbf{k}). \qquad \text{(M.10)}$$

We thus see that $\psi(\mathbf{r})$ is written as the sum of an incident wave and an outgoing spherical wave. This wavefunction has a similar structure to the continuum eigenfunction of a one-dimensional system with a potential barrier: an incident wave e^{ikx} on which we have superimposed a reflected wave Re^{-ikx} for $x \to -\infty$, and a transmitted wave Te^{ikx} for $x \to \infty$. In three dimensions the reflected and transmitted wave are both incorporated into the outgoing spherical wave. The probability of being reflected (transmitted) is obtained by choosing $\mathbf{r}$, or equivalently $\mathbf{k}'$, in the opposite (same) direction to the incident wavevector $\mathbf{k}$. In three dimensions, however, there are infinitely many more possibilities and the outgoing wave yields the probability of measuring a relative momentum $\mathbf{k}'$ for two particles scattering with relative momentum $\mathbf{k}$.

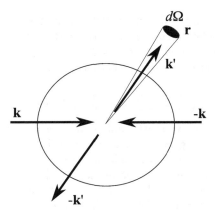

In an experiment we can measure the flux of incident particles as well as how many particles per unit time hit a detector in $\mathbf{r}$ with surface area $r^2 d\Omega$. The flux of incident particles is just the modulus of the current density of the initial plane-wave, which in our case is k. The current density of the scattered particles in the direction $\mathbf{k}'$ is, to lowest order in $1/r$, given by $k|f(\mathbf{k}',\mathbf{k})/r|^2$. Therefore, the number of particles which hit the detector per unit time is $k|f(\mathbf{k}',\mathbf{k})|^2 d\Omega$. The ratio between these two quantities has the physical dimension of an area and is called the *differential cross-section*:

$$d\sigma \equiv \frac{k|f(\mathbf{k}',\mathbf{k})|^2 d\Omega}{k} \qquad \Rightarrow \qquad \frac{d\sigma}{d\Omega} = |f(\mathbf{k}',\mathbf{k})|^2. \tag{M.11}$$

We can calculate f by iterating (M.10), i.e., by replacing ψ in the r.h.s. with the whole r.h.s. *ad infinitum*. The first iteration corresponds to the second Born approximation and gives

$$f(\mathbf{k}',\mathbf{k}) = -\frac{m}{4\pi}\tilde{v}_{\mathbf{k}'-\mathbf{k}} = -\frac{m}{4\pi}\tilde{v}_{\mathbf{k}-\mathbf{k}'}, \tag{M.12}$$

where in the last equality we use $v(\mathbf{r}) = v(-\mathbf{r})$ and hence its Fourier transform is real and symmetric.

So far we have considered two distinguishable particles. If the particles are identical their orbital wavefunction is either symmetric or antisymmetric. This means that the initial ket is $|\mathbf{k}\rangle \pm |-\mathbf{k}\rangle$. Consequently we must replace (M.10) with

$$\psi(\mathbf{r}) = \left(e^{i\mathbf{k}\cdot\mathbf{r}} \pm e^{-i\mathbf{k}\cdot\mathbf{r}}\right) - \frac{m}{4\pi}\frac{e^{ikr}}{r}\int d\mathbf{r}' e^{-i\mathbf{k}'\cdot\mathbf{r}'} v(\mathbf{r}')\psi(\mathbf{r}')$$

$$= \left(e^{i\mathbf{k}\cdot\mathbf{r}} \pm e^{-i\mathbf{k}\cdot\mathbf{r}}\right) + \frac{e^{ikr}}{r}f(\mathbf{k}',\mathbf{k}). \tag{M.13}$$

In the second Born approximation we then have $f(\mathbf{k}',\mathbf{k}) = -\frac{m}{4\pi}(\tilde{v}_{\mathbf{k}-\mathbf{k}'} \pm \tilde{v}_{\mathbf{k}+\mathbf{k}'})$ and the differential cross-section becomes

$$\frac{d\sigma_\pm}{d\Omega} = \left(\frac{m}{4\pi}\right)^2 \left(\tilde{v}_{\mathbf{k}-\mathbf{k}'}^2 + \tilde{v}_{\mathbf{k}+\mathbf{k}'}^2 \pm 2\tilde{v}_{\mathbf{k}-\mathbf{k}'}\tilde{v}_{\mathbf{k}+\mathbf{k}'}\right).$$

We can rewrite this formula using the momentum of the particles in the original reference frame. We use the same notation as in Section 13.3 and call $\mathbf{p} = \mathbf{K}/2 + \mathbf{k}$, $\mathbf{p}' = \mathbf{K}/2 - \mathbf{k}$ the momenta of the incident particles, and $\bar{\mathbf{p}} = \mathbf{K}/2 + \mathbf{k}'$, $\bar{\mathbf{p}}' = \mathbf{K}/2 - \mathbf{k}'$ the momenta of the particles after the scattering. Then we see that $\mathbf{k} - \mathbf{k}' = \mathbf{p} - \bar{\mathbf{p}}$ and $\mathbf{k} + \mathbf{k}' = \mathbf{p} - \bar{\mathbf{p}}'$ and hence

$$\frac{d\sigma_\pm}{d\Omega} = \left(\frac{m}{4\pi}\right)^2 \left(\tilde{v}^2_{\mathbf{p}-\bar{\mathbf{p}}} + \tilde{v}^2_{\mathbf{p}-\bar{\mathbf{p}}'} \pm 2\tilde{v}_{\mathbf{p}-\bar{\mathbf{p}}}\tilde{v}_{\mathbf{p}-\bar{\mathbf{p}}'}\right). \tag{M.14}$$

To complete the discussion we must combine this result with the spin degrees of freedom. Let us start by considering two bosons of spin S (hence S is integer). In the basis of the total spin they can form multiplets of spin $2S$, $2S - 1, \ldots, 0$. The multiplets with even total spin are symmetric under an exchange of the particles while those with odd total spin are antisymmetric. We define N_+ and N_- as the total number of states in the symmetric and antisymmetric multiplets respectively. We then have $N_+ + N_- = (2S + 1)^2$. Furthermore, since the multiplet with spin $S' \leq 2S$ always has two states more than the multiplet with spin $S' - 1$ (the eigenvalues of the z component of the total spin go from $-S'$ to S') we also have $N_+ = N_- + 2S + 1$. These two relations allow us to express N_+ and N_- in terms of S, and what we find is

$$N_+ = \frac{(2S + 1)(2S + 2)}{2}, \qquad N_- = \frac{(2S + 1)2S}{2}. \tag{M.15}$$

In the absence of any information about the spin of the particles we must calculate the differential cross-section using a statistical weight $N_+/(2S + 1)^2$ for the symmetric orbital part and $N_-/(2S + 1)^2$ for the antisymmetric orbital part, i.e.,

$$\begin{aligned}
\frac{d\sigma}{d\Omega} &= \frac{N_+}{(2S + 1)^2}\frac{d\sigma_+}{d\Omega} + \frac{N_-}{(2S + 1)^2}\frac{d\sigma_-}{d\Omega} \\
&= \left(\frac{m}{4\pi}\right)^2 \left(\tilde{v}^2_{\mathbf{p}-\bar{\mathbf{p}}} + \tilde{v}^2_{\mathbf{p}-\bar{\mathbf{p}}'} + \frac{2}{2S + 1}\tilde{v}_{\mathbf{p}-\bar{\mathbf{p}}}\tilde{v}_{\mathbf{p}-\bar{\mathbf{p}}'}\right),
\end{aligned} \tag{M.16}$$

which is proportional to the bosonic $B(\mathbf{p}, \bar{\mathbf{p}}, \bar{\mathbf{p}}')$ in (13.20). In a similar way we can proceed for two fermions. It is easy to show that N_+ and N_- are still given by (M.15). This time, however, we must combine the symmetric multiplets with the antisymmetric wave function and the antisymmetric multiplets with the symmetric wave function. We find

$$\begin{aligned}
\frac{d\sigma}{d\Omega} &= \frac{N_-}{(2S + 1)^2}\frac{d\sigma_+}{d\Omega} + \frac{N_+}{(2S + 1)^2}\frac{d\sigma_-}{d\Omega} \\
&= \left(\frac{m}{4\pi}\right)^2 \left(\tilde{v}^2_{\mathbf{p}-\bar{\mathbf{p}}} + \tilde{v}^2_{\mathbf{p}-\bar{\mathbf{p}}'} - \frac{2}{2S + 1}\tilde{v}_{\mathbf{p}-\bar{\mathbf{p}}}\tilde{v}_{\mathbf{p}-\bar{\mathbf{p}}'}\right),
\end{aligned} \tag{M.17}$$

which is proportional to the fermionic $B(\mathbf{p}, \bar{\mathbf{p}}, \bar{\mathbf{p}}')$ in (13.20).

N

Why the name Random Phase Approximation?

In Chapter 15 we discussed the density response function χ in the time-dependent Hartree approximation and we called this approximation the Random Phase Approximation (RPA). The origin of this name is based on an idea of Bohm and Pines who derived the RPA response function without using any diagrammatic technique [136, 178]. In this appendix we do not present the original derivation of Bohm and Pines since it is rather lengthy (a pedagogical discussion of the original derivation can be found in Ref. [179]). We instead use the idea of Bohm and Pines in a derivation based on the equation of motion for χ, thus still providing a justification of the name RPA. A similar derivation can be found in Refs. [46, 75].

Let us consider the Hamiltonian of an electron gas in a large three-dimensional box of volume $V = L^3$:

$$\hat{H} = -\frac{1}{2} \int d\mathbf{x}\, \hat{\psi}^\dagger(\mathbf{x}) \nabla^2 \hat{\psi}(\mathbf{x}) + \frac{1}{2} \int d\mathbf{x} d\mathbf{x}'\, v(\mathbf{r} - \mathbf{r}') \hat{\psi}^\dagger(\mathbf{x}) \hat{\psi}^\dagger(\mathbf{x}') \hat{\psi}(\mathbf{x}') \hat{\psi}(\mathbf{x})$$

$$= -\frac{1}{2} \int d\mathbf{x}\, \hat{\psi}^\dagger(\mathbf{x}) \left(\nabla^2 + v(0) \right) \hat{\psi}(\mathbf{x}) + \frac{1}{2} \int d\mathbf{x} d\mathbf{x}'\, v(\mathbf{r} - \mathbf{r}') \hat{n}(\mathbf{x}) \hat{n}(\mathbf{x}'),$$

where the space integral is restricted to the box. Imposing the BvK periodic boundary conditions along x, y, and z we can expand the field operators according to

$$\hat{\psi}(\mathbf{x}) = \frac{1}{V} \sum_{\mathbf{p}} e^{i\mathbf{p}\cdot\mathbf{r}}\, \hat{d}_{\mathbf{p}\sigma}, \qquad \Rightarrow \qquad \hat{\psi}^\dagger(\mathbf{x}) = \frac{1}{V} \sum_{\mathbf{p}} e^{-i\mathbf{p}\cdot\mathbf{r}}\, \hat{d}_{\mathbf{p}\sigma}^\dagger,$$

where $\mathbf{p} = \frac{2\pi}{L}(n_x, n_y, n_z)$. The inverse relations read

$$\hat{d}_{\mathbf{p}\sigma} = \int d\mathbf{r}\, e^{-i\mathbf{p}\cdot\mathbf{r}}\, \hat{\psi}(\mathbf{x}) \qquad \Rightarrow \qquad \hat{d}_{\mathbf{p}\sigma}^\dagger = \int d\mathbf{r}\, e^{i\mathbf{p}\cdot\mathbf{r}}\, \hat{\psi}^\dagger(\mathbf{x}).$$

It is easy to check that the $\hat{d}$-operators satisfy the anticommutation relations

$$\left[\hat{d}_{\mathbf{p}\sigma}, \hat{d}_{\mathbf{p}'\sigma'}^\dagger \right]_- = V \delta_{\sigma\sigma'} \delta_{\mathbf{p}\mathbf{p}'} \tag{N.1}$$

and that in the limit $V \to \infty$ they become the creation/annihilation operators of the momentum–spin kets, see (1.65). In terms of the $\hat{d}$-operators the density operator reads

$$\hat{n}(\mathbf{x}) = \frac{1}{V^2} \sum_{\mathbf{pp}'} e^{-i(\mathbf{p}-\mathbf{p}')\cdot\mathbf{r}} \, \hat{d}_{\mathbf{p}\sigma}^{\dagger} \hat{d}_{\mathbf{p}'\sigma} = \frac{1}{V^2} \sum_{\mathbf{pq}} e^{i\mathbf{q}\cdot\mathbf{r}} \, \hat{d}_{\mathbf{p}\sigma}^{\dagger} \hat{d}_{\mathbf{p}+\mathbf{q}\sigma}. \qquad (N.2)$$

We define the density fluctuation operator as

$$\hat{\rho}_{\mathbf{q}} \equiv \frac{1}{V} \sum_{\mathbf{p}\sigma} \hat{d}_{\mathbf{p}\sigma}^{\dagger} \hat{d}_{\mathbf{p}+\mathbf{q}\sigma}. \qquad (N.3)$$

This operator can be seen as the Fourier transform of $\hat{n}(\mathbf{r})$, since from (N.2)

$$\hat{n}(\mathbf{r}) = \sum_{\sigma} \hat{n}(\mathbf{x}) = \frac{1}{V} \sum_{\mathbf{q}} e^{i\mathbf{q}\cdot\mathbf{r}} \hat{\rho}_{\mathbf{q}} \qquad \Rightarrow \qquad \hat{\rho}_{\mathbf{q}} = \int d\mathbf{r} \, e^{-i\mathbf{q}\cdot\mathbf{r}} \, \hat{n}(\mathbf{r}).$$

The $\hat{d}$-operators and the density fluctuation operator can be used to rewrite the Hamiltonian of the electron gas in the following form:

$$\hat{H} = \frac{1}{V} \sum_{\mathbf{p}\sigma} \left(\epsilon_{\mathbf{p}} - \frac{v(0)}{2} \right) \hat{d}_{\mathbf{p}\sigma}^{\dagger} \hat{d}_{\mathbf{p}\sigma} + \frac{1}{2V} \sum_{\mathbf{q}} \tilde{v}_{\mathbf{q}} \, \hat{\rho}_{\mathbf{q}} \hat{\rho}_{-\mathbf{q}},$$

where $\epsilon_{\mathbf{p}} = p^2/2$.

Let us now consider the density response function

$$\chi^{\mathrm{R}}(\mathbf{r}, t; \mathbf{r}', t') = \sum_{\sigma\sigma'} -i\theta(t - t') \langle \, [\hat{n}_H(\mathbf{x}, t), \hat{n}_H(\mathbf{x}', t')]_- \rangle$$

$$= \frac{1}{V^2} \sum_{\mathbf{qq}'} -i\theta(t - t') e^{i\mathbf{q}\cdot\mathbf{r}+i\mathbf{q}'\cdot\mathbf{r}'} \langle \, [\hat{\rho}_{\mathbf{q},H}(t), \hat{\rho}_{\mathbf{q}',H}(t')]_- \rangle, \qquad (N.4)$$

where $\langle \ldots \rangle$ denotes the equilibrium ensemble average. Due to translational invariance the response function depends only on the difference $\mathbf{r} - \mathbf{r}'$ and hence the only nonvanishing averages in (N.4) are those for which $\mathbf{q}' = -\mathbf{q}$. Thus if we define

$$\chi_{\mathbf{p}}^{\mathrm{R}}(\mathbf{q}, t - t') \equiv \frac{1}{V} \sum_{\sigma} -i\theta(t - t') \langle \, [\hat{d}_{\mathbf{p}\sigma,H}^{\dagger}(t)\hat{d}_{\mathbf{p}+\mathbf{q}\sigma,H}(t), \hat{\rho}_{-\mathbf{q},H}(t')]_- \rangle$$

we can rewrite the density response function as

$$\chi^{\mathrm{R}}(\mathbf{r}, t; \mathbf{r}', t') = \frac{1}{V^2} \sum_{\mathbf{q}} \sum_{\mathbf{p}} e^{i\mathbf{q}\cdot(\mathbf{r}-\mathbf{r}')} \chi_{\mathbf{p}}^{\mathrm{R}}(\mathbf{q}, t - t'). \qquad (N.5)$$

We now derive the equation of motion for $\chi_{\mathbf{p}}^{\mathrm{R}}(\mathbf{q}, t)$. The derivative of this quantity with respect to t contains a term in which we differentiate the Heaviside function and a term in which we differentiate the operators in the Heisenberg picture. We have

$$i\frac{d}{dt} \chi_{\mathbf{p}}^{\mathrm{R}}(\mathbf{q}, t) = \frac{1}{V} \sum_{\sigma} \delta(t) \langle \, [\hat{d}_{\mathbf{p}\sigma,H}^{\dagger}(t)\hat{d}_{\mathbf{p}+\mathbf{q}\,\sigma,H}(t), \hat{\rho}_{-\mathbf{q},H}(t)]_- \rangle$$

$$+ \frac{1}{V} \sum_{\sigma} -i\theta(t) \langle \, [[\hat{d}_{\mathbf{p}\sigma,H}^{\dagger}(t)\hat{d}_{\mathbf{p}+\mathbf{q}\,\sigma,H}(t), \hat{H}_H(t)]_-, \hat{\rho}_{-\mathbf{q},H}(0)]_- \rangle. \qquad (N.6)$$

To evaluate the r.h.s. of this equation we need the basic commutator

$$\sum_{\sigma\sigma'} [\hat{d}^\dagger_{\mathbf{p}\sigma}\hat{d}_{\mathbf{p}+\mathbf{q}\,\sigma}, \hat{d}^\dagger_{\mathbf{p}'\sigma'}\hat{d}_{\mathbf{p}'+\mathbf{q}'\,\sigma'}]_- = V \sum_\sigma \left(\delta_{\mathbf{p}+\mathbf{q}\,\mathbf{p}'}\, \hat{d}^\dagger_{\mathbf{p}\sigma}\hat{d}_{\mathbf{p}+\mathbf{q}+\mathbf{q}'\,\sigma} - \delta_{\mathbf{p}\,\mathbf{p}'+\mathbf{q}'}\, \hat{d}^\dagger_{\mathbf{p}-\mathbf{q}'\sigma}\hat{d}_{\mathbf{p}+\mathbf{q}\,\sigma}\right),$$

where we have explicitly used the anticommutation rules (N.1). The first commutator in the r.h.s. of (N.6) is then

$$\sum_\sigma [\hat{d}^\dagger_{\mathbf{p}\sigma}\hat{d}_{\mathbf{p}+\mathbf{q}\,\sigma}, \hat{\rho}_{-\mathbf{q}}]_- = \sum_\sigma \left(\hat{d}^\dagger_{\mathbf{p}\sigma}\hat{d}_{\mathbf{p}\sigma} - \hat{d}^\dagger_{\mathbf{p}+\mathbf{q}\,\sigma}\hat{d}_{\mathbf{p}+\mathbf{q}\,\sigma}\right). \tag{N.7}$$

The commutator with the Hamiltonian is the sum of two terms. The first term involves the one-body part of $\hat{H}$ and yields

$$\sum_\sigma [\hat{d}^\dagger_{\mathbf{p}\sigma}\hat{d}_{\mathbf{p}+\mathbf{q}\,\sigma}, \frac{1}{V}\sum_{\mathbf{p}'\sigma'}(\epsilon_{\mathbf{p}'} - \frac{1}{2}v(0))\hat{d}^\dagger_{\mathbf{p}'\sigma'}\hat{d}_{\mathbf{p}'\sigma'}]_- = (\epsilon_{\mathbf{p}+\mathbf{q}} - \epsilon_\mathbf{p}) \sum_\sigma \hat{d}^\dagger_{\mathbf{p}\sigma}\hat{d}_{\mathbf{p}+\mathbf{q}\,\sigma}.$$

We thus see that the dependence on $v(0)$ disappears. The second term involves the commutator with $\hat{\rho}_\mathbf{q}\hat{\rho}_{-\mathbf{q}}$ and must be approximated. Here is where the idea of Bohm and Pines comes into play. It is easy to find

$$\sum_\sigma [\hat{d}^\dagger_{\mathbf{p}\sigma}\hat{d}_{\mathbf{p}+\mathbf{q}\,\sigma}, \frac{1}{2V}\sum_{\mathbf{q}'}\tilde{v}_{\mathbf{q}'}\,\hat{\rho}_{\mathbf{q}'}\hat{\rho}_{-\mathbf{q}'}]_-$$

$$= \frac{1}{2V}\sum_{\mathbf{q}'\sigma}\tilde{v}_{\mathbf{q}'}\left[\left(\hat{d}^\dagger_{\mathbf{p}\sigma}\hat{d}_{\mathbf{p}+\mathbf{q}+\mathbf{q}'\,\sigma} - \hat{d}^\dagger_{\mathbf{p}-\mathbf{q}'\,\sigma}\hat{d}_{\mathbf{p}+\mathbf{q}\sigma}\right), \hat{\rho}_{-\mathbf{q}'}\right]_+. \tag{N.8}$$

To cast the r.h.s. of this equation in terms of an anticommutator we use $\tilde{v}_\mathbf{q} = \tilde{v}_{-\mathbf{q}}$. Now we observe that the low energy states of an electron gas are states with very delocalized electrons. This means that the expansion

$$|\Psi\rangle = \frac{1}{N!}\int d\mathbf{x}_1 \ldots d\mathbf{x}_N \Psi(\mathbf{x}_1, \ldots, \mathbf{x}_N)|\mathbf{x}_1 \ldots \mathbf{x}_N\rangle \tag{N.9}$$

of a low-energy state has a wavefunction which is a smooth function of the coordinates. Let us consider the action of $\sum_\sigma \hat{d}^\dagger_{\mathbf{p}\sigma}\hat{d}_{\mathbf{p}+\mathbf{q}\sigma}$ on a position–spin ket

$$\sum_\sigma \hat{d}^\dagger_{\mathbf{p}\sigma}\hat{d}_{\mathbf{p}+\mathbf{q}\sigma}|\mathbf{x}_1 \ldots \mathbf{x}_N\rangle = \int d\mathbf{r}d\mathbf{r}' e^{i\mathbf{p}\cdot(\mathbf{r}-\mathbf{r}')}e^{-i\mathbf{q}\cdot\mathbf{r}'}\,\hat{\psi}^\dagger(\mathbf{r}\sigma)\hat{\psi}(\mathbf{r}'\sigma)|\mathbf{x}_1 \ldots \mathbf{x}_N\rangle$$

$$= \int d\mathbf{r} \sum_{i=1}^N e^{i\mathbf{p}\cdot(\mathbf{r}-\mathbf{r}_i)}e^{-i\mathbf{q}\cdot\mathbf{r}_i}|\mathbf{x}_1 \ldots \mathbf{x}_{i-1}\,\mathbf{r}\sigma_i\,\mathbf{x}_{i+1}\mathbf{x}_N\rangle,$$

where we use (1.45). Multiplying this equation by the wavefunction of a low-energy state and integrating over all coordinates we see that the dominant contribution of the integral over $\mathbf{r}$ comes from the region $\mathbf{r} \sim \mathbf{r}_i$, since for large $|\mathbf{r} - \mathbf{r}_i|$ the exponential $e^{i\mathbf{p}\cdot(\mathbf{r}-\mathbf{r}_i)}$

becomes a highly oscillating function. Under the integral over $\mathbf{x}_1 \ldots \mathbf{x}_N$ we then make the approximation

$$\sum_\sigma \hat{d}^\dagger_{\mathbf{p}\sigma} \hat{d}_{\mathbf{p}+\mathbf{q}\,\sigma} |\mathbf{x}_1 \ldots \mathbf{x}_N\rangle \sim \Delta V \sum_{i=1}^N e^{-i\mathbf{q}\cdot\mathbf{r}_i} |\mathbf{x}_1 \ldots \mathbf{x}_N\rangle, \qquad (N.10)$$

where ΔV is some volume element proportional to $1/p$. The r.h.s. in (N.10) is the position-spin ket multiplied by the sum of exponentials with "randomly" varying phases. In the large N limit, or equivalently in the limit $r_s \to 0$, this sum approaches zero for all $\mathbf{q} \neq 0$. The idea of Bohm and Pines was then to say that every operator $\hat{d}^\dagger_{\mathbf{p}\sigma} \hat{d}_{\mathbf{p}+\mathbf{q}\,\sigma}$ appearing in (N.8) generates small contributions unless $\mathbf{q} = 0$. This means that the dominant contributions in (N.8) are those with either $\mathbf{q}' = -\mathbf{q}$ (in this case the operator in parenthesis becomes $\hat{d}^\dagger_{\mathbf{p}\sigma} \hat{d}_{\mathbf{p}\sigma} - \hat{d}^\dagger_{\mathbf{p}+\mathbf{q}\,\sigma} \hat{d}_{\mathbf{p}+\mathbf{q}\sigma}$) or those with $\mathbf{q}' = 0$, since according to (N.3) the main contribution to the density fluctuation operator comes from $\hat{\rho}_{\mathbf{q}'}$ with $\mathbf{q}' = 0$ (in this case $\hat{\rho}_0 = \frac{1}{V} \sum_{\mathbf{p}'\sigma} \hat{d}^\dagger_{\mathbf{p}'\sigma} \hat{d}_{\mathbf{p}'\sigma}$). For $\mathbf{q}' = 0$, however, the r.h.s. of (N.8) vanishes and therefore the Bohm–Pines approximation reduces to

$$\sum_\sigma \left[\hat{d}^\dagger_{\mathbf{p}\sigma} \hat{d}_{\mathbf{p}+\mathbf{q}\,\sigma}, \frac{1}{2V} \sum_{\mathbf{q}'} \tilde{v}_{\mathbf{q}'}\, \hat{\rho}_{\mathbf{q}'} \hat{\rho}_{-\mathbf{q}'} \right]_- \sim \frac{1}{2V} \sum_\sigma \tilde{v}_{\mathbf{q}} \left[\left(\hat{d}^\dagger_{\mathbf{p}\sigma} \hat{d}_{\mathbf{p}\sigma} - \hat{d}^\dagger_{\mathbf{p}+\mathbf{q}\,\sigma} \hat{d}_{\mathbf{p}+\mathbf{q}\sigma} \right), \hat{\rho}_{\mathbf{q}} \right]_+ .$$

Next we observe that $\hat{d}^\dagger_{\mathbf{p}\sigma} \hat{d}_{\mathbf{p}\sigma}$ is the occupation operator for electrons of momentum $\mathbf{p}$ and spin σ. The noninteracting ground state of the electron gas has N-electrons in the lowest energy $\mathbf{p}$-levels and therefore is an eigenstate of $\hat{d}^\dagger_{\mathbf{p}\sigma} \hat{d}_{\mathbf{p}\sigma}$ with eigenvalue $V f_{\mathbf{p}}$ where $f_{\mathbf{p}} = 1$ if $\mathbf{p}$ is occupied and zero otherwise. The second approximation of Bohm and Pines was to assume that things are not so different in the interacting case and hence that we can replace $\hat{d}^\dagger_{\mathbf{p}\sigma} \hat{d}_{\mathbf{p}\sigma}$ with $V f_{\mathbf{p}}$. In conclusion

$$\sum_\sigma \left[\hat{d}^\dagger_{\mathbf{p}\sigma} \hat{d}_{\mathbf{p}+\mathbf{q}\,\sigma}, \frac{1}{2V} \sum_{\mathbf{q}'} \tilde{v}_{\mathbf{q}'}\, \hat{\rho}_{\mathbf{q}'} \hat{\rho}_{-\mathbf{q}'} \right]_- \sim 2(f_{\mathbf{p}} - f_{\mathbf{p}+\mathbf{q}})\, \tilde{v}_{\mathbf{q}}\, \hat{\rho}_{\mathbf{q}},$$

where the factor of 2 comes from spin. Putting together all these results the equation of motion (N.6) becomes

$$i\frac{d}{dt} \chi^R_{\mathbf{p}}(\mathbf{q}, t) = 2\delta(t)(f_{\mathbf{p}} - f_{\mathbf{p}+\mathbf{q}}) + (\epsilon_{\mathbf{p}+\mathbf{q}} - \epsilon_{\mathbf{p}}) \chi^R_{\mathbf{p}}(\mathbf{q}, t) + 2\tilde{v}_{\mathbf{q}}(f_{\mathbf{p}} - f_{\mathbf{p}+\mathbf{q}})\frac{1}{V}\sum_{\mathbf{p}'} \chi^R_{\mathbf{p}'}(\mathbf{q}, t).$$

Fourier transforming as

$$\chi^R_{\mathbf{p}}(\mathbf{q}, t) = \int \frac{d\omega}{2\pi} e^{-i\omega t} \chi^R_{\mathbf{p}}(\mathbf{q}, \omega)$$

and rearranging the terms, the equation of motion for $\chi^R_{\mathbf{p}}(\mathbf{q}, t)$ yields the following solution:

$$\chi^R_{\mathbf{p}}(\mathbf{q}, \omega) = 2\frac{f_{\mathbf{p}} - f_{\mathbf{p}+\mathbf{q}}}{\omega - \epsilon_{\mathbf{p}+\mathbf{q}} - \epsilon_{\mathbf{p}} + i\eta} \left(1 + \tilde{v}_{\mathbf{q}} \frac{1}{V} \sum_{\mathbf{p}'} \chi^R_{\mathbf{p}'}(\mathbf{q}, \omega) \right), \qquad (N.11)$$

where we add the infinitesimal $i\eta$ to ensure that $\chi_{\mathbf{p}}^{R}(\mathbf{q}, t < 0) = 0$. From (N.5) we see that the Fourier transform of the density response function is

$$\chi^{R}(\mathbf{q}, \omega) = \frac{1}{V} \sum_{\mathbf{p}} \chi_{\mathbf{p}}^{R}(\mathbf{q}, \omega).$$

Thus, summing (N.11) over $\mathbf{p}$ and dividing by the volume we get

$$\chi^{R}(\mathbf{q}, \omega) = \frac{2}{V} \sum_{\mathbf{p}} \frac{f_{\mathbf{p}} - f_{\mathbf{p}+\mathbf{q}}}{\omega - \epsilon_{\mathbf{p}+\mathbf{q}} - \epsilon_{\mathbf{p}} + i\eta} \left(1 + \tilde{v}_{\mathbf{q}} \chi^{R}(\mathbf{q}, \omega)\right).$$

In the limit $V \to \infty$ we have $\frac{1}{V} \sum_{\mathbf{p}} \to \int \frac{d\mathbf{p}}{(2\pi)^3}$, and we recognize in this equation the noninteracting response function (15.66). Thus

$$\chi^{R}(\mathbf{q}, \omega) = \chi_{0}^{R}(\mathbf{q}, \omega) \left(1 + \tilde{v}_{\mathbf{q}} \chi^{R}(\mathbf{q}, \omega)\right). \tag{N.12}$$

Solving (N.12) for χ^{R} we recover the RPA response function (15.78).

O

Kramers–Kronig relations

Let us consider a function $A(z)$ of the complex variable $z = x + iy$ which is analytic in the upper-half plane and with the property that

$$\lim_{z \to \infty} z A(z) = 0, \qquad \text{with } \mathrm{Arg}(z) \in (0, \pi). \tag{0.1}$$

For z in the upper-half plane we can use the Cauchy residue theorem to write this function as

$$A(z) = \oint_C \frac{dz'}{2\pi i} \frac{A(z')}{z' - z}, \tag{0.2}$$

where the integral is along an anti-clockwise oriented curve C that is entirely contained in the upper-half plane and inside which there is the point z. Let C now be the curve in the figure below and $z = x + i\eta$ a point infinitesimally above the real axis.

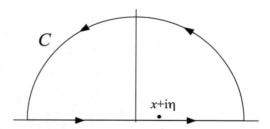

Due to property (0.1) the contribution of the integral along the arc vanishes in the limit of infinite radius and therefore (0.2) implies that

$$A(x + i\eta) = \int_{-\infty}^{\infty} \frac{dx'}{2\pi i} \frac{A(x')}{x' - (x + i\eta)} = P \int_{-\infty}^{\infty} \frac{dx'}{2\pi i} \frac{A(x')}{x' - x} + \frac{1}{2} A(x), \tag{0.3}$$

where in the second equality we use the Cauchy relation

$$\frac{1}{x' - x - i\eta} = P \frac{1}{x' - x} + i\pi \delta(x' - x), \tag{0.4}$$

with P the principal part. Since η is an infinitesimal positive constant we can safely replace $A(x + i\eta)$ with $A(x)$ in the l.h.s. of (O.3), thereby obtaining the following important identity:

$$A(x) = \frac{i}{\pi} P \int_{-\infty}^{\infty} dx' \frac{A(x')}{x - x'}.$$

If we separate $A(x) = A_1(x) + iA_2(x)$ into a real and an imaginary part, then we see that A_1 and A_2 are related through a Hilbert transformation

$$\boxed{A_1(x) = -\frac{1}{\pi} P \int_{-\infty}^{\infty} dx' \frac{A_2(x')}{x - x'}}$$

$$\boxed{A_2(x) = \frac{1}{\pi} P \int_{-\infty}^{\infty} dx' \frac{A_1(x')}{x - x'}} \qquad (O.5)$$

These relations are known as the *Kramers–Kronig relations* and allow us to express the real/imaginary part of the function A in terms of its imaginary/real part. Consequently, the full function A can be written solely in terms of its real or imaginary part since

$$A(x) = A_1(x) + i\frac{1}{\pi} P \int_{-\infty}^{\infty} dx' \frac{A_1(x')}{x - x'} = \frac{i}{\pi} \int_{-\infty}^{\infty} dx' \frac{A_1(x')}{x - x' + i\eta},$$

and also

$$A(x) = -\frac{1}{\pi} P \int_{-\infty}^{\infty} dx' \frac{A_2(x')}{x - x'} + iA_2(x) = -\frac{1}{\pi} \int_{-\infty}^{\infty} dx' \frac{A_2(x')}{x - x' + i\eta}.$$

P

Algorithm for solving the Kadanoff–Baym equations

In this appendix we describe a practical propagation algorithm to solve the four Kadanoff–Baym equations of Section 16.3.1. This algorithm has been applied to finite systems [98, 147] as well as to open systems [165, 180]. As usual we start by writing Σ_{tot} as the sum of the Hartree–Fock self-energy and a nonlocal (in time) self-energy which is the sum of the embedding self-energy and the correlation self-energy,

$$\Sigma_{\text{tot}} = \Sigma_{\text{HF}} + \tilde{\Sigma}_{\text{tot}}.$$

Then, the four Kadanoff–Baym equations become

$$\left[i\frac{d}{dt} - h_{\text{HF}}(t)\right] G^{\rceil}(t, \tau) = \tilde{I}_L^{\rceil}(t, \tau), \tag{P.1}$$

$$\left[i\frac{d}{dt} - h_{\text{HF}}(t)\right] G^{>}(t, t') = \tilde{I}_L^{>}(t, t'), \tag{P.2}$$

$$G^{<}(t, t')\left[-i\frac{\overleftarrow{d}}{dt'} - h_{\text{HF}}(t')\right] = \tilde{I}_R^{<}(t, t'), \tag{P.3}$$

$$i\frac{d}{dt}G^{<}(t, t) - \left[h_{\text{HF}}(t), G^{<}(t, t)\right]_{-} = \tilde{I}_L^{<}(t, t) - \tilde{I}_R^{<}(t, t), \tag{P.4}$$

where $h_{\text{HF}}(t) = h(t) + \Sigma_{\text{HF}}(t)$ is the Hartree–Fock Hamiltonian and the collision integrals with a tilde are the original collision integrals in which $\Sigma_{\text{tot}} \to \tilde{\Sigma}_{\text{tot}}$. Since the integral of $h_{\text{HF}}(t)$ can attain large values, it is favorable to eliminate this term from the time-stepping equations. For each time step $T \to T + \Delta_t$ we therefore absorb the term in a time evolution operator of the form

$$U(t) = e^{-i\bar{h}_{\text{HF}}(T)t}, \qquad 0 \leq t \leq \Delta_t$$

where $\bar{h}_{\text{HF}}(T) = h(T + \Delta_t/2) + \Sigma_{\text{HF}}(T)$. The Hamiltonian $h(t)$ is explicitly known as a function of time and can be evaluated at half the time step. The term Σ_{HF} is only known at time T and is recalculated in the repeated time step (predictor corrector). In terms of the operator $U(t)$ we define new Green's functions g^{x} ($\text{x} = \rceil, \lceil, >, <$) as

$$G^{\rceil}(T+t,\tau) = U(t)g^{\rceil}(t,\tau),$$

$$G^{>}(T+t,t') = U(t)g^{>}(t,t'),$$

$$G^{<}(t',T+t) = g^{<}(t',t)U^{\dagger}(t),$$

$$G^{<}(T+t,T+t) = U(t)g^{<}(t,t)U^{\dagger}(t),$$

where $t' < T$ and $0 < t < \Delta_t$. We can now transform the Kadanoff–Baym equations into equations for g. For instance $g^{\rceil}(t,\tau)$ satisfies the equation

$$i\frac{d}{dt}g^{\rceil}(t,\tau) = U^{\dagger}(t)\left[h_{\mathrm{HF}}(T+t) - \bar{h}_{\mathrm{HF}}(T)\right]G^{\rceil}(T+t,\tau) + U^{\dagger}(t)\tilde{I}_L^{\rceil}(T+t,\tau).$$

Since $\bar{h}_{\mathrm{HF}}(T) \sim h_{\mathrm{HF}}(T+t)$ for times $0 \leq t \leq \Delta_t$, we can neglect for these times the first term on the r.h.s.. We then find

$$G^{\rceil}(T+\Delta_t,\tau) = U(\Delta_t)\left[G^{\rceil}(T,\tau) + \int_0^{\Delta_t} dt\,\frac{d}{dt}g^{\rceil}(t,\tau)\right]$$

$$\sim U(\Delta_t)G^{\rceil}(T,\tau) - iU(\Delta_t)\left[\int_0^{\Delta_t} dt\,e^{i\bar{h}_{\mathrm{HF}}(T)t}\right]\tilde{I}_L^{\rceil}(T,\tau)$$

$$= U(\Delta_t)G^{\rceil}(T,\tau) - V(\Delta_t)\tilde{I}_L^{\rceil}(T,\tau), \tag{P.5}$$

where $V(\Delta_t)$ is defined according to

$$V(\Delta_t) = \frac{1 - e^{-i\bar{h}_{\mathrm{HF}}(T)\Delta_t}}{\bar{h}_{\mathrm{HF}}(T)}.$$

In a similar way we can use (P.2) and (P.3) to propagate the greater and lesser Green's functions and we find

$$G^{>}(T+\Delta_t,t') = U(\Delta_t)G^{>}(T,t') - V(\Delta_t)\tilde{I}_L^{>}(T,t'), \tag{P.6}$$

$$G^{<}(t',T+\Delta_t) = G^{<}(t',T)U^{\dagger}(\Delta_t) - \tilde{I}_R^{<}(t',T)V^{\dagger}(\Delta_t). \tag{P.7}$$

For time-stepping along the diagonal the equation for $g^{<}(t,t)$ reads

$$i\frac{d}{dt}g^{<}(t,t) = U^{\dagger}(t)\left[\tilde{I}_L^{<}(t,t) - \tilde{I}_R^{<}(t,t)\right]U(t),$$

where again we approximate the difference $\bar{h}_{\mathrm{HF}}(T) - h_{\mathrm{HF}}(T+t) \sim 0$. Integrating over t between 0 and Δ_t we then find

$$G^{<}(T+\Delta_t,T+\Delta_t) \sim U(\Delta_t)G^{<}(T,T)U^{\dagger}(\Delta_t)$$

$$- iU(\Delta_t)\left[\int_0^{\Delta_t} dt\,U^{\dagger}(t)\left(\tilde{I}_L^{<}(T,T) - \tilde{I}_R^{<}(T,T)\right)U(t)\right]U^{\dagger}(\Delta_t).$$

By using the operator expansion

$$e^A B e^{-A} = B + [A, B]_- + \frac{1}{2!} \left[A, [A, B]_- \right]_- + \frac{1}{3!} \left[A, \left[A, [A, B]_- \right]_- \right]_- + \dots$$

it follows that

$$-\mathrm{i} \int_0^{\Delta_t} dt \, U^\dagger(t) \left(\tilde{I}_L^<(T, T) - \tilde{I}_R^<(T, T) \right) U(t) = \sum_{n=0}^{\infty} C_n,$$

where

$$C_{n+1} = \frac{\mathrm{i}\Delta_t}{n+2} \left[\bar{h}_{\mathrm{HF}}(T), C_n \right]_-,$$

and $C_0 = -\mathrm{i}\Delta_t \left(\tilde{I}_L^<(T, T) - \tilde{I}_R^<(T, T) \right)$. Inserting this result into the formula for the equal time $G^<$ we finally obtain

$$G^<(T + \Delta_t, T + \Delta_t) = U(\Delta_t) \left[G^<(T, T) + \sum_{n=0}^{\infty} C_n \right] U^\dagger(\Delta_t). \qquad \text{(P.8)}$$

For the systems considered in Chapter 16 we have found that keeping terms for $n \leq 3$ yields sufficient accuracy. The equations (P.5), (P.6), (P.7), and (P.8) together with the symmetry properties discussed in Section 16.3.1 form the basis of the time-stepping algorithm. At each time step, it requires the construction of the operators $U(\Delta_t)$ and $V(\Delta_t)$ and therefore the diagonalization of $\bar{h}_{\mathrm{HF}}(T)$. The implementation of the predictor corrector consists in carrying out the time step $T \to T + \Delta_t$ as many times as needed to get the desired accuracy. According to our experience, repeating every time step twice is already enough to obtain accurate results (provided that the time step Δ_t is small enough). To summarize, the procedure is as follows:

- The collision integrals and $\bar{h}_{\mathrm{HF}}$ at time T are calculated from the Green's functions with real times up to T.

- A step in the Green's function is taken according to the equations (P.5), (P.6), (P.7), and (P.8).

- A new $\bar{h}_{\mathrm{HF}}$ and new collision integrals $\tilde{I}_L^<(T + \Delta_t, \tau)$, $I_L^>(T + \Delta_t, t')$, and $I_R^<(t', T + \Delta_t)$ are calculated from the new Green's functions with real times up to $T + \Delta_t$.

- The arithmetic average of $\bar{h}_{\mathrm{HF}}$ and of the collision integrals for times T and $T + \Delta_t$ is calculated.

- The time step $T \to T + \Delta_t$ is then repeated using the arithmetic average values of $\bar{h}_{\mathrm{HF}}$ and of the collision integrals according to (P.5), (P.6), (P.7), and (P.8).

This concludes the general time-stepping procedure for the Green's function.

References

[1] P. A. M. Dirac, *The Principles of Quantum Mechanics* (Oxford University Press, Oxford, 1999).

[2] J. J. Sakurai, *Modern Quantum Mechanics* (Addison-Wesley, Reading MA, 1994).

[3] A. Bohm, *Quantum Mechanics: Foundations and Applications* (Springer, Berlin, 1994).

[4] R. McWeeny, *Rev. Mod. Phys.* **32**, 335 (1960).

[5] R. Pariser and R. G. Parr, *J. Chem. Phys.* **21**, 466 (1953).

[6] J. A. Pople, *Trans. Faraday Soc.* **42**, 1375 (1953).

[7] J. Linderberg and Y. Öhrn, *Propagators in Quantum Chemistry* (Wiley, New York, 2000).

[8] P. Atkins, *Physical Chemistry* (W. H. Freeman, New York, 1998).

[9] K. S. Novoselov *et al.*, *Science* **306**, 666 (2004).

[10] A. H. C. Neto, F. Guinea, N. M. R. Peres, K. S. Novoselov, and A. K. Geim, *Rev. Mod. Phys.* **81**, 109 (2009).

[11] A. Mielke, *J. Phys. A* **24**, L73 (1991).

[12] H. Tasaki, *Phys. Rev. Lett.* **69**, 1608 (1992).

[13] A. Mielke and H. Tasaki, *Commun. Math. Phys.* **158**, 341 (1993).

[14] U. Fano, *Phys. Rev.* **124**, 1866 (1961).

[15] J. Hubbard, *Proc. Roy. Soc. A* **276**, 238 (1963).

[16] H. Tasaki, *J. Phys.: Condens. Matter* **10**, 4353 (1998).

[17] *The Hubbard Model: Its Physics and Mathematical Physics*, edited by D. Baeriswyl, D. K. Campbell, J. M. P. Carmelo, F. Guinea, and E. Louis, NATO ASI Series, Series B: Physics vol. 343 (1995).

[18] *The Hubbard Model:* a reprint volume, edited by A. Montorsi (World Scientific Publ. Co., Singapore, 1992).

[19] P. W. Anderson, *Phys. Rev.* **115**, 2 (1959).

[20] J. Bardeen, L. N. Cooper, and J. R. Schrieffer, *Phys. Rev.* **106**, 162 (1957).

[21] D. C. Ralph, C. T. Black, and M. Tinkham, *Phys. Rev. Lett.* **76**, 688 (1996).

[22] R. W. Richardson, *Phys. Lett.* **3**, 277 (1963).

[23] L. N. Cooper, *Phys. Rev.* **104**, 1189 (1956).

[24] H. Frölich, *Advances in Physics* **3**, 325 (1954).

[25] R. P. Feynman, *Phys. Rev.* **97**, 660 (1955).

[26] W. P. Su, J. R. Schrieffer, and A. J. Heeger, *Phys. Rev. B* **22**, 2099 (1980).

[27] T. Holstein, *Annals of Physics* **8**, 325 (1959).

[28] P. A. M. Dirac, *Proc. Roy. Soc. Lond. A* **117**, 610 (1928).

[29] C. Itzykson and J.-B. Zuber, *Quantum Field Theory* (McGraw-Hill, Singapore, 1980).

[30] L. H. Ryder, *Quantum Field Theory* (Cambridge University Press, Cambridge, 1985).

[31] F. Mandl and G. Shaw, *Quantum Field Theory* (J. Wiley & Sons, New York, 1984).

[32] S. Weinberg, *The Quantum Theory of Fields Ñ*, Volume I: *Foundations* (Cambridge University Press, Cambridge, 1995).

[33] P. Ramond, *Field Theory: A Modern Primer* (Benjamin/Cummings, Inc., London, 1981).

[34] M. E. Peskin and D. V. Schroeder, *An Introduction to Quantum Field Theory* (Addison-Wesley, USA, 1995).

[35] R. E. Peierls, *Quantum Theory of Solids* (Clarendon, Oxford, 1955).

[36] I. G. Lang and Y. A. Firsov, *Zh. Eksp. Teor. Fiz.* **43**, 1843 (1962), [*Sov. Phys. JETP* **16**, 1301 (1962)].

[37] L. V. Keldysh, *Sov. Phys. JETP* **20**, 1018 (1965).

[38] J. Schwinger, *J. Math. Phys.* **2**, 407 (1961).

[39] L. V. Keldysh, *Progress in Nonequilibrium Green's Functions II*, editors M. Bonitz and D. Semkat (World Scientific, London, 2003).

[40] O. V. Konstantinov and V. I. Perel', *Sov. Phys. JETP* **12**, 142 (1961).

[41] P. Danielewicz, *Ann. Physics* **152**, 239 (1984).

[42] M. Wagner, *Phys. Rev. B* **44**, 6104 (1991).

[43] P. Gaspard and M. Nagaoka, *J. Chem. Phys.* **111**, 5676 (1999).

[44] M. Gell-Mann and F. Low, *Phys. Rev.* **84**, 350 (1951).

[45] A. L. Fetter and J. D. Walecka, *Quantum Theory of Many-Particle Systems* (McGraw-Hill, New York, 1971).

[46] H. Bruus and K. Flensberg, *Many-Body Quantum Theory in Condensed Matter Physics: An Introduction* (Oxford University Press, Oxford, 2004).

[47] P. C. Martin and J. Schwinger, *Phys. Rev.* **115**, 1342 (1959).

[48] R. Kubo, *J. Phys. Soc. Jpn.* **12**, 570 (1957).

[49] R. Haag, N. M. Hugenholtz, and M. Winnink, *Comm. Math. Phys.* **5**, 215 (1967).

[50] G. C. Wick, *Phys. Rev.* **80**, 268 (1950).

[51] D. Kaiser, *Phil. Mag.* **43**, 153 (1952).

[52] J. Olsen, P. Jørgensen, T. Helgaker, and O. Christiansen, *J. Chem. Phys.* **112**, 9736 (2000).

[53] D. C. Langreth, in *Linear and Nonlinear Electron Transport in Solids*, edited by J. T. Devreese and E. van Doren (Plenum, New York, 1976), pp. 3–32.

[54] G. Stefanucci and C.-O. Almbladh, *Phys. Rev. B* **69**, 195318 (2004).

[55] G. Stefanucci, *Phys. Rev. B* **75**, 195115 (2007).

[56] M. Knap, E. Arrigoni, and W. von der Linden, *Phys. Rev. B* **81**, 024301 (2010).

[57] S. Ejima, H. Fehske, and F. Gebhard, *Europhys. Lett.* **93**, 30002 (2011).

[58] W. Schattke, M. A. Van Hove, F. J. García de Abajo, R. Díez Muiño, and N. Mannella, *Solid-State Photoemission and Related Methods: Theory and Experiment*, edited by Wolfgang Schattke and Michel A. Van Hove (Wiley-VCH Verlag GmbH & Co. KGaA, Weinheim, 2003).

[59] P. Nozières and C. T. De Dominicis, *Phys. Rev.* **178**, 1097 (1969).

[60] K. S. Thygesen and A. Rubio, *Phys. Rev. Lett.* **102**, 046802 (2009).

[61] J. B. Neaton, M. S. Hybertsen, and S. G. Louie, *Phys. Rev. Lett.* **97**, 216405 (2006).

[62] P. Myöhänen, R. Tuovinen, T. Korhonen, G. Stefanucci, and R. van Leeuwen, *Phys. Rev. B* **85**, 075105 (2012).

[63] D. R. Hartree, *Proc. Cambridge Phil. Soc.* **24**, 89 (1928).

[64] E. P. Gross, *Il Nuovo Cimento* **20**, 454 (1961).

[65] L. P. Pitaevskii, *JETP* **13**, 451 (1961).

[66] F. Dalfovo, S. Giorgini, L. P. Pitaevskii, and S. Stringari, *Rev. Mod. Phys.* **71**, 463 (1999).

[67] G. F. Giuliani and G. Vignale, *Quantum Theory of the Electron Liquid* (Cambridge University, Canbridge, 2005).

[68] L. P. Kadanoff and G. Baym, *Quantum Statistical Mechanics* (W. A. Benjamin, Inc. New York, 1962).

[69] M. A. Reed, C. Zhou, C. J. Muller, T. P. Burgin, and J. M. Tour, *Science* **278**, 252 (1997).

[70] A. I. Yanson, G. R. Bollinger, H. E. van den Brom, N. Agrait, and J. M. van Ruitenbeek, *Nature* **395**, 783 (1998).

[71] N. Bushong, N. Sai, and M. Di Ventra, *Nano Lett.* **5**, 2569 (2005).

[72] R. Rajaraman, *Solitons and Instantons* (North-Holland Publishing Company, Amsterdam, 1982).

[73] J. C. Slater, *Phys. Rev.* **35**, 210 (1930).

[74] V. Fock, *Z. Physik* **61**, 126 (1930).

[75] G. D. Mahan, *Many-Particle Physics*, 2nd ed. (Plenum Press, New York, 1990).

[76] E. Lipparini, *Modern Many-Particle Physics: Atomic Gases, Quantum Dots and Quantum Fluids* (World Scientific, London, 2003).

[77] G. Baym and L. P. Kadanoff, *Phys. Rev.* **124**, 287 (1961).

[78] G. Baym, *Phys. Rev.* **171**, 1391 (1962).

[79] L. Hedin, *Phys. Rev.* **139**, A796 (1965).

[80] F. Aryasetiawan and O. Gunnarsson, *Rep. Prog. Phys.* **61**, 237 (1998).

[81] J. M. Luttinger and J. C. Ward, *Phys. Rev.* **118**, 1417 (1960).

[82] N. E. Dahlen, R. van Leeuwen, and U. von Barth, *Phys. Rev. A* **73**, 012511 (2006).

[83] N. E. Dahlen and U. von Barth, *Phys. Rev. B* **69**, 195102 (2004); *J. Chem. Phys.* **120**, 6826 (2004).

[84] U. von Barth, N. E. Dahlen, R. van Leeuwen, and G. Stefanucci, *Phys. Rev. B* **72**, 235109 (2005).

[85] A. Klein, *Phys. Rev.* **121**, 950 (1961).

[86] C.-O. Almbladh, U. von Barth, and R. van Leeuwen, *Int. J. Mod. Phys. B* **13**, 535 (1999).

[87] C. T. De Dominicis, *J. Math. Phys.* **4**, 255 (1963).

[88] C. T. De Dominicis and P. C. Martin, *J. Math. Phys.* **5**, 14 (1964); *J. Math. Phys.* **5**, 31 (1964).

[89] M. Hindgren, *Model Vertices Beyond the GW Approximation*, PhD thesis, University of Lund, Lund, Sweden (1997). ISBN: 91-628-2555-0. ISRN: LUNFD6 / (NTFTF-1034) / 1-33 / (1997).

[90] A. C. Phillips, *Phys. Rev.* **142**, 984 (1966).

[91] M. Cini, E. Perfetto, G. Stefanucci, and S. Ugenti, *Phys. Rev. B* **76**, 205412 (2007).

[92] L. D. Faddeev, *JETP* **12**, 1014 (1961).

[93] G. Grosso and G. Pastori Parravicini, *Solid State Physics* (Academic Press, London, 2000).

[94] W. Hanke and L. J. Sham, *Phys. Rev. Lett.* **33**, 582 (1974); *Phys. Rev. B* **12**, 4501 (1975); *Phys Rev. B* **21**, 4656 (1980).

[95] G. Onida, L. Reining, and A. Rubio, *Rev. Mod. Phys.* **74**, 601 (2002).

[96] D. Sangalli, P. Romaniello, G. Onida, and A. Marini, *J. Chem. Phys.* **134**, 034115 (2011).

[97] N.-H. Kwong and M. Bonitz, *Phys. Rev. Lett.* **84**, 1768 (2000).

[98] N. E. Dahlen and R. van Leeuwen, *Phys. Rev. Lett.* **98**, 153004 (2007).

[99] N. Säkkinen, M. Manninen, and R. van Leeuwen, *New J. Phys.* **14**, 013032 (2012).

[100] A. Schindlmayr and R. W. Godby, *Phys. Rev. Lett.* **80**, 1702 (1998).

[101] J. M. Luttinger, *Phys. Rev.* **121**, 942 (1961).

[102] T. Giamarchi, *Quantum Physics in One Dimension* (Clarendon, Oxford, 2004).

[103] M. Puig von Friesen, C. Verdozzi, and C.-O. Almbladh, *Phys. Rev. Lett.* **103**, 176404 (2009); *Phys. Rev. B* **82**, 155108 (2010).

[104] M. Gell-Mann and K. Brueckner, *Phys. Rev.* **106**, 364 (1957).

[105] T. Endo, M. Horiuchi, Y. Takada, and H. Yasuhara, *Phys. Rev. B* **59**, 7367 (1999).

[106] J. Sun, J. P. Perdew, and M. Seidl, *Phys. Rev. B* **81**, 085123 (2010).

[107] D. M. Ceperly and B. J. Alder, *Phys. Rev. Lett.* **45**, 566 (1980).

[108] W. M. C. Foulkes, L. Mitas, R. J. Needs, and G. Rajagopal, *Rev. Mod. Phys.* **73**, 33 (2001).

[109] K. Binder and D. W. Heermann, *Monte Carlo Simulations in Statistical Physics: An Introduction* (Springer-Verlag, Berlin/Heidelberg, 2010).

[110] B. Holm and U. von Barth, *Phys. Rev. B* **57**, 2108 (1998).

[111] G. D. Mahan and B. E. Sernelius, *Phys. Rev. Lett.* **62**, 2718 (1989).

[112] S. Hong and G. D. Mahan, *Phys. Rev. B* **50**, 8182 (1994).

[113] P. A. Bobbert and W. van Haeringen, *Phys. Rev. B* **49**, 10326 (1994).

[114] R. Del Sole, L. Reining, and R. W. Godby, *Phys. Rev. B* **49**, 8024 (1994).

[115] M. Holzmann *et al.*, *Phys. Rev. Lett.* **107**, 110402 (2011).

[116] D. J. Thouless, *Ann. Phys.* **10**, 553 (1960).

[117] S. Schmitt-Rink, C. M. Varma, and A. E. Ruckenstein, *Phys. Rev. Lett.* **63**, 445 (1989).

[118] L. P. Kadanoff and P. C. Martin, *Phys. Rev.* **124**, 670 (1961).

[119] A.-M. Uimonen *et al.*, *Phys. Rev. B* **84**, 115103 (2011).

[120] C. Cohen-Tannoudji, B. Diu, and F. Laloë, *Quantum Mechanics* (Wiley, New York, 1977).

[121] P. A. M. Dirac, *Proc. Roy. Soc. Lond. A* **114**, 243 (1927).

[122] E. Fermi, *Nuclear Physics* (University of Chicago, Chicago IL, 1950).

[123] T. D. Visser, *Am. J. Phys.* **77**, 487 (2009).

[124] G. Strinati, *Riv. Nuovo Cim.* **11**, 1 (1986).

[125] G. D. Mahan, *Phys. Rev. B* **2**, 4334 (1970).

[126] W. L. Schaich and N. W. Ashcroft, *Phys. Rev. B* **3**, 2452 (1970).

[127] C. Caroli, D. Lederer-Rozenblatt, B. Roulet, and D. Saint-James, *Phys. Rev. B* **8**, 4552 (1973).

[128] W. Thomas, *Naturwissenschaften* **13**, 627 (1925).

[129] W. Kuhn, *Z. Phys.* **33**, 408 (1925).

[130] F. Reiche and W. Thomas, *Z. Phys.* **34**, 510 (1925).

[131] J. C. Ward, *Phys. Rev.* **79**, 182 (1950).

[132] Y. Takahashi, *Nuovo Cim.* **6**, 371 (1957).

[133] P. Nozières, *Theory of Interacting Fermi Systems* (Benjamin, New York, NY, 1964).

[134] M. Hellgren and U. von Barth, *J. Chem. Phys.* **131**, 044110 (2009).

[135] J. Lindhard, *Det Kgl. Danske Vid. Selskab, Matematisk-fysiske Meddelelser* **28** (1954).

[136] D. Bohm and D. Pines, *Phys. Rev.* **92**, 609 (1953).

[137] L. H. Thomas, *Proc. Cambridge Phil. Soc.* **23**, 542 (1927).

[138] E. Fermi, *Rend. Accad. Naz. Lincei* **6**, 602 (1927).

[139] G. S. Canright, *Phys. Rev. B* **38**, 1647 (1988).

[140] J. S. Langer and S. H. Vosko, *J. Phys. Chem. Solids* **12**, 196 (1960).

[141] J. Friedel, *American Scientist* **93**, 156 (2005).

[142] F. Aryasetiawan, L. Hedin, and K. Karlsson, *Phys. Rev. Lett.* **77**, 2268 (1996).

[143] L. Hedin, *Phys. Rev.* **139**, A796 (1963).

[144] L. Landau, *Soviet Physics JETP* **3**, 920 (1957); *Soviet Physics JETP* **5**, 101 (1957); *Soviet Physics JETP* **8**, 70 (1959).

[145] B. I. Lundqvist, *Phys. Kondens. Materie* **6**, 193 (1967); *Phys. Kondens. Materie* **6**, 206 (1967); *Phys. Kondens. Materie* **7**, 117 (1968).

[146] B. I. Lundqvist and V. Samathiyakanit, *Phys. Kondens. Materie* **9**, 231 (1969).

[147] A. Stan, N. E. Dahlen, and R. van Leeuwen, *J. Chem. Phys.* **130**, 224101 (2009).

[148] Y. Meir and N. S. Wingreen, *Phys. Rev. Lett.* **68**, 2512 (1992).

[149] A.-P. Jauho, N. S. Wingreen, and Y. Meir, *Phys. Rev. B* **50**, 5528 (1994).

[150] E. Khosravi, G. Stefanucci, S. Kurth, and E. K. U. Gross, *App. Phys. A* **93** 355 (2008); *Phys. Chem. Chem. Phys.* **11**, 4535 (2009).

[151] E. Khosravi *et al.*, *Phys. Rev. B* **85**, 075103 (2012).

[152] M. Cini, *Phys. Rev. B* **22**, 5887 (1980).

[153] E. Perfetto, G. Stefanucci, and M. Cini, *Phys. Rev. B* **78**, 155301 (2008).

[154] J. A. Gupta, D. D. Awschalom, X. Peng, and A. P. Alivisatos, *Phys. Rev. B* **59**, R10421 (1999).

[155] A. Greilich *et al.*, *Phys. Rev. Lett.* **96**, 227401 (2006).

[156] F. M. Souza, *Phys. Rev. B* **76**, 205315 (2007).

[157] R. Landauer, *IBM J. Res. Dev.* **1**, 233 (1957).

[158] M. Büttiker, *Phys. Rev. Lett.* **57**, 1761 (1989).

[159] E. Perfetto, G. Stefanucci, and M. Cini, *Phys. Rev. Lett.* **105**, 156802 (2010).

[160] M. A. Cazalilla, *Phys. Rev. Lett.* **97**, 076401 (2006).

[161] E. Perfetto, *Phys. Rev. B* **74**, 205123 (2006).

[162] B. Dóra, M. Haque, and G. Zaránd, *Phys. Rev. Lett.* **106**, 156406 (2011).

[163] E. Perfetto and G. Stefanucci, *Europhys. Lett.* **95**, 10006 (2011).

[164] M. A. Cazalilla, R. Citro, T. Giamarchi, E. Orignac, and M. Rigol, *Rev. Mod. Phys.* **83**, 1405 (2011).

[165] P. Myöhänen, A. Stan, G. Stefanucci, and R. van Leeuwen, *Europhys. Lett.* **84**, 67001 (2008).

[166] K. S. Thygesen and A. Rubio, *Phys. Rev. B* **77**, 115333 (2008).

[167] K. S. Thygesen, *Phys. Rev. Lett.* **100**, 166804 (2008).

[168] T. J. Park and J. C. Light, *J. Chem. Phys.* **85**, 5870 (1986).

[169] H. Q. Lin and J. E. Gubernatis, *Comput. Phys.* **7**, 400 (1993).

[170] R. van Leeuwen and G. Stefanucci, *Phys. Rev. B* **85**, 115119 (2012).

[171] M. Born and H. S. Green, *Proc. Roy. Soc. Lond. A* **191**, 168 (1947).

[172] N. N. Bogoliubov, *The Dynamical Theory in Statistical Physics* (Hindustan Pub. Corp., Delhi, 1965).

[173] J. Kirkwood, *J. Chem. Phys.* **14**, 180 (1946).

[174] J. Yvon, *Nucl. Phys.* **4**, 1 (1957).

[175] A. Akbari, M. J. Hashemi, R. M. Nieminen, R. van Leeuwen, and A. Rubio, *Phys. Rev. B* **85**, 235121 (2012).

[176] N. E. Dahlen and R. van Leeuwen, *J. Chem. Phys.* **122**, 164102 (2005).

[177] L. Hedin and S. Lundqvist, Effects of Electron–Electron and Electron–Phonon Interactions on the One-Electron States of Solids, *Solid State Phys.* **23**, 1 (1970).

[178] D. Pines, *Elementary Excitations in Solids*, (Benjamin, New York, NY, 1964).

[179] E. K. U. Gross, E. Runge and O. Heinonen, *Many-Particle Theory* (Adam Hilger, Bristol, 1991).

[180] P. Myöhänen, A. Stan, G. Stefanucci, and R. van Leeuwen, *Phys. Rev. B* **80**, 115107 (2009).

Index

图书在版编目（CIP）数据

量子系统的非平衡多体理论 = Nonequilibrium Many–Body Theory of Quantum
Systems: A Modern Introduction：英文 /（意）G.斯蒂芬尼茨，（德）R.冯·莱
文著 . — 影印本 . — 北京：世界图书出版有限公司北京分公司，2019.7（2023.3重印）
ISBN 978–7–5192–6419–2

Ⅰ．①量… Ⅱ．① G… ② R… Ⅲ．①非平衡统计理论—英文 Ⅳ．① O414.22

中国版本图书馆 CIP 数据核字（2019）第 137036 号

中文书名　量子系统的非平衡多体理论
英文书名　Nonequilibrium Many–Body Theory of Quantum Systems: A Modern Introduction
著　　者　Gianluca Stefanucci, Robert van Leeuwen
责任编辑　刘　慧　高　蓉

出版发行　世界图书出版有限公司北京分公司
地　　址　北京市东城区朝内大街 137 号
邮　　编　100010
电　　话　010–64038355（发行）　　64033507（总编室）
网　　址　http://www.wpcbj.com.cn
邮　　箱　wpcbjst@vip.163.com
销　　售　新华书店
印　　刷　北京建宏印刷有限公司
开　　本　710 mm×1000 mm　1/16
印　　张　38.75
字　　数　620 千字
版　　次　2019 年 9 月第 1 版
印　　次　2023 年 3 月第 3 次印刷
版权登记　01–2019–3163
国际书号　ISBN 978–7–5192–6419–2
定　　价　129.00 元
